NEW DIRECTIONS IN QUANTUM CHROMODYNAMICS

New Directions in
Quantum Chromodynamics

The APCTP Topical Meeting and
The 11th International Light-Cone School
May 26-June 18, 1999

supported by
the Asia Pacific Center for Theoretical Physics

The 12th Nuclear Physics Summer School and
Symposium and
The 11th International Light Cone Workshop
June 21-25, 1999

supported by
the Center for Theoretical Physics in Seoul
National University, the Daewoo Foundation, the
Korea Science and Engineering Foundation and
the Deutsche Forschungsgemeinschaft

NEW DIRECTIONS
IN QUANTUM
CHROMODYNAMICS

Seoul and Kyungju, Korea   May-June 1999

EDITORS
Chueng-Ryong Ji
North Carolina State University

Dong-Pil Min
Seoul National University

American Institute of Physics

**AIP CONFERENCE
PROCEEDINGS 494**

Melville, New York

Editors:

Chueng-Ryong Ji
Department of Physics
North Carolina State University
Raleigh, NC 27695-8202
USA

E-mail: crji@unity.ncsu.edu

Dong-Pil Min
Department of Physics
Seoul National University
Seoul 151-742
KOREA

E-mail: dpmin@phya.snu.ac.kr

L.C. Catalog Card No. 99-067662
ISBN 1-56396-908-4
ISSN 0094-243X
Printed in the United States of America

CONTENTS

SCHOOL LECTURES

WORKSHOP TALKS

LIGHT-CONE FIELD THEORY

NEW METHODS IN NONPERTURBATIVE QCD

PHENOMENOLOGY CONFRONTING QCD

EFFECTIVE FIELD THEORY IN HADRONS

APPENDIX

PREFACE

These Proceedings contain the lectures presented during the 11th International Light Cone (ILC) School held May 26-June 18, 1999, at the Asia Pacific Center for Theoretical Physics (APCTP) in Seoul, Korea and the talks presented during the 12th Nuclear Physics Summer School and Symposium (NuSS) and the 11th ILC Workshop held June 21-25, 1999, in Kyungju, Korea. The main purpose of this conference was to survey the new directions for solving QCD strong interaction problems and apply various approaches to hadron physics where abundant experimental data are available.

The 11th ILC School was held as a Topical Meeting of APCTP and a series of lectures were presented for three weeks by the invited lecturers Stan Brodsky, Simon Dalley, Jerry Miller, Chris Pauli and Steve Pinsky. A couple of seminars were also given by Nicolai Kochelev. While most of the lectures were presented at APCTP, some lectures and seminars were given at the newly constructed Sangsan Math Science Building at Seoul National University (SNU).

The 12th NuSS (NuSS'99) and the 11th ILC Workshop were held at the Concord Hotel in Kyungju for the subsequent one week and attracted more than 80 participants. The world experts in theory and experiment presented the most recent developments in the six topics; Light-Cone Field Theory, New Methods in Nonperturbative QCD, Phenomenology Confronting QCD, Effective Field Theory in Hadrons, High Temperatures and Density QCD and Critical Experiments Testing QCD. We are very grateful to the twelve convenors who arranged such a nice program in each topic. Including the talks contributed by the young active researchers in these fields, about 45 talks were presented altogether.

Both the School and the Workshop were useful especially to the young participants in the beginning stages of their research careers. They had an opportunity to see the variety of available new directions in the research of QCD and hadron physics. It was a great occasion for all the participants to gain some insight in pursuing their future researches in these fields.

We would like to thank all the speakers for their dedicated presentation of lectures and talks and all others for their continued participation in the stimulated discussions that made the whole conference a great success. We also want to thank the APCTP, the Center for Theoretical Physics at SNU, the Daewoo Foundation, the Korea Science and Engineering Foundation (KOSEF) and the Deutsche Forschungsgemeinschaft (DFG) for the financial support which made this meeting possible.

Chueng-Ryong Ji
Dong-Pil Min

Advisory Committee

A. Bassetto (INFN), S. J. Brodsky (SLAC),
I. T. Cheon (Yonsei),Y. M. Cho (APCTP),
D. I. Choi (KAIST), J. R. Hiller (Minnesota),
B. T. Kim (Sungkyunkwan), C. W. Kim (KIAS),
J. K. Kim (KAIST), J. W. Kim (Seoul),
Y. S. Koh (Seoul), H.-C. Pauli (MPI),
S. Pinsky (Ohio), S. A. Shin (Ewha),
K. S. Soh (Seoul), H. S. Song (CTP, Seoul),
H. Toki (RCNP), J. D. Vary (Iowa), and
E. Werner (Regensburg)

Local Organizing Committee

Co-Chair
  Chueng-Ryong Ji (North Carolina, ji@ncsu.edu)
  Dong-Pil Min   (Seoul, dpmin@phya.snu.ac.kr)

Members
J.-B. Choi (Chonbuk), D. S. Hwang (Sejong),
H. Jung (Sookmyung), J. H. Kang (Yonsei),
C. S. Kim (Yonsei), Y. Kim (Sungkyunkwan),
P. Ko (KAIST), B.-H. Lee (Sogang),
S. H. Lee (Yonsei), W. Namgung (Dongkuk),
Y. Oh (CTP, Seoul, Secretary), and
M. Oka (TIT, Tokyo)

International Light-Cone Advisory Committee (ILCAC)

A. Bassetto (INFN), S. J. Brodsky (SLAC),
M. Burkardt (New Mexico), S. Dalley (Oxford),
Y. Frishman (Weizmann), S. D. Glazek (Warsaw),
J. R. Hiller (Minnesota-Duluth), C.-R. Ji (North Carolina),
M. Karliner (Tel-Aviv), L. Lipatov (St. Petersburg),
G. McCartor (SMU, Dallas), G. A. Miller (Washington, Seattle),
H.-C. Pauli (Max-Planck, Heidelberg), S. Pinsky (Ohio),
J. D. Vary (Iowa), E. Werner (Regensburg), and
A. Williams (Adelaide)

Topics	Convenors
Light-Cone Field Theory	Gary McCartor (mccartor@mail.physics.amu.edu) Yoonbai Kim (yoonbai@cosmos.skku.ac.kr)
New Methods in Nonperturbative QCD	Koichi Yamawaki (yamawaki@eken.phys.nagoya-u.ac.jp) Hong Jung (jung@nt2.sookmyung.ac.kr)
Phenomenology Confronting QCD	Carl Carlson (carlson@physics.wm.edu) Jong-Bum Choi (jbchoi@moak.chonbuk.ac.kr)
Effective Field Theory in Hadrons	Matthias Lutz (mlutz@clri6a.gsi.de) Bum-Hoon Lee (bhl@ccs.sogang.ac.kr)
High Temperature and Density QCD	Frieder Lenz (flenz@theorie3.physik.uni-erlangen.de) Su Houng Lee (shlee@tpri6n.gsi.de)
Critical Experiments Testing QCD	Gerald Garvey (garvey@lanl.gov) Ju-Hwan Kang (jhkang@npl1.yonsei.ac.kr)

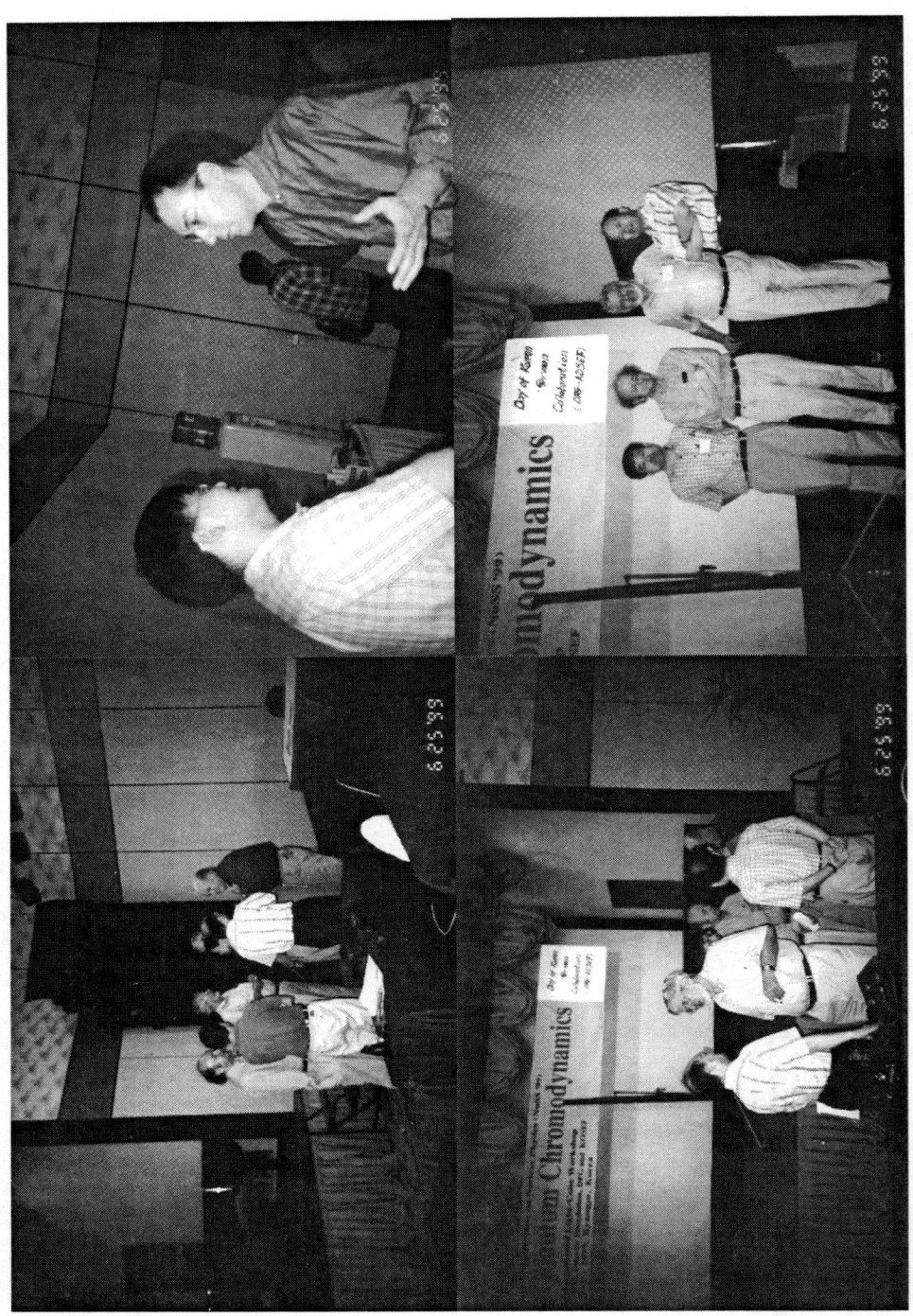

SCHOOL LECTURES

QCD Technology:
Light-Cone Quantization and
Commensurate Scale Relations*

Stanley J. Brodsky

Stanford Linear Accelerator Center
Stanford University, Stanford, California 94309

Abstract. I discuss several theoretical tools which are useful for analyzing perturbative and non-perturbative problems in quantum chromodynamics, including (a) the light-cone Fock expansion, (b) the effective charge α_V, (c) conformal symmetry, and (d) commensurate scale relations. Light-cone Fock-state wavefunctions encode the properties of a hadron in terms of its fundamental quark and gluon degrees of freedom. Given the proton's light-cone wavefunctions, one can compute not only the quark and gluon distributions measured in deep inelastic lepton-proton scattering, but also the multi-parton correlations which control the distribution of particles in the proton fragmentation region and dynamical higher twist effects. Light-cone wavefunctions also provide a systematic framework for evaluating exclusive hadronic matrix elements, including timelike heavy hadron decay amplitudes and form factors. The α_V coupling, defined from the QCD heavy quark potential, provides a physical expansion parameter for perturbative QCD with an analytic dependence on the fermion masses which is now known to two-loop order. Conformal symmetry provides a template for QCD predictions, including relations between observables which are present even in a theory which is not scale invariant. Commensurate scale relations are perturbative QCD predictions based on conformal symmetry relating observable to observable at fixed relative scale. Such relations have no renormalization scale or scheme ambiguity.

INTRODUCTION

A primary goal of both high energy and nuclear physics is to unravel the structure and dynamics of nucleons and nuclei in terms of their fundamental quark and gluon degrees of freedom. Our present empirical knowledge of the quark and gluon distributions of the proton has revealed a remarkably complex substructure. It is helpful to categorize the parton distributions as "intrinsic" –pertaining to the composition of the target hadron, and "extrinsic", reflecting the substructure of the individual quarks and gluons themselves. For example, the $\overline{u}(x)$ and $\overline{d}(x)$ antiquark distributions of the proton at $Q^2 \sim 10$ GeV^2 to be quite different in

---

*) Work supported by the Department of Energy, contract DE–AC03–76SF00515.

CP494, *New Directions in Quantum Chromodynamics*, edited by C.-R. Ji and D.-P. Min
© 1999 American Institute of Physics 1-56396-908-4/99/$15.00

shape [1] and thus must reflect dynamics intrinsic to the proton's structure. If the sea quarks were generated solely by perturbative QCD evolution via gluon splitting, the anti-quark distributions would be isospin symmetric. Evidence for a difference between the $\bar{s}(x)$ and $s(x)$ distributions has also been claimed. [2] Gluons carry a significant fraction of the proton's spin as well as its momentum. Since gluon exchange between valence quarks contributes to the $p - \Delta$ mass splitting, it follows that the gluon distributions must be cannot be solely accounted for by gluon bremsstrahlung from individual quarks, the process responsible for DGLAP evolutions of the structure functions. Similarly, in the case of heavy quarks, $s\bar{s}$, $c\bar{c}$, $b\bar{b}$, the diagrams in which the sea quarks are multiply connected to the valence quarks are intrinsic to the proton structure itself. Thus neither gluons nor sea quarks are solely generated by DGLAP evolution, and one cannot define a resolution scale Q_0 where the sea or gluon degrees of freedom can be neglected. There have also been surprises associated with the chirality distributions $\Delta q = q_{\uparrow/\uparrow} - q_{\downarrow/\uparrow}$ of the valence quarks which again show that a simple valence quark approximation to nucleon spin structure functions is far from the actual dynamical situation. For a recent discussion and references, see Ref. [3].

A traditional focus of QCD has been on hard inclusive processes and jet physics where perturbative methods and leading-twist factorization provide predictions up to next-to-next-to leading order (NNLO) with very good precision. More recently, the domain of reliable perturbative QCD predictions has been extended to much more complex phenomena, such as a fundamental understanding of the hard QCD BFKL pomeron in deep inelastic scattering at small x_{bj} and hard diffractive processes, such as $\gamma^* p \to \rho^0 p$. In these lectures I will discuss applications of QCD where the non-perturbative composition of hadrons in terms of their quark and gluon degrees of freedom play a crucial role, for example the x_{bj}-dependence of structure functions measured in deep inelastic scattering, exclusive and semi-exclusive processes such as form factors, two-photon processes, elastic scattering at fixed θ_{cm}, as well as the semi-leptonic decays of heavy hadrons. The analysis of QCD processes at the amplitude level is a challenging problem, mixing issues involving non-perturbative and perturbative dynamics. However, a number of tools are available:

1. The Light-Cone Fock expansion provides a frame-independent representation of a hadrons in terms of a set of wavefunctions $\{\psi_{n/H}(x_i, \vec{k}_{\perp i}, \lambda_i)\}$ describing its composition into relativistic quark and gluon constituents. The light-cone wavefunctions can be derived from the eigensolutions of the QCD Hamiltonian defined at fixed light-cone time $\tau = t + z/c$. Structure functions are obtained from the sum over absolute squares of the light-cone wavefunctions. Spacelike form factors and semi-leptonic decay amplitudes can be written as exact identities in terms of the convolution of the light-cone wavefunctions.

2. Factorization theorems for hard exclusive, semi-exclusive, and diffractive processes allow a rigorous separation of soft non-perturbative dynamics of the bound state hadrons from the hard dynamics of the perturbatively-calculable quark-gluon

scattering amplitude $T_H^{(\Lambda)}$. The key non-perturbative input is the gauge and frame independent hadron distribution amplitude [4] defined as the integral over transverse momenta of the valence (lowest particle number) Fock wavefunction; *e.g.* for the pion

$$\phi_\pi(x_i, Q) \equiv \int d^2 k_\perp \, \psi_{q\bar{q}/\pi}^{(Q)}(x_i, \vec{k}_{\perp i}, \lambda) \tag{1}$$

where the global cutoff Λ is identified with the resolution Q. The distribution amplitude controls leading-twist exclusive amplitudes at high momentum transfer, and it can be related to the gauge-invariant Bethe-Salpeter wavefunction at equal light-cone time $\tau = x^+$. Thus hard exclusive hadronic amplitudes such as quarkonium decay, heavy hadron decay, and scattering amplitudes where the hadrons are scattered with momentum transfer can be factorized as the convolution of the light-cone Fock state wavefunctions with quark-gluon matrix elements [4]

$$\mathcal{M}_{\text{Hadron}} = \prod_H \sum_n \int \prod_{i=1}^n d^2 k_\perp \prod_{i=1}^n dx_i \delta\left(1 - \sum_{i=1}^n x_i\right) \delta\left(\sum_{i=1}^n \vec{k}_{\perp i}\right)$$

$$\times \psi_{n/H}^{(\Lambda)}(x_i, \vec{k}_{\perp i}, \lambda_i) \, T_H^{(\Lambda)} \; . \tag{2}$$

Here $T_H^{(\Lambda)}$ is the underlying quark-gluon subprocess scattering amplitude, where the (incident or final) hadrons are replaced by quarks and gluons with momenta $x_i p^+$, $x_i \vec{p}_\perp + \vec{k}_{\perp i}$ and invariant mass above the separation scale $\mathcal{M}_n^2 > \Lambda^2$.

3. The logarithmic evolution of hadron distribution amplitudes $\phi_H(x_i, Q)$ can be derived from the perturbatively-computable tail of the valence light-cone wavefunction in the high transverse momentum regime. [4]

4. Conformal symmetry provides a template for QCD predictions, leading to relations between observables which are present even in a theory which is not scale invariant. Thus an important guide in QCD analyses is to identify the underlying conformal relations of QCD which are manifest if we drop quark masses and effects due to the running of the QCD couplings. In fact, if QCD has an infrared fixed point (vanishing of the Gell Mann-Low function at low momenta), the theory will closely resemble a scale-free conformally symmetric theory in many applications.

5. Commensurate scale relations are perturbative QCD predictions which relate observable to observable at fixed relative scale, such as the "generalized Crewther relation", which connects the Bjorken and Gross-Llewellyn Smith deep inelastic scattering sum rules to measurements of the e^+e^- annihilation cross section. The relations have no renormalization scale or scheme ambiguity. The coefficients in the perturbative series for commensurate scale relations are identical to those of conformal QCD; thus no infrared renormalons are present. All non-conformal effects are absorbed by fixing the ratio of the respective momentum transfer and energy scales. In the case of fixed-point theories, commensurate scale relations relate both the ratio of couplings and the ratio of scales as the fixed point is approached.

6. α_V Scheme. A natural scheme for defining the QCD coupling in exclusive and other processes is the α_V scheme defined from heavy quark interactions. All vacuum polarization corrections due to fermion pairs are then automatically and analytically incorporated into the Gell Mann-Low function, thus avoiding the problem of explicitly computing and resumming quark mass corrections related to the running of the coupling.

7. The Abelian Correspondence Principle. One can consider QCD predictions as analytic functions of the number of colors N_C and flavors N_F. In particular, one can show at all orders of perturbation theory that PQCD predictions reduce to those of an Abelian theory at $N_C \rightarrow 0$ with $\hat{\alpha} = C_F \alpha_s$ and $\widehat{N}_F = N_F/T C_F$ held fixed. [93] There is thus a deep connection between QCD processes and their corresponding QED analogs.

THE LIGHT-CONE FOCK EXPANSION IN QCD

In a relativistic collision, an incident hadron projectile presents itself as an ensemble of coherent states containing various numbers of quark and gluon quanta. Thus when a laser beam crosses a proton at fixed "light-cone" time $\tau = t + z/c = x^0 + x^z$, it encounters a baryonic state with a given number of quarks, anti-quarks, and gluons in flight with $n_q - n_{\bar{q}} = 3$. The natural formalism for describing these hadronic components in QCD is the light-cone Fock representation obtained by quantizing the theory at fixed τ. [5] For example, the proton state has the Fock expansion

$$
\begin{aligned}
|p\rangle &= \sum_n \langle n \,|\, p \rangle \,|\, n \rangle \\
&= \psi_{3q/p}^{(\Lambda)}(x_i, \vec{k}_{\perp i}, \lambda_i) \,|\, uud \rangle \\
&\quad + \psi_{3qg/p}^{(\Lambda)}(x_i, \vec{k}_{\perp i}, \lambda_i) \,|\, uudg \rangle + \cdots
\end{aligned}
\tag{3}
$$

representing the expansion of the exact QCD eigenstate on a non-interacting quark and gluon basis. The probability amplitude for each such n-particle state of on-mass shell quarks and gluons in a hadron is given by a light-cone Fock state wavefunction $\psi_{n/H}(x_i, \vec{k}_{\perp i}, \lambda_i)$, where the constituents have longitudinal light-cone momentum fractions

$$
x_i = \frac{k_i^+}{p^+} = \frac{k^0 + k_i^z}{p^0 + p^z}, \quad \sum_{i=1}^n x_i = 1 ,
\tag{4}
$$

relative transverse momentum

$$
\vec{k}_{\perp i}, \quad \sum_{i=1}^n \vec{k}_{\perp i} = \vec{0}_\perp ,
\tag{5}
$$

and helicities λ_i. The effective lifetime of each configuration in the laboratory frame is $\frac{2P_{lab}}{\mathcal{M}_n^2 - M_p^2}$ where

$$\mathcal{M}_n^2 = \sum_{i=1}^{n} \frac{k_\perp^2 + m^2}{x} < \Lambda^2 \qquad (6)$$

is the off-shell invariant mass and Λ is a global ultraviolet regulator. The form of $\psi_{n/H}^{(\Lambda)}(x_i, \vec{k}_{\perp i}, \lambda_i)$ is invariant under longitudinal boosts; *i.e.*, the light-cone wavefunctions expressed in the relative coordinates x_i and $k_{\perp i}$ are independent of the total momentum P^+, $\vec{P}_\perp$ of the hadron.

The interactions of the proton reflects an average over the interactions of its fluctuating states. For example, a valence state with small impact separation, and thus a small color dipole moment, would be expected to interact weakly in a hadronic or nuclear target reflecting its color transparency. The nucleus thus filters differentially different hadron components. [6,7] The ensemble $\{\psi_{n/H}\}$ of such light-cone Fock wavefunctions is a key concept for hadronic physics, providing a conceptual basis for representing physical hadrons (and also nuclei) in terms of their fundamental quark and gluon degrees of freedom. Given the $\psi_{n/H}^{(\Lambda)}$, we can construct any spacelike electromagnetic or electroweak form factor from the diagonal overlap of the LC wavefunctions. [8] Similarly, the matrix elements of the currents that define quark and gluon structure functions can be computed from the integrated squares of the LC wavefunctions. [4,9] In general the LC ultraviolet regulators provide a factorization scheme for elastic and inelastic scattering, separating the hard dynamical contributions with invariant mass squared $\mathcal{M}^2 > \Lambda^2$ from the soft physics with $\mathcal{M}^2 \leq \Lambda^2$ which is incorporated in the nonperturbative LC wavefunctions. (Similarly, the DGLAP evolution of quark and gluon distributions can be derived by computing the variation of the Fock expansion with respect to Λ^2. [4])

The light-cone Fock formalism is derived in the following way: one first constructs the light-cone time evolution operator $P^- = P^0 - P^z$ and the invariant mass operator $H_{LC} = P^- P^+ - P_\perp^2$ in light-cone gauge $A^+ = 0$ from the QCD Lagrangian. The total longitudinal momentum $P^+ = P^0 + P^z$ and transverse momenta $\vec{P}_\perp$ are conserved, *i.e.* are independent of the interactions. The matrix elements of H_{LC} on the complete orthonormal basis $\{|n\rangle\}$ of the free theory $H_{LC}^0 = H_{LC}(g = 0)$ can then be constructed. The matrix elements $\langle n|H_{LC}|m\rangle$ connect Fock states differing by 0, 1, or 2 quark or gluon quanta, and they include the instantaneous quark and gluon contributions imposed by eliminating dependent degrees of freedom in light-cone gauge.

It is thus important to not only compute the spectrum of hadrons and gluonic states, but also to determine the wavefunction of each QCD bound state in terms of its fundamental quark and gluon degrees of freedom. If we could obtain such nonperturbative solutions of QCD, then we could compute the quark and gluon structure functions and distribution amplitudes which control hard-scattering inclusive and exclusive reactions as well as calculate the matrix elements of currents which underlie electroweak form factors and the weak decay amplitudes of the light and heavy hadrons. The light-cone wavefunctions also determine the multi-parton correlations which control the distribution of particles in the proton fragmentation region as well as dynamical higher twist effects. Thus one can analyze not only

7

the deep inelastic structure functions but also the fragmentation of the spectator system. Knowledge of hadron wavefunctions would also open a window to a deeper understanding of the physics of QCD at the amplitude level, illuminating exotic effects of the theory such as color transparency, intrinsic heavy quark effects, hidden color, diffractive processes, and the QCD van der Waals interactions.

Solving a quantum field theory such as QCD is clearly not easy. However, highly nontrivial, one-space one-time relativistic quantum field theories which mimic many of the features of QCD, have already been completely solved using light-cone Hamiltonian methods. [5] Virtually any (1+1) quantum field theory can be solved using the method of Discretized Light-Cone-Quantization (DLCQ). [10,11] where the matrix elements $\langle n \mid H_{LC}^{\Lambda} \mid m \rangle$, are made discrete in momentum space by imposing periodic or antiperiodic boundary conditions in $x^- = x^0 - x^z$ and $\vec{x}_\perp$. Upon diagonalization of H_{LC}, the eigenvalues provide the invariant mass of the bound states and eigenstates of the continuum. In DLCQ, the Hamiltonian H_{LC}, which can be constructed from the Lagrangian using light-cone time quantization, is completely diagonalized, in analogy to Heisenberg's solution of the eigenvalue problem in quantum mechanics. The quantum field theory problem is rendered discrete by imposing periodic or antiperiodic boundary conditions. The eigenvalues and eigensolutions of collinear QCD then give the complete spectrum of hadrons, nuclei, and gluonium and their respective light-cone wavefunctions. A beautiful example is "collinear" QCD: a variant of $QCD(3+1)$ defined by dropping all of interaction terms in H_{LC}^{QCD} involving transverse momenta. [12] Even though this theory is effectively two-dimensional, the transversely-polarized degrees of freedom of the gluon field are retained as two scalar fields. Antonuccio and Dalley [13] have used DLCQ to solve this theory. The diagonalization of H_{LC} provides not only the complete bound and continuum spectrum of the collinear theory, including the gluonium states, but it also yields the complete ensemble of light-cone Fock state wavefunctions needed to construct quark and gluon structure functions for each bound state. Although the collinear theory is a drastic approximation to physical $QCD(3+1)$, the phenomenology of its DLCQ solutions demonstrate general gauge theory features, such as the peaking of the wavefunctions at minimal invariant mass, color coherence and the helicity retention of leading partons in the polarized structure functions at $x \to 1$. The solutions of the quantum field theory can be obtained for arbitrary coupling strength, flavors, and colors.

In practice it is essential to introduce an ultraviolet regulator in order to limit the total range of $\langle n \mid H_{LC} \mid m \rangle$, such as the "global" cutoff in the invariant mass of the free Fock state. One can also introduce a "local" cutoff to limit the change in invariant mass $|\mathcal{M}_n^2 - \mathcal{M}_m^2| < \Lambda_{\text{local}}^2$ which provides spectator-independent regularization of the sub-divergences associated with mass and coupling renormalization. Recently, Hiller, McCartor, and I have shown [14] that the Pauli-Villars method has advantages for regulating light-cone quantized Hamitonian theory. We show that Pauli-Villars fields satisfying three spectral conditions will regulate the interactions

8

in the ultraviolet, while at same time avoiding spectator-dependent renormalization and preserving chiral symmetry. Although gauge theories are usually quantized on the light-cone in light-cone gauge $A^+ = 0$, it is also possible and interesting to quantize the theory in Feynman gauge [15]. Covariant gauges are advantageous since they preserve the rotational symmetry of the gauge interactions.

The natural renormalization scheme for the QCD coupling is $\alpha_V(Q)$, the effective charge defined from the scattering of two infinitely-heavy quark test charges. This is discussed in more detail below. The renormalization scale can then be determined from the virtuality of the exchanged momentum, as in the BLM and commensurate scale methods. [16–19] Similar effective charges have been proposed by Watson [20] and Czarnecki et $al.$ [21]

In principle, we could also construct the wavefunctions of QCD(3+1) starting with collinear QCD(1+1) solutions by systematic perturbation theory in ΔH, where ΔH contains the terms which produce particles at non-zero $k_\perp$, including the terms linear and quadratic in the transverse momenta $\vec{k}_{\perp i}$ which are neglected in the Hamilton H_0 of collinear QCD. We can write the exact eigensolution of the full Hamiltonian as

$$\psi_{(3+1)} = \psi_{(1+1)} + \frac{1}{M^2 - H + i\epsilon} \Delta H \, \psi_{(1+1)} \, ,$$

where

$$\frac{1}{M^2 - H + i\epsilon} = \frac{1}{M^2 - H_0 + i\epsilon} + \frac{1}{M^2 - H + i\epsilon} \Delta H \frac{1}{M^2 - H_0 + i\epsilon}$$

can be represented as the continued iteration of the Lippmann Schwinger resolvant. Note that the matrix $(M^2 - H_0)^{-1}$ is known to any desired precision from the DLCQ solution of collinear QCD.

ELECTROWEAK MATRIX ELEMENTS AND LIGHT-CONE WAVEFUNCTIONS

Dae Sung Hwang and I have recently shown that exclusive semileptonic B-decay amplitudes, such as $B \to A\ell\bar{\nu}$ can be evaluated exactly in the light-cone formalism. [22] These timelike decay matrix elements require the computation of the diagonal matrix element $n \to n$ where parton number is conserved, and the off-diagonal $n + 1 \to n - 1$ convolution where the current operator annihilates a $q\bar{q}'$ pair in the initial B wavefunction. See Fig. 1. This term is a consequence of the fact that the time-like decay $q^2 = (p_\ell + p_{\bar{\nu}})^2 > 0$ requires a positive light-cone momentum fraction $q^+ > 0$. Conversely for space-like currents, one can choose $q^+ = 0$, as in the Drell-Yan-West representation of the space-like electromagnetic form factors. However, as can be seen from the explicit analysis of the form factor in a perturbation model, the off-diagonal convolution can yield a nonzero q^+/q^+ limiting form as $q^+ \to 0$.

9

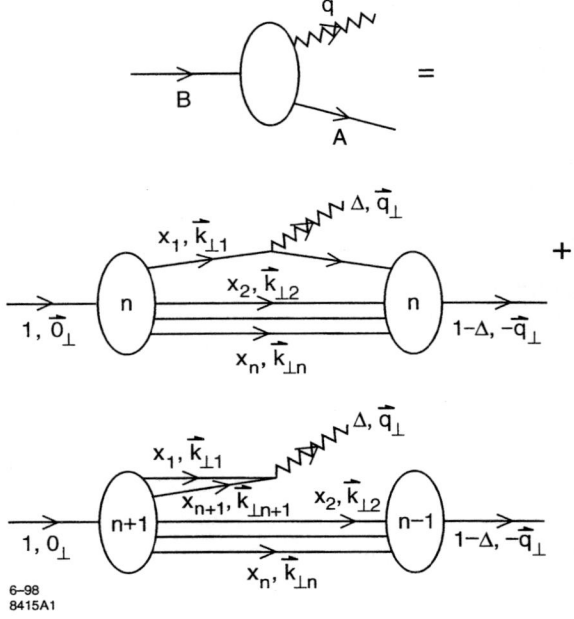

FIGURE 1. Exact representation of electroweak decays and time-like form factors in the light-cone Fock representation.

This extra term appears specifically in the case of "bad" currents such as J^- in which the coupling to $q\bar{q}$ fluctuations in the light-cone wavefunctions are favored. In effect, the $q^+ \to 0$ limit generates $\delta(x)$ contributions as residues of the $n+1 \to n-1$ contributions. The necessity for such "zero mode" $\delta(x)$ terms has been noted by Chang, Root and Yan, [23],Burkardt, [24] and Ji and Choi. [25]

The off-diagonal $n + 1 \to n - 1$ contributions give a new perspective for the physics of B-decays. A semileptonic decay involves not only matrix elements where a quark changes flavor, but also a contribution where the leptonic pair is created from the annihilation of a $q\bar{q}'$ pair within the Fock states of the initial B wavefunction. The semileptonic decay thus can occur from the annihilation of a nonvalence quark-antiquark pair in the initial hadron. This feature will carry over to exclusive hadronic B-decays, such as $B^0 \to \pi^- D^+$. In this case the pion can be produced from the coalescence of a $d\bar{u}$ pair emerging from the initial higher particle number Fock wavefunction of the B. The D meson is then formed from the remaining quarks after the internal exchange of a W boson.

In principle, a precise evaluation of the hadronic matrix elements needed for B-decays and other exclusive electroweak decay amplitudes requires knowledge of all of the light-cone Fock wavefunctions of the initial and final state hadrons. In the case of model gauge theories such as QCD(1+1) [26] or collinear QCD [13] in one-space and one-time dimensions, the complete evaluation of the light-cone wavefunction is possible for each baryon or meson bound-state using the DLCQ method. It would be interesting to use such solutions as a model for physical B-decays.

The existence of an exact formalism for electroweak matrix elements gives a basis for systematic approximations and a control over neglected terms. For example, one can analyze exclusive semileptonic B-decays which involve hard internal momentum transfer using a perturbative QCD formalism patterned after the analysis of form factors at large momentum transfer. [4] The hard-scattering analysis proceeds by writing each hadronic wavefunction as a sum of soft and hard contributions

$$\psi_n = \psi_n^{\text{soft}}(\mathcal{M}_n^2 < \Lambda^2) + \psi_n^{\text{hard}}(\mathcal{M}_n^2 > \Lambda^2), \tag{7}$$

where $\mathcal{M}_n^2$ is the invariant mass of the partons in the n-particle Fock state and Λ is the separation scale. The high internal momentum contributions to the wavefunction ψ_n^{hard} can be calculated systematically from QCD perturbation theory by iterating the gluon exchange kernel. The contributions from high momentum transfer exchange to the B-decay amplitude can then be written as a convolution of a hard scattering quark-gluon scattering amplitude T_H with the distribution amplitudes $\phi(x_i, \Lambda)$, the valence wavefunctions obtained by integrating the constituent momenta up to the separation scale $\mathcal{M}_n < \Lambda < Q$. This is the basis for the perturbative hard scattering analyses. [27–30] In the exact analysis, one can identify the hard PQCD contribution as well as the soft contribution from the convolution of the light-cone wavefunctions. Furthermore, the hard scattering contribution can be systematically improved. For example, off-shell effects can be retained in the evaluation of T_H by utilizing the exact light-cone energy denominators.

Given the solution for the hadronic wavefunctions $\psi_n^{(\Lambda)}$ with $\mathcal{M}_n^2 < \Lambda^2$, one can construct the wavefunction in the hard regime with $\mathcal{M}_n^2 > \Lambda^2$ using projection operator techniques. [4] The construction can be done perturbatively in QCD since only high invariant mass, far off-shell matrix elements are involved. One can use this method to derive the physical properties of the LC wavefunctions and their matrix elements at high invariant mass. Since $\mathcal{M}_n^2 = \sum_{i=1}^n \left(\frac{k_\perp^2 + m^2}{x}\right)_i$, this method also allows the derivation of the asymptotic behavior of light-cone wavefunctions at large $k_\perp$, which in turn leads to predictions for the fall-off of form factors and other exclusive matrix elements at large momentum transfer, such as the quark counting rules for predicting the nominal power-law fall-off of two-body scattering amplitudes at fixed θ_{cm}. [9] The phenomenological successes of these rules can be understood within QCD if the coupling $\alpha_V(Q)$ freezes in a range of relatively small momentum transfer. [19]

11

OTHER APPLICATIONS OF LIGHT-CONE QUANTIZATION TO QCD PHENOMENOLOGY

Diffractive vector meson photoproduction. The light-cone Fock wavefunction representation of hadronic amplitudes allows a simple eikonal analysis of diffractive high energy processes, such as $\gamma^*(Q^2)p \to \rho p$, in terms of the virtual photon and the vector meson Fock state light-cone wavefunctions convoluted with the $gp \to gp$ near-forward matrix element. [31] One can easily show that only small transverse size $b_\perp \sim 1/Q$ of the vector meson distribution amplitude is involved. The hadronic interactions are minimal, and thus the $\gamma^*(Q^2)N \to \rho N$ reaction can occur coherently throughout a nuclear target in reactions without absorption or shadowing. The $\gamma^* A \to VA$ process thus is a laboratory for testing QCD color transparency. [32] This is discussed further in the next section.

Regge behavior of structure functions. The light-cone wavefunctions $\psi_{n/H}$ of a hadron are not independent of each other, but rather are coupled via the equations of motion. Antonuccio, Dalley and I [33] have used the constraint of finite "mechanical" kinetic energy to derive "ladder relations" which interrelate the light-cone wavefunctions of states differing by one or two gluons. We then use these relations to derive the Regge behavior of both the polarized and unpolarized structure functions at $x \to 0$, extending Mueller's derivation of the BFKL hard QCD pomeron from the properties of heavy quarkonium light-cone wavefunctions at large N_C QCD. [34]

Structure functions at large x_{bj}. The behavior of structure functions where one quark has the entire momentum requires the knowledge of LC wavefunctions with $x \to 1$ for the struck quark and $x \to 0$ for the spectators. This is a highly off-shell configuration, and thus one can rigorously derive quark-counting and helicity-retention rules for the power-law behavior of the polarized and unpolarized quark and gluon distributions in the $x \to 1$ endpoint domain. It is interesting to note that the evolution of structure functions is minimal in this domain because the struck quark is highly virtual as $x \to 1$; *i.e.* the starting point Q_0^2 for evolution cannot be held fixed, but must be larger than a scale of order $(m^2 + k_\perp^2)/(1-x)$. [4,9,35]

Intrinsic gluon and heavy quarks. The main features of the heavy sea quark-pair contributions of the Fock state expansion of light hadrons can also be derived from perturbative QCD, since $\mathcal{M}_n^2$ grows with m_Q^2. One identifies two contributions to the heavy quark sea, the "extrinsic" contributions which correspond to ordinary gluon splitting, and the "intrinsic" sea which is multi-connected via gluons to the valence quarks. The intrinsic sea is thus sensitive to the hadronic bound state structure. [36] The maximal contribution of the intrinsic heavy quark occurs at $x_Q \simeq m_{\perp Q}/\sum_i m_\perp$ where $m_\perp = \sqrt{m^2 + k_\perp^2}$; *i.e.* at large x_Q, since this minimizes the invariant mass $\mathcal{M}_n^2$. The measurements of the charm structure function by the EMC experiment are consistent with intrinsic charm at large x in the nucleon with a probability of order $0.6 \pm 0.3\%$. [37] Similarly, one can distinguish intrinsic gluons which are associated with multi-quark interactions and extrinsic gluon contributions associated with quark substructure. [38] One can also use this framework to isolate the physics of the anomaly contribution to the Ellis-Jaffe sum rule.

Materialization of far-off-shell configurations. In a high energy hadronic collisions, the highly-virtual states of a hadron can be materialized into physical hadrons simply by the soft interaction of any of the constituents. [39] Thus a proton state with intrinsic charm $|uudc\bar{c}\rangle$ can be materialized, producing a J/ψ at large x_F, by the interaction of a light-quark in the target. The production occurs on the front-surface of a target nucleus, implying an $A^{2/3}$ J/ψ production cross section at large x_F, which is consistent with experiment, such as Fermilab experiments E772 and E866.

Rearrangement mechanism in heavy quarkonium decay. It is usually assumed that a heavy quarkonium state such as the J/ψ always decays to light hadrons via the annihilation of its heavy quark constituents to gluons. However, as Karliner and I [40] have recently shown, the transition $J/\psi \to \rho\pi$ can also occur by the rearrangement of the $c\bar{c}$ from the J/ψ into the $|q\bar{q}c\bar{c}\rangle$ intrinsic charm Fock state of the ρ or π. On the other hand, the overlap rearrangement integral in the decay $\psi' \to \rho\pi$ will be suppressed since the intrinsic charm Fock state radial wavefunction of the light hadrons will evidently not have nodes in its radial wavefunction. This observation gives a natural explanation of the long-standing puzzle why the J/ψ decays prominently to two-body pseudoscalar-vector final states, whereas the ψ' does not.

Asymmetry of intrinsic heavy quark sea. The higher Fock state of the proton $|uuds\bar{s}\rangle$ should resemble a $|K\Lambda\rangle$ intermediate state, since this minimizes its invariant mass $\mathcal{M}$. In such a state, the strange quark has a higher mean momentum fraction x than the $\bar{s}$. [41–43] Similarly, the helicity intrinsic strange quark in this configuration will be anti-aligned with the helicity of the nucleon. [41,43] This $Q \leftrightarrow \overline{Q}$ asymmetry is a striking feature of the intrinsic heavy-quark sea.

Comover phenomena. Light-cone wavefunctions describe not only the partons that interact in a hard subprocess but also the associated partons freed from the projectile. The projectile partons which are comoving (*i.e.*, which have similar rapidity) with final state quarks and gluons can interact strongly producing (a) leading particle effects, such as those seen in open charm hadroproduction; (b) suppression of quarkonium [44] in favor of open heavy hadron production, as seen in the E772 experiment; (c) changes in color configurations and selection rules in quarkonium hadroproduction, as has been emphasized by Hoyer and Peigne. [45] All of these effects violate the usual ideas of factorization for inclusive reactions. Further, more than one parton from the projectile can enter the hard subprocess, producing dynamical higher twist contributions, as seen for example in Drell-Yan experiments. [46,47]

Jet hadronization in light-cone QCD. One of the goals of nonperturbative analysis in QCD is to compute jet hadronization from first principles. The DLCQ solutions provide a possible method to accomplish this. By inverting the DLCQ solutions, we can write the "bare" quark state of the free theory as $|q_0\rangle = \sum |n\rangle \langle n|q_0\rangle$ where now $\{|n\rangle\}$ are the exact DLCQ eigenstates of H_{LC}, and $\langle n|q_0\rangle$ are the DLCQ projections of the eigensolutions. The expansion in automatically infrared and ultraviolet regulated if we impose global cutoffs on the DLCQ basis: $\lambda^2 < \Delta \mathcal{M}_n^2 < \Lambda^2$ where $\Delta \mathcal{M}_n^2 = \mathcal{M}_n^2 - (\Sigma \mathcal{M}_i)^2$. It would be interesting to study jet hadronization at the amplitude level for the existing DLCQ solutions to QCD (1+1) and collinear QCD.

Hidden Color. The deuteron form factor at high Q^2 is sensitive to wavefunction configurations where all six quarks overlap within an impact separation $b_{\perp i} < \mathcal{O}(1/Q)$; the leading power-law fall off predicted by QCD is $F_d(Q^2) = f(\alpha_s(Q^2))/(Q^2)^5$, where, asymptotically, $f(\alpha_s(Q^2)) \propto \alpha_s(Q^2)^{5+2\gamma}$. [48] The derivation of the evolution equation for the deuteron distribution amplitude and its leading anomalous dimension γ is given in Ref. [49] In general, the six-quark wavefunction of a deuteron is a mixture of five different color-singlet states. The dominant color configuration at large distances corresponds to the usual proton-neutron bound state. However at small impact space separation, all five Fock color-singlet components eventually acquire equal weight, *i.e.*, the deuteron wavefunction evolves to 80% "hidden color." The relatively large normalization of the deuteron form factor observed at large Q^2 points to sizable hidden color contributions. [50]

Spin-Spin Correlations in Nucleon-Nucleon Scattering and the Charm Threshold. One of the most striking anomalies in elastic proton-proton scattering is the large spin correlation A_{NN} observed at large angles. [51] At $\sqrt{s} \simeq 5$ GeV, the rate for scattering with incident proton spins parallel and normal to the scattering plane is four times larger than that for scattering with anti-parallel polarization. This strong polarization correlation can be attributed to the onset of charm production in the intermediate state at this energy. [52] The intermediate state $|uuduudc\bar{c}\rangle$ has odd intrinsic parity and couples to the $J = S = 1$ initial state, thus strongly enhancing scattering when the incident projectile and target protons have their spins parallel and normal to the scattering plane. The charm threshold can also explain the anomalous change in color transparency observed at the same energy in quasi-elastic pp scattering. A crucial test is the observation of open charm production near threshold with a cross section of order of 1μb.

HARD EXCLUSIVE REACTIONS

Exclusive hard-scattering reactions and hard diffractive reactions are now giving a valuable window into the structure and dynamics of hadronic amplitudes. Recent measurements of the photon-to-pion transition form factor at CLEO, [53] the diffractive dissociation of pions into jets at Fermilab, [54] diffractive vector meson leptoproduction at Fermilab and HERA, and the new program of experiments on exclusive proton and deuteron processes at Jefferson Laboratory are now yielding fundamental information on hadronic wavefunctions, particularly the distribution amplitude of mesons. Such information is also critical for interpreting exclusive heavy hadron decays and the matrix elements and amplitudes entering CP-violating processes at the B factories.

14

There has been much progress analyzing exclusive and diffractive reactions at large momentum transfer from first principles in QCD. Rigorous statements can be made on the basis of asymptotic freedom and factorization theorems which separate the underlying hard quark and gluon subprocess amplitude from the nonperturbative physics incorporated into the process-independent hadron distribution amplitudes $\phi_H(x_i, Q)$, [4] the valence light-cone wavefunctions integrated over $k_\perp^2 < Q^2$. An important new application is the recent analysis of hard exclusive B decays by Beneke, *et al.* [55] Key features of such analyses are: (a) evolution equations for distribution amplitudes which incorporate the operator product expansion, renormalization group invariance, and conformal symmetry; [4,56–60] (b) hadron helicity conservation which follows from the underlying chiral structure of QCD; [61] (c) color transparency, which eliminates corrections to hard exclusive amplitudes from initial and final state interactions at leading power and reflects the underlying gauge theoretic basis for the strong interactions; [32,62] and (d) hidden color degrees of freedom in nuclear wavefunctions, which reflects the color structure of hadron and nuclear wavefunctions. [49] There have also been recent advances eliminating renormalization scale ambiguities in hard-scattering amplitudes via commensurate scale relations [63–65] which connect the couplings entering exclusive amplitudes to the α_V coupling which controls the QCD heavy quark potential. [66] The postulate that the QCD coupling has an infrared fixed-point can explain the applicability of conformal scaling and dimensional counting rules to physical QCD processes. [67,68,66] The field of analyzable exclusive processes has recently been expanded to a new range of QCD processes, such as electroweak decay amplitudes, highly virtual diffractive processes such as $\gamma^* p \to \rho p$, [31,69] and semi-exclusive processes such as $\gamma^* p \to \pi^+ X$ [70–72] where the π^+ is produced in isolation at large p_T.

The natural renormalization scheme for the QCD coupling in hard exclusive processes is $\alpha_V(Q)$, the effective charge defined from the scattering of two infinitely-heavy quark test charges. The renormalization scale can then be determined from the virtuality of the exchanged momentum of the gluons, as in the BLM and commensurate scale methods. [16,63–65]

The main features of exclusive processes to leading power in the transferred momenta are:

(1) The leading power fall-off is given by dimensional counting rules for the hard-scattering amplitude: $T_H \sim 1/Q^{n-1}$, where n is the total number of fields (quarks, leptons, or gauge fields) participating in the hard scattering. [67,68] Thus the reaction is dominated by subprocesses and Fock states involving the minimum number of interacting fields. The hadronic amplitude follows this fall-off modulo logarithmic corrections from the running of the QCD coupling, and the evolution of the hadron distribution amplitudes. In some cases, such as large angle $pp \to pp$ scattering, pinch contributions from multiple hard-scattering processes must also be included. [73] The general success of dimensional counting rules implies that the effective coupling $\alpha_V(Q^*)$ controlling the gluon exchange propagators in T_H are frozen in the infrared, *i.e.*, have an infrared fixed point, since the effective momentum transfers Q^* exchanged by the gluons are often a small fraction of the

overall momentum transfer. [66] The pinch contributions are suppressed by a factor decreasing faster than a fixed power. [67]

(2) The leading power dependence is given by hard-scattering amplitudes T_H which conserve quark helicity. [61,74] Since the convolution of T_H with the light-cone wavefunctions projects out states with $L_z = 0$, the leading hadron amplitudes conserve hadron helicity; *i.e.*, the sum of initial and final hadron helicities are conserved.

(3) Since the convolution of the hard scattering amplitude T_H with the light-cone wavefunctions projects out the valence states with small impact parameter, the essential part of the hadron wavefunction entering a hard exclusive amplitude has a small color dipole moment. This leads to the absence of initial or final state interactions among the scattering hadrons as well as the color transparency. of quasi-elastic interactions in a nuclear target. [32,62] For example, the amplitude for diffractive vector meson photoproduction $\gamma^*(Q^2)p \to \rho p$, can be written as convolution of the virtual photon and the vector meson Fock state light-cone wave-functions the $gp \to gp$ near-forward matrix element. [31] One can easily show that only small transverse size $b_\perp \sim 1/Q$ of the vector meson distribution amplitude is involved. The sum over the interactions of the exchanged gluons tend to cancel reflecting its small color dipole moment. Since the hadronic interactions are minimal, the $\gamma^*(Q^2)N \to \rho N$ reaction at large Q^2 can occur coherently throughout a nuclear target in reactions without absorption or final state interactions. The $\gamma^*A \to VA$ process thus provides a natural framework for testing QCD color transparency. Evidence for color transparency in such reactions has been found by Fermilab experiment E665. [75]

Diffractive multi-jet production in heavy nuclei provides a novel way to measure the shape of the LC Fock state wavefunctions and test color transparency. For example, consider the reaction [6,7,76] $\pi A \to \text{Jet}_1 + \text{Jet}_2 + A'$ at high energy where the nucleus A' is left intact in its ground state. The transverse momenta of the jets have to balance so that $\vec{k}_{\perp i} + \vec{k}_{\perp 2} = \vec{q}_\perp < R^{-1}{}_A$, and the light-cone longitudinal momentum fractions have to add to $x_1 + x_2 \sim 1$ so that $\Delta p_L < R_A^{-1}$. The process can then occur coherently in the nucleus. Because of color transparency, *i.e.*, the cancelation of color interactions in a small-size color-singlet hadron, the valence wavefunction of the pion with small impact separation will penetrate the nucleus with minimal interactions, diffracting into jet pairs. [6] The $x_1 = x$, $x_2 = 1 - x$ dependence of the di-jet distributions will thus reflect the shape of the pion distribution amplitude; the $\vec{k}_{\perp 1} - \vec{k}_{\perp 2}$ relative transverse momenta of the jets also gives key information on the underlying shape of the valence pion wavefunction. [7,76] The QCD analysis can be confirmed by the observation that the diffractive nuclear amplitude extrapolated to $t = 0$ is linear in nuclear number A, as predicted by QCD color transparency. The integrated diffractive rate should scale as $A^2/R_A^2 \sim A^{4/3}$. A diffractive dissociation experiment of this type, E791, is now in progress at Fermilab using 500 GeV incident pions on nuclear targets. [54] The preliminary results from E791 appear to be consistent with color transparency. The momentum fraction distribution of the jets is consistent with a valence light-cone wavefunction of the pion consistent with the shape of the asymptotic distribution amplitude, $\phi_\pi^{\text{asympt}}(x) = \sqrt{3} f_\pi x(1-x)$. As discussed below, data from CLEO [53] for the $\gamma\gamma^* \to \pi^0$ transition form factor also favor a form for the pion distribution amplitude close to the asymptotic solution [4] to the perturbative QCD evolution equation. [77,78,66,79,80] It will also be interesting to study diffractive tri-jet production using proton beams $pA \to \text{Jet}_1 + \text{Jet}_2 + \text{Jet}_3 + A'$ to determine the fundamental shape of the 3-quark structure of the valence light-cone wavefunction of the nucleon at small transverse separation. [7] One interesting possibility is that the distribution amplitude of the $\Delta(1232)$ for $J_z = 1/2, 3/2$ is close to the asymptotic form $x_1 x_2 x_3$, but that the proton distribution amplitude is more complex. This would explain why the $p \to \Delta$ transition form factor appears to fall faster at large Q^2 than the elastic $p \to p$ and the other $p \to N^*$ transition form factors. [81] Conversely, one can use incident real and virtual photons: $\gamma^* A \to \text{Jet}_1 + \text{Jet}_2 + A'$ to confirm the shape of the calculable light-cone wavefunction for transversely-polarized and longitudinally-polarized virtual photons. Such experiments will open up a direct window on the amplitude structure of hadrons at short distances.

There are a large number of measured exclusive reactions in which the empirical power law fall-off predicted by dimensional counting and PQCD appears to be accurate over a large range of momentum transfer. These include processes such as the proton form factor, time-like meson pair production in e^+e^- and $\gamma\gamma$ annihilation, large-angle scattering processes such as pion photoproduction $\gamma p \to \pi^+ p$, and nuclear processes such as the deuteron form factor at large momentum transfer and deuteron photodisintegration. [48] A spectacular example is the recent measurements at CESR of the photon to pion transition form factor in the reaction $e\gamma \to e\pi^0$. [53] As predicted by leading twist QCD [4] $Q^2 F_{\gamma\pi^0}(Q^2)$ is essentially constant for $1 \text{ GeV}^2 < Q^2 < 10 \text{ GeV}^2$. Further, the normalization is consistent with QCD at NLO if one assumes that the pion distribution amplitude takes on the form $\phi_\pi^{\text{asympt}}(x) = \sqrt{3} f_\pi x(1-x)$ which is the asymptotic solution [4] to the evolution equation for the pion distribution amplitude. [77,78,66,80]

The measured deuteron form factor and the deuteron photodisintegration cross section appear to follow the leading-twist QCD predictions at large momentum transfers in the few GeV region. [82,83] The normalization of the measured deuteron form factor is large compared to model calculations [50] assuming that the deuteron's six-quark wavefunction can be represented at short distances with the color structure of two color singlet baryons. This provides indirect evidence for the presence of hidden color components as required by PQCD. [49]

If the pion distribution amplitude is close to its asymptotic form, then one can predict the normalization of exclusive amplitudes such as the spacelike pion form factor $Q^2 F_\pi(Q^2)$. Next-to-leading order predictions are now becoming available which incorporate higher order corrections to the pion distribution amplitude as well as the hard scattering amplitude. [58,84,85] However, the normalization of the PQCD prediction for the pion form factor depends directly on the value of the effective coupling $\alpha_V(Q^*)$ at momenta $Q^{*2} \simeq Q^2/20$. Assuming $\alpha_V(Q^*) \simeq 0.4$, the QCD

17

LO prediction appears to be smaller by approximately a factor of 2 compared to the presently available data extracted from the original pion electroproduction experiments from CEA. [86] A definitive comparison will require a careful extrapolation to the pion pole and extraction of the longitudinally polarized photon contribution of the $ep \to \pi^+ n$ data.

An important debate has centered on whether processes such as the pion and proton form factors and elastic Compton scattering $\gamma p \to \gamma p$ might be dominated by higher twist mechanisms until very large momentum transfers. [87–89] For example, if one assumes that the light-cone wavefunction of the pion has the form $\psi_{\text{soft}}(x, k_\perp) = A \exp(-b \frac{k_\perp^2}{x(1-x)})$, then the Feynman endpoint contribution to the overlap integral at small $k_\perp$ and $x \simeq 1$ will dominate the form factor compared to the hard-scattering contribution until very large Q^2. However, the above form of $\psi_{\text{soft}}(x, k_\perp)$ has no suppression at $k_\perp = 0$ for any x; i.e., the wavefunction in the hadron rest frame does not fall-off at all for $k_\perp = 0$ and $k_z \to -\infty$. Thus such wavefunctions do not represent well soft QCD contributions. Furthermore, endpoint contributions will be suppressed by the QCD Sudakov form factor, reflecting the fact that a near-on-shell quark must radiate if it absorbs large momentum. If the endpoint contribution dominates proton Compton scattering, then both photons will interact on the same quark line in a local fashion and the amplitude is real, in strong contrast to the QCD predictions which have a complex phase structure. The perturbative QCD predictions [90] for the Compton amplitude phase can be tested in virtual Compton scattering by interference with Bethe-Heitler processes. [91] It should be noted that there is no apparent endpoint contribution which could explain the success of dimensional counting in large angle pion photoproduction.

It is interesting to compare the corresponding calculations of form factors of bound states in QED. The soft wavefunction is the Schrödinger-Coulomb solution $\psi_{1s}(\vec{k}) \propto (1 + \vec{p}^2/(\alpha m_{\text{red}})^2)^{-2}$, and the full wavefunction, which incorporates transversely polarized photon exchange, only differs by a factor $(1 + \vec{p}^2/m_{\text{red}}^2)$. Thus the leading twist dominance of form factors in QED occurs at relativistic scales $Q^2 > m_{\text{red}}^2$. [92] Furthermore, there are no extra relative factors of α in the hard-scattering contribution. If the QCD coupling α_V has an infrared fixed point, then the fall-off of the valence wavefunctions of hadrons will have analogous power-law forms, consistent with the Abelian correspondence principle. [93] If power-law wavefunctions are indeed applicable to the soft domain of QCD then, the transition to leading-twist power law behavior will occur in the nominal hard perturbative QCD domain where $Q^2 \gg \langle k_\perp^2 \rangle, m_q^2$.

SEMI-EXCLUSIVE PROCESSES: NEW PROBES OF HADRON STRUCTURE

A new class of hard "semi-exclusive" processes of the form $A + B \to C + Y$, have been proposed as new probes of QCD. [72,70,71] These processes are characterized by a large momentum transfer $t = (p_A - p_C)^2$ and a large rapidity gap between the final state particle C and the inclusive system Y. Here A, B and C can be hadrons or (real or virtual) photons. The cross sections for such processes factorize in terms of the distribution amplitudes of A and C and the parton distributions in the target B. Because of this factorization semi-exclusive reactions provide a novel array of generalized currents, which not only give insight into the dynamics of hard scattering QCD processes, but also allow experimental access to new combinations of the universal quark and gluon distributions.

QCD scattering amplitude for deeply virtual exclusive processes like Compton scattering $\gamma^* p \rightarrow \gamma p$ and meson production $\gamma^* p \rightarrow M p$ factorizes into a hard subprocess and soft universal hadronic matrix elements. [94,69,31] For example, consider exclusive meson electroproduction such as $ep \rightarrow e\pi^+ n$ (Fig. 2a). Here one takes (as in DIS) the Bjorken limit of large photon virtuality, with $x_B = Q^2/(2m_p\nu)$ fixed, while the momentum transfer $t = (p_p - p_n)^2$ remains small. These processes involve 'skewed' parton distributions, which are generalizations of the usual parton distributions measured in DIS. The skewed distribution in Fig. 2a describes the emission of a u-quark from the proton target together with the formation of the final neutron from the d-quark and the proton remnants. As the subenergy $\hat{s}$ of the scattering process $\gamma^* u \rightarrow \pi^+ d$ is not fixed, the amplitude involves an integral over the u-quark momentum fraction x.

An essential condition for the factorization of the deeply virtual meson production amplitude of Fig. 2a is the existence of a large rapidity gap between the produced meson and the neutron. This factorization remains valid if the neutron is replaced with a hadronic system Y of invariant mass $M_Y^2 \ll W^2$, where W is the c.m. energy of the $\gamma^* p$ process. For $M_Y^2 \gg m_p^2$ the momentum k' of the d-quark in Fig. 2b is large with respect to the proton remnants, and hence it forms a jet. This jet hadronizes independently of the other particles in the final state if it is not in the direction of the meson, $i.e.$, if the meson has a large transverse momentum $q'_\perp = \Delta_\perp$ with respect to the photon direction in the $\gamma^* p$ c.m. Then the cross section for an inclusive system Y can be calculated as in DIS, by treating the d-quark as a final state particle.

The large $\Delta_\perp$ furthermore allows only transversally compact configurations of the projectile A to couple to the hard subprocess H of Fig. 2b, as it does in exclusive processes. [4] Hence the above discussion applies not only to incoming virtual photons at large Q^2, but also to real photons ($Q^2 = 0$) and in fact to any hadron projectile.

Let us then consider the general process $A + B \rightarrow C + Y$, where B and C are hadrons or real photons, while the projectile A can also be a virtual photon. In the semi-exclusive kinematic limit $\Lambda_{QCD}^2, M_B^2, M_C^2 \ll M_Y^2, \Delta_\perp^2 \ll W^2$ we have a large rapidity gap $|y_C - y_d| = \log \frac{W^2}{\Delta_\perp^2 + M_Y^2}$ between C and the parton d produced in the hard scattering (see Fig. 2c). The cross section then factorizes into the form

$$\frac{d\sigma}{dt\,dx_S}(A + B \rightarrow C + Y)$$

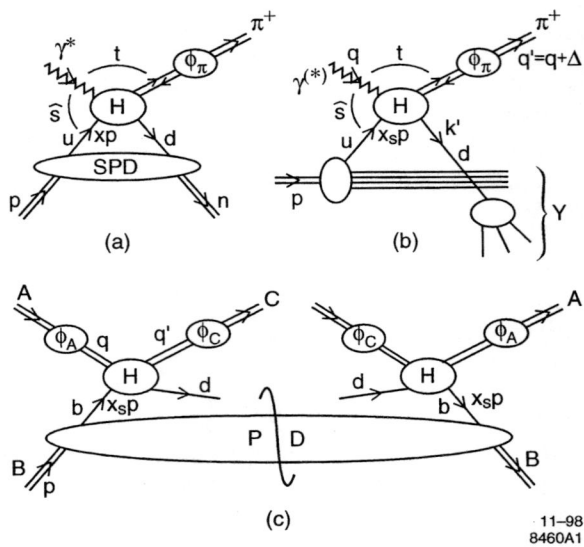

(a)

(b)

(c)

11–98
8460A1

FIGURE 2. (a): Factorization of $\gamma^* p \to \pi^+ n$ into a skewed parton distribution (SPD), a hard scattering H and the pion distribution amplitude ϕ_π. (b): Semi-exclusive process $\gamma^{(*)} p \to \pi^+ Y$. The d-quark produced in the hard scattering H hadronizes independently of the spectator partons in the proton. (c): Diagram for the cross section of a generic semi-exclusive process. It involves a hard scattering H, distribution amplitudes ϕ_A and ϕ_C and a parton distribution (PD) in the target B.

$$= \sum_b f_{b/B}(x_S, \mu^2) \frac{d\sigma}{dt}(Ab \to Cd) , \qquad (8)$$

where $t = (q - q')^2$ and $f_{b/B}(x_S, \mu^2)$ denotes the distribution of quarks, antiquarks and gluons b in the target B. The momentum fraction x_S of the struck parton b is fixed by kinematics to the value $x_S = \frac{-t}{M_Y^2 - t}$ and the factorization scale μ^2 is characteristic of the hard subprocess $Ab \to Cd$.

It is conceptually helpful to regard the hard scattering amplitude H in Fig. 2c as a generalized current of momentum $q - q' = p_A - p_C$, which interacts with the target parton b. For $A = \gamma^*$ we obtain a close analogy to standard DIS when particle C is removed. With $q' \to 0$ we thus find $-t \to Q^2$, $M_Y^2 \to W^2$, and see that x_S goes over into $x_B = Q^2/(W^2 + Q^2)$. The possibility to control the value of q' (and hence the momentum fraction x_S of the struck parton) as well as the quantum numbers of particles A and C should make semi-exclusive processes a versatile tool for studying hadron structure. The cross section further depends on the distribution amplitudes ϕ_A, ϕ_C (*c.f.* Fig. 2c), allowing new ways of measuring these quantities.

CONFORMAL SYMMETRY AS A TEMPLATE

Testing quantum chromodynamics to high precision is not easy. Even in processes involving high momentum transfer, perturbative QCD predictions are complicated by questions of the convergence of the series, particularly by the presence of "renormalon" terms which grow as $n!$, reflecting the uncertainty in the analytic form of the QCD coupling at low scales. Virtually all QCD processes are complicated by the presence of dynamical higher twist effects, including power-law suppressed contributions due to multi-parton correlations, intrinsic transverse momentum, and finite quark masses. Many of these effects are inherently nonperturbative in nature and require knowledge of hadron wavefunction themselves. The problem of interpreting perturbative QCD predictions is further compounded by theoretical ambiguities due to the apparent freedom in the choice of renormalization schemes, renormalizations scales, and factorization procedures.

A central principle of renormalization theory is that predictions which relate physical observables to each other cannot depend on theoretical conventions. For example, one can use any renormalization scheme, such as the modified minimal subtraction dimensional regularization scheme, and any choice of renormalization scale μ to compute perturbative series observables A and B. However, all traces of the choices of the renormalization scheme and scale must disappear when we algebraically eliminate the $\alpha_{\overline{MS}}(\mu)$ and directly relate A to B. This is the principle underlying "commensurate scale relations" (CSR) [17], which are general leading-twist QCD predictions relating physical observables to each other. For example, the "generalized Crewther relation", which is discussed in more detail below, provides a scheme-independent relation between the QCD corrections to the Bjorken (or Gross Llewellyn-Smith) sum rule for deep inelastic lepton-nucleon scattering, at a given momentum transfer Q, to the radiative corrections to the annihilation cross section $\sigma_{e^+e^- \to \text{hadrons}}(s)$, at a corresponding "commensurate" energy scale $\sqrt{s}$. [17,95] The specific relation between the physical scales Q and $\sqrt{s}$ reflects the fact that the radiative corrections to each process have distinct quark mass thresholds.

Any perturbatively calculable physical quantity can be used to define an effective charge [96–98] by incorporating the entire radiative correction into its definition. For example, the $e^+e^-\gamma^* \to$ hadrons annihilation to muon pair cross section ratio can be written

$$R_{e^+e^-}(s) \equiv R^0_{e^+e^-}(s)[1 + \frac{\alpha_R(s)}{\pi}], \tag{9}$$

where $R^0_{e^+e^-}$ is the prediction at Born level. Similarly, we can define the entire radiative correction to the Bjorken sum rule as the effective charge $\alpha_{g_1}(Q^2)$ where Q is the corresponding momentum transfer:

$$\int_0^1 dx \left[g_1^{ep}(x, Q^2) - g_1^{en}(x, Q^2) \right] \equiv \frac{1}{6} \left| \frac{g_A}{g_V} \right| C_{\text{Bj}}(Q^2) = \frac{1}{6} \left| \frac{g_A}{g_V} \right| \left[1 - \frac{3}{4} C_F \frac{\alpha_{g_1}(Q^2)}{\pi} \right].$$

(10)

By convention, each effective charge is normalized to α_s in the weak coupling limit. One can define effective charges for virtually any quantity calculable in perturbative QCD; e.g. moments of structure functions, ratios of form factors, jet observables, and the effective potential between massive quarks. In the case of decay constants of the Z or the τ, the mass of the decaying system serves as the physical scale in the effective charge. In the case of multi-scale observables, such as the two-jet fraction in e^+e^- annihilation, the multiple arguments of the effective coupling $\alpha_{2jet}(s, y)$ correspond to the overall available energy s variables such as $y = \max_{ij}(p_i + p_j)^2/s$ representing the maximum jet mass fraction.

Commensurate scale relations take the general form

$$\alpha_A(Q_A) = C_{AB}[\alpha_B(Q_B)] .$$

(11)

The function $C_{AB}(\alpha_B)$ relates the observables A and B in the conformal limit; i.e., C_{AB} gives the functional dependence between the effective charges which would be obtained if the theory had zero β function. The conformal coefficients can be distinguished from the terms associated with the β function at each order in perturbation theory from their color and flavor dependence, or by an expansion about a fixed point.

The ratio of commensurate scales is determined by the requirement that all terms involving the β function are incorporated into the arguments of the running couplings, as in the original BLM procedure. Physically, the ratio of scales corresponds to the fact that the physical observables have different quark threshold and distinct sensitivities to fermion loops. More generally, the differing scales are in effect relations between mean values of the physical scales which appear in loop integrations. Commensurate scale relations are transitive; i.e., given the relation between effective charges for observables A and C and C and B, the resulting between A and B is independent of C. In particular, transitivity implies $\Lambda_{AB} = \Lambda_{AC} \times \Lambda_{CB}$.

One can consider QCD predictions as functions of analytic variables of the number of colors N_C and flavors N_F. For example, one can show at all orders of perturbation theory that PQCD predictions reduce to those of an Abelian theory at $N_C \to 0$ with $\hat{\alpha} = C_F \alpha_s$ and $\widehat{N}_F = N_f/T C_F$ held fixed. In particular, CSRs obey the "Abelian correspondence principle" in that they give the correct Abelian relations at $N_c \to 0$.

Similarly, commensurate scale relations obey the "conformal correspondence principle": the CSRs reduce to correct conformal relations when N_C and N_F are tuned to produce zero β function. Thus conformal symmetry provides a *template* for QCD predictions, providing relations between observables which are present even in theories which are not scale invariant. All effects of the nonzero beta function are encoded in the appropriate choice of relative scales $\Lambda_{AB} = Q_A/Q_B$.

The scale Q which enters a given effective charge corresponds to a physical momentum scale. The total logarithmic derivative of each effective charge effective charge $\alpha_A(Q)$ with respect to its physical scale is given by the Gell Mann-Low equation:

$$\frac{d\alpha_A(Q,m)}{d\log Q} = \Psi_A(\alpha_A(Q,m), Q/m), \tag{12}$$

where the functional dependence of Ψ_A is specific to its own effective charge. Here m refers to the quark's pole mass. The pole mass is universal in that it does not depend on the choice of effective charge. The Gell Mann-Low relation is reflexive in that ψ_A depends on only on the coupling α_A at the same scale. It should be emphasized that the Gell Mann-Low equation deals with physical quantities and is independent of the renormalization procedure and choice of renormalization scale. A central feature of quantum chromodynamics is asymptotic freedom; *i.e.*, the monotonic decrease of the QCD coupling $\alpha_A(\mu^2)$ at large spacelike scales. The empirical test of asymptotic freedom is the verification of the negative sign of the Gell Mann-Low function at large momentum transfer, which must be true for any effective charge.

In perturbation theory,

$$\Psi_A = -\psi_A^{\{0\}} \frac{\alpha_A^2}{\pi} - \psi_A^{\{1\}} \frac{\alpha_A^3}{\pi^2} - \psi_A^{\{2\}} \frac{\alpha_A^4}{\pi^3} + \cdots \tag{13}$$

At large scales $Q^2 \gg m^2$, the first two terms are universal and identical to the first two terms of the β function $\psi_A^{\{0\}} = \beta_0 = \frac{11N_C}{3} - \frac{2}{3}N_F, \psi_A^{\{1\}} = \beta_1$, whereas $\psi_A^{\{n\}}$ for $n \geq 2$ is process dependent. The quark mass dependence of the ψ function is analytic, and in the case of α_V scheme is known to two loops.

The commensurate scale relation between α_A and α_B implies an elegant relation between their conformal dependence C_{AB} and their respective Gell Mann Low functions:

$$\psi_B = \frac{dC_{BA}}{d\alpha_A} \times \psi_A. \tag{14}$$

Thus given the result for $N_{F,V}(m/Q)$ in α_V scheme we can use the CSR to derive $N_{F,A}(m/Q)$ for any other effective charge, at least to two loops. The above relation also shows that if one effective charge has a fixed point $\psi_A[\alpha_A(Q_A^{FP})] = 0$, then all effective charges B have a corresponding fixed point $\psi_B[\alpha_B(Q_B^{FP})] = 0$ at the corresponding commensurate scale and value of effective charge.

In quantum electrodynamics, the running coupling $\alpha_{QED}(Q^2)$, defined from the Coulomb scattering of two infinitely heavy test charges at the momentum transfer $t = -Q^2$, is taken as the standard observable. Is there a preferred effective charge which we should use to characterize the coupling strength in QCD? In the case of QCD, the heavy-quark potential $V(Q^2)$ is defined via a Wilson loop from the interaction energy of infinitely heavy quark and antiquark at momentum transfer

$t = -Q^2$. The relation $V(Q^2) = -4\pi C_F \alpha_V(Q^2)/Q^2$ then defines the effective charge $\alpha_V(Q)$. As in the corresponding case of Abelian QED, the scale Q of the coupling $\alpha_V(Q)$ is identified with the exchanged momentum. Thus there is never any ambiguity in the interpretation of the scale. All vacuum polarization corrections due to fermion pairs are incorporated in α_V through the usual vacuum polarization kernels which depend on the physical mass thresholds. Other observables could be used to define the standard QCD coupling, such as the effective charge defined from heavy quark radiation. [99]

Commensurate scale relations between α_V and the QCD radiative corrections to other observables have no scale or scheme ambiguity, even in multiple-scale problems such as multi-jet production. As is the case in QED, the momentum scale which appears as the argument of α_V reflect the mean virtuality of the exchanged gluons. Furthermore, we can write a commensurate scale relation between α_V and an analytic extension of the $\alpha_{\overline{MS}}$ coupling, thus transferring all of the unambiguous scale-fixing and analytic properties of the physical α_V scheme to the $\overline{MS}$ coupling.

An elegant example is the relation between the rate for semi-leptonic B-decay and α_V:

$$\Gamma(b \to X_u \ell \nu) = \frac{G_F^2 |V_{ub}|^2 M_b^2}{192\pi^3} \left[1 - 2.41 \frac{\alpha_V(0.16M_b)}{\pi} - 1.43 \frac{\alpha_V(0.16M_b)}{\pi}^2 \right], \quad (15)$$

where M_b is the scheme independent b-quark pole mass. The coefficient of $\alpha_{\overline{MS}}^2(\mu)$ in the usual expansion with $\mu = m_b$ is 26.8.

Some other examples of CSR's at NLO:

$$\alpha_R(\sqrt{s}) = \alpha_{g_1}(0.5\sqrt{s}) - \frac{\alpha_{g_1}^2(0.5\sqrt{s})}{\pi} + \frac{\alpha_{g_1}^3(0.5\sqrt{s})}{\pi^2} \quad (16)$$

$$\alpha_R(\sqrt{s}) = \alpha_V(1.8\sqrt{s}) + 2.08\frac{\alpha_V^2(1.8\sqrt{s})}{\pi} - 7.16\frac{\alpha_V^3(1.8\sqrt{s})}{\pi^2} \quad (17)$$

$$\alpha_\tau(\sqrt{s}) = \alpha_V(0.8\sqrt{s}) + 2.08\frac{\alpha_V^2(0 - .8\sqrt{s})}{\pi} - 7.16\frac{\alpha_V^3(0.8\sqrt{s})}{\pi^2} \quad (18)$$

$$\alpha_{g1}(\sqrt{s}) = \alpha_V(0.8Q) + 1.08\frac{\alpha_V^2(0.8Q)}{\pi} - 10.3\frac{\alpha_V^3(0.8Q)}{\pi^2} \quad (19)$$

For numerical purposes in each case we have used $N_F = 5$ and $\alpha_V = 0.1$ to compute the NLO correction to the CSR scale.

Commensurate scale relations thus provide fundamental and precise scheme-independent tests of QCD, predicting how observables track not only in relative normalization, but also in their commensurate scale dependence.

THE GENERALIZED CREWTHER RELATION

The generalized Crewther relation can be derived by calculating the QCD radiative corrections to the deep inelastic sum rules and $R_{e^+e^-}$ in a convenient renormalization scheme such as the modified minimal subtraction scheme $\overline{MS}$. One then algebraically eliminates $\alpha_{\overline{MS}}(\mu)$. Finally, BLM scale-setting [16] is used to eliminate the β-function dependence of the coefficients. The form of the resulting relation between the observables thus matches the result which would have been obtained had QCD been a conformal theory with zero β function. The final result relating the observables is independent of the choice of intermediate $\overline{MS}$ renormalization scheme.

More specifically, consider the Adler function [100] for the e^+e^- annihilation cross section

$$D(Q^2) = -12\pi^2 Q^2 \frac{d}{dQ^2}\Pi(Q^2), \quad \Pi(Q^2) = -\frac{Q^2}{12\pi^2}\int_{4m_\pi^2}^\infty \frac{R_{e^+e^-}(s)ds}{s(s+Q^2)}. \tag{20}$$

The entire radiative correction to this function is defined as the effective charge $\alpha_D(Q^2)$:

$$D\left(Q^2/\mu^2, \alpha_{\rm s}(\mu^2)\right) = D\left(1, \alpha_{\rm s}(Q^2)\right) \tag{21}$$

$$\equiv 3\sum_f Q_f^2\left[1 + \frac{3}{4}C_F\frac{\alpha_D(Q^2)}{\pi}\right] + (\sum_f Q_f)^2 C_{\rm L}(Q^2)$$

$$\equiv 3\sum_f Q_f^2 C_D(Q^2) + (\sum_f Q_f)^2 C_{\rm L}(Q^2),$$

where $C_F = \frac{N_C^2-1}{2N_C}$. The coefficient $C_{\rm L}(Q^2)$ appears at the third order in perturbation theory and is related to the "light-by-light scattering type" diagrams. (Hereafter $\alpha_{\rm s}$ will denote the $\overline{MS}$ scheme strong coupling constant.)

It is straightforward to algebraically relate $\alpha_{g_1}(Q^2)$ to $\alpha_D(Q^2)$ using the known expressions to three loops in the $\overline{MS}$ scheme. If one chooses the renormalization scale to resum all of the quark and gluon vacuum polarization corrections into $\alpha_D(Q^2)$, then the final result turns out to be remarkably simple [95] ($\hat\alpha = 3/4\,C_F\,\alpha/\pi$):

$$\hat\alpha_{g_1}(Q) = \hat\alpha_D(Q^*) - \hat\alpha_D^2(Q^*) + \hat\alpha_D^3(Q^*) + \cdots, \tag{22}$$

where

$$\ln\left(\frac{Q^{*2}}{Q^2}\right) = \frac{7}{2} - 4\zeta(3) + \left(\frac{\alpha_D(Q^*)}{4\pi}\right)\left[\left(\frac{11}{12} + \frac{56}{3}\zeta(3) - 16\zeta^2(3)\right)\beta_0\right.$$

$$\left. + \frac{26}{9}C_{\rm A} - \frac{8}{3}C_{\rm A}\zeta(3) - \frac{145}{18}C_{\rm F} - \frac{184}{3}C_{\rm F}\zeta(3) + 80C_{\rm F}\zeta(5)\right]. \tag{23}$$

where in QCD, $C_{\rm A} = N_C = 3$ and $C_{\rm F} = 4/3$. This relation shows how the coefficient functions for these two different processes are related to each other at their respective commensurate scales. We emphasize that the $\overline{MS}$ renormalization scheme is

25

used only for calculational convenience; it serves simply as an intermediary between observables. The renormalization group ensures that the forms of the CSR relations in perturbative QCD are independent of the choice of an intermediate renormalization scheme.

The Crewther relation was originally derived assuming that the theory is conformally invariant; *i.e.*, for zero β function. In the physical case, where the QCD coupling runs, all non-conformal effects are resummed into the energy and momentum transfer scales of the effective couplings α_R and α_{g_1}. The general relation between these two effective charges for non-conformal theory thus takes the form of a geometric series

$$1 - \widehat{\alpha}_{g_1} = [1 + \widehat{\alpha}_D(Q^*)]^{-1} \ . \tag{24}$$

We have dropped the small light-by-light scattering contributions. This is again a special advantage of relating observable to observable. The coefficients are independent of color and are the same in Abelian, non-Abelian, and conformal gauge theory. The non-Abelian structure of the theory is reflected in the expression for the scale Q^*.

Is experiment consistent with the generalized Crewther relation? Fits [101] to the experimental measurements of the R-ratio above the thresholds for the production of $c\bar{c}$ bound states provide the empirical constraint: $\alpha_R(\sqrt{s} = 5.0 \text{ GeV})/\pi \simeq 0.08 \pm 0.03$. The prediction for the effective coupling for the deep inelastic sum rules at the commensurate momentum transfer Q is then $\alpha_{g_1}(Q = 12.33 \pm 1.20 \text{ GeV})/\pi \simeq \alpha_{\mathrm{GLS}}(Q = 12.33 \pm 1.20 \text{ GeV})/\pi \simeq 0.074 \pm 0.026$. Measurements of the Gross-Llewellyn Smith sum rule have so far only been carried out at relatively small values of Q^2 [102,103]; however, one can use the results of the theoretical extrapolation [104] of the experimental data presented in [105]: $\alpha_{\mathrm{GLS}}^{\mathrm{extrapol}}(Q = 12.25 \text{ GeV})/\pi \simeq 0.093 \pm 0.042$. This range overlaps with the prediction from the generalized Crewther relation. It is clearly important to have higher precision measurements to fully test this fundamental QCD prediction.

GENERAL FORM OF COMMENSURATE SCALE RELATIONS

In general, commensurate scale relations connecting the effective charges for observables A and B have the form

$$\alpha_A(Q_A) = \alpha_B(Q_B)\left(1 + r_{A/B}^{(1)}\frac{\alpha_B(Q_B)}{\pi} + r_{A/B}^{(2)}\frac{\alpha_B(Q_B)^2}{\pi} + \cdots\right), \tag{25}$$

where the coefficients $r_{A/B}^n$ are identical to the coefficients obtained in a conformally invariant theory with $\beta_B(\alpha_B) \equiv (d/d\ln Q^2)\alpha_B(Q^2) = 0$. The ratio of the scales Q_A/Q_B is thus fixed by the requirement that the couplings sum all of the effects of the non-zero β function. In practice the NLO and NNLO coefficients and relative scales can be identified from the flavor dependence of the perturbative series; *i.e.* by shifting scales such that the N_F-dependence associated with $\beta_0 = 11/3C_A - 4/3T_F N_F$ and $\beta_1 = -34/3C_A^2 + \frac{20}{3}C_A T_F N_F + 4C_F T_F N_F$ does not appear in the coefficients. Here $C_A = N_C$, $C_F = (N_C^2 - 1)/2N_C$ and $T_F = 1/2$. The shift in scales which gives conformal coefficients in effect pre-sums the large and strongly divergent terms in the PQCD series which grow as $n!(\beta_0 \alpha_s)^n$, *i.e.*, the infrared renormalons associated with coupling-constant renormalization. [106,34,107,108]

The renormalization scales Q^* in the BLM method are physical in the sense that they reflect the mean virtuality of the gluon propagators. This scale-fixing procedure is consistent with scale fixing in QED, in agreement with in the Abelian limit, $N_C \to 0$. [16,93,109–111] The ratio of scales $\lambda_{A/B} = Q_A/Q_B$ guarantees that the observables A and B pass through new quark thresholds at the same physical scale. One can also show that the commensurate scales satisfy the transitivity rule $\lambda_{A/B} = \lambda_{A/C}\lambda_{C/B}$, which ensures that predictions are independent of the choice of an intermediate renormalization scheme or intermediate observable C.

In general, we can write the relation between any two effective charges at arbitrary scales μ_A and μ_B as a correction to the corresponding relation obtained in a conformally invariant theory:

$$\alpha_A(\mu_A) = C_{AB}[\alpha_B(\mu_B)] + \beta_B[\alpha_B(\mu_B)]F_{AB}[\alpha_B(\mu_B)] \qquad (26)$$

where

$$C_{AB}[\alpha_B] = \alpha_B + \sum_{n=1} C_{AB}^{(n)}\alpha_B^n \qquad (27)$$

is the functional relation when $\beta_B[\alpha_B] = 0$. In fact, if α_B approaches a fixed point $\overline{\alpha}_B$ where $\beta_B[\overline{\alpha}_B] = 0$, then α_A tends to a fixed point given by

$$\alpha_A \to \overline{\alpha}_A = C_{AB}[\overline{\alpha}_B]. \qquad (28)$$

The commensurate scale relation for observables A and B has a similar form, but in this case the relative scales are fixed such that the non-conformal term F_{AB} is zero. Thus the commensurate scale relation $\alpha_A(Q_A) = C_{AB}[\alpha_B(Q_B)]$ at general commensurate scales is also the relation connecting the values of the fixed points for any two effective charges or schemes. Furthermore, as $\beta \to 0$, the ratio of commensurate scales Q_A^2/Q_B^2 becomes the ratio of fixed point scales $\overline{Q}_A^2/\overline{Q}_B^2$ as one approaches the fixed point regime.

IMPLEMENTATION OF α_V SCHEME

The effective charge $\alpha_V(Q)$ provides a physically-based alternative to the usual modified minimal subtraction ($\overline{\text{MS}}$) scheme. All vacuum polarization corrections due to fermion pairs are incorporated in α_V through the usual vacuum polarization kernels which depend on the physical mass thresholds. When continued to time-like momenta, the coupling has the correct analytic dependence dictated by the production thresholds in the crossed channel. Since α_V incorporates quark mass effects exactly, it avoids the problem of explicitly computing and resumming quark mass corrections which are related to the running of the coupling. Thus the effective number of flavors $N_F(Q/m)$ is an analytic function of the scale Q and the quark masses m. The effects of finite quark mass corrections on the running of the strong coupling were first considered by De Rújula and Georgi [112] within the momentum subtraction schemes (MOM) (see also Georgi and Politzer [113], Shirkov and collaborators [114], and Chýla [115]).

One important advantage of the physical charge approach is its inherent gauge invariance to all orders in perturbation theory. This feature is not manifest in massive β-functions defined in non-physical schemes such as the MOM schemes. A second, more practical, advantage is the automatic decoupling of heavy quarks according to the Appelquist-Carazzone theorem [116].

By employing the commensurate scale relations other physical observables can be expressed in terms of the analytic coupling α_V without scale or scheme ambiguity. This way the quark mass threshold effects in the running of the coupling are taken into account by utilizing the mass dependence of the physical α_V scheme. In effect, quark thresholds are treated analytically to all orders in m^2/Q^2; i.e., the evolution of the physical α_V coupling in the intermediate regions reflects the actual mass dependence of a physical effective charge and the analytic properties of particle production. Furthermore, the definiteness of the dependence in the quark masses automatically constrains the scale Q in the argument of the coupling. There is thus no scale ambiguity in perturbative expansions in α_V.

In the conventional $\overline{MS}$ scheme, the coupling is independent of the quark masses since the quarks are treated as either massless or infinitely heavy with respect to the running of the coupling. Thus one formulates different effective theories depending on the effective number of quarks which is governed by the scale Q; the massless β-function is used to describe the running in between the flavor thresholds. These different theories are then matched to each other by imposing matching conditions at the scale where the effective number of flavors is changed (normally the quark masses). The dependence on the matching scale can be made arbitrarily small by calculating the matching conditions to high enough order. For physical observables one can then include the effects of finite quark masses by making a higher-twist expansion in m^2/Q^2 and Q^2/m^2 for light and heavy quarks, respectively. These higher-twist contributions have to be calculated for each observables separately, so that in principle one requires an all-order resummation of the mass corrections to the effective Lagrangian to give correct results.

The specification of the coupling and renormalization scheme also depends on the definition of the quark mass. In contrast to QED where the on-shell mass provides a natural definition of lepton masses, an on-shell definition for quark masses is complicated by the confinement property of QCD. In this paper we will use the pole mass $m(p^2 = m^2) = m$ which has the advantage of being scheme and renormalization-scale invariant. Furthermore, when combined with the α_V scheme, the pole mass gives predictions which are free of the leading renormalon ambiguity.

A technical complication of massive schemes is that one cannot easily obtain analytic solutions of renormalization group equations to the massive β function, and the Gell-Mann Low function is scheme-dependent even at lowest order.

In a recent paper we have presented a two-loop analytic extension of the α_V-scheme based on the recent results of Ref. [117]. The mass effects are in principle treated exactly to two-loop order and are only limited in practice by the uncertainties from numerical integration. The desired features of gauge invariance and decoupling are manifest in the form of the two-loop Gell-Mann Low function, and we give a simple fitting-function which interpolates smoothly the exact two-loop results obtained by using the adoptive Monte Carlo integrator VEGAS [118]. Strong consistency checks of the results are performed by comparing the Abelian limit to the well known QED results in the on-shell scheme. In addition, the massless as well as the decoupling limit are reproduced exactly, and the two-loop Gell-Mann Low function is shown to be renormalization scale independent.

The results of our numerical calculation of $N_{F,V}^{(1)}$ in the V-scheme for QCD and QED are shown in Fig. 3. The decoupling of heavy quarks becomes manifest at small Q/m, and the massless limit is attained for large Q/m. The QCD form actually becomes negative at moderate values of Q/m, a novel feature of the antiscreening non-Abelian contributions. This property is also present in the (gauge dependent) MOM results. In contrast, in Abelian QED the two-loop contribution to the effective number of flavors becomes larger than 1 at intermediate values of Q/m. We also display the one-loop contribution $N_{F,V}^{(0)}\left(\frac{Q}{m}\right)$ which monotonically interpolates between the decoupling and massless limits. The solid curves displayed in Fig. 3 shows that the parameterizations which we used for fitting the numerical results are quite accurate.

The relation of $\alpha_V(Q^2)$ to the conventional $\overline{MS}$ coupling is now known to NNLO, [119] but for clarity in this section only the NLO relation will be used. The commensurate scale relation is given by [120]

$$\begin{aligned}
\alpha_{\overline{MS}}(Q) &= \alpha_V(Q^*) + \frac{2}{3}N_C\frac{\alpha_V^2(Q^*)}{\pi} \\
&= \alpha_V(Q^*) + 2\frac{\alpha_V^2(Q^*)}{\pi} ,
\end{aligned} \tag{29}$$

which is valid for $Q^2 \gg m^2$. The coefficients in the perturbation expansion have their conformal values, *i.e.*, the same coefficients would occur even if the theory had been conformally invariant with $\beta = 0$. The commensurate scale is given by

$$Q^* = Q\exp\left[\frac{5}{6}\right] . \tag{30}$$

The scale in the $\overline{MS}$ scheme is thus a factor ~ 0.4 smaller than the physical scale. The coefficient $2N_C/3$ in the NLO coefficient is a feature of the non-Abelian

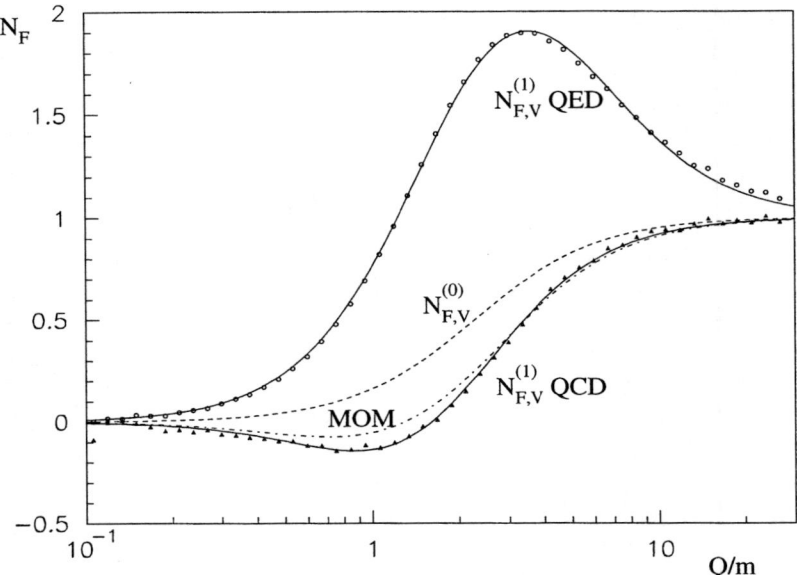

FIGURE 3. The numerical results for the gauge-invariant $N_{F,V}^{(1)}$ in QED (open circles) and QCD (triangles) with the best χ^2 fits superimposed respectively. The dashed line shows the one-loop $N_{F,V}^{(0)}$ function . For comparison we also show the gauge dependent two-loop result obtained in MOM schemes (dash-dot) [121,122]. At large $\frac{Q}{m}$ the theory becomes effectively massless, and both schemes agree as expected. The figure also illustrates the decoupling of heavy quarks at small $\frac{Q}{m}$.

couplings of QCD; the same coefficient occurs even if the theory were conformally invariant with $\beta_0 = 0$.

Using the above QCD results, we can transform any NLO prediction given in $\overline{MS}$ scheme to a scale-fixed expansion in $\alpha_V(Q)$. We can also derive the connection between the $\overline{MS}$ and α_V schemes for Abelian perturbation theory using the limit $N_C \to 0$ with $C_F \alpha_s$ and N_F/C_F held fixed. [93]

The use of α_V and related physically defined effective charges such as α_p (to NLO the effective charge defined from the (1,1) plaquette, α_p is the same as α_V) as expansion parameters has been found to be valuable in lattice gauge theory, greatly increasing the convergence of perturbative expansions relative to those using the bare lattice coupling. [109] Recent lattice calculations of the Υ- spectrum [123] have been used with BLM scale-fixing to determine a NLO normalization of the static heavy quark potential: $\alpha_V^{(3)}(8.2\text{GeV}) = 0.196(3)$ where the effective number of light flavors is $n_f = 3$. The corresponding modified minimal subtraction coupling evolved to the Z mass and five flavors is $\alpha_{\overline{MS}}^{(5)}(M_Z) = 0.1174(24)$. Thus a high precision value for $\alpha_V(Q^2)$ at a specific scale is available from lattice gauge theory. Predictions for other QCD observables can be directly referenced to this value without the scale or scheme ambiguities, thus greatly increasing the precision of QCD tests.

One can also use α_V to characterize the coupling which appears in the hard scattering contributions of exclusive process amplitudes at large momentum transfer, such as elastic hadronic form factors, the photon-to-pion transition form factor at large momentum transfer [16,19] and exclusive weak decays of heavy hadrons. [124] Each gluon propagator with four-momentum k^μ in the hard-scattering quark-gluon scattering amplitude T_H can be associated with the coupling $\alpha_V(k^2)$ since the gluon exchange propagators closely resembles the interactions encoded in the effective potential $V(Q^2)$. [In Abelian theory this is exact.] Commensurate scale relations can then be established which connect the hard-scattering subprocess amplitudes which control exclusive processes to other QCD observables.

We can anticipate that eventually nonperturbative methods such as lattice gauge theory or discretized light-cone quantization will provide a complete form for the heavy quark potential in QCD. It is reasonable to assume that $\alpha_V(Q)$ will not diverge at small space-like momenta. One possibility is that α_V stays relatively constant $\alpha_V(Q) \simeq 0.4$ at low momenta, consistent with fixed-point behavior. There is, in fact, empirical evidence for freezing of the α_V coupling from the observed systematic dimensional scaling behavior of exclusive reactions. [19] If this is in fact the case, then the range of QCD predictions can be extended to quite low momentum scales, a regime normally avoided because of the apparent singular structure of perturbative extrapolations.

There are a number of other advantages of the V-scheme:

1. Perturbative expansions in α_V with the scale set by the momentum transfer cannot have any β-function dependence in their coefficients since all running coupling effects are already summed into the definition of the potential. Since coefficients involving β_0 cannot occur in an expansions in α_V, the divergent infrared renormalon series of the form $\alpha_V^n \beta_0^n n!$ cannot occur. The general convergence properties of the scale Q^* as an expansion in α_V is not known. [34]

2. The effective coupling $\alpha_V(Q^2)$ incorporates vacuum polarization contributions with finite fermion masses. When continued to time-like momenta, the coupling has the correct analytic dependence dictated by the production thresholds in the t channel. Since α_V incorporates quark mass effects exactly, it avoids the problem of explicitly computing and resumming quark mass corrections.

3. The α_V coupling is the natural expansion parameter for processes involving non-relativistic momenta, such as heavy quark production at threshold where the Coulomb interactions, which are enhanced at low relative velocity v as

$\pi\alpha_V/v$, need to be re-summed. [125–127] The effective Hamiltonian for non-relativistic QCD is thus most naturally written in α_V scheme. The threshold corrections to heavy quark production in e^+e^- annihilation depend on α_V at specific scales Q^*. Two distinct ranges of scales arise as arguments of α_V near threshold: the relative momentum of the quarks governing the soft gluon exchange responsible for the Coulomb potential, and a high momentum scale, induced by hard gluon exchange, approximately equal to twice the quark mass for the corrections. [126] One thus can use threshold production to obtain a direct determination of α_V even at low scales. The corresponding QED results for τ pair production allow for a measurement of the magnetic moment of the τ and could be tested at a future τ-charm factory. [125,126]

We also note that computations in different sectors of the Standard Model have been traditionally carried out using different renormalization schemes. However, in a grand unified theory, the forces between all of the particles in the fundamental representation should become universal above the grand unification scale. Thus it is natural to use α_V as the effective charge for all sectors of a grand unified theory, rather than in a convention-dependent coupling such as $\alpha_{\overline{MS}}$.

THE ANALYTIC EXTENSION OF THE $\overline{MS}$ SCHEME

The standard $\overline{MS}$ scheme is not an analytic function of the renormalization scale at heavy quark thresholds; in the running of the coupling the quarks are taken as massless, and at each quark threshold the value of N_F which appears in the β function is incremented. Thus Eq. (29) is technically only valid far above a heavy quark threshold. However, we can use this commensurate scale relation to define an extended $\overline{MS}$ scheme which is continuous and analytic at any scale. The new modified scheme inherits all of the good properties of the α_V scheme, including its correct analytic properties as a function of the quark masses and its unambiguous scale fixing. [120] Thus we define

$$\widetilde{\alpha}_{\overline{MS}}(Q) = \alpha_V(Q^*) + \frac{2N_C}{3}\frac{\alpha_V^2(Q^{**})}{\pi} + \cdots, \tag{31}$$

for all scales Q. This equation not only provides an analytic extension of the $\overline{MS}$ and similar schemes, but it also ties down the renormalization scale to the physical masses of the quarks as they enter into the vacuum polarization contributions to α_V.

The modified scheme $\widetilde{\alpha}_{\overline{MS}}$ provides an analytic interpolation of conventional $\overline{MS}$ expressions by utilizing the mass dependence of the physical α_V scheme. In effect, quark thresholds are treated analytically to all orders in m^2/Q^2; i.e., the evolution of the analytically extended coupling in the intermediate regions reflects the actual mass dependence of a physical effective charge and the analytic properties of particle production. Just as in Abelian QED, the mass dependence of the effective potential and the analytically extended scheme $\widetilde{\alpha}_{\overline{MS}}$ reflects the analyticity of the physical thresholds for particle production in the crossed channel. Furthermore, the definiteness of the dependence in the quark masses automatically constrains the renormalization scale. There is thus no scale ambiguity in perturbative expansions in α_V or $\tilde{\alpha}_{\overline{MS}}$.

In leading order the effective number of flavors in the modified scheme $\tilde{\alpha}_{\overline{MS}}$ is given to a very good approximation by the simple form [120]

$$\widetilde{N}_{F,\overline{MS}}^{(0)}\left(\frac{m^2}{Q^2}\right) \simeq \left(1+\frac{5m^2}{Q^2\exp(\frac{5}{3})}\right)^{-1} \simeq \left(1+\frac{m^2}{Q^2}\right)^{-1}. \tag{32}$$

Thus the contribution from one flavor is $\simeq 0.5$ when the scale Q equals the quark mass m_i. The standard procedure of matching $\alpha_{\overline{MS}}(\mu)$ at the quark masses serves as a zeroth-order approximation to the continuous N_F.

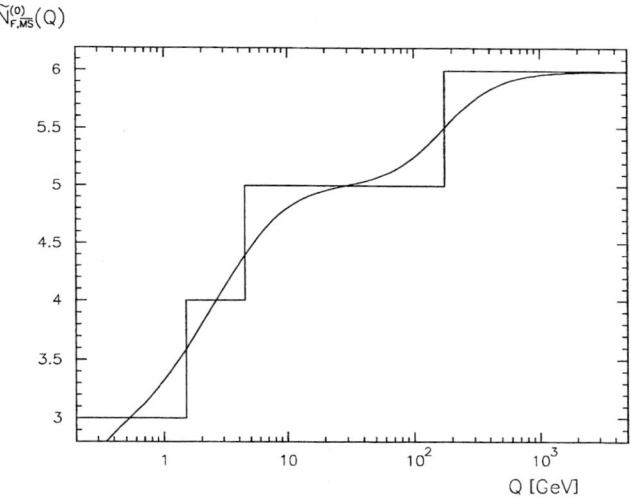

FIGURE 4. The continuous $\widetilde{N}_{F,\overline{MS}}^{(0)}$ in the analytic extension of the $\overline{MS}$ scheme as a function of the physical scale Q. (For reference the continuous N_F is also compared with the conventional procedure of taking N_F to be a step-function at the quark-mass thresholds.)

Adding all flavors together gives the total $\widetilde{N}_{F,\overline{MS}}^{(0)}(Q)$ which is shown in Fig. 4. For reference, the continuous N_F is also compared with the conventional procedure of taking N_F to be a step-function at the quark-mass thresholds. The figure shows clearly that there are hardly any plateaus at all for the continuous $\widetilde{N}_{F,\overline{MS}}^{(0)}(Q)$ in between the quark masses. Thus there is really no scale below 1 TeV where $\widetilde{N}_{F,\overline{MS}}^{(0)}(Q)$ can be approximated by a constant; for all Q below 1 TeV there is always one quark with mass m_i such that $m_i^2 \ll Q^2$ or $Q^2 \gg m_i^2$ is not true. We also note that if one would use any other scale than the BLM-scale for $\widetilde{N}_{F,\overline{\text{MS}}}^{(0)}(Q)$, the result would be to increase the difference between the analytic N_F and the standard procedure of using the step-function at the quark-mass thresholds.

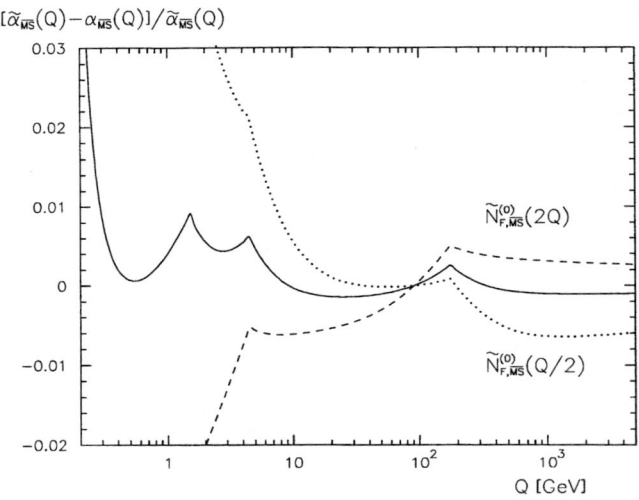

FIGURE 5. The solid curve shows the relative difference between the solutions to the 1-loop renormalization group equation using continuous N_F, $\widetilde{\alpha}_{\overline{\text{MS}}}(Q)$, and conventional discrete theta-function thresholds, $\alpha_{\overline{\text{MS}}}(Q)$. The dashed (dotted) curves shows the same quantity but using the scale $2Q$ ($Q/2$) in $\widetilde{N}_{F,\overline{\text{MS}}}^{(0)}$. The solutions have been obtained numerically starting from the world average [128] $\alpha_{\overline{\text{MS}}}(M_Z) = 0.118$.

Figure 5 shows the relative difference between the two different solutions of the 1-loop renormalization group equation, $i.e.$ $(\widetilde{\alpha}_{\overline{\text{MS}}}(Q) - \alpha_{\overline{\text{MS}}}(Q))/\widetilde{\alpha}_{\overline{\text{MS}}}(Q)$. The solutions have been obtained numerically starting from the world average [128] $\alpha_{\overline{\text{MS}}}(M_Z) = 0.118$. The figure shows that taking the quark masses into account in the running leads to effects of the order of one percent which are most especially pronounced near thresholds.

The extension of the $\overline{\text{MS}}$-scheme proposed here provides a coupling which is an analytic function of both the scale and the quark masses. The new modified coupling $\widetilde{\alpha}_{\overline{\text{MS}}}(Q)$ inherits most of the good properties of the α_V scheme, including its correct analytic properties as a function of the quark masses and its unambiguous scale fixing [120]. However, the conformal coefficients in the commensurate scale relation between the α_V and $\overline{\text{MS}}$ schemes does not preserve one of the defining criterion of the potential expressed in the bare charge, namely the non-occurrence of color factors corresponding to an iteration of the potential. This is probably an effect of the breaking of conformal invariance by the $\overline{\text{MS}}$ scheme. The breaking of conformal symmetry has also been observed when dimensional regularization is used as a factorization scheme in both exclusive [57,129,130] and inclusive [131] reactions. Thus, it does not turn out to be possible to extend the modified scheme $\tilde{\alpha}_{\overline{MS}}$ beyond leading order without running into an intrinsic contradiction with conformal symmetry.

APPLICATION OF COMMENSURATE SCALE RELATIONS TO THE HARD QCD POMERON

The observation of rapidly increasing structure functions in deep inelastic scattering at small-x_{bj} and the observation of rapidly increasing diffractive processes such as $\gamma^* p \rightarrow \rho p$ at high energies at HERA is in agreement with the expectations of the BFKL [132] QCD high-energy limit. The highest eigenvalue, ω^{max}, of the leading order (LO) BFKL equation [132] is related to the intercept of the Pomeron which in turn governs the high-energy asymptotics of the cross sections: $\sigma \sim s^{\alpha_{IP}-1} = s^{\omega^{max}}$. The BFKL Pomeron intercept in LO turns out to be rather large: $\alpha_{IP} - 1 = \omega_L^{max} = 12 \ln 2 \, (\alpha_S/\pi) \simeq 0.55$ for $\alpha_S = 0.2$; hence, it is very important to know the next-to-leading order (NLO) corrections.

Recently the NLO corrections to the BFKL resummation of energy logarithms were calculated [133,134] by employing the modified minimal subtraction scheme ($\overline{MS}$) [135] to regulate the ultraviolet divergences with arbitrary scale setting. The NLO corrections [133,134] to the highest eigenvalue of the BFKL equation turn out to be negative and even larger than the LO contribution for $\alpha_s > 0.157$. It is thus important to analyze the NLO BFKL resummation of energy logarithms [133,134] in physical renormalization schemes and apply the BLM-CSR method. In fact, as shown in a recent paper [136], the reliability of QCD predictions for the intercept of the BFKL Pomeron at NLO when evaluated using BLM scale setting [16] within non-Abelian physical schemes, such as the momentum space subtraction (MOM) scheme [137,138] or the Υ-scheme based on $\Upsilon \rightarrow ggg$ decay, is significantly improved compared to the $\overline{MS}$-scheme.

The renormalization scale ambiguity problem can be resolved if one can optimize the choice of scales and renormalization schemes according to some sensible criteria. In the BLM optimal scale setting [16], the renormalization scales are chosen such that all vacuum polarization effects from the QCD β-function are resummed into the running couplings. The coefficients of the perturbative series are thus identical to the perturbative coefficients of the corresponding conformally invariant theory with $\beta = 0$.

In the present case one can show that within the V-scheme (or the $\overline{MS}$-scheme) the BLM procedure does not change significantly the value of the NLO coefficient $r(\nu)$. This can be understood since the V-scheme, as well as $\overline{MS}$-scheme, are adjusted primarily to the case when in the LO there are dominant QED (Abelian) type contributions, whereas in the BFKL case there are important LO gluon-gluon (non-Abelian) interactions. Thus one can choose for the BFKL case the MOM-scheme [137,138] or the Υ-scheme based on $\Upsilon \to ggg$ decay.

Adopting BLM scale setting, the NLO BFKL eigenvalue in the MOM-scheme is

$$\omega_{BLM}^{MOM}(Q^2, \nu) = N_C \chi_L(\nu) \frac{\alpha_{MOM}(Q_{BLM}^{MOM\,2})}{\pi} \left[1 + r_{BLM}^{MOM}(\nu) \frac{\alpha_{MOM}(Q_{BLM}^{MOM\,2})}{\pi} \right], \quad (33)$$

$$r_{BLM}^{MOM}(\nu) = r_{MOM}^{conf}(\nu).$$

The β-dependent part of the $r_{MOM}(\nu)$ defines the corresponding BLM optimal scale

$$Q_{BLM}^{MOM\,2}(\nu) = Q^2 \exp\left[-\frac{4r_{MOM}^{\beta}(\nu)}{\beta_0}\right] = Q^2 \exp\left[\frac{1}{2}\chi_L(\nu) - \frac{5}{3} + 2\left(1 + \frac{2}{3}I\right)\right]. \quad (34)$$

At $\nu = 0$ we have $Q_{BLM}^{MOM\,2}(0) = Q^2(4\exp[2(1 + 2I/3) - 5/3]) \simeq Q^2\,127$. Note that $Q_{BLM}^{MOM\,2}(\nu)$ contains a large factor, $\exp[-4T_{MOM}^{\beta}/\beta_0] = \exp[2(1 + 2I/3)] \simeq 168$, which reflects a large kinematic difference between MOM- and $\overline{MS}$- schemes [139,16].

One of the striking features of this analysis is that the NLO value for the intercept of the BFKL Pomeron, improved by the BLM procedure, has a very weak dependence on the gluon virtuality Q^2. This agrees with the conventional Regge-theory where one expects an universal intercept of the Pomeron without any Q^2-dependence. The minor Q^2-dependence obtained, on one side, provides near insensitivity of the results to the precise value of Λ, and, on the other side, leads to approximate scale and conformal invariance. Thus one may use conformal symmetry [140,141] for the continuation of the present results to the case $t \neq 0$.

The NLO corrections to the BFKL equation for the QCD Pomeron thus become controllable and meaningful provided one uses physical renormalization scales and schemes relevant to non-Abelian gauge theory. BLM optimal scale setting automatically sets the appropriate physical renormalization scale by absorbing the non-conformal β-dependent coefficients. The strong renormalization scheme dependence of the NLO corrections to BFKL resummation then largely disappears. This is in contrast to the unstable NLO results obtained in the conventional $\overline{MS}$-scheme with arbitrary choice of renormalization scale. A striking feature of the NLO BFKL Pomeron intercept in the BLM approach is its very weak Q^2-dependence, which provides approximate conformal invariance.

The new results presented here open new windows for applications of NLO BFKL resummation to high-energy phenomenology.

Recently the $L3$ collaboration at LEPL3 has presented new results for the virtual photon cross section $\sigma(\gamma^*(Q_A)\gamma^*(Q_b)) \to$ hadrons using double tagged $e^+e^- \to e^+e^-$hadrons. This process provides a remarkably clean possible test of the perturbative QCD pomeron since there are no initial hadrons. [142] The calculation of $\sigma(\gamma^*\gamma^*)$ and is discussed in detail in references [142]. We note here some important features:

i) for large virtualities, $\sigma(\gamma^*\gamma^*)$ the longitudinal cross section σ_{LL} dominates and scales like $1/Q^2$, where $Q^2 \sim \max\{Q_A^2, Q_B^2\}$. This is characteristic of the perturbative QCD prediction. Models based on Regge factorization (which work well in the soft-interaction regime dominating $\gamma\gamma$ scattering near the mass shell) would predict a higher power in $1/Q$.

ii) $\sigma(\gamma^*\gamma^*)$ is affected by logarithmic corrections in the energy s to all orders in α_s. As a result of the BFKL summation of these contributions, the cross section rises like a power in s, $\sigma \propto s^\lambda$. The Born approximation to this result (that is, the $\mathcal{O}(\alpha_s^2)$ contribution, corresponding to single gluon exchange gives a constant cross section, $\sigma_{Born} \propto s^0$. A fit to photon-photon sub-energy dependence measured by L3 at $\sqrt{s}_{e^+e^-} = 91$ GeV and $< Q_A^2 >=< Q_A^2 >= 3.5$ GeV^2 gives $\alpha_P - 1 = 0.28 \pm 0.05$. The L3 data at $\sqrt{s}_{e^+e^-} = 183$ GeV and $< Q_A^2 >=< Q_A^2 >= 14$ GeV^2, gives $\alpha_P - 1 = 0.40 \pm 0.07$ which shows a rise of the virtual photon cross section much stronger than single gluon or soft pomeron exchange, but it is compatible with the expectations from the NLO scale- and scheme-fixed BFKL predictions. It will be crucial to measure the Q_A^2 and Q_B^2 scaling and polarization dependence and compare with the detailed predictions of PQCD [142].

SUMMARY ON COMMENSURATE SCALE RELATIONS

Commensurate scale relations have a number of attractive properties:

1. The ratio of physical scales Q_A/Q_B which appears in commensurate scale relations reflects the relative position of physical thresholds, *i.e.* quark antiquark pair production.

2. The functional dependence and perturbative expansion of the CSR are identical to those of a conformal scale-invariant theory where $\beta_A(\alpha_A) = 0$ and $\beta_B(\alpha_B) = 0$.

3. In the case of theories approaching fixed-point behavior $\beta_A(\overline{\alpha}_A) = 0$ and $\beta_B(\overline{\alpha}_B) = 0$, the commensurate scale relation relates both the ratio of fixed point couplings $\overline{\alpha}_A/\overline{\alpha}_B$, and the ratio of scales as the fixed point is approached.

4. Commensurate scale relations satisfy the Abelian correspondence principle [93]; *i.e.* the non-Abelian gauge theory prediction reduces to Abelian theory for $N_C \to 0$ at fixed $C_F\alpha_s$ and fixed N_F/C_F.

5. The perturbative expansion of a commensurate scale relation has the same form as a conformal theory, and thus has no $n!$ renormalon growth arising from the β-function. [143] It is an interesting conjecture whether the perturbative expansion relating observables to observable are in fact free of all $n!$ growth. The generalized Crewther relation, where the commensurate relation's perturbative expansion forms a geometric series to all orders, has convergent behavior.

Virtually any perturbative QCD prediction can be written in the form of a commensurate scale relation, thus eliminating any uncertainty due to renormalization scheme or scale dependence. Recently it has been shown [144] how the commensurate scale relation between the radiative corrections to τ-lepton decay and $R_{e^+e^-}(s)$ can be generalized and empirically tested for arbitrary τ mass and nearly arbitrarily functional dependence of the τ weak decay matrix element.

An essential feature of the $\alpha_V(Q)$ scheme is the absence of any renormalization scale ambiguity, since Q^2 is, by definition, the square of the physical momentum transfer. The α_V scheme naturally takes into account quark mass thresholds, which is of particular phenomenological importance to QCD applications in the mass region close to threshold. As we have seen, commensurate scale relations provide an analytic extension of the conventional $\overline{MS}$ scheme in which many of the advantages of the α_V scheme are inherited by the $\tilde{\alpha}_{\overline{MS}}$ scheme, but only minimal changes have to be made. Given the commensurate scale relation connecting $\tilde{\alpha}_{\overline{MS}}$ to α_V expansions in $\tilde{\alpha}_{\overline{MS}}$ are effectively expansions in α_V to the given order in perturbation theory at a corresponding commensurate scale.

The calculation of $\psi_V^{(1)}$, the two-loop term in the Gell-Mann Low function for the α_V scheme, with massive quarks gives for the first time a gauge invariant and renormalization scheme independent two-loop result for the effects of quarks masses in the running of the coupling. Renormalization scheme independence is achieved by using the pole mass definition for the "light" quarks which contribute to the scale dependence of the static heavy quark potential. Thus the pole mass and the V-scheme are closely connected and have to be used in conjunction to give reasonable results.

It is interesting that the effective number of flavors in the two-loop coefficient of the Gell-Mann Low function in the α_V scheme, $N_{F,V}^{(1)}$, becomes negative for intermediate values of Q/m. This feature can be understood as anti-screening from the non-Abelian contributions and should be contrasted with the QED case where the effective number of flavors becomes larger than 1 for intermediate Q/m. For small Q/m the heavy quarks decouple explicitly as expected in a physical scheme, and for large Q/m the massless result is retained.

The analyticity of the α_V coupling can be utilized to obtain predictions for any perturbatively calculable observables including the finite quark mass effects associated with the running of the coupling. By employing the commensurate scale relation method, observables which have been calculated in the $\overline{MS}$ scheme can be related to the analytic V-scheme without any scale ambiguity. The commensurate scale relations provides the relation between the physical scales of two effective charges where they pass through a common flavor threshold. We also note the utility of the α_V effective charge in supersymmetric and grand unified theories, particularly since the unification of couplings and masses would be expected to occur in terms of physical quantities rather than parameters defined by theoretical convention.

38

As an example we have showed in Ref. [120] how to calculate the finite quark mass corrections connected with the running of the coupling for the non-singlet hadronic width of the Z-boson compared with the standard treatment in the $\overline{\text{MS}}$ scheme. The analytic treatment in the V-scheme gives a simple and straightforward way of incorporating these effects for any observable. This should be contrasted with the $\overline{\text{MS}}$ scheme where higher twist corrections due to finite quark mass threshold effects have to be calculated separately for each observable. The V-scheme is especially suitable for problems where the quark masses are important such as for threshold production of heavy quarks and the hadronic width of the τ lepton.

It has also been shown that the NLO corrections to the BFKL equation for the QCD Pomeron become controllable and meaningful provided one uses physical renormalization scales and schemes relevant to non-Abelian gauge theory. BLM optimal scale setting automatically sets the appropriate physical renormalization scale by absorbing the non-conformal β-dependent coefficients. The strong renormalization scheme dependence of the NLO corrections to BFKL resummation then largely disappears. This is in contrast to the unstable NLO results obtained in the conventional $\overline{\text{MS}}$-scheme with arbitrary choice of renormalization scale. A striking feature of the NLO BFKL Pomeron intercept in the BLM/CSR approach is its very weak Q^2-dependence, which provides approximate conformal invariance. The new results presented here open new windows for applications of NLO BFKL resummation to high-energy phenomenology, particularly virtual photon-photon scattering.

Acknowledgments

Many of the results presented here are based on collaborations with a number of colleagues, including Victor Fadin, Gregory Gabadadze, Mandeep Gill, John Hiller, Dae Sung Hwang, Chueng Ji, Andrei Kataev, Victor Kim, Peter Lepage, Lev Lipatov, Hung Jung Lu, Gary McCartor, Michael Melles, Chris Pauli, Grigorii B. Pivovarov. and Johan Rathsman. I thank S. Dalley, Yitzhak Frishman, Einan Gardi, Georges Grunberg, Paul Hoyer, Marek Karliner, Carlos Merino, and Jose Pelaez for helpful conversations. The sections on commensurate scale relations are based on a review written in collaboration with Johan Rathsman. [65] I also wish to thank Chueng Ji and Dong-Pil Min for their outstanding hospitality at the Asia Pacific Center for Theoretical Physics in Seoul.

REFERENCES

1. J.P. Nasalski [New Muon Collaboration], Nucl. Phys. **A577**, 325C (1994).
2. V. Barone, C. Pascaud and F. Zomer, hep-ph/9907512.
3. M. Karliner and H.J. Lipkin, hep-ph/9906321.
4. G. P. Lepage and S. J. Brodsky, *Phys. Rev.* **D22**, 2157 (1980); *Phys. Lett.* **B87**, 359 (1979); *Phys. Rev. Lett.* **43**, 545, 1625(E) (1979).
5. S. J. Brodsky, H. Pauli and S. S. Pinsky, *Phys. Rept.* **301**, 299 (1998) hep-ph/9705477.

6. G. Bertsch, S. J. Brodsky, A. S. Goldhaber, and J. F. Gunion, *Phys. Rev. Lett.* **47**, 297 (1981).

7. L. Frankfurt, G. A. Miller, and M. Strikman, *Phys. Lett.* **B304**, 1 (1993), hep-ph/9305228.

8. S. J. Brodsky and S. D. Drell, *Phys. Rev.* **D22**, 2236 (1980).

9. S. J. Brodsky and G. P. Lepage, in *Perturbative Quantum Chromodynamics*, A. H. Mueller, Ed. (World Scientific, 1989).

10. H. C. Pauli and S. J. Brodsky, *Phys. Rev.* **D32**, 1993 (1985); *Phys. Rev.* **D32**, 2001 (1985).

11. S.J. Brodsky and H.C. Pauli, *lectures given at 30th Schladming Winter School in Particle Physics: Field Theory, Schladming, Austria, Feb 27 - Mar 8, 1991.*

12. S. Dalley, and I. R. Klebanov, *Phys. Rev.* **D47**, 2517 (1993).

13. F. Antonuccio and S. Dalley, *Phys. Lett.* **B348**, 55 (1995); *Phys. Lett.* **B376**, 154 (1996); *Nucl. Phys.* **B461**, 275 (1996).

14. S.J. Brodsky, J.R. Hiller and G. McCartor, Phys. Rev. **D58**, 025005 (1998) hep-th/9802120.

15. P.P. Srivastava and S.J. Brodsky, hep-ph/9906423.

16. S. J. Brodsky, G. P. Lepage and P. B. Mackenzie, Phys. Rev. **D28**, 228 (1983).

17. S.J. Brodsky, H.J. Lu, Phys. Rev. **D51**, 3652 (1995); hep-ph/9506322.

18. S. J. Brodsky, G. T. Gabadadze, A. L. Kataev and H. J. Lu, *Phys. Lett.* **372B**, 133, (1996).

19. S. J. Brodsky, C.-R. Ji, A. Peng and D. G. Robertson, Phys. Rev. **D57**, 345 (1998).

20. N.J. Watson, Nucl. Phys. **B494**, 388 (1997) hep-ph/9606381.

21. A. Czarnecki, K. Melnikov and N. Uraltsev, Phys. Rev. Lett. **80**, 3189 (1998) hep-ph/9708372.

22. S.J. Brodsky and D.S. Hwang, Nucl. Phys. **B543**, 239 (1999) hep-ph/9806358.

23. S. J. Chang, R.G. Root and T. M. Yan, *Phys. Rev.* **D7**, 1133 (1973).

24. M. Burkardt, *Nucl. Phys.* **A504**, 762 (1989); *Nucl. Phys.* **B373**, 613 (1992); *Phys. Rev.* **D52**, 3841 (1995).

25. H. Choi and C. Ji, Phys. Rev. **D58**, 071901 (1998) hep-ph/9805438.

26. K. Hornbostel, S. J. Brodsky, and H. C. Pauli, *Phys. Rev.* **D41** 3814 (1990).

27. A. Szczepaniak, E. M. Henley and S. J. Brodsky, *Phys. Lett.* **B243**, 287 (1990).

28. A. Szczepaniak, *Phys. Rev.* **D54**, 1167 (1996).

29. P. Ball, JHEP **09**, 005 (1998) hep-ph/9802394.

30. P. Ball and V. M. Braun, *Phys. Rev.* **D58**, 094016 (1998) hep-ph/9805422.

31. S. J. Brodsky, L. Frankfurt, J. F. Gunion, A. H. Mueller, and M. Strikman, *Phys. Rev.* **D50**, 3134 (1994), hep-ph/9402283.

32. S. J. Brodsky and A. H. Mueller, *Phys. Lett.* **206B**, 685 (1988). L. Frankfurt and M. Strikman, *Phys. Rept.* **160** , 235 (1988); P. Jain, B. Pire and J. P. Ralston, *Phys. Rept.* **271**, 67(1996).

33. F. Antonuccio, S. J. Brodsky, and S. Dalley, SLAC-PUB-7472, *Phys. Lett.* **B412** 104 (1997), hep-ph/9705413.

34. A. H. Mueller, Phys. Lett. **B308**, 355 (1993).

35. D. Mueller, SLAC-PUB-6496, May 1994, hep-ph/9406260.

36. S. J. Brodsky, P. Hoyer, C. Peterson, and N. Sakai, *Phys. Lett.* **93B**, 451 (1980).

37. B. W. Harris, J. Smith, and R. Vogt, *Nucl. Phys.* **B461**, 181 (1996), hep-ph/9508403.
38. S. J. Brodsky and I. A. Schmidt, *Phys. Lett.* **B234**, 144 (1990).
39. S. J. Brodsky, P. Hoyer, A. H. Mueller, W.-K. Tang, *Nucl. Phys.* **B369**, 519 (1992).
40. S. J. Brodsky and M. Karliner, *Phys. Rev. Lett.* **78**, 4682 (1997) hep-ph/9704379.
41. M. Burkardt and Brian Warr, *Phys. Rev.* **D45**, 958 (1992).
42. A. I. Signal and A. W. Thomas, *Phys. Lett.* **191B**, 205 (1987).
43. S. J. Brodsky and B-Q Ma, *Phys. Lett.* **B381**, 317 (1996), hep-ph/9604393.
44. S. J. Brodsky and A. Mueller, *Phys. Lett.* **206B**, 685 (1988). R. Vogt, S. J. Brodsky, and P. Hoyer, SLAC-PUB-5421,*Nucl. Phys.* **B360**, 67 (1991); SLAC-PUB-5827, *Nucl. Phys.* **B383**, 643 (1992).
45. P. Hoyer and S. Peigne, *Phys. Rev.* **D59**, 034011 (1999) hep-ph/9806424.
46. E. L. Berger and S. J. Brodsky, *Phys. Rev. Lett.* **42**, 940 (1979).
47. A. Brandenburg, S. J. Brodsky, V.V. Khoze, and D. Mueller, *Phys. Rev. Lett.* **73**, 939 (1994), hep-ph/9403361.
48. S. J. Brodsky and B. T. Chertok, *Phys. Rev.* **D14**, 3003 (1976).
49. S. J. Brodsky, C.-R. Ji, and G. P. Lepage, *Phys. Rev. Lett.* **51**, 83 (1983).
50. G. R. Farrar, K. Huleihel and H. Zhang, *Phys. Rev. Lett.* **74**, 650 (1995).
51. A. D. Krisch, *Nucl. Phys. B (Proc. Suppl.)* **25**, 285 (1992).
52. S. J. Brodsky and G. F. de Teramond, *Phys. Rev. Lett.* **60**, 1924 (1988).
53. J. Gronberg *et al.* [CLEO Collaboration], *Phys. Rev.* **D57**, 33 (1998) hep-ex/9707031.
54. D. F. Ashery *et al.*, Fermilab E791 Collaboration, to be published.
55. M. Beneke, G. Buchalla, M. Neubert and C.T. Sachrajda, hep-ph/9905312.
56. S. J. Brodsky, Y. Frishman, G. P. Lepage and C. Sachrajda, *Phys. Lett.* **91B**, 239 (1980).
57. S. J. Brodsky, Y. Frishman and G. P. Lepage, *Phys. Lett.* **167B**, 347 (1986).
58. D. Müller, *Phys. Rev.* **D49**, 2525 (1994).
59. P. Ball and V. M. Braun, *Nucl. Phys.* **B543**, 201 (1999) hep-ph/9810475.
60. V.M. Braun, S.E. Derkachov, G.P. Korchemsky and A.N. Manashov, hep-ph/9902375.
61. S. J. Brodsky and G. P. Lepage, *Phys. Rev.* **D24**, 2848 (1981).
62. L. Frankfurt, G. A. Miller and M. Strikman, *Comments Nucl. Part. Phys.* **21**, 1 (1992).
63. S. J. Brodsky and H. J. Lu, *Phys. Rev.* **D51**, 3652 (1995) hep-ph/9405218.
64. S. J. Brodsky, G. T. Gabadadze, A. L. Kataev and H. J. Lu, *Phys. Lett.* **B372**, 133 (1996) hep-ph/9512367.
65. S. J. Brodsky and J. Rathsman, hep-ph/9906339.
66. S.J. Brodsky, C. Ji, A. Pang and D.G. Robertson, Phys. Rev. **D57**, 245 (1998) hep-ph/9705221.
67. S. J. Brodsky and G. R. Farrar, *Phys. Rev.* **D11**, 1309 (1975).
68. V. A. Matveev, R. M. Muradian and A. N. Tavkhelidze, *Nuovo Cim. Lett.* **7**, 719 (1973).
69. J.C. Collins, L. Frankfurt and M. Strikman, Phys. Rev. **D56**, 2982 (1997) hep-ph/9611433.

70. C. E. Carlson and A. B. Wakely, *Phys. Rev.* **D48**, 2000 (1993); A. Afanasev, C. E. Carlson and C. Wahlquist, *Phys. Lett.* **B398**, 393 (1997), hep-ph/9701215, and *Phys. Rev.* **D58**, 054007 (1998), hep-ph/9706522.

71. S. J. Brodsky, M. Diehl, P. Hoyer and S. Peigne, *Phys. Lett.* **B449**, 306 (1999) hep-ph/9812277.

72. J. F. Gunion, S. J. Brodsky and R. Blankenbecler, *Phys. Rev.* **D6**, 2652 (1972); R. Blankenbecler and S. J. Brodsky, *Phys. Rev.* **D10**, 2973 (1974).

73. P.V. Landshoff, *Phys. Rev.* **D10**, 1024 (1974).

74. V. Chernyak, hep-ph/9906387.

75. M.R. Adams *et al.* [E665 Collaboration], Z. Phys. **C74**, 237 (1997).

76. L. Frankfurt, G. A. Miller and M. Strikman, hep-ph/9907214.

77. P. Kroll and M. Raulfs, *Phys. Lett.* **B387**, 848 (1996).

78. I. V. Musatov and A. V. Radyushkin, *Phys. Rev.* **D56**, 2713 (1997).

79. T. Feldmann, hep-ph/9907226.

80. A. Schmedding and O. Yakovlev, hep-ph/9905392.

81. P. Stoler, Few Body Syst. Suppl. **11**, 124 (1999).

82. R. J. Holt, *Phys. Rev.* **C41**, 2400 (1990).

83. C. Bochna *et al.* [E89-012 Collaboration], *Phys. Rev. Lett.* **81**, 4576 (1998) nucl-ex/9808001.

84. B. Melic, B. Nizic and K. Passek, hep-ph/9903426.

85. A. Szczepaniak, A. Radyushkin and C. Ji, *Phys. Rev.* **D57**, 2813 (1998) hep-ph/9708237.

86. C. J. Bebek *et al.*, *Phys. Rev.* **D13**, 25 (1976).

87. N. Isgur and C. H. Llewellyn Smith, *Phys. Lett.* **B217**, 535 (1989).

88. A. V. Radyushkin, *Phys. Rev.* **D58**, 114008 (1998) hep-ph/9803316.

89. J. Bolz and P. Kroll, *Z. Phys.* **A356**, 327 (1996) hep-ph/9603289.

90. A.S. Kronfeld and B. Nizic, Phys. Rev. **D44**, 3445 (1991).

91. S. J. Brodsky, F. E. Close and J. F. Gunion, *Phys. Rev.* **D6**, 177 (1972).

92. For reviews and further references see S. J. Brodsky and G. P. Lepage, SLAC-PUB-4947. Published in 'Perturbative Quantum Chromodynamics', Ed. by A.H. Mueller, World Scientific Publ. Co. (1989), p. 93-240 (QCD161:M83); V. L. Chernyak and A. R. Zhitnitsky, *Phys. Rept.* **112**, 173 (1984).

93. S.J. Brodsky and P. Huet, Phys. Lett. **B417**, 145 (1998) hep-ph/9707543.

94. X. Ji, *Phys. Rev.* **D55**, 7114 (1997), hep-ph/9609381; X. Ji and J. Osborne, *Phys. Rev.***D58**, 094018 (1998), hep-ph/9801260; A.V. Radyushkin, *Phys. Rev.***D56**, 5524 (1997), hep-ph/9704207.

95. S. J. Brodsky , G. T. Gabadadze, A. L. Kataev, and H. J. Lu. Phys. Lett. **B 372**, 133 (1996).

96. G. Grunberg, Phys. Lett. **B85**. 70 (1980); Phys. Lett. **B110**, 501 (1982); Phys. Rev. **D29**, 2315 (1984).

97. A. Dhar and V. Gupta, Phys. Rev. **D29**, 2822 (1984).

98. V. Gupta, D. V. Shirkov and O. V. Tarasov, Int. J. Mod. Phys. **A6**, 3381 (1991).

99. A. Czarnecki, K. Melnikov, and N. Uraltsev, Phys. Rev. Lett. **80**, 3189 (1998). Yu. L. Dokshitser and B. R. Webber, Phys. Lett. **B404**, 321 (1997).

100. S. Adler, Phys. Rev. **182**, 1517 (1969).

101. A. C. Mattingly and P. M. Stevenson, Phys. Rev. **D49**, 437 (1994).

102. CCFR Collaboration, W.C. Leung, *et al.*, Phys. Lett. **B317**, 655 (1993).

103. CCFR and NuTeV Collaboration, presented by D. Harris at XXX Recontre de Moriond, 1995, presented by J. H. Kim at the European Conference on High Energy Physics, Brussels, July 1995.

104. A. L. Kataev, A.V. Sidorov, Phys. Lett. **B331**, 179 (1994).

105. CCFR Collaboration, P.Z. Quinta, *et al.*, Phys. Rev. Lett. **71**, 1307 (1993).

106. G. 't Hooft, in the Proceedings of the International School, Erice, Italy, 1977, edited by A. Zichichi, Subnuclear Series Vol. 15 (Plenum, New York, 1979).

107. H. J. Lu, Phys. Rev. **D45**, 1217 (1992)..

108. M. Beneke and V. M. Braun, Phys. Lett. **B348**, 513 (1995).

109. G. Peter Lepage and P. B. Mackenzie, Phys. Rev. **D48**, 2250 (1993).

110. M. Neubert, Phys. Rev. **D51**, 5924 (1995).

111. P. Ball, M. Beneke and V. M. Braun, Nucl. Phys. **B452**, 563 (1995).

112. A. De Rújula and H. Georgi, Phys. Rev. **D13**, 1296 (1976).

113. H. Georgi and H.D. Politzer, Phys. Rev. **D14**, 1829 (1976).

114. D. V. Shirkov, Teor. Mat. Fiz. **98**, 500 (1992) [Theor. Math. Phys. **93**, 1403 (1992)]; D. V. Shirkov and S. V. Mikhailov, Zeit. Phys. **C63**, 463 (1994).

115. J. Chýla, Phys. Lett. **B 351**, 325 (1995).

116. T. Appelquist, J. Carazzone, Phys. Rev. **D11**, 2856 (1975).

117. M. Melles, hep-ph/9805216, Phys. Rev. **D58**:114004, 1998.

118. G.P. Lepage, J. Comp. Phys. **27**, 192 (1978); G.P. Lepage, Cornell preprint, CLNS-80/447, March 1980.

119. M. Peter, Nucl. Phys. **B501**, 471 (1997).

120. S. J. Brodsky, M. S. Gill, M. Melles, and J. Rathsman, Phys. Rev. **D58**, 116006 (1998).

121. T. Yoshino, K. Hagiwara, Z.Phys. **C 24**, 185 (1984).

122. F. Jegerlehner, O.V. Tarasov, hep-ph/9809485 and DESY 98-093.

123. C.T.H. Davies, *et al.*, Phys. Lett. **B345**, 42 (1995); Phys. Rev. **D56**, 2755 (1997).

124. A. Szczepaniak, E. M. Henley, S. J. Brodsky, Phys. Lett. **B243**, 287 (1990).

125. B. H. Smith, M. B. Voloshin, Phys. Lett. **B324**, 117 (1994). Erratum-ibid. **B333**, 564 (1994).

126. S.J. Brodsky, A. H. Hoang, J. H. Kuhn, and T. Teubner, Phys. Lett. **B359**, 355 (1995).

127. V. S. Fadin, V. A. Khoze, A. D. Martin, and W. J. Stirling, Phys. Lett. **B363**, 112 (1995).

128. P. N. Burrows, Acta Phys. Polon. **B28**, 701 (1997).

129. S. J. Brodsky, P. Damgaard, Y. Frishman and G. P. Lepage, Phys. Rev. **D33**, 1881 (1986).

130. D. Muller, Phys. Rev. **D59**, 116001 (1999); A. V. Belitsky and D. Muller, Nucl. Phys. **B537**, 397 (1999); D. Muller, Phys. Rev. **D49**, 2525 (1994).

131. J. Blumlein, V. Ravindran and W. L. van Neerven, hep-ph/9812450.

132. V. S. Fadin, E. A. Kuraev and L. N. Lipatov, Phys. Lett. **60B**, 50 (1975); L. N. Lipatov, Yad. Fiz. **23**, 642 (1976) [Sov. J. Nucl. Phys. **23**, 338 (1976)]; E. A. Kuraev, L. N. Lipatov and V. S. Fadin, Zh. Eksp. Teor. Fiz. **71**, 840 (1976) [Sov. JETP **44**,

443 (1976)]; *ibid.* **72**, 377 (1977) [**45**, 199 (1977)]; Ya. Ya. Balitskiĭ and L. N. Lipatov, Yad. Fiz. **28**, 1597 (1978) [Sov. J. Nucl. Phys. **28**, 822 (1978)].

133. V. S. Fadin and L. N. Lipatov, Phys. Lett. **429B**, 127 (1998).
134. G. Camici and M. Ciafaloni, Phys. Lett. **430B**, 349 (1998).
135. W. A. Bardeen, A. J. Buras, D. W. Duke and T. Muta, Phys. Rev. **D18**, 3998 (1978).
136. S.J. Brodsky, V.S. Fadin, V.T. Kim, L.N. Lipatov and G.B. Pivovarov, hep-ph/9901229.
137. W. Celmaster and R. J. Gonsalves, Phys. Rev. **D20**, 1420 (1979); Phys. Rev. Lett. **42**, 1435 (1979).
138. P. Pascual and R. Tarrach, Nucl. Phys. **B174**, 123 (1980); (E) **B181**, 546 (1981).
139. W. Celmaster and P. M. Stevenson, Phys. Lett. **125B**, 493 (1983).
140. L. N. Lipatov, Phys. Rept. **C286**, 131 (1997).
141. L. N. Lipatov, Zh. Eksp. Teor. Fiz. **90**, 1536 (1986) [Sov. JETP **63**, 904 (1986)]; in *Perturbative Quantum Chromodynamics*, ed. A.H. Mueller (World Scientific, Singapore, 1989) p. 411; R. Kirschner and L. Lipatov, Zeit. Phys. **C45**, 477 (1990).
142. S. J. Brodsky, F. Hautmann and D.E. Soper, Phys. Rev. **D56**, 6957 (1997); S. J. Brodsky, F. Hautmann and D. E. Soper, Phys. Rev. Lett. **78**, 803 (1997); F. Hautmann, talk at ICHEP96 (Warsaw, July 1996), preprint OITS 613/96, in Proceedings of the XXVIII International Conference on High Energy Physics, eds. Z. Ajduk and A. K. Wroblewski, World Scientific, p.705; J. Bartels, A. De Roeck and H. Lotter, Phys. Lett. **B389**, 742 (1996).
143. S. J. Brodsky, E. Gardi, G. Grunberg, and J. Rathsman (in preparation).
144. S.J. Brodsky, J.R. Pelaez and N. Toumbas, Phys. Rev. **D60**, 037501 (1999) hep-ph/9810424.

Introduction to
Transverse Lattice Gauge Theory

Simon Dalley

Department of Applied Mathematics and Theoretical Physics
Silver Street, Cambridge CB3 9EW, England

Abstract. I review ideas behind the light-front quantization of lattice gauge theories and give some illustrative results: [I] Transverse Lattice Gauge Theory; [II] Pure Glue; [III] Heavy Sources and Winding Modes; [IV] An Example – Large-N QCD in $2 + 1$ Dimensions. (Very recent results for $3 + 1$-dimensions can be found in the presentation of B. van de Sande in the workshop proceedings.)

I TRANSVERSE LATTICE GAUGE THEORY

A Light-Front Co-ordinates

In the constituent quark model, hadrons are composed of a few quarks moving about in empty space and bound by non-relativistic potentials. It is an *extremely* successful model. Before the advent of QCD as the quantum field theory of hadron structure, there was little reason to think of hadrons in any other way. But while there is consensus that QCD is fundamentally correct, the low-energy dynamics of this gauge theory are thought to involve a complicated vacuum with arbitrary numbers of relativistic quarks and gluons. How could these pictures be reconciled? Indeed, how can one efficiently tackle the problem of relativistic strongly bound states at all?

There is a popular yet poorly understood formulation of quantum field theory — Light-Front quantisation — where something like the constituent picture of bound-states arises naturally. In fact, it offers much more than that, since it furnishes a Schrodinger equation suitable for relativistic many-body systems.

In the co-ordinate system $\{x^0, x^1, x^2, x^3\}$, where x^0 is time (I set $c = \hbar = 1$), one usually initializes the wavefunction $\Psi(x^1, x^2, x^3)$ on a hyper-plane $x^0 = $ constant. If the four-momentum of the system is $\{P^0, P^1, P^2, P^3\}$, the energy-operator $\hat{P}^0$ is the equal-x^0 hamiltonian that evolves the wavefunction in x^0 according to Shrodinger's equation

$$i\frac{\partial \Psi}{\partial x^0} = \hat{P}^0 \Psi. \tag{1}$$

CP494, *New Directions in Quantum Chromodynamics*, edited by C.-R. Ji and D.-P. Min
© 1999 American Institute of Physics 1-56396-908-4/99/$15.00

This formulation has manifest rotational invariance and is tractable for most non-relativistic problems in physics. It is not so suitable when the momenta $P^{i=1,2,3} \sim P^0$ (recall that P^0 contains the rest-energy).

In light-front (LF) quantisation [1] on the other hand, we (conventionally) define

$$x^{\pm} = \frac{x^0 \pm x^3}{\sqrt{2}} \ , \quad P^{\pm} = \frac{P^0 \pm P^3}{\sqrt{2}} \ . \tag{2}$$

$P^{\pm}$ is canonically conjugate to $x^{\mp}$, and we interpret (x^-, P^+) as space and momentum, while (x^+, P^-) as time and energy. $\mathbf{x} = \{x^1, x^2\}$ and $\mathbf{P} = \{P^1, P^2\}$ are the transverse co-ordinates. The Minkowski metric in the new co-ordinate system is such that the invariant length

$$ds^2 = 2dx^+ dx^- - (dx^1)^2 - (dx^2)^2 \tag{3}$$

and $x^+ \equiv x_-$. The wavefunction Ψ_{lc} is initialized on a null hyper-plane $x^+ = $ constant and the LF Schrodinger equation that evolves $\Psi_{\mathrm{lc}}(x^-, x^1, x^2)$ in LF time x^+ is

$$i\frac{\partial \Psi_{\mathrm{lc}}}{\partial x^+} = \hat{P}^- \Psi_{\mathrm{lc}}. \tag{4}$$

This equation is now suitable for relativistic problems. For example, Ψ_{lc} is manifestly Lorentz-boost invariant; a detailed discussion of the advantages and historical survey of LF ideas can be found in ref. [2].

It is easy to see why LF co-ordinates may help with the constituent question, if we consider the energy-momentum relation for a (free) particle of mass μ

$$P^- = \frac{(P^1)^2 + (P^2)^2 + \mu^2}{2P^+} \ . \tag{5}$$

With the conventional interpretation of positive energy $(P^0 \geq 0)$ for particles and anti-particles, it follows that light-front momentum $P^+ \geq 0$. Because P^+ is conserved and the total P^+ of the vacuum must be zero (by translation invariance), each particle contributing to the vacuum state can have only $P^+ = 0$. But according to to (5), these 'zero modes' have infinite light-front energy. If we introduce a high energy cut-off, as is always necessary for defining a quantum field theory, these zero modes and the vacuum structure they carry will be explicitly removed. Their effects on observables below the cut-off should be felt through renormalisation of the hamiltonian. If this renormalisation entails quarks getting a constituent mass, and the gluon degrees of freedom getting at least a mass gap, if not constituent masses themselves, then we have arrived at the basic picture of the constituent quark model.

The trouble with the above idea is that the LF renormalisation procedure is still quite poorly understood for theories like QCD, though interesting efforts are being made in this direction [8–11]. A formulation of quantum field theory where non-perturbative effects normally associated with the vacuum have to appear via explicit counter-terms in the hamiltonian is also alien to many people. For these reasons there have been some reservations about the consistency of light-front quantisation. Nevertheless, the ideas are physically appealing and a number of examples are known where one can derive the renormalisation resulting from the removal of certain non-dynamical zero modes; see Burkardt's lectures at this school in a previous year [12]. The attitude I will take in these lectures is that the potential applications of light-front quantisation are far too important for it to be simply dismissed. I will describe a practical framework for doing calculations in LF QCD with a high-energy cut-off, pioneered by Bardeen and Pearson [13,14], which uses symmetry to guide renormalisation of the theory. The symmetries that define QCD are gauge, Lorentz, and chiral invariance. In these lectures, I will only discuss pure gauge theories and infinitely heavy quarks, so chiral invariance will not be treated.

The method will initially allow all operators in the LF hamiltonian that respect symmetries unviolated by the cut-off — allowed operators. After systematically truncating to a subset of these allowed operators, one then tunes the remaining couplings *a posteriori* to restore the violated symmetries in observables as best one can. This is a physically motivated, systematically improvable framework which does not use any fits to experimental data, i.e. it uses only first principles. At present, the tests for violations of Lorentz covariance are done in a rather inefficient way, simply by solving a whole bunch of hamltonians to find the correct one. Nevertheless, the recent results obtained in this way [5–7] are surprisingly accurate. Once a more efficient treatment of Lorentz covariance is developed, or non-perturbative LF hamiltonian renormalisation group ideas are better understood, vast new areas of non-perturbative QCD will be opened up.

B   Transverse Lattices

Light-front quantisation already has more manifest Lorentz symmetry than any other hamiltonian quantisation scheme [1], but typically one cannot avoid breaking rotational invariance; the choice x^3 in (2), rather than x^1 or x^2, is arbitrary. On the other hand, we would like to preserve gauge invariance (later this will also greatly facilitate the treatment of confinement). This can be done with a lattice cut-off [15,16]. In a hamiltonian formulation time remains continous and infinite, of course. For a LF hamiltonian $\hat{P}^-$, whose configuration space is $\{x^-, x^1, x^2\}$, only lattice discretization of the transverse directions **x** is appropriate. Discretizing the longitudinal co-ordinate x^- is not appropriate since it would cut-off large values of the conjugate variable P^+; but it's small values of P^+ that correspond to high energy. To remove the high-energy region in a gauge-invariant way we can impose, say, anti-periodic boundary conditions on x^- [17]. In the following we use indices, $\mu, \nu \in \{0, 1, 2, 3, \}$, $\alpha, \beta \in \{+, -\}$ and $r \in \{1, 2\}$, and sum over repeated indices. We introduce a transverse lattice spacing a and a longitudinal period $\mathcal{L}$.

To construct an $SU(N)$ transverse lattice (pure) gauge theory on this spacetime,

47

we introduce gauge potentials A_α in the Lie algebra of $SU(N)$ for the continuum directions, and lattice variables for the transverse directions. The particular choice of lattice variables which will be convenient for LF quantisation are complex NxN matrices $M_r(x^+, x^-, \mathbf{x})$ associated with the transverse link from $\mathbf{x}$ to $\mathbf{x} + a\hat{\mathbf{r}}$, at position (x^+, x^-) — colour-dielectric link variables. Because they are linear variables, it is easy to identify the independent degrees of freedom. It is sometimes helpful to think of them as an average over paths $\mathcal{C}$ between these points, with some weight $\rho(\mathcal{C})$, of the short-distance continuum gauge potentials

$$M_r = \sum_\mathcal{C} \rho(\mathcal{C}) \mathrm{P} \, \exp\left\{ \mathrm{i} \int_{\mathbf{x}}^{\mathbf{x}+a\hat{\mathbf{r}}} A_\mu dx^\mu \right\} \qquad (6)$$

Near the continuum limit $a = 0$, the potentials must change only very slowly over many lattice spacings, and M must be forced to lie in the $SU(N)$ group. At larger a there is no such restriction however. In fact, for non-abelian gauge theories, it makes more physical sense to use disordered complex variables M when the lattice cut-off a is quite large.

Continuum gauge transformations induce the following lattice gauge transformations on these variables

$$A_\alpha(\mathbf{x}) \rightarrow V(\mathbf{x}) A_\alpha(\mathbf{x}) V^\dagger(\mathbf{x}) + \mathrm{i} \left(\partial_\alpha V(\mathbf{x}) \right) V^\dagger(\mathbf{x}) , \qquad (7)$$
$$M_r(\mathbf{x}) \rightarrow V(\mathbf{x}) M_r(\mathbf{x}) V^\dagger(\mathbf{x} + a\hat{\mathbf{r}}) , \qquad (8)$$

where $V \in SU(N)$. The strategy will be to construct LF hamiltonians invariant under these gauge transformations and any Lorentz symmetries that have not be violated by the cut-offs; this include boosts along x^3 and discrete Z_4 rotations about x^3. We will seek to enhance the remaining Lorentz symmetries, generically broken by the cut-offs, by tuning the couplings of these hamiltonians.

The continuum QCD action is

$$- \int dx^4 \frac{1}{2g^2} \mathrm{Tr}\{F_{\mu\nu} F^{\mu\nu}\} . \qquad (9)$$

The following gauge-invariant transverse lattice action is the simplest that reduces to (9) in the naive continuum limit $a \rightarrow 0$, $\mathcal{L} \rightarrow \infty$, [14]

$$S = \int dx^+ dx^- \sum_{\mathbf{x}} \mathrm{Tr}\left\{ \overline{D}_\alpha M_r(\mathbf{x}) \left(\overline{D}^\alpha M_r(\mathbf{x}) \right)^\dagger \right\}$$
$$- \frac{a^2}{2g^2} \mathrm{Tr}\{F_{\alpha\beta} F^{\alpha\beta}\} + \frac{\beta}{Na^2} \mathrm{Tr}\{M_{\mathrm{plaq}} + M^\dagger_{\mathrm{plaq}}\} - U[M] \qquad (10)$$

where

$$\overline{D}_\alpha M_r(\mathbf{x}) = (\partial_\alpha + i A_\alpha(\mathbf{x})) M_r(\mathbf{x}) - i M_r(\mathbf{x}) A_\alpha(\mathbf{x} + a\hat{\mathbf{r}}) , \qquad (11)$$

M_{plaq} is the product of link matrices around a transverse plaquette, and $U[M]$ is any gauge-invariant potential that forces M into $SU(N)$ as $a \rightarrow 0$, such as

$$U[M] = \frac{N}{\lambda} \left(\mathrm{Tr}\{(1 - M_r^\dagger(\mathbf{x})M_r(\mathbf{x}))^2\} + (\det M - 1)^2 \right) , \qquad (12)$$

where $\lambda \to 0$ as $a \to 0$.

More generally, $S[M]$ can consist of all the combinations of $M, F^{\alpha\beta}, \bar{D}^\alpha M$ which are invariant under gauge and residual Lorentz symmetries. Their couplings are to be chosen so as to restore the symmetries violated by the cut-offs.

II PURE GLUE

A Colour-Dielectric Expansion

$S[M]$ contains an infinite number of allowed operators, so to begin calculations we need to truncate them to a finite set in some physically reasonable way. Of the two cut-offs, a and $\mathcal{L}$, it will be possible to extrapolate the latter. It makes sense therefore to restrict allowed operators on dimensional grounds w.r.t. $\{0, 3\}$ co-ordinates. The divergences that appear as $\mathcal{L} \to \infty$ are of normal-ordering type, and thus easily dealt with. We will also demand an interaction-independent (kinetic) momentum operator P^+ and naive LF parity $x^+ \leftrightarrow x^-$. The main further restriction will be to expand the LF hamiltonian $\hat{P}^-$, that results from $S[M]$, in powers of M; let us call it the 'colour-dielectric expansion'. This only makes sense if, for a given a, the P^- that best recovers Lorentz covariance is analytic about $M = 0$ and (classically) minimized there. (This could not be the case in a neighborhood of $a = 0$, where M should be forced into $SU(N)$). We are not guaranteed to find anything at all. There is, however, circumstantial evidence from Euclidean lattice work [18] that a Lorentz covariant scaling trajectory in the space of couplings flows from the continuum limit to a large a region where $M = 0$ is the classical minimum. Our job is to explictly verify the existence of such a trajectory for the transverse lattice by testing this region for signs of Lorentz covariance restoration.

B Gauge-Fixing

Light-front quantization is greatly simpled by the LF gauge choice $A_- = 0$.[1] This still leaves x^--independent gauge transformations unfixed. Suppose we start with an action

$$S = \int dx^0 dx^3 \sum_{\mathbf{x}} \left(\mathrm{Tr}\{\overline{D}_\alpha M_r(\mathbf{x}) \left(\overline{D}^\alpha M_r(\mathbf{x})\right)^\dagger\} \right.$$
$$\left. - \frac{1}{2G^2} \mathrm{Tr}\{F_{\alpha\beta}F^{\alpha\beta}\} - V_{\mathbf{x}}[M] \right) , \qquad (13)$$

---

[1] Because we use anti-periodic x^- boundaries, the $\int dx^- A_-$ zero mode has been removed from the theory. Its omission can be shown to renormalise the mass of the M field.

where $V[M]$ contains all gauge-invariant products of M up $O(M^4)$ and $G^2 \to g^2/a^2$ as $a \to 0$. This is still not yet a finite number of operators, but we can obtain a finite set by further requiring a degree of *transverse locality* for operators that are products of gauge-invariant operators (see later).

This action in LF gauge reduces to

$$
S(A_- = 0) = \int dx^0 dx^3 \sum_{\mathbf{x}} \left(\text{Tr}\{\partial_+ M_r(\mathbf{x})\partial_- M_r(\mathbf{x})^\dagger\} + \text{c.c.} \right.
$$
$$
\left. + \frac{1}{G^2} \text{Tr}\{(\partial_- A_+)^2\} + \text{Tr}\{A_+ J^+(\mathbf{x})\} - V_{\mathbf{x}}[M] \right) , \tag{14}
$$

where

$$
J^+ = i \left(M_r(\mathbf{x}) \overset{\leftrightarrow}{\partial}_- M_r^\dagger(\mathbf{x}) + M_r^\dagger(\mathbf{x} - a\hat{\mathbf{r}}) \overset{\leftrightarrow}{\partial}_- M_r(\mathbf{x} - a\hat{\mathbf{r}}) \right) . \tag{15}
$$

We note that the equation of motion for A_+ is a constraint (no x^+-derivatives)

$$
\frac{2}{G^2}(\partial_-)^2 A_+ = J^+ - \frac{1}{N} \text{Tr}\, J^+ , \tag{16}
$$

which can be inverted for A_+

$$
A_+ = \frac{G^2}{2} \frac{1}{\partial_-^2} \left(J^+ - \frac{1}{N} \text{Tr}\, J^+ \right) . \tag{17}
$$

C   Canonical LF Quantisation

The canonical momenta, including the LF hamiltonian P^-, can be deduced from the usual definition in terms of the energy momentum tensor $P^\alpha = \int dx^- \sum_{\mathbf{x}} T^{+\alpha}$;

$$
P^+ = \int dx^- \sum_{\mathbf{x}} \text{Tr}\{\partial_- M_r(\mathbf{x})\partial_- M_r(\mathbf{x})^\dagger\} \tag{18}
$$
$$
P^- = \int dx^- \sum_{\mathbf{x}} V[M] - \frac{G^2}{4} \text{Tr}\left\{J^+ \frac{1}{\partial_-^2} J^+\right\} + \frac{G^2}{4N} \text{Tr}\, J^+ \frac{1}{\partial_-^2} \text{Tr}\, J^+ . \tag{19}
$$

The transverse momenta $\mathbf{P}$ and boost-rotations $M^{\mu\nu}$ will be discussed shortly. M_r is conjugate to $\partial_- M_r^\dagger$, and we impose equal-x^+ commutation relations

$$
\left[M_{r,ij}(x^-, \mathbf{x}), \left(\partial_- M_{s,kl}(y^-, \mathbf{y})\right)^\dagger \right] = \frac{1}{2}\delta_{il}\, \delta_{jk}\, \delta(x^- - y^-)\, \delta_{\mathbf{x},\mathbf{y}}\, \delta_{r,s} . \tag{20}
$$

where $i, j \in \{1, \cdots N\}$ are colour indices.

It is useful to make a Fourier decomposition at $x^+ = 0$ with respect to x^- in terms of free fields

$$M_r(x^+ = 0, x^-, \mathbf{x}) = \frac{1}{\sqrt{4\pi}} \int_0^\infty \frac{dk^+}{\sqrt{k^+}} \left(a_{-r}(k^+, \mathbf{x}) \, e^{-ik^+x^-} + a_r^\dagger(k^+, \mathbf{x}) \, e^{ik^+x^-} \right) \quad (21)$$

that induces

$$\left[a_{\lambda,ij}(k^+, \mathbf{x}), a_{\rho,kl}^*(\tilde{k}^+, \mathbf{y}) \right] = \delta_{ik} \, \delta_{jl} \, \delta_{\lambda\rho} \, \delta_{\mathbf{x},\mathbf{y}} \, \delta(k^+ - \tilde{k}^+) , \quad (22)$$

$$\left[a_{\lambda,ij}(k^+, \mathbf{x}), a_{\rho,kl}(\tilde{k}^+, \mathbf{y}) \right] = 0 , \quad (23)$$

$\lambda, \rho \in \{\pm 1, \pm 2\}$, $a_{\lambda,ij}^* = (a_\lambda^\dagger)_{ji}$. Acting on a Fock vacuum $|0>$ defined by

$$a_{\lambda,ij}(k^+, \mathbf{x})|0 >= 0 \quad \forall \; k^+, \mathbf{x}, i, j, \lambda , \quad (24)$$

$a_{\lambda,ij}^*(k^+, \mathbf{x})$ creates a link-parton on $(\mathbf{x}, \mathbf{x}+a\hat{\lambda})$ on the transverse lattice and carrying longitudinal momentum k^+. (This mixed momentum-coordinate representation will be useful for displaying confinement shortly.) The general Fock space then splits into blocks of definite total P^+

$$\{a^*(P^+)|0 >, a^*(P^+ - k^+)a^*(k^+)|0 >, \cdots\} . \quad (25)$$

States of definite $\mathbf{P}$ will be constructed shorty. The above Fock basis can be used to write down a matrix representation of $\hat{P}^-$, which can then be diagonalised to obtain LF wavefunctions Ψ_{lc}.

D   Physical States

The first key point to note at this stage is that, because of positivity $k^+ > 0$, P^- and P^+ contain no terms with only creation operators a^*. This means that the Fock vacuum satisfies : $P^- : |0 >=: P^+ : |0 >= 0$. Of course, this is only true so long as M is a massive degree of freedom; this is the region of couplings we will be exploring for Lorentz covariance. Provided no tachyons appear in the physical spectrum of boundstates, $|0 >$ is the physical vacuum of the theory (in the presence of our cutoffs).

Fock space is further simplified by confinement [13]. Since the operator corresponding to k^+ is ∂_-, a glance at (19) shows that eliminating A_+ has introduced small-k^+ singularities. These are cut-off by $\mathcal{L}$, but will blow up as $\mathcal{L} \to \infty$ for fixed a in general. The divergence is avoided if the LF charge Q associated to the current J^+ (15) vanishes

$$\lim_{k^+ \to 0} \int dx^- e^{ik^+x^-} J^+ = 0 . \quad (26)$$

Only certain Fock states satisfy this condition, : Q : $|$Fock $>= 0$. They are the states invariant under the residual x^--independent gauge transformations

$$M_r(\mathbf{x}) \to V(\mathbf{x})M_r(\mathbf{x})V^\dagger(\mathbf{x} + a\hat{\mathbf{r}}) , \quad \partial_- V = 0 . \quad (27)$$

51

This will include, for example, closed loops of link-partons on the transverse lattice

$\mathrm{Tr}\{a_\lambda^\dagger(P^+ - k^+, \mathbf{x})a_{-\lambda}^\dagger(k^+, \mathbf{x})\}|0>$,

$\mathrm{Tr}\{a_\lambda^\dagger(P^+ - k_1^+ - k_2^+ - k_3^+, \mathbf{x})a_\lambda^\dagger(k_1^+, \mathbf{x} + a\hat{\lambda})a_{-\lambda}^\dagger(k_2^+, \mathbf{x} + 2a\hat{\lambda})a_{-\lambda}^\dagger(k_3^+, \mathbf{x} + a\hat{\lambda})\}|0>$,

etc..

This simple geometrical picture of the finite-energy, gauge-singlets is why we chose to work in transverse configuration space but longitudinal momentum space initially.

A remarkable property of the Fock space of connected closed loops of links[2] is that it is finite-dimensional in a sector of fixed P^+ and $\mathbf{P}$ for given cut-off $\mathcal{L}$. If we write the momentum of the ith parton as

$$k_i^+ = \frac{n_i P^+}{K} \; , \tag{28}$$

where $n_i \in \{1/2, 3/2, \ldots, K - 1/2\}$ and K is an integer that plays the role of dimensionless cut-off, the Fock space is finite-dimensional because individual momenta are discrete and bounded and the number of partons is bounded (by $2K$). This is the basis of the numerical treatment known as DLCQ [19]. Matrices of finite dimension can be calculated and diagonalised, and the results extrapolated to large K.

E   Momentum Eigenstates

To test Lorentz covariance it will be important to have explicit forms for bound-state wavefunctions at non-zero momentum. So far we have constructed a Fock basis diagonal in $\hat{P}^+$, with modes localised on transverse links. To obtain states of definite non-zero transverse momentum $\mathbf{P}$, we can take a gauge-singlet p-link shape, say, and translate it over the transverse lattice sites $\mathbf{y}$ according to

$$\sum_{\mathbf{y}} e^{i\mathbf{P}\cdot(\mathbf{y}+\bar{\mathbf{x}})} \mathrm{Tr}\left\{a_{\lambda_1}^\dagger(k_1^+, \mathbf{x}_1 + \mathbf{y})\, a_{\lambda_2}^\dagger(k_2^+, \mathbf{x}_2 + \mathbf{y}) \cdots a_{\lambda_p}^\dagger(k_p^+, \mathbf{x}_p + \mathbf{y})\right\}|0\rangle \; . \tag{29}$$

We must have

$$\sum_{i=1}^{p} \hat{\lambda}_i = 0$$

$$\mathbf{x}_i = \mathbf{x}_{i-1} + a\,\hat{\lambda}_{i-1} \, , \quad i < 1 \le p \tag{30}$$

$$\sum_{i=1}^{p} k_i^+ = P^+$$

---

[2] These are the relevant ones in the $N = \infty$ limit for example.

and it is convenient to set the phase convention by adopting a momentum-weighted 'centre-of-mass'

$$\bar{\mathbf{x}} = \frac{1}{P+} \sum_{i=1}^{p} k_i^+ \left(\mathbf{x}_i + \frac{a \hat{\lambda}_i}{2} \right) . \tag{31}$$

In this case, the M_{-r} component of the boost–rotation tensor gives

$$e^{-ib^r M_{-r}} a_\lambda^\dagger (k^+, \mathbf{x}) e^{ib^r M_{-r}} = a_\lambda^\dagger (k^+, \mathbf{x}) e^{-ik^+ \mathbf{b} \cdot (\mathbf{x} + a\hat{\lambda}/2)} \tag{32}$$

$$e^{-ib^r M_{-r}} |\Psi_{\text{lc}}(P^+, \mathbf{P})\rangle = |\Psi_{\text{lc}}(P^+, \mathbf{P} - \mathbf{b}P^+)\rangle , \tag{33}$$

and non-zero $\mathbf{P}$ states in the Brillouin zone $(-\pi/a, \pi/a)$ can be conveniently obtained from those at $\mathbf{P} = 0$. Longitudinal boosts are even more trivial, being simply a rescaling

$$e^{isM_{+-}} |\Psi_{\text{lc}}(P^+, \mathbf{P})\rangle = |\Psi_{\text{lc}}(e^{-s} P^+, \mathbf{P})\rangle . \tag{34}$$

Thus, longitudinal momentum fractions $x_i = k_i^+/P^+$ are Lorentz-boost invariant. A perfectly relativistic state should have a dispersion of the form

$$\mathcal{M}^2 = 2P^+ P^- - \mathbf{P}^2 \tag{35}$$

and so diagonalising $\hat{P}^-$ in a basis of fixed P^+ and $\mathbf{P}$ is equivalent to finding the masses $\mathcal{M}$ of boundstates. Because of the transverse lattice however, there will be corrections to (35) (see later) which we will try to minimize as part of the optimization of Lorentz covariance.

III   HEAVY SOURCES AND WINDING MODES

Pure QCD is characterised by a single scale in terms of which all dimensionful quantities, such as hadron masses, can be expressed. A convenient measure of this scale is the string tension σ, the asymptotic slope of the linearly-rising confining potential between heavy sources. Moreover, by measuring this slope in both transverse lattice directions $\mathbf{x}$ and the continuum space direction x^3, and requiring it to be rotationally invariant, we fix the lattice spacing a in dimensionful units.

A   Winding Modes

In the transverse directions, a more accurate measurement of σ can be made by examining winding modes rather than heavy sources. By making the transverse lattice compact in direction $\hat{\lambda}$ say

$$\mathbf{x} \equiv \mathbf{x} + a\hat{\lambda}D_\lambda , \tag{36}$$

where D_λ is the number of transverse links in direction λ, we can construct a basis of Fock states that wind around these directions, e.g.

$$\text{Tr}\left\{a^\dagger_{\lambda_1}(k_1^+, \mathbf{x})\, a^\dagger_{\lambda_2}(k_2^+, \mathbf{x} + a\hat{\lambda}_1) \cdots a^\dagger_{\lambda_p}(k_p^+, \mathbf{x} + a\hat{\lambda}D_\lambda - a\hat{\lambda}_p)\right\}|0\rangle \tag{37}$$

has winding number 1 in direction $\hat{\lambda}$. The mass spectrum of such winding modes should rise linearly with D_λ with slope σ as $D_\lambda \to \infty$. Unlike the potential between heavy sources, there are no 'endpoint' effects, and so the asymptotic linear rise should set in more quickly, especially for the lowest mass eigenvalue

$$\mathcal{M} \to \sigma_\text{T}|\mathbf{n}|a \quad, \quad |\mathbf{n}| \to \infty\,, \tag{38}$$

where $\mathbf{n} = (D_1 n_1, D_2 n_2)$ and n_1, n_2 are winding numbers in directions $(1,0)$ and $(0,1)$. (Note that we cannot construct winding modes around a compactified x^3 as this clashes with the choice of LF coordinates). The suffix 'T' indicates a transverse measurement of σ.

B Heavy Sources

To measure the potential between two heavy sources in transverse lattice gauge theory [20], we will start with a heavy scalar field $\phi(x^+, x^-, \mathbf{x})$ of large mass ρ in the fundamental representation of $SU(N)$. In addition to the pure glue 'link-link' interactions, we must now also consider interactions with ϕ's. The simplest are

$$\left(\partial_\alpha \phi^\dagger - i\phi^\dagger A_\alpha\right)\left(\partial_\alpha \phi + iA_\alpha \phi\right) \quad, \quad -\rho^2 \phi^\dagger \phi \tag{39}$$

such that ϕ couples to M via A_α.[3] In LF gauge $A_- = 0$ this leads to the substitution

$$J^+ \to J^+_\text{pure glue}[M] + J^+_\text{source}[\phi] \tag{40}$$

$$(J^+_\text{source})_{ij} = -i\partial_-\phi_i\phi_j^* + i\phi_i\partial_-\phi_j^* \tag{41}$$

Now for some kinematics: Let P^α_full represent the full 2-momentum of a system containing h heavy particles. It is convenient to split the full momentum into a "heavy" part plus a "residual" part P^α due to interactions,

$$P^\alpha_\text{full} = \rho h v^\alpha + P^\alpha\,, \tag{42}$$

where v^α is the covariant velocity of the heavy sources, $v^\alpha v_\alpha = 1$. Note that P^α, as defined here, is not positive definite. The full invariant mass-squared (at $\mathbf{P} = 0$) is

$$\mathcal{M}^2 = 2P^+_\text{full}P^-_\text{full} = (h\rho)^2 + 2h\rho v^+ P^- + 2P^+\left(P^- + h\rho v^-\right) \tag{43}$$

The choice of v^+ is arbitrary and it is convenient to choose it such that $P^+ = 0$. Consequently, $v^+ P^-$ is just the shift of the full invariant mass $\mathcal{M}$ due to the interactions:

$$\mathcal{M} = h\rho + v^+ P^- + O\left(1/\rho\right)\,. \tag{44}$$

Thus, $v^+ P^-$ is the usual energy associated with the heavy quark potential.

---

[3] The full set of operators allowed at leading order of the colour-dielectric expansion is described in ref. [7], but we skip the details here.

C   Fock Space

We define creation-annihilation operators associated with the heavy field:

$$\phi_i(x^+ = 0, x^-, \mathbf{x}) = \frac{1}{\sqrt{4\pi}} \int_{-\infty}^{\infty} \frac{dk}{\sqrt{\rho v^+ + k}} \left(b_i(k, \mathbf{x})\, e^{-i(v^+\rho + k)x^-} + d_i^*(k, \mathbf{x})\, e^{i(v^+\rho + k)x^-} \right)$$

(45)

$$\left[b_i(k, \mathbf{x}), b_j^*(\tilde{k}, \mathbf{y}) \right] = \delta_{ij}\, \delta_{\mathbf{x},\mathbf{y}}\, \delta(k - \tilde{k}) , \quad \text{etc.}$$

(46)

The $e^{i\rho v_\alpha x^\alpha}$ term removes an overall ρv^α from the 2-momentum. $b_i^*(k, \mathbf{x})$ creates a source particle at site $\mathbf{x}$ carrying residual momentum k, while d^* creates an antiparticle. The residual gauge invariance in $A_- = 0$ gauge again leads to confinement into singlet states. An example of a gauge-invariant Fock state with two co-moving sources at $\mathbf{x}$ and $\mathbf{y}$ respectively, maintaining a fixed a x^3-separation L is

$$\int_{-\infty}^{\infty} dl\, e^{2iLlv^-} b_i^*(l - P_{\text{link}}/2, \mathbf{x})\, a_{ij}^*(k_1, \mathbf{x}) \cdots a_{mn}^*(k_p, \mathbf{y})\, d_n^*(-l - P_{\text{link}}/2, \mathbf{y})|0>$$

(47)

where $\sum_{i=1}^{p} k_i = P_{\text{link}}$. To obtain LF eigenfunctions of $\hat{P}^-$ we must take linear combinations of all possible $P_{\text{link}} > 0$. In practice this requires us to introduce a high-energy cut-off on P_{link}, in addition to the DLCQ one, K. $v^+ \hat{P}^-$ is then a finite dimensional matrix (at least for the connected Fock states that dominate in the large-N limit) whose eigenvalues may be extrapolated to infinite cutoff.

The lowest eigenvalue of $v^+ \hat{P}^-$ can be compared with the popular phenomenological form of the potential between heavy sources of spatial separation R

$$V(R) = \sigma R + c_1 + \frac{c_2}{R}$$

(48)

$$R = \sqrt{a^2|\mathbf{n}|^2 + L^2} \quad \text{Source Separation,}$$

where $a\mathbf{n} = \mathbf{y} - \mathbf{x}$ and $L = y^3 - x^3$. For generic couplings in $\hat{P}^-$ the eigenvalues will not simply be a function of R, i.e. σ, c_1, c_2 extracted from (48) will depend upon L and $|n|$ separately. Once rotational invariance is restored, the value of σ can be set from experiment by the masses and decays of heavy mesons using $V(R)$ in a conventional Schrodinger equation. A typical value is $\sqrt{\sigma} \sim 440 MeV$, though one must remember that this will includes effects of light-quarks which are neglected in pure gauge theory. Higher eigenvalues of $v^+ \hat{P}^-$ should correspond to hybrid heavy mesons.

D   Setting the Scales

We now have everything necessary for performing a first-principles calculation. Let us now combine results from each sector — pure glue, winding, heavy sources.

Pure Glue

For generic couplings in $\hat{P}^-$, the glueball eigenstates at fixed momenta $(P^+, \mathbf{P})$ have eigenvalues which we may expand in powers of transverse momentum thus

$$2P^+P^- = G^2N \left(\mathcal{M}_0^2 + \mathcal{M}_1^2 a^2 |\mathbf{P}|^2 + 2\overline{\mathcal{M}}_1^2 a^2 P^1 P^2 + \mathcal{M}_2^2 a^4 |\mathbf{P}|^4 + \cdots \right) . \tag{49}$$

Here we have chosen to factor out the coupling G^2 as an overall mass scale (G has units of energy). $\mathcal{M}_0, \mathcal{M}_1, \overline{\mathcal{M}}_2, \cdots$ are then dimensionless numbers which we calculate when diagonalising $\hat{P}^-$. Eq.(49) is not, of course, in the form of a relativistic dispersion relation (35) in general. The coefficients, however, are functions of the couplings at our disposal and may be adjusted to regain a relativistic form.

Winding Modes

Let us write the lowest winding eigenvalue (38) as

$$2P^+P^- = a^2 \sigma_T^2 |\mathbf{n}|^2 = G^2 N \mathcal{W} |\mathbf{n}|^2 \ , \quad |\mathbf{n}| \to \infty \ . \tag{50}$$

Again, we factor out G^2 to set the scale, and $\mathcal{W}$ is a dimensionless measured quantity dependent on the couplings.

Heavy Sources

Similarly, at fixed sources separation R

$$v^+P^- = \sigma R = G^2 N \mathcal{S} R \ , \quad R \to \infty \ . \tag{51}$$

$\mathcal{S}$ is the dimensionless measured quantity that we can adjust with the couplings in $\hat{P}^-$.

The following conditions are necessary for Lorentz covariance of eigenstates

$$\sigma_T = \sigma = \text{constant} \tag{52}$$
$$\mathcal{M}_1^2 a^2 G^2 N - 1 = 0 \tag{53}$$
$$\overline{\mathcal{M}}_1 = 0 \tag{54}$$

Of course, there are further conditions that one could impose on solutions. But as we shall see in the next lecture, in practice these seem to be the most significant when trying to remove lattice discretization errors to a first approximation.

To test for (53) we need the dimensionless combination $a^2 G^2 N = \mathcal{W}/\mathcal{S}^2$ that follows from (50)(51). Combining (49)(51) allows us to express boundstate masses in units of the string tension

$$2P^+P^- (\mathbf{P} = 0) = \mathcal{M}^2 = \frac{\mathcal{M}_0^2 \sigma}{\mathcal{S}} \tag{55}$$

and also the lattice spacing itself $a = \sqrt{\mathcal{W}/\mathcal{S}\sigma}$.

IV   EXAMPLE – LARGE-N QCD IN $2+1$ DIMENSIONS

We need to do a test calculation that can be accurately compared with known results and is relevant to non-abelian gauge theories in 3+1 dimensions. $SU(\infty)$ gauge theory in $2+1$ dimensions provides an ideal test. It is physically very similar to $SU(3)$ pure gauge theory in $3+1$ dimensions, having dynamically-generated linear confinement and a discrete spectrum of glueballs; it is even quantitatively similar. But factorization of colour Traces in the Fock space states in the large-N limit and having only one transverse dimension make an accurate calculation feasible. Moreover, good Euclidean lattice Monte Carlo (ELMC) data has recently become available for this problem [21], which we can take as 'the right answer'.[4] The results presented here have been done with B. van de Sande and are new or improvements on our previously published work [5].

The discussion of the previous lectures is trivially adapted to $2+1$ dimensions. Here $x^{\pm} = (x^0 \pm x^2)/\sqrt{2}$ etc. and x^1 is the single transverse lattice dimension. Up order order M^4, we have in the large N limit with G^2N finite (suppressing transverse indices for clarity)

$$P^- = \int dx^- -\frac{G^2}{4}\,\mathrm{Tr}\left\{J^+\frac{1}{\partial_-^2}J^+\right\} + \frac{G^2}{4N}\,\mathrm{Tr}J^+\frac{1}{\partial_-^2}\,\mathrm{Tr}J^+ + \mu^2\,\mathrm{Tr}\left\{MM^\dagger\right\}$$
$$+\frac{\lambda_1}{aN}\,\mathrm{Tr}\left\{MM^\dagger MM^\dagger\right\} + \frac{\lambda_2}{aN}\,\mathrm{Tr}\left\{MMM^\dagger M^\dagger\right\} + \frac{\lambda_3}{aN^2}\left(\mathrm{Tr}\left\{MM^\dagger\right\}\right)^2 \quad (56)$$

where, in anticipation of always taking the $\mathcal{L} \to \infty$ limit, we have also used power-counting in longitudinal coordinates to limit the number of terms. There are in principle an infinite number of different products of individually gauge-invariant operators (like the λ_3 term), obtained by separating the operators in the transverse direction. We have also assumed a degree of transverse locality by keeping only the most local product to a first approximation. This approximation, like the colour-dielectric expansion in gauge-invariant powers of M, can of course be systematically investigated.

To maintain a trivial LF vacuum, we shall be considering the region $\mu^2 > 0$ with tachyon-free physical spectrum. G has dimensions of energy and will be used to express dimensionful quantities; it is convenient to form dimensionless versions of all the other couplings

$$m^2 = \frac{\mu^2}{G^2N}\,, \qquad l_i = \frac{\lambda_i}{aG^2N}\,. \quad (57)$$

The basic technique we follow is to search the space $\{m, l_1, l_2, l_3\}$ for a one parameter trajectory on which observables show enhancement of Lorentz covariance — a

---

[4]   The calculations of Teper are based on the traditional Wilson Euclidean lattice path integral formulation of gauge theory. The large N calculations were in fact performed about the same time as the light-front work, and to some extent motivated by it.

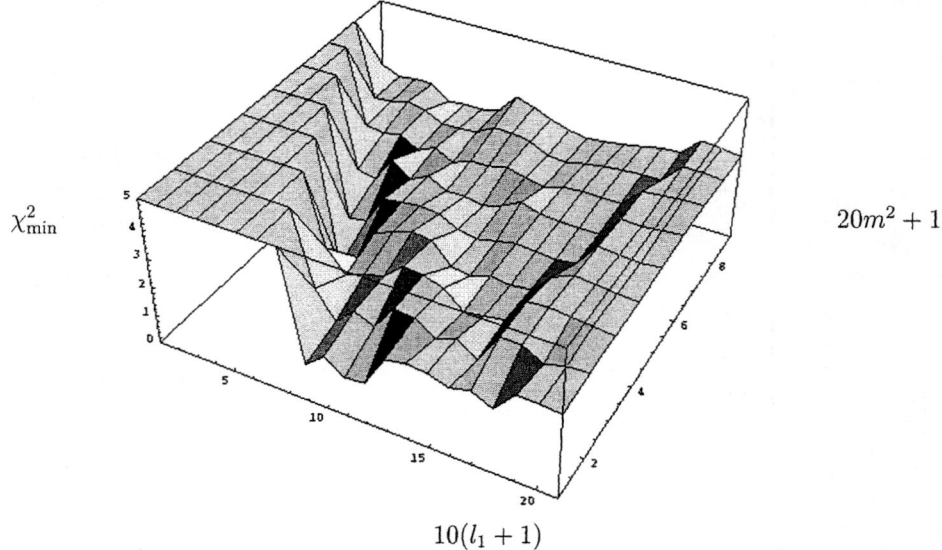

$$\chi^2_{\min} \qquad\qquad\qquad\qquad\qquad\qquad\qquad 20m^2 + 1$$

$$10(l_1 + 1)$$

FIGURE 1. χ^2-chart for l_1.

'Lorentz trajectory'. Moving along this trajectory should correspond to changing the spacing a, eventually taking us to the continuum limit. We will, in fact, be prevented from reaching $a = 0$ by our self-imposed restriction $m^2 > 0$. Nevertheless, assuming gauge and Lorentz covariance *define* pure QCD, observables should be invariant along the (exact) Lorentz trajectory.[5] Of course, we can only obtain some approximation to such a trajectory with a finite number of couplings.

To get some intuition for the space of couplings, let us construct charts that display the 'size' of violations of Lorentz covariance. We can use a χ^2-test based on the variables (52))(53) (condition (54) is absent in $2+1$-dimensions). Condition (53) in fact provides a number of variables, one for each boundstate (glueball) in the low-lying spectrum. To begin with, let us set the lowest glueball mass to the known value $\mathcal{M} = 4.05\sqrt{\sigma}$ [21], and try to get an isotropic speed of light (53) for this and the heavier low-lying glueballs. With a particular choice of variance assignments in the χ^2-test, we can make plots of $\{\chi^2_{\min}, m^2, l_i\}$ for each l_i, where $\chi^2_{\min}$ is the minimum value of χ^2 with respect to the couplings except m^2 and l_i. For technical reasons l_3 almost completely decouples (we can set it to almost any fixed large value, see ref. [5]) so we only display charts for l_1 (fig.1) and l_2 (fig.2).

---

[5] The idea is similar to the renormalised trajectory in the renormalisation group associated to a fixed point with one relevant direction. However we are not performing any explicit RG transformations here!

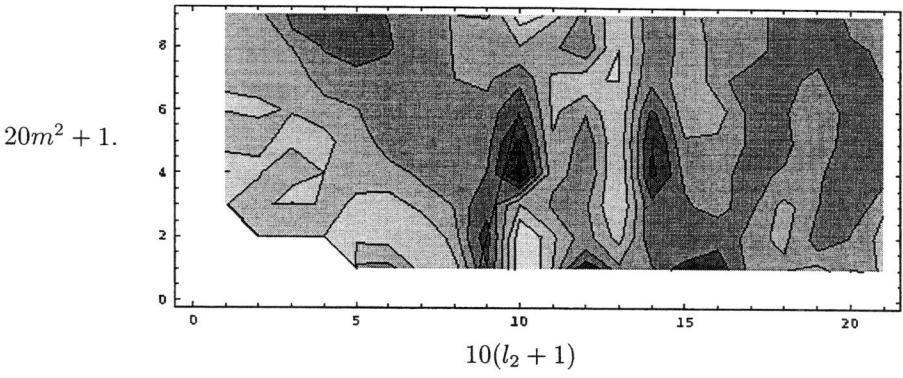

$20m^2 + 1.$

$10(l_2 + 1)$

FIGURE 2. χ^2-chart for l_2. Darker shades correspond to lower $\chi^2_{\min}$.

They each show an unique, narrow valley, running from small to large m^2, at the bottom of which Lorentz covariance is optimised. One finds that changing the details of the χ^2 test changes these charts somewhat, *except* in the neighborhood of the valley in each case. Thus one finds a degree of universality in the results. If there were no Lorentz trajectory, one would not expect to obtain a robust valley. Having roughly located the candidate Lorentz trajectory, one can increase the resolution in its neighborhood, and perform more efficient iterative searches for the bottom of the valley.

In a first-principles calculation we cannot use the result for the lightest glueball mass in σ units — this should be predicted by the method. Performing a new χ^2 search in the neighborhood of the valley, but now including a variable to test (52), one finds a similar position for the valley bottom, without having to input the lightest glueball mass. The χ^2 per degree of freedom is always about 1 near the Lorentz trajectory, as it should be for reasonable criteria. Results as one moves along the valley bottom are displayed in fig.3 and fig.4. We see that the lattice spacing gradually decreases with m^2 as one moves along the Lorentz trajectory, but never becomes zero for $m^2 > 0$. Despite fluctuations, the lightest glueball mass scales in a way roughly consistent with the correct continuum answer.

Although the lightest glueball (a $J^{PC} = 0^{++}$) is behaving covariantly along the Lorentz trajectory — the speed of light deduced from the 0^{++}'s dispersion is isotropic to with % 2-3 — fluctuations are still present in fig.4. These are mostly due to the difficulty in accurately establishing the scale σ. Most of the fluctuations tend to cancel if we consider ratios of glueball masses, and good scaling is obtained along the Lorentz trajectory [5]. For higher glueballs with nearly-covariant wavefunctions, such as the 0^{--} and its excited state 0^{--}_*, this provides rather accurate mass ratio determinations in the large-N limit; see figure 5. Finite-N results from conventional

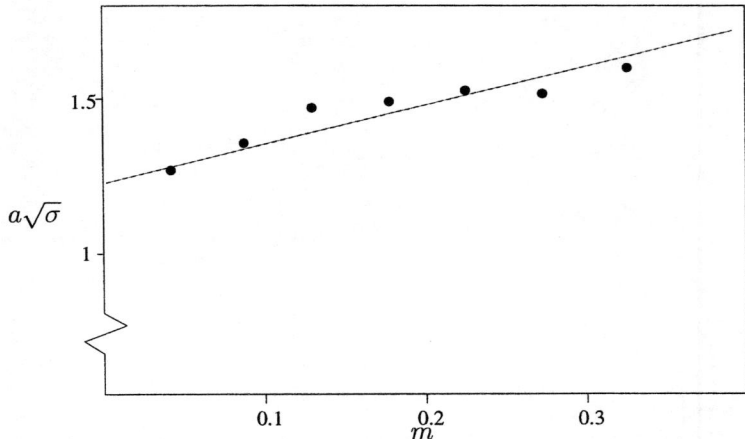

FIGURE 3. Variation of the transverse lattice spacing along the Lorentz trajectory. The fit is $1.275m + 1.23$.

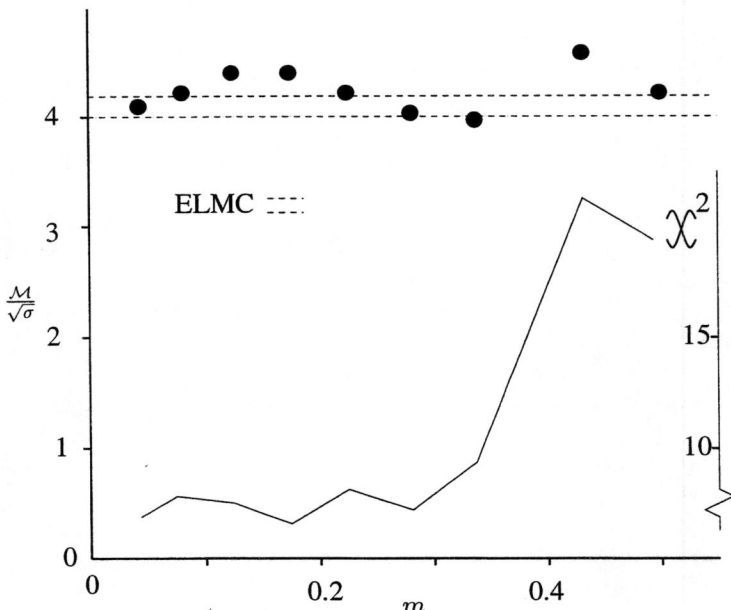

FIGURE 4. The variation of the lightest glueball mass along the Lorentz trajectory (together with the associated variation of the χ^2).

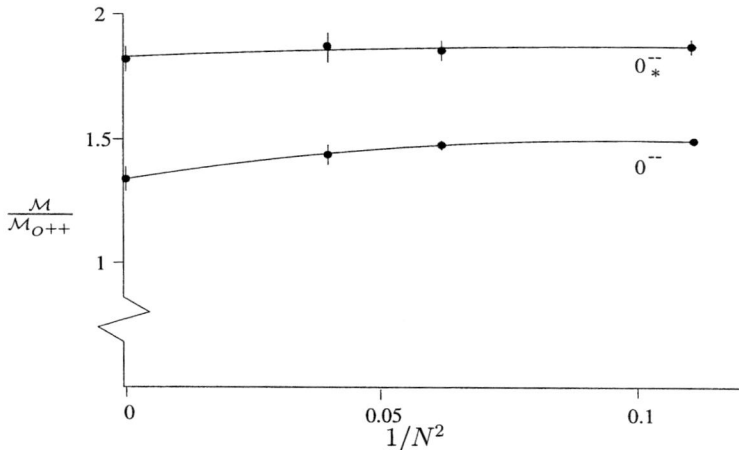

FIGURE 5. Variation (or lack of it!) of the glueball mass ratios $0^{--}/0^{++}$ and $0^{--}_*/0^{++}$ with N. The fits are $1.349 + 2.914/N^2 - 13.968/N^4$ for 0^{--} and $1.824 + 1.274/N^2 - 7.143/N^4$ for 0^{--}_*. The error on the $N = \infty$ results is from extrapolation to $\mathcal{L} = \infty$.

ELMC in ref. [21] were fit to $A + B/N^2$, at confidence levels of order $\%50 - 70$. Including our $N = \infty$ data, and fitting to $A + B/N^2 + C/N^4$, improves this to $\%95$ for $0^{--}_*/0^{++}$ and $> \%99$ for $0^{--}/0^{++}$ glueball mass ratios!

The heavy-source potential calculated at a point on the Lorentz trajectory is displayed in fig.6. It shows good restoration of spatial symmetry. The potential in the continuum spatial direction x^2 is a fit to

$$v^+ P^- = 0.154 L G^2 N + 0.183\sqrt{G^2 N} - \frac{0.178}{L} \tag{58}$$

at the point on the Lorentz trajectory of lowest overall χ^2. One must be careful when interpreting (58) since the Coulomb potential in $2 + 1$ dimensions is logarithmic. The form (58) should be appropriate except at the very smallest L, where Coulomb corrections are expected. The $1/L$ term is a universal correction expected on the grounds of models of flux-string oscillations [22]. Universality implies that its coefficient should be invariant along the Lorentz trajectory. The fact that, in reality, it drifts slowly, is a symptom that our approximation to the Lorentz trajectory is not an exact scaling trajectory and/or the form (58) is not sufficient to fit the potential. Nevertheless, the coefficient 0.178 is close to the theoretical value $d\pi/24$, where d is the number of transverse dimensions in which an ideal thin-string can oscillate ($d = 1$ here). Interestingly, it is too large (0.178 gives $d \sim 1.4$) by roughly the same amount as that given by another method to measure the 'central charge' $d + 2$, using the density of glueball states [4]. In ref. [4] we observed that this excess in the central charge comes from longitudinal degrees of freedom in the large-N QCD flux string, a mode of oscillation which an ideal thin-string does not possess.

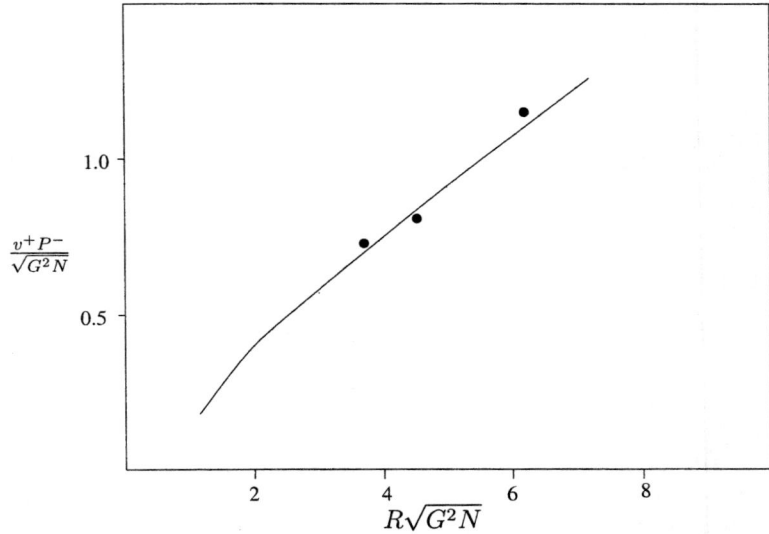

FIGURE 6. The heavy-source potential. Solid line is fit to potential for sources with x^2-separation only; data points are values at one-link transverse separation and x^2-separation $L\sqrt{G^2 N} = 0, 2.5, 5$.

One may wonder whether results really do improve as more operators are added to the hamiltonian. For a given χ^2 test, adding more operators to the hamiltonian will inevitably bring one closer to the Lorentz trajectory as measured by this test. We see from fig. 7 that the masses of glueballs, which are not part of the test, rapidly approach the correct values as more operators are added and covariance is improved.[6]

A   Future Work

The transverse lattice idea is already quite old [13]. The breakthrough, that has been made over the last year or so, was to show that in $2 + 1$ and $3 + 1$ dimensions large-N gauge theory exhibit a unique Lorentz covariant scaling trajectory on coarse transverse lattices [5–7]. Glueball masses on this trajectory are consistent with known results. Thus there is good reason to believe that the Lorentz-invariant wavefunctions obtained with this method are accurate also. They are completely new results, essentially unobtainable within any other quantisation scheme. It will be interesting to see what implications they have for experiment. The next step towards this goal is to couple these pure gauge theories to propagating quarks [13,23,7], and then enforce the Lorentz and chiral symmetries that define QCD to

---

[6] The 2^{-+} state appears to overshoot the correct mass. It is the only state whose wavefunction actually becomes less covariant when λ_1 is turned on.

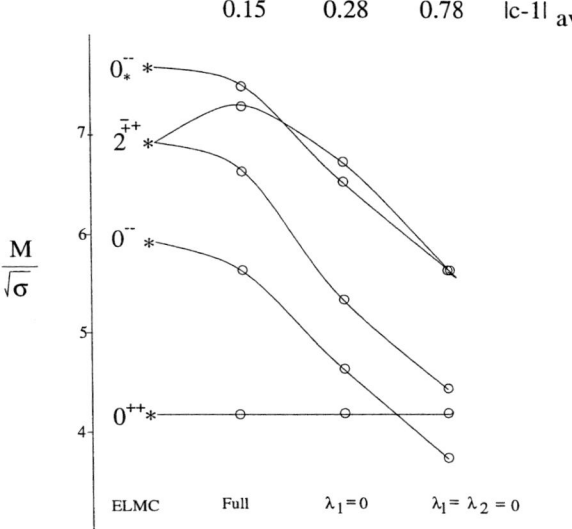

FIGURE 7. Variation of glueball mass ratios to the lightest mass (0^{++}), as more operators are included in the hamiltonian. $|c-1|_{av}$ measures the average deviation from 1 of the speed of light in the transverse direction, for the displayed glueballs.

determine the couplings of the hamiltonian. We also need to understand better how to extrapolate results into the small a region. This is important not only formally for the existence of the continuum limit, but to compare calculations of hadronic structure with hard process experiments where the usual factorization theorems simplify the analysis. We are only at the beginning of the development of this subject, and it is difficult to know how far it will progress, but there does appear to be too much truth in it to ignore.

Acknowledgments I would like to thank Prof. Ji and Prof. Min for inviting me to give these lectures and for their hospitality in Korea.

REFERENCES

1. P. A. M. Dirac, *Rev. Mod. Phys.* **21**, (1949) 392.
2. S. J. Brodsky, H.-C. Pauli, and S. Pinsky, *Phys. Rep.* **301**, (1998) 299.
3. S. Dalley and B. van de Sande, *Nucl. Phys.* **B53** (Proc. Suppl.) (1997) 827.
4. S. Dalley and B. van de Sande, *Phys. Rev.* **D56**, (1997) 7917.
5. S. Dalley and B. van de Sande, *Phys. Rev.* **D59**, (1999) 065008.
6. S. Dalley and B. van de Sande, *Phys. Rev. Lett.* **82**, (1999) 1088.
7. S. Dalley and B. van de Sande, preprint DAMTP-99-107.
8. S. Glazek and K. G. Wilson, *Phys. Rev.* **D47**, (1993) 4657.
9. M. Brisudova, R. J. Perry, and K. G. Wilson, *Phys. Rev. Lett* **78**, (1997) 1227.
10. K. G. Wilson *et. al.*, *Phys. Rev.* **D49**, (1994) 6720.
11. R. J. Perry, Lectures at *Hadrons 94*, Gramado, Brasil (1994) hep-th/9407056; B. H. Allen and R. J. Perry, *Phys. Rev.* **D58**, (1998) 125017.
12. M. Burkardt, Lectures at NuSS97, Seoul, (1997), hep-ph/9709421.
13. W. A. Bardeen and R. B. Pearson, *Phys. Rev.* **D14**, (1976) 547.
14. W. A. Bardeen, R. B. Pearson, and E. Rabinovici, *Phys. Rev.* **D21**, (1980) 1037.
15. K. G. Wilson, *Phys. Rev.* **D10**, (1974) 2445.
16. J. B. Kogut and L. Susskind, *Phys. Rev.* **D11**, (1975) 395.
17. A. Casher, *Phys. Rev.* **D14**, (1976) 452.
18. H.-J. Pirner, *Prog. Part. Nucl. Phys.* **29**, (1992) 33.
19. H.-C. Pauli and S. J. Brodsky, *Phys. Rev.* **D32**, (1985) 1993 and 2001; S. Dalley and I. R. Klebanov, *Phys. Lett.* **B298**, (1993) 79; *Phys. Rev.* **D47**, (1993) 2517.
20. M. Burkardt and B. Klindworth, *Phys. Rev.* **D55**, (1997) 1001; in *Confinement III*, Newport News VA, (June 1998), hep-ph/9809283.
21. M. Teper, *Phys. Rev.* **D59**, (1998) 014512.
22. M. Luscher, *Nucl. Phys.* **B180**, (1981) 317.
23. M. Burkardt and H. El-Khozondar, *Phys. Rev.* **D60**, (1999) 054504.

Nuclear Physics on the Light Front–a new old way to do an old new problem

Gerald A. Miller*

*Department of Physics
University of Washington
Seattle, Washington 98195-1560[1]

Abstract. A brief introduction to light front techniques is presented. This is followed by a review of recent attempts to perform realistic, relativistic nuclear physics with those techniques.

MOTIVATION

This lecture series is aimed at describing our recent attempts to derive the properties of nuclei using the light-front formalism. Nuclear properties are very well handled within existing conventional nuclear theory, so it is necessary to explain the motivation. It seems to me that understanding experiments involving high energy nuclear reactions requires that light-front dynamics and light cone variables be used. Consider the EMC experiment [1], which showed that there is a significant difference between the parton distributions of free nucleons and nucleons in a nucleus. This difference can interpreted as a shift in the momentum distribution of valence quarks towards smaller values of the Bjorken variable x. This variable is a ratio of the plus-momentum $k^+ = k^0 + k^3$ of a quark to that of the target. If one uses k^+ as a momentum variable, the corresponding canonical spatial variable is $x^- = x^0 - x^3$ and the time variable is $x^+ = x^0 + x^3$. To do calculations in this framework is to use light front dynamics.

Light front dynamics applies to nucleons within the nucleus as well as to partons of the nucleons, and this is a useful approach whenever the momentum of initial or final state nucleons is large compared to their mass [2]. For example, this technique can be used for $(e, e'p)$ and $(p, 2p)$ reactions at sufficiently high energies. The use of light-front variables for nucleons in a nucleus is not sufficient. It is also necessary to include all the relevant features of conventional nuclear dynamics. Combining these two aspects provides the technical challenge which we have been addressing.

---

[1] This work is partially supported by the USDOE

CP494, *New Directions in Quantum Chromodynamics*, edited by C.-R. Ji and D.-P. Min
© 1999 American Institute of Physics 1-56396-908-4/99/$15.00

I'd like to begin by describing how using the light-front approach leads to important simplifications. Consider high energy electron scattering from nucleons in nuclei. Let the four-momentum q of the exchanged virtual photon be given by $\left(\nu, 0, 0, -\sqrt{Q^2 + \nu^2}\right)$, with $Q^2 = -q^2$, and Q^2 and ν^2 are both very large but Q^2/ν is finite (the Bjorken limit). Use the light-cone variables $q^\pm = q^0 \pm q^3$ in which $q^+ \approx Q^2/2\nu = Mx$, $q^- \approx 2\nu - Q^2/2\nu$, so that $q^- \gg q^+$. Here M is the mass of a nucleon. We neglect q^- in comparison to q^+; corrections to this can be handled in a systematic fashion. Then the scattering cross section for $e + A \to e' + (A-1)_f + p$, where f represents the final nuclear eigenstate of P^-, and p the four-momentum of the final proton, takes the form

$$d\sigma \sim \sum_f \int \frac{d^3 p_f}{E_f} \int d^4 p \, \delta(p^2 - M^2)\delta^{(4)}(q + p_i - p_f - p)|\langle p, f \mid J(q) \mid i \rangle|^2, \quad (1)$$

with the operator $J(q)$ as a schematic representation of the electromagnetic current. Performing the four-dimensional integral over p leads to the expression

$$d\sigma \sim \sum_f \int \frac{d^2 p_f dp_f^+}{p_f^+} \delta \left((p_i - p_f + q)^2 - m^2\right) \mid \langle p, f \mid J(q) \mid i \rangle \mid^2. \quad (2)$$

The argument of the delta function $(p_i - p_f + q)^2 - M^2 \approx -Q^2 + 2q^-(p_i - p_f)^+$. Thus p_f^- does not appear in the argument of the delta function, or anywhere else, so that we can replace the sum over intermediate states by unity. In the usual equal-time representation, the argument of the delta function is $-Q^2 + 2\nu(E_i - E_f)$. The energy of the final state appears, and one can not do the sum over states. It is useful to define $\boldsymbol{p_B} \equiv \boldsymbol{p_i} - \boldsymbol{p_f}$, because $p_B^+ = Q^2/2\nu \equiv Mx$ (as demanded by the delta function). Then one can re-express Eq. (2) as

$$d\sigma \sim d^2 p_{B\perp} n(Mx, p_{B\perp}), \quad (3)$$

where $n(Mx, p_{B\perp})$ is the probability for a nucleon in the ground state to have a momentum $(Mx, p_{B\perp})$. Integration in Eq. (3) leads to

$$\sigma \sim \int d^2 p_\perp \, n(Mx, p_\perp) \equiv f(Mx), \quad (4)$$

with $f(Mx)$ as the probability for a nucleon in the ground state to have a plus momentum of Mx. The use of light-front dynamics to compute nuclear wave functions should allow us to compute $f(Mx)$ from first principles. We also claim that using light-front dynamics incorporates the experimentally relevant kinematics from the beginning, and therefore is the most efficient way to compute the cross sections for nuclear deep inelastic scattering and nuclear quasi-elastic scattering.

Since much of this work is motivated by the desire to understand nuclear deep inelastic scattering and related experiments, it is worthwhile to review some of the features of the EMC effect [1,3]. One key experimental result is the suppression of the structure function for $x \sim 0.5$. This means that the valence quarks of bound nucleons carry less plus-momentum than those of free nucleons. This may be understood by postulating that mesons carry a larger fraction of the plus-momentum in the nucleus than in free space. While such a model explains the shift in the valence distribution, one obtains at the same time a meson (i.e. anti-quark) distribution in the nucleus, which is strongly enhanced compared to free nucleons and which should be observable in Drell-Yan experiments [4]. However, no such enhancement has been observed experimentally [5], and the implications are analyzed in Ref. [6].

The use of light-front dynamics should allow us to compute the necessary nuclear meson distribution functions using variables which are experimentally relevant. The need for a computation of such functions in a manner consistent with generally known properties of nuclei led us to begin this program. There are other motivations for using the light front formalism that have been emphasized in many reviews [7]. One key feature is that the vacuum of the theory is trivial because it can not create pairs. Another is that the theory is a Hamiltonian theory and the many-body techniques of equal time theory can be used here too. I also like to say: Ask not what the light front can do for nuclear physics; instead ask what nuclear physics can do for the light front. This is to provide a set of non-trivial four dimensional examples with real physics content. Finally I quote the review by Geesaman et al. "In light front dynamics LFD, the particles are on mass-shell, and there are no off-shell ambiguities. However, ... we have little or no experience in calculating the wave function of a realistic nucleus in LFD". The aim here is to provide such wave functions.

Outline

We shall begin with a simple description of what is light front dynamics. Then the formal procedures of light front quantization of a hadronic Lagrangian $\mathcal{L}$ will be discussed. The first application is a study of infinite nuclear matter within the mean field approximation [8]. The distribution functions $f(y)$ for nucleons and mesons will be computed. The above topics comprise the first lecture. The next lecture is devoted to a study of finite nuclei [9] using the mean field approximation. Here one must confront a difficulty. The use of $x^- = t - z$ as a spatial variable violates manifest rotational invariance because x^- and $x_\perp$ are different variables. We show that rotational invariance re-emerges after one does the appropriate dynamical calculation. It is necessary to go beyond the mean field approximation, and the third lecture deals with that [10]. Nucleon-nucleon scattering is studied first and used in the many-body calculation. The influence of nucleon-nucleon correlations on the properties of nuclear matter is studied by making the necessary light front calculations. Applications are to compute the nuclear pionic content and to nuclear deep inelastic scattering and Drell-Yan processes. The goal is to provide a series of examples showing that the light front can be used for high energy realistic and relativistic nuclear physics.

WHAT IS LIGHT FRONT DYNAMICS?

This is a relativistic treatment of dynamics in which the fields are quantized at a fixed "time" $\tau = t + z = x^0 + x^3 \equiv x^+$. This means that the orthogonal spatial variable must be $x^- \equiv t - z$ so that the canonical momentum is $p^0 + p^3 \equiv p^+$. The remainder of the spatial variables are given by: $\vec{x}_\perp, \vec{p}_\perp$.

The consequence of using τ as a "time" variable is that the canonical energy is $p^- = p^0 - p^3$. In general our notation is given by

$$A^\pm \equiv A^0 \pm A^3, \tag{5}$$

with

$$A \cdot B = A^\mu B_\mu = \frac{1}{2}\left(A^+ B^- + A^- B^+\right) - \vec{A}_\perp \cdot \vec{B}_\perp. \tag{6}$$

The key reason for using such unusual coordinates is phenomenological. For a particle with $\vec{v} \approx c\hat{e}_3$, the quantity p^+ is BIG. Thus experiments tend to measure quantities associated with p^+.

Another important feature is the relativistic dispersion relation $p^\mu p_\mu = m^2$, which in light front dynamics takes the form:

$$p^- = \frac{p_\perp^2 + m^2}{p^+}. \tag{7}$$

Thus one has a form of relativistic kinematics that avoids using a square root.

The main formal consequence of using light front dynamics is that the minus component of the total momentum, P^-, is used as a Hamiltonian operator, and the plus component P^+ is used as a momentum operator. The procedures to obtain these operators are discussed in the next section.

LIGHT FRONT QUANTIZATION

My intent here is to discuss the basic aspects in as informal way as possible. For more details see the reviews and the references. I'll start by considering one free field at a time. These will be the scalar meson ϕ, the Dirac fermion ψ and the massive vector meson V^μ.

Free Scalar field

Consider the Lagrangian

$$\mathcal{L}_\phi = \frac{1}{2}(\partial^+ \phi \partial^- \phi - \boldsymbol{\nabla}_\perp \phi \cdot \boldsymbol{\nabla}_\perp \phi - m_s^2 \phi^2). \tag{8}$$

The notation is such that $\partial^\pm = \partial^0 \pm \partial^3 = 2\frac{\partial}{\partial x^\mp}$. The Euler-Lagrange equation leads to the wave equation

$$i\partial^- \phi = \frac{-\nabla_\perp^2 + m_s^2}{i\partial^+} \phi. \tag{9}$$

The most general solution is a superposition of plane waves:

$$\phi(x) = \int \frac{d^2 k_\perp dk^+ \theta(k^+)}{(2\pi)^{3/2} \sqrt{2k^+}} \left[a(\boldsymbol{k}) e^{-ik\cdot x} + a^\dagger(\boldsymbol{k}) e^{ik\cdot x} \right], \tag{10}$$

where $k \cdot x = \frac{1}{2}(k^- x^+ + k^+ x^-) - \boldsymbol{k}_\perp \cdot \boldsymbol{x}_\perp$ with $k^- = \frac{k_\perp^2 + m_s^2}{k^+}$, and $\boldsymbol{k} \equiv (k^+, \boldsymbol{k}_\perp)$. The θ function restricts k^+ to positive values. Note that

$$i\partial^+ e^{-ik\cdot x} = k^+ e^{-ik\cdot x}. \tag{11}$$

The value of x^+ that appears in Eq. (10) can be set to zero, but only after taking necessary derivatives.

Deriving the equal x^+ commutation relations for the fields is a somewhat obscure procedure [11], but the result can be stated in terms of familiar commutation relations:

$$[a(\boldsymbol{k}), a^\dagger(\boldsymbol{k}')] = \delta(\boldsymbol{k}_\perp - \boldsymbol{k}'_\perp)\delta(k^+ - k'^+) \tag{12}$$

with $[a(\boldsymbol{k}), a(\boldsymbol{k}')] = 0$.

The next step is compute the Hamiltonian P^- for this system. The conserved energy-momentum tensor is given in terms of the Lagrangian:

$$T_\phi^{\mu\nu} = -g^{\mu\nu}\mathcal{L}_\phi + \frac{\partial \mathcal{L}_\phi}{\partial(\partial_\mu \phi)}\partial^\nu \phi.. \tag{13}$$

This brings us to the question of what is $g^{\mu\nu}$? This is straightforward, although the results (viewed for the first time) can be surprising:

$$g^{+\nu} = g^{0\nu} + g^{3\nu} \tag{14}$$

Thus

$$g^{++} = g^{00} + g^{03} + g^{30} + g^{33} = 1 + 0 + 0 - 1 = 0$$
$$g^{ij} = -\delta_{i,j}(i = 1, 2, j = 1, 2); \quad g^{+-} = g^{-+} = 2. \tag{15}$$

Then one finds that

$$T_\phi^{+-} = \frac{1}{2}\boldsymbol{\nabla}_\perp \phi \cdot \boldsymbol{\nabla}_\perp \phi + \frac{1}{2}m_s^2 \phi^2. \tag{16}$$

The term T^{+-} is the density for the operator P^-:

$$P^- = \frac{1}{2} \int d^2 x_\perp dx^- T^{+-}. \tag{17}$$

The use of the field expansion (10), along with normal ordering followed by integration leads to the result:

$$P_\phi^- = \int d^2 k_\perp dk^+ \theta(k^+) a^\dagger(\boldsymbol{k}) a(\boldsymbol{k}) \frac{k_\perp^2 + m_s^2}{k^+}. \tag{18}$$

One defines a vacuum state $| 0 \rangle$ such that $a(\boldsymbol{p}) | 0 \rangle = 0$. Then the creation operators acting on the vacuum give the usual single particle states:

$$P_\phi^- a^\dagger(\boldsymbol{p}) | 0 \rangle = \frac{p_\perp^2 + m_s^2}{p^+} a^\dagger(\boldsymbol{p}) | 0 \rangle. \tag{19}$$

The momentum operator P^+ is constructed by integrating T^{++}:

$$P_\phi^+ = \int d^2 k_\perp dk^+ \theta(k^+) a^\dagger(\boldsymbol{k}) a(\boldsymbol{k}) k^+. \tag{20}$$

Interactions and Light Front Simplification

Suppose we take the Lagrangian

$$\mathcal{L} = \frac{1}{2}(\partial_\mu \phi \partial^\mu \phi - m_s^2 \phi^2) + \lambda \phi^4. \tag{21}$$

The operator ϕ creates or destroys a particles of plus-momenta $k^+ > 0$. Thus a possible term in which $\lambda \phi^4$ term converts the vacuum $| 0 \rangle$ into a four particle state vanishes by virtue of the conservation of plus-momentum. The vacuum of $p^+ = 0$ can not be connected to four particles, each having a positive k^+. This vanishing simplifies Hamiltonian (x^+-ordered perturbation) calculations.

Free Dirac Field

Consider the Lagrangian

$$\mathcal{L}_\psi = \bar{\psi}(\gamma^\mu \frac{i}{2} \overset{\leftrightarrow}{\partial}_\mu - M)\psi, \tag{22}$$

and its equation of motion:

$$(i\gamma^\mu \partial_\mu - M)\psi = 0. \tag{23}$$

A fermion has spin $1/2$, so there can only be two independent degrees of freedom. The standard Dirac spinor has four components, so two of these must represent dependent degrees of freedom. In the light front formalism one separates the independent and dependent degrees of freedom by using projection operators: $\Lambda_\pm \equiv \frac{1}{2}\gamma^0 \gamma^\pm$. Then the independent field is $\psi_+ = \Lambda_+ \psi$ and the dependent one is $\psi_- = \Lambda_- \psi$

The Dirac equation (23) is re-written as

$$\left(\frac{i}{2}\gamma^+\partial^- + \frac{i}{2}\gamma^-\partial^+ + i\boldsymbol{\gamma}_\perp \cdot \boldsymbol{\nabla}_\perp - M\right)\psi = 0. \tag{24}$$

Equations for $\psi_\pm$ can be obtained by multiplying Eq. (24) on the left by $\Lambda_\pm$:

$$i\partial^-\psi_+ = (\boldsymbol{\alpha}_\perp \cdot \frac{\boldsymbol{\nabla}_\perp}{i} + \beta M)\psi_-$$

$$i\partial^+\psi_- = (\boldsymbol{\alpha}_\perp \cdot \frac{\boldsymbol{\nabla}_\perp}{i} + \beta M)\psi_+, \tag{25}$$

so that the equation of motion of ψ_+ becomes

$$i\partial^-\psi_+ = (\boldsymbol{\alpha}_\perp \cdot \frac{\boldsymbol{\nabla}_\perp}{i} + M)\frac{1}{i\partial^+}(\boldsymbol{\alpha}_\perp \cdot \frac{\boldsymbol{\nabla}_\perp}{i} + M)\psi_+. \tag{26}$$

One can make the field expansion and determine the momenta in a manner similar to the previous section. The key results are

$$T_\psi^{+-} = \psi_+^\dagger \left(\alpha_\perp \cdot \frac{\boldsymbol{\nabla}_\perp}{i} + \beta M\right) \frac{1}{i\partial^+} \left(\alpha_\perp \cdot \frac{\boldsymbol{\nabla}_\perp}{i} + \beta M\right) \psi_+, \tag{27}$$

$$P_\psi^- = \sum_\lambda \int d^2p_\perp dp^+ \theta(p^+) \frac{p_\perp^2 + M^2}{p^+} \left[b^\dagger(\boldsymbol{p}, \lambda)b(\boldsymbol{p}, \lambda) + d^\dagger(\boldsymbol{p}, \lambda)d(\boldsymbol{p}, \lambda)\right], \tag{28}$$

where $b(\boldsymbol{p}, \lambda), d(\boldsymbol{p}, \lambda)$ are nucleon and anti-nucleon destruction operators.

Free Vector Meson

The formalism for massive vector mesons was worked out by Soper [12] and later by Yan [13] using a different formulation. I generally follow Yan's approach. The formalism is lengthy and detailed in the references, so I only state the minimum. There are three independent degrees of freedom, even though the Lagrangian depends on V^μ and $V^{\mu\nu} = \partial^\mu V^\nu - \partial^\nu V^\mu$. These are chosen to be V^+ and V^{+i}. The other terms V^-, V^i, V^{-i} and V^{ij} can be written in terms of V^+ and V^{+i}.

We need a Lagrangian, no matter how bad

It seems to me that one can not do complete dynamical calculations using the light front formalism without specifying some Lagrangian. One starts [8] with $\mathcal{L}$ and derives field equations. These are used to express the dependent degrees of freedom in terms of independent ones. One also uses $\mathcal{L}$ to derive $T^{\mu\nu}$ (as a function of independent degrees of freedom) which is used to obtain the total momentum operators $P^\pm$. It is P^- that acts as a Hamiltonian operator in the light front x^+-ordered perturbation theory.

We start with a Lagrangian containing scalar and vector mesons and nucleons ψ'. This is the minimal Lagrangian for obtaining a caricature of nuclear physics because the exchange of scalar mesons provides a medium range attraction which can bind the nucleons and the exchange of vector mesons provides the short-range repulsion which prevents a collapse. Thus we take

$$\mathcal{L} = \frac{1}{2}(\partial_\mu \phi \partial^\mu \phi - m_s^2 \phi^2) - \frac{1}{4}V^{\mu\nu}V_{\mu\nu} + \frac{m_v^2}{2}V^\mu V_\mu$$
$$+ \bar{\psi}'\left(\gamma^\mu(\frac{i}{2}\overset{\leftrightarrow}{\partial}_\mu - g_v V_\mu) - M - g_s\phi\right)\psi', \tag{29}$$

with the effects of other mesons included elsewhere and below. The equations of motion are

$$\partial_\mu V^{\mu\nu} + m_v^2 V^\nu = g_v \bar{\psi}' \gamma^\nu \psi' \tag{30}$$
$$\partial_\mu \partial^\mu \phi + m_s^2 \phi = -g_s \bar{\psi}' \psi', \tag{31}$$
$$(i\partial^- - g_v V^-)\psi'_+ = (\boldsymbol{\alpha}_\perp \cdot (\boldsymbol{p}_\perp - g_v \boldsymbol{V}_\perp) + \beta(M + g_s\phi))\psi'_- \tag{32}$$
$$(i\partial^+ - g_v V^+)\psi'_- = (\boldsymbol{\alpha}_\perp \cdot (\boldsymbol{p}_\perp - g_v \boldsymbol{V}_\perp) + \beta(M + g_s\phi))\psi'_+. \tag{33}$$

The presence of the interaction term V^+ on the left-hand side of the second equation presents a problem because one can not easily solve for ψ_- in terms of ψ_+. This difficulty is handled by using the Soper-Yan transformation:

$$\psi' = e^{-ig_v \Lambda(x)}\psi, \qquad \partial^+ \Lambda = V^+. \tag{34}$$

Using this in Eqs. (32)-(33) leads to the more usable form

$$(i\partial^- - g_v \bar{V}^-)\psi_+ = (\boldsymbol{\alpha}_\perp \cdot (\boldsymbol{p}_\perp - g_v \bar{\boldsymbol{V}}_\perp) + \beta(M + g_s\phi))\psi_-$$
$$i\partial^+ \psi_- = (\boldsymbol{\alpha}_\perp \cdot (\boldsymbol{p}_\perp - g_v \bar{\boldsymbol{V}}_\perp) + \beta(M + g_s\phi))\psi_+. \tag{35}$$

The cost of the transformation is that one gets new terms resulting from taking derivatives of $\Lambda(x)$. One uses $\bar{V}^\mu$ with $\bar{V}^\mu = V^\mu - \frac{1}{\partial^+}\partial^\mu V^+$, and $\bar{V}^\mu$ enters in the nucleon field equations, but V^μ enters in the meson field equations.

NUCLEAR MATTER MEAN FIELD THEORY

The philosophy [14] is that the nucleonic densities which are mesonic sources are large enough to generate a large number of mesons to enable a classical treatment (replacing an operator by an expectation value). In infinite nuclear matter, the volume is taken as infinity so that all positions are equivalent. Thus we make the replacement:

$$g_s\bar{\psi}(x)\psi(x) \rightarrow g_s\langle\bar{\psi}(0)\psi(0)\rangle, \quad \phi = \frac{-g_s}{m_s^2}\langle\bar{\psi}(0)\psi(0)\rangle, \tag{36}$$

in which the expectation value is in the ground state, and second part of the equation is obtained from the field equation (31) with a constant source. Similarly $g_v \psi(x) \gamma^\mu \psi(x) \to g_v \langle \psi(0) \gamma^\mu \psi(0) \rangle \delta_{\mu,0}$, in which the notion that there is no special direction in space is used. (The nucleus is taken to be at rest.) Again the source is constant, so that the solution of the field equation (30) is

$$\bar{V}^- = V^- = V^0 = \frac{g_v}{m_v^2} \langle \psi^\dagger(0) \psi(0) \rangle; \quad \bar{V}^{+,i} = 0. \tag{37}$$

Since the potentials entering the light-front Dirac equation (35) are constant, the nucleon modes are plane waves $\psi \sim e^{ik \cdot x}$, and the many-body system is a kind of Fermi gas. The solutions of Eq. (35) are

$$i \partial^- \psi_+ = g_v \bar{V}^- \psi_+ + \frac{k_\perp^2 + (M + g_s \phi)^2}{k^+} \psi_+. \tag{38}$$

Solving the equations (36),(37) and (38) yields a self-consistent solution.

Nuclear Momentum Content

The expectation value of $T^{+\mu}$ is used to obtain the total momentum:

$$P^\mu = \frac{1}{2} \int d^2 x_\perp dx^- \langle T^{+\mu} \rangle. \tag{39}$$

The expectation value is constant so that the volume $\Omega = \frac{1}{2} \int d^2 x_\perp dx^-$ will enter as a factor. A straightforward evaluation leads to the results

$$\frac{P^-}{\Omega} = m_s^2 \phi^2 + \frac{4}{(2\pi)^3} \int_F d^2 k_\perp dk^+ \frac{k_\perp^2 + (M + g_s \phi)^2}{k^+} \tag{40}$$

$$\frac{P^+}{\Omega} = m_v^2 (V^-)^2 + \frac{4}{(2\pi)^3} \int_F d^2 k_\perp dk^+ \, k^+. \tag{41}$$

To proceed further one needs to define the Fermi surface F. The use of a transformation $k^+ \equiv \sqrt{(M + g_s \phi)^2 + \vec{k}^2} + k^3 \equiv E(k) + k^3$ to define a new variable k^3 enables one to simplify the integrals. One replaces the integral over k^+ by one over k^3 (including the Jacobian factor $\frac{\partial k^+}{\partial k^3} = \frac{k^+}{E}$) leads to:

$$\int_F d^2 k_\perp dk^+ \cdots \equiv \int d^3 k \, \theta(k_F - |\vec{k}|) \cdots. \tag{42}$$

The nuclear energy E is the average of P^+ and P^-: $E \equiv \frac{1}{2}(P^- + P^+)$ and one gets the very same expression as in the original Walecka model. This provides a useful check on the algebra.

There is a potential problem: for nuclear matter in its rest frame we need to have $P^+ = P^- = M_A$. If one looks at the expressions for $P^\pm$ this result does not seem likely. However, the value of the fermi momentum has not yet been determined. There is one more condition to be satisfied:

$$\left(\frac{\partial(E/A)}{\partial k_F}\right)_\Omega = 0. \tag{43}$$

Satisfying this equation determines k_f and for the value so obtained the values of P^+ and P^- turn are the same.

Thus we see that our light front procedure reproduces standard results for energy and density. We use the parameters of Chin and Walecka [15] $g_v^2 M^2/m_v^2 = 195.9$ and $g_s^2 M^2/m_s^2 = 267.1$ to obtain first numerical results. Then $k_F = 1.42 \quad \text{fm}^{-1}$, the binding energy per nucleon is 15.75 MeV and $M + g_s\phi = 0.56M$. The last number corresponds to a huge attraction that is nearly cancelled by the huge repulsion. Then one may use Eq. (41) to obtain the separate contributions of the vector mesons and nucleons, with spectacular results. Nucleons carry only 65% of the plus-momentum. Thus is much less than the 90% needed to explain the EMC effect for infinite nuclear matter [16]. Furthermore, vector mesons carry 35% of the plus-momentum, which is an amazingly large number.

The distribution of this vector meson plus-momentum is an interesting quantity. The mean fields ϕ, V^μ are constants in space and time. Thus V^- has support only for $k^+ = 0$. The physical interpretation of this is that ∞ number of mesons carry a vanishingly small ϵ of the plus-momentum, but the product is 35%. One can also show [17] that

$$k^+ f_v(k^+) = 0.35M\delta(k^+). \tag{44}$$

There is an important phenomenological consequence the value $k^+ = 0$ corresponds to $x_{Bj} = 0$ which can not be reached in experiments. This means one can't use the momentum sum rule as a phenomenological tool to analyze deep inelastic scattering data to determine the different contributions to the plus-momentum.

Of course this result is caused by solving a simple model for a simple system with a simple mean field approximation. It is necessary to ask if any of the qualitative features of the present results will persist in more detailed treatments.

MEAN FIELD THEORY FOR FINITE-SIZED NUCLEI

It is important to make calculations for finite nuclei because all laboratory experiments are done for such targets or projectiles. The most basic feature of all of nuclear physics is that the shell model is able to explain the magic numbers. Rotational invariance causes the $2j + 1$ degeneracy of the single particle orbitals, and full occupation leads to increased binding. But light front dynamics does not make rotational invariance manifest because the different components of the spatial variable are treated differently: $x^-, \boldsymbol{x}_\perp$. However, the final results must respect rotational invariance. Therefore, the challenge of making successful calculations of the properties of finite nuclei is important to us.

Let's discuss, in a general way, how it is that we will be able to find spectra which have the correct number of degenerate states. Suppose we try to determine eigenstates of a LF Hamiltonian by means of a variational calculation. Simply minimizing the LF energy leads to nonsensical results because $P^- = M_A^2/P^+$. One can easily reach zero energy by letting P^+ be infinite. This is not a problem if one is able to use a Fock space basis in which the total plus and $\perp$ momentum of each component are fixed. But in calculations involving many particles, the Fock state approach cannot be used in practical calculations. One needs to find a sensible variational procedure. One such is to perform a constrained variation, in which the total LF momentum is fixed by including a Lagrange multiplier term proportional to the total momentum in the LF Hamiltonian. We minimize the expectation value of P^+ subject to the condition that the expectation values of P^- and P^+ are equal. This is the same as minimizing the expectation value of the average of P^- and P^+.

The need to include the plus-momentum along with the minus momentum can be seen in a simple example. Consider a nucleus of A nucleons of momentum $P_A^+ = M_A$, $\boldsymbol{P}_{A\perp} = 0$, which consists of a nucleon of momentum $(p^+, \boldsymbol{p}_\perp)$, and a residual $(A-1)$ nucleon system which must have momentum $(P_A^+ - p^+, -\boldsymbol{p}_\perp)$. The kinetic energy K is given by the expression

$$K = \frac{p_\perp^2 + M^2}{p^+} + \frac{p_\perp^2 + M_{A-1}^2}{P_A^+ - p^+}. \tag{45}$$

In the second expression, one is tempted to neglect the term p^+ in comparison with $P_A^+ \approx M_A$. This would be a mistake. Instead make the expansion

$$K \approx \frac{p_\perp^2 + M^2}{p^+} + \frac{M_{A-1}^2}{P_A^+}\left(1 + \frac{p^+}{P_A^+}\right)$$
$$\approx \frac{p_\perp^2 + M^2}{p^+} + p^+ + M_{A-1}, \tag{46}$$

because for large A, $M_{A-1}^2/P_A^2 \approx 1$. For free particles, of ordinary three momentum $\boldsymbol{p}$ one has $E^2(p) = \boldsymbol{p}^2 + m^2$ and $p^+ = E(p) + p^3$, so that

$$K \approx \frac{(E^2(p) - (p^3)^2)}{E(p) + p^3} + E(p) + p^3 + M_{A-1} = 2E(p) + M_{A-1}. \tag{47}$$

We see that K depends only on the magnitude of a three-momentum and rotational invariance is restored. The physical mechanism of this restoration is the inclusion of the recoil kinetic energy of the residual nucleus.

Results

The formalism is described in recent papers [9], so I simply summarize the results. If our solutions are to have any relevance, they should respect rotational invariance.

The success in achieving this is examined in Tables I and II of [9] which give our results for the spectra of ^{16}O and ^{40}Ca, respectively. Scalar and vector meson parameters are taken from Horowitz and Serot [18], and we have assumed isospin symmetry. We see that the $J_z = \pm 1/2$ spectrum contains the eigenvalues of all states, since all states must have a $J_z = \pm 1/2$ component. Furthermore, the essential feature that the expected degeneracies among states with different values of J_z are reproduced numerically.

The obtained eigenvalues of the nucleon mode equation are essentially the same as the single particle energies of the ET formalism, to within the expected numerical accuracy of our program. This equality is not mandated by spherical symmetry alone because the solutions in the equal-time framework have non-vanishing components with negative values of p^+. Table III of [9] gives the contributions to the total P^+ momentum from the nucleons, scalar mesons, and vector mesons for ^{16}O, ^{40}Ca, and ^{80}Zr, as well as the nuclear matter limit. The vector mesons carry approximately 30% of the nuclear plus-momentum. The technical reason for the difference with the scalar mesons (which have negligible effect) is that the evaluation of $a^\dagger(\boldsymbol{k}, \omega) a(\boldsymbol{k}, \omega)$ counts vector mesons "in the air" and the resulting expression contains polarization vectors that give a factor of $\frac{1}{k^+}$ which enhances the distribution of vector mesons of low k^+. The results for the vector meson distribution are shown in Fig. 2 of [9]. As the size of the nucleus increases the enhancement of the distribution at lower values of k^+ becomes more evident.

Lepton-nucleus deep inelastic scattering

It is worthwhile to see how the present results are related to lepton-nucleus deep inelastic scattering experiments. We find that the nucleons carry only about 70% of the plus-momentum. The use of our f_N in standard convolution formulae lead to a reduction in the nuclear structure function that is far too large ($\sim$95% is needed [3]) to account for the reduction observed [3] in the vicinity of $x \sim 0.5$. The reason for this is that the quantity $M + g_s\phi$ acts as a nucleon effective mass of about 670 MeV, which is very small. A similar difficulty occurs in the (e, e') reaction [19] when the mean field theory is used for the initial and final states. The use of a small effective mass and a large vector potential enables a simple reproduction of the nuclear spin orbit force [14,18]. Furthermore, the use of other Lagrangians [21,22] will lead to improved results. We also expect that including effects beyond the mean field would lead to a significant effective tensor coupling of the isoscalar vector meson [20], and to an increased value of the effective mass. Such effects are incorporated in Brueckner theory, and a light-front version [10] could be applied to finite nuclei with better success in reproducing the data. This is discussed in the next sections.

CORRELATED INFINITE NUCLEAR MATTER

The first step is to derive a light front version of the nucleon-nucleon interaction. This is most easily done within the framework of the one boson exchange approximation. The formalism and philosophy are discussed in [8], and the calculation is discussed in [10]. The nucleon-nucleon potential $V(NN)$ describes phase shifts reasonably well. The corresponding density is $\mathcal{V}(NN)$. The basic Lagrangian density contains a free nucleon term $\mathcal{L}_0(N)$, a free meson term $\mathcal{L}_0(\text{mesons})$ and an interaction term $\mathcal{L}_I(N, \text{mesons})$ but does not contain $\mathcal{V}(NN)$. Thus one adds this term and subtracts it:

$$\mathcal{L} = \mathcal{L}_0(N) - \mathcal{V}(NN) + \mathcal{L}_m \qquad (48)$$
$$\mathcal{L}_m = \mathcal{L}_I(N, \text{mesons}) + \mathcal{L}_0(\text{mesons}) + \mathcal{V}(NN). \qquad (49)$$

We use the term $\mathcal{L}_0(N) - \mathcal{V}(NN)$ to obtain a first solution $\mid \Phi \rangle$ to the many-body problem. The term $\mathcal{L}_m$ accounts for mesonic content of Fock space, and we present [10] a scheme to incorporate the effects of $\mathcal{L}_m$ and calculate the full wave function $\mid \Psi \rangle$. Our procedure allows us to assess whether or not $\mathcal{V}(NN)$ has been chosen well. If it has, the effects of $\mathcal{L}_m$ can be treated perturbatively.

Solving for $\mid \Phi \rangle$ is no easy task –it demands a separate non-perturbative treatment. One introduces a mean field U_{MF} which acts on single nucleons.

$$\mathcal{L}_0(N) - \mathcal{V}(NN) = \mathcal{L}_0(N) - U_{MF} + (U_{MF} - \mathcal{V}(NN)). \qquad (50)$$

The operator U_{MF} is chosen to minimize the effects of $\langle \Psi | U_{MF} - \mathcal{V}(NN) | \Psi \rangle$. There is a well-known procedure, called Brueckner theory, which is used to determine U_{MF}. In schematic terms:

$$U_{MF} \sim G \times \rho, \qquad (51)$$

in which G is a nucleon-nucleon scattering matrix, as modified by the Pauli principle, ρ is the nuclear density, and the $\times$ represents a convolution.

The result [10] is a rather complete theory in which the full wave function $|\Psi\rangle$ includes the effects of both NN correlations and explicit mesons.

Results

The trivial nature of the vacuum in the light front formalism was exploited in deriving [10] the necessary equations. Applying our light front OBEP, the nuclear matter saturation properties are reasonably well reproduced. The binding energy per nucleon is 14.71 MeV with a value of k_F of 1.35 fm^{-1}. This is good considering that we have no three-body force. The computed value of the compressibility, 180 MeV, is smaller than that of alternative relativistic approaches to nuclear matter in which the compressibility usually comes out too large. The replacement of meson degrees of freedom by a NN interaction was shown to be a reasonable approximation, and that the formalism allows one to calculate corrections to this approximation in a well-organized manner. The mesonic Fock space components of the nuclear wave function are studied we find that there are about 0.05 excess pions per nucleon.

The magnitudes of the scalar and vector potentials are far smaller than found in the mean field approximation. Our first calculation neglected the influence of two-particle-two-hole states to obtain an approximate version of $f(k^+)$ the nucleons carry 81% (as opposed to the 65% of mean field theory) of the nuclear plus momentum. This is a vast improvement in the description of nuclear deep inelastic scattering as the minimum value of the ratio F_{2A}/F_{2N} is increased by a factor of twenty towards the data. This is not enough to provide a satisfactory description, but it is an excellent start. I am optimistic about future results because including nucleons with momentum greater than k_F can be expected to substantially increase the computed ratio F_{2A}/F_{2N} [10].

Let me discuss the observational aspects, concentrating on the experimental information about the nuclear pionic content. The Drell-Yan experiment on nuclear targets [5] showed no enhancement of nuclear pions within an error of about 5%-10% for their heaviest target. Understanding this result is an important challenge to the understanding of nuclear dynamics [6]. Here we have a good description of nuclear dynamics, and our 5% enhancement is consistent, within errors, with the Drell-Yan data.

SUMMARY

The light front approach has been applied, within the mean field approximation, to both infinite and finite nuclear matter. Furthermore, LF studies of πN and NN scattering have been made. This is input to LF calculations of correlated nucleons in infinite nuclear matter. One can use light front dynamics to compute nuclear energies, wave functions and the experimentally observable plus-momentum distributions for a wide variety of Lagrangians. There are indications that the computed quantities will ultimately be in good agreement with experiment. The use of light front dynamics in nuclear physics is only in its infancy, but it seems to be a tool that can be used for any problem in high energy nuclear physics.

ACKNOWLEDGMENTS

These lectures are based on work performed in collaboration with P.G. Blunden, M. Burkardt, and R. Machleidt.

REFERENCES

1. J. Aubert *et al.*, Phys. Lett. **123B**, 275-278 (1982); R.G. Arnold *et al.*, Phys. Rev. Lett. **52**, 727-730 (1984); A. Bodek *et al.*, Phys. Rev. Lett. **51**, 534-537 (1983).
2. L.L. Frankfurt and M.I. Strikman, Phys. Rep. **76**, 215-347 (1981).
3. R.L. Jaffe, in *Relativistic Dynamics and Quark-Nuclear Physics*, pp. 537-618 edited by M.B. Johnson and A. Picklesimer (Wiley, New York, 1985); L.L. Frankfurt and M.I. Strikman, Phys. Rep. **160**, 235-427 (1988); M. Arneodo, Phys. Rep. **240**, 301-393 (1994); D.F. Geesaman, K. Saito, A.W. Thomas, Ann. Rev. Nucl. Part. Sci. **45**, 337-390 (1995).
4. R.P. Bickerstaff, M.C. Birse, and G.A. Miller, Phys. Rev. Lett. **53**, 2532-2535 (1984); M. Ericson and A.W. Thomas, Phys. Lett. **148B**, 191-193 (1984).
5. D.M. Alde *et al.*, Phys. Rev. Lett. **64**, 2479-2482 (1990).
6. G.F. Bertsch, L. Frankfurt, and M. Strikman, Science **259**, 773-774 (1993).
7. S. J. Brodsky, H-C Pauli, S.S. Pinsky, Phys. Rep. **301**, 299-486 (1998); *Theory of hadrons and light-front QCD*, edited by S.D. Glazek, (World Scientific, Singapore, 1994).
8. G.A. Miller, Phys. Rev. C **56**, R8-11 (1997); **56**, 2789-2805 (1997).
9. P.G. Blunden, M. Burkardt, and G.A. Miller, Phys. Rev. **C59**, R2998-3001 (1999);
10. G.A. Miller and R. Machleidt, Phys. Lett. **B455**, 19-24 (1999); Phys. Rev. **C60**, 035202-1-23 (1999).
11. S.-J. Chang, R.G. Root, and T.-M. Yan, Phys. Rev. D **7**, 1133-1146 (1973); **7**, 1147-1161 (1973).
12. D.E. Soper, SLAC pub-137 (1971); D.E. Soper, Phys. Rev. D **4**, 1620-1625 (1971).
13. T.-M. Yan, Phys. Rev. D **7**, 1760-1778 (1974); **7**, 1780-1800 (1974).
14. B.D. Serot and J.D. Walecka, Adv. Nucl. Phys. **16**, 1-320 (1986); Int. J. Mod. Phys. **E6**, 515-631 (1997).
15. S.A. Chin and J.D. Walecka, Phys. Lett. **52B**, 24-28 (1974).
16. I. Sick, and D. Day, Phys. Lett. **B274**, 16-20 (1992).
17. M. Burkardt and G.A. Miller, Phys. Rev. C **58**, 2450-2458 (1998).
18. C.J. Horowitz and B.D. Serot, Nucl. Phys. **A368**, 503-528 (1981).
19. H. Kim, C.J. Horowitz, and M.R. Frank, Phys.Rev. C **51**, 792-796 (1995).
20. R.J. Furnstahl, J.J. Rusnak, and B.D. Serot, Nucl. Phys. **A632**, 607-623 (1998).
21. J. Zimanyi, S.A. Moszkowski, Phys. Rev. C **42**, 1416-1421 (1990); N.K. Glendenning, F. Weber, and S.A. Moszkowski, Phys. Rev. C **45**, 844-855 (1992).
22. K. Saito and A.W. Thomas, Phys. Lett. **327B**, 9-15 (1994); P.G. Blunden and G.A. Miller, Phys. Rev. C **54**, 359-370 (1996).

Discretized light-cone quantization and the effective interaction in hadrons

Hans-Christian Pauli

Max-Planck-Institut für Kernphysik
D-69029 Heidelberg

Abstract. Light-cone quantization of gauge theories is discussed from two perspectives: as a calculational tool for representing hadrons as QCD bound-states of relativistic quarks and gluons, and as a novel method for simulating quantum field theory on a computer. A general non-perturbative method for numerically solving quantum field theories, 'discretized light-cone quantization', is outlined. Both the bound-state spectrum and the corresponding relativistic wavefunctions can be obtained by matrix diagonalization and related techniques. Emphasis is put on the construction of the light-cone Fock basis and on how to reduce the many-body problem to an effective Hamiltonian. The usual divergences are avoided by cut-offs and subsequently removed by the renormalization group. For the first time, this programme is carried out within a Hamiltonian approach, from the beginning to the end. Starting with the QCD-Lagrangian, a regularized effective interaction is derived and renormalized, ending up with an almost solvable integral equation. Its eigenvalues yield the mass spectrum of physical mesons, its eigenfunctions yield their wavefunctions including the higher Fock-space components. An approximate but analytic mass formula is derived for all physical mesons.

INTRODUCTION

One of the outstanding problems in particle physics is the determination of the structure of hadrons such as the proton and neutron in terms of their fundamental quark and gluon degrees of freedom. Over the past twenty years two fundamentally different pictures have developed. One, the constituent quark model is closely related to experimental observation. The other, quantum chromodynamics is based on a covariant non-abelian quantum field theory. The front form (also known as light-cone quantization) appears to be the only hope of reconciling these two. This elegant approach to quantum field theory is a Hamiltonian gauge-fixed formulation that avoids many of the most difficult problems in the conventional equal-time formulation of the theory.

The natural gauge for light-cone Hamiltonian theories is the light-cone gauge $A^+ = 0$. In this physical gauge the gluons have only the two physical transverse

CP494, *New Directions in Quantum Chromodynamics*, edited by C.-R. Ji and D.-P. Min

degrees of freedom. One imagines that there is an expansion in multi-particle occupation number Fock states. But even in the case of the simpler abelian quantum theory of electrodynamics very little is known about the nature of the bound state solutions in the strong-coupling domain. In the non-abelian quantum theory of chromodynamics a calculation of bound-state structure has to deal with many difficult aspects simultaneously. Confinement, vacuum structure and chiral symmetry inter-twine with the difficulties of describing a (relativistic) many-body system and the non-perturbative renormalization of a Hamiltonian.

In the conventional approach based on equal-time quantization the Fock state expansion becomes quickly intractable because of the complexity of the vacuum. Furthermore, boosting such a wavefunction from the hadron's rest frame to a moving frame is as complex a problem as solving the bound state problem itself. The presence of the square root operator in the equal-time Hamiltonian approach presents severe mathematical difficulties.

Fortunately 'light-cone quantization' offers an elegant avenue of escape. It can be formulated independent of the Lorentz frame. The square root operator does not appear, and the vacuum structure is relatively simple. There is no spontaneous creation of massive fermions in the light-cone quantized vacuum.

In fact, there are many reasons to quantize relativistic field theories at fixed light-cone time. Dirac [1] showed, in 1949, that in this so called 'front form' of Hamiltonian dynamics a maximum number of Poincaré generators become independent of the interaction, including certain Lorentz boosts. In fact, unlike the traditional equal-time Hamiltonian formalism, quantization on a plane tangential to the light-cone, on the 'null plane', can be formulated without reference to a specific frame. One can construct an operator whose eigenvalues are the invariant masses of the composite physical particles. The eigenvectors describe bound states of arbitrary four-momentum and invariant mass, and allow the computation of scattering amplitudes and other dynamical quantities. In many field theories the vacuum state of the free Hamiltonian is also an eigenstate of the light-cone Hamiltonian. The Fock expansion built on this vacuum state provides a complete relativistic many-particle basis for diagonalizing the full theory.

The main thrust of these lectures will be to discuss the complexities that are unique to this formulation of QCD, in varying degrees of detail. The goal is to present a self-consistent framework rather than trying to cover the subject exhaustively. A review all of the successes or applications will, however, not be undertaken.

One of the reasons is, that the subject was reviewed recently [2]. Another is that other lecturers in this school emphasize complementary aspects. Stan Brodsky shows how the knowledge of the the light-cone wavefunction has impact on hadronic physics and exclusive processes. Steve Pinsky demonstrates how the method of discretized light-cone quantization is constructive for analyzing supersymmetric string and M(atrix) theories. Last not least, Simon Dalley expands on the transverse-lattice calculations within the light-cone approach.

Comparatively little space will be devoted to canonical field theory, just so much as to plausibilize that a light-cone Hamiltonian exists. This so called 'naive Hamiltonian' will be derived and written down explicitly as a Fock-space operator in the light-cone gauge and disregarding zero modes. For historical and paedagogical reasons these notes will be rather outspoken for one space and one time dimension, mostly to show that periodic boundary conditions (discretized light-cone quantization) are helpful, indeed, for solving the bound-state problem in a relativistic theory. The attempt is made to be complementary to the review [2] and some new material on the Schwinger model is included.

The bulk of these notes deals with the many-body aspects of a gauge field theory in the real world of 3+1 dimensions. A hadron not only contains the valence quarks but also an infinite amount of gluons and sea-quarks. Progress often comes with new technologies. The presentation of the method of iterated resolvents therefore takes broad room. It allows to derive a well-defined effective interaction. Some thus far unpublished work is included, in particular more instructive examples. The presentation is separated into two parts. In the first considerations are essentially exact. In the second some approximations and simplifications are admitted and well marked in the text. One arrives at the effective interaction in the form of a tractable and solvable integral equation. Its solutions allow to construct explicitly the many-body amplitudes corresponding to see-quarks and gluons by comparatively simple quadratures, without the need of solving another bound-state problem. Explicit equations for that are given. Moreover, some new research work will be included in the section on renormalization.

What is not included in these notes, however, is a complete survey of the literature. Mostly for the reasons of space, it is refered to the some 469 items of [2]. I apologize with my colleagues whose work is not mentioned. But I will be careful to cite the work which I need for the present discussion and presentation. Some selected monographies which I found useful to consult are [3–7], some selected review articles or conference proceedings might be [8–11].

FORMS OF HAMILTONIAN DYNAMICS

Dirac defines the Hamiltonian H of a closed system as that operator whose action on the state vector $\mid t \, \rangle$ has the same effect as taking the partial derivative with respect to time t, $i.e.$

$$H \mid t \, \rangle = i \frac{\partial}{\partial t} \mid t \, \rangle. \tag{1}$$

The concept of an Hamiltonian is applicable irrespective of whether one deals with the motion of a non-relativistic particle in classical mechanics or with a non-relativistic wave function in the Schrödinger equation, and it generalizes almost unchanged to a relativistic and covariant field theory.

In a covariant theory, the very notion of 'time' becomes, however, questionable since the time is only one component of four-dimensional space-time. But the concept of usual space and of usual time can be generalized. One can define 'space' as that hypersphere in four-space on which one chooses the initial conditions. The remaining fourth coordinate can be understood as 'time'.

More formally, one conveniently introduces generalized coordinates $\tilde{x}^\nu$. Starting from a baseline parametrization of space-time like the instant form in Figure 1, one parametrizes space-time by a coordinate transformation $\tilde{x}^\nu = \tilde{x}^\nu(x^\mu)$. The metric tensors for the two parametrizations are then related by

$$\tilde{g}_{\kappa\lambda} = \left(\frac{\partial x^\mu}{\partial \tilde{x}^\kappa}\right) g_{\mu\nu} \left(\frac{\partial x^\nu}{\partial \tilde{x}^\lambda}\right).$$

The physical content of the theory can not depend on such a re-parametrization.

But the raising and the lowering of Lorentz indices are then non-trivial operations. As an example consider the front form parametrization in Figure 1. Interpret $\tilde{x}^3 = x^0 - x^3$ as the new space coordinate and denote it by $x^- = ct - z$. Then $\tilde{x}^0 = x^0 + x^3$ must be interpreted as the new time coordinate denoted by $x^+ = ct + z$, or by the 'light-cone time' $\tau = t + z/c$. Of course, one also could exchange the two. Since the lowering operation is $x_\mu = g_{\mu\nu} x^\nu$, both $x_+ = \frac{1}{2} x^-$ are space, and $x_- = \frac{1}{2} x^+$ are time coordinates. The new space derivative is therefore $\partial_- = \frac{1}{2} \partial^+$, while $\partial_+ = \frac{1}{2} \partial^-$ is a time-derivative. The Lorentz indices '+' and '-' have a different physical meaning, depending on whether they occur up- or down-stairs. Co-variant and contra-variant vectors are different objects. The Hamiltonian is only one component of a four-vector P^μ, particularly its time-like component. Taking the partial time derivative of the state vector

$$P_+ \mid x^+ \rangle = i\frac{\partial}{\partial x^+} \mid x^+ \rangle,$$

defines then $P_+ = \frac{1}{2} P^-$ as the Hamiltonian in the transformed coordinates, in line with Eq.(1), and $P_- = \frac{1}{2} P^+$ as the longitudinal momentum.

Following Dirac [1] there are no more than three different parametrisations of space-time. They are illustrated in Figure 1, and cannot be mapped onto each other by a Lorentz transform. They differ by the hypersphere on which the fields are initialized. They have thus different 'times' and different 'Hamiltonians'. Dirac [1] speaks of the three *forms of Hamiltonian dynamics*: The *instant form* is the familiar one, with its hypersphere given by $t = 0$. In the *front form* the hypersphere is a tangent plane to the light cone. In the *point form* the time-like coordinate is identified with the eigentime of a physical system and the hypersphere has a shape of a hyperboloid.

Which of the three forms should be prefered, is an ill-posed question, since all three forms must yield the same physical results. Comparatively little work has been done in the point form. The bulk of research on field theory implicitly uses the instant form. Although it is the conventional choice for quantizing field theory, it has many practical disadvantages. For example, given the wavefunctions of an n-electron atom at an initial time $t = 0$, $\psi_n(\vec{x}_i, t = 0)$, one can use the Hamiltonian H to evolve $\psi_n(\vec{x}_i, t)$ to later times t. However, an experiment which specifies the

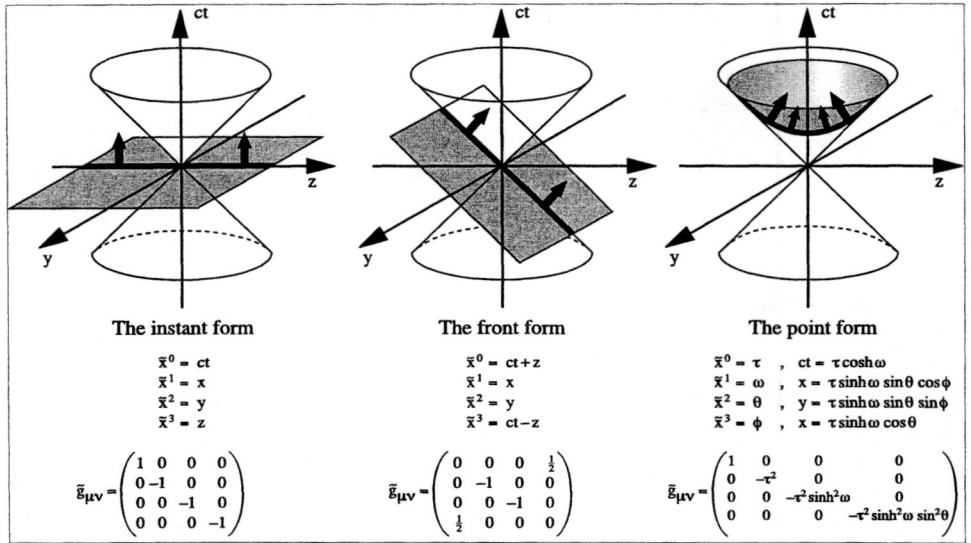

FIGURE 1. Dirac's three forms of Hamiltonian dynamics.

initial wave function would require the simultaneous measurement of the positions of all of the bounded electrons. In contrast, determining the initial wave function at fixed light-cone time $\tau = 0$ only requires an experiment which scatters one plane-wave laser beam, since the signal reaching each of the n electrons, along the light front, at the same light-cone time $\tau = t_i + z_i/c$.

Dirac's legacy had been forgotten and was re-invented several times. The front form approach carries therefore names as different as *Infinite-Momentum Frame*, *Null-Plane Quantization*, *Light-Cone Quantization*, or most recently *Light-Front Quantization*. In the essence they are all the same. The infinite-momentum frame is a misnomer since the total momentum is finite and since the front form is *frame-independent* and covariant. Light-cone quantization is also unfortunate, since the inital data are set one a plane tangential to but not on the light cone, and since both equal-time or equal light-cone-time quantization stand both for the same quantization, for a quantum as opposed to a classical theory. We propose to stay with Dirac's different 'forms of Hamiltonians'.

The canonical Hamiltonian for gauge theory

The prototype of a field theory is Faraday's and Maxwell's electrodynamics. Every field theory has its own canonical Hamiltonian and is governed by the action density. The Lagrangian is the subject of the canonical calculus of variation, given in many text books [4]. Its essentials shall be recalled briefly.

The Lagrangian, in general, is a function of a finite number of fields $\phi_r(x)$ and their first space-time derivatives $\partial_\mu \phi_r(x)$, thus $\mathcal{L} = \mathcal{L}[\phi_r, \partial_\mu \phi_r]$. Independent variation of $\mathcal{L}$ with respect to ϕ_r and $\partial_\mu \phi_r$ results in the equations of motion,

$$\partial_\kappa \pi_r^\kappa - \delta \mathcal{L}/\delta \phi_r = 0 , \quad \text{with} \quad \pi_r^\kappa[\phi] \equiv \frac{\delta \mathcal{L}}{\delta (\partial_\kappa \phi_r)},$$

canonically refered to as the Euler equations. The canonical formalism is particularly suited for discussing the symmetries of a field theory. Every continuous symmetry of the Lagrangian is associated with a vanishing four-divergence of a current and a conserved charge. Since $\mathcal{L}$ does not explicitly depend on the coordinates, every field theory in 3+1 dimensions has ten conserved four-currents. The four-divergences of the energy-momentum tensor $T^{\lambda\nu}$ and of the boost-angular-momentum stress tensor $J^{\lambda,\mu\nu}$ vanish,

$$\partial_\lambda T^{\lambda\nu} = 0, \quad \partial_\lambda J^{\lambda,\mu\nu} = 0.$$

As a consequence the Lorentz group has ten 'conserved charges', the 4 components total momentum P^ν and the 6 components of boost-angular momentum $M^{\mu\nu}$,

$$P^\nu = \int_\Omega d\omega_0 T^{0\nu} , \quad \text{and} \quad M^{\mu\nu} = \int_\Omega d\omega_0 J^{0,\mu\nu}. \tag{2}$$

With the totally antisymmetric tensor in four dimensions $\epsilon_{\lambda\mu\nu\rho}$ ($\epsilon_{0123} = 1$), the three-dimensional surface elements of a hypersphere are $d\omega_\lambda = \epsilon_{\lambda\mu\nu\rho} dx^\mu dx^\nu dx^\rho/3!$. A finite volume is thus $\Omega = \int d\omega_0 = \int dx^1 dx^2 dx^3$, and correspondingly in the front form $d\omega_+ = \int dx_+ d^2 x_\perp$. The time-like components of P_μ is the Hamiltonian, i.e. P_0 for the instant form and P_+ for the front form. The transition from instant to front form is thus simple: substitute '0' by '+'.

Working out the canonical procedure for QED with its Lagrangian,

$$\mathcal{L} = -\frac{1}{4} F^{\mu\nu} F_{\mu\nu} + \frac{1}{2} \left[\overline{\Psi} \left(i\gamma^\mu D_\mu - m \right) \Psi + \text{h.c.} \right], \tag{3}$$

where $F^{\mu\nu}$ is the electro-magnetic field tensor and D_μ the covariant derivative,

$$F^{\mu\nu} = \partial^\mu A^\nu - \partial^\nu A^\mu, \qquad D_\mu = \partial_\mu - ig A_\mu,$$

one ends up with a manifestly gauge-invariant total momentum,

$$P_\nu = \int_\Omega d\omega_0 \left(F^{0\kappa} F_{\kappa\nu} + \frac{1}{4} g_\nu^0 F^{\kappa\lambda} F_{\kappa\lambda} + \frac{1}{2} \left[i\overline{\Psi}\gamma^0 D_\nu \Psi + \text{h.c.} \right] \right),$$

$$P_\nu = \int_\Omega d\omega_+ \left(F^{+\kappa} F_{\kappa\nu} + \frac{1}{4} g_\nu^+ F^{\kappa\lambda} F_{\kappa\lambda} + \frac{1}{2} \left[i\overline{\Psi}\gamma^+ D_\nu \Psi + \text{h.c.} \right] \right),$$

in both the instant and the front form, respectively. The boost angular momenta will not be used explicitly.

The gauge invariant Lagrangian density for QCD is

$$\mathcal{L} = -\frac{1}{2}\text{Tr}(\mathbf{F}^{\mu\nu}\mathbf{F}_{\mu\nu}) + \frac{1}{2}[\overline{\Psi}(i\gamma^{\mu}\mathbf{D}_{\mu} - \mathbf{m})\Psi + \text{ h.c.}],$$

$$= -\frac{1}{4}F_a^{\mu\nu}F_{\mu\nu}^a + \frac{1}{2}[\overline{\Psi}(i\gamma^{\mu}\mathbf{D}_{\mu} - \mathbf{m})\Psi + \text{ h.c.}],$$

The color-electro-magnetic fields and the covariant derivative are now

$$\mathbf{F}^{\mu\nu} \equiv \partial^{\mu}\mathbf{A}^{\nu} - \partial^{\nu}\mathbf{A}^{\mu} + ig[\mathbf{A}^{\mu}, \mathbf{A}^{\nu}] = \mathbf{T}^a\left(\partial^{\mu}A_a^{\nu} - \partial^{\nu}A_a^{\mu} - gf^{ars}A_r^{\mu}A_s^{\nu}\right),$$

$$\mathbf{D}_{cc'}^{\mu} = \delta_{cc'}\partial^{\mu} + ig\mathbf{A}_{cc'}^{\mu} = \delta_{cc'}\partial^{\mu} + igT_{cc'}^a A_a^{\mu}.$$

As compared to QED, each local gauge field $A^{\mu}(x)$ is replaced by the 3×3 matrix $\mathbf{A}^{\mu}(x)$. All such matrices can be parametrized $\mathbf{A}^{\mu} \equiv T_{cc'}^a A_a^{\mu}$. More generally for SU(N), the vector potentials $\mathbf{A}^{\mu}$ are hermitian and traceless $N \times N$ matrices. The color index c (or c') runs now from 1 to n_c, and correspondingly the gluon index a (or r, s, t) from 1 to $n_c^2 - 1$. No distinction will be made between raising or lowering them. In order to make sense of expressions like $\overline{\Psi}\mathbf{A}^{\mu}\Psi$ the quark fields $\Psi_{c,\alpha}(x)$ must carry a color index c. They, as well as the Dirac indices, are usually suppressed. The color matrices $T_{cc'}^a$ obey

$$\left[T^r, T^s\right]_{cc'} = if^{rsa}T_{cc'}^a \qquad \text{and} \qquad \text{Tr}\,(T^r T^s) = \frac{1}{2}\delta_r^s. \tag{4}$$

For SU(2) the color matrices are $T^a = \frac{1}{2}\sigma^a$, with σ^a being the Pauli matrices. The structure constants f^{rst} are therefore the totally antisymmetric tensor ϵ_{rst}. For SU(3), $T^a = \frac{1}{2}\lambda^a$ with the Gell-Mann matrices λ^a, with the corresponding structure constants tabulated in the literature [6]. Everything proceeds in analogy with QED. The energy-momentum vector,

$$P_{\nu} = \int_{\Omega} d\omega_0 \left(F_a^{0\kappa}F_{\kappa\nu}^a + \frac{1}{4}g_{\nu}^0 F_a^{\kappa\lambda}F_{\kappa\lambda}^a + \frac{1}{2}\left[i\overline{\Psi}\gamma^0 T^a D_{\nu}^a\Psi + \text{ h.c.}\right]\right),$$

$$P_{\nu} = \int_{\Omega} d\omega_+ \left(F_a^{+\kappa}F_{\kappa\nu}^a + \frac{1}{4}g_{\nu}^+ F_a^{\kappa\lambda}F_{\kappa\lambda}^a + \frac{1}{2}\left[i\overline{\Psi}\gamma^+ T^a D_{\nu}^a\Psi + \text{ h.c.}\right]\right), \tag{5}$$

is manifestly gauge-invariant both in the instant and the front form.

The Poincaré symmetries in the front form

The ten constants of motion P^{μ} and $M^{\mu\nu}$ are observables with real eigenvalues. It is advantageous to construct representations in which the constants of motion are diagonal. But one cannot diagonalize all ten constants of motion simultaneously because they do not commute. The algebra of the four-energy-momentum $P^{\mu} = p^{\mu}$ and four-angular-momentum $M^{\mu\nu} = x^{\mu}p^{\nu} - x^{\nu}p^{\mu}$ for free particles with the basic commutator $\frac{1}{i\hbar}[x^{\mu}, p_{\nu}] = \delta_{\nu}^{\mu}$ is

$$\frac{1}{i\hbar}\left[P^{\rho}, M^{\mu\nu}\right] = g^{\rho\mu}P^{\nu} - g^{\rho\nu}P^{\mu}, \qquad \frac{1}{i\hbar}\left[P^{\rho}, P^{\mu}\right] = 0,$$

$$\text{and} \qquad \frac{1}{i\hbar}\left[M^{\rho\sigma}, M^{\mu\nu}\right] = g^{\rho\nu}M^{\sigma\mu} + g^{\sigma\mu}M^{\rho\nu} - g^{\rho\mu}M^{\sigma\nu} - g^{\sigma\nu}M^{\rho\mu}.$$

It is postulated that the generalized momentum operators satisfy the same commutator relations as a single particle. They form thus a group, the Poincaré group.

It is convenient to discuss the structure of the Poincaré group in terms of the Pauli-Lubansky vector $V^\kappa \equiv \epsilon^{\kappa\lambda\mu\nu} P_\lambda M_{\mu\nu}$. V is orthogonal to the generalized momenta, $P_\mu V^\mu = 0$. The two group invariants are the operator for the invariant mass-squared $M^2 = P^\mu P_\mu$ and the operator for intrinsic spin-squared $V^2 = V^\mu V_\mu$. They are Lorentz scalars and commute with all generators P^μ and $M^{\mu\nu}$, as well as with all V^μ. A convenient choice of six mutually commuting operators is: The invariant mass squared $M^2 = P^\mu P_\mu$, the three space-like momenta P^+ and $\vec{P}_\perp$, the total spin squared $S^2 = V^\mu V_\mu$, and one component of V, say $V^+ \equiv S_z$.

Inspecting the definition of boost-angular momentum $M_{\mu\nu}$ in Eq.(2) one identifies which components are dependent on the interaction and which are not. Dirac [1] calls them complicated and simple, or dynamic and kinematic, or Hamiltonians and Momenta, respectively. In the instant form, the three components of the boost vector $K_i = M_{i0}$ are dynamic, and the three components of angular momentum $J_i = \epsilon_{ijk} M_{jk}$ are kinematic. As noted already by Dirac, the front form is special in having four kinematic components of $M_{\mu\nu}$ ($M_{+-}, M_{12}, M_{1-}, M_{2-}$) and only two dynamic ones (M_{+1} and M_{+2}). In the front form one deals thus with seven mutually commuting operators

$$(M_{+-}, M_{12}, M_{1-}, M_{2-}), \quad \text{and all } P^\mu,$$

instead of the six in the instant form. These symmetries imply the very important aspect of the front form that both the Hamiltonian and all amplitudes obtained in light-cone perturbation theory are manifestly invariant under a large class of Lorentz transformations:

$$
\begin{aligned}
p^+ &\to C_\parallel\, p^+, & \vec{p}_\perp &\to \vec{p}_\perp, & p^- &\to C_\parallel^{-1} p^-, \\
p^+ &\to p^+, & \vec{p}_\perp &\to \vec{p}_\perp + p^+ \vec{C}_\perp, & p^- &\to p^- + 2\vec{p}_\perp \cdot \vec{C}_\perp + p^+ \vec{C}_\perp^2, \\
p^+ &\to p^+, & \vec{p}_\perp^{\,2} &\to \vec{p}_\perp^{\,2},
\end{aligned}
$$

i.e. for parallel boosts, transverse boosts, and rotations, respectively. All of these hold for every single particle momentum p^μ, and for any set of dimensionless c-numbers $C_\parallel$ and $\vec{C}_\perp$.

The light-cone Hamiltonian operator

The four-vector of energy-momentum for gauge theory in Eq.(5) contains time-derivatives and other constraint field components. They will be eliminated in this section using the equations of motion with the goal to express P^ν in terms of the free fields and to isolate the dependence on the coupling constant.

The color-Maxwell equations are four equations for determining the four functions A_a^μ. One of the equations of motion is identically fullfilled by choosing the light-cone gauge $A_a^+ = 0$. Two of the equations give expressions the time derivatives of the two transversal components $\vec{A}_\perp^a$. The fourth is the Gauss' law in the front form, $\partial_\mu F_a^{\mu+} = g J_a^+$, or explicitly

$$-\partial^+\partial_- A_a^- - \partial^+\partial_i A_{\perp a}^i = g J_a^+. \tag{6}$$

It contains only space-derivatives and is a constrained equation for the components of A_a^μ. For the free case ($g = 0$), A_a^μ reduces to $\tilde{A}_a^\mu$ and therefore to

$$\tilde{A}_a^\mu = \left(\tilde{A}_a^+, \vec{A}_{\perp a}, \tilde{A}_a^-\right) = \left(0, \vec{A}_{\perp a}, -\frac{1}{\partial_-}\left[\partial_i A_{\perp a}^i\right]\right).$$

As a consequence, $\tilde{A}_a^\mu$ is purely transverse. The formal inversion of Eq.(6) is therefore

$$A_a^- = \tilde{A}_a^- + \frac{g}{(i\partial_-)^2} J_a^+, \tag{7}$$

which must be substituted everywhere. The inverse space derivatives $(i\partial^+)^{-1}$ and $(i\partial^+)^{-2}$, used here and below, are actually Green's functions. Since they depend only on x^-, they are comparatively simple.

The color-Dirac equations $(i\gamma^\mu \mathbf{D}_\mu - m)\Psi = 0$ can be used to express the time derivatives $\partial_+ \Psi$ as function of the other fields. After multiplication with γ^0 one gets first

$$(i\gamma^0\gamma^+ T^a D_+^a + i\gamma^0\gamma^- T^a D_-^a + i\alpha_\perp^i T^a D_{\perp i}^a)\Psi = m\beta\Psi,$$

with the usual Dirac matrices $\beta = \gamma^0$ and $\alpha^k = \gamma^0\gamma^k$. In order to isolate the time derivative one introduces the projectors $\Lambda_\pm$ and projected spinors $\Psi_\pm$ by

$$\Lambda_\pm = \frac{1}{2}(1 \pm \alpha^3) \qquad \text{and} \qquad \Psi_\pm = \Lambda_\pm \Psi .$$

Multiplying the color-Dirac equation once with Λ_+ and once with Λ_- gives

$$2i\partial_+ \Psi_+ = \left(m\beta - i\alpha_\perp^i T^a D_{\perp i}^a\right)\Psi_- + 2gA_+^a T^a\Psi_+,$$
$$\text{and} \quad 2i\partial_- \Psi_- = \left(m\beta - i\alpha_\perp^i T^a D_{\perp i}^a\right)\Psi_+ + 2gA_-^a T^a\Psi_-, \tag{8}$$

a coupled set of spinor equations. The first of them contains a time derivative. The second contains a space derivative and is a constraint equation. The component

$$\Psi_- = \frac{1}{2i\partial_-}\left(m\beta - i\alpha_\perp^i T^a D_{\perp i}^a\right)\Psi_+ \tag{9}$$

must therefore be substituted everywhere. The time derivative becomes then

$$2i\partial_+ \Psi_+ = 2gA_+^a T^a\Psi_+ + \left(m\beta - i\alpha_\perp^j T^a D_{\perp j}^a\right)\frac{1}{2i\partial_-}\left(m\beta - i\alpha_\perp^i T^a D_{\perp i}^a\right)\Psi_+ .$$

88

In analogy to the color-Maxwell case one defines free spinors by

$$\tilde{\Psi} = \Psi_+ + \left(m\beta - i\alpha^i_\perp \partial_{\perp i}\right) \frac{1}{2i\partial_-} \Psi_+ .$$

Contrary to the full spinor, $\tilde{\Psi}$ is independent of the interaction.

Inserting the above expressions into Eq.(5), the space-like components of P^ν become

$$P_k = \int dx_+ d^2 x_\perp \left(\overline{\tilde{\Psi}}\, \gamma^+ i\partial_k \tilde{\Psi} + \tilde{A}^\mu_a\, \partial^+ \partial_k \tilde{A}^a_\mu\right), \qquad \text{for } k = 1, 2, -. \tag{10}$$

Inserting them into P_+ gives rather lengthy expressions, which are conveniently written as a sum of five terms

$$P_+ = T + V + W_1 + W_2 + W_3. \tag{11}$$

Only the first term survives the limit $g \to 0$, and therefore is called the free part of the Hamiltonian, or its 'kinetic energy'

$$T = \frac{1}{2} \int dx_+ d^2 x_\perp \left(\overline{\tilde{\Psi}}\gamma^+ \frac{m^2 + (i\nabla_\perp)^2}{i\partial^+} \tilde{\Psi} + \tilde{A}^\mu_a (i\nabla_\perp)^2 \tilde{A}^a_\mu\right).$$

The vertex interaction

$$V = g \int dx_+ d^2 x_\perp \; \tilde{J}^\mu_a \tilde{A}^a_\mu, \qquad \text{with} \quad \tilde{J}^\nu_a(x) = \overline{\tilde{\Psi}}\gamma^\nu T^a \tilde{\Psi} + f^{abc}\partial^\mu \tilde{A}^\nu_b \tilde{A}^c_\nu, \tag{12}$$

is linear in the coupling constant and is the light-cone analogue of the conventional $J_\mu A^\mu$-structures in the instant form. Note that the current $\tilde{J}^\mu_a$ has contributions from both quarks and gluons. The four-point gluon interaction

$$W_1 = \frac{g^2}{4} \int dx_+ d^2 x_\perp \; \tilde{B}^{\mu\nu}_a \tilde{B}^a_{\mu\nu}, \qquad \text{with} \quad B^{\mu\nu}_a = f^{abc}\tilde{A}^\mu_b \tilde{A}^\nu_c,$$

describes the four-point gluon-vertices which is quadratic in g. The remainders are the 'instantaneous interactions'. They are characterized by the inverse derivatives. The instantaneous gluon interaction arises from the Gauss equation,

$$W_2 = \frac{g^2}{2} \int dx_+ d^2 x_\perp \; \tilde{J}^+_a \frac{1}{(i\partial^+)^2} \tilde{J}^+_a,$$

and is the light-cone analogue of the Coulomb energy. The instantaneous fermion interaction originates from the light-cone specific decomposition of Dirac's equation

$$W_3 = \frac{g^2}{2} \int dx_+ d^2 x_\perp \; \overline{\tilde{\Psi}}\gamma^\mu T^a \tilde{A}^a_\mu \frac{\gamma^+}{i\partial^+} \left(\gamma^\nu T^b \tilde{A}^b_\nu \tilde{\Psi}\right).$$

It has no analogue in the instant form.

Most remarkable, however, is that the relativistic Hamiltonian is additive in the 'kinetic' and the 'potential' energy, very much like a non-relativistic Hamiltonian $H = T + U$. In this respect the front form is distinctly different from the conventional instant form.

The free field solutions

The free solutions of the Dirac and the Maxwell equations are in the front form

$$\tilde{\Psi}_{acf}(x) = \sum_\lambda \int \frac{dp^+ d^2 p_\perp}{\sqrt{2p^+(2\pi)^3}} \left(b(q)u_\alpha(p,\lambda)e^{-ipx} + d^\dagger(q)v_\alpha(p,\lambda)e^{+ipx} \right),$$

$$\tilde{A}_\mu^a(x) = \sum_\lambda \int \frac{dp^+ d^2 p_\perp}{\sqrt{2p^+(2\pi)^3}} \left(a(q)\epsilon_\mu(p,\lambda)e^{-ipx} + a^\dagger(q)\epsilon_\mu^\star(p,\lambda)e^{+ipx} \right).$$

The properties of the Dirac spinors u_α and v_α, and of the polarization vectors ϵ_μ, are given for example in [2]. The single particle state are specified by the quantum numbers $q = (p^+, p_{\perp x}, p_{\perp y}, \lambda, c, f)$. Their creation and destruction operators are subject to the usual relations

$$\left[a(q), a^\dagger(q') \right] = \left\{ b(q), b^\dagger(q') \right\} = \left\{ d(q),\ d^\dagger(q') \right\} = \delta(p^+ - p^{+\prime})\delta^{(2)}(\vec{p}_\perp - \vec{p}_\perp')\delta_\lambda^{\lambda'}\delta_c^{c'}\delta_f^{f'},$$

which carry the operator structure of the theory. When inserting the free fields into P_μ one can integrate out the dependence on x^μ, producing essentially Dirac delta-functions in the single particle momenta, which reflect momentum conservation:

$$\int \frac{dx_+}{2\pi} e^{ix_+(\sum_j p_j^+)} = \delta\left(\sum_j p_j^+\right), \qquad \int \frac{d^2 x_\perp}{(2\pi)^2} e^{-i\vec{x}_\perp(\sum_j \vec{p}_{\perp j})} = \delta^{(2)}\left(\sum_j \vec{p}_{\perp j}\right).$$

In detail this can be quite laborious, as shown by the example with the fermionic contribution to the vertex interaction

$$V_f = g \int dx_+ d^2 x_\perp\ \overline{\tilde{\Psi}}(x)\gamma^\mu T^a \tilde{\Psi}(x)\tilde{A}_\mu^a(x)\bigg|_{x^+=0}$$

$$= \frac{g}{\sqrt{(2\pi)^3}} \sum_{\lambda_1,\lambda_2,\lambda_3} \sum_{c_1,c_2,a_3} \int \frac{dp_1^+ d^2 p_{\perp 1}}{\sqrt{2p_1^+}} \int \frac{dp_2^+ d^2 p_{\perp 2}}{\sqrt{2p_2^+}} \int \frac{dp_3^+ d^2 p_{\perp 3}}{\sqrt{2p_3^+}}$$

$$\times \int \frac{dx_+ d^2 x_\perp}{(2\pi)^3} \left[\left(b^\dagger(q_1)\overline{u}_\alpha(p_1,\lambda_1)e^{+ip_1 x} + d(q_1)\overline{v}_\alpha(p_1,\lambda_1)e^{-ip_1 x} \right) T_{c_1,c_2}^{a_3} \right.$$

$$\times \quad \gamma_{\alpha\beta}^\mu \left(d^\dagger(q_2)v_\beta(p_2,\lambda_2)e^{+ip_2 x} + b(q_2)u_\beta(p_2,\lambda_2)e^{-ip_2 x} \right)$$

$$\times \quad \left. \left(a^\dagger(q_3)\epsilon_\mu^\star(p_3,\lambda_3)e^{+ip_3 x} + a(q_3)\epsilon_\mu(p_3,\lambda_3)e^{-ip_3 x} \right). \right.$$

Note that the sum of these single particle momenta is essentially the sum of the particle momenta minus the sum of the hole momenta. Consequently, if a particular term has only creation or only destruction operators as in

$$b^\dagger(q_1)d^\dagger(q_2)a^\dagger(q_3)\ \delta\left(p_1^+ + p_2^+ + p_3^+\right) \simeq 0,$$

its contribution vanishes since the light-cone longitudinal momenta p^+ are all positive and can not add to zero. As a consequence, all energy diagrams which generate the vacuum fluctuations in the usual formulation of quantum field theory are absent in the front form.

TABLE 1. *The vertex interaction in terms of Dirac spinors. The matrix elements V_n are displayed on the right, the corresponding (energy) graphs on the left. All matrix elements are proportional to $\Delta_V = \hat{g}\delta(k_1^+|k_2^++3)\delta^{(2)}(\vec{k}_{\perp,1}|\vec{k}_{\perp,2} + \vec{k}_{\perp,3})$, with $\hat{g} = g/\sqrt{2(2\pi)^3}$. For the periodic boundary conditions one uses $\hat{g} = g/\sqrt{\Omega}$.*

$$V_1 \quad + \quad \frac{\Delta_V}{\sqrt{k_1^+ k_2^+ k_3^+}} \quad (\bar{u}_1 \slashed{\epsilon}_3 T^{a_3} u_2)$$

$$V_3 \quad + \quad \frac{\Delta_V}{\sqrt{k_1^+ k_2^+ k_3^+}} \quad (\bar{v}_2 \slashed{\epsilon}_1^\star T^{a_1} u_3)$$

$$V_4 = \quad \frac{iC_{a_2 a_3}^{a_1} \, \Delta_V}{\sqrt{k_1^+ k_2^+ k_3^+}} \quad (\epsilon_1^\star k_3)\,(\epsilon_2 \epsilon_3) \quad -$$

$$+ \quad \frac{iC_{a_2 a_3}^{a_1} \, \Delta_V}{\sqrt{k_1^+ k_2^+ k_3^+}} \quad (\epsilon_3 k_1)\,(\epsilon_1^\star \epsilon_2)$$

$$+ \quad \frac{iC_{a_2 a_3}^{a_1} \, \Delta_V}{\sqrt{k_1^+ k_2^+ k_3^+}} \quad (\epsilon_3 k_2)\,(\epsilon_1^\star \epsilon_2)$$

The Hamiltonian as a Fock-space operator

The *kinetic energy* T becomes a sum of 3 diagonal operators

$$T = \int dk_- d^2\vec{k}_\perp \sum_{\lambda,c,f} \left(\frac{m^2 + \vec{k}_\perp^2}{k_-}\right)_q \left(b_q^\dagger b_q + d_q^\dagger d_q + a_q^\dagger a_q\right)$$

$$\equiv \sum_q \left(\frac{m^2 + \vec{k}_\perp^2}{k_-}\right)_q \left(b_q^\dagger b_q + d_q^\dagger d_q + a_q^\dagger a_q\right).$$

Here and in the sequel it is convenient to abbreviate the integration and summation over the single particle coordinates by the symbol $\sum$ to replace for instance $b(q)$ with b_q.

The *vertex interaction* V becomes a sum of 4 operators

$$V = V_1 + V_2 + V_3 + V_4$$
$$= \sum_{1,2,3} \left[b_1^\dagger b_2 a_3 \, V_1(1;2,3) + \text{h.c.}\right] + \sum_{1,2,3} \left[d_1^\dagger d_2 a_3 \, V_2(1;2,3) + \text{h.c.}\right] +$$
$$+ \sum_{1,2,3} \left[a_1^\dagger d_2 b_3 \, V_3(1;2,3) + \text{h.c.}\right] + \sum_{1,2,3} \left[a_1^\dagger a_2 a_3 \, V_4(1;2,3) + \text{h.c.}\right].$$

It connects Fock states whose particle number differs by 1. The *matrix elements* $V_n(1;2,3) = V_n(q_1; q_2, q_3)$ are simple functions of the three single-particle states q_i, which are given in Table 1.

TABLE 2. *The fork interaction in terms of Dirac spinors. The matrix elements $F_{n,j}$ are displayed on the right, the corresponding (energy) graphs on the left. All matrix elements are proportional to $\Delta = \tilde{g}^2 \delta(k_1^+ | k_2^+ + k_3^+ + k_4^+) \delta^{(2)}(\vec{k}_{\perp,1} | \vec{k}_{\perp,2} + \vec{k}_{\perp,3} + \vec{k}_{\perp,4})$, with $\hat{g} = g/\sqrt{2(2\pi)^3}$. For the periodic boundary conditions one uses $\hat{g} = g/\sqrt{\Omega}$.*

$F_1 = + \dfrac{2\Delta}{\sqrt{k_1^+ k_2^+ k_3^+ k_4^+}}$	$\dfrac{(\bar{u}_1 T^a \gamma^+ u_2)\,(\bar{v}_3 \gamma^+ T^a u_4)}{(k_1^+ - k_2^+)^2}$
$F_{3,1} = + \dfrac{\Delta}{\sqrt{k_1^+ k_2^+ k_3^+ k_4^+}}$	$\dfrac{(\bar{u}_1 T^{a_4} \slashed{\epsilon}_4 \gamma^+ \slashed{\epsilon}_3 T^{a_2} u_2)}{(k_1^+ - k_4^+)}$
$F_{3,2} = - \dfrac{2k_3^+ \Delta}{\sqrt{k_1^+ k_2^+ k_3^+ k_4^+}}$	$\dfrac{(\bar{u}_1 T^a \gamma^+ u_2)\,(\epsilon_3 i C^a \epsilon_4)}{(k_1^+ - k_2^+)^2}$
$F_{5,1} = + \dfrac{\Delta}{\sqrt{k_1^+ k_2^+ k_3^+ k_4^+}}$	$\dfrac{(\bar{v}_3 T^{a_1} \slashed{\epsilon}_1^{\,*} \gamma^+ \slashed{\epsilon}_2 T^{a_2} u_4)}{(k_1^+ - k_3^+)}$
$F_{5,2} = - \dfrac{\Delta}{\sqrt{k_1^+ k_2^+ k_3^+ k_4^+}}$	$\dfrac{(\bar{v}_3 T^{a_2} \slashed{\epsilon}_2 \gamma^+ \slashed{\epsilon}_1^{\,*} T^{a_1} u_4)}{(k_1^+ - k_4^+)}$
$F_{5,3} = + \dfrac{2(k_1^+ + k_2^+)\Delta}{\sqrt{k_1^+ k_2^+ k_3^+ k_4^+}}$	$\dfrac{(\bar{v}_3 T^a \gamma^+ u_4)\,(\epsilon_1^* i C^a \epsilon_2)}{(k_1^+ - k_2^+)^2}$
$F_{6,1} = + \dfrac{2k_3^+ (k_1^+ + k_2^+)\Delta}{\sqrt{k_1^+ k_2^+ k_3^+ k_4^+}}$	$\dfrac{(\epsilon_1^* C^a \epsilon_2)\,(\epsilon_3 C^a \epsilon_4)}{(k_1^+ - k_2^+)^2}$
$F_{6,2} = + \dfrac{2\Delta}{\sqrt{k_1^+ k_2^+ k_3^+ k_4^+}}$	$(\epsilon_1^* \epsilon_3)\,(\epsilon_2 \epsilon_4)\, C^a_{a_1 a_2} C^a_{a_3 a_4}$

The four-point interactions are broken up conveniently into fork and seagull interactions, F and S, depending on whether they have an odd or an even number of creation operators, thus

$$ P_+ = T + V + F + S. $$

The *fork interaction* F becomes then a sum of 6 operators,

$$
\begin{aligned}
F &= F_1 + F_2 + F_3 + F_4 + F_5 + F_6 \\
&= \sum_{1,2,3,4} \left[b_1^\dagger b_2 d_3 b_4 \, F_1(1;2,3,4) + \text{h.c.} \right] + \left[d_1^\dagger d_2 b_3 d_4 \, F_2(1;2,3,4) + \text{h.c.} \right] \\
&\quad + \sum_{1,2,3,4} \left[b_1^\dagger b_2 a_3 a_4 \, F_3(1;2,3,4) + \text{h.c.} \right] + \left[d_1^\dagger d_2 a_3 a_4 \, F_4(1;2,3,4) + \text{h.c.} \right] \\
&\quad + \sum_{1,2,3,4} \left[a_1^\dagger a_2 d_3 b_4 \, F_5(1;2,3,4) + \text{h.c.} \right] + \left[a_1^\dagger a_2 a_3 a_4 \, F_6(1;2,3,4) + \text{h.c.} \right]. \quad (13)
\end{aligned}
$$

It changes the particle number by 2. The matrix elements are given in Table 2.

The *seagull interaction* S, finally, becomes a sum of 7 operators

$$S = S_1 + S_2 + S_3 + S_4 + S_5 + S_6 + S_7$$
$$= \sum_{1,2,3,4} b_1^\dagger b_2^\dagger b_3 b_4 \, S_1(1,2;3,4) + \sum_{1,2,3,4} d_1^\dagger d_2^\dagger d_3 d_4 \, S_2(1,2;3,4)$$
$$+ \sum_{1,2,3,4} b_1^\dagger d_2^\dagger b_3 d_4 \, S_3(1,2;3,4) + \sum_{1,2,3,4} b_1^\dagger a_2^\dagger b_3 a_4 \, S_4(1,2;3,4)$$
$$+ \sum_{1,2,3,4} d_1^\dagger a_2^\dagger d_3 a_4 \, S_5(1,2;3,4) + \sum_{1,2,3,4} (b_1^\dagger d_2^\dagger a_3 a_4 \, S_6(1,2;3,4) + \text{h.c.})$$
$$+ \sum_{1,2,3,4} a_1^\dagger a_2^\dagger a_3 a_4 \, S_7(1,2;3,4).$$

Its matrix elements are given in Table 3. It can act only between Fock states with the same particle number.

The above results are are quite generally applicable: They hold for arbitrary non-abelian gauge theory $SU(N)$. They hold for abelian gauge theory (QED), formally by replacing the color-matrices $T_{c,c'}^a$ with the unit matrix and by setting to zero the structure constants f^{abc}, thus $B^{\mu\nu} = 0$ and $\chi^\mu = 0$. They hold for 1 time dimension and arbitrary $d + 1$ space dimensions, with $i = 1, \ldots, d$. All what has to be adjusted is the volume integral $\int dx_+ dx_{\perp,1} \ldots dx_{\perp,d}$.

THE HADRONIC BOUND-STATE PROBLEM

One has to find a language in which one can represent hadrons in terms of relativistic confined quarks and gluons. As reviewed in [12], the Bethe-Salpeter formalism has been the central method for analyzing hydrogenic atoms in QED and provides a completely covariant procedure for obtaining bound state solutions. However, calculations using this method are extremely complex and appear to be intractable much beyond the ladder approximation. It also appears impractical to extend this method to systems with more than a few constituent particles.

An intuitive approach for solving relativistic bound-state problems would be to solve the instant form Hamiltonian eigenvalue problem

$$H \, |\Psi\rangle = \sqrt{M^2 + \vec{P}^2} \, |\Psi\rangle$$

for the hadron's mass and wave function. Here, one imagines that $|\Psi\rangle$ is an expansion in multi-particle occupation number Fock states, and that the operators H and $\vec{P}$ are second-quantized Heisenberg operators. Unfortunately, this method is complicated by its non-covariant structure and the necessity to first understand its complicated vacuum eigenstate over all space and time. The presence of the square root operator presents severe mathematical difficulties. Even if these problems could be solved, the eigensolution is only determined in its rest system $(\vec{P} = 0)$; determining the boosted wave function is as complicated as diagonalizing H itself.

TABLE 3. *The seagull interaction in terms of Dirac spinors. The matrix elements $S_{n,j}$ are displayed on the right, the corresponding (energy) graphs on the left. All matrix elements are proportional to $\Delta = \widehat{g}^2 \delta(k_1^+ + k_2^+ | k_3^+ + k_4^+) \delta^{(2)}(\vec{k}_{\perp,1} + \vec{k}_{\perp,2} | \vec{k}_{\perp,3} + \vec{k}_{\perp,4})$, with $\widehat{g} = g/\sqrt{2(2\pi)^3}$. For periodic boundary conditions one uses $\widehat{g} = g/\sqrt{\Omega}$.*

$$S_1 \;=\; -\frac{\Delta}{\sqrt{k_1^+ k_2^+ k_3^+ k_4^+}} \frac{(\bar{u}_1 T^a \gamma^+ u_3)\,(\bar{u}_2 \gamma^+ T^a u_4)}{(k_1^+ - k_3^+)^2}$$

$$S_{3,1} \;=\; +\frac{2\Delta}{\sqrt{k_1^+ k_2^+ k_3^+ k_4^+}} \frac{(\bar{u}_1 T^a \gamma^+ u_3)\,(\bar{v}_2 \gamma^+ T^a v_4)}{(k_1^+ - k_3^+)^2}$$

$$S_{3,2} \;=\; -\frac{2\Delta}{\sqrt{k_1^+ k_2^+ k_3^+ k_4^+}} \frac{(\bar{v}_2 T^a \gamma^+ u_1)\,(\bar{v}_4 \gamma^+ T^a u_3)}{(k_1^+ + k_2^+)^2}$$

$$S_{4,1} \;=\; +\frac{\Delta}{\sqrt{k_1^+ k_2^+ k_3^+ k_4^+}} \frac{(\bar{u}_1 T^{a_4}\!\!\not{\epsilon}_4 \gamma^+ \not{\epsilon}_2^\star T^{a_2} u_3)}{(k_1^+ - k_4^+)}$$

$$S_{4,2} \;=\; +\frac{\Delta}{\sqrt{k_1^+ k_2^+ k_3^+ k_4^+}} \frac{(\bar{u}_1 T^{a_2}\!\!\not{\epsilon}_2^\star \gamma^+ \not{\epsilon}_4 T^{a_4} u_3)}{(k_1^+ + k_2^+)}$$

$$S_{4,3} \;=\; +\frac{2(k_2^+ + k_4^+)\Delta}{\sqrt{k_1^+ k_2^+ k_3^+ k_4^+}} \frac{(\bar{u}_1 T^a \gamma^+ u_3)\,(\epsilon_2^\star i C^a \epsilon_4)}{(k_1^+ - k_3^+)^2}$$

$$S_{6,1} \;=\; +\frac{\Delta}{\sqrt{k_1^+ k_2^+ k_3^+ k_4^+}} \frac{(\bar{u}_1 T^{a_3}\!\!\not{\epsilon}_3 \gamma^+ \not{\epsilon}_4 T^{a_4} v_2)}{(k_1^+ - k_3^+)}$$

$$S_{6,2} \;=\; -\frac{(k_3^+ - k_4^+)\Delta}{\sqrt{k_1^+ k_2^+ k_3^+ k_4^+}} \frac{(\bar{u}_1 T^a \,\gamma^+ \, v_2)\,(\epsilon_3 i C^a \epsilon_4)}{(k_1^+ + k_2^+)^2}$$

$$S_{7,1} \;=\; -\frac{(k_1^+ + k_3^+)(k_2^+ + k_4^+)\Delta}{\sqrt{k_1^+ k_2^+ k_3^+ k_4^+}} \frac{(\epsilon_1^\star C^a \epsilon_3)\,(\epsilon_2^\star C^a \epsilon_4)}{(k_1^+ - k_3^+)^2}$$

$$S_{7,2} \;=\; +\frac{2 k_3^+ k_4^+ \Delta}{\sqrt{k_1^+ k_2^+ k_3^+ k_4^+}} \frac{(\epsilon_1^\star C^a \epsilon_2^\star)\,(\epsilon_3 C^a \epsilon_4)}{(k_1^+ + k_2^+)^2}$$

$$S_{7,3} \;=\; +\frac{\Delta}{\sqrt{k_1^+ k_2^+ k_3^+ k_4^+}} (\epsilon_1^\star \epsilon_3)\,(\epsilon_2^\star \epsilon_4)\, C^a_{a_1 a_2} C^a_{a_3 a_4}$$

$$S_{7,4} \;=\; +\frac{\Delta}{\sqrt{k_1^+ k_2^+ k_3^+ k_4^+}} (\epsilon_1^\star \epsilon_3)\,(\epsilon_2^\star \epsilon_4)\, C^a_{a_1 a_4} C^a_{a_3 a_2}$$

$$S_{7,5} \;=\; +\frac{\Delta}{\sqrt{k_1^+ k_2^+ k_3^+ k_4^+}} (\epsilon_1^\star \epsilon_2^\star)\,(\epsilon_3 \epsilon_4)\, C^a_{a_1 a_3} C^a_{a_2 a_4}$$

This is why instant form wave function cannot be applied in practice to scattering problems. Structure functions for example cannot be calculated.

In principle, the front form approach works in the same way. One aims at solving the Hamiltonian eigenvalue problem

$$H\,|\Psi\rangle = \frac{M^2 + \vec{P}_\perp^2}{P^+}\,|\Psi\rangle\,,\qquad(14)$$

which for several reasons is easier: Contrary to P_z the operator P^+ is positive, having only positive eigenvalues. The square-root operator is absent. The boost operators are kinematic. Having determined the wave function in a particular frame with fixed total momenta P^+ and $\vec{P}_\perp$ the kinematic boost operators allow to covariantly transcribe to any other frame. In fact, as discussed below, one can formulate the theory frame-independently.

The ket $|\Psi\rangle$ can be calculated in terms of a complete set of functions $|\mu\rangle$ or $|\mu_n\rangle$,

$$\int d[\mu]\,|\mu\rangle\,\langle\mu| = \sum_n \int d[\mu_n]\,|\mu_n\rangle\,\langle\mu_n| = \mathbf{1}.$$

The transformation between the complete set of eigenstates $|\Psi\rangle$ and the complete set of basis states $|\mu_n\rangle$ are then $\langle\mu_n|\Psi\rangle$ and usually called the *wavefunctions* $\Psi_{n/h)}(\mu) \equiv \langle\mu_n|\Psi\rangle$. In addition to the quantum numbers of the Lorentz group, the eigenfunction is labeled by quantum numbers like charge, parity, or baryon number which specify a particular hadron h, thus

$$|\Psi\rangle = \sum_n \int d[\mu_n]\,|\mu_n\rangle\,\Psi_{n/h}(\mu).$$

One constructs the complete basis of Fock states $|\mu_n\rangle$ in the usual way by applying products of free field creation operators to the vacuum state $|0\rangle$:

$$
\begin{aligned}
n = 0: &\quad |0\rangle\,, \\
n = 1: &\quad \left|q\bar{q} : k_i^+, \vec{k}_{\perp i}, \lambda_i\right\rangle &=&\quad b^\dagger(q_1)\,d^\dagger(q_2)\,|0\rangle\,, \\
n = 2: &\quad \left|q\bar{q}g : k_i^+, \vec{k}_{\perp i}, \lambda_i\right\rangle &=&\quad b^\dagger(q_1)\,d^\dagger(q_2)\,a^\dagger(q_3)\,|0\rangle\,, \\
n = 3: &\quad \left|gg : k_i^+, \vec{k}_{\perp i}, \lambda_i\right\rangle &=&\quad a^\dagger(q_1)\,a^\dagger(q_2)\,|0\rangle\,, \\
\vdots &\quad \vdots &&\quad \vdots\;\; |0\rangle\,.
\end{aligned}
$$

The operators $b^\dagger(q)$, $d^\dagger(q)$ and $a^\dagger(q)$ create bare leptons (electrons or quarks), bare anti-leptons (positrons or antiquarks) and bare vector bosons (photons or gluons). All of these particles are 'on-shell', $(k^\mu k_\mu)_i = m_i^2$. The various Fock-space classes are conveniently labeled with a running index n. Each Fock state $|\mu_n\rangle = |n : k_i^+, \vec{k}_{\perp i}, \lambda_i\rangle$ is an eigenstate of P^+ and $\vec{P}_\perp$ and the free part of the energy P_0^-, with eigenvalues

$$P^+ = \sum_{i\in n} k_i^+\,,\quad \vec{P}_\perp = \sum_{i\in n} \vec{k}_{\perp i}\,,\quad P_0^- = \sum_{i\in n} \frac{m_i^2 + k_{\perp i}^2}{k_i^+}\,.$$

95

The free invariant mass square of a Fock-state is $M_0^2 = (p_1 + p_2 + \ldots + p_{n_i})^2$, thus

$$M_0^2 = P_0^\mu P_{0,\mu} = P^+ P_0^- - \vec{P}_\perp^2 = P^+ \left(\sum_{i \in n} \frac{m_i^2 + k_{\perp i}^2}{k_i^+} \right) - \vec{P}_\perp^2. \qquad (15)$$

The Fock and the physical vacuum have eigenvalues 0.

The restriction to $k^+ > 0$ is a key difference between light-cone quantization and ordinary equal-time quantization. In equal-time quantization, the state of a parton is specified by its ordinary three-momentum $\vec{k} = (k_x, k_y, k_z)$. Since each component of $\vec{k}$ can be either positive or negative, there exist zero total momentum Fock states of arbitrary particle number, and these will mix with the zero-particle state to build up the ground state, the physical vacuum. However, in light-cone quantization each of the particles forming a zero-momentum state must have vanishingly small k^+. The free or Fock space vacuum $|0\rangle$ is then an exact eigenstate of the full front form Hamiltonian H, in stark contrast to the quantization at equal usual-time. However, the vacuum in QCD is undoubtedly more complicated due to the possibility of color-singlet states with $P^+ = 0$ built on zero-mode massless gluon quanta, but the physical vacuum in the front form is *still far simpler* than in the usual instant form.

Since $k_i^+ > 0$ and $P^+ > 0$, one can define boost-invariant longitudinal momentum fractions

$$x_i = \frac{k_i^+}{P^+}, \qquad \text{with} \quad 0 < x_i < 1,$$

and boost-invariant intrinsic transverse momenta $\vec{k}_{\perp i}$ Their values are constrained,

$$\sum_{i \in n} x_i = 1 \quad \text{and} \quad \sum_{i \in n} \vec{k}_{\perp i} = \vec{0}, \qquad (16)$$

corresponding to the intrinsic frame $\vec{P}_\perp = \vec{0}$. All particles in a Fock state $|\mu_n\rangle = |n : x_i, \vec{k}_{\perp i}, \lambda_i\rangle$ have a boosted four-momentum

$$p_i^\mu \equiv (p^+, \vec{p}_\perp, p^-)_i = \left(x_i P^+, \vec{k}_{\perp i} + x_i \vec{P}_\perp, \frac{m_i^2 + (\vec{k}_{\perp i} + x_i \vec{P}_\perp)^2}{x_i P^+} \right).$$

The free invariant mass square of the Fock state, Eq.(15), is therefore

$$M_0^2 = \sum_{i \in n} \left(\frac{m_i^2 + (\vec{k}_{\perp i} + x_i \vec{P}_\perp)^2}{x_i} \right) - \vec{P}_\perp^2 = \sum_{i \in n} \left(\frac{m^2 + \vec{k}_\perp^2}{x} \right)_i,$$

as a direct consequence of the transverse boost properties.

The phase-space differential $d[\mu_n]$ depends on how one normalizes the single particle states. In the convention where commutators are normalized to a Dirac-delta function, the phase space integration is

$$\int d[\mu_n] \ldots = \sum_{\lambda_i \in n} \int \left[dx_i d^2 k_{\perp i} \right] \ldots , \qquad \text{with}$$

$$\left[dx_i d^2 k_{\perp i} \right] = \delta \Big(1 - \sum_{j \in n} x_j \Big) \delta^{(2)} \Big(\sum_{j \in n} \vec{k}_{\perp j} \Big) \, dx_1 \ldots dx_{N_n} \, d^2 k_{\perp 1} \ldots d^2 k_{\perp N_n},$$

where N_n is the number of particles in Fock state μ_n. The additional Dirac δ-functions account for the constraints (16). The eigenvalue equation (14) stands then for an infinite set of coupled integral equations

$$\sum_{n'} \int [d\mu'_{n'}] \, \langle n : x_i, \vec{k}_{\perp i}, \lambda_i | H | n' : x'_i, \vec{k}'_{\perp i}, \lambda'_i \rangle \, \Psi_{n'/h}(x'_i, \vec{k}'_{\perp i}, \lambda'_i)$$

$$= \frac{M^2 + \vec{P}_\perp^2}{P^+} \Psi_{n/h}(x_i, \vec{k}_\perp, \lambda_i), \qquad \text{for} \quad n = 1, \ldots, \infty. \tag{17}$$

Since P^+ and $\vec{P}_\perp$ are diagonal operators one can rewrite this equation as

$$\sum_{n'} \int [d\mu'_{n'}] \, \langle n : x_i, \vec{k}_{\perp i}, \lambda_i | H P^+ - \vec{P}_\perp^2 | n' : x'_i, \vec{k}'_{\perp i}, \lambda'_i \rangle \, \Psi_{n'/h}(x'_i, \vec{k}'_{\perp i}, \lambda'_i)$$

$$= M^2 \Psi_{n/h}(x_i, \vec{k}_\perp, \lambda_i). \tag{18}$$

It is therefore possible to define a 'light-cone Hamiltonian' as the operator

$$H_{LC} = H P^+ - \vec{P}_\perp^2 = P^\mu P_\mu, \tag{19}$$

so that its eigenvalues correspond to the invariant mass spectrum M_i of the theory. Eq.(18) thus stands for

$$H_{LC} |\Psi\rangle = M^2 |\Psi\rangle , \tag{20}$$

in analogy to Eq.(14).

The Lorentz invariance of H_{LC} and the boost invariance of the wave functions reflects the fact that the boost operators are kinematical. In fact one can boost the system to an 'intrinsic frame' in which the transversal momentum vanishes $\vec{P}_\perp = \vec{0}$, thus $H_{LC} = P^- P^+$. The transformation to an arbitrary frame with finite values of $\vec{P}_\perp$ is then trivially performed. Consider a pion in QCD with momentum $P = (P^+, \vec{P}_\perp)$ as an example. It is described by

$$|\pi : P\rangle = \sum_{n=1}^{\infty} \int d[\mu_n] \big| n : x_i P^+, \vec{k}_{\perp i} + x_i \vec{P}_\perp, \lambda_i \big\rangle \, \Psi_{n/\pi}(x_i, \vec{k}_{\perp i}, \lambda_i),$$

where the sum is over all Fock space sectors of Eq.(15). The ability to specify wavefunctions simultaneously in any frame is a special feature of light-cone quantization. The light-cone wavefunctions $\Psi_{n/\pi}$ do not depend on the total momentum, since x_i is the longitudinal momentum fraction carried by the ith parton and $\vec{k}_{\perp i}$ is

FIGURE 2. *The Hamiltonian matrix for a meson. The matrix elements are represented by energy diagrams. Within each block they are all of the same type: either vertex, fork or seagull diagrams. Zero matrices are denoted by a dot (·). The single gluon is absent since it cannot be color neutral.*

n	Sector	1 $q\bar{q}$	2 gg	3 $q\bar{q}\,g$	4 $q\bar{q}\,q\bar{q}$	5 $gg\,g$	6 $q\bar{q}\,gg$	7 $q\bar{q}\,q\bar{q}\,g$	8 $q\bar{q}\,q\bar{q}\,q\bar{q}$	9 $gg\,gg$	10 $q\bar{q}\,gg\,g$	11 $q\bar{q}\,q\bar{q}\,gg$	12 $q\bar{q}\,q\bar{q}\,q\bar{q}\,g$	13 $q\bar{q}\,q\bar{q}\,q\bar{q}\,q\bar{q}$
1	$q\bar{q}$					·		·	·	·	·	·	·	·
2	gg				·			·	·		·	·	·	·
3	$q\bar{q}\,g$								·	·		·	·	·
4	$q\bar{q}\,q\bar{q}$		·			·				·	·		·	·
5	$gg\,g$	·			·			·	·			·	·	·
6	$q\bar{q}\,gg$								·				·	·
7	$q\bar{q}\,q\bar{q}\,g$	·	·			·				·				·
8	$q\bar{q}\,q\bar{q}\,q\bar{q}$	·	·	·		·	·			·	·			
9	$gg\,gg$	·		·	·			·	·			·	·	·
10	$q\bar{q}\,gg\,g$	·	·		·				·				·	·
11	$q\bar{q}\,q\bar{q}\,gg$	·	·	·		·				·				·
12	$q\bar{q}\,q\bar{q}\,q\bar{q}\,g$	·	·	·	·	·	·			·	·			
13	$q\bar{q}\,q\bar{q}\,q\bar{q}\,q\bar{q}$	·	·	·	·	·	·	·		·	·	·		

its momentum "transverse" to the direction of the meson; both of these are frame-independent quantities. They are the probability amplitudes to find a Fock state of bare particles in the physical pion. But given these light-cone wavefunctions $\Psi_{n/h}(x_i, \vec{k}_{\perp i}, \lambda_i)$, one can compute any hadronic quantity by convolution with the appropriate quark and gluon matrix elements, see for example [2].

In addressing to solve Eq.(20) one faces several major difficulties, among them that the above equations are ill-defined for very large values of the transversal momenta ('ultraviolet singularities') and for values of the longitudinal momenta close to the endpoints $x \sim 0$ or $x \sim 1$ ('endpoint singularities'). One has to introduce cut-offs Λ to regulate the theory in some convenient way. Subsequently one has to remove the cut-off dependence by renormalization group analysis. Renormalization theory is known however only for perturbation theory, for example when calculating Feynman scattering amplitudes in a certain order. Renormalization theory is not available for the bound-state problem. – But even if one has found a convenient regularization scheme, one faces the large (infinite) number of coupled integral equations. Their nature resides in the complicated structure of the kernel

$$\langle n|H|n'\rangle \equiv \langle n : x_i, \vec{k}_{\perp i}, \lambda_i |H| n' : x_i', \vec{k}_{\perp i}', \lambda_i'\rangle.$$

In analyzing the structure of this very complicated many-body problem, as done in Figure 2, one realizes that most of its matrix elements are zero by nature of the operator structure of the Hamiltonian. The Hamiltonian is zero for all sectors whose particle number difference is larger than 2. As an example, consider the fork interaction given in Eq.(13) particularly its matrix element F_3. It scatters a quark into an other momentum state and distroys two gluons. In the block matrix element $\langle 1|H|6\rangle = \langle q\bar{q}|F_3|q\bar{q}\,gg\rangle$ one has many ways to realize that, without that the anti-quark $\bar{q}$ changes its momentum. In Fig. 2, all matrix elements in the block $\langle 1|H|6\rangle$ are therefore represented diagrammatically by the same energy diagram as in Table 2. Usually, the typical seagull matrix elements occur for the diagonal blocs of this matrix, with one exception: In the graph S_6 of Table 3 a $q\bar{q}$-pair is annihilated and scatteredg instantaneously into a gg-pair. Correspondingly, not all entries of the block matrix element $\langle 5|H|5\rangle = \langle q\bar{q}\,g|S_6|gg\,g\rangle$ can vanish. In order to cope with the formidable many-body problem imposed by Eq.(20), the method of discretized light-cone quantization is very useful.

Perturbation theory in the front form

Let us devote a section to the perturbative treatment of front-form gauge theory. Light-cone perturbation theory is really Hamiltonian perturbation theory, and we give the complete set of rules which are the analogues of the Feynman rules. We shall demonstrate in a selected example, that one gets the same covariant and gauge-invariant scattering amplitude as in Feynman theory, see also [2].

The Green's functions $\hat{G}_{fi}(x^+)$ are the probability amplitudes that a state starting in Fock state $|i\rangle$ ends up in Fock state $|f\rangle$ at a later time x^+

$$\left\langle f \mid \hat{G}(x^+) \mid i \right\rangle = \langle f|e^{-iP_+ x^+}|i\rangle = i \int \frac{d\epsilon}{2\pi} e^{-i\epsilon x^+} \langle f|G(\epsilon)|i\rangle.$$

Its Fourier transform $\langle f \mid G(\epsilon) \mid i\rangle$ is called the resolvent of the Hamiltonian H, i.e.

$$\langle f \mid G(\epsilon) \mid i \rangle = \left\langle f \left| \frac{1}{\epsilon - H + i0_+} \right| i \right\rangle = \left\langle f \left| \frac{1}{\epsilon - H_0 - U + i0_+} \right| i \right\rangle.$$

Separating the Hamiltonian $H = H_0 + U$ into a free part H_0 and an interaction U, one can expand the resolvent into the series

$$\langle f|G(\epsilon)|i\rangle = \langle f|\tilde{G}(\epsilon) + \tilde{G}(\epsilon)U\tilde{G}(\epsilon) + \tilde{G}(\epsilon)U\tilde{G}(\epsilon)U\tilde{G}(\epsilon) + \ldots |i\rangle.$$

The rules for x^+-ordered perturbation theory follow when the resolvent of the free Hamiltonian $\tilde{G}(\epsilon) = 1/(\epsilon - H_0 + i0_+)$ is replaced by its spectral decomposition

$$\tilde{G}(\epsilon) = \sum_{n=1}^{\infty} \tilde{G}_n(\epsilon), \quad \tilde{G}_n(\epsilon) = \int d[\mu_n] \left| n : k_i^+, \vec{k}_{\perp i}, \lambda_i \right\rangle \frac{1}{\Delta_n} \left\langle n : k_i^+, \vec{k}_{\perp i}, \lambda_i \right|, \quad (21)$$

with the energy denominator $\Delta_n = \epsilon - \sum_{i \in n} \left((k_\perp^2 + m^2)/k^+ \right)_i + i0_+$. The sum runs over all Fock states n intermediate between two interactions U.

To calculate then $\langle f|(\epsilon)|i\rangle$ perturbatively, all x^+-ordered diagrams must be considered, the contribution from each graph computed according to the rules of old-fashioned Hamiltonian perturbation theory [13,14]:

1. Draw all topologically distinct x^+-ordered diagrams.

2. Assign to each line a single particle momentum k^μ, a helicity λ, as well as color and flavor. With fermions (electrons or quark) associate a spinor $u_\alpha(k, \lambda)$, with antifermions $v_\alpha(k, \lambda)$, and with vector bosons (photons or gluons) a polarization vector $\epsilon_\mu(k, \lambda)$.

3. For each vertex include the matrix element $\langle n|V|n'\rangle$ between Fock state n and n' as given in Table 1.

4. For each intermediate state there is an energy denominator $1/\Delta_n$ in which $\epsilon = P_{0,\text{in}}^-$ is the incident free light-cone energy.

5. To account for three-momentum conservation include for each intermediate state the delta-functions $\delta\left(P^+ - \sum_i k_i^+\right)$ and $\delta^{(2)}\left(\vec{P}_\perp - \sum_i \vec{k}_{\perp i}\right)$.

6. Sum over all internal helicities (and colors for gauge theories) and integrate over each internal k with the weight $\int d^2 k_\perp dk^+ \theta(k^+)(2\pi)^{-3/2}$.

7. Include a factor -1 for each closed fermion loop, for each fermion line that both begins and ends in the initial state, and for each diagram in which fermion lines are interchanged in either of the initial or final states.

8. Imagine that every internal line is a sum of a 'dynamic' and an 'instantaneous' line, and draw all diagrams with $1, 2, 3, \ldots$ instantaneous lines.

9. Two consecutive instantaneous interactions give a vanishing contribution.

10. For each instantaneous line insert a factor $\langle n|W|n'\rangle$, with the matrix element given in Table 2 and 3.

The light-cone Fock state representation can thus be used advantageously in perturbation theory. The sum over intermediate Fock states is equivalent to summing all x^+-ordered diagrams and integrating over the transverse momentum and light-cone fractions x. Because of the restriction to positive x, diagrams corresponding to vacuum fluctuations or those containing backward-moving lines are eliminated.

The $q\bar{q}$-scattering amplitude

The simplest application of the above rules is the calculation of the electron-muon scattering amplitude to lowest non-trivial order. But the quark-antiquark scattering is only marginally more difficult. We thus imagine an initial $(q, \bar{q})$-pair with different flavors $f \neq \bar{f}$ to be scattered off each other by exchanging a gluon.

Let us treat this problem as a pedagogical example to demonstrate the rules. Rule 1: There are two time-ordered diagrams associated with this process. In the first one the gluon is emitted by the quark and absorbed by the antiquark, and in the second it is emitted by the antiquark and absorbed by the quark. For the first diagram, we assign the momenta required in rule 2 by giving explicitly the initial Fock state $|q, \bar{q}\rangle = \frac{1}{\sqrt{n_c}} \sum_{c=1}^{n_c} b_{cf}^\dagger(k_q, \lambda_q) d_{c\bar{f}}^\dagger(k_{\bar{q}}, \lambda_{\bar{q}})|0\rangle$. Note that it is invariant under $SU(n_c)$. The final Fock state is $|q', \bar{q}'\rangle = \frac{1}{\sqrt{n_c}} \sum_{c=1}^{n_c} b_{cf}^\dagger(k_q', \lambda_q') d_{c\bar{f}}^\dagger(k_{\bar{q}}', \lambda_{\bar{q}}')|0\rangle$. The intermediate state

$$|q', \bar{q}, g\rangle = \sqrt{\frac{2}{n_c^2 - 1}} \sum_{c=1}^{n_c} \sum_{c'=1}^{n_c} \sum_{a=1}^{n_c^2-1} T_{c,c'}^a b_{cf}^\dagger(k_q', \lambda_q') d_{c'\bar{f}}^\dagger(k_{\bar{q}}, \lambda_{\bar{q}}) a_a^\dagger(k_g, \lambda_g)|0\rangle , \qquad (22)$$

has 'a gluon in flight'. Since the gluons longitudinal momentum is positive, the diagram allows only for $k_q'^+ < k_q^+$. Rule 3 requires at each vertex the factors

$$\langle q, \bar{q}| V |q', \bar{q}, g\rangle = \frac{g}{(2\pi)^{\frac{3}{2}}} \sqrt{\frac{n_c^2 - 1}{2n_c}} \frac{\left[\bar{u}(k_q, \lambda_q) \gamma^\mu \epsilon_\mu(k_g, \lambda_g) u(k_q', \lambda_q')\right]}{\sqrt{2k_q^+} \sqrt{2k_g^+} \sqrt{2k_q'^+}} , \qquad (23)$$

$$\langle q', \bar{q}, g| V |q', \bar{q}'\rangle = \frac{g}{(2\pi)^{\frac{3}{2}}} \sqrt{\frac{n_c^2 - 1}{2n_c}} \frac{\left[\bar{v}(k_{\bar{q}}', \lambda_{\bar{q}}') \gamma^\nu \epsilon_\nu^*(k_g, \lambda_g) v(k_{\bar{q}}, \lambda_{\bar{q}})\right]}{\sqrt{2k_{\bar{q}}^+} \sqrt{2k_g^+} \sqrt{2k_{\bar{q}}'^+}} , \qquad (24)$$

respectively, wich are found by means of Table 1. Working with color neutral Fock states, all color structure reduces to an overall factor C, with $C^2 = (n_c^2 - 1)/2n_c$. Rule 4 requires the energy denominator $1/\Delta_3$. It is useful to work the four-momentum transfers of the quark $Q^2 = -(k_q - k_q')^2 = k_g^+(k_g + k_q' - k_q)^-$. The anti-quark has $\overline{Q}^2 = (k_{\bar{q}} - k_{\bar{q}}')^2 = k_g^+(k_g + k_{\bar{q}} - k_{\bar{q}}')^-$. With the initial energy $\epsilon = \widetilde{P}_+ = (k_q + k_{\bar{q}})_+ = \frac{1}{2}(k_q + k_{\bar{q}})^-$, the energy denominator becomes then

$$\Delta_3 = (k_q + k_{\bar{q}})^- - (k_g + k_q' + k_{\bar{q}})^- = -\frac{Q^2}{k_g^+}$$

Rule 5 requires two Dirac-delta functions, one at each vertex, to account for conservation of three-momentum. One of them is removed by the requirement of rule 6, namely to integrate over all intermediate internal momenta and the other remains in the final equation (26). The gluon momentum is thus fixed by the external legs of the graph. The polarization sum over the gluon helicity gives

$$d_{\mu\nu}(k_g) \equiv \sum_{\lambda_g} \epsilon_\mu(k_g, \lambda_g) \epsilon_\nu^*(k_g, \lambda_g) = -g_{\mu\nu} + \frac{k_{g,\mu}\eta_\nu + k_{g,\nu}\eta_\mu}{k_g^\kappa \eta_\kappa}.$$

The null vector has the components $\eta^\mu = (\eta^+, \vec{\eta}_\perp, \eta^-) = (0, \vec{0}_\perp, 2)$ and thus the properties $\eta^2 \equiv \eta^\mu \eta_\mu = 0$ and $k\eta = k^+$. As shown explicitly in [2] one gets for the second order diagram $\langle q, \bar{q}|V\widetilde{G}_3 V|q', \bar{q}'\rangle$ after some non-trivial steps

101

$$\langle q, \bar{q} | V \widetilde{G}_3 V | q', \bar{q}' \rangle = \frac{g^2 C^2}{(2\pi)^3} \frac{\left[\overline{u}(k_q, \lambda_q) \gamma^\mu u(k'_q, \lambda'_q) \right]}{\sqrt{4 k_q^+ k_q'^+}} \frac{1}{Q^2} \frac{\left[\overline{u}(k_{\bar{q}}, \lambda_{\bar{q}}) \gamma_\mu u(k'_{\bar{q}}, \lambda'_{\bar{q}}) \right]}{\sqrt{4 k_{\bar{q}}^+ k_{\bar{q}}'^+}}$$

$$- \frac{g^2 C^2}{(2\pi)^3} \frac{\left[\overline{u}(k_q, \lambda_q) \gamma^+ u(k'_q, \lambda'_q) \right]}{\sqrt{4 k_q^+ k_q'^+}} \frac{1}{(k_g^+)^2} \frac{\left[\overline{u}(k_{\bar{q}}, \lambda_{\bar{q}}) \gamma^+ u(k'_{\bar{q}}, \lambda'_{\bar{q}}) \right]}{\sqrt{4 k_{\bar{q}}^+ k_{\bar{q}}'^+}}. \quad (25)$$

The delta-functions and a step function $\Theta(k_q'^+ \leq k_q^+)$ are omitted for simplicity. One proceeds with rule 8, by including the instantaneous lines. Table 3 gives

$$\langle q, \bar{q} | S | q', \bar{q}' \rangle = \frac{g^2 C^2}{(2\pi)^3} \frac{[\overline{u}(k, \lambda) \gamma^+ u(k', \lambda')]_q}{\sqrt{4 k_q^+ k_q'^+}} \frac{1}{(k_q^+ - k_q'^+)^2} \frac{[\overline{u}(k, \lambda) \gamma^+ u(k', \lambda')]_{\bar{q}}}{\sqrt{4 k_{\bar{q}}^+ k_{\bar{q}}'^+}}.$$

The $q\bar{q}$-scattering amplitude, the sum $\langle q, \bar{q} | S + V \widetilde{G}_3 V | q', \bar{q}' \rangle$ has then the correct gauge-invariant result known from Feynman theory up to second order

$$\langle q, \bar{q} | S + V \widetilde{G}_3 V | q', \bar{q}' \rangle = \frac{(-1)}{(k_q - k'_{\bar{q}})^2} \frac{1}{\sqrt{k_q^+ k_{\bar{q}}^+ k_q'^+ k_{\bar{q}}'^+}} \delta(P^+ - P'^+) \delta^{(2)}(\vec{P}_\perp - \vec{P}'_\perp)$$

$$\times \frac{(gC)^2}{(2\pi)^3} [\overline{u}(k, \lambda) \gamma^\mu u(k', \lambda')]_q [\overline{u}(k, \lambda) \gamma_\mu u(k', \lambda')]_{\bar{q}}. \quad (26)$$

The instantaneous diagram is cancelled exactly.

DISCRETIZED LIGHT-CONE QUANTIZATION

The infinitely many coupled integral equations in the preceeding section a very difficult to cope with in practice. Because of the many integrations and summations it is very difficult to even write them down. But field theory becomes much more transparent when one works with periodic boundary conditions. The coupled integral equations become then coupled matrix equations. Since rows and colums of a matrix can be denumerated, one can keep track of the necessary manipulations much easier. In fact, working with periodic boundary conditions, or with 'Discretized Light-Cone Quantization' (DLCQ), one has obtained the first total solutions to non-trivial quantum field theories in 1+1 dimensions. In 3+1 dimensions the method has the ambitious goal to calculate the spectra and wavefunctions of physical hadrons from a covariant gauge field theory. The ingredients of the method shall be reviewed in short in this section.

The non-relativistic A-body problem in one dimension

Let us first briefly review the difficulties for a conventional non-relativistic many-body theory. One starts with a many-body Hamiltonian $H = T + U$. The kinetic

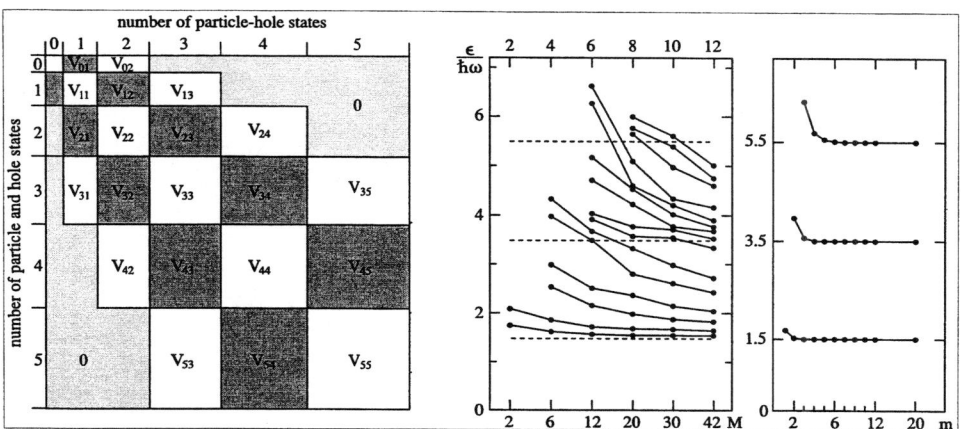

FIGURE 3. *Non-relativistic many-body theory.*

energy T is usually a one-body operator and thus simple. The potential energy U is at least a two-body operator and thus complicated. One has solved the problem if one has found the eigenvalues and eigenfunctions of the Hamiltonian equation, $H\Psi = E\Psi$. One always can expand the eigenstates in terms of products of single particle states $\langle \vec{x}|m \rangle$ belonging to a complete set of ortho-normal functions. When antisymmetrized, one refers to them as 'Slater-determinants'. All Slater-determinants with a fixed particle number form a complete set.

One can proceed as follows. In the first step one chooses a complete set of single particle wave functions. These single particle wave functions are solutions of an arbitrary single particle Hamiltonian. In a second step, one selects one Slater determinat as a reference state. All Slater determinants can be classified relative to this reference state as 1-particle-1-hole (1-ph) states, 2-particle-2-hole (2-ph) states, and so on. The Hilbert space is truncated at some level. In a third step, one calculates the Hamiltonian matrix within this Hilbert space.

In Figure 3, the Hamiltonian matrix for a two-body interaction is displayed schematically. Most of the matrix-elements vanish, since a 2-body Hamiltonian changes the state by up to 2 particles. Therefore the structure of the Hamiltonian is a finite penta-diagonal block matrix. The dimension within a block is made finite by an artificial cut-off on the kinetic energy, *i.e.* on the single particle quantum numbers m. A finite matrix can be diagonalized on a computer. At the end one must verify that the physical results are reasonably insensitive to the cut-off(s) and other formal parameters.

This procedure was actually carried out in one space dimension [15] with two different sets of single-particle functions,

$$\langle x|m \rangle = N_m H_m\left(\frac{x}{L}\right) \exp\left\{-\frac{1}{2}\left(\frac{x}{L}\right)^2\right\} \quad \text{and} \quad \langle x|m \rangle = N_m \exp\left\{im\frac{x}{L}\pi\right\}. \quad (27)$$

The two sets are the eigenfunctions of the harmonic oscillator ($L \equiv \hbar/m\omega$) with its

Hermite polynomials H_m, and the eigenfunctions of the momentum of a free particle with periodic boundary conditions. Both are suitably normalized (N_m), and both depend parametrically on a characteristic length parameter L. The calculations are particularly easy for particle number 2, and for a harmonic two-body interaction. The results are displayed in Figure 3, and surprisingly different. For the plane waves, the results converge rapidly to the exact eigenvalues $E = \frac{3}{2}, \frac{7}{2}, \frac{11}{2}, \ldots$, as shown in the right part of the figure. Opposed to this, the results with the oscillator states converge extremely slowly. Obviously, the larger part of the Slater determinants is wasted on building up the plane wave states of center of mass motion from the Slater determinants of oscillator wave functions. It is obvious, that the plane waves are superior, since they account for the symmetry of the problem, namely Galilean covariance. The approach was successfully applied for getting the exact eigenvalues and eigenfunctions for up to 30 particles.

From these calculations, one should conclude that discretized plane waves are a useful tool for many-body problems, and that the generate good wavefunctions even for a 'confining' potential like the harmonic oscillator.

QED in 1+1 dimension

DLCQ had been applied first to Yukawa theory [16] in 1-space and 1-time dimensions followed by an application to QED [17] and to QCD [18], but the advantages of working with periodic boundary conditions in the front form, particularly when discussing the 'zero modes' in the ϕ^4-theory, had been noted also by Maskawa and Yamawaki [26] in 1976.

In one space dimension are no rotations — hence no spin. The Dirac spinor has two components and the Dirac matrices are 2 by 2 matrices. The gauge field A^μ field has two components. One of them is eliminated by fixing the gauge, and the other determined by Gauss' law. The gauge field carries no dynamical degree of freedom. The theory confines quarks (or electrons) because the Poisson equation in 1 space dimension gives rise to a linearly rising potential.

Quantum electrodynamics in 1+1 dimension has played an important role in field theory because its massless version, the Schwinger model, is analytically soluable. The fundamental solution corresponds to a composite $e\bar{e}$-state – the Schwinger-boson – with invariant mass $m_B \equiv g/\sqrt{\pi}$. Note that the coupling constant g in 1+1 dimension has the dimension of a mass.

Consider first the massive Schwinger model. The Lagrangian for the theory takes the same form as in Eq.(3). Again one works in the light-cone gauge $A^+ = 0$, and uses the same projection operators $\Lambda_\pm$. The Dirac equation decomposes again into a time-derivative $2i\partial^+\Psi_+ = m\beta\Psi_- + gA^-\Psi_+$ and a constraint $2i\partial^-\Psi_- = m\beta\Psi_+$. One is left with only one independent field, Ψ_+, which is canonically quantized at $x^+ = 0$,

$$\{\Psi_+(x^-, x^+), \Psi_+^\dagger(y^-, y^+)\}_{x^+=y^+=0} = \Lambda_+\delta(x^- - y^-). \tag{28}$$

TABLE 4. *Fock-space sectors and block matrix structure for QED in 1+1 dimension. Diagonal blocs (D=T+S) refer to seagull, off-diagonal (F) to fork matrix elements. A zero block matrix is marked by (·).*

Sector	n	1	2	3	4	5	6	7	8
$q\bar{q}$	1	D	F	·	·	·	·	·	·
$q\bar{q}\,q\bar{q}$	2	F	D	F	·	·	·	·	·
$q\bar{q}\,q\bar{q}\,q\bar{q}$	2	·	F	D	F	·	·	·	·
$q\bar{q}\,q\bar{q}\,q\bar{q}\,q\bar{q}$	4	·	·	F	D	F	·	·	·
5 $q\bar{q}$-pairs	5	·	·	·	F	D	F	·	·
6 $q\bar{q}$-pairs	6	·	·	·	·	F	D	F	·
7 $q\bar{q}$-pairs	7	·	·	·	·	·	F	D	F
8 $q\bar{q}$-pairs	8	·	·	·	·	·	·	F	D

The formalism gets somewhat simplified if one uses the chiral representation [18]

$$\gamma^0 = \begin{pmatrix} 0 & 1 \\ 1 & 0 \end{pmatrix}, \quad \gamma^1 = \begin{pmatrix} 0 & 1 \\ -1 & 0 \end{pmatrix}, \quad \gamma^5 = \gamma^0\gamma^1 = \begin{pmatrix} -1 & 0 \\ 0 & 1 \end{pmatrix}.$$

The $\Lambda_\pm = (1 \pm \gamma^0\gamma^1)/2$ are then diagonal and project on the chiral components

$$\Psi = \begin{pmatrix} \Psi_L \\ \Psi_R \end{pmatrix}, \quad \Psi_+ = \begin{pmatrix} 0 \\ \Psi_R \end{pmatrix}, \quad \Psi_- = \begin{pmatrix} \Psi_L \\ 0 \end{pmatrix}.$$

The (light-cone) momentum and energy operators become

$$P^+ = \int_{-L}^{+L} dx^- \Psi_R^\dagger \partial_- \Psi_R,$$

$$P^- = m^2 \int_{-L}^{+L} dx^- \Psi_R^\dagger \frac{1}{i\partial_-} \Psi_R + \frac{g^2}{2} \int_{-L}^{+L} dx^- \Psi_R^\dagger \Psi_R \frac{1}{(i\partial_-)^2} \Psi_R^\dagger \Psi_R.$$

One can expand Ψ_R (or Ψ_+) with periodic boundary conditions [17], but anti-periodic boundary conditions [18] avoid the zero mode:

$$\Psi_R(x^-) = \frac{1}{\sqrt{2L}} \sum_{n=\frac{1}{2},\frac{3}{2},\ldots}^{\infty} \left(b_n e^{-i\frac{n\pi}{L}x^-} + d_n^\dagger e^{i\frac{n\pi}{L}x^-} \right).$$

The creation and destruction operators obey $\left\{ b_n^\dagger, b_m \right\} = \left\{ d_n^\dagger, d_m \right\} = \delta_{n,m}$, consistent with Eq.(28). One redefines units by

$$P^+ = \frac{2\pi}{L} K, \quad P_0^- = \frac{L}{2\pi} \left(m^2 H_0 + \frac{g^2}{\pi} U \right).$$

The length L drops out in $H_{\text{LC}} = P^+ P^-$. Inserting the above fields gives

TABLE 5. *The 5 Fock states for* $K = 4$.

1	$\frac{1}{2}; \frac{7}{2}$
2	$\frac{3}{2}; \frac{5}{2}$
2	$\frac{5}{2}; \frac{3}{2}$
4	$\frac{7}{2}; \frac{1}{2}$
5	$\frac{1}{2}, \frac{3}{2}; \frac{1}{2}, \frac{3}{2}$

TABLE 6. *Number of Fock states versus harmonic resolution.*

K	1 $q\bar{q}$	2 $q\bar{q}$	3 $q\bar{q}$	4 $q\bar{q}$	Total
1	1	-	-	-	1
4	4	1	-	-	5
9	9	20	1	-	30
16	16	140	74	1	231

$$K = \sum_{n=\frac{1}{2},\frac{3}{2},\ldots}^{\infty} n \left(b_n^\dagger b_n + d_n^\dagger d_n \right), \quad H_0 = \sum_{n=\frac{1}{2},\frac{3}{2},\ldots}^{\infty} \frac{1}{n} \left(b_n^\dagger b_n + d_n^\dagger d_n \right),$$

$$U = :U: + \sum_{n=\frac{1}{2},\frac{3}{2},\ldots}^{\infty} \frac{I_n}{n} \left(b_n^\dagger b_n + d_n^\dagger d_n \right), \quad \text{with} \quad I_n = -\frac{1}{2n} + \sum_{m=1}^{n+\frac{1}{2}} \frac{1}{m^2},$$

$$:U: = \sum_{n_1,n_2,n_3,n_4} \frac{\delta(n_1 + n_2 | n_3 + n_4)}{2(n_1 - n_3)^2} \left(b_1^\dagger b_2^\dagger b_3 b_4 + d_1^\dagger d_2^\dagger d_3 d_4 \right)$$

$$+ \sum_{n_1,n_2,n_3,n_4} \delta(n_1 + n_2 | n_3 + n_4) \left(\frac{1}{(n_1 - n_3)^2} - \frac{1}{(n_1 + n_2)^2} \right) b_1^\dagger d_2^\dagger b_3 d_4$$

$$+ \sum_{n_1,n_2,n_3,n_4} \frac{\delta(n_1 | n_2 + n_3 + n_4)}{(n_1 - n_3)^2} \left(b_1^\dagger b_2 b_3 d_4 + d_1^\dagger d_2 d_3 b_4 + h.c. \right).$$

The symbols $\delta(n|m) = \delta_{n,m}$ are Kronecker delta's and $b_1 \equiv b_{n_1}$. In 1+1 dimensions it is very important to keep the 'self-induced inertias' I_n from the normal ordering. They are needed to cancel the infrared singularity in the interaction term in the continuum limit.

The next step is to actually solve the equations of motions in the discretized space. Typically one proceeds as follows: One constructs the Fock space

$$|\mu_n\rangle = |q_1, \ldots, q_n; \bar{q}_1, \ldots, \bar{q}_n\rangle = b_{q_1}^\dagger, \ldots, b_{q_n}^\dagger d_{\bar{q}_1}^\dagger, \ldots, d_{\bar{q}_n}^\dagger |0\rangle,$$

in the same way as above in Eq.(15), and arranges it in denumerated Fock space sectors, as illustrated in Table 4. Each Fock state must be an eigenstate to P^+ and thus of the *harmonic resolution* K [17], with eigenvalue $K = \sum_{i \in \mu_n} n_i$. One selects

106

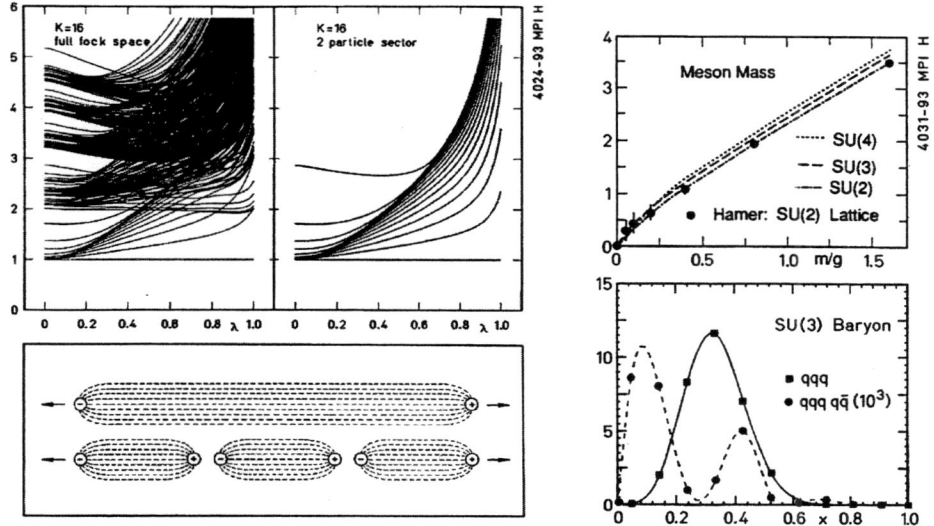

FIGURE 4. Spectra and wavefunctions in 1+1 dimensions, taken from [17,18].

now one value for K and constructs all Fock states. Both the number of Fock space sectors and the number of Fock states within each sector are finite for every finite K, due to the positivity condition on the light-cone momenta. For $K = 1$ one has only one Fock-space class with one Fock state $|\frac{1}{2}; \frac{\bar{1}}{2}\rangle = b^\dagger_{\frac{1}{2}} d^\dagger_{\frac{1}{2}} |0\rangle$, for $K = 4$ one has the 5 Fock states given in Table 5. The numbers of Fock states increase with given K, as shown in Table 6, but less than exponentially due to the exclusion principle. Next, one calculates the matrix elements of $H = P^-$. In the last step one diagonalizes H. Any of its eigenvalues $E(K)$ depends on K and corresponds to an invariant mass $M^2(K) \equiv P^+ P^- = K E(K)$. Notice that one gets a spectrum of invariant mass-squares for any value of K.

The eigenvalue spectrum of QED1 + 1 was given first by Eller [17], for periodic boundary conditions on the fermion fields. The plot likeon the left side in Figure 4 was calculated [22] with anti-periodic boundary conditions. It shows the full mass spectrum of QED in the charge zero sector for all values of the coupling constant and the fermion mass, parametrized by $\lambda = (1 + \pi(m/g)^2)^{-\frac{1}{2}}$. The eigenvalues M_i are plotted in units where the mass of the lowest 'positronium' state has the numerical value 1. All states with $M > 2$ are unbound. The plot includes the free case $\lambda = 0$ $(g = 0)$ and the the Schwinger limit $M = 1$ for $\lambda = 1$ $(m = 0)$, where DLCQ generates the exact eigenvalue. – The lower left part of the figure illustrates the following point. The rich complexity of the spectrum allows for multi-particle Fock states *at the same invariant mass* as the 'simple $q\bar{q}$-states' shown in the figure as the '2 particle sector'. The spectrum includes not only the simple bound state

107

spectrum, but also the associated discretized continuum of the same particles in relative motion. One can identify the simple bound states as two quarks connected by a confining string as displayed in the figure. The smallest residual interaction mixes the simple configuration with the large number of 'continuum states' at the same mass. The few simple states have a much smaller statistical weight, and it looks as if the long string 'breaks' into several pieces of smaller strings. Loosely speaking one can interpret such a process as the decay of an excited pion into multi-pion configurations $\pi^\star \to \pi\pi\pi$.

In 1+1 dimensions quantum electro dynamics [17] and quantum chromo dynamics [18] show many similarities, both from the technical and from the phenomenological point of view. In the right part of Figure 4 some of the results of Hornbostel [18] on the spectrum and the wavefunctions for QCD are displayed. Fock states in non-abelian gauge theory SU(N) can be made color singlets for any order of the gauge group and thus one can calculate mass spectra for mesons and baryons for almost arbitrary values of N. In the upper right part of the figure the lowest mass eigenvalue of a meson is given for $N = 2, 3, 4$. Lattice gauge calculations are available only for $N = 2$ and for the lowest two eigenstates [19]. In general the agreement is very good. In the left lower part of the figure the structure function of a baryon is plotted versus (Bjørken-)x for $m/g = 1.6$. With DLCQ it is possible to calculate also higher Fock space components. As an example, the figure includes the probability distribution to find a quark in a $qqq\,q\bar{q}$-state.

Fermion condensates and the small mass limit

Based on the low energy theorems from times prior to QCD, it is believed that the square of the pion mass is linear in the quark mass m for sufficiently small m. The proportionality constant has to have a dimension of mass, and since there is no other scale in the problem except the *quark condensates in the vacuum* $\langle 0|\overline{\Psi}\Psi|0\rangle$, one believes that the square of the pion mass is approximatively given by $m_\pi^2 \sim 2\langle 0|\overline{\Psi}\Psi|0\rangle m$, a theorem which was succesful in many phenomenological applications.

The Schwinger model has played an important role as a paradigm for our understanding of hadronic physics. Among other aspects it has the desired feature that the invariant mass square of the $e\bar{e}$-boson is linear in the electron mass m

$$M^2 = m_B^2 + 2m\,\langle 0|\overline{\Psi}\Psi|0\rangle, \quad \text{with} \quad \langle 0|\overline{\Psi}\Psi|0\rangle = e^\gamma\, m_B, \tag{29}$$

where $m_B \equiv g/\sqrt{\pi}$ is the invariant mass in the Schwinger limit. Euler's constant γ was understood as a signal for non-perturbative physics. In his analysis of the Schwinger model, Bergknoff [20] showed that the take-off from the Schwinger limit obeys

$$\langle 0|\overline{\Psi}\Psi|0\rangle = \frac{\pi}{\sqrt{3}}\, m_B. \tag{30}$$

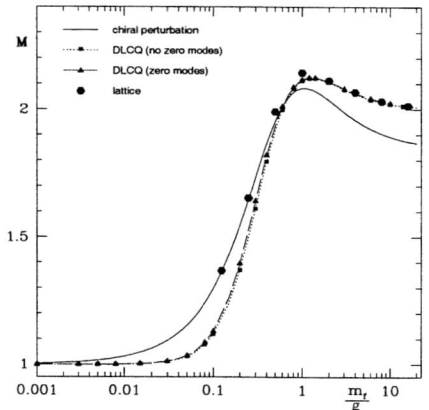

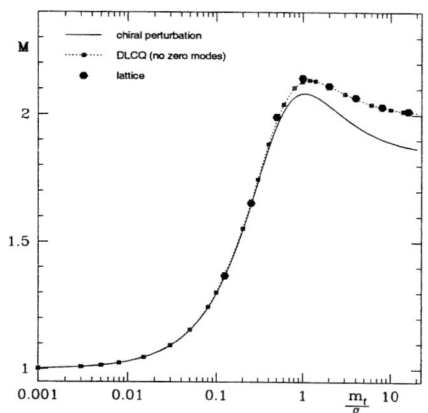

FIGURE 5. *The lowest mass eigenvalue of* QED_{1+1} *in units of* m_B *is plotted versus the fermion mass* m_f *in units of the coupling constant* g*, as calculated with 'conventional DLCQ' ($K = 16$). Taken from Ref. [23]. See also discussion in the text.*

FIGURE 6. *The lowest mass eigenvalue of* QED_{1+1} *in units of* m_B *is plotted versus the fermion mass* m_f *in units of the coupling constant* g*, as calculated with 'improved DLCQ' ($K = 20$), as proposed in Ref. [25]. Taken from Ref. [23]. See also discussion in the text.*

The result of this 'chiral perturbation theory' to first order, $\frac{\pi}{\sqrt{3}} \sim 1.81$, is numerically very close to $e^{\gamma} \sim 1.78$.

Its is actually quite easy to derive the Bergknoff equation from DLCQ [21]. For a fixed resolution K, a $q\bar{q}$-state is fixed by the quantum number n of the electron. The eigenvalue equation is then given by

$$M^2 \langle n|\Psi\rangle = \sum_{n'=\frac{1}{2}}^{[K]} \langle n|H_{LC}|n'\rangle \, \langle n'|\Psi\rangle, \quad \text{for} \quad n = \frac{1}{2}, \frac{3}{2}, \ldots, [K],$$

with $[K] \equiv K - \frac{1}{2}$. Using the Hamiltonian matrix elements are given above, one introduces $x = p^+/P^+ = n/K$ and goes to the continuum limit. After a few steps [21] one gets the integral equation of Bergknoff [20],

$$M^2 \langle x|\Psi\rangle = \frac{m^2}{x(1-x)} \langle x|\Psi\rangle + m_B^2 \int_0^1 dx' \langle x'|\Psi\rangle + m_B^2 \fint_0^1 dx' \frac{\langle x|\Psi\rangle - \langle x'|\Psi\rangle}{(x-x')^2}.$$

For $m = 0$, the solution $\langle x|\Psi\rangle = 1$ has the eigenvalue $M^2 = m_B^2$, the Schwinger boson.

Eqs.(29) and (30) state that the mass-squared of the Schwinger boson is *linear in the fermion mass* m, in the limit when m goes to zero. For a long time, this result was in conflict with the explicit DLCQ-calulations [17,22,23]. The shortcoming was taken as a hint [24] that light-cone quantization was in failure because of the trivial vacuum structure which does not allow for condensates.

The most recent results by Völlinger [23] for DLCQ are displayed in Figure 5. They are compared with the lattice calculations of Crewther and Hamer [19] and with chiral perturbation theory up to second order [24]. The lattice results and DLCQ show some discrepancy for very small m which however fades away for larger m. On the other hand, the lattice results and chiral perturbation theory agree perfectly at small and deviate for large m, where chiral perturbation theory is not suposed to work. The figure show aslo that a correct inclusion of the (gauge field) zero modes has no significant impact. The re-solution of this puzzle came by van de Sande [25]. He realized that conventional DLCQ has difficulties to reproduce the wave function $\langle x|\Psi \rangle$ near the endpoints $x \to 0$ and $x \to 1$ very close to the Schwinger limit. His 'improved DLCQ' accounts for that, and indeed when properly included as shown in Figure 6 all discrepancies fade away.

What should be learned from this exercise is that not everything what is called a 'vacuum condensate' in the literature deserves this name in a physical sense: In naive light-cone quantization particularly DLCQ the vacuum is trivial and can not have condensates.

Φ^4 in 1+1 dim's: Zero modes and phase transitions

The naive front-form vacuum is simple. However, one commonly associates important long range properties of a field theory with the vacuum like spontaneous symmetry breaking, the Goldstone pion, or color confinement. If one cannot associate long range phenomena with the vacuum state itself, then the only alternative is the zero momentum components of the field, the 'zero modes'. In some cases, the zero mode operator is not an independent degree of freedom but obeys a constraint equation. Consequently, it is a complicated operator-valued function of all the other modes of the field. Zero modes of this type have been investigated first by Maskawa and Yamawaki as early as in 1976 [26]. An analysis of the zero mode constraint equation for $(1+1)$–dimensional ϕ^4 field theory, by van de Sande and Pinsky [27], shows how spontaneous symmetry breaking occurs within the context of this model.

The model represents a new paradigm for spontaneous symmetry breaking and shall be reviewed shortly. The Lagrangian in one space and one time dimension is $\mathcal{L} = \partial_+\phi\partial_-\phi - \frac{\mu^2}{2}\phi^2 - \frac{\lambda}{4!}\phi^4$. Imposing periodic boundary conditions with a length parameter $d = 2L$ gives $\phi(x) = \frac{1}{\sqrt{d}}\sum_{n=-\infty}^{\infty} q_n(x^+)e^{i\frac{\pi}{L}nx^-}$. The field integral $\Sigma_n = \int dx^- \, \phi(x)^n - (zero\ modes)$ is convenient to discuss the problem and becomes

$$\Sigma_n = \frac{1}{n!} \sum_{i_1,i_2,\ldots,i_n \neq 0} q_{i_1} q_{i_2} \cdots q_{i_n} \, \delta_{i_1+i_2+\ldots+i_n,0}.$$

110

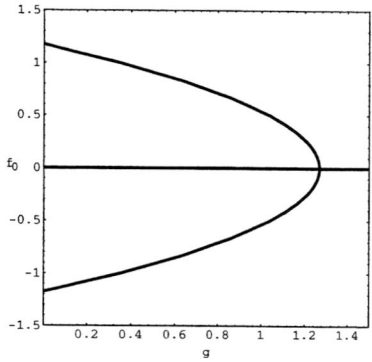

FIGURE 7. *The vacuum expectation value of ϕ, $f_0 = \sqrt{4\pi}\langle 0|\phi|0\rangle$ is plotted versus $g = 24\pi\mu^2/\lambda$, which is the inverse of the coupling constant λ. Taken from [27]. — In the front form, the vacuum state $|0\rangle$ is simple, but the operator ϕ, or a_0, is complicated. In the conventional instant form, the vacuum $|0\rangle$ is complicated but the operator ϕ is simple.*

Also the canonical Hamiltonian can be disentangled into Σ_n and the zero mode q_0

$$P^- = \frac{\mu^2}{2}q_0^2 + \mu^2\Sigma_2 + \frac{\lambda}{4!d}q_0^4 + \frac{\lambda}{2!d}q_0^2\Sigma_2 + \frac{\lambda}{d}q_0\Sigma_3 + \frac{\lambda}{d}\Sigma_4.$$

One can apply canonical quantization. Following the Dirac-Bergman prescription, described in [2], one identifies first-class constraints which define the conjugate momenta $0 = p_n - ik_n^+ q_{-n}$,, where $[q_m, p_n] = \delta_{n,m}/2$ and $m, n \neq 0$. The secondary constraint

$$0 = \mu^2 q_0 + \frac{\lambda}{3!d}q_0^3 + \frac{\lambda}{d}q_0\Sigma_2 + \frac{\lambda}{d}\Sigma_3 \tag{31}$$

determines the zero mode q_0. This result can also be obtained by integrating the equations of motion, $\partial_+\partial_-\phi + \mu^2\phi = \frac{\lambda}{3!}\phi^3$. To quantize the system one replaces the Dirac bracket by a commutator. One must choose a regularization and an operator-ordering prescription in order to make the system well-defined. One begins by defining creation and annihilation operators $a_k^\dagger$ and a_k,

$$q_k = \sqrt{\frac{d}{4\pi|k|}}\,a_k\,, \quad a_k = a_{-k}^\dagger\,, \quad k \neq 0\,,$$

which satisfy the usual commutation relations $\left[a_k, a_l^\dagger\right] = \delta_{k,l}$. Likewise, one defines the zero mode operator $q_0 = a_0\sqrt{d/4\pi}$. In the quantum case, one normal orders the operator Σ_n. One redefines this the fields in terms of operators

$$\phi(x) = \frac{1}{\sqrt{d}}\left(a_0 + \sum_{n=1}^{\infty}\frac{1}{n}\left(a_n e^{-i\frac{\pi}{L}nx^-} + a_n^\dagger e^{i\frac{\pi}{L}nx^-}\right)\right), \tag{32}$$

and notes that they obey the canonical commutation relations

$$\left[\phi(x^+, x^-), \partial_-\phi(y^+, y^-)\right]_{x^+=y^+=0} = \delta(x^- - y^-), \tag{33}$$

111

for a boson field in the front form, see also Eq.(28).

The solution of the constraint Eq.(31) is very difficult. Van de Sande and Pinsky [27] have shown that the zero mode a_0 could acquire finite values,

$$\langle 0|\phi|0\rangle = a_0 \neq 0,$$

depending on the coupling constant.

One finds the following general behavior: for small coupling (large g, where $g \propto 1/\text{coupling}$) the constraint equation has a single solution and the field has no vacuum expectation value (VEV). As one increases the coupling (decreases g) to the "critical coupling" g_{critical}, two additional solutions which give the field a nonzero VEV appear. These solutions differ only infinitesimally from the first solution near the critical coupling, indicating the presence of a second order phase transition. Above the critical coupling ($g < g_{\text{critical}}$), there are three solutions: one with zero VEV, the "unbroken phase," and two with nonzero VEV, the "broken phase". The "critical curves" shown in Figure 7, is a plot of the VEV as a function of g.

DLCQ IN 3+1 DIMENSIONS

Periodic boundary conditions on $\mathcal{L}$ can be realized by periodic boundary conditions on the vector potentials A_μ and anti-periodic boundary conditions on the spinor fields, since $\mathcal{L}$ is bilinear in the Ψ_α. In momentum representation one expands these fields into plane wave states $e^{-ip_\mu x^\mu}$, and satisfies the boundary conditions by *discretized momenta*

$$p_- = \begin{cases} \frac{\pi}{L}n, & \text{with } n = \frac{1}{2}, \frac{3}{2}, \ldots, \infty \text{ for fermions,} \\ \frac{\pi}{L}n, & \text{with } n = 1, 2, \ldots, \infty \text{ for bosons,} \end{cases}$$

$$\text{and } \vec{p}_\perp = \frac{\pi}{L_\perp}\vec{n}_\perp, \quad \text{with } n_x, n_y = 0, \pm 1, \pm 2, \ldots, \pm\infty \quad \text{for both.}$$

As an expense, one has to introduce two artificial length parameters, L and $L_\perp$. They also define the normalization volume $\Omega \equiv 2L(2L_\perp)^2$. More explicitly, the free fields are expanded as

$$\tilde{\Psi}_\alpha(x) = \frac{1}{\sqrt{\Omega}}\sum_q \frac{1}{\sqrt{p^+}}\left(b_q u_\alpha(p, \lambda)e^{-ipx} + d_q^\dagger v_\alpha(p, \lambda)e^{ipx}\right),$$

$$\text{and } \tilde{A}_\mu(x) = \frac{1}{\sqrt{\Omega}}\sum_q \frac{1}{\sqrt{p^+}}\left(a_q \epsilon_\mu(p, \lambda)e^{-ipx} + a_q^\dagger \epsilon_\mu^\star(p, \lambda)e^{ipx}\right), \quad (34)$$

particularly for the two transverse vector potentials $\tilde{A}^i \equiv \tilde{A}_\perp^i$, $(i = 1, 2)$. As above, each single particle state 'q' is specified by six quantum numbers, the three discrete momenta n, n_x, n_y, helicity, color and flavor. The creation and destruction operators like $a_q^\dagger$ and a_q create and destroy single particle states q, and obey (anti-) commutation relations like

$$[a_q, a_{q'}^\dagger] = \{b_q, b_{q'}^\dagger\} = \{d_q, d_{q'}^\dagger\} = \delta_{q,q'}.$$

The Kronecker symbol is unity only if all six quantum numbers coincide.

Inserting the free fields of Eq.(34) into the Hamiltonian, one performs the space-like integrations and ends up with the light-cone energy-momenta $P^{\nu} = P^{\nu}(a_q, a_q^{\dagger}, b_q, b_q^{\dagger}, d_q, d_q^{\dagger})$ as operators acting in Fock space. The spatial components of P^k are simple and diagonal, the temporal component P^- is complicated and off-diagonal. The integrals over the coordinates x^- are conveiently expressed in Kronecker delta functions

$$\delta(k^+|p^+) = \frac{1}{2L} \int_{-L}^{+L} dx^- e^{+i(k_- - p_-)x^-} = \frac{1}{2L} \int_{-L}^{+L} dx^- e^{+i(n-m)\frac{\pi x^-}{L}} = \delta_{n,m},$$

and correspondingly for the transversal integrations. In the tables given above they appear typically in the overall factor

$$\Delta(q_1; q_2, q_3, q_4) = \frac{g^2}{2\Omega} \delta(k_1^+|k_2^+ + k_3^+ + k_4^+) \delta^{(2)}(\vec{k}_{\perp 1}|\vec{k}_{\perp 2} + \vec{k}_{\perp 3} + \vec{k}_{\perp 4}).$$

Retrieving the continuum formulation

The continuum formulation of the Hamiltonian problem in gauge field theory with its endless multiple integrals is usually cumbersome and untransparent. In DLCQ, the continuum limit corresponds to harmonic resolution $K \to \infty$. The compactified formulation with its simple multiple sums is straightforward. The key relation is the connection between sums and integrals

$$\int dk^+ f(k^+, \vec{k}_{\perp}) \Longleftrightarrow \frac{\pi}{2L} \sum_n f(k^+, \vec{k}_{\perp}),$$

$$\int d^2\vec{k}_{\perp} f(k^+, \vec{k}_{\perp}) \Longleftrightarrow \frac{\pi^2}{L_{\perp}^2} \sum_{n_{\perp}} f(k^+, \vec{k}_{\perp}).$$

Combined they yield

$$\int dk^+ d^2\vec{k}_{\perp} f(k^+, \vec{k}_{\perp}) \Longleftrightarrow \frac{2(2\pi)^3}{\Omega} \sum_{n, n_{\perp}} f(k^+, \vec{k}_{\perp}).$$

Similarly, Dirac delta and Kronecker delta functions are related by

$$\delta(k^+) \delta^{(2)}(\vec{k}_{\perp}) \Longleftrightarrow \frac{\Omega}{2(2\pi)^3} \delta(k^+|0) \delta^{(2)}(\vec{k}_{\perp}|\vec{0}).$$

Because of that, in order to satisfy the respective commutation relations, one must modify also the creation and destruction operators. Denoting the single boson operators in the continuum by $\tilde{a}$ and in the discretized case by a, they must be related by

$$\tilde{a}(q) \Longleftrightarrow \sqrt{\frac{\Omega}{2(2\pi)^3}} a_q.$$

and correspondingly for fermion operators. Of course, one has formally to replace sums by integrals, Kronecker delta by Dirac delta functions, and single particle operators by their tilded versions. In practice, it suffices to replace the tilded coupling constant

$$\tilde{g}^2 = \frac{g^2}{2\Omega} \qquad \text{by} \qquad \tilde{g}^2 = \frac{g^2}{4(2\pi)^3}$$

in order to convert the discretized expressions in Tables 1-3 to the continuum formulation.

The DLCQ method can be considered a general framework for solving problems such as relativistic many-body theories or approximate models. The general procedure is: (1) Phrase the physics problem in DLCQ; (2) Apply approximation and simplifications; (3) Derive the final result; (4) At the end convert th so obtained expressions to the continuum.

Fock-space and vertex regularization

The finite number of Fock-space sectors is a consequence of the positivity of the longitudinal light-cone momentum p^+. The transversal momenta $\vec{p}_\perp$ can take either sign, and the number of Fock states within each sector can be arbitrarily large. In order to face a finite dimensional Hamiltonian matrix one must have a finite number of Fock states, and this is achieved by *Fock space regularization*: Following Lepage and Brodsky [14], a Fock state with n particles is included only if its free invariant mass does not exceed a certain threshold

$$(p_1 + p_2 + \ldots p_n)^2 - (m_1 + m_2 + \ldots m_n)^2 \le \Lambda_0^2.$$

The sum extends over all n particles in a Fock state. The lowest possible value of M_0^2 is taken when all particles are at rest relative to each other, *i.e.* $(M_0^2)_{min} = (m_1 + m_2 + \ldots m_n)^2$. This frozen invariant mass should be removed from the cut-off The mass scale Λ_0 is a Lorentz scalar and one of the parameters of the theory.

However, it was not realized in the past [8], that Fock-space regularization is almost irrelevant in the continuum theory. *Vertex regularization* seem to be a better alternative. At each vertex, a particle with four-momentum p^μ is scattered into two particles with respective four-momentum p_1^μ and p_2^μ. In order to avoid potential singularities one can *regulate the interaction* by setting the matrix element to zero if the off-shell mass $(p_1 + p_2)^2$ exceeds a certain scale Λ. The condition

$$R(\Lambda) = \Theta\left((p_1 + p_2)^2 - (m_1 + m_2)^2 - \Lambda^2\right) \tag{35}$$

will be referred to as the sharp cut-off for *vertex regularization*.

The dependence on the regularization parameter Λ must be removed by the renormalization group. Renormalization looks like a terrible problem in the context of non-perturbative theory, but with the above regularization scheme it could be simple in principle: The eigenvalues may not depend on the regulator scale(s) Λ. To require this is easier than to find a practical realization.

THE MANY-BODY PROBLEM IN GAUGE THEORY

In principle one proceeds in 3+1 like in 1+1 dimensions: One selects a particular value of the harmonic resolution K and the cut-off Λ and diagonalizes the finite dimensional Hamiltonian matrix by numerical methods. But the bottle neck of any Hamiltonian approach is that the dimension of the Hamiltonian matrix increases exponentially fast with the cut-off. As a concrete example consider the matrix structure as given for $K = 4$ in Fig. 2. Suppose the regularization procedure allows for 10 discrete momentum states in each direction. For every single particle one has about 10^3 possibilities to define a momentum state. A Fock-space sector with n particles and fixed total momentum has then roughly 10^{n-1} different Fock states. Sector 13 in Fig. 2 alone, with its 8 particles, has thus about 10^{21} Fock states. Chemists are able to handle matrices with some 10^7 dimensions, but 10^{21} dimensions exceeds the calculational capacity of any computer in the foreseeable future.

For 3+1 dimensions one is thus confronted with a a similar problem as in conventional many-body physics, displayed in Fig. 3. One has to diagonalize finite matrices with exponentially large dimensions (typically $> 10^6$). In fact, the problem in quantum field theory is even more difficult since the particle number is unlimited. One needs an effective interaction which acts in smaller matrix spaces and which has a well defined relation to the full interaction. One needs it also for the physical understanding. The effective interaction between two electrons, for example, is the Coulomb interaction, at least to lowest order of approximation.

The goal is therefore to develop an exact effective interaction between a quark and an anti-quark in a meson. To achieve this goal, the Hamiltonian DLCQ-matrix is discussed in terms of block matrices, since the Fock-space sectors appear quite naturally in a gauge theory, see also Eq.(18). Each Fock-space sector has a finite number of Fock states, which are kept track of collectively. Eq.(20) is therefore rewritten as a block matrix equation, with $H \equiv H_{LC}$ and $E \equiv M^2$,

$$\sum_{j=1}^{N} \langle i|H|j \rangle \langle i|\Psi \rangle = E \langle i|\Psi \rangle \qquad \text{for all } i = 1, 2, \ldots, N. \qquad (36)$$

Rows and columns are denumerated in the same convention as in Table 7. As to be shown, it can be mapped identically on a matrix equation which acts only in sector $|1\rangle$. Once this is achieved, one can go to the continuum limit.

The approach of Tamm and Dancoff

Effective interactions are a well known tool in many-body physics [5]. In field theory the method is known as the Tamm-Dancoff-approach, applied first by Tamm [28] and by Dancoff [29], which shall be reviewed it in short.

The rows and columns of a matrix are split into the P- and the Q-space. In terms of the sector numbers of Eq.(36), $P = \sum_{j=1}^{n} |j\rangle\langle j|$ and $Q = \sum_{j=n+1}^{N} |j\rangle\langle j|$, where $1 \leq n < N$. Eq.(36) can thus be written as a 2 by 2 block matrix equation

$$\langle P|H|P\rangle \langle P|\Psi\rangle + \langle P|H|Q\rangle \langle Q|\Psi\rangle = E \langle P|\Psi\rangle, \tag{37}$$

$$\langle Q|H|P\rangle \langle P|\Psi\rangle + \langle Q|H|Q\rangle \langle Q|\Psi\rangle = E \langle Q|\Psi\rangle. \tag{38}$$

If one can invert the quadratic matrix $\langle Q|E - H|Q\rangle$ one could express the Q-space in terms of the P-space wavefunction. But here is a problem: The eigenvalue E is unknown at this point. But one can replace it by another number, *the starting point energy* ω, which is at first a free parameter. The matrix inverse to $\langle Q|\omega - H|Q\rangle$ is called the resolvent of the Hamiltonian in the Q-space and is denoted by $G_Q(\omega)$. The Q-space wave function becomes then

$$\langle Q|\Psi(\omega)\rangle = G_Q(\omega)\langle Q|H|P\rangle \langle P|\Psi\rangle, \qquad G_Q(\omega) = \frac{1}{\langle Q|\omega - H|Q\rangle}. \tag{39}$$

Substituting it in Eq.(37) produces an eigenvalue equation in the P-space,

$$H_{\text{eff}}(\omega)|P\rangle \langle P|\Psi_k(\omega)\rangle = E_k(\omega) |\Psi_k(\omega)\rangle, \tag{40}$$

and defines the effective P-space Hamiltonian

$$H_{\text{eff}}(\omega) = H + H|Q\rangle G_Q(\omega) \langle Q|H.$$

Varying ω one generates a set of *energy functions* $E_k(\omega)$. Every solution of the *fix-point equation*

$$E_k(\omega) = \omega,$$

generates one of the eigenvalues H, in fact, it generates all of them.

If one identifies the P- with the $q\bar{q}$-space one seems to have found the effective interaction which acts in the Fock space of a single quark and a single anti-quark. It looks as if one has mapped a difficult problem, the diagonalization of a big matrix onto a simpler problem, the diagonalization of a small matrix. But the price to pay is to invert a matrix. Matrix inversion takes about the same numerical effort as its diagonalization. In view of having to vary ω, the numerical work is therefore rather larger than smaller as compared to a direct diagonalization. The advantage of working with a resolvent is of analytical nature to the extent that resolvents can be approximated systematically. The two resolvents

$$G_Q(\omega) = \frac{1}{\langle Q|\omega - T - U|Q\rangle}, \quad \text{and} \quad \tilde{G}(\omega) = \frac{1}{\langle Q|\omega - T|Q\rangle}, \tag{41}$$

defined once with and once without the off-diagonal interaction U, are identically related by $G_Q = \tilde{G} + \tilde{G}UG_Q$, or by the infinite series of (Tamm-Dancoff) perturbation theory

TABLE 7. *The Fock-space sectors and the Hamiltonian block matrix structure for QCD. Diagonal blocs are marked by D. Off-diagonal blocks are labeled by V, F and S_6, corresponding to vertex, fork and seagull interactions, respectively. Zero-matrices are denoted by dots. Taken from [32]. See also Fig. 2.*

Sector	n	1	2	3	4	5	6	7	8	9	10	11	12	13
$q\bar{q}$	1	D	S	V	F	.	F	.	.	.	.	.	.	.
$g\ g$	2	S	D	V	.	V	F	.	.	F	.	.	.	.
$q\bar{q}\ g$	3	V	V	D	V	S	V	F	.	.	F	.	.	.
$q\bar{q}\ q\bar{q}$	4	F	.	V	D	.	S	V	F	.	.	F	.	.
$g\ g\ g$	5	.	V	S	.	D	V	.	.	V	F	.	.	.
$q\bar{q}\ g\ g$	6	F	F	V	S	V	D	V	.	S	V	F	.	.
$q\bar{q}\ q\bar{q}\ g$	7	.	.	F	V	.	V	D	V	.	S	V	F	.
$q\bar{q}\ q\bar{q}\ q\bar{q}$	8	.	.	.	F	.	.	V	D	.	.	S	V	F
$g\ g\ g\ g$	9	.	F	.	.	V	S	.	.	D	V	.	.	.
$q\bar{q}\ g\ g\ g$	10	.	.	F	.	F	V	S	.	V	D	V	.	.
$q\bar{q}\ q\bar{q}\ g\ g$	11	.	.	.	F	.	F	V	S	.	V	D	V	.
$q\bar{q}\ q\bar{q}\ q\bar{q}\ g$	12	.	.	.	.	.	.	F	V	.	.	V	D	V
$q\bar{q}\ q\bar{q}\ q\bar{q}\ q\bar{q}$	13	.	.	.	.	.	.	.	F	.	.	.	V	D

$$G_Q = \tilde{G} + \tilde{G}U\tilde{G} + \tilde{G}U\tilde{G}U\tilde{G} + \ldots, \qquad (42)$$

see also Eq.(21). The free resolvent $\tilde{G}$ can be obtained trivially since the kinetic energy T is diagonal. Conceptually, it is the same object as the one in Eq.(21), except for two aspects: The present $\tilde{G}$ acts only in the Q-space, and the starting point energy ω is a *constant* (one of the eigenvalues); the free energy ϵ in (21), however, is a *function of the incoming momenta*. The starting point energy not being a kinetic energy creates problems allover the place, since non-integrable singularities appear in every order of (Tamm-Dancoff) perturbation theory. Essentially two conclusions are possible: Either gauge theory has no bound-state solution, or the series in Eq.(42) has to be resumed to all orders of perturbation theory before the singularities begin cancel eachother.

In practice, Tamm and Dancoff [28,29] have restricted themselves to the first non-trivial order. In order to make things work, they have replaced the energy denominator with the eigenvalue ω by the energy denominator with the function ϵ, as it appears in the perturbative scattering amplitudes. The same trick was applied in the later work with the front form [30,31].

Perturbation theory within a bound state problem is known to be a very difficult question. It has motivated the formal work in the next few sections. What we are after is the impossible, some kind of non-perturbative perturbation theory!

117

The method of iterated resolvents

The Tamm-Dancoff approach can be interpreted as the reduction of a block matrix dimension from 2 to 1. But having a matrix with block matrix dimension $N = 13$ as in Table 7, it could be interpreted as the reduction from $N \to N - 1$, simply by choosing the Q-space appropriately. But then the procedure can be iterated, one can reduce the block matrix dimension from $N - 1 \to N - 2$, and so on until one arrives at $2 \to 1$. This method of 'iterated of resolvents' has certain advantages, which will be discussed as the formalism develops.

First, one needs a reasonable and compact notation. One of them is to denumerate the Fock space sectors as in Table 7. The full Hamiltonian will be denoted by $H \equiv H_N$, since by the definition in Eq.(36) it has N blocks. Suppose, during the reduction one has arrived at block matrix dimension n, with $1 < n \le N$. The eigenvalue problem corresponding to Eq.(40) reads then

$$\sum_{j=1}^{n} \langle i|H_n(\omega)|j\rangle\langle j|\Psi(\omega)\rangle = E(\omega)\,\langle i|\Psi(\omega)\rangle, \qquad \text{for } i = 1, 2, \ldots, n.$$

In analogy to Eq.(39), define the n-space resolvent, and get

$$\langle n|\Psi(\omega)\rangle = G_n(\omega)\sum_{j=1}^{n-1}\langle n|H_n(\omega)|j\rangle\,\langle j|\Psi(\omega)\rangle, \quad G_n(\omega) = \frac{1}{\langle n|\omega - H_n(\omega)|n\rangle}. \qquad (43)$$

The effective interaction in the $(n - 1)$-space becomes then

$$H_{n-1}(\omega) = H_n(\omega) + H_n(\omega)G_n(\omega)H_n(\omega) \qquad (44)$$

for every block matrix element. It is unpleasant but unavoidable, that the symbol n denotes both the block matrix dimension and the number of the last sector. Else one proceeds like for Tamm-Dancoff, including the fixed point equation $E(\omega) = \omega$. But one has achieved much more: Eq.(44) is a *recursion relation*!

A few comments seem to be in order. The method of iterated resolvents [32,33] is particularly suited for gauge theory with its many zero block matrices, see Table 8. The zero matrices remove many of the multiplications in Eq.(43). The Tamm-Dancoff procedure cannot make use of them. – Both the Tamm-Dancoff procedure and the iterated resolvents can be put on a computer in model studies. The iterated resolvents require to invert several smaller matrices insted of one big one. Since matrix diagonalization (and inversion) grows with power 3 in the dimension, the technique of iterated resolvents might thus even be faster. – The iterated and the Tamm-Dancoff resolvents are distinctly different in the following aspect: G_n conserves particle number, but G_Q does not. Both however conserve the incoming momentum. – The notation in Eq.(44) is very compact and will be explained further below. – The method applies also equally well to conventional many-body problems since a pair-interaction generates for example the bloc matrix structure shown in

Figure 3. – Finally, it should be emphasized that the higher sector wavefunctions can be retrieved by matrix multiplications from the eigenfunction in the lowest sector, from $\langle 1|\Psi\rangle$. No additional matrix diagonalizations or inversions are required. To show this, consider Eq.(43) for $n = 1$, *i.e.*

$$\langle 2|\Psi\rangle = \langle 2|G_2 H_2|1\rangle \langle 1|\Psi\rangle.$$

Both G_2 and H_2 were calculated as a matrix for getting down to the effective interaction in the 1-space. Required is thus one additional matrix multiplication. Next get $\langle 3|\Psi\rangle = \langle 3|G_3 H_3|1\rangle \langle 1|\Psi\rangle + \langle 3|G_3 H_3|2\rangle \langle 2|\Psi\rangle$. Substituting $\langle 2|\Psi\rangle$ gives

$$\langle 3|\Psi\rangle = \langle 3|G_3 H_3 (1 + G_2 H_2)|1\rangle \langle 1|\Psi\rangle.$$

The general case,

$$\langle n|\Psi\rangle = \langle n|G_n H_n (1 + G_{n-1} H_{n-1}) \ldots (1 + G_2 H_2)|1\rangle \langle 1|\Psi\rangle, \tag{45}$$

can be proven by induction.

A simple numerical example

It might be interesting to study a Hamiltonian with a tridigonal band structure, such as was given for example in Table 4:

$$\begin{pmatrix} \langle 1|1 + S|1\rangle & \langle 1|F|2\rangle & . & . \\ \langle 2|F|1\rangle & \langle 2|1 + S|2\rangle & \langle 2|F|3\rangle & . \\ . & \langle 3|F|2\rangle & \langle 3|1 + S|3\rangle & \langle 3|F|4\rangle \\ . & . & \langle 4|F|3\rangle & \langle 4|1 + S|4\rangle \end{pmatrix}. \tag{46}$$

According to the rules it develops the structure a of continued fraction, since with $G_n(\omega) = 1/(\omega - H_n)$ one has explicitly

$$\begin{aligned} H_4 &= \langle 4|1 + S|4\rangle, & \langle 4|\Psi\rangle &= \langle 4|G_4 F G_3 F G_2 F|1\rangle \langle 1|\Psi\rangle, \\ H_3 &= \langle 3|1 + S + F G_4 F|3\rangle, & \langle 3|\Psi\rangle &= \langle 3|G_3 F G_2 F|1\rangle \langle 1|\Psi\rangle, \\ H_2 &= \langle 2|1 + S + F G_3 F|2\rangle, & \langle 2|\Psi\rangle &= \langle 2|G_2 F|1\rangle \langle 1|\Psi\rangle, \\ H_1 &= \langle 1|1 + S + R G_2 R|1\rangle, & \langle 1|\Psi\rangle. \end{aligned} \tag{47}$$

Here and above a very compact notation is used, which is explained by the example

$$\langle 3|D + V G_4 V|3\rangle = \langle 3|D|3\rangle + \langle 3|F|4\rangle \langle 4|G_4|4\rangle \langle 4|F|3\rangle.$$

By reasons of space, the abbrevation $\langle i|D|i\rangle = \langle i|T + S|i\rangle$ is used sometimes in the diagonal blocks. The kinetic energies T are diagonal matrix elements, and the seagulls do not change particle number by definition. Even that the above is very compact. Here is what it means in terms of matrix operations:

119

The 4 × 4 matrix

$$\begin{pmatrix} 0 & 1 & . & . \\ 1 & 2 & 3 & . \\ . & 3 & 4 & 5 \\ . & . & 5 & 6 \end{pmatrix}$$

has eigenvalues E_i = -1.872, -0,01518, 3.333, and 10.55. Its energy function is

$$E(\omega) = \cfrac{1 \cdot 1}{\omega - 2 - \cfrac{3 \cdot 3}{\omega - 4 - \cfrac{5 \cdot 5}{\omega - 6}}}.$$

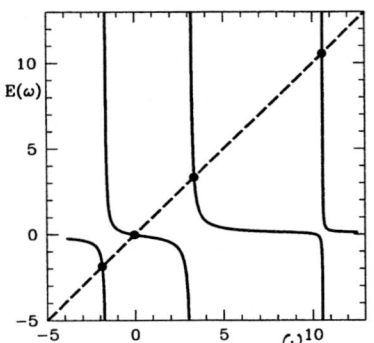

FIGURE 8. *The the energy function $E(\omega)$ is plotted versus ω. The dotted line represents $E = \omega$.*

The solutions of $E(\omega) = \omega$ agree with E_i to within computer accuracy. The are marked with • in the figure.

$$\langle 3, i|D + FG_4F|3, i'\rangle = \langle 3, i|D|3, i'\rangle$$
$$+ \sum_{j,k} \langle 3, i|F|4, j\rangle \langle 4, j|G_4|4, k\rangle \langle 4, k|F|3, i'\rangle.$$

Within each sector $|i\rangle$, the number of basis or Fock states is finite, *i.e.* $|i\rangle \equiv |i; j\rangle$ with $j = 1, 2, \ldots, N_i$. The steps to be taken are then: Take the reactangular block matrix $\langle 4, k|F|1, i'\rangle$; matrix-multiply it from the left with the inverse matrix of $\langle 4|H|4\rangle$ which is $\langle 4, j|G_4|4, k\rangle$; matrix-multiply the result with the transpose (and complex conjugated) matrix $\langle 4, j|F^+|1, i\rangle$; the result is the quadratic matrix $\langle 3, i|FG_4F|3, i'\rangle$; add this to the quadratic matrix $\langle 3, i|D|3, i'\rangle$. As a net result, the old matrix $\langle 3, i|D|3, i'\rangle$ in Eq.(48) is replaced the effective (and ω-dependent) matrix $\langle 1, i|\overline{D}(\omega)|1, i'\rangle$.

To play the game a little further, one can reduce the number of states within a block to 1. Eq.(46) stands then for a tri-diagonal matrix and Eq.(47) stands literally for a continued fraction. Such one is analyzed in Figure 8, with a perfect agreement between the method of iterated resolvents and a direct numerical diagonalization.

The 4 by 4 block matrix for gauge theory

In the front form, the Fock-space for harmonic resolution $K = 2$ has only 4 Fock-space sectors, see Table 7. $K = 2$ provides thus a good example for all 4 by 4 hermitean bloc matrices, in which the block matrix element $\langle 2|H|4\rangle$ vanishes. The case is considered as an exercise and a non-trivial example.

Since $N = 4$ one starts out with the block matrix

$$
\begin{array}{cccc}
& 1=(q\bar{q}) & 2=(gg) & 3=(q\bar{q}\,g) & 4=(q\bar{q}\,q\bar{q}) \\
\begin{array}{l}1=(q\bar{q})\\2=(gg)\\3=(q\bar{q}\,g)\\4=(q\bar{q}\,q\bar{q})\end{array}
&
\left(\begin{array}{cccc}
\langle 1|D|1\rangle & \langle 1|S|2\rangle & \langle 1|V|3\rangle & \langle 1|F|4\rangle \\
\langle 2|S|1\rangle & \langle 2|D|2\rangle & \langle 2|V|3\rangle & 0 \\
\langle 3|V|1\rangle & \langle 3|V|2\rangle & \langle 3|D|3\rangle & \langle 3|V|4\rangle \\
\langle 4|F|1\rangle & 0 & \langle 4|V|3\rangle & \langle 4|D|4\rangle
\end{array}\right).
\end{array}
\tag{48}
$$

The block matrix element $\langle 2|H_{\mathrm{LC}}|4\rangle$ is a rectangular zero-matrix, since the light-cone operator has no matrix elements between two gluons gg and two $q\bar{q}$-pairs. The notation keeps track of the field theoretic property, that each block matrix has only one type of interactions, namely instaneneous seagull (S) and fork (F), or dynamic vertex (V) interactions.

The Hamiltonian in the 4-sector is quadratic and has the resolvent

$$
G_4(\omega) = \frac{1}{\langle 4|\omega - H_4|4\rangle}, \qquad \langle 4|H_4|4\rangle = \langle 4|T + S|4\rangle.
\tag{49}
$$

Using Eq.(44) to reduce the block matrix dimension from 4 to 3 gives

$$
\begin{array}{cccc}
& 1=(q\bar{q}) & 2=(gg) & 3=(q\bar{q}\,g) \\
\begin{array}{l}1=(q\bar{q})\\2=(gg)\\3=(q\bar{q}\,g)\end{array}
&
\left(\begin{array}{ccc}
\langle 1|D+FG_4F|1\rangle & \langle 1|S|2\rangle & \langle 1|(1+FG_4)V|3\rangle \\
\langle 2|S|1\rangle & \langle 2|D|2\rangle & \langle 2|V|3\rangle \\
\langle 3|V(1+G_4F)|1\rangle & \langle 3|V|2\rangle & \langle 3|D+VG_4V|3\rangle
\end{array}\right)
\end{array}
$$

Almost every block matrix element is replaced by an 'effective' element.

Now continue to reduce from 3 to 2: The resolvent of $\langle 3|H|3\rangle$ is now

$$
G_3(\omega) = \frac{1}{\langle 3|\omega - H_3|3\rangle}, \qquad \langle 3|H_3|3\rangle = \langle 3|T + S + VG_4V|3\rangle.
$$

With $\langle 1|A|1\rangle \equiv \langle 1|T + S + FG_4F + (1 + FG_4)VG_3V(1 + G_4F)|1\rangle$ one gets

$$
\begin{array}{ccc}
& 1=(q\bar{q}) & 2=(gg) \\
\begin{array}{l}1=(q\bar{q})\\2=(gg)\end{array}
&
\left(\begin{array}{cc}
\langle 1|A|1\rangle & \langle 1|S+(1+FG_4)VG_3V|2\rangle \\
\langle 2|S+VG_3V(1+G_4F)|1\rangle & \langle 2|T+S+VG_3V|2\rangle
\end{array}\right)
\end{array}
$$

Finally, evaluate the last resolvent

$$
G_2(\omega) = \frac{1}{\langle 2|\omega - H_2|2\rangle}, \qquad H_2 = \langle 2|T + S + VG_3V|2\rangle,
$$

to end up with the 1 by 1 block matrix

$$
\begin{aligned}
\langle 1|H_1|1\rangle = {}& \langle 1|D + FG_4F + (1 + FG_4)VG_3V(1 + G_4F)|1\rangle \\
& + \langle 1|(S + VG_3V + FG_4VG_3V)\,G_2\,(S + VG_3V + VG_3VG_4F)|1\rangle.
\end{aligned}
\tag{50}
$$

Here the procedure stops. For gauge theory one can order the result with the power of the coupling constant

$$\langle 1|H_1|1\rangle = \langle 1|T + \overline{VG_3V} + \overline{VG_3VG_2VG_3V}|1\rangle$$
$$+ \langle 1|\, SG_2VG_3VG_4F + FG_4VG_3VG_2S +$$
$$VG_3VG_2VG_3VG_4F + FG_4VG_3VG_2VG_3V +$$
$$FG_4VG_3VG_4F + FG_4VG_3VG_2VG_3VG_4F\,|1\rangle. \tag{51}$$

with the abbreviations

$$\overline{VG_3V} = VG_3V + S,$$
$$\overline{VG_3VG_2VG_3V} = VG_3VG_2VG_3V + SG_2VG_3V + VG_3VG_2S$$
$$+ FG_4VG_3V + VG_3VG_4F + SG_2S + FG_4F. \tag{52}$$

Finally, for completeness, the higher sector wave function in Eq.(45) are written out explicitly

$$\langle 2|\Psi\rangle = G_2\left[\overline{VG_3V} + VG_3VG_4F\right]|1\rangle\,\langle 1|\Psi\rangle, \tag{53}$$

$$\langle 3|\Psi\rangle = G_3\left[V + \overline{VG_2VG_3V} + VG_2VG_3VG_4F\right]|1\rangle\,\langle 1|\Psi\rangle, \tag{54}$$

$$\langle 4|\Psi\rangle = G_4\left[\overline{VG_3V} + \overline{VG_3VG_2VG_3V} + VG_3VG_2VG_3VG_4F\right]|1\rangle\,\langle 1|\Psi\rangle, \tag{55}$$

with the additional abbreviations

$$\overline{\overline{VG_3V}} = VG_3V + F$$
$$\overline{VG_2VG_3V} = VG_2VG_3V + VG_4F + VG_2S$$
$$\overline{VG_3VG_2VG_3V} = VG_3VG_2VG_3V + VG_3VG_4F + VG_3VG_2S. \tag{56}$$

These abbreviations are not very relevant for the general case. But for gauge theory they have a deeper physical meaning as to be discussed further below.

Discussion and gauge invariant interactions

All effective interactions have a <u>finite</u> number of <u>finite</u> strings like $V\,GV\ldots GV$. This is a colossal simplification as compared infinite number of terms of Tamm-Dancoff perturbation theory. Iterated resolvents resume them in a systematic way. Both kinds of series can be identified term by term if one expands the iterated resolvents also about the kinetic energies. Denoting the free resolvent by $\tilde{G}_n$ it is related to the Tamm-Dancoff resolvent $\tilde{G}$ by

$$\tilde{G}_n = \sum_{i=1}^{N_n}|n,i\rangle\frac{1}{\omega - T_{n,i}}\langle n,i| \quad , \qquad \tilde{G}_n = \sum_{n=1}^{N}\tilde{G}_n$$

If one then expands G_n using $H_n = T_n + U_n$ one gets

$$G_n = \tilde{G}_n + \tilde{G}_n U_n \tilde{G}_n + \tilde{G}_n U_n \tilde{G}_n U_n \tilde{G}_n + \ldots,$$

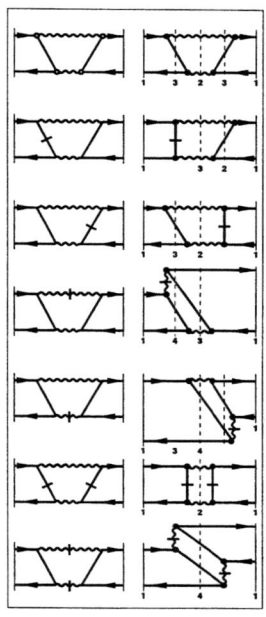

FIGURE 9. *A particular time-ordering of the two-gluon-annihilation interaction is drawn in the upper left of the figure. If all intrinsic lines are understood as a sum of a 'dynamical' and an 'instantaneous' line, one gets the seven diagrams in the figure, drawn in two different conventions. One observes that the very same strings appear in $\overline{VG_3VG_2VG_3V}$ as defined by Eq.(52).*

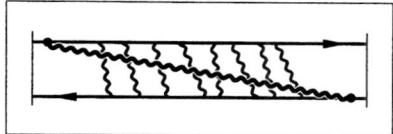

FIGURE 10. *Each propagator G_n respresents an infinite resummation of Tamm-Dancoff perturbative diagrams. The graph $\langle 1|VG_3V|1\rangle$ respresents a resummation of all Tamm-Dancoff graphs with at least one $q\bar{q}$ pair, and where a first gluons is emitted by the quark and a (perhaps other) gluon is arbsorbed by the anti-quark.*

thus infinite series of infinite series, which after reordering become identical with the Tamm-Dancoff series.

A look back to Eq.(51) is helpful. The various term have been arranged there by order of the coupling constant. In the first line one finds the combination $\overline{VG_3VG_2VG_3V}$, defined in Eq.(52). In the caption to Figure 9 it is explained how one can simplify the formal procedures, if one associates with each intrisic line a dynamic plus an instantaneous line. Doing that replaces the above sum of six terms into a single one namely $VG_3VG_2VG_3V$, and similarily one replaces the term in the first line by $VG_3V + S \rightarrow VG_3V$.

There is a physical reason behind that: When calculating scattering diagrams to a fixed order in perturbation theory, one obtains the gauge invariant Feynman amplitudes only, if all instantaneous graphs of the same order are included. An example for that was given above, when calculating the scattering graph up to second order. Obviously, the method of iterated resolvents accounts for this aspect automatically, in a formal way.

Since the remaining terms in Eq.(51) are of order 6 or 8 in the coupling constant, one concludes that the restriction to the phase space for $K = 2$ satisfies gauge invariance only up to terms of order 6 in the coupling constant. One can verify, that the order of violating gauge invariance is pushed up if one includes more Fock-space sectors. This rule was checked explicitly in many rather laborious calculations in [32].

Theorem 1 *An easy trick to achieve gauge invariance is the following: Set to zero* *formally all instantaneous interactions, perform the steps required by the resolvents. At the end, when having calculated the effective interaction in terms of the vertex interaction V, one replaces each internal line by the sum of a dynamical and a instantaneous line, according to the general rules of light-cone perturbation theory.*

If one follows thes rules for the 4 by 4 matrix given in Eq.(48) one obtains for the sector Hamiltonians and wavefunctions simply

$$H_4 = T_4, \quad \langle 4|\Psi\rangle = \langle 4|G_4 V G_3 V + G_4 V G_3 V G_2 V G_3 V|1\rangle \langle 1|\Psi\rangle,$$
$$H_3 = T_3 + V G_4 V, \quad \langle 3|\Psi\rangle = \langle 3|G_3 V + G_3 V G_2 V G_3 V|1\rangle \langle 1|\Psi\rangle,$$
$$H_2 = T_2 + V G_3 V, \quad \langle 2|\Psi\rangle = \langle 2|V G_3 V|1\rangle \langle 1|\Psi\rangle,$$
$$H_1 = T_1 + V G_3 V + V G_3 V G_2 V G_3 V. \tag{57}$$

These expressions are more transparent than the full equations; they are also gauge invariant.

The iterated resolvents for arbitrary K

When one substitutes formally all instantaneous interactions by block matrices with only zeros, keeping only the vertex interactions V, one gets a block matrix structure as displayed in Table 8. Because of the many zero matrices, the construction of the the sector Hamiltonians is now rather straightforward and simple. Here they are for the first 12 sectors:

$$H_{12} = T_{12} + V G_{17} V + V G_{17} V G_{16} V G_{12} V + V G_{13} V, \tag{58}$$
$$H_{11} = T_{11} + V G_{16} V + V G_{16} V G_{15} V G_{16} V + V G_{12} V, \tag{59}$$
$$H_{10} = T_{10} + V G_{15} V + V G_{15} V G_{14} V G_{15} V + V G_{11} V, \tag{60}$$
$$H_9 = T_9 + V G_{10} V + V G_{14} V, \tag{61}$$
$$H_8 = T_8 + V G_{12} V + V G_{12} V G_{11} V G_{12} V, \tag{62}$$
$$H_7 = T_7 + V G_{11} V + V G_{11} V G_{10} V G_{11} V + V G_8 V, \tag{63}$$
$$H_6 = T_6 + V G_{10} V + V G_{10} V G_9 V G_{10} V + V G_7 V, \tag{64}$$
$$H_5 = T_5 + V G_6 V + V G_9 V, \tag{65}$$
$$H_4 = T_4 + V G_7 V + V G_7 V G_6 V G_7 V, \tag{66}$$
$$H_3 = T_3 + V G_6 V + V G_6 V G_5 V G_6 V + V G_4 V, \tag{67}$$
$$H_2 = T_2 + V G_3 V + V G_5 V, \tag{68}$$
$$H_1 = T_1 + V G_3 V + V G_3 V G_2 V G_3 V, \tag{69}$$

Actually, for $K = 4$, the exact expressions corresponding to Table 8 are obtained by setting $G_{13} = 1/(\omega - T_{13})$ and $G_n = 0$ for $n \geq 14$. But it is easy to write down the expressions in Eqs.(58)-(60) valid for a higher value of K. One also notes that they give all sector Hamiltonians for $K = 2$ in Eqs.(57), by setting formally $G_n = 0$ for $n \geq 5$.

TABLE 8. *The Fock space and the Hamiltonian matrix $H' = T + V$ for a meson at fixed value of $K = 4$. — See discussion in the text. The diagonal blocks are denoted by T. Most of the block matrices are zero matrices, marked by a dot (·). The block matrices marked by V are potentiall non-zero due to the vertex interaction.*

Sector	n	1	2	3	4	5	6	7	8	9	10	11	12	13
$q\bar{q}$	1	D	·	V	·	·	·	·	·	·	·	·	·	·
$g\ g$	2	·	D	V	·	V	·	·	·	·	·	·	·	·
$q\bar{q}\ g$	3	V	V	D	V	·	V	·	·	·	·	·	·	·
$q\bar{q}\ q\bar{q}$	4	·	·	V	D	·	·	V	·	·	·	·	·	·
$g\ g\ g$	5	·	V	·	·	D	V	·	·	V	·	·	·	·
$q\bar{q}\ g\ g$	6	·	·	V	·	V	D	V	·	·	V	·	·	·
$q\bar{q}\ q\bar{q}\ g$	7	·	·	·	V	·	V	D	V	·	·	V	·	·
$q\bar{q}\ q\bar{q}\ q\bar{q}$	8	·	·	·	·	·	·	V	D	·	·	·	V	·
$g\ g\ g\ g$	9	·	·	·	·	V	·	·	·	D	V	·	·	·
$q\bar{q}\ g\ g\ g$	10	·	·	·	·	·	V	·	·	V	D	V	·	·
$q\bar{q}\ q\bar{q}\ g\ g$	11	·	·	·	·	·	·	V	·	·	V	D	V	·
$q\bar{q}\ q\bar{q}\ q\bar{q}\ g$	12	·	·	·	·	·	·	·	V	·	·	V	D	V
$q\bar{q}\ q\bar{q}\ q\bar{q}\ q\bar{q}$	13	·	·	·	·	·	·	·	·	·	·	·	V	D

Theorem 2 *For sufficiently large harmonic resolution K the formal expressions for effective Hamiltonian in sufficiently low sectors become independent of K.*

One therefore can go to the limit $K \to \infty$ and thus to the continuum limit.

Propagation in medium

Here seems to be a problem: For calculating G_3 one needs G_6, G_5 and G_4, for calculating G_6 one needs G_{10}, G_9 and G_7, and so on. In the next section will be shown how the hierarchy can be broken in a rather effective way. That final step will be comparatively simple if one has analyzed the propagators for the sectors with one $q\bar{q}$-pair and arbitrarily many gluons, as follows next.

Consider first the the effective interaction in the space of one $q\bar{q}$-pair and one gluon as given by Eq.(67). The corresponding diagrams can be grouped into two topologically distinct classes, displayed in Figs. 12 and 13. The the diagrams in Fig. 12 are obtained by adding a non-interacting gluon line to the former diagrams in Fig. 11. The gluon does not change quantum numbers under impact of the interaction and acts like a spectator. Therefore, the graphs in Fig. 12 will be refered to as the 'spectator interaction' $\overline{U}_3$. In the graphs of Fig. 13 the gluons are scattered by the interaction, and correspondingly these graphs will be refered to as the 'participant interaction' $\widetilde{U}_3$. Thus, $U_3 = \overline{U}_3 + \widetilde{U}_3$. The separation into a graph with only one interacting quark-pair Fock-space sectors

125

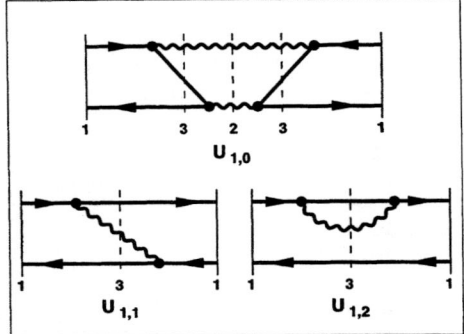

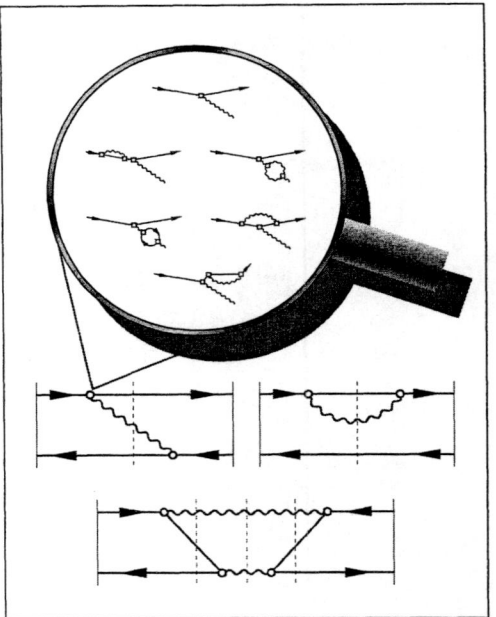

FIGURE 11. *The three graphs of the effective interaction in the $q\bar{q}$-space. The vertical lines denote the propagators G_n. On the right, the vertical lines denote the propagators $\overline{G}_n$. The coupling function at the vertices is symbolized by graphs as they would appear in a perturbative analysis.*

$$U_n = \overline{U}_n + \tilde{U}_n, \tag{70}$$

except in those with only gluons. More explicitly, the spectator and participant interactions in the lowest sectors with one quark-pair become

$$
\begin{aligned}
\overline{U}_3 &= V G_6 V + V G_6 V G_5 V G_6 V, & \tilde{U}_3 &= V G_4 V + V G_6 V, \\
\overline{U}_6 &= V G_{10} V + V G_{10} V G_9 V G_{10} V, & \tilde{U}_6 &= V G_7 V + V G_{10} V.
\end{aligned}
\tag{71}
$$

The same operators can appear in both $\overline{U}$ and $\tilde{U}$, but they refer to different graphs. Since the Hamiltonian is additive in spectator and participant interactions, $\overline{U}_n$ can be associated with its own resolvent

$$\overline{G}_n = \frac{1}{\omega - T_n - \overline{U}_n}, \qquad \left(\text{while } G_n \equiv \frac{1}{\omega - H_n} = \frac{1}{\omega - T_n - \overline{U}_n - \tilde{U}_n}\right). \tag{72}$$

The relation of $\overline{G}_n$ to the full resolvent is $G_n = \overline{G}_n + \overline{G}_n \tilde{U}_n G_n$, or

$$G_n = \overline{G}_n + \overline{G}_n \tilde{U}_n \overline{G}_n + \overline{G}_n \tilde{U}_n \overline{G}_n \tilde{U}_n \overline{G}_n + \ldots, \tag{73}$$

writing it as an infinite series. Note that the propagator $\overline{G}_n$ contains the interaction $\tilde{U}_n$ in contrast to the free Tamm-Dancoff propagator $\tilde{G}$ in Eq.(41). One deals here therefore with 'perturbation theory in medium'. But since it is driven to all orders one better speaks of 'propagation in medium'.

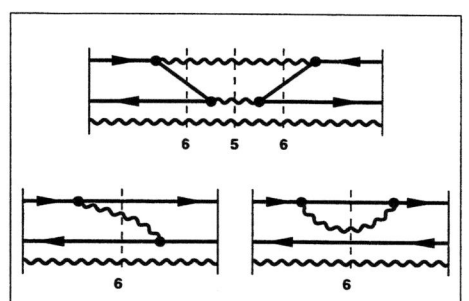

FIGURE 12. *The three graphs of the specta-tor interaction in the $q\bar{q}\,g$-space. Note the role of the gluon as a spectator.*

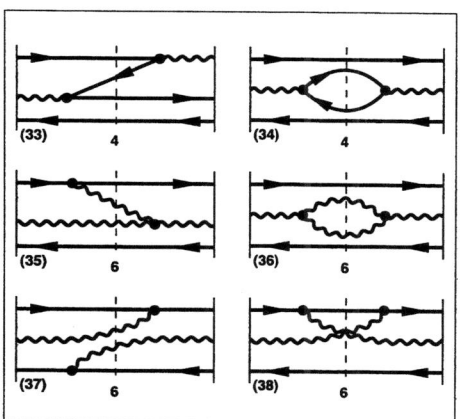

FIGURE 13. *Some six graphs of the partici-pant interaction in the $q\bar{q}\,g$-space.*

The series can be identically written as

$$G_n = C_n\,\overline{G}_n\,C_n^\dagger, \qquad \text{with} \quad C_n = \sqrt{\frac{1}{1-\overline{G}_n\tilde{U}_n}}. \tag{74}$$

One can verify this order by order in perturbation theory or by the identity

$$(\omega - \overline{H}_n)(1 - \overline{G}_n\tilde{U}_n) = (\omega - \overline{H}_n)\left(1 - \frac{1}{\omega - \overline{H}_n}\tilde{U}_n\right) = \omega - H_n. \tag{75}$$

The inverse gives $C_n^2\,\overline{G}_n = G_n$, and with the identity $C_n\,\overline{G}_n = \overline{G}_n\,C_n^\dagger$, one gets Eq.(74). This remarkable property is peculiar to the method of iterated resolvents. Whenever a quark-pair-glue resolvent is sandwiched in between two vertex interactions, it allows to define a modified vertex interaction $\overline{V}$, since

$$V\,G_n\,V^\dagger = V\,C_n\,\overline{G}_n\,C_n^\dagger\,V^\dagger = \overline{V}G_n\,\overline{V}^\dagger, \quad \text{with} \quad \overline{V} = VC_n. \tag{76}$$

One can thus rewrite the sector Hamiltonians like for example

$$\overline{U}_6 = \overline{V}\,\overline{G}_{10}\overline{V} + \overline{V}\,\overline{G}_{10}\overline{V}G_9\overline{V}\,\overline{G}_{10}\overline{V}, \tag{77}$$

$$\overline{U}_3 = \overline{V}\,\overline{G}_6\overline{V} + \overline{V}\,\overline{G}_6\overline{V}G_5\overline{V}\,\overline{G}_6\overline{V}, \tag{78}$$

$$U_1 = \overline{V}\,\overline{G}_3\overline{V} + \overline{V}\,\overline{G}_3\overline{V}G_2\overline{V}\,\overline{G}_3\overline{V}. \tag{79}$$

Now they have all essentially the same structure, contrary to Eqs.(69) where they were different. Note that G_2, G_5 and G_9 are pure gluon propagators.

Generally spoken, the rectangular block matrix V is multiplied with the square matrix C_n. Below it will be shown that C_n is approximately diagonal and independent of the spin. Each vertex matrix element is multiplied with a simple function which depends on the momentum transfer Q across the vertex. Equivalently one replaces the coupling constant g by $\bar{g} = gC_n(Q)$, such that C_n can be interpreted as a *coupling function*.

The exact effective interaction

The most important result of this section is that gauge theory particularly QCD has only two structurally different contributions to the effective interaction in the $q\bar{q}$-space, see Eq.(79). It scatters a quark with helicity λ_q and four-momentum $p = (xP^+, x\vec{P}_\perp + \vec{k}_\perp, p^-)$ into a state with λ_q' and four-momentum $p' = (x'P^+, x'\vec{P}_\perp + \vec{k}_\perp', p'^-)$, and correspondingly the antiquark. In the *continuum limit*, the resolvents are replaced by propagators and the matrix eigenvalue problem $H_{\text{eff}}|\psi\rangle = M^2|\psi\rangle$ becomes an integral equation with a rather transparent structure:

$$M_i^2 \langle x, \vec{k}_\perp; \lambda_q, \lambda_{\bar{q}}|\psi_i\rangle = \left[\frac{\overline{m}_q^2 + \vec{k}_\perp^2}{x} + \frac{\overline{m}_{\bar{q}}^2 + \vec{k}_\perp^2}{1-x}\right] \langle x, \vec{k}_\perp; \lambda_q, \lambda_{\bar{q}}|\psi_i\rangle \qquad (80)$$

$$+ \sum_{\lambda_q', \lambda_{\bar{q}}'} \int dx' d^2\vec{k}_\perp' \ R(x', k_\perp') \ \langle x, \vec{k}_\perp; \lambda_q, \lambda_{\bar{q}}|U_{\text{OGE}}|x', \vec{k}_\perp'; \lambda_q', \lambda_{\bar{q}}'\rangle \ \langle x', \vec{k}_\perp'; \lambda_q', \lambda_{\bar{q}}'|\psi_i\rangle$$

$$+ \sum_{\lambda_q', \lambda_{\bar{q}}'} \int dx' d^2\vec{k}_\perp' \ R(x', k_\perp') \ \langle x, \vec{k}_\perp; \lambda_q, \lambda_{\bar{q}}|U_{\text{TGA}}|x', \vec{k}_\perp'; \lambda_q', \lambda_{\bar{q}}'\rangle \ \langle x', \vec{k}_\perp'; \lambda_q', \lambda_{\bar{q}}'|\psi_i\rangle.$$

The domain of integration is set by the cut-off function R given in Eq.(35). The effective one-gluon exchange

$$U_{\text{OGE}} = \overline{V}G_3\overline{V} \qquad (81)$$

conserves the flavor along the quark line. As illustrated in Fig. 11 the vertex interaction V creates a gluon and scatters the system virtually into the $q\bar{q}\,g$-space. As indicated in the figure by the vertical line with subscript '3', the three particles propagate there under impact of the full Hamiltonian before the gluon is absorbed. The gluon can be absorbed either by the antiquark or by the quark. If it is absorbed by the quark, it contributes to the effective quark mass $\overline{m}$. The second term in Eq.(79) is the effective two-gluon-annihilation interaction,

$$U_{\text{TGA}} = \overline{V}G_3\overline{V}G_2\overline{V}G_3\overline{V}, \qquad (82)$$

as represented by the graph $U_{1,0}$ in Fig. 11. The virtual annihilation of the $q\bar{q}$-pair into two gluons can generate an interaction between different quark flavors. – The eigenvalue M_i^2 is one of the invaraint-mass squared eigenvalues of the full light-cone Hamiltonian, and the wavefunction $\langle x, \vec{k}_\perp; \lambda_q, \lambda_{\bar{q}}|\psi_i\rangle$ gives the probability amplitudes for finding in the $q\bar{q}$-state a flavored quark with momentum fraction x, intrinsic transverse momentum $\vec{k}_\perp$ and helicity λ_q, and correspondingly an antiquark with $1-x$, $-\vec{k}_\perp$ and $\lambda_{\bar{q}}$.

THE BREAKING OF THE PROPAGATOR HIERARCHY

The content of the preceeding sections is exact but rather formal. In the sequel, rigor will be given up by the aim to obtain a solvable equation. In order to spot easier the essential approximations, they will be dressed as 'theorems', which can – or cannot – be proven later.

All sector Hamiltonians can be diagonalized on their own merit, for example

$$M_{1;i}^2 \langle q; \bar{q} | \Psi_{1;i} \rangle = \sum_{q', \bar{q}'} \langle q; \bar{q} | H_1 | q'; \bar{q}' \rangle \langle q'; \bar{q}' | \Psi_{1;i} \rangle,$$

$$M_{3;i}^2 \langle q; \bar{q}; g | \Psi_{3;i} \rangle = \sum_{q', \bar{q}', g'} \langle q; \bar{q}; g | \overline{H}_3 | q'; \bar{q}'; g' \rangle \langle q'; \bar{q}'; g' | \Psi_{3;i} \rangle. \tag{83}$$

The spectra might be continuous, but the eigenvalues are denumerated for simplicity by i. By construction, one knows the relation between these two sets of eigenvalues, by the following reason: Because of the separation into spectators and participants the gluon in the $q\bar{q}\,g$-equation is a non-interacting particle. it moves freely relative to a $q\bar{q}$-bound state. If its four-momentum is parametrized as $q^\mu = (yP^+, y\vec{P}_\perp + \vec{q}_\perp, q_g^-)$ one has

$$M_{3;i}^2 = \frac{M^2 + \vec{q}_\perp^2}{(1-y)} + \frac{\vec{q}_\perp^2}{y}. \tag{84}$$

Every $q\bar{q}$ bound state M is a band-head in the $q\bar{q}\,g$ spectrum. With $\omega = M_{1,0}^2(\omega) \equiv M^2$ one gets

$$\omega - M_{3;i}^2 = M^2 - \left(\frac{M_{1;0}^2 + \vec{q}_\perp^2}{(1-y)} + \frac{\vec{q}_\perp^2}{y} \right) = -\frac{y^2 M^2 + \vec{q}_\perp^2}{y(1-y)}. \tag{85}$$

Knowing the eigenvalues and the eigenfunctions of the $q\bar{q}$ bound state one alos know the spectrum and the eigenfunctions of $\overline{H}_3$. Knowing the eigenfunctions, the exact resolvent can be calculated.

Theorem 3 *The exact propagators can be approximated by closure.*

The substitution of the exact propagators by closure is a widely used approximation in many-body theory and in chemistry. In principle, the exact propagator is non-diagonal. Performing closure, it becomes diagonal, which is reflected by the Dirac-delta function $\langle q; \bar{q}; g | q'; \bar{q}'; g' \rangle$ in the single particle momenta and helicities

$$\langle q; \bar{q}; g | \overline{G}_3 | q'; \bar{q}'; g' \rangle = \langle q; \bar{q}; g | q'; \bar{q}'; g' \rangle \overline{G}_3(q; \bar{q}; g). \tag{86}$$

The Dirac-delta function is multiplied with the function

$$\overline{G}_3(q; \bar{q}; g) = -\frac{y(1-y)}{y^2 M^2 + \vec{q}_\perp^2}. \tag{87}$$

The in-medium propagator $\overline{G}_3$ ceases to be a functional of the higher resolvents. *The hierarchy of the iterated resolvents is broken.*

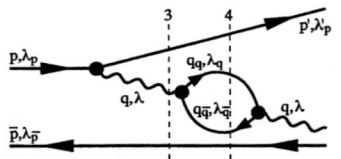

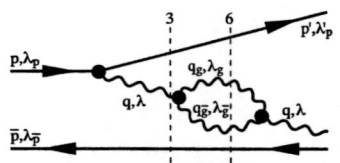

FIGURE 14. *The $q\bar{q}$ vacuum polarization graph.*

FIGURE 15. *The gg vacuum polarization graph.*

Theorem 4 *In the solution, the interacting particles propagate free particles. The in-medium propagators can be replaced by the free propagators.*

The free propagator in the $q\bar{q}g$-space can be written

$$G_{3,\text{free}} = \frac{1}{P^+(p^- - p'^- - q^-)} = -\frac{y}{Q^2} = -\frac{y}{y^2(2\overline{m})^2 + \vec{q}_\perp^2}, \tag{88}$$

where $Q^2 = (p - p')^2$ is the four-momentum transfer along the quark line, in the notation of Fig. 14. For sufficiently small y holds $(p - p')^2 = -[y^2(2\overline{m})^2 + \vec{q}_\perp^2]$. If one substitutes $M \simeq 2\overline{m}$, which holds to rather good approximation, the momentum transfer in Eq.(88) is the same as in Eq.(87).

Similar considerations also hold all other propagators with at least one $q\bar{q}$-pair.

Having replaced the in-medium-propagators $\overline{G}_n$ by the free propagators, one can calculate the graphs like in the usual (light-cone-) time-ordered perturbation theory. The diagram $U_{1,2}$ in Fig. 11 yields the effective quark mass

$$\overline{m}_f^2 = m_f^2 + m_f^2 \frac{\alpha}{\pi} \frac{n_c^2 - 1}{2n_c} \ln \frac{\Lambda^2}{m_g^2}. \tag{89}$$

Correspondingly one can calculate the coupling function

$$C_3 = \sqrt{\frac{1}{1 - \overline{G}_3 \tilde{U}_3}} = 1 + \frac{1}{2}\overline{G}_3 \tilde{U}_3 + \frac{3}{8}\overline{G}_3 \tilde{U}_3 \, \overline{G}_3 \tilde{U}_3 + \dots, \tag{90}$$

as defined by Eq.(74). With $\tilde{U}_3 = VG_4V + VG_6V$ from Eq.(71), and thus $\tilde{U}_3 = \overline{VG_4V} + \overline{VG_6V}$ from Eq.(76), the first two terms in the expansion are

$$\overline{V} \simeq V + \frac{1}{2}(V\overline{G}_3\overline{VG_4}\overline{V} + V\overline{G}_3\overline{VG_6}\overline{V}) \simeq V + \frac{1}{2}(V\overline{G}_3VG_4V + V\overline{G}_3VG_6V). \tag{91}$$

In the last step $R_4 = R_6 = 1$ was set. In general, they contribute terms of higher orders in g, which must be suppressed for consistency. Figs. 14 and 15 give only 2 representative graphs. According to the rules of time ordered perturbation theory, there are about 22 different time-ordered and instantaneous diagrams. All of them were re-calculated by Raufeisen [34], for all values of Q. As shown to some detail in [33] one gets for sufficiently large Λ

$$\overline{G}_3 V(\overline{G}_6 + \overline{G}_4)V = \frac{11\alpha n_c}{12\pi} \ln\left(\frac{\Lambda^2}{4m_g^2 + Q^2}\right) - \frac{\alpha}{6\pi} \sum_f \ln\left(\frac{\Lambda^2}{4m_f^2 + Q^2}\right),$$

$$= \alpha\, b_0 \ln\left(\Lambda^2/\kappa^2\right) - b(Q), \quad b_0 = \frac{11 n_c - 2 n_f}{12\pi}, \tag{92}$$

with an arbitrary κ and a $b(Q)$ independent of Λ given in Eq.(93). One concludes that the coupling function C_3 is identical with the familiar vertex correction. In the limit where Q is larger than all mass scales, this result agrees with the calculations of both, Thorn [35] and Perry [36], having been done previously.

What happens if one substitutes the free propagators in Eq.(90) by the non-perturbative propagators $\overline{G}_4 = (\omega - \overline{H}_{q\bar{q}q\bar{q}})^{-1}$ and $\overline{G}_6 = (\omega - \overline{H}_{q\bar{q}gg})^{-1}$, at least in an approximate fashion? — There are additional graphs. In the fermion loop of vacuum polarization appear two graphs in addition to Fig. 14. In one of them, a gluon is emitted and absorbed on the same quark line which changes the bare quark mass m_f into the physical quark mass $\overline{m}_f$. In the other graph, the gluon is emitted from the quark and absorbed by the anti-quark which represents an interaction. In consequence one has a bound state with a physical mass scale μ_f. We replace therefore $2m_f \Longrightarrow 2\overline{m}_f \Longrightarrow \mu_f$. Similar considerations hold for the gluon loop in Fig. 15 and lead to the substitution $2m_g \Longrightarrow 2\overline{m}_g \Longrightarrow \mu_g$. Both μ_g and μ_f are interpreted as physical mass scales. The physical gluon mass $\overline{m}_g$ vanishes of course due to gauge invariance. This is not in conflict with f.e. Cornwall's suggestion of a finite effective gluon mass [37] since one can define $(\overline{m}_g)_{\text{eff}} \equiv \mu_g/2$.

As a consequence of the approximation in Eq.(86), the propagators $\overline{G}_3$ and $\overline{G}_4$ are diagonal in Fock space, which in turn leads to diagonal products $\overline{G}_3 V \overline{G}_4 V$. They can be resumed therefore to all orders according to Eq.(90). With $\overline{\alpha} \equiv g^2 C_3^2/4\pi$, one gets

$$\overline{\alpha}(Q;\Lambda) = \frac{1}{1/\alpha - b_0 \ln\left(\Lambda^2/\kappa^2\right) + b(Q)} \quad \text{with}$$

$$b(Q) = \frac{11 n_c}{12\pi} \ln\left(\frac{\kappa^2}{\mu_g^2 + Q^2}\right) - \frac{1}{6\pi} \sum_f \ln\left(\frac{\kappa^2}{\mu_f^2 + Q^2}\right). \tag{93}$$

The effective fine structure constant depends on the momentum transfer Q across the vertex and on the cut-off Λ. Now, all the pieces are together which are needed for a further discussion of the effective interaction as defined in Eq.(80). Finally, the the instantaneous interaction in the effective one-gluon-exchange interaction $U_{OGE} = \overline{V G_3 V}$ is restored according to $\overline{V G_3 V} \longrightarrow C_3^2(V\overline{G}_3 V + S)$, with the expression given in Eq.(26).

STATUS AND DISCUSSION

What have we reached? – The general expression for the effective Hamiltonian in the $q\bar{q}$-sector was given in Eq.(80). If one restricts to consider mesons in which the quark and the anti-quark have different flavors, $f_q \neq f_{\bar{q}}$, the two-gluon annihilation diagram can not get active and one remains with

$$M_n^2 \langle x, \vec{k}_\perp; \lambda_q, \lambda_{\bar{q}} | \psi_n \rangle = \left[\frac{\overline{m}_q^2(\Lambda) + \vec{k}_\perp^2}{x} + \frac{\overline{m}_{\bar{q}}^2(\Lambda) + \vec{k}_\perp^2}{1-x} \right] \langle x, \vec{k}_\perp; \lambda_q, \lambda_{\bar{q}} | \psi_n \rangle$$

$$+ \sum_{\lambda_q', \lambda_{\bar{q}}'} \int dx' d^2 \vec{k}_\perp' \ \langle x, \vec{k}_\perp; \lambda_q, \lambda_{\bar{q}} | U_{\mathrm{OGE}} | x', \vec{k}_\perp'; \lambda_q', \lambda_{\bar{q}}' \rangle \ \langle x', \vec{k}_\perp'; \lambda_q', \lambda_{\bar{q}}' | \psi_n \rangle. \qquad (94)$$

The essential achievement of the two preceeding sections is that the kernel of this integral equation can be written down in the very explicit form

$$\langle \lambda_q, \lambda_{\bar{q}} | U_{\mathrm{OGE}} | \lambda_q', \lambda_{\bar{q}}' \rangle = -\frac{\overline{\alpha}(Q; \Lambda)}{3\pi^2 Q^2} \ \langle \lambda_q, \lambda_{\bar{q}} | S | \lambda_q', \lambda_{\bar{q}}' \rangle \frac{R(x', k_\perp'; \Lambda)}{\sqrt{x(1-x)x'(1-x')}}. \qquad (95)$$

All many-body effects reside in the effective mass $\overline{m}_q(\Lambda)$ and in the effective coupling constant $\overline{\alpha}(Q; \Lambda)$. They are given in Eqs.(89) and (93), respectively. The mean of the four-momentum transfers of quark and anti-quark, $Q_q^2 = -(k_q - k_q')^2$ and $\overline{Q}_{\bar{q}}^2 = (k_{\bar{q}} - k_{\bar{q}}')^2$, respectively, is denoted by Q^2. The spinor factor is

$$\langle \lambda_q, \lambda | S | \lambda_q', \lambda' \rangle = [\overline{u}(k, \lambda) \gamma^\mu u(k', \lambda')]_q \ [\overline{u}(k, \lambda) \gamma_\mu u(k', \lambda')]_{\bar{q}} .$$

The cut-off function R sets the domain of integration and is defined in Eq.(35). The approximations made have been carefully enumerated in Section 7.

One has thus reached the goal of reducing the field theoretical many-body problem of $M^2 = H_{\mathrm{LC}}$ to a one-body problem with an effective interaction. For any value of the Lagrangian coupling constant g, the flavor masses m_f, and the cut-off Λ, one can generate the eigenvalues M_n^2 on a computer, very much in analogy to QED [30,31]. With the known eigenfunction in the $q\bar{q}$-sector one an generate all higher Fock-space amplitudes, according to the explicit formulas in Section 6. It looks as if one has solved the problem.

But one has solved the problem only in a superficial way, since the solutions depend on the unphysical parameter Λ. One of the central issues in gauge-field theory is to remove this dependence by a renormalization group analysis, see below.

The eigenvalues of the integral equation are the invariant masses squared. Very often it is more convenient to think in terms of an analogue of a non-relativistic Hamiltonian H_S,

$$H_{\mathrm{LC}} = (\overline{m}_{\bar{q}} + \overline{m}_{q'})^2 + 2(\overline{m}_{\bar{q}} + \overline{m}_{q'}) \ H_S, \qquad (96)$$

whose eigenvalues $E = H_S$ are the more convential binding energies, while the eigenfunctions are the same. As shown in [2], the so defined 'Schrödinger operator' has much in common with a non-relativistic Hamiltonian for the Coulomb problem.

132

Renormalization group analysis

All eigenvalues M_n^2 depend thus on the cut-off Λ, the bare masses m and the bare α. In line with modern interpretation of quantum field theory [7], they are unphysical parameters and can be replaced by $(m_\Lambda, \alpha_\Lambda)$, by functions of Λ, such that the eigenvalues are cut-off independent, $i.e.$

$$\frac{d}{d\Lambda} M_n^2(\Lambda, \alpha_\Lambda, m_\Lambda) = 0. \tag{97}$$

This fundamental equation of the renormalization group must hold for all eigenvalues.

The effective Hamiltonian $H = H_{LC,\text{eff}}$ depends on Λ only through $\overline{m}_f$, $\overline{\alpha}$ and the regulator function R. The renormalization group equations can be written therefore in the operator form

$$\delta \overline{m}_f \frac{\delta H}{\delta \overline{m}_f} |\psi_n\rangle + \delta \overline{\alpha} \frac{\delta H}{\delta \overline{\alpha}} |\psi_n\rangle + \delta R \frac{\delta H}{\delta R} |\psi_n\rangle = 0. \tag{98}$$

The simultaneous variation of all three terms is a very difficult problem. But one can replace this equation by three independent ones,

$$\frac{d}{d\Lambda} \overline{m}_f = 0, \tag{99}$$

$$\frac{d}{d\Lambda} \overline{\alpha} = 0, \tag{100}$$

$$\frac{dR}{d\Lambda} \frac{\delta H}{\delta R} |\psi_n\rangle = 0. \tag{101}$$

This is possible since one still solves Eq.(98).

Eqs.(99) and (100) are two equations for the two unknown functions m_Λ and α_Λ. The first equation says that the effective flavor masses $\overline{m}_f = \overline{m}_f(\Lambda, \alpha_\Lambda, m_\Lambda)$ are renormalization group invariants. The numerical value of $\overline{m}_f$ must be fixed by experiment. – Eq.(100) is then considered as an equation for α_Λ at fixed values of $\overline{m}_f$. Using the expression for $\overline{\alpha}(Q; \Lambda)$ given in Eq.(93) yields then [33]

$$\alpha_\Lambda = \frac{1}{b_0 \ln(\Lambda^2/\kappa^2)}, \quad \text{thus} \quad \overline{\alpha}(Q) = \frac{1}{b(Q)}, \tag{102}$$

with $b(Q)$ given in Eq.(93). All Λ-dependence cancels exactly in favor of the renormalization group invariant κ, which is sometimes called the QCD-scale Λ_{QCD}. The scale κ must be determind from experiment. – Once α_Λ is known, Eq.(99) is an equation for m_Λ. This function can be determined from Eq.(89), but we renounce to do that here, since m_Λ is needed nowhere in the present context. – Having fixed the functions α_Λ and m_Λ one has exhausted all freedom provided by conventional renormalization anaysis of field theory [7]. Since the cutoff-dependence was removed by renormalization, the cut-off can be driven to infinity, thus

$$R(x, \vec{k}_\perp; \Lambda) = 1.$$

Formally, this solves Eq.(101).

The remaining divergence

After renormalization, the kernel of the one-body equation is independent of Λ. Let us discuss its relevant part

$$\mathcal{K} = \frac{1}{b(Q)} \frac{S}{Q^2}.$$

At very small momentum transfers $Q^2 \to 0$, where $x \sim x'$ and $\vec{k}_\perp \sim \vec{k}'_\perp$, the effective coupling constant locks into the finite value $b(0)$, see Eq.(93). The spinor function also goes to a constant, $S \sim 4\overline{m}_q \overline{m}_{\bar{q}}$, see [31] or [2]. The remaining 'Coulomb singularity' Q^{-2} is square-integrable and can be dealt with by convenient numerical means [30,31]. For very large momentum transfers $Q^2 \to \infty$, the behaviour is quite different. Both S and Q^2 tend to diverge, $S \propto (\vec{k}'_\perp)^2$ and $Q^2 \propto (\vec{k}'_\perp)^2$, but such that the ratio S/Q^2 tends to a *finite constant*. Disregarding the very slow logarithmic increase of $b(Q)$, the kernel of the integral equation is therefore essentially a dimensionless constant

$$\mathcal{K} \sim constans, \qquad \text{for} \quad Q^2 \gg 0.$$

One has sufficient evidence [31] that this behaviour creates all kinds of problems, among them a diverging eigenvalue.

One can separate the kernel of the integral equation identically into two pieces $\mathcal{K} = \mathcal{K}_1 + \mathcal{K}_2$, *i.e.*

$$\mathcal{K} = \frac{1}{b(Q)} \frac{S}{Q^2} \frac{\mu^2 + Q^2}{\mu^2 + Q^2} = \frac{1}{b(Q)} \frac{S}{Q^2(1 + Q^2/\mu^2)} + \frac{S}{b(Q)(\mu^2 + Q^2)},$$

with an arbitrary mass parameter μ. The first part $\mathcal{K}_1$ is well behaved and vanishes at least like Q^{-2} in the limit of large Q. The second part $\mathcal{K}_2$ is a strange object: it is a constant almost everywhere, in particular

$$\mathcal{K}_2 \sim constans, \qquad \text{since} \quad \frac{S}{Q^2} \sim constans, \qquad \text{for} \quad Q^2 \gg 0.$$

One should emphasize the following aspect. The typical field theoretical divergences like the divergence of the effective coupling constant lead to a *divergence* of the kernel, but adding a *finite constant* to the kernel leads also to a divergence. Thus far one does not understand the reason for the latter.

Perhaps one can get further insight by studying the following eigenvalue in (usual) three-momentum space $(\vec{k})$

$$E\psi(\vec{k}) = \frac{\vec{k}^2}{2m}\psi(\vec{k}) - \frac{\alpha}{\pi^2} \int d^3k' \left[\frac{1}{(\vec{k} - \vec{k}')^2} + \frac{1}{\mu^2} \right] \psi(\vec{k}'). \tag{103}$$

TABLE 9. *The empirical masses of the flavor-off-diagonal physical mesons in MeV. Vector mesons are given in the upper, pseudo-scalar mesons in the lower triangle. Their physical nomenclature is given on the right.*

	$\bar{u}$	$\bar{d}$	$\bar{s}$	$\bar{c}$	$\bar{b}$		$\bar{u}$	$\bar{d}$	$\bar{s}$	$\bar{c}$	$\bar{b}$
u		768	892	2007	5325	u		ρ^+	K^{*+}	$\overline{D}^{*0}$	B^{*+}
d	140		896	2010	5325	d	π^-		K^{*0}	D^{*-}	B^{*0}
s	494	498		2110	—	s	K^-	$\overline{K}^0$		D_s^{*-}	B_s^{*0}
c	1865	1869	1969		—	c	D^0	D^+	D_s^+		B_c^{*+}
b	5278	5279	5375	—		b	B^-	$\overline{B}^0$	$\overline{B}_s^0$	B_c^-	

TABLE 10. *The calculated masses of the flavor-off-diagonal mesons in MeV, as obtained from a fit to Eq.(105) on the left and to Eq.(104) on the right. Vector mesons are given in the upper, pseudo-scalar mesons in the lower triangle. The *upperscript marks mesons where the shift and mass parameters have be fitted to.*

	$\bar{u}$	$\bar{d}$	$\bar{s}$	$\bar{c}$	$\bar{b}$		$\bar{u}$	$\bar{d}$	$\bar{s}$	$\bar{c}$	$\bar{b}$
u		*768	902	2012	5331	u		*768	1002	2301	5696
d	*140		902	2012	5331	d	*140		1002	2301	5696
s	*494	494		2155	5476	s	*494	494		2535	5829
c	*1865	1865	2108		6617	c	*1865	1865	2102		7227
b	*5278	5278	5423	6573		b	*5278	5278	5512	6811	

In the spirit of Eq.(96), the right-hand side is a conventional Schrödinger equation for the Coulomb problem with the reduced mass $1/m = 1/m_q + 1/m_{\bar{q}}$, but where an arbitrary constant has been added in the kernel. In Heidelberg, we are presently working on the question whether its eigenvalue E diverges. Preliminary studies with the Fourier transform of Eq.(103), where the constant works its way into a weird Dirac-delta function $\delta^{(3)}(\vec{x})$ as a potential,

$$E\phi(\vec{x}) = -\frac{\vec{\nabla}^2}{2m}\phi(\vec{x}) - \left[\frac{\alpha}{|\vec{x}|} + \frac{\alpha}{\mu^2}\delta^{(3)}(\vec{x})\right]\phi(\vec{x}'),$$

indicate indeed that the eigenvalue of the $1s$-state in particular diverges.

If these preliminary results materialize, one must regularize and subsequently renormalize Eq.(103). The energy of the $1s$-state will then be essentially a renormalization constant $E_{1s} = \bar{s}$. Given that to be true, all of us have been riding the wrong horse. It is not the regular part of the interaction kernel $\mathcal{K}_1$, but it is the 'constant' $\mathcal{K}_2$ which determines the mass in the first place. Since $\mathcal{K}_2$ involves a renormalization and since the singlet can have an energy shift $(\bar{s}_-)$ different from the triplet $(\bar{s}_+)$, one has at most two additional renormalization constants, subject to be determined by experiment.

135

TABLE 11. *The quark mass parameters $\overline{m}_q$ and the energy shifts $\overline{s}_\pm$, all in MeV, as obtained from a fit to the physical meson masses.*

method	$\overline{m}_u$	$\overline{m}_d$	$\overline{m}_s$	$\overline{m}_c$	$\overline{m}_b$	$\overline{s}_-$	$\overline{s}_+$
Eq.(104)	350	350	583	1881	5275	-336	71
Eq.(105)	350	350	495	1637	4972	-336	71

A mass formula

Is there some physical evidence for the above considerations? – Inspired by Eq.(96), one can seek the masses of the physical mesons in the form [38]

$$M^2 = (\overline{m}_{\bar{q}} + \overline{m}_{q'})^2 + 2(\overline{m}_{\bar{q}} + \overline{m}_{q'})\, \overline{s}_\pm, \tag{104}$$

and fit the energy shifts $\overline{s}_\pm$ and the physical quark masses $\overline{m}_q$ to the physical meson masses, compiled in Table 9 from the data of the particle data group [39] which do not include yet the topped mesons.

The following procedure was applied. First, the up and the down mass were chosen equal. Then the empirical masses of π^+ and ρ^+ were used to determine the mass shifts for the singlet and the triplet. The remaining quark masses are obtained from the pseudo-scalar mesons with an up quark, which exhausts all freedom in determining physical parameters. The remaining 13 off-diagonal pseudo-scalar and vector meson masses are calculated straightforwardly and compiled in Table 10. In comparing the so obtained numbers with the experiment, one notes with great surprise that the discrepancy exceeds only rarely an estimated error of 10%. The resulting quark masses are given in Table 11.

Actually, there is no reason why one should stick to Eq.(104). Equivalently, one can replace $\overline{s}_\pm$ by $\overline{s}_\pm \, a/(\overline{m}_{\bar{q}} + \overline{m}_{q'})$. Choosing $a = \overline{m}_{\bar{u}} + \overline{m}_d = 700$ MeV, one gets

$$M^2 = (\overline{m}_{\bar{q}} + \overline{m}_{q'})^2 + 2(\overline{m}_{\bar{u}} + \overline{m}_d)\, \overline{s}_\pm. \tag{105}$$

A form like this was suggested to me by Dae Sung Hwang during the lectures. Redoing the fit for the heavy quarks gives the results compiled in Table 10. The discrepancy with the experimental numbers is now on the level of 1%. This excellent agreement should not be over-emphasized. In any case it does not falsify the above considerations.

A SHORT SUMMARY

In these lecture notes the attempt was made to phrase and discuss a relativistic quantum theory such as QCD from the point of view of Dirac's front form of hamiltonian dynamics. Sometimes refered to as light-cone quantization, it has the goal to describe and understand the bound-state structure of hadrons in terms of their constituents from a covariant theory. As the lectures show, this goal has not been reached yet.

As reviewed in [2], there might be alternative roads to reach the same goal, like for example the Bethe-Salpeter equations, lattice gauge theory, Hamiltonian flow equations, just to name a few. Each of these approaches have their own inherent adventages or disadvantages. By reasons of space their discussion is left out in these notes. By the same reason the light-front renormalizations approach of Wilson and collaborators [40,41] is omitted. The many articles have thus far not provided conclusive evidence for success, apart from the fact that virtually every Langrangian symmetry has been violated in the approach. It is much too early to draw definitive conclusions.

The front form is however useful to study the structure of arbitrary field theories. The theories considered are often not physical, but are selected to help in the understanding of a particular non-perturbative phenomenon. The relatively simple vacuum properties of light-front field theories underly many of these 'analytical' approaches. The relative simplicity of the light-cone vacuum provides a firm starting point to attack many non perturbative issues. As we have seen, the problems in two dimensions, in one space and one time dimension, not only are they tractable from the outset but in many cases like the Schwinger model can they be solved numerically by 'discretized light-cone quantization (DLCQ)' to almost arbitrary accuracy. This solution gives a unique insight and understanding. The Schwinger particle indeed has the simple parton structure that one hopes to see in QCD.

Unfortunately the same cannot be said for the physical world in 3 space and 1 time dimensions. The essential problem is that the number of degrees of freedom needed to specify each Fock state even in a discrete basis grows much too quickly. As discussed in this review, the basic procedure is to diagonalize the full light-cone Hamiltonian in the free light-cone Hamiltonian basis. The eigenvalues are the invariant mass squared of the discrete and continuum eigenstates of the spectrum. The projection of the eigenstate on the free Fock basis are the light-cone wavefunctions and provide a rigorous relativistic many-body representation in terms of its particle degrees of freedom. Given the light-cone wavefunction one can compute the structure functions and distribution amplitudes. More generally, the light-cone wavefunctions provide the interpolation between hadron scattering amplitudes and the underlying parton subprocesses. The method of iterated resolvents can be a useful intermediary step to generate these wavefunctions. The unique property of light-cone quantization that makes the calculations of light cone wavefunctions particularly useful is that they are independent of the reference frame and that the same wavefunction can be use in many different problems.

Finally, let us highlight the intrinsic advantages of light-cone field theory:

- The light-cone wavefunctions are independent of the momentum of the bound state – only relative momentum coordinates appear.

- The vacuum state is simple and in many cases trivial.

- Fermions and fermion derivatives are treated exactly; there is no fermion doubling problem.

- The minimum number of physical degrees of freedom are used because of the light-cone gauge. No Gupta-Bleuler or Faddeev-Popov ghosts occur and unitarity is explicit.

- The output is the full color-singlet spectrum of the theory, both bound states and continuum, together with their respective wavefunctions.

REFERENCES

1. P.A.M. Dirac, Rev. Mod. Phys. **21**, 392 (1949).
2. S.J.Brodsky, H.C. Pauli, S.S. Pinsky, Physics Reports **301**, 299-486 (1998).
3. A. Bassetto, G.Nardelli and R. Soldati, *Yang-Mills Theories in Algebraic Noncovariant Gauges* (World Scientific,Singapore, 1991).
4. J.D. Bjørken and S.D. Drell, *Relativistic Quantum Mechanics* (McGraw-Hill, New York, 1964); J.D.Bjørken and S. D. Drell, *Relativistic Quantum Fields* (McGraw-Hill,NewYork, 1965).
5. P.M. Morse and H. Feshbach, *Methods in Theoretical Physics*, 2 Vols, (Mc Graw-Hill, New York, N.Y., 1953).
6. T. Muta, *Foundations of Quantum Chromodynamic*. Lecture notes in Physics, Vol.5 (World Scientific, Singapore, 1987).
7. S. Weinberg, *The quantum theory of fields*, 2 Vols, (Cambridge University Press, New York, 1995).
8. S.J. Brodsky and H.C. Pauli, in *Recent Aspect of Quantum Fields*, H. Mitter and H. Gausterer, Eds.; Lecture Notes in Physics,Vol. 396 (Springer-Verlag, Berlin, Heidelberg, 1991).
9. F. Coester, Prog. Nuc. Part. Phys. **29**, 1 (1992).
10. *Theory of Hadrons and Light-front QCD*, S. Glazek, Ed., (World Scientific Publishing Co., Singapore, 1995).
11. *New non-perturbative methods and quantization on the light cone*, P. Grangé, A. Neveu, H.C. Pauli, S. Pinsky and E. Werner, Eds., (Springer Verlag, Berlin, Heidelberg, 1997).
12. W. Lucha, F.F. Schöberl, and D. Gromes, Physics Reports **200**, 127 (1991).
13. J.B. Kogut and D.E. Soper, Phys. Rev. **D1**, 2901 (1970).
14. G.P. Lepage and S.J. Brodsky, Phys. Rev. **D22**, 2157 (1980).
15. H.C. Pauli, Z. Phys. **A319**, 303 (1984).
16. H.C. Pauli and S.J. Brodsky, Phys. Rev. **D32**, 1993 (1985); **D32**, 2001 (1985).
17. T. Eller, H.C. Pauli and S.J. Brodsky, Phys. Rev. **D35**, 1493 (1987).
18. K. Hornbostel, S. J. Brodsky and H. C. Pauli, Phys. Rev. **D41**, 3814 (1990).
19. D.P. Crewther and C.J. Hamer, Nucl. Phys. **B170**, 353 (1980).
20. H. Bergknoff, Nucl. Phys. **B122**, 215 (1977).

21. T. Eller and H.C. Pauli, Z. Physik **C42**, 59 (1989).
22. S. Elser, Proceedings of *Hadron structure '94*, Kosice, Slowakia (1994); and Diplomarbeit, U. Heidelberg (1994).
23. M. Völlinger, Diplomarbeit, U. Heidelberg (1996).
24. J.P. Vary, T.J. Fields and H.-J. Pirner, Phys. Rev. **D53**, 7231-7238 (1996).
25. B. van de Sande, Phys. Rev. **D54**, 6347 (1996). hep-ph/9605409.
26. T. Maskawa and K. Yamawaki, Prog. Theor. Phys. **56**, 270 (1976).
27. S.S. Pinsky and B. van de Sande, Phys. Rev. **D49**, 2001 (1994).
28. I. Tamm, J. Phys. (USSR) **9**, 449 (1945).
29. S.M. Dancoff, Phys. Rev. **78**, 382 (1950).
30. M. Krautgärtner, H.C. Pauli and F. Wölz, Phys. Rev. **D45**, 3755 (1992).
31. U. Trittmann and H.C. Pauli, "Quantum electrodynamics at strong coupling", Heidelberg preprint MPI H-V4-1997, Jan. 1997. hep-th/9704215
32. H.C. Pauli, in: *Neutrino mass, Monopole Condensation, Dark matter, Gravitational waves and Light-Cone Quantization*, B.N. Kursunoglu, S. Mintz and A. Perlmutter, Eds., (Plenum Press, New York, 1996) p.183-204; hep-th/9608035, hep-th/9707361, Heidelberg preprint MPIH-V9-1998, Mar. 1998.
33. H.C. Pauli, Eur.Phys.J.**C7**, 289 (1998); hep-th/9809005, hep-th/9608035.
34. J. Raufeisen, Diplomarbeit (in German), Universität Heidelberg (1997).
35. C.B. Thorn, Phys. Rev. **D19**, 639 (1979); Phys. Rev. **D20**, 1934 (1979).
36. R.J. Perry, A. Harindranath, and W.M. Zhang, Phys. Lett. **B300**, 8 (1993).
37. J.M. Cornwall and A. Soni, Phys.Lett. **120B**, 431 (1983).
38. H.C. Pauli, Preprint MPIH-V15-1999, Preprint MPIH-V16-1999.
39. C. Caso *et al.* (Particle Data Group), Eur.Phys.J. **C3**, 1 (1998).
40. K.G. Wilson, T.S. Walhout, A. Harindranath, W.M. Zhang, R.J. Perry, S.D. Głazek, Phys. Rev.**D49**, 6720 (1994).
41. S.D. Glazek and K.G. Wilson, Phys. Rev.**D57**, 3558 (1998).

SDLCQ
Supersymmetric Discrete Light Cone Quantization

Oleg Lunin and Stephen Pinsky

Department of Physics
The Ohio State University
Columbus Ohio 43210

Abstract. In these lectures we discuss the application of discrete light cone quantization (DLCQ) to supersymmetric field theories. We will see that it is possible to formulate DLCQ so that supersymmetry is exactly preserved in the discrete approximation. We call this formulation of DLCQ, SDLCQ and it combines the power of DLCQ with all of the beauty of supersymmetry. In these lecture we will review the application of SDLCQ to several interesting supersymmetric theories. We will discuss two dimensional theories with (1,1), (2,2) and (8,8) supersymmetry, zero modes, vacuum degeneracy, massless states, mass gaps, theories in higher dimensions, and the Maldacena conjecture among other subjects.

Introduction.

In the last decade there have been significant improvements in our understanding of gauge theories and important breakthroughs in the nonperturbative description of supersymmetric gauge theories [73,74]. In the last few year various relations between string theory, brane theory and gauge fields [34,2] have also emerged. While these developments give us some insight into strong coupled gauge theories [74], they do not offer a direct method for non-perturbative calculations. In these lectures we discuss some recent development in the light cone quantization approaches to non-perturbative problems. We will see that these methods have the potential to expand our understanding of strongly coupled gauge theories in directions not previously available.

The original idea was formulated half of a century ago [31], but apart from several technical clarifications [62] it remained mostly undeveloped. The first change came in the mid 80–th when the Discrete Light Cone Quantization (DLCQ) was suggested as a practical way for calculating the masses and wavefunctions of hadrons [67]. Although the direct application of the method to realistic problems meets

CP494, *New Directions in Quantum Chromodynamics,* edited by C.-R. Ji and D.-P. Min
© 1999 American Institute of Physics 1-56396-908-4/99/$15.00

some difficulties (for review see [24]), DLCQ has been successful in studying various two dimensional models. Given the importance of supersymmetric theories, it is not surprising thatlight cone quantization was ultimately applied to such models [56,19,22]. Even in this early work the mass spectrum was shown to be supersymmetric in the continuum and a great deal of information about the properties of bound states in supersymmetric theories was extracted. However the straightforward application of DLCQ to the supersymmetric systems had one disadvantage: the supersymmetry was lost in the discrete formulation. The way to solve this problem was suggested in [63], where the alternative formulation of DLCQ was introduced. Namely it was noted that since the supercharge is the "square root" of Hamiltonian one can define the new DLCQ procedure based on the supercharge. We will study this formulation (called SDLCQ) in these lectures.

The lectures have the following organization. In section I we introduce the basic concepts of DLCQ and SDLCQ. We also define the systems to be studied in the remaining part of the lectures. We will concentrate our attention on the two dimensional models with adjoint matter, several examples of such systems can be constructed from supersymmetric Yang–Mills theories in higher dimensions using the dimensional reduction. In section II we address the problem of the DLCQ vacuum. In the continuum theory the light cone vacuum is very simple: it coincides with the usual Fock vacuum. This property is related to the decoupling between positive and negative frequency modes on the light cone and it is not true for equal time quantization. In DLCQ however one encounters the problem of zero modes which complicates the structure of vacuum and allows us to reproduce the correct vacuum degeneracy in certain theories. We continue to analyze zero modes in the section III and they are shown to play an important role in the explaining the difference between DLCQ and SDLCQ regularization procedures.

Section IV is devoted to the studying of massless states. The numerical analysis [63,5,10,8] shows an important property of the mass spectrum for some supersymmetric theories. Unlike the QCD–like models [45], such systems appear to have a lot of massless bound states. In fact the supersymmetric $SU(\infty)$ gauge theory seems to have an infinite number of such states in the continuum limit. Since the states with zero mass dominate the partition function for low enough temperatures they deserve to be studied very carefully and in section IV we analyze the structure of such states.

As we already mentioned in the beginning of this introduction, the relations between string theory and gauge fields attracted a lot of attention over the recent years. In particular it was conjectured [60] that one can extract some information about strong coupled gauge theory from supergravity calculations. The problem however is that in the relevant regime the usual field theoretic methods do not work, so it is hard to really test the conjecture. However for two dimensional systems DLCQ gives the solution of bound state problem which is valid beyond perturbation theory, so the results can be used to test the conjecture. We report the results of this first test in section V. For realistic systems with eight supersymmetries we still don't have enough computer power to compare the results with the supergravity predictions. The general techniques described in this section can also be used to calculate other correlation functions in nonperturbative regime.

Finally in section VI we make a first attempt to move beyond two dimensions. We present the general ideas for formulating SDLCQ in more than two dimension. As an example we present the numerical results of SYM for the simplest case when only one transverse momentum mode is introduced.

I   SUPERSYMMETRIC YANG–MILLS THEORY IN THE LIGHT–CONE GAUGE.

A   DLCQ and Its Supersymmetric Version.

In this work we will study the bound state problem for various supersymmetric matrix models in two dimensions. The examples of such models may be constructed by dimensional reduction of supersymmetric Yang–Mills theory in higher dimensions. In this subsection we will consider such reduction for three dimensional SYM. Before we begin the detailed analysis of bound state problem for the specific systems it is worthwhile to summarize some basic ideas of Discrete Light Cone Quantization, for a complete review see [24].

Let us consider general relativistic system in two dimensions. Usual canonical quantization of such system means that one imposes certain commutation relations between coordinates and momenta at equal time. However as was pointed out by Dirac long ago [31] this is not the only possibility. Another scheme of quantization treats the light like coordinate $x^+ = \frac{1}{\sqrt{2}}(x^0 + x^1)$ as a new time and then the system is quantized canonically. This scheme (called light cone quantization) has both positive and negative sides. The main disadvantage of light cone quantization is the presence of constraints, even the system as simple as free bosonic field has one: from the action

$$S = \int d^2x \, \partial_+ \phi \partial_- \phi \tag{1.1}$$

one can derive the constraint relating coordinate and momentum:

$$\pi = \partial_- \phi. \tag{1.2}$$

For more complicated systems the constraints are also present and in general they are hard to resolve.

The main advantage of the light cone is the decoupling between positive and negative momentum modes. This property is crucial for DLCQ. In the Discrete Light Cone Quantization one considers the theory on the finite circle along the x^- axis: $-L < x^- < L$. Then all the momenta become quantized and the integer number measuring the total momentum in terms of "elementary momentum" is called the harmonic resolution K. Due to decoupling property one may work only

142

in the sector with positive momenta then there is a finite number of states for any finite value of resolution. Of course the full quantum field theory in the continuum corresponds to the limit $L \to \infty$, in this limit the elementary bit of momentum goes to zero, as the harmonic resolution goes to infinity and the infinite number of degrees of freedom is restored. However it is believed that the "quantum mechanical" approximation is suitable for describing the lowest states in the spectrum. Note that the problem of constraints in DLCQ is a quantum mechanical one and thus it is easier to solve. Usually this problem can be reformulated in terms of zero modes and the solution can be found for any value of the resolution.

DLCQ is mainly used in order to solve the bound state problem. Let us formulate this problem for general two dimensional theory. The theory in the continuum has the full Poincare symmetry, thus the states are naturally labeled by the eigenvalues of Casimir operators of the Poincare algebra. One such Casimir is the mass operator: $M^2 = P^\mu P_\mu$. Another Casimir is related to the spin of the particle and we will not use it. After compactifying the x^- direction one looses Lorenz symmetry, but not the translational invariance in x^+ and x^- directions. Thus P^+ and P^- are still conserved charges, but now the mass operator is not the only Casimir of the symmetry group: the states are characterized by both P^+ and P^-. However if we consider DLCQ as an approximation to the continuum theory we anticipate that in the limit of infinite harmonic resolution (or $L \to \infty$) the Poincare symmetry is restored and the mass will be the only quantity having invariant meaning. Thus the aim would be to study the value of M^2 as function of K and to extrapolate the results to the $K = \infty$.

The usual way to define M^2 in DLCQ is based on separate calculation of P^+ and P^- in matrix form and then bringing them together:

$$M^2 = 2P^+ P^-. \qquad (1.3)$$

Usually one works in the sector with fixed P^+, but calculation of light cone Hamiltonian P^- is a nontrivial problem. An important simplifications occur for supersymmetric theories [63].

Supersymmetry is the only nontrivial extension of Poincare algebra compatible with the existence of S matrix [79]. Namely in addition to usual bosonic generators of symmetries, fermionic ones are allowed and the full (super)algebra in two dimensions reads:

$$\{Q_\alpha^I, Q_\beta^J\} = 2\delta^{IJ}\gamma_{\alpha\beta}^\mu P_\mu + \varepsilon_{\alpha\beta} Z^{IJ}, \qquad (1.4)$$

$$[P_\mu, P_\nu] = 0, \qquad [P_\mu, Q_\alpha^I] = 0. \qquad (1.5)$$

In this expression ε is antisymmetric 2×2 matrix, $\varepsilon_{12} = 1$ and Z^{IJ} is the set of c–numbers called central charges. In these lectures we will put them equal to zero. It is convenient to choose two dimensional gamma matrices in the form: $\gamma^0 = \sigma^2$, $\gamma^1 = i\sigma^1$, then one can rewrite (1.4) in terms of light cone components:

$$\{Q_I^+, Q_J^+\} = 2\sqrt{2}\delta^{IJ}P^+, \tag{1.6}$$

$$\{Q_I^-, Q_J^-\} = 2\sqrt{2}\delta^{IJ}P^-, \tag{1.7}$$

$$\{Q_I^+, Q_J^-\} = 2Z_{IJ}. \tag{1.8}$$

As we mentioned before, in DLCQ diagonalization of P^+ is trivial and the construction of Hamiltonian is the main problem. The last set of equations suggests an alternative way of dealing with this problem: one can first construct the matrix representation for the supercharge Q^- and then just square it. This version of DLCQ first suggested in [63] appeared to be very fruitful. First of all it preserves supersymmetry at finite resolution, while the conventional DLCQ applied to supersymmetric theories doesn't (we will consider the relation between these two approaches in section III). The supersymmetric version of DLCQ (SDLCQ) also provides the better numerical convergence.

To summarize, in this subsection we defined two procedures for studying the bound state spectrum: DLCQ and SDLCQ. To implement the first one we should construct the light cone Hamiltonian and diagonalize it, while the second approach requires the construction of supercharge. Of course the SDLCQ method is appropriate only for the theories with supersymmetries, although it can be modified to study models with soft supersymmetry breaking (see section VI).

B   Reduction from Three Dimensions.

Let us start by the defining a simple supersymmetric system in two dimensions. It can be constructed by dimensional reduction of SYM from three dimensions to two dimension. The more general case can be found in the next subsection.

Our starting point is the action for SYM in $2 + 1$ dimensions:

$$S = \int d^3x \operatorname{tr}\left(-\frac{1}{4}F_{AB}F^{AB} + i\bar{\Psi}\gamma^A D_A\Psi\right). \tag{1.9}$$

The system consists of gauge field A_A and two–component Majorana fermion Ψ, both transforming according to adjoint representation of gauge group. We assume that this group is either $U(N)$ or $SU(N)$ and thus matrices A_{ij}^A and Ψ_{ij} are hermitian. Studying dimensional reduction of SYM_D we introduce the following conventions for the indices: the capital latin letters correspond to D dimensional spacetime, greek indices label two dimensional coordinates and the lower case letters are used as matrix indices. According to this conventions the indexes in (1.9) go from zero to two, the field strength F_{AB} and covariant derivative D_A are defined in the usual way:

$$F_{AB} = \partial_A A_B - \partial_B A_A + ig[A_A, A_B],$$
$$D_A\Psi = \partial_A\Psi + ig[A_A, \Psi]. \tag{1.10}$$

Dimensional reduction to $1+1$ means that we require all fields to be independent on coordinate x^2, in other words we place the system on the cylinder with radius

$L_\perp$ along the x^2 axis and consider only zero modes of the fields. The possible improvement of this approximation will be suggested in section VI, here we consider this reduction as a formal way of getting two dimensional matrix model. In the reduced theory it is convenient to introduce two dimensional indices and treat A^2 component of gauge field as two dimensional scalar ϕ. The action for the reduced theory has the form:

$$S = \int d^2 x \, \mathrm{tr} \left(-\frac{1}{4} F_{\mu\nu} F^{\mu\nu} + \frac{1}{2} D_\mu \phi D^\mu \phi + i \bar\Psi \gamma^\mu D_\mu \Psi - 2ig\phi \bar\Psi \gamma_5 \Psi \right), \qquad (1.11)$$

We also could choose the special representation of three dimensional gamma matrices:

$$\gamma^0 = \sigma^2, \qquad \gamma^1 = i\sigma^1, \qquad \gamma^2 = i\sigma^3, \qquad (1.12)$$

then it would be natural to write the spinor Ψ in terms of its components:

$$\Psi = (\psi, \chi)^T. \qquad (1.13)$$

Taking all these definitions into account one can rewrite the dimensional reduction of (1.9) as:

$$S = L_\perp \int d^2 x \left(\frac{1}{2} D_\mu \phi D^\mu \phi + i\sqrt{2} \psi D_+ \psi + i\sqrt{2} \chi D_- \chi + \right.$$
$$\left. + 2g\psi\{\psi, \chi\} - \frac{1}{4} F_{\mu\nu} F^{\mu\nu} \right). \qquad (1.14)$$

The covariant derivatives here are taken with respect to the light cone coordinates:

$$x^\pm = \frac{x^0 \pm x^1}{\sqrt{2}}. \qquad (1.15)$$

Note that by rescaling the fields and coupling constant g we can make the constant $L_\perp$ to be equal to one, so below we simply omit this constant.

The bound state problem for the system (1.14) was first studied in [63]. The supersymmetric version of the discrete light cone quantization was used in order to find the mass spectrum. However the zero modes were neglected by authors of [63], so we spend some time studying this problem in the next section. As we will see, while zero modes are not very important for calculations of massive spectrum, they play crucial role in the description of the vacuum of the theory.

Let us consider (1.14) as the theory in the continuum. In this case one can choose the light cone gauge:

$$A^+ = 0, \qquad (1.16)$$

then equations of motion for A^- and χ give constraints:

145

$$-\partial_-^2 A^- = g J^+, \tag{1.17}$$

$$\sqrt{2} i \partial_- \chi = g[\phi, \psi], \tag{1.18}$$

$$J^+(x) = \frac{1}{i}[\phi(x), \partial_- \phi(x)] - \frac{1}{\sqrt{2}}\{\psi(x), \psi(x)\}. \tag{1.19}$$

Solving this constraints and substituting the result back into the action one determines the Lagrangian as function of physical fields ϕ and ψ only. Then using the usual Noether technique we can construct the conserved charges corresponding to the translational invariance:

$$P^+ = \int dx^- \mathrm{tr}\left((\partial_- \phi)^2 + i\sqrt{2}\psi\partial_- \psi\right), \tag{1.20}$$

$$P^- = \int dx^- \mathrm{tr}\left(-\frac{g^2}{2}J^+ \frac{1}{\partial_-^2}J^+ + \frac{ig^2}{2\sqrt{2}}[\phi, \psi]\frac{1}{\partial_-}[\phi, \psi]\right). \tag{1.21}$$

We can also construct the Noether charges corresponding to the supersymmetry transformation. However the naive SUSY transformations break the gauge fixing condition $A^+ = 0$, so they should be accompanied by compensating gauge transformation:

$$\delta A_\mu = \frac{i}{2}\bar{\varepsilon}\gamma_\mu \Psi - D_\mu \frac{i}{2}\bar{\varepsilon}\gamma_- \frac{1}{\partial_-}\Psi, \tag{1.22}$$

$$\delta\Psi = \frac{1}{4}F_{\mu\nu}\gamma^{\mu\nu}\varepsilon - \frac{g}{2}[\bar{\varepsilon}\gamma_- \frac{1}{\partial_-}\Psi, \Psi].$$

The resulting supercharges are:

$$Q^+ = 2\int dx^- \mathrm{tr}\left(\psi\partial_- \phi\right), \tag{1.23}$$

$$Q^- = -2g \int dx^- \mathrm{tr}\left(J^+ \frac{1}{\partial_-}\psi\right). \tag{1.24}$$

Finally we make a short comment on supersymmetry in the pure fermionic system. As one can see the expression for Q^- contains the term cubic in fermions, so if we formally put $\phi = 0$ this supercharge will not vanish. One may ask what kind of supersymmetric system this supercharge corresponds to. This answer was found by Kutasov [56] who discovered the supersymmetry in the system of adjoint fermions, namely the square of supercharge including fermions only gives Hamiltonian:

$$P^- = \int dx^- \mathrm{tr}\left(-\frac{im^2}{2}\psi\frac{1}{\partial_-}\psi - \frac{g^2}{2}J^+\frac{1}{\partial_-^2}J^+\right), \tag{1.25}$$

$m^2 = g^2 N/\pi$. This P^- corresponds to the system of adjoint fermions in two dimensions with the special value of mass m. We will consider this system in details in section III.

C   Reduction from Higher Dimensions.

In this subsection we consider the general reduction of SYM$_D$ to two dimensions. By counting the fermionic and bosonic degrees of freedom one can see that the SYM can be defined only in limited number of spacetime dimensions, namely D can be equal to 2, 3, 4, 6 or 10. The last case is the most general one: all other system can be obtained by dimensional reduction and appropriate truncation of degrees of freedom. So in this subsection we will concentrate on reduction $10 \to 2$, and the comments on four and six dimensional cases will be made in the end.

As in the last subsection we start from ten dimensional action:

$$S = \int d^3x \mathrm{tr} \left(-\frac{1}{4} F_{AB} F^{AB} + i \bar{\Psi} \gamma^A D_A \Psi \right). \tag{1.26}$$

According to our general conventions the indexes in (1.26) go from zero to nine, Ψ is the ten dimensional Majorana–Weyl spinor. A general spinor in ten dimensions has $2^{10/2} = 32$ complex components, if the appropriate basis of gamma matrices is chosen then Majorana condition makes all the components real. Since all the matrices in such representation are real, the Weyl condition

$$\Gamma_{11} \Psi = \Psi \tag{1.27}$$

is compatible with the reality of Ψ and thus it eliminates half of its components. In the special representation of Dirac matrices:

$$\Gamma^0 = \sigma_2 \otimes \mathbf{1}_{16}, \tag{1.28}$$
$$\Gamma^I = i\sigma_1 \otimes \gamma^I, \quad I = 1,\ldots,8; \tag{1.29}$$
$$\Gamma^9 = i\sigma_1 \otimes \gamma^9, \tag{1.30}$$

the $\Gamma_{11} = \Gamma^0 \cdots \Gamma^9$ has very simple form: $\Gamma_{11} = \sigma_3 \otimes \mathbf{1}_{16}$. Then the Majorana spinor of positive chirality can be written in terms of 16–component real object ψ:

$$\Psi = 2^{1/2} \begin{pmatrix} \psi \\ 0 \end{pmatrix}. \tag{1.31}$$

Let us return to the expressions for Γ matrices. The ten dimensional Dirac algebra

$$\{\Gamma_\mu, \Gamma_\nu\} = 2g_{\mu\nu}$$

is equivalent to the spin(8) algebra for γ matrices: $\{\gamma_I, \gamma_J\} = 2\delta_{IJ}$ and the ninth matrix can be chosen to be $\gamma^9 = \gamma^1 \ldots \gamma^8$. Note that the 16 dimensional representation of spin(8) is the reducible one: it can be decomposed as $\mathbf{8}_s + \mathbf{8}_c$

$$\gamma^I = \begin{pmatrix} 0 & \beta_I \\ \beta_I^T & 0 \end{pmatrix}, \quad I = 1,\ldots,8. \tag{1.32}$$

The explicit expressions for the β_I satisfying $\{\beta_I, \beta_J\} = 2\delta_{IJ}$ can be found in [35]. Such choice leads to the convenient form of γ^9:

$$\gamma^9 = \begin{pmatrix} \mathbf{1}_8 & 0 \\ 0 & -\mathbf{1}_8 \end{pmatrix}. \tag{1.33}$$

So far we have found nonzero components of the spinor given by (1.31). However as we saw in the last subsection not all such components are physical in the light cone gauge, so it is useful to perform the analog of decomposition (1.13). In ten dimension it is related with breaking the sixteen component spinor ψ on the left and right–moving components using the projection operators

$$P_L = \frac{1}{2}(1 - \gamma^9), \qquad P_R = \frac{1}{2}(1 + \gamma^9). \tag{1.34}$$

After introducing the light–cone coordinates $x^{\pm} = \frac{1}{\sqrt{2}}(x^0 \pm x^9)$ the action (1.26) can be rewritten as

$$\begin{aligned} S_{9+1}^{LC} = \int dx^+ dx^- d\mathbf{x}^{\perp}\, \mathrm{tr} \Big(&\frac{1}{2}F_{+-}^2 + F_{+I}F_{-I} - \frac{1}{4}F_{IJ}^2 \\ &+ i\sqrt{2}\psi_R^T D_+ \psi_R + i\sqrt{2}\psi_L^T D_- \psi_L + 2i\psi_L^T \gamma^I D_I \psi_R \Big), \end{aligned} \tag{1.35}$$

where the repeated indices I, J are summed over $(1, \ldots, 8)$. After applying the light–cone gauge $A^+ = 0$ one can eliminate nonphysical degrees of freedom using the Euler–Lagrange equations for ψ_L and A^-:

$$\partial_- \psi_L = -\frac{1}{\sqrt{2}}\gamma^I D_I \psi_R, \tag{1.36}$$

$$\partial_-^2 A_+ = \partial_- \partial_I A_I + g J^+ \tag{1.37}$$

$$J^+ = i[A_I, \partial_- A_I] + 2\sqrt{2}\psi_R^T \psi_R. \tag{1.38}$$

Performing the reduction to two dimensions means that all fields are assumed to be independent on the transverse coordinates: $\partial_I \Phi = 0$. Then as in previous subsection one can construct the conserved momenta $P^{\pm}$ in terms of physical degrees of freedom:

$$P^+ = \int dx^- \mathrm{tr} \left((\partial_- A_I)^2 + i\sqrt{2}\psi_R \partial_- \psi_R \right), \tag{1.39}$$

$$P^- = \int dx^- \mathrm{tr} \left(-\frac{g^2}{2}J^+ \frac{1}{\partial_-^2}J^+ + \frac{ig^2}{2\sqrt{2}}[A_I, \psi_R^T]\beta_I^T \frac{1}{\partial_-}\beta_J[A_J, \psi_R] \right) - \\ - \frac{1}{4}\int dx^- \mathrm{tr} \left([A_I A_J]^2 \right). \tag{1.40}$$

We can also construct the Noether charges corresponding to the supersymmetry transformation (1.22). As in the three dimensional case it is convenient to decompose the supercharge in two components:

$$Q^+ = P_L Q, \qquad Q^- = P_R Q.$$

The resulting eight component supercharges are given by

$$Q^+ = 2 \int dx^- \text{tr} \left(\beta_I^T \psi_R \partial_- A_I \right), \tag{1.41}$$

$$Q^- = -2g \int dx^- \text{tr} \left(J^+ \frac{1}{\partial_-} \psi_R + \frac{i}{4} [A_I A_J](\beta_I \beta_J^T - \beta_J \beta_I^T) \psi_R \right). \tag{1.42}$$

Finally we make a short comment on dimensional reduction of SYM_{3+1} and SYM_{5+1}. These systems can be constructed repeating the procedure just described. However there is an easier way to construct the Hamiltonian and supercharges for the dimensionally reduced theories, namely one has to truncate the unwanted degrees of freedom in the ten dimensional expressions. This is especially easy for the bosonic coordinates: one simply considers indices I and J running from one to two (for $D = 4$) or to four (for $D = 6$). The fermionic truncation can also be performed by requiring the spinor ψ_R to be 2– or 4–component. Then the only problem is the choice of 2×2 or 4×4 beta matrices satisfying

$$\{\beta_I, \beta_J\} = 2\delta_{IJ}, \tag{1.43}$$

that can be done easily.

II   ZERO MODES AND LIGHT CONE VACUUM.

The results of this section are based on the paper [9]

A   Gauge Fixing in DLCQ

We consider the supersymmetric Yang-Mills theory in 1+1 dimensions [32] which is described by the action (1.11):

$$S = \int d^2x \, \text{tr} \left(-\frac{1}{4} F_{\mu\nu} F^{\mu\nu} + \frac{1}{2} D_\mu \phi D^\mu \phi + i\bar{\Psi}\gamma^\mu D_\mu \Psi - 2ig\phi\bar{\Psi}\gamma_5\Psi \right). \tag{2.1}$$

A convenient representation of the gamma matrices is $\gamma^0 = \sigma^2$, $\gamma^1 = i\sigma^1$ and $\gamma^5 = \sigma_3$ where σ^a are the Pauli matrices. In this representation the Majorana spinor is real. We use the matrix notation for $SU(N)$ so that A_{ij}^μ and Ψ_{ij} are $N \times N$ traceless matrices. We now introduce the light-cone coordinates $x^\pm = \frac{1}{\sqrt{2}}(x^0 \pm x^1)$. The longitudinal coordinate x^- is compactified on a finite interval $x^- \in [-L, L]$ [62,67] and we impose periodic boundary conditions on all fields to ensure unbroken supersymmetry.

The light-cone gauge $A^+ = 0$ can not be used in a finite compactification radius, but the modified condition $\partial_- A^+ = 0$ [51] is consistent with the light-like compactification. We can make a global rotation in color space so that the zero mode is diagonalized $A_{ij}^+(x^+) = v_i(x^+)\delta_{ij}$ with $\sum_i v_i = 0$ [51]. The gauge zero modes

149

correspond to a (quantized) color electric flux loops around the compactified space. The modified light-cone gauge is not a complete gauge fixing. We still have large gauge transformations preserving the gauge condition $\partial_- A^+ = 0$. There are two types of such transformations [57,58]: displacements T_D and central conjugations T_C. Their actions on the physical fields of the theory and complete gauge fixing will be discussed in the end of this subsection. Now we just mention that being discrete transformations, T_D and T_C don't affect quantization procedure.

The quantization in the light–cone gauge with or without dynamical A^+ is widely explored in the literature [70,64,63,5,24], here we provide only the results which are useful for later purposes. The quantization proceeds in two steps. First, we must resolve the constraints to eliminate the redundant degrees of freedom. There are two constraints in the theory,

$$-D_-^2 A^- = g J^+, \tag{2.2}$$

$$\sqrt{2} i D_- \chi = g[\phi, \psi], \tag{2.3}$$

where $\Psi \equiv (\psi, \chi)^T$ and the current operator is

$$J^+(x) = \frac{1}{i}[\phi(x), D_-\phi(x)] - \frac{1}{\sqrt{2}}\{\psi(x), \psi(x)\}. \tag{2.4}$$

Different components of (2.2), (2.3) play different roles in the theory. First we look at diagonal zero modes of these equations. The diagonal zero mode of (2.3) gives us constraints on the physical fields:

$$[\phi, \psi]_{ii}^0 = 0. \tag{2.5}$$

There is no sum over i in above expression. As one can see this constraint leads to decoupling of $\overset{0}{\chi}_{ii}$, this field plays the role of Lagrange multiplier for above condition. The same is true for $\overset{0}{A}_{ii}^-$, the corresponding constraint is $\overset{0}{J}_{ii} = 0$. The reason we treated the diagonal zero modes of (2.2) and (2.3) separately is that for all other modes the D_- operator is invertible and instead of constraints on physical fields ψ and ϕ one gets expressions for non-dynamical ones:

$$A^- = -\frac{g}{D_-^2} J^+, \qquad \chi = \frac{g}{i\sqrt{2}} \frac{1}{D_-}[\phi, \psi]. \tag{2.6}$$

The next step is to derive the commutation relations for the physical degrees of freedom. As in the ordinary quantum mechanics, the zero mode v_i has a conjugate momentum $p = 2L\partial_+ v_i$ and the commutation relation is $[v_i, p_j] = i\delta_{ij}$. The off–diagonal components of the scalar field are complex valued operators with $\phi_{ij} = (\phi_{ji})^\dagger$. The canonical momentum conjugate to ϕ_{ij} is $\pi_{ij} = (D_-\phi)_{ji}$. They satisfy the canonical commutation relations [70,63]

$$[\phi_{ij}(x), \pi_{kl}(y)]_{x^+=y^+} = [\phi_{ij}(x), D_-\phi_{lk}(y)]_{x^+=y^+} = \frac{i}{2}(\delta_{ik}\delta_{jl} - \frac{1}{N}\delta_{ij}\delta_{kl})\delta(x^- - y^-).$$

$$(2.7)$$

On the other hand, the quantization of the diagonal component ϕ_{ii} needs care. As mentioned in [70], the zero mode of ϕ_{ii}, the mode independent of x^-, is not an independent degree of freedom but obeys a certain constrained equation [62,70,50]. Except the zero mode, the commutation relation is canonical

$$[\phi_{ii}(x), \partial_-\phi_{jj}(y)]_{x^+=y^+} = \frac{i}{2}(1 - \frac{1}{N})\delta_{ij}\left[\delta(x^- - y^-) - \frac{1}{2L}\right].$$

$$(2.8)$$

The commutator of diagonal and non-diagonal elements of ϕ vanishes. The canonical anti-commutation relations for fermion fields are [63]

$$\{\psi_{ij}(x), \psi_{kl}(y)\}_{x^+=y^+} = \frac{1}{\sqrt{2}}\delta(x^- - y^-)(\delta_{il}\delta_{jk} - \frac{1}{N}\delta_{ij}\delta_{kl}).$$

$$(2.9)$$

There are two differences between this expression and one from [63]. First one is technical: we consider commutators for $SU(N)$ group, this gives $1/N$ term. Second difference is that unlike [63] we include zero modes in the expansion of ψ, we also include such modes in non-diagonal elements of ϕ.

Finally we return to the problem of complete gauge fixing. The actions of T_D and T_C on physical fields are given by [57,58]:

$$T_D : v_i(x^+) \to v_i(x^+) + \frac{n_i\pi}{gL}, \qquad n_i \in \mathbb{Z}, \qquad \sum n_i = 0, \qquad (2.10)$$

$$\psi_{ij} \to \exp(\frac{\pi i(n_i - n_j)x^-}{L})\psi_{ij}, \qquad \phi_{ij} \to \exp(\frac{\pi i(n_i - n_j)x^-}{L})\phi_{ij};$$

$$T_C : v_i(x^+) \to v_i(x^+) + \frac{\nu_i\pi}{gL}, \qquad \nu_i = n(1/N - \delta_{iN}), \qquad (2.11)$$

$$\psi_{ij} \to \exp(\frac{\pi i(\nu_i - \nu_j)x^-}{L})\psi_{ij}, \qquad \phi_{ij} \to \exp(\frac{\pi i(\nu_i - \nu_j)x^-}{L})\phi_{ij}.$$

There are also permutations of the color basis $i \to P(i)$ which leave the theory invariant. These symmetries preserve the gauge condition $\partial_-A^+ = 0$, but two configurations related by T_D, T_C or P are equivalent. To fix the gauge completely one therefore considers v_i only in the fundamental domain, other regions related with this domain by T_D, T_C or P give gauge "copies" of it [36]. The easiest thing to do is to describe the boundaries of fundamental domain imposed by displacements T_D: $-\frac{\pi}{2gL} < v_i < \frac{\pi}{2gL}$. The invariance under T_C limits this region even more, but since we will not need the explicit form of fundamental domain, we do not discuss such limits for $SU(N)$. For the simplest case of $SU(2)$ the fundamental domain is given by $0 < v_1 = -v_2 < \frac{\pi}{2gL}$, the result for $SU(3)$ can be found in [58]. The P symmetries do not respect the fundamental domain, so they are not symmetries of

151

gauge fixed theory. However there is one special transformation among P which being accompanied with combination of T_D and T_C leaves fundamental domain invariant. Namely if R is cyclic permutation of color indexes then there exists a combination T of T_D and T_C such that $S = TR$ is the symmetry of gauge fixed theory. The explicit form of T depends on the rank of the group, for $SU(2)$ and $SU(3)$ it may be found in [58]. The operator S satisfies the condition $S^N = 1$ and it was used in classifying the vacua [58,69].

B   Current Operators

The resolution of the Gauss-law constraint (2.2) is a necessary step for obtaining the light-cone Hamiltonian. The expression for the current operator is, however, ill–defined unless an appropriate definition is specified, since the operator products are defined at the same point. We shall use the point–splitting regularization which respects the symmetry of the theory under the large gauge transformation.

To simplify notation it is convenient to introduce the dimensionless variables $z_i = Lgv_i/\pi$ instead of quantum mechanical coordinates v_i describing A^+. The mode–expanded fields at the light-cone time $x^+ = 0$ are

$$\phi_{ij}(x) = \frac{1}{\sqrt{4\pi}} \left(\sum_{n=0}^{\infty} a_{ij}(n) u_{ij}(n) e^{-ik_n x^-} + \sum_{n=1}^{\infty} a_{ji}^{\dagger}(n) u_{ij}(-n) e^{ik_n x^-} \right), \quad i \neq j,$$

$$\phi_{ii}(x) = \frac{1}{\sqrt{4\pi}} \sum_{n=1}^{\infty} \frac{1}{\sqrt{n}} \left(a_{ii}(n) e^{-ik_n x^-} + a_{ii}^{\dagger}(n) e^{ik_n x^-} \right),$$

$$\psi_{ij}(x) = \frac{1}{2^{\frac{1}{4}} \sqrt{2L}} \left(\sum_{n=0}^{\infty} b_{ij}(n) e^{-ik_n x^-} + \sum_{n=1}^{\infty} b_{ji}^{\dagger}(n) e^{ik_n x^-} \right), \quad (2.12)$$

where $k_n = n\pi/L$, $u_{ij}(n) = 1/\sqrt{|n - z_i + z_j|}$ [1]. The (anti)commutation relations for Fourier modes are found in [70,64] and in our notation they take the form

$$[a_{ij}(n), a_{kl}^{\dagger}(m)] = \operatorname{sgn}(n + z_j - z_i) \delta_{n,m} (\delta_{ik} \delta_{jl} - \frac{1}{N} \delta_{ij} \delta_{kl}),$$

$$\{b_{ij}(n), b_{kl}^{\dagger}(m)\} = \delta_{n,m} (\delta_{ik} \delta_{jl} - \frac{1}{N} \delta_{ij} \delta_{kl}) \quad (2.13)$$

The zero modes in above relations deserve special consideration. Although we formally wrote them as $a_{ij}(0)$ and $b_{ij}(0)$, these modes also act as creation operators because the conjugation of zero mode gives another zero mode:

$$a_{ij}^{\dagger}(0) = a_{ji}(0), \qquad b_{ij}^{\dagger}(0) = b_{ji}(0). \quad (2.14)$$

---

[1] $u_{ij}(n)$ is well-defined in the fundamental domain. Similarly, $(D_-)^2$ in the Gauss-law constraint have no zero modes in this domain.

In particular the diagonal components of fermionic zero mode are real and we will use them later to describe the degeneracy of vacua. Now we concentrate our attention on non-diagonal zero modes. In the fundamental domain all z_i are different, then one can always make take them to satisfy the inequality $z_N < z_{N-1} < \ldots < z_1$ in this domain. Such condition together with (2.13) leads to interpretation of $a_{ij}(0)$ as creation operator if $i < j$ and as annihilation operator otherwise. The situation for fermions is more ambiguous. One can consider $b_{ij}(0)$ as creation operator either when $i < j$ or when $i > j$, both assumptions are consistent with (2.13). Later we will explore each of these situations.

Let us now discuss the definition of singular operator products in the current (2.4). We define the current operator by point splitting:

$$J^+ \equiv \lim_{\epsilon \to 0} \left(J^+{}_\phi(x; \epsilon) + J^+{}_\psi(x; \epsilon) \right), \tag{2.15}$$

where the divided pieces are given by

$$J^+{}_\phi(x; \epsilon) = \frac{1}{i} \left[e^{-i\frac{\pi\epsilon}{2L} M} \phi(x^- - \epsilon) e^{i\frac{\pi\epsilon}{2L} M}, D_- \phi(x^-) \right] \tag{2.16}$$

$$J^+{}_\psi(x; \epsilon) = -\frac{1}{\sqrt{2}} \left\{ e^{-i\frac{\pi\epsilon}{2L} M} \psi(x^- - \epsilon) e^{i\frac{\pi\epsilon}{2L} M}, \psi(x^-) \right\}. \tag{2.17}$$

Here M is diagonal matrix: $M = diag(z_1, \ldots, z_N)$. An advantage of this regularization is that the current transforms covariantly under the large gauge transformation.

To evaluate (2.16) and (2.17) we will generalize the approach used in [70,64] to the SU(N) case. First let us calculate the vacuum average of bosonic current. Taking into account the interpretation of zero modes as creation–annihilation operators we obtain:

$$\langle 0|J_{ij}^+{}_\phi(x; \epsilon)|0\rangle = \frac{1}{i} \langle 0|e^{-i\frac{\pi\epsilon}{2L}(z_i-z_k)} \phi_{ik}(x^- - \epsilon) D_- \phi(x^-)_{kj} -$$

$$-e^{-i\frac{\pi\epsilon}{2L}(z_k-z_j)} \phi_{kj}(x^- - \epsilon) D_- \phi(x^-)_{ik}|0\rangle =$$

$$= \frac{1}{4L} \sum_k \sum_{m>0} \left(e^{-i\frac{\pi\epsilon}{2L}(z_i-z_k)} - e^{-i\frac{\pi\epsilon}{2L}(z_k-z_j)} \right) e^{-ik_m\epsilon} \left(\delta_{ij} - \frac{1}{N} \delta_{ik}\delta_{jk} \right) +$$

$$+\frac{1}{4L} \sum_{k<j} e^{-i\frac{\pi\epsilon}{2L}(z_i-z_k)} \delta_{ij} - \frac{1}{4L} \sum_{k>i} e^{-i\frac{\pi\epsilon}{2L}(z_k-z_j)} \delta_{ij}. \tag{2.18}$$

Evaluating the sum and taking the limit one finds:

$$\lim_{\epsilon \to 0} J_{ij}^+{}_\phi(x; \epsilon) =: J_{ij}^+{}_\phi(x) : +\frac{1}{4L} \left(z_i - (N+1-2i) \right) \delta_{ij}, \tag{2.19}$$

where $: J^+{}_\phi :$ is the naive normal ordered currents. To be more precise, we have omitted the zero modes of the diagonal color sectors in which the notorious constrained zero mode [62] appears.

The result for fermionic current depends on our interpretation of zero modes as creation–annihilation operators and it is given by

$$\lim_{\epsilon \to 0} J^+_{ij\,\psi}(x; \epsilon) =: J^+_{ij\,\psi}(x) : -\frac{1}{4L} \left(z_i \mp (N + 1 - 2i)\right) \delta_{ij}. \tag{2.20}$$

The minus sign here corresponds to the case where $b_{ij}(0)$ is a creation operator if $i < j$ (i.e. the convention is the same as for the bosons) and plus corresponds to the opposite situation. As can be seen, $J^+{}_\phi$ and $J^+{}_\psi$ acquire extra z dependent terms, so called gauge corrections. Integrating these charges over x^-, one finds that the charges are time dependent. Of course this is an unacceptable situation, and implies the need to impose special conditions to single out 'physical states' to form a sensible theory. The important simplification of the supersymmetric model is that these time dependent terms cancel, and the full current (2.15) becomes

$$J^+_{ij}(x) =: J^+_{ij\,\phi} : + : J^+_{ij\,\psi} : + C_i \delta_{ij}. \tag{2.21}$$

Depending on the convention for fermionic zero modes the z *independent* constants C_i either vanish or they are given by

$$C_i = -\frac{1}{2L}(N + 1 - 2i). \tag{2.22}$$

The regularized current is thus equivalent to the naive normal ordered current up to an irrelevant constant. Similarly, one can show that P^+ picks up gauge correction when the adjoint scalar or adjoint fermion are considered separately but in the supersymmetric theory it is nothing more than the expected normal ordered contribution of the matter fields.

In one sense these results are a consequence of the well known fact that the normal ordering constants in a supersymmetric theory cancel between fermion and boson contributions. The important point here is that these normal ordered constants are not actually constants, but rather quantum mechanical degrees of freedom. It is therefore not obvious that they should cancel. Of course, this property profoundly effects the dynamics of the theory.

C   Vacuum Energy

The wave function of the vacuum state for the supersymmetric Yang-Mills theory in 1+1 dimensions has already been discussed in the equal-time formulation [66]. An effective potential is computed in a weak coupling region as a function of the gauge zero mode by using the adiabatic approximation. Here we analyze the vacuum structure of the same theory in the context of the DLCQ formulation.

The presence of zero modes renders the light-cone vacuum quite nontrivial, but the advantage of the light-cone quantization becomes evident: the ground state is the Fock vacuum for a fixed gauge zero mode and therefore our ground state may be written in the tensor product form

$$|\Omega\rangle \equiv \Phi[z] \otimes |0\rangle, \qquad (2.23)$$

where we have taken the Schrödinger representation for the quantum mechanical degree of freedom z which is defined in the fundamental domain. In contrast, to find the ground state of the fermion and boson for a fixed value of the gauge zero mode turns out to be a highly nontrivial task in the equal-time formulation [66].

Our next task is to derive an effective Hamiltonian acting on $\Phi[z]$. The light-cone Hamiltonian $H \equiv P^-$ is obtained from energy momentum tensors, or through the canonical procedure:

$$H = -\frac{g^2 L}{4\pi^2} \frac{1}{K(z)} \sum_i \frac{\partial}{\partial z_i} K(z) \frac{\partial}{\partial z_i} +$$

$$+ \int_{-L}^{L} dx^- \mathrm{tr} \left(-\frac{g^2}{2} J^+ \frac{1}{D_-^2} J^+ + \frac{ig^2}{2\sqrt{2}} [\phi, \psi] \frac{1}{D_-} [\phi, \psi] \right), \qquad (2.24)$$

$$K(z) = \prod_{i>j} \sin^2\left(\frac{\pi(z_i - z_j)}{2}\right), \qquad (2.25)$$

where the first term is the kinetic energy of the gauge zero mode, and in the second term the zero modes of D_- are understood to be removed. Note that the kinetic term of the gauge zero mode is not the standard form $-d^2/dz^2$ but acquires a nontrivial Jacobian K which is nothing but the Haar measure of SU(N). The Jacobian originates from the unitary transformation of the variable from A^+ to v, and can be derived by explicit evaluation of a functional determinant [57,58]. In the present context it is found in [50]. Also we mention that Hamiltonian (2.24) seems to contain terms quadratic in diagonal zero modes $\overset{0}{\psi}_{ii}$. However using constraint equations one can show that the total contribution of all such term vanishes. This also can be seen by using the fact that Hamiltonian is proportional to the square of supercharge (2.34).

Projecting the light-cone Hamiltonian onto the Fock vacuum sector we obtain the quantum mechanical Hamiltonian

$$H_0 = -\frac{g^2 L}{4\pi^2} \frac{1}{K(z)} \sum_i \frac{\partial}{\partial z_i} K(z) \frac{\partial}{\partial z_i} + V_{JJ} + V_{\phi\psi}, \qquad (2.26)$$

where the reduced potentials are defined by

$$V_{JJ} \equiv -\frac{g^2}{2} \int_{-L}^{L} dx^- \langle \mathrm{tr} J^+ \frac{1}{D_-^2} J^+ \rangle, \qquad (2.27)$$

$$V_{\phi\psi} \equiv \frac{ig^2}{2\sqrt{2}} \int_{-L}^{L} dx^- \langle \mathrm{tr}[\phi, \psi] \frac{1}{D_-} [\phi, \psi] \rangle, \qquad (2.28)$$

respectively. As stated in the previous subsection, the gauge invariantly regularized current turns out to be precisely the normal ordered current in the absence of the zero modes. It is now straightforward to evaluate V_{JJ} and $V_{\phi\psi}$ in terms of modes. One finds that they cancel among themselves as expected from the supersymmetry:

$$
V_{JJ} = -V_{\phi\psi} = \frac{g^2 L}{16\pi^2} \left[\sum_{n,m=1}^{\infty} \sum_{ijk} \frac{1}{(n - z_i + z_k)(m + z_j - z_k)} - \sum_{n,m=1}^{\infty} \frac{N}{mn} + \right.
$$
$$
+ \sum_{n=1}^{\infty} \sum_{ij} \left(\sum_{k>j} \frac{1}{(n - z_i + z_k)(z_j - z_k)} + \sum_{k<i} \frac{1}{(n + z_j - z_k)(z_k - z_i)} \right) +
$$
$$
\left. + \sum_{ij} \sum_{i>k>j} \frac{1}{(z_k - z_i)(z_j - z_k)} \right]. \tag{2.29}
$$

This cancellation was found as the result of formal manipulations with divergent series like ones in the right hand side of the last formula. Such transformations are not well defined mathematically and as the result they may lead to the finite "anomalous" contribution. The famous chiral anomaly initially was found as the result of careful analysis of transformations analogous to ones we just performed [1]. However if one considers derivatives of V_{JJ} or $V_{\phi\psi}$ with respect to any z_i then all the sums become convergent, the order of summations becomes interchangeable and as the result the derivatives of $V_{JJ} + V_{\phi\psi}$ vanish. Thus if there is any anomaly in the expression above it is given by z–independent constant. Such constant in the Hamiltonian would correspond to the shift of energy levels and usually it is ignored. However in supersymmetric case there is a natural choice for such constant: in order for vacuum to be supersymmetric it should be zero. Below we assume that SUSY is not broken, then we expect that (2.29) is true.

Thus we arrive at

$$
H_0 = -\frac{g^2 L}{4\pi^2} \frac{1}{K(z)} \sum_i \frac{\partial}{\partial z_i} K(z) \frac{\partial}{\partial z_i}. \tag{2.30}
$$

The relevant solutions of this equation should be finite in the fundamental domain, this requirement leads to discrete spectrum due to the fact that Jacobian vanishes on the boundary of this domain. However the operator H_0 is elliptic, and therefore it can't have negative eigenvalues. If the eigenvalue problem

$$
H_0 \Phi(z) = E \Phi(z) \tag{2.31}
$$

has a solution for $E = 0$, this solution corresponds to the ground state of the theory. It is easy to see that such solution exists and it is given by $\Phi(z) = const$ [2]. We have thus found that the ground state has a vanishing vacuum energy, suggesting that the supersymmetry is not broken spontaneously.

---

[2] some authors prefer to rewrite this to include the measure in the definition of the wave function and then in SU(2) for example the ground state wave function is a sin

D   Supersymmetry and Degenerate Vacua.

As we saw in the previous subsection supersymmetry leads to the cancellation of the anomaly terms in current operator. However these terms played an important role in the description of Z_N degeneracy of vacua [69], so we should find another explanation of this fact here. It appears that fermionic zero modes give a natural framework for such treatment.

First we will generalize the supersymmetry transformation given in [63] to the present case, i.e. we include A^+ and the zero modes of fermions. The naive SUSY transformations spoil the gauge fixing condition, so we combine them with compensating gauge transformation following [63]. In three dimensional notation (spinors have two components and indices go from 0 to 2) the result reads:

$$\delta A_\mu = \frac{i}{2}\bar{\varepsilon}\gamma_\mu\Psi - D_\mu\frac{i}{2}\bar{\varepsilon}\gamma_-\frac{1}{D_-}\tilde{\Psi},$$

$$(2.32)$$

$$\delta\Psi = \frac{1}{4}F_{\mu\nu}\gamma^{\mu\nu}\varepsilon - \frac{g}{2}[\bar{\varepsilon}\gamma_-\frac{1}{D_-}\tilde{\Psi}, \Psi].$$

The difference between above expression and those in [63] is that we include the zero modes. Namely we defined Ψ as the complete field with all the zero modes included and $\tilde{\Psi}$ as fermion without diagonal zero modes. The introducing of $\tilde{\Psi}$ is necessary, because diagonal zero modes form the kernel of operator D_-, so $\frac{1}{D_-}$ is not defined on this subspace.

In particular we are interested in supersymmetry transformations for A^+ and fermionic zero modes. Performing a mode expansion one can check that diagonal elements of matrix $[\frac{1}{D_-}\psi, \psi]^0$ vanish, then from (2.32) we get:

$$\delta A_{ii}^+ = \frac{i}{\sqrt{2}}\varepsilon_+^T \overset{0}{\psi}_{ii},$$

$$\delta \overset{0}{\psi}_{ii} = -2\partial_+ A_{ii}^+\varepsilon_+.$$

$$(2.33)$$

This expression is written in two component notation and the decomposition of spinor ε: $\varepsilon = (\varepsilon_+, \varepsilon_-)^T$ is used. Note that since $\bar{\varepsilon}Q = \sqrt{2}(\varepsilon_+Q^- + \varepsilon_-Q^+)$ the fields involved in transformations (2.33) don't contribute to Q^+, this is consistent to the fact that being x^- independent they don't contribute to P^+. The equations (2.33) look like supersymmetry transformation for the quantum mechanical system built from free bosons and free fermions. In fact as one can see the supercharge Q^- is the sum of supercharge for the quantum mechanical system and from the QFT without diagonal zero modes:

$$Q^- = -2g\int dx^-\text{tr}(J^+\frac{1}{D_-}\psi) + 4L\text{tr}(\partial_+ A^+ \overset{0}{\psi}).$$

$$(2.34)$$

Calculating $(Q^-)^2$ and writing the momentum conjugate to A^+ as differential operator [3] we reproduce Hamiltonian (2.24). Note that ψ there has all the zero modes in it. The square of another supercharge

$$Q^+ = 2 \int dx^- \mathrm{tr}(\psi D_- \phi) \tag{2.35}$$

gives P^+ while the anti-commutator of Q^- with Q^+ is proportional to the constraint (2.5) and thus vanishes.

One can check that although $[\overset{0}{\psi}_{ii}, H]$ does not vanish, this commutator annihilates Fock vacuum $|0\rangle$, then it also annihilates $|\Omega\rangle$. In subsection 1 we mentioned that $\overset{0}{\chi}_{ii}$ decouples from the theory, and therefore it commutes with Hamiltonian. Thus acting on the vacuum state $|\Omega\rangle$ by diagonal elements of either $\overset{0}{\psi}$ or $\overset{0}{\chi}$ we get states annihilated by P^- and P^+ (the latter statement is obvious since zero modes commute with momentum). Not all such states however may be considered as vacua. Although we fixed the gauge in subsection 1, the theory still has residual symmetry P, corresponding to permutations of the color basis. Physical states are constructed from operator acting on the physical vacuum $|\Omega\rangle$ and both the operators and the physical vacuum must be invariant under P. Such objects can always be written as combinations of traces. The candidates for the vacuum state may have any combination of $\overset{0}{\psi}$ and $\overset{0}{\chi}$ inside the trace, here and below we consider only diagonal components of zero modes. Since $\overset{0}{\chi}$ is not dynamical we have the usual c–number relation

$$\{\overset{0}{\chi}_{ii}, \overset{0}{\chi}_{jj}\} = 0 \tag{2.36}$$

instead of canonical anti-commutator, so $\overset{0}{\chi}\overset{0}{\chi} = 0$. From the relations (2.13) one finds:

$$\overset{0}{\psi}\overset{0}{\psi} = \frac{1}{4L\sqrt{2}}\left(1 - \frac{1}{N}\right), \tag{2.37}$$

also we have $\overset{0}{\chi}\overset{0}{\psi} = -\overset{0}{\psi}\overset{0}{\chi}$. Using all these relations and the $SU(N)$ conditions $\mathrm{tr}(\overset{0}{\psi}) = 0$ and $\mathrm{tr}(\overset{0}{\chi}) = 0$ we find that the only nontrivial trace involving only zero modes is $\mathrm{tr}(\overset{0}{\psi}\overset{0}{\chi})$. Then the family of vacua is given by:

$$\left(\mathrm{tr}(\overset{0}{\psi}\overset{0}{\chi})\right)^n |\Omega\rangle, \qquad 0 \le n \le N - 1. \tag{2.38}$$

---

[3] using Schrödinger coordinate representation for quantum mechanical degree of freedom - note that the QFT term has non-trivial dependence on the quantum mechanical coordinate.

The region for n is determined taking into account the fact that $\overset{0}{\chi}$ is anti-commuting field with $N - 1$ independent components. Thus we explained the Z_N degeneracy of vacua first mentioned in [80].

In addition to this discrete vacuum degeneracy supersymmetric theories also have a continuum space of vacua called moduli space. In DLCQ approach the moduli space is easy to understand. Let us suppose that scalar field ϕ developed a VEV. To have a consistent theory this VEV should commute with the Wilson loop in the compact direction, which in our case happened to be $\exp(i \int dx^- A^+)$. Since A^+ is a general diagonal matrix this leads to the condition for the VEV: $\langle \phi_{ij} \rangle = w_i \delta_{ij}$. Now we can make the substitution $\phi \rightarrow \phi + \langle \phi \rangle$ in the supercharges (2.34) to find the correction in Q^- due to the scalar VEV:

$$\delta Q^- = -2ig\text{tr}\left(w \int dx^- [\phi \psi]\right). \tag{2.39}$$

We used integration by part and the equation $D_- w = 0$. Taking into account the fermionic constraint (2.5) we conclude that for any diagonal w: $\delta Q^- = 0$, i.e. we can choose the state with arbitrary VEV $\langle \phi_{ij} \rangle$ as the new vacuum. This is precisely the moduli space of the theory: the models constructed starting from different vacua are not coupled with each other.

E   Solving for Massive Bound States.

As we saw the zero modes play an important role in the description of the vacuum. However solving for massive bound states one usually neglect the zero mode contribution. Does this lead to errors in the mass spectrum? The answer depends on the problem we are solving. If one is interested in the spectrum of the theory at the finite value of resolution then zero modes are important, but as we will show their contribution becomes smaller and smaller as the resolution goes to infinity, so they may be neglected if one is interested only in the large K extrapolation.

First let us formulate the DLCQ problem with zero modes precisely. We will use the Hamiltonian formulation, but the consideration for SDLCQ formalism is the same. The space of states of the theory is the direct product of Fock space and quantum mechanical Hilbert space for zero modes: a general state can be written as

$$|state\rangle = \Phi(z) \otimes |FockState\rangle, \tag{2.40}$$

the Hamiltonian has the form:

$$H = K(z) + V(z, a, a^\dagger, b, b^\dagger). \tag{2.41}$$

Here $K(z)$ is some differential operator, while V is some function of zero modes z and creation–annihilation operators (see for example (2.24)). In general one

159

should solve the bound state problem $H|\Psi\rangle = E|\Psi\rangle$ in two steps: first one should determine the effective potential $\tilde{V}$:

$$V(z, a, a^{\dagger}, b, b^{\dagger})|FockState\rangle = \tilde{V}(z)\||FockState\rangle \qquad (2.42)$$

and then solve the Schrödinger equation for zero modes:

$$(K(z) + \tilde{V}(z))\Psi(z) = E\Psi(z). \qquad (2.43)$$

However in practice this is hard to carry out. Fortunately, solving the Schrödinger equation is not important for calculating the continuum limit of mass spectrum. The reason for this is the following.

Studying the large L limit in DLCQ one is usually interested in situation when the total momentum $P^{+} = \sum n_i/L$ is kept fixed. Then most of the terms in V (and thus in $\tilde{V}$) are of order L^0, while $K(z)$ scales like L. Assume for a moment that the whole $\tilde{V}$ is of order one, then one can consider $\tilde{V}$ as perturbation and use the standard expression for the eigenvalue:

$$E_i = E_i^{(0)} + \int dz \Psi_i^{\dagger}(z)\tilde{V}(z)\Psi_i(z), \qquad (2.44)$$

where $E_i^{(0)}$ and $\Psi_i(z)$ are eigenvalue and eigenfunction of unperturbed system. To get finite masses in the continuum limit only the ground state of $K(z)$ should be considered: $i = 0$ and $E_0^{(0)} = 0$ in the last expression. Introducing the averaging procedure as

$$\langle A \rangle = \int dz \Psi_0^{\dagger}(z) A(z) \Psi_0(z) \qquad (2.45)$$

we find: $E = \langle \tilde{V} \rangle$ and thus the continuum eigenvalues are just solutions of the z–independent equation:

$$\langle V(a, a^{\dagger}, b, b^{\dagger}) \rangle |FockState\rangle = E|FockState\rangle. \qquad (2.46)$$

The assumption of L^0 scaling for $\tilde{V}$ is not the trivial one. Namely it is responsible for the difference in the constraint equations in DLCQ and continuum cases. For example looking at the Hamiltonian (2.24) one can see that $V(z)$ includes a term linear in L:

$$\frac{g^2 L}{2} \frac{1}{(z_i - z_j)^2} \tilde{J}_{ij}^{+}(0)\tilde{J}_{ji}^{+}(0), \qquad (2.47)$$

so the assumption being false for V may be satisfied for $\tilde{V}$ only dynamically. One can make this specific term vanish if instead of DLCQ constraint $\int dx^- J_{ii}(x) = 0$ its continuum version

$$\int dx^- J_{ij}(x) = 0 \qquad (2.48)$$

is used. Of course imposing this condition is not enough to make all the terms in $\tilde{V}$ to be of order L^0, but following the usual path in DLCQ calculations we choose not to impose other conditions explicitly. In our numerical study we rather perform calculations with Hamiltonian $\langle V(a, a^\dagger, b, b^\dagger) \rangle$ in the sector satisfying (2.48) and then concentrate our attention only on states whose masses can be extrapolated to finite value. This way we make sure that our assumption $\tilde{V} \sim 1$ holds and thus the z dependence is not important.

To summarize, we have shown that zero modes of gauge field and diagonal zero modes of fermions play an important role in the description of vacuum structure. However if studying the bound state problem for the states with nonzero total momentum P^+ one is interested only in the extrapolation to the continuum limit, the zero mode of A^+ can be omitted from the theory. This fact leads to significant simplifications in the numerical procedure. As soon as A^+ is excluded from the theory one also has to exclude the bosonic zero modes (otherwise the expression $1/0$ is encountered in the (2.12)). What about the fermionic zero modes? In principle we can either keep them or disregard them. However in the latter case one should be very careful: as we will see in the next section such modes play an important role in the ensuring of supersymmetry.

III   FERMIONIC ZERO MODES AND EXACT SUPERSYMMETRY.

In this section we study the relation between conventional DLCQ and its supersymmetric version. Since usual DLCQ is formulated for the Hamiltonian we should rewrite SDLCQ in the same form. Here one encounters the first difference between two schemes: in DLCQ the fermions can be chosen to be either periodic or antiperiodic on x^-, but in the Hamiltonian formulation of SDLCQ they must be periodic due to supersymmetry. Then one encounters the problem of fermionic zero modes. However the boundary conditions is not the only difference between the two approaches. Even after we choose periodic fermions, DLCQ still has an ambiguity emerging from the choice of regularization scheme. Taking the simplest SUSY system as an example we will show that supersymmetry dictates the unique regularization and we study the relation between this prescription and the principal value scheme, which is usually used in the DLCQ calculations. We show that fermionic zero modes play an important role in deriving this relation.

As we already mentioned in section one the simplest supersymmetric system in two dimension is the one involving only gauge fields and adjoint fermions [56]. We derive all the relations for this particular system.

A   Zero Modes and Supersymmetric Regularization.

We consider the $1 + 1$ dimensional $SU(N)$ gauge theory coupled to an adjoint Majorana fermion. The light-cone quantization of this model in the light-cone gauge and large N limit has been dealt with explicitly before [28,19]. The expressions for the light-cone momentum P^+ and light-cone Hamiltonian P^- for this model are

$$P^+ = \int dx^- \mathrm{tr}(i\sqrt{2}\psi\partial_-\psi), \tag{3.1}$$

$$P^- = \int dx^- \mathrm{tr}\left(-\frac{im^2}{\sqrt{2}}\psi\frac{1}{\partial_-}\psi - \frac{g^2}{2}J^+\frac{1}{\partial_-^2}J^+\right). \tag{3.2}$$

Here $J_{ij}^+ = -\sqrt{2}\psi_{ik}\psi_{kj}$ is the longitudinal current. It is well known that at a special value of fermionic mass (namely $m^2 = g^2 N/\pi$) this system is supersymmetric [56]. This special value of the fermion mass will be denoted by m_{SUSY}. At this supersymmetric point, the supercharge is given by

$$Q^- = \sqrt{2}g \int dx^- \mathrm{tr}(\psi\psi\frac{1}{\partial_-}\psi) \tag{3.3}$$

which satisfies the supersymmetry relation $\{Q^-, Q^-\} = 2\sqrt{2}P^-$. This may be checked explicitly by using the anticommutator at equal x^+:

$$\{\psi_{ij}(x^-), \psi_{kl}(y^-)\} = \frac{1}{2}\delta_{il}\delta_{jk}\delta(x^- - y^-). \tag{3.4}$$

In the DLCQ formulation, the theory is regularized by a light-like compactification, and either periodic or antiperiodic boundary conditions may be imposed for fermions. If P^+ denotes the total light-cone momentum, light-like compactification is equivalent to restricting the light-cone momentum of partons to be non-negative integer multiples of P^+/K, where K is some positive integer that is sent to infinity in the decompactified limit[4]. Anti-periodic boundary conditions will in general explicitly break the supersymmetry in the discretized theory, although supersymmetry will be restored in the decompactification limit $K \to \infty$ [19]. If we wish to maintain supersymmetry at any finite K, we must at least impose periodic boundary conditions for the fermions. This, however, leads to the notorious "zero-mode problem"[5]. ¿From a numerical perspective, omitting zero-momentum modes in our analysis is absolutely necessary, since it guarantees a *finite* Fock basis for each finite resolution K. The mass spectrum of the continuum theory may be obtain by extrapolating from a sequence of finite mass matrices $M^2 = 2P^+P^-$. But are we really justified in omitting the zero-momentum modes? To date, the general consensus is that omitting zero momentum modes in a two dimensional interacting field theory does not affect the spectrum of the decompactified theory, where

---

[4] K is sometimes called the 'harmonic resolution', or just 'resolution'.
[5] For anti-periodic boundary conditions, the light-cone momentum of partons is restricted to *odd* integer multiples of P^+/K, and so there are no zero-momentum modes in such a formulation.

$K \to \infty$. Actually, the numerical results of the next subsection are consistent with this viewpoint.

However, the goal of this work is to understand the structure of a supersymmetric theory at finite resolution. As we will see shortly, understanding why the DLCQ and SDLCQ prescriptions differ involves studying certain intermediate zero-momentum processes. But first, we need to be more precise about the form of the light-cone operators of the theory. If we expand the fermion field ψ_{ij} in terms of its Fourier components, we may express the uncompactified light-cone supercharge and Hamiltonian in a momentum space representation involving fermion creation and annihilation operators: ([56,28,19]):

$$Q^- = \frac{i 2^{-1/4} g}{\sqrt{\pi}} \int_0^\infty dk_1 dk_2 dk_3 \delta(k_1 + k_2 - k_3) \left(\frac{1}{k_1} + \frac{1}{k_2} - \frac{1}{k_3}\right) \times$$
$$\times \left(b_{ik}^\dagger(k_1) b_{kj}^\dagger(k_2) b_{ij}(k_3) + b_{ij}^\dagger(k_3) b_{ik}(k_1) b_{kj}(k_2)\right),$$

(3.5)

$$P^- = \frac{m^2}{2} \int_0^\infty \frac{dk}{k} b_{ij}^\dagger(k) b_{ij}(k) + \frac{g^2 N}{\pi} \int_0^\infty \frac{dk}{k} \int_0^k dp \frac{k}{(p-k)^2} b_{ij}^\dagger(k) b_{ij}(k) +$$
$$+ \frac{g^2}{2\pi} \int_0^\infty dk_1 dk_2 dk_3 dk_4 \left[\delta(k_1 + k_2 - k_3 - k_4) A(k) b_{kj}^\dagger(k_3) b_{ji}^\dagger(k_4) b_{kl}(k_1) b_{li}(k_2) + \right.$$
$$+ \delta(k_1 + k_2 + k_3 - k_4) B(k) \times$$
$$\left. \times \left(b_{kj}^\dagger(k_4) b_{kl}(k_1) b_{li}(k_2) b_{ij}(k_4) - b_{kj}^\dagger(k_1) b_{jl}^\dagger(k_2) b_{li}^\dagger(k_3) b_{ki}(k_4)\right) \right]$$

(3.6)

with

$$A(k) = \frac{1}{(k_4 - k_2)^2} - \frac{1}{(k_1 + k_2)^2},$$
$$B(k) = \frac{1}{(k_3 + k_2)^2} - \frac{1}{(k_1 + k_2)^2}.$$

(3.7)

As we mentioned earlier, the continuum theory is supersymmetric for a special value of fermion mass. We will therefore consider only the case $m = m_{SUSY}$. In the DLCQ formulation, one simply restricts integration of the light-cone momenta k_i in expression (3.6) for P^- above to be *positive* integer multiples of P^+/K. i.e. one simply drops the zero-momentum mode. The DLCQ mass spectrum is then obtained by diagonalizing the mass operator $M^2 = 2P^+P^-$. Similarly, in the SDLCQ formulation, the light-cone momenta k_i in expression (3.5) for Q^- are restricted to *positive* integer multiples of P^+/K. One then simply *defines* P^- to be the square of the supercharge: $2\sqrt{2} P^- = \{Q^-, Q^-\}$. The mass operator $M^2 = 2P^+P^-$ is then easily constructed and diagonalized to obtain the SDLCQ spectrum.

In general, the following observations are made; at finite resolution, the DLCQ spectrum of a supersymmetric theory is not supersymmetric. However, supersymmetry is restored after extrapolating to the continuum limit $K \to \infty$ (see [19], for example). In contrast, for any finite resolution, the SDLCQ spectrum is supersymmetric. The DLCQ and SDLCQ spectra agree only in the decompactified limit $K \to \infty$.

Not surprisingly, the difference in the DLCQ and SDLCQ prescriptions at finite resolution may be understood as a zero-mode contribution. What is surprising is that we can encode the effect of these zero-mode contributions into a simple well defined operator. The main result here is the the precise operator form of this contribution at finite K.

In order to motivate our argument, note that the anticommutator for the supercharge Q^- in the continuum theory involves products of terms of the form $b^\dagger(k)b^\dagger(0)b(k)$ and $b^\dagger(p)b(0)b(p)$, and these provide contributions to P^- that may be expressed in terms of non-zero momentum modes. The problem is exacerbated by the fact that the coefficients of these terms behave singularly. To examine this more closely, we consider the discretized theory where the light-cone momenta k_i in the expression for Q^- [eqn(3.5)] are restricted to positive integer multiples of P^+/K. We also include the effects of zero-momentum modes by introducing an 'ϵ regulated zero mode', which are modes with momentum $k_i = \epsilon$, where ϵ is much less than P^+/K, and is sent to zero at the end of the calculation. Then the anticommutator of two Q^- gives contributions of the following form:

$$
\left\{ (\frac{1}{\epsilon} + \frac{\epsilon}{k(k+\epsilon)})b^\dagger(k)b^\dagger(\epsilon)b(k+\epsilon), (\frac{1}{\epsilon} + \frac{\epsilon}{p(p+\epsilon)})b^\dagger(p+\epsilon)b(\epsilon)b(p) \right\} =
$$
$$
= b^\dagger(k)b(k+\epsilon)b^\dagger(p+\epsilon)b(p) \left[\frac{1}{\epsilon^2} + (\frac{1}{p(p+\epsilon)} + \frac{1}{k(k+\epsilon)}) + \frac{\epsilon^2}{pk(p+\epsilon)(k+\epsilon)} \right],
$$
(3.8)

where any terms involving an ϵ regularized zero mode on the right-hand-side are dropped and zero modes are omitted from P^-. We have suppressed all matrix indices in this expression. In the limit $\epsilon \to 0$ the last term on the right-hand-side in the brackets vanishes, while the first term is the pure momentum–independent divergence that was identified in an earlier study of this model [19], and is canceled if we adopt a principal value prescription for singular amplitudes in the definition of P^-. The second term however, is clearly a finite contribution to P^-, although it arises from the ϵ regulated zero modes in Q^-, which are not present in the SDLCQ prescription for defining Q^-. Consequently, in order to ensure the supersymmetry relation $\{Q^-, Q^-\} = 2\sqrt{2}P^-$ in the discretized formulation, we must include an ϵ regularization of the zero modes in the definition for Q^-, and then apply a principal value prescription in the presence of any singular processes to eliminate $1/\epsilon$ divergences.

Stated slightly differently, we may decompose the supercharge into a part without zero modes Q^-_{SDLCQ} (i.e. $k_i = nP^+/K, n = 1, 2, \ldots$), and terms with ϵ regularized zero modes, Q_ϵ^-. The anti-commutator $\{Q_{SDLCQ}^-, Q_\epsilon^-\}$ contains only terms with ϵ regulated zero-modes. Since $Q^- = Q_{SDLCQ}^- + Q_\epsilon^-$ one finds

$$\{Q_{SDLCQ}^-, Q_{SDLCQ}^-\} = 2\sqrt{2}P_{SDLCQ}^- = 2\sqrt{2}P_{DLCQ}^- - \{Q_\epsilon^-, Q_\epsilon^-\}_{PV}, \qquad (3.9)$$

after dropping any ϵ regulated zero-mode terms in the calculated expression for $\{Q^-, Q^-\}$. Note that the first equality above is just the definition for the light-cone Hamiltonian P^- in the SDLCQ prescription. The PV abbreviation on the right hand side indicates a principal value regularization prescription, which is tantamount to dropping all $1/\epsilon$ terms as $\epsilon \to 0$. The procedure is well known in the context of the present model [19]. It is clear that our definition for P_{SDLCQ}^- gives rise to the supersymmetry relation $[Q_{SDLCQ}^-, P_{SDLCQ}^-] = 0$, which yields a supersymmetric spectrum for any finite resolution K. Moreover, we know that P_{SDLCQ}^- and P_{DLCQ}^- yield the same spectrum in the continuum limit $K \to \infty$, so it remains to calculate the difference at finite resolution K. We will write this difference in terms of their respective mass operators: $M^2 = 2P^+P^-$. A straightforward calculation of the anticommutator on the right-hand-side of (3.9) leads to the result:

$$M_{SDLCQ}^2 - M_{DLCQ}^2 = M_\Delta^2 = -\frac{g^2 NK}{\pi} \sum_n \frac{1}{n^2} B_{ij}^\dagger(n) B_{ij}(n)$$

$$-\frac{g^2 NK}{\pi} \sum_{mn} (\frac{1}{m^2} + \frac{1}{n^2}) \frac{1}{N} B_{kj}^\dagger(m) B_{ji}^\dagger(n) B_{kl}(m) B_{li}(n). \qquad (3.10)$$

We also write down the expression for M_{DLCQ}^2 in the theory with periodic fermions:

$$M_{DLCQ}^2 = \frac{g^2 NK}{\pi} \sum_n B_{ij}^\dagger(n) B_{ij}(n) (\frac{x}{n} + \sum_m^{n-1} \frac{2}{(n-m)^2}) +$$

$$\frac{g^2 K}{\pi} {\sum_{n_i}}' \left\{ \delta_{n_1+n_2}^{n_3+n_4} \left[\frac{1}{(n_2-n_4)^2} - \frac{1}{(n_1+n_2)^2} \right] B_{kj}^\dagger(n_3) B_{ji}^\dagger(n_4) B_{kl}(n_1) B_{li}(n_2) \right.$$

$$+ \delta_{n_1+n_2+n_3,n_4} \left[\frac{1}{(n_2+n_3)^2} - \frac{1}{(n_1+n_2)^2} \right] \times \qquad (3.11)$$

$$\left. \left(B_{kj}^\dagger(n_4) B_{kl}(n_1) B_{li}(n_2) B_{ij}(n_3) - B_{kj}^\dagger(n_1) B_{jl}^\dagger(n_2) B_{li}^\dagger(n_3) B_{ki}(n_4) \right) \right\}.$$

In this expression the variable $x = \frac{\pi m^2}{g^2 N}$ is a dimensionless mass parameter, and for the supersymmetric point we have $x = 1$. The sums are performed over positive integers, $0 < n_i < K$, and we employ a principal value prescription in sums labeled as $\sum'$, which implies that terms of the form $1/(k-k)^2$ are dropped. In the SDLCQ procedure we calculate Q^- which is non-singular and requires no principal value prescription.

The term M_Δ^2 appears to be non-trivial due to the presence of $B^\dagger B^\dagger BB$ terms on the right hand side of (3.10). However, the action of this term on any SU(N) Fock state turns out to be equivalent to the first term, although with opposite sign, and twice the magnitude. Thus the action of the right hand side of (3.10) is equivalent to the single quadratic operator:

$$M_\Delta^2 = \frac{g^2 N K}{\pi} \sum_n \frac{1}{n^2} B_{ij}^\dagger(n) B_{ij}(n). \tag{3.12}$$

Fortunately, we are able to test this analytical result by performing direct numerical simulations of this model using both prescriptions, and comparing the differences observed with the above prediction. Interestingly, although this result was derived for large N, agreement turns out to be perfect for both finite and large N, which was verified using the finite N DLCQ algorithms developed in [12]. We discuss this further in the next subsection.

B Soft SUSY Breaking and Numerical Results.

In this subsection we compare the numerical results for different regularization schemes. Although in the continuum limit both PV and SUSY prescription ns should give the same results, the convergence of the masses as $K \to \infty$ might be different. So if at a given value of K one wants to get a better approximation to continuum masses, one scheme might work better than other. In previous subsection we described two regularization schemes and found the operator M_Δ^2 describing the difference between them. It is convenient to introduce the family of regularizations labeled by parameter Y:

$$P_Y^- = P_{PV}^- + Y M_\Delta^2. \tag{3.13}$$

Then at two special values of Y we get the PV and SUSY prescriptions: $P_{PV}^- = P_{Y=0}^-$, $P_{SUSY}^- = P_{Y=1}^-$. Since P_{PV}^- is defined for arbitrary value of fermionic mass m (not only for supersymmetric one) the last equation also defines the family of regularizations beyond the SUSY point. On the other hand shifting the fermionic mass from its supersymmetric value is equivalent to introducing additional fermionic mass, i.e. to the soft SUSY breaking. Below we will give a numerical results for bound state masses in the theory with two new "coupling constants": $X = \frac{\pi m^2}{g^2 N}$ and Y which determine the value of fermionic mass and regularization scheme accordingly.

Our investigation of this theory indicates that at $X = 1$ (the supersymmetric value of the fermion mass) the lightest fermionic and bosonic bound states are degenerate with continuum masses approximately $M^2 = 26$ [19,5]. Using P_{SUSY}^- we arrive at the same conclusion for any value of Y.

Boorstein and Kutasov [22] have investigated 'soft' supersymmetry breaking for small values of this difference, $X - 1$ and they found that the degeneracy between the fermion and boson bound state masses is broken according to

$$M_F^2(X) - M_B^2(X) = (1 - X) M_B(1) + O((X-1)^3). \tag{3.14}$$

They calculated these masses using the PV prescription ($Y = 0$) with anti-periodic BC and found very good agreement with the theoretical prediction. We have compared this theoretical prediction at $Y = 1$ and we find that eq (3.14) is very well satisfied. At resolution $K = 5$, for example, the slope is 4.76 and the predicted slope $M_B(1)$ is 4.76. The indication is that this result is true for any value of Y.

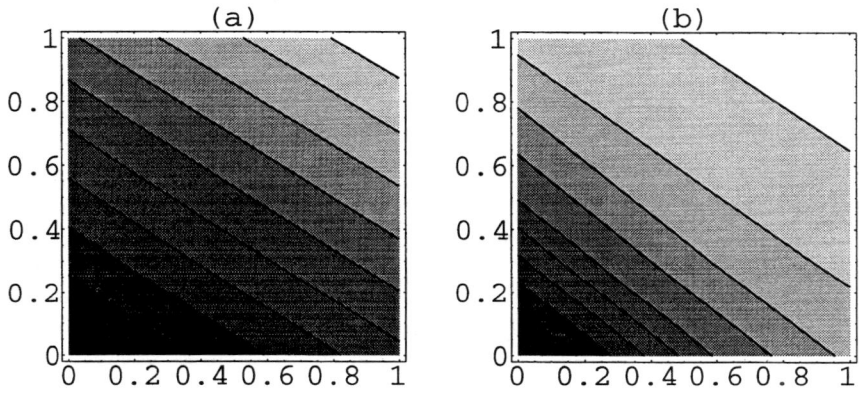

FIGURE 1. (a) The contour plots of $Y = Y(X)$ for the mass squared of the lowest bound state in units of g^2N/π as a function of $X = m\pi/g^2N$ and Y (b)The contour plots of $Y = Y(X)$ for the mass squared of the second lowest bound state in units of g^2N/π as a function of $X = m\pi/g^2N$ and Y (b)

In Fig. 1 we show the contour plots of the mass squared M^2 of the two lightest bosonic bound states as a function of X and Y at resolution $K = 10$. These contours are lines of constant mass squared. Selecting a particular value of the mass of the first bound state then fixes a particular contour in Fig. 1a as a contour of fixed mass, which we can write as $Y = Y_p(X)$.

Interestingly, constructing the same contour plot for the next to lightest bosonic bound state – see Fig. 1b – yields contours that have approximately the same functional dependence implied by Fig. 1a. In fact, one obtains approximately the same contour plots for the next twenty bound states (which is as far as we checked). The simple conclusion is that the coupling Y which represents the strength of the additional operator affects all bound state masses more or less equally. This in turn suggests that at finite resolution, we can smoothly interpolate between different values of fermion mass X, and different prescriptions specified by the coupling Y, without affecting too much the actual numerical spectrum. Of course, in the decompactification limit $K \to \infty$, such a dependence on Y disappears, due to scheme independence.

Since the lightest bosonic bound state is primarily a two particle state it is reasonable to truncate the Fock basis to two particle states. This will permit very high resolutions, which will be needed to carefully scrutinize any possible discrepancies between the two versions of 'soft' symmetry breaking presented here. In fact, we are able to study the theory for K up to 800. The mass of the lowest state as a function of the resolution for various values of X and Y are shown in Fig. 2. Each converging pair of lines – which extrapolate the actual data points – in Fig. 2 corresponds to different values of fermion mass X. The top upper curve in each pair runs through data points that were calculated via SDLCQ (i.e. $Y = 1$), while the lower corresponds to the PV (i.e. $Y = 0$) prescription commonly adopted in the literature. We find that each pair of curves converge to the same point at infinite resolution, although this may not be completely obvious for the lowest pair in the figure (corresponding to the critical mass $X = 0$).

Away from $X = 0$, the SDLCQ formulation is fitted with a linear function of $1/K$, while the PV formulation is fit with a polynomial of $1/K^{2\beta}$, where β is the solution of $1 - X/2 = \pi\beta Cot(\pi\beta)$ [78]. It now appears that SDLCQ not only provides more rapid convergence for supersymmetric models, but also for the massive t'Hooft model, which is not supersymmetric. For the massless case, the situation is reversed; the SDLCQ formulation converges slower. It is fit by a polynomial in $1/Log(K)$ and gives the same mass at infinite resolution as the PV formulation. This behavior may be understood from the observation that the wave function of this state does not vanish at $x = 0$. We have looked closely at 'small' masses, such as $X = .1$, and one finds that both PV and SDLCQ vary as a polynomial in $1/K^{2\beta}$ at large resolution. Thus careful extrapolation schemes must be adopted at small masses.

We therefore conclude that the continuum of regularization schemes that interpolate smoothly between the SDLCQ and PV prescriptions – which we characterized by the parameter Y – yield the same continuum bound state masses, although the rate of convergence of the DLCQ spectrum may be altered significantly. This implies that the contour plots observed in Fig. 1 eventually approach lines parallel to the Y axis, and the sole dependence on the parameter X is recovered.

Interestingly, since the two-body equation studied here for the adjoint fermion model is simply the t'Hooft equation with a rescaling of coupling constant, we have arrived at an alternative prescription for regulating the Coulomb singularity in the massive t'Hooft model that improves the rate of convergence towards the actual continuum mass. Thus, a prescription that arises naturally in the study of supersymmetric theories is also applicable in the study of a theory without supersymmetry. We believe that this idea deserves to be exploited further in a wider context of theories. In particular, it is an open question whether this procedure could provide a sensible approach to regularizing softly broken gauge theories with bosonic degrees of freedom, and in higher dimensions.

In any case, it appears that the special cancellations afforded by supersymmetry – especially in the context of DLCQ bound state calculations – might have scope beyond the domain of supersymmetric field theory. This would be a crucial first step

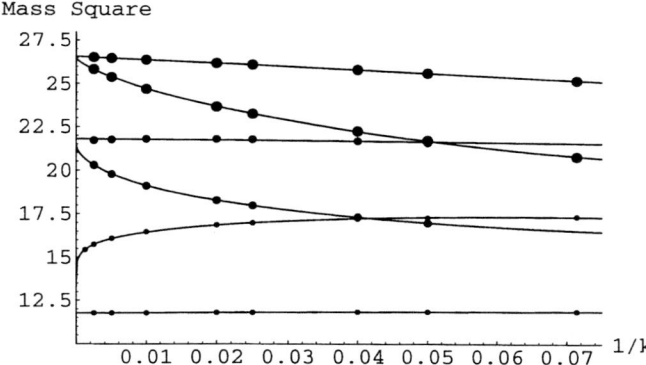

Mass Square

FIGURE 2. Mass of the of the lowest bound state in units of $g^2 N/\pi$ calculated in the t'Hooft model. The top pair is at $X = 1$, the second is at $X = .5$, and the bottom pair is at $X = 0$

towards a serious non-perturbative study of theories with broken supersymmetry.

IV MASSLESS STATES IN TWO DIMENSIONAL MODELS.

In this section we will study the structure of bound states for two dimensional supersymmetric models defined in section 1. We will concentrate most of the attention on the model obtained by dimensional reduction from SYM_{2+1}. For this theory we will prove that any normalizable bound state in the continuum must include a contribution with arbitrarily large number of partons. By generalizing this proof to the theories with extended SUSY we show that this is the general property of supersymmetric matrix models. This scenario is to be contrasted with the simple bound states discovered in a number of $1 + 1$ dimensional theories with complex fermions, such as the Schwinger model, the t'Hooft model, and a dimensionally reduced theory with complex adjoint fermions [11,68]. We also study the massless states of SYM_{2+1} in DLCQ. Some of them are constructed explicitly and the general formula for the number of massless states as function of harmonic resolution is derived for the large N case. This section is based in part on the results of [4].

A   Formulation of the bound state problem.

The light-cone formulation of the supersymmetric matrix model obtained by dimensionally reducing $\mathcal{N} = 1$ SYM_{2+1} to $1 + 1$ dimensions was initially given in [63], and it was summarized in the section 2 of these lectures. We simply note here that the light-cone Hamiltonian P^- is given in terms of the supercharge Q^- via the supersymmetry relation $\{Q^-, Q^-\} = 2\sqrt{2}P^-$, where

$$Q^- = 2^{3/4}g \int dx^- \text{tr}\left\{ (i[\phi, \partial_-\phi] + 2\psi\psi)\frac{1}{\partial_-}\psi \right\}. \qquad (4.1)$$

In the above, $\phi_{ij} = \phi_{ij}(x^+, x^-)$ and $\psi_{ij} = \psi_{ij}(x^+, x^-)$ are $N \times N$ Hermitian matrix fields representing the physical boson and fermion degrees of freedom (respectively) of the theory, and are remnants of the physical transverse degrees of freedom of the original $2 + 1$ dimensional theory. This is a special feature of light-cone quantization in light-cone gauge: all unphysical degrees of freedom present in the original Lagrangian may be explicitly eliminated. There are no ghosts.

In order to quantize ϕ and ψ on the light-cone, we first introduce the following expansions at fixed light-cone time $x^+ = 0$ (the continuum counterpart of (2.12):

$$\phi_{ij}(x^-, 0) = \frac{1}{\sqrt{2\pi}} \int_0^\infty \frac{dk^+}{\sqrt{2k^+}} \left(a_{ij}(k^+)e^{-ik^+x^-} + a_{ji}^\dagger(k^+)e^{ik^+x^-} \right); \qquad (4.2)$$

$$\psi_{ij}(x^-, 0) = \frac{1}{2\sqrt{\pi}} \int_0^\infty dk^+ \left(b_{ij}(k^+)e^{-ik^+x^-} + b_{ji}^\dagger(k^+)e^{ik^+x^-} \right). \qquad (4.3)$$

We then specify the commutation relations

$$\left[a_{ij}(p^+), a_{lk}^\dagger(q^+)\right] = \left\{b_{ij}(p^+), b_{lk}^\dagger(q^+)\right\} = \delta(p^+ - q^+)\delta_{il}\delta_{jk} \qquad (4.4)$$

for the gauge group U(N), or

$$\left[a_{ij}(p^+), a_{lk}^\dagger(q^+)\right] = \left\{b_{ij}(p^+), b_{lk}^\dagger(q^+)\right\} = \delta(p^+ - q^+)\left(\delta_{il}\delta_{jk} - \frac{1}{N}\delta_{ij}\delta_{kl}\right) \qquad (4.5)$$

for the gauge group SU(N)[6].

For the bound state eigen-problem $2P^+P^-|\Psi> = M^2|\Psi>$, we may restrict to the subspace of states with fixed light-cone momentum P^+, on which P^+ is diagonal, and so the bound state problem is reduced to the diagonalization of the light-cone Hamiltonian P^-. Since P^- is proportional to the square of the supercharge Q^-, any eigenstate $|\Psi>$ of P^- with mass squared M^2 gives rise to a natural four-fold degeneracy in the spectrum because of the supersymmetry algebra—all four states below have the same mass:

$$|\Psi>, \quad Q^+|\Psi>, \quad Q^-|\Psi>, \quad Q^+Q^-|\Psi>. \qquad (4.6)$$

Although this four-fold degeneracy is realized in the continuum formulation of the theory, this property will not necessarily survive if we choose to discretize the theory

---

[6] We assume the normalization $tr[T^aT^b] = \delta^{ab}$, where the T^a's are the generators of the Lie algebra of SU(N).

in an arbitrary manner. However, a nice feature of SDLCQ is that it does preserve the supersymmetry (and hence the *exact* four-fold degeneracy) for any resolution.

Focusing attention on zero mass eigenstates, we simply note that a massless eigenstate of P^- must also be annihilated by the supercharge Q^-, since P^- is proportional to $(Q^-)^2$. Thus the relevant eigen-equation is $Q^-|\Psi> = 0$. We wish to study this equation. However, first we need to state the explicit equation for Q^-, in the momentum representation, which is obtained by substituting the quantized field expressions (4.2) and (4.3) directly into the the definition of the supercharge (4.1). The result is:

$$
Q^- = \frac{i2^{-1/4}g}{\sqrt{\pi}} \int_0^\infty dk_1 dk_2 dk_3 \delta(k_1 + k_2 - k_3) \Big\{
$$

$$
\frac{1}{2\sqrt{k_1 k_2}} \frac{k_2 - k_1}{k_3} [a_{ik}^\dagger(k_1) a_{kj}^\dagger(k_2) b_{ij}(k_3) - b_{ij}^\dagger(k_3) a_{ik}(k_1) a_{kj}(k_2)]
$$

$$
\frac{1}{2\sqrt{k_1 k_3}} \frac{k_1 + k_3}{k_2} [a_{ik}^\dagger(k_3) a_{kj}(k_1) b_{ij}(k_2) - a_{ik}^\dagger(k_1) b_{kj}^\dagger(k_2) a_{ij}(k_3)]
$$

$$
\frac{1}{2\sqrt{k_2 k_3}} \frac{k_2 + k_3}{k_1} [b_{ik}^\dagger(k_1) a_{kj}^\dagger(k_2) a_{ij}(k_3) - a_{ij}^\dagger(k_3) b_{ik}(k_1) a_{kj}(k_2)]
$$

$$
(\frac{1}{k_1} + \frac{1}{k_2} - \frac{1}{k_3})[b_{ik}^\dagger(k_1) b_{kj}^\dagger(k_2) b_{ij}(k_3) + b_{ij}^\dagger(k_3) b_{ik}(k_1) b_{kj}(k_2)] \Big\}.
$$

$$(4.7)$$

In order to implement the DLCQ formulation [67,62] of the theory, we simply restrict the momenta k_1, k_2 and k_3 appearing in the above equation to the following set of allowed momenta: $\{\frac{P^+}{K}, \frac{2P^+}{K}, \frac{3P^+}{K}, \ldots\}$. Here, K is some arbitrary positive integer, and must be sent to infinity if we wish to recover the continuum formulation of the theory. The integer K is called the *harmonic resolution*, and $1/K$ measures the coarseness of our discretization. Physically, $1/K$ represents the smallest unit of longitudinal momentum fraction allowed for each parton. As soon as we implement the DLCQ procedure, which is specified unambiguously by the harmonic resolution K, the integrals appearing in the definition of Q^- are replaced by finite sums, and the eigen-equation $Q^-|\Psi> = 0$ is reduced to a finite matrix problem. For sufficiently small values of K (in this case for $K \leq 4$) this eigen-problem may be solved analytically. For values $K > 5$, we may still compute the DLCQ supercharge analytically as a function of N, but the diagonalization procedure must be performed numerically.

For now, we concentrate on the structure of the zero mass eigenstates for the continuum theory. Firstly, note that for the U(N) bound state problem, massless states appear automatically because of the decoupling of the U(1) and SU(N) degrees of freedom that constitute U(N). More explicitly, we may introduce the U(1) operators

$$
\alpha(k^+) = \frac{1}{N} \text{tr}[a(k^+)] \quad \text{and} \quad \beta(k^+) = \frac{1}{N} \text{tr}[b(k^+)], \tag{4.8}
$$

which allow us to decompose any U(N) operator into a sum of U(1) and SU(N) operators:

$$a(k^+) = \alpha(k^+) \cdot \mathbf{1}_{N \times N} + \tilde{a}(k^+) \quad \text{and} \quad b(k^+) = \beta(k^+) \cdot \mathbf{1}_{N \times N} + \tilde{b}(k^+), \qquad (4.9)$$

where $\tilde{a}(k^+)$ and $\tilde{b}(k^+)$ are traceless $N \times N$ matrices. If we now substitute the operators above into the expression for the supercharge (4.7), we find that all terms involving the U(1) factors $\alpha(k^+), \beta(k^+)$ vanish – only the SU(N) operators $\tilde{a}(k^+), \tilde{b}(k^+)$ survive. i.e. starting with the definition of the U(N) supercharge, we end up with the definition of the SU(N) supercharge. In addition, the (anti)commutation relations $[\tilde{a}_{ij}(k_1), \alpha^\dagger(k_2)] = 0$ and $\{\tilde{b}_{ij}(k_1), \beta^\dagger(k_2)\} = 0$ imply that this supercharge acts only on the SU(N) creation operators of a fock state - the U(1) creation operators only introduce degeneracies in the SU(N) spectrum. Clearly, since Q^- has no U(1) contribution, any fock state made up of only U(1) creation operators must have zero mass. The non-trivial problem here is to determine whether there are massless states for the SU(N) sector. We will address this topic next.

B   The Proof for (1,1) Model.

It was pointed out in the previous subsection that a zero mass eigenstate is annihilated by the light-cone supercharge (4.7):

$$Q^-|\Psi\rangle = 0 \qquad (4.10)$$

We wish to show that if such an SU(N) eigenstate is normalizable, then it must involve a superposition of an *infinite* number of Fock states. The basic strategy is quite simple: normalizability will impose certain conditions on the light-cone wave functions as one or several momentum variables vanish. Moreover, if we assume a given eigenstate $|\Psi\rangle$ has at most n partons, then the terms in $Q^-|\Psi\rangle$ consisting of $n+1$ partons must sum to zero, providing relations between the n parton wave functions only. We then show these wave functions must all vanish by studying various zero momentum limits of these relations. Interestingly, the utility of studying light-cone wave functions at small momenta also appears in the context of light-front QCD$_{3+1}$ [3].

In order to proceed with a systematic presentation of the proof, we start by considering the large N limit case. This simply means that we consider Fock states that are made from a *single* trace of a product of boson or fermion creation operators acting on the light-cone Fock vacuum $|0\rangle$. Multiple trace states correspond to $1/N$ corrections to the theory, and are therefore ignored. In this limit, a general state $|\Psi\rangle$ is a superposition of Fock states of any length, and may be written in the form

$$|\Psi\rangle = \sum_{n=2}^{\infty} \sum_{r=0}^{n} \sum_P \int_0^{P^+} \frac{dq_1 \dots dq_n}{\sqrt{q_1 \dots q_n}} \delta(q_1 + \dots + q_n - P^+) \times$$
$$f_P^{(n,r)}(q_1, \dots, q_n) \text{tr}[c^\dagger(q_1) \dots c^\dagger(q_n)]|0\rangle, \qquad (4.11)$$

172

where $c^\dagger(q^+)$ represents either a boson or fermion creation operator carrying light-cone momentum q^+, and $f_P^{(n,r)}$ denotes the wave function of an n parton Fock state containing r fermions in a particular arrangement P. It is implied that we sum over all such arrangements, which may not necessarily be distinct with respect to cyclic symmetry of the trace.

At this point, we simply remark that normalizability of a general state $|\Psi\rangle$ above implies

$$\int_0^{P^+} \frac{dq_1 \ldots dq_n}{q_1 \ldots q_n} \delta(q_1 + \cdots + q_n - P^+) |f_P^{(n,r)}(q_1, \ldots, q_n)|^2 < \infty \qquad (4.12)$$

for any particular wave function $f_P^{(n,r)}$. Therefore, any wave function vanishes if one or several of its momenta are made to vanish.

We are now ready to carry out the details of the proof. But first a little notation. We will write $|\Psi_{(n,m)}\rangle$ to denote a superposition of all Fock states – as in (4.11) – with precisely n partons, m of which are fermions. Such a Fock expansion involves only the wave functions $f_P^{(n,m)}$, and the number of them is enumerated by the index P. For the special case $|\Psi_{(n,0)}\rangle$ (i.e. no fermions), there is only one wave function, which we denote by $f^{(n,0)}$ for brevity:

$$|\Psi_{(n,0)}\rangle = \int_0^{P^+} \frac{dq_1 \ldots dq_n}{\sqrt{q_1 \ldots q_n}} \delta(q_1 + \ldots + q_n - P^+) \, f^{(n,0)}(q_1 \ldots q_n) \mathrm{tr}[a^\dagger(q_1) \ldots a^\dagger(q_n)]|0\rangle.$$

$$(4.13)$$

There is another special case we wish to consider; namely, the state $|\Psi_{(n,2)}\rangle$ consisting of n parton Fock states with precisely two fermions. If we place one of the fermions at the beginning of the trace, then there are $n-1$ ways of positioning the second fermion, yielding $n-1$ possible wave functions. We will enumerate such wave functions by the subscript index k, as in $f_k^{(n,2)}$, where $k = 2, 3, \ldots, n$. The subscript k denotes the location of the second fermion. Explicitly, we have

$$|\Psi_{(n,2)}\rangle = \sum_{k=2}^n \int_0^{P^+} \frac{dq_1 \ldots dq_n}{\sqrt{q_1 \ldots q_n}} \delta(q_1 + \cdots + q_n - P^+) \times$$

$$f_k^{(n,2)}(q_1, \ldots, q_k, \ldots, q_n) \mathrm{tr}[b^\dagger(q_1) a^\dagger(q_2) \ldots b^\dagger(q_k) \ldots a^\dagger(q_n)]|0\rangle.$$

$$(4.14)$$

Of course, depending upon the symmetry, the $n-1$ Fock states enumerated in this way need not be distinct with respect to the cyclic properties of the trace. This provides us with additional relations between wave functions – a fact we will make use of later on.

Now let us assume that $|\Psi\rangle$ is a normalizable SU(N) zero mass eigenstate with at most n partons. Glancing at the form of (4.7), we see that the $n+1$ parton Fock states containing a single fermion in each of the combinations $Q^- |\Psi_{(n,0)}\rangle$

173

and $Q^-|\Psi_{(n,2)}\rangle$ must cancel each other to guarantee a massless eigenstate. This immediately gives rise to the following wave function relation:

$$\frac{q_1 + 2q_2}{q_1 + q_2} f^{(n,0)}(q_1 + q_2, q_3, \ldots, q_{n+1}) - \frac{q_1 + 2q_{n+1}}{q_1 + q_{n+1}} f^{(n,0)}(q_{n+1} + q_1, q_2, \ldots, q_n) =$$

$$= 2\frac{\sqrt{q_1}}{n} \sum_{k=2}^{n} \frac{q_{k+1} - q_k}{(q_{k+1} + q_k)^{3/2}} f_k^{(n,2)}(q_1, \ldots, q_{k-1}, q_k + q_{k+1}, q_{k+2}, \ldots, q_{n+1}).$$

$$(4.15)$$

In the limit $q_i \to 0$, for $3 \le i \le n$, this last equation is reduced to

$$\frac{1}{\sqrt{q_{i+1}}} f_i^{(n,2)}(q_1, \ldots, q_{i-1}, q_{i+1}, \ldots, q_{n+1})$$

$$- \frac{1}{\sqrt{q_{i-1}}} f_{i-1}^{(n,2)}(q_1, \ldots, q_{i-1}, q_{i+1}, \ldots, q_{n+1}) = 0. \qquad (4.16)$$

An immediate consequence is that any wave function $f_i^{(n,2)}$ for $i = 3, 4, \ldots, n$, may be expressed in terms of $f_2^{(n,2)}$. Explicitly, we have

$$f_i^{(n,2)}(q_1, q_2, \ldots, q_n) = \sqrt{\frac{q_i}{q_2}} f_2^{(n,2)}(q_1, q_2, \ldots, q_n), \qquad i = 3, 4, \ldots, n. \qquad (4.17)$$

Moreover, the limit $q_2 \to 0$ of equation (4.15) yields the further relation after a suitable change of variables:

$$f^{(n,0)}(q_1, q_2, q_3, \ldots, q_n) = \frac{2}{n} \sqrt{\frac{q_1}{q_2}} f_2^{(n,2)}(q_1, q_2, q_3, \ldots, q_n). \qquad (4.18)$$

Finally, because of the cyclic properties of the trace, there is an additional relation between wave functions:

$$f_i^{(n,2)}(q_1, q_2, \ldots, q_i, \ldots, q_n) = -f_{n-i+2}^{(n,2)}(q_i, q_{i+1}, \ldots, q_n, q_1, q_2, \ldots, q_{i-1}). \qquad (4.19)$$

Setting $i = 2$ in the above equation, and $i = n$ in equation (4.17), we deduce

$$f_2^{(n,2)}(q_1, q_2, \ldots, q_n) = -\sqrt{\frac{q_1}{q_2}} f_2^{(n,2)}(q_2, q_3, \ldots, q_n, q_1). \qquad (4.20)$$

Combining this with equation (4.18), we conclude $\left(\frac{\sqrt{q_2}}{q_1} + \frac{\sqrt{q_3}}{q_2}\right) f^{(n,0)}(q_1, \ldots, q_n) = 0$, where we use the fact that the wave functions $f^{(n,0)}$ are cyclically symmetric. Thus $f^{(n,0)}$ must vanish. It immediately follows that $f_i^{(n,2)}$ vanish for all i as well.

To summarize, we have shown that if $|\Psi\rangle$ is a normalizable zero mass eigenstate, where each Fock state in its Fock state expansion has no more than n partons, the contributions $|\Psi_{(n,0)}\rangle$ and $|\Psi_{(n,2)}\rangle$ in this Fock state expansion must vanish. Since we may assume $|\Psi\rangle$ is bosonic, the only other contributions involve Fock states with an even number of fermions: $|\Psi_{(n,4)}\rangle$, $|\Psi_{(n,6)}\rangle$, and so on. We claim that all such contributions vanish. To see this, first observe that the $n+1$ parton Fock states with three fermions in the combinations $Q^-|\Psi_{(n,2)}\rangle$ and $Q^-|\Psi_{(n,4)}\rangle$ must cancel each other, in order to guarantee a zero eigenstate mass. But our previous analysis demonstrated that $|\Psi_{(n,2)}\rangle \equiv 0$, and thus the $n+1$ parton Fock states with three fermions in $Q^-|\Psi_{(n,4)}\rangle$ alone must sum to zero.

We are now ready to perform an induction procedure. Namely, we assume that for some positive integer k the state $|\Psi_{(n,2[k-1])}\rangle$ vanishes. Then the $n+1$ parton Fock states in $Q^-|\Psi\rangle$ which contain $2k-1$ fermions receive contributions only from $Q^-|\Psi_{(n,2k)}\rangle$ in which a fermion is replaced by two bosons. This has to sum to zero. We therefore obtain a relation among the wave functions $f_P^{(n,2k)}$ by considering the action of the supercharge (4.7) in which a fermion is replaced by two bosons. Keeping in mind that we are free to renormalize any wave function by a constant, we end up with the following relation:

$$\sum_P f_P^{(n,2k)}(s_1,\ldots,s_{i-1},s_i+s_{i+1},s_{i+2},\ldots,s_{n+1})\frac{s_{i+1}-s_i}{(s_{i+1}+s_i)^{3/2}} = 0. \qquad (4.21)$$

It is now an easy task to show that the wave functions $f_P^{(n,2k)}$ appearing in equation (4.21) must vanish; one simply considers various limits $s_j \to 0$ as we did before. This completes our proof by induction. Namely, there can be no non-trivial normalizable massless state with an upper limit on the number of allowed partons. Of course, this proof is valid only in the large N limit. We now turn our attention to the finite N case.

Let us define Q_{lead}^- to be that part of the supercharge Q^- that replaces a fermion with two bosons, or replaces a boson with a boson and fermion pair. As in the large N case we begin by assuming that we have a normalizable zero mass eigenstate $|\Psi\rangle$ which is a sum of Fock states that have at most n partons. The proof for finite N consists of two parts. First, we consider bosonic states consisting of only n parton Fock states that have at most two fermions, and show the wave functions must vanish. We then invoke an induction argument to consider n parton wave functions involving an even number of fermions, and show they must vanish as well.

The additional complication introduced by the assumption that N is finite is that a given Fock state may involve more than just a single trace. However, note that Q_{lead}^- cannot decrease the number of traces; it can either increase the number of traces by one, or leave the number unchanged. Thus we have a natural induction procedure in the number of traces as well. Since the terms in Q_{lead}^- have only one annihilation operator, it acts on a given product of traces according to the Leibniz rule:

$$Q_{lead}^-\left(\mathrm{tr}[A]\mathrm{tr}[B]\ldots\right)|0\rangle = \left(Q_{lead}^-\mathrm{tr}[A]\right)\mathrm{tr}[B]\ldots|0\rangle +$$
$$(-1)^{F(A)}\mathrm{tr}[A]Q_{lead}^-\left(\mathrm{tr}[B]\ldots\right)|0\rangle. \qquad (4.22)$$

175

Schematically, the general structure of an arbitrary Fock state with k traces has the form

$$f_P^{(n,i_1,i_2,\ldots,i_k)} \mathrm{tr}[(b^\dagger)^{i_1} a^\dagger \ldots a^\dagger] \ldots \mathrm{tr}[(b^\dagger)^{i_k} a^\dagger \ldots a^\dagger]|0\rangle, \qquad (4.23)$$

where n denotes the total number of partons in each Fock state, and the integers $i_1, i_2, \ldots$ denote the number of fermions in the first trace, second trace, and so on. We will always order the traces so that the number of fermions in each trace decreases to the right. The index P labels a particular arrangement of fermions.

We now consider the $n+1$ parton Fock states of $Q_{lead}^-|\Psi\rangle$ that have precisely one fermion. The only possible contributions involve three types of wave functions; $f^{(n,0)}$, $f_P^{(n,2)}$ and $f^{(n,1,1)}$ (we only include the permutation index P if there is more than one distinct arrangement). If these three wave functions contribute to the same one fermion Fock state, then the distribution of bosons in the Fock state corresponding to $f_P^{(n,2)}$ determines the distribution of bosons for $f^{(n,0)}$ and $f^{(n,1,1)}$. We allow Q_{lead}^- to act only on the first trace in both $f^{(n,0)}$ and $f_P^{(n,2)}$, and only on the second one in $f^{(n,1,1)}$. If there are more than two traces in these states they must be identical in all the components, and so don't play a role in the calculation. Thus, it is sufficient to consider states with two traces only. Such a state has the form

$$|\Phi\rangle = \int_0^{P^+} \frac{d^{m+n}q}{\sqrt{q_1 \cdots q_{n+m}}} \delta(q_1 + \cdots + q_{n+m} - P^+)$$

$$f^{(n+m,0)}(q_1, \ldots, q_m | q_{m+1}, \ldots, q_{m+n})$$

$$\times \mathrm{tr}\left[a^\dagger(q_1) \ldots a^\dagger(q_m)\right] \mathrm{tr}\left[a^\dagger(q_{m+1}) \ldots a^\dagger(q_{m+n})\right]|0\rangle$$

$$+ \int_0^{P^+} \frac{d^{m+n-2}q\, dp_1 dp_2}{\sqrt{q_1 \cdots q_{n+m-2} p_1 p_2}} \delta(q_1 + \cdots + q_{n+m-2} + p_1 + p_2 - P^+)\Bigg\{$$

$$f^{(n+m,1,1)}(p_1, q_1, \ldots, q_m | p_2, q_{m+3}, \ldots, q_{m+n}) \times$$

$$\times \mathrm{tr}\left[b^\dagger(p_1) a^\dagger(q_1) \ldots a^\dagger(q_m)\right]\mathrm{tr}[b^\dagger(p_2) a^\dagger(q_{m+3}) \ldots a^\dagger(q_{m+n})] +$$

$$+ \sum_P f_P^{(n+m,2)}(p_1, P[q_1 \ldots q_{m-2}; p_2] | q_{m+1} \ldots q_{m+n}) \times$$

$$\times \mathrm{tr}\left(b^\dagger(p_1) P[a^\dagger(q_1) \ldots a^\dagger(q_{m-2}); b^\dagger(p_2)]\right) \mathrm{tr}\left[a^\dagger(q_{m+1}) \ldots a^\dagger(q_{m+n})\right]\Bigg\}|0\rangle,$$

$$(4.24)$$

where we have summed over the index P representing all possible permutation arrangements of bosons and fermions that contribute. We then find:

$$F(p, q_1, \ldots, q_m | q_{m+1}, q_{m+2}, \ldots, q_{m+n}) + \qquad (4.25)$$

$$+ \frac{q_{m+2} - q_{m+1}}{(q_{m+2} + q_{m+1})^{3/2}} f^{(n+m,1,1)}(p, q_1 \ldots q_m | q_{m+1} + q_{m+2}, q_{m+3} \ldots q_{m+n}) = 0,$$

where F is the contribution from $f^{(n+m,0)}$ and $f_P^{(n+m,2)}$. Now we see that the limit $q_{m+1} \to 0$ gives: $f^{(n+m,1,1)} \equiv 0$. Thus if (4.24) represents a contribution to the massless eigenstate state $|\Psi\rangle$, then $|\Phi\rangle$ takes the form

$$
|\Phi\rangle = \int_0^{P^+} \frac{d^{m+n-2}qdK^+}{\sqrt{q_1 \cdots q_{n+m-2}}} \delta(q_1 + \cdots + q_{n+m-2} - (P^+ - K^+)) \Bigg[
$$

$$
\int_0^{P^+} \frac{dq_{m-1}dq_m}{\sqrt{q_{m-1}q_m}} \delta(q_{m-1} + q_m - K^+)
$$

$$
f^{(n+m,0)}(q_1, \ldots, q_m | q_{m+1}, \ldots, q_{m+n}) \mathrm{tr}(a^\dagger(q_1) \ldots a^\dagger(q_m))
$$

$$
+ \int_0^{P^+} \frac{dp_1 dp_1}{\sqrt{p_1 p_2}} \delta(p_1 + p_2 - K^+)
$$

$$
\sum_P f_P^{(n+m,2)}(p_1, P[q_1, \ldots, q_{m-2}; p_2] | q_{m+1}, \ldots, q_{m+n}) \tag{4.26}
$$

$$
\mathrm{tr}(b^\dagger(p_1) P[a^\dagger(q_1) \ldots a^\dagger(q_{m-2}); b^\dagger(p_2)]) \mathrm{tr}(a^\dagger(q_{m+1}) \ldots a^\dagger(q_{m+n})) \Bigg] |0\rangle
$$

and Q_{lead}^- acts only on the terms in the square brackets. All these terms have only one trace, which is a scenario we already encountered in the large N limit case. Using the results of that discussion, we find that the only massless solution of the form (4.26) is the trivial one. This is the starting point of the induction procedure for finite N.

As explained earlier, we look for n parton Fock states in the expansion for $|\Psi\rangle$ that have $2k$ fermions ($k > 1$), To finish the proof we need to show that for any k the only allowed wave function is the trivial one. ¿From the large N result we know there are no such one trace states. We now consider the state with an arbitrary number of traces,

$$
|\Psi_{(n,2k)}\rangle = \sum_P \int_0^{P^+} \frac{ds_1 \ldots ds_n}{\sqrt{s_1 \ldots s_n}} \delta(s_1 + \cdots + s_n - P^+) \tag{4.27}
$$

$$
f_P^{(n,2k)}(s_1 \ldots s_{i_1} | \ldots | \ldots s_n) \mathrm{tr}\left(c^\dagger(s_1) \ldots c^\dagger(s_{i_1})\right) \mathrm{tr}(\ldots) \mathrm{tr}\left(\ldots c^\dagger(s_n)\right) |0\rangle,
$$

then the analog of (4.21) for such states reads:

$$
\sum_i' f_{P_i}^{(n,2k)}(s_1 \ldots | s_{j_a} \ldots s_{i-1}, s_i + s_{i+1}, s_{i+2} \ldots s_{j_a+k_a} | \ldots s_{n+1}) \frac{s_{i+1} - s_i}{(s_{i+1} + s_i)^{3/2}} = 0.
$$

$$
\tag{4.28}
$$

Here, $\sum_i'$ means that for each trace we should include one additional term with "i" $= j_a + k_a$, "$i+1$" $= j_a$ if c corresponding to both $j_a + k_a$ and j_a is a. If the number of traces is a, we introduce

$$
j_a = \sum_{b=1}^{a-1} k_b.
$$

If any of the blocks tr(...) in the state for which (4.28) is written contains two or more fermions, then, as in the large N case, all the corresponding wave functions $f_P^{(n,2k)}$ vanish. So we only need to consider the states of the form:

$$|\Psi_{(n,k_1+1,...)}\rangle = \sum_P \int dpdq f_P^{(n,k_1+1,...)}(p_1, q_1, \ldots, q_{k_1}|p_2, q_{k_1+1}, \ldots, q_{k_1+k_2}|\ldots) \times$$
$$\text{tr}\left(b^\dagger(p_1)a^\dagger(q_1)\ldots a^\dagger(q_{k_1})\right) \text{tr}\left(b^\dagger(p_2)a^\dagger(q_{k_1+1})\ldots a^\dagger(q_{k_1+k_2})\right)\ldots|0\rangle.$$

$$(4.29)$$

Let $\tilde{Q}$ denote that part of the supercharge Q^- which replaces a fermion with two bosons. Let us consider the result of such a change in the first trace. Suppose there are a traces having the same form as the first trace. Then without loss of generality, we may assume they are the first a traces. Then using the symmetries of the wave functions we find:

$$\tilde{Q}|\Psi_{(n,k_1+1,...)}\rangle = -\frac{1}{2\sqrt{2\pi}} \sum_P \int_0^{P^+} dkdpdq$$

$$f_P^{(n,k_1+1,...)}(p_1, q_1, \ldots, q_{k_1}|p_2, q_{k_1+1}, \ldots, q_{2k_1}|\ldots) \sum_{b=1}^a \frac{p_b - 2k}{p_b} \frac{1}{\sqrt{k(p_b-k)}}(-1)^{b+1} \times$$

$$\text{tr}\left(b^\dagger(p_1)a^\dagger(q_1)\ldots a^\dagger(q_{k_1})\right)\ldots \text{tr}\left(a^\dagger(k)a^\dagger(p_b-k)a^\dagger(q_{(b-1)k_1+1})\ldots a^\dagger(q_{bk_1})\right)\ldots|0\rangle$$

$$= -\frac{1}{2\sqrt{2\pi}} \sum_P \int_0^{P^+} dkdpdq \frac{p_1-2k}{p_1}\frac{1}{\sqrt{k(p_1-k)}}\text{tr}\left(a^\dagger(k)a^\dagger(p_1-k)a^\dagger(q_1)\ldots a^\dagger(q_{k_1})\right) \times$$

$$\text{tr}\left(b^\dagger(p_2)a^\dagger(q_{k_1+1})\ldots a^\dagger(q_{k_1+k_2})\right)\ldots|0\rangle \sum_{b=1}^a (-1)^{b+1}(-1)^{b+1} \times$$

$$f_P^{(n,k_1+1,...)}(p_1, q_1, \ldots, q_{k_1}|p_2, q_{k_1+1}, \ldots, q_{k_1+k_2}|\ldots).$$

If the above expression vanishes then the only solution is the trivial one in which all wave functions vanish. This finishes the proof of the induction procedure for the finite N case.

The extension of the proof to massive bound states is straightforward. Firstly, assume $|\Psi\rangle$ is a normalizable eigenstate of $2P^+P^-$ with mass squared $M^2 \neq 0$. Then, since $P^- = \frac{1}{\sqrt{2}}(Q^-)^2$, the state

$$|\tilde{\Psi}\rangle \equiv |\Psi\rangle + \alpha Q^-|\Psi\rangle \tag{4.30}$$

for $\alpha^2 = \sqrt{2}P^+/M^2$ is a normalizable eigenstate of the supercharge Q^-, with eigenvalue $1/\alpha$. We therefore study the eigen-problem $Q^-|\tilde{\Psi}\rangle = \frac{1}{\alpha}|\tilde{\Psi}\rangle$. The resulting constraints on the wave functions may be obtained by modifying our original expressions by including a wave function multiplied by a finite constant. However, in our analysis, we always need to let some of the momenta vanish, and therefore this additional contribution vanishes. The analysis (and therefore the conclusions) remains unchanged.

We therefore conclude that any normalizable SU(N) bound state (massless or massive) that exists in the model must be a superposition of an infinite number of Fock states.

C   Higher Dimensional Theories.

In this subsection we extend our theorem to the two–dimensional supersymmetric theories obtained as the result of dimensional reduction from $D > 3$ dimensions. The most important cases are $D = 4$, 6 and 10 which have 2, 4 and 8 supersymmetries in two dimensions. Below we consider only large N case, the generalization to arbitrary $SU(N)$ group is trivial repetition of the arguments given in previous subsection.

Again our starting point is the fact that if there is normalizable eigenstate of Hamiltonian having finite length than its main symbol satisfies the condition:

$$Q^-_{lead}|\Psi >= 0, \tag{4.31}$$

where Q^-_{lead} is the part of supercharge increasing the number of partons. In three dimensional case we had only one supercharge Q^-, for general D dimensional SYM reduced in $1 + 1$ there are $D - 2$ supercharges, each of them squared gives P^- and (4.31) should be true for all of them. In general different supercharges are not anticommute with each other, but since we consider quantization near trivial classical configuration (with no monopoles and no external charges) then they do. It is easy to derive the general form of supercharge:

$$Q^-_\alpha = \int_o^\infty \frac{dk}{k}(b^{\alpha\dagger}_{ij}(k)J_{ij}(-k) - (J_{ij}(-k))^\dagger b^\alpha_{ij}(k)) + \tag{4.32}$$
$$+ \frac{\mu}{2\sqrt{2\pi}} \int_{-\infty}^\infty dk M^{\alpha\beta}_{IJ}[A_I, A_J]_{ij}(k)\Psi^\beta_{ji}(-k),$$

$$J_{ij}(-k) = \frac{1}{2\sqrt{2\pi}} \int_o^\infty dp \frac{2p+k}{\sqrt{p(p+k)}} \left(a^{I\dagger}_{ki}(p)a^I_{kj}(k+p) - a^{I\dagger}_{jk}(p)a^I_{ik}(k+p)\right) + \tag{4.33}$$
$$+ \frac{1}{2\sqrt{2\pi}} \int_o^k dp \frac{k-2p}{\sqrt{p(k-p)}} a^I_{ik}(p)a^I_{kj}(k-p) + \frac{1}{\sqrt{2\pi}} \int_o^k dp\, b^\alpha_{ik}(p)b^\alpha_{kj}(k-p) +$$
$$+ \frac{1}{\sqrt{2\pi}} \int_o^\infty dp \left(b^{\alpha\dagger}_{ki}(p)b^\alpha_{kj}(k+p) - b^{\alpha\dagger}_{jk}(p)b^\alpha_{ik}(k+p)\right).$$

In the above expression we introduced $d = D-2$ kinds of bosons ($I = 1, ..., d$) and d kinds of fermions ($\alpha = 1, ..., d$) which we get as the result of compactification. The μ is nonzero constant depending on D and M are combinations of d dimensional Dirac matrices:

$$M_{IJ}^{\alpha\beta} = (\gamma_I \gamma_J^T - \gamma_J \gamma_I^T)_{\alpha\beta}. \tag{4.34}$$

As before our proof is based on the induction on the number of fermionic operators in the state. First we consider main symbol being superposition of purely bosonic states and ones containing two fermionic operators. Now we have d types of bosons and d types of fermions so some additional indices should be included in the wavefunctions. Defining bosonic indexes to be capital letters A, B... and fermionic ones to be Greek letters we write:

$$|\Psi, 0> = \int_0^{P^+} \frac{dq_1...dq_n}{\sqrt{q_1...q_n}} \delta(q_1 + ... + q_n - P^+) \sum_A f_{[A_1...A_n]}^{(0)}(q_1...q_n) \times$$
$$\times \, tr[a_{A_1}^\dagger(q_1)...a_{A_n}^\dagger(q_n)]|0>, \tag{4.35}$$

$$|\Psi, 2> = \sum_{k=1}^{n-1} \int_0^{P^+} \frac{dq_1...dq_n}{\sqrt{q_1...q_n}} \delta(q_1 + ... + q_n - P^+) \sum_{A,\alpha} f_{[A_1...A_{k-1}\alpha_1 A_k...A_{n-2}\alpha_2]}^{(2)k}(q_1...q_n) \times$$
$$\times \, tr[a_{A_1}^\dagger(q_1)...a_{A_{k-1}}^\dagger(q_{k-1})b_{\alpha_1}^\dagger(q_k)a_{A_k}^\dagger(q_{k+1})...a_{A_{n-2}}^\dagger(q_{n-1})b_{\alpha_2}^\dagger(q_n)]|0>. \tag{4.36}$$

It is now easy to find the one fermionic part of the result of action by (4.32) on the main symbol of the state. The vanishing of this contribution leads to the generalization of the equation (4.15):

$$\delta_{\alpha\beta} \frac{n}{p} \left(\frac{2q_n + p}{q_n + p} f_{A_1...A_n}^{(0)}(q_1...q_{n-1}, q_n + p) - \frac{2q_1 + p}{q_1 + p} f_{A_1...A_n}^{(0)}(q_1 + p...q_{n-1}, q_n) \right) -$$

$$\frac{2}{\sqrt{p}} \sum_{k=1}^{n-1} \frac{q_{k+1} - q_k}{(q_{k+1} + q_k)^{3/2}} \delta_{A_k A_{k+1}} f_{[A_1...A_{k-1}\alpha A_{k+2}...A_n\beta]}^{(2)k}(q_1...q_{k-1}, q_k + q_{k+1}, q_{k+2}...q_n, p) +$$

$$n\mu \left(\frac{M_{A_n B}^{\alpha\beta}}{q_n + p} f_{A_1...A_{n-1}B}^{(0)}(q_1...q_{n-1}, q_n + p) - \frac{M_{A_1 B}^{\alpha\beta}}{q_1 + p} f_{BA_2...A_n}^{(0)}(q_1 + p...q_{n-1}, q_n) \right) +$$

$$\frac{2\mu}{\sqrt{p}} \sum_{k=1}^{n-1} \frac{M_{A_{k+1} A_k}^{\alpha\gamma}}{\sqrt{q_{k+1} + q_k}} f_{[A_1...A_{k-1}\gamma A_{k+2}...A_n\beta]}^{(2)k}(q_1...q_{k-1}, q_k + q_{k+1}, q_{k+2}...q_n, p) = 0 \tag{4.37}$$

This equation should be true for any possible $A_1...A_N$, α and β. We will show that the only solution of such system of equations is trivial one so all the $f_{[...]}^{(0)}$ and $f_{[...]}^{(2)k}$ vanish. This will be proven by induction. First we note that if $A_1 = ... = A_n$ and $\alpha = \beta$ then equation (4.37) is reduced to (4.15) written for $f_{[A...A]}^{(0)}$ and $f_{[A...A\alpha A...A\alpha]}^{(2)k}$ and as we saw this leads to

$$f_{[A...A]}^{(0)} = 0, \qquad f_{[A...A\alpha A...A\alpha]}^{(2)k} = 0 \tag{4.38}$$

for arbitrary A and α. The next case to consider is $A_1 = ... = A_n$, $\alpha \neq \beta$. Using relation just found the (4.37) for this case again gives us (4.15), but this time correspondence reads:

$$f^{(0)} \to p\mu M_{AB}^{\alpha\beta} f_{[BA...A]}^{(0)},$$

$$f^{(2)k} \to f_{[A...A\alpha A...A\beta]}^{(0)}. \tag{4.39}$$

The proven property of (4.15) together with trivial identity

$$\sum_{\alpha\beta} M_{CA}^{\beta\alpha} M_{AB}^{\alpha\beta} = 4d\delta_{BC}(1 - \delta_{AC}) \tag{4.40}$$

leads to

$$f_{[BA...A]}^{(0)} = 0, \qquad f_{[A...A\alpha A...A\beta]}^{(2)k} = 0 \tag{4.41}$$

for any A, B, α, β (A could be equal to B and α to β). We use this equation as starting point of the induction procedure.

Let us introduce one useful function. For each set $\{A_1...A_k\}$ we define $\bar{n}(\{A_1...A_k\})$ to be the maximal number of identical A in the set:

$$\bar{n}(\{A_1...A_k\}) = \max_{I \le d} \left(\sum_{i=1}^{k} \theta(A_i - I) \right). \tag{4.42}$$

In terms of this new function our result (4.41) can be rewritten as

$$f_{[A_1...A_n]}^{(0)} = 0 \quad if \quad \bar{n}(\{A_1...A_n\}) \ge n - 1$$

$$f_{[A_1...A_{k-1}\alpha A_k...A_{n-2}]}^{(2)k} = 0 \quad if \quad \bar{n}(\{A_1...A_n\}) = n - 2. \tag{4.43}$$

This condition will be used as starting point of induction then the assumption of induction procedure is:

$$f_{[A_1...A_n]}^{(0)} = 0 \quad if \quad \bar{n}(\{A_1...A_n\}) \ge m$$

$$f_{[A_1...A_{k-1}\alpha A_k...A_{n-2}\beta]}^{(2)k} = 0 \quad if \quad \bar{n}(\{A_1...A_n\}) \ge m - 1 \tag{4.44}$$

and we checked it for $m = n - 1$. In our induction procedure we will decrease parameter m instead of increasing it. To perform the proof, we start with writing (4.37) for the set $\{A_1...A_k\}$ with $\bar{n} = m$:

$$\frac{2}{\sqrt{p}} \sum_{k=1}^{n-1} \frac{q_{k+1} - q_k}{(q_{k+1} + q_k)^{3/2}} \delta_{A_k A_{k+1}} f_{[A_1...A_{k-1}\alpha A_{k+2}...A_n\beta]}^{(2)k}(q_1...q_{k-1}, q_k + q_{k+1}, q_{k+2}...q_n, p) =$$

$$n\mu \left(\frac{M_{A_n B}^{\alpha\beta}}{q_n + p} f_{A_1..A_{n-1}B}^{(0)}(q_1..q_{n-1}, q_n + p) - \frac{M_{A_1 B}^{\alpha\beta}}{q_1 + p} f_{BA_2..A_n}^{(0)}(q_1 + p..q_{n-1}, q_n) \right). \tag{4.45}$$

If $\alpha = \beta$ then the right hand side is zero and we have a recurrent relations for $f_{[A_1...A_{k-1}\alpha A_{k+2}...A_n\alpha]}^{(2)k}$ with different k. Due to the presence of δ symbol these relations would connect only $f^{(2)k}$ inside some clusters (for $m < n$) and the boundary elements of such clusters should be zero. Thus we deduce that if $\bar{n}(\{A_1...A_{n-2}\}) = m - 2$ then

$$f^{(2)k}_{[A_1...A_{k-1}\alpha A_k...A_{n-2}\alpha]} = 0. \tag{4.46}$$

For $\alpha \neq \beta$ we consider different limits $q_i \to 0$ in (4.45):

$$\frac{2\sqrt{p}}{\sqrt{q_2}} \delta_{A_1 A_2} f^{(2)1}_{[\alpha A_3...A_n\beta]}(q_2...q_n, p) = -n\mu M^{\alpha\beta}_{A_1 B} f^{(0)}_{B A_2...A_n}(p...q_{n-1}, q_n) \tag{4.47}$$

for $i = 1$ and

$$\frac{1}{\sqrt{q_{i+1}}} \delta_{A_i A_{i+1}} f^{(2)i}_{[A_1...A_{i-1}\alpha A_{i+2}...A_n\beta]}(q_1...q_{i-1}, q_{i+1}...q_n, p) =$$

$$= \frac{1}{\sqrt{q_{i-1}}} \delta_{A_i A_{i-1}} f^{(2)i}_{[A_1...A_{i-2}\alpha A_{i+1}...A_n\beta]}(q_1...q_{i-1}, q_{i+1}...q_n, p) \tag{4.48}$$

for $1 < i < n$. Since $\bar{n} < n - 1$ then there exist $i < n$: $\delta_{A_i A_{i-1}} = 0$ then from above equations we deduce for $\alpha \neq \beta$:

$$f^{(0)}_{[A_1...A_n]} = 0, \quad \bar{n}(\{A_1...A_n\}) = m - 1$$

$$f^{(2)k}_{[A_1...A_{k-1}\alpha A_k...A_{n-2}\beta]} = 0, \quad \bar{n}(\{A_1...A_n\}) = m - 2. \tag{4.49}$$

This finishes the proof by induction. Thus we have proven that equation (4.37) doesn't have any normalizable solutions.

To show that there are no finite length bound states we now turn to the analog of equation (4.21). This analog reads:

$$\sum_i A_i f^{(2k)}_{P_i}(s_1...s_{i-1}, s_i + s_{i+1}, s_{i+2}...s_{n+1}) \times$$

$$\times \left(\delta^{\alpha\alpha_i}_{A_i A_{i+1}} \frac{s_{i+1} - s_i}{(s_{i+1} + s_i)^{3/2}} + \mu M^{\alpha\alpha_i}_{A_i A_{i+1}} \frac{1}{\sqrt{s_{i+1} + s_i}} \right) = 0, \tag{4.50}$$

where P_i describes different permutations of A and α. This equation gives linear relations between wavefunctions inside the block of a in

$$tr(...b^\dagger a^\dagger...a^\dagger b^\dagger...)|0 >, \tag{4.51}$$

"boundary" elements (when index i corresponds to fermions) vanish, so as in the case $3 \to 2$ all the $f^{(2k)}$ are zero. This completes the proof for general compactification.

D   Bound States in DLCQ.

In the previous subsection we proved that the continuum formulation of the theory does not have any normalizable bound states with a finite number of partons. Our proof used the behavior of wave functions at small momenta arising from the normalizability assumption. Neither of these properties can be used in DLCQ, however. Here we consider some simple examples of massless DLCQ solutions with n bosons to help shed some light on the relation between DLCQ solutions and the solutions of the continuum theory. For simplicity, we work in the large N limit case.

We write the momentum of a state in DLCQ in terms of the momentum fraction q_i where $q_i = \frac{r_i}{r} P^+$, and the r_i are positive integers. The wave function of such a state is $f^{(n,0)}(r_1, \ldots, r_n)$. There are two conditions that must be satisfied to show that it is massless. One is the that the coefficient of the term with one additional fermion that is produced by the action of Q^- is zero. This condition gives the relation,

$$\frac{2r_n + t}{r_n + t} f^{(n,0)}(r_1, \ldots, r_{n-1}, r_n + t) - \frac{2r_{n-1} + t}{r_{n-1} + t} f^{(n,0)}(r_1, \ldots, r_{n-1} + t, r_n) = 0. \quad (4.52)$$

where t correspond to the momentum fraction of the one fermion. The second is that the coefficient of the state with two fewer bosons and one additional fermion which is also produced by the action of Q^- is zero. This condition gives the relation,

$$\sum_{k,t} \frac{t - 2k}{k(t - k)} f^{(n,0)}(r_1, \ldots, r_{n-2}, k, t - k)\delta_{(r_{n-1}+r_n,t)} = 0. \quad (4.53)$$

For the case where all $r_i = 1$, and the total harmonic resolution is n, it is trivial that eqn(4.52) is satisfied since there is not enough resolution to increase the number of particles in the state. It is also easy to see from eqn (4.53) since the coefficient of the one term in the sum is zero. Thus the wave function $f^{(n,0)}(1, 1, \ldots.1)$ is a massless state for every resolution

To discuss additional solutions it is useful to start by considering eqn(4.52). The case $t = 1$, gives the equation

$$f^{(n,0)}(r_1, \ldots, r_{n-2}, r_{n-1}, r_n + 1) = \frac{2r_{n-1} + 1}{2r_n + 1} \frac{r_n + 1}{r_{n-1} + 1} f^{(n,0)}(r_1, \ldots, r_{n-2}, r_{n-1} + 1, r_n).$$

$$(4.54)$$

This equation is trivial to satisfy if $r_i = 1$ for all i. The contributions in eqn(4.53) come from the two terms in the sum, $k = 1$, $t = 3$ and $k = 2$, $t = 3$. Each term has the same coefficient but of opposite sign and cancel. Therefore the state $f^{(n,0)}(1, \ldots 1, 2)$ is a massless state for all resolutions

The next case $t = 2$ in eqn(4.52) gives,

$$f^{(n,0)}(r_1, \ldots, r_{n-2}, r_{n-1}, r_n + 2) = \frac{2r_{n-1} + 2}{2r_n + 2} \frac{r_n + 2}{r_{n-1} + 2} f^{(n,0)}(r_1, \ldots, r_{n-2}, r_{n-1} + 2, r_n).$$

$$(4.55)$$

Using (4.54) twice we find:

$$f^{(n,0)}(r_1 \ldots r_{n-2}, r_{n-1}, r_n + 2) = \frac{2r_{n-1} + 1}{2r_n + 3} \frac{r_n + 2}{r_{n-1} + 1} f^{(n.0)}(r_1 \ldots r_{n-2}, r_{n-1} + 1, r_n + 1) =$$

$$= \frac{2r_{n-1} + 1}{2r_n + 3} \frac{r_n + 2}{r_{n-1} + 1} \frac{2r_{n-1} + 3}{2r_n + 1} \frac{r_n + 1}{r_{n-1} + 2} f^{(n,0)}(r_1 \ldots r_{n-2}, r_{n-1} + 2, r_n). \qquad (4.56)$$

Comparing with (4.55) we have:

$$f^{(n,0)}(r_1 \ldots r_{n-2}, r_{n-1} + 2, r_n) \left(\frac{(r_n + 1)^2}{(2r_n + 3)(2r_n + 1)} - \frac{(r_{n-1} + 1)^2}{(2r_{n-1} + 3)(2r_{n-1} + 1)} \right) = 0.$$

$$(4.57)$$

Using relation (4.52) several times we can always express an arbitrary wave function in the following form:

$$f^{(n,0)}(r_1 \ldots r_n) = C(r_1 \ldots r_n) f^{(n,0)}(1 \ldots 1, L + 1, 1) \qquad (4.58)$$

where $L = r_1 + \ldots + r_n - n$ and $C(r_1 \ldots r_n)$ is some nonzero coefficient. The two massless states we found above correspond to $L = 0$ and $L = 1$. Choosing $r_1 = \ldots = r_{n-2} = r_n = 1$ in (4.57) we find,

$$f^{(n,0)}(1 \ldots 1, (L - 1) + 2, 1) = 0 \quad for \quad L > 2 \qquad (4.59)$$

due to monotonic behavior of the function in the parenthesis. Then using (4.58) we conclude that all the wave functions with $L > 2$ vanish. So the only case we need consider is $L = 2$. In this case (4.52) has only two nontrivial cases: $t = 1$ and $t = 2$ which are given by (4.54) and (4.55). In the second of these equations we can only have $r_1 = \ldots = r_n = 1$ so it is trivially satisfied. Equation (4.54) however gives a nontrivial relation for the wave function:

$$f^{(n,0)}(1, \ldots, 1, 2, 2) = f^{(n,0)}(1, \ldots, 2, 1, 2) = \ldots =$$

$$f^{(n,0)}(2, \ldots, 1, 1, 2) = \frac{10}{9} f^{(n,0)}(1, \ldots, 1, 3). \qquad (4.60)$$

finally we must show that eqn(4.53) is satisfied which is straight forward.

These are only a few examples of massless states, and there are in fact many more in DLCQ [4]. But the results of our numerical analysis show that the states we just described are closely connected with the massless states in the continuum. Let us formulate this relation for $N = \infty$. In this case only single trace states should be kept in the spectrum and DLCQ massless states have the following structure.

The state first appear as $tr(a^\dagger(1) \ldots a^\dagger(1))$ at resolution P, then one can trace it to resolutions $P + 1$ and $P + 2$ as states with wavefunctions (4.59) and (4.60). As we just proved at higher resolutions there are no massless states containing exactly K partons, however at any resolution $K \geq P$ there is exactly one massless state whose wavefunction is localized predominantly in the sector with P partons $a^\dagger$. So it is natural to collect all such states in the single sequence and to call the limits of this sequence "continuum massless state with P bosons", although as we saw the wavefunction of continuum state has contributions from sectors with different number of partons. The interesting feature of this theory is that such "continuum massless states with P bosons" are the only bosonic massless states seen by DLCQ (in principle the theory in the continuum might have massless state whose wavefunction is localized in sector with infinite number of partons, but we will ignore this possibility). Thus one can easily count bosonic massless states at any resolution P: they are just images of states with P bosons for all $P \leq K$, thus there are $K - 1$ such states. Acting on any of such states by Q^+ we can get the fermionic massless state (then there are also $K - 1$ of them), while acting by Q^- doesn't give any new state (the result is zero). We will do the counting for finite N case in the next subsection.

In the continuum limit we have proven that there are no massless normalizable states with a finite number of particles. However, at each finite value of the harmonic resolution, one obtains an exactly massless bound state, but as the harmonic resolution is sent to infinity, the number of Fock states required to keep the bound state massless must also be infinite.

E   Counting of Massless States in DLCQ.

Finally in this section we will count the massless states in DLCQ as function of resolution, keeping the number of colors N finite. However we will assume that N is not too small, so that the relation between N and resolution K: $K < N^2 - 1$ is satisfied. We will need this condition in order to insure all states of the form $tr(c^\dagger \ldots c^\dagger) \ldots tr(c^\dagger \ldots c^\dagger)|0\rangle$ are linearly independent unless they are related by either cyclic permutations in one of the traces or permutations of traces themselves. The simplest example of violation of this condition is the state

$$tr(a^\dagger(1)a^\dagger(1)a^\dagger(1))|0\rangle = 0$$

for $SU(2)$ (here $3 = 4 - 1$). Although for $K > N^2 - 1$ some conclusions can be made, the different N and K requires special consideration and we are not going to proceed in this direction. From the numerical perspective we should mention that in our calculation $K < 11$, so the only excluded values of N are 2 and 3.

As soon as the condition $K < N^2 - 1$ is satisfied the DLCQ Fock spaces for $SU(N)$ and $SU(\infty)$ are the same if all multitrace states are taken into account. Moreover our numerical analysis strongly suggests that the number of massless states is the same for all $N > \sqrt{K + 1}$, while wavefunctions depend on N. However

185

talking about $SU(\infty)$ Fock space one usually considers only single trace states as fundamental ones, while multitrace states are though of as the system of free bound states. Let us explain the reason for this. The light–cone Hamiltonian can be written in the following schematic form:

$$p^- = \frac{1}{N}P^- = \alpha c_{ij}^\dagger c_{ij} + \frac{1}{N}\beta c_{ij}^\dagger(ccc)_{ij} + \frac{1}{N}\beta(c^\dagger c^\dagger)_{ij}(cc)_{ij} + \frac{1}{N}\beta(c^\dagger c^\dagger c^\dagger)_{ij}(c)_{ij}. \quad (4.61)$$

The $\frac{1}{N}$ is introduced in order to make the eigenvalues of p^- finite as $N \to \infty$. Let us consider two eigenstates of p^-, which are chosen to be combination of single traces in the large N limit:

$$p^- A|0\rangle = m_A A|0\rangle, \quad (4.62)$$
$$p^- B|0\rangle = m_A B|0\rangle. \quad (4.63)$$

This is equivalent to the following commutation relations:

$$[p^-, A] = m_A A + \frac{1}{N}\sum \mu_A \mathrm{ntr}(c^\dagger \ldots c^\dagger c) + \frac{1}{N}\sum \nu_A \mathrm{tr}(c^\dagger \ldots c^\dagger cc) + O\left(\frac{1}{N^2}\right),$$
$$[p^-, B] = m_B B + \frac{1}{N}\sum \mu_B \mathrm{ntr}(c^\dagger \ldots c^\dagger c) + \frac{1}{N}\sum \nu_B \mathrm{ntr}(c^\dagger \ldots c^\dagger cc) + O\left(\frac{1}{N^2}\right).$$
$$(4.64)$$

We introduced a convenient notation for the normalized trace here:

$$\mathrm{ntr}(\underbrace{c^\dagger \ldots c^\dagger c \ldots c}_{n}) = \frac{1}{N^{n/2}}(c^\dagger \ldots c^\dagger)_{ij}(c \ldots c)_{ij}. \quad (4.65)$$

This way the state $\mathrm{ntr}(c^\dagger \ldots c^\dagger)|0\rangle$ has a finite norm in the large N limit. ¿From the equations (4.64) one can easily see that

$$p^- AB|0\rangle = (m_A + m_B)AB|0\rangle + O\left(\frac{1}{N}\right), \quad (4.66)$$

i.e. we indeed have a combination of two free states in the large N limit. In particular this fact may be applied toward the classification of DLCQ massless states at $N = \infty$: the multitrace state is massless if and only if all of the traces involved correspond to massless states. We also mention the trivial fact that if state A has resolution K_A and B has K_B then AB is the state at resolution $K_A + K_B$.

Let us summarize what we have learned so far. The number of massless states in $SU(N)$ theory at resolution $K < N^2 - 1$ is the same as one for $SU(\infty)$ theory if the multitrace states are included in the latter. On the other hand due to the fact that multitrace massless states in $SU(\infty)$ have special structure (namely any single trace in them is massless state itself), their number can be calculated from the known number of massless states written as linear combination of single traces. The remaining part of this subsection is devoted to such calculation.

As we found in the end of the last subsection there are $2(K-1)$ single trace massless states at resolution K. We will show how this information can be used in order to count the total number of massless states. Let us introduce some notation first. The value we want to calculate is N_k — the number of massless states at resolution k. We also define $N_k^{(m)}$ as number of such massless states at resolution k that the resolution of any single traces in them is greater or equal to m. Then for example $N_k = N_k^{(2)}$ and $N_k^{(k)} = 2(k-1)$. Finally we define $f_n(m)$ to be the number of different massless states containing n traces, each of which corresponds has resolution *equal* to m. Then one can derive the recurrent relation:

$$N_k^{(m)} = N_k^{(m+1)} + \sum_{n=1}^{[k/m]} f_n(m) \left(N_{k-mn}^{(m+1)} + \delta_{mn}^{k} \right). \tag{4.67}$$

The starting point of the recurrent procedure are the relations $N_k^{(k)} = 2(k-1)$ and $N_k^{(m)} = 0$ for $m > k$. In order to apply (4.67) we only have to evaluate $f_n(m)$. This will be our next task.

To calculate $f_n(m)$ we first assume that we have only bosonic traces at our disposal. Let us count the states which contain p such traces. After combining identical traces together one can reduce the problem further by considering only states with special trace structure. Namely we will concentrate our attention on the massless states having n_i traces of type i, for different values of $i = 1 \ldots r$. Without the loss of generality we can require $n_1 \geq \ldots \geq n_r \geq 1$, one can also see that the relation $\sum n_i = p$ holds. If all n_i are different then the number of massless states with fixed structure is given by simple formula:

$$g(m) \, (g(m) - 1) \ldots (g(m) - r + 1) \,,$$

where $g(m) = m-1$ is the number of bosonic single trace massless states. In general one gets additional combinatoric coefficient $C(n_1, \ldots, n_r)$ in the last expression, it is defined by the following rules:

$$C(n_1, \ldots, n_i, n_{i+1}, \ldots, n_r) = C(n_1, \ldots, n_i) C(n_{i+1}, \ldots, n_r), \quad \text{if} \quad n_i > n_{i+1},$$

$$C(\underbrace{n, \ldots, n}_{a}) = \frac{1}{a!}. \tag{4.68}$$

Now let us include fermionic traces in the picture. The only difference between them and bosons is the Pauli principle, so considering the product of q fermionic traces one has to choose all of them to be different. Thus the coefficient C for this case is

$$C_F(q) = C(\underbrace{1, \ldots, 1}_{q}) = \frac{1}{q!}. \tag{4.69}$$

Collecting all the information together we finally get:

187

$$f_n(m) = \sum_{q=0}^{n} \frac{1}{q!} g(m) \left(g(m) - 1\right) \ldots \left(g(m) - q + 1\right) \times \qquad (4.70)$$

$$\times \sum_{r=0}^{n-q} \mathcal{F}(n - q, r) g(m) \left(g(m) - 1\right) \ldots \left(g(m) - r + 1\right),$$

$$\mathcal{F}(p, r) = \sum_{\{n_1, \ldots, n_r\}} C(n_1, \ldots, n_r), \qquad \begin{array}{l} n_1 + \ldots + n_r = p \\ n_1 \geq \ldots \geq n_r \geq 1. \end{array} \qquad (4.71)$$

Now we can use the relations (4.70), (4.71) and (4.68) to determine all the coefficients $f_n(m)$ and then substituting them to the recurrent relation (4.67) one can find all the $N_k^{(m)}$. This in turn leads to the results for $N_k = N_k^{(2)}$. Although we were not able to find an analytic expression for N_k as function of k, the number of states can be evaluated numerically for arbitrary k using the procedure we just described. We performed such calculations using Mathematica and the results for lowest resolutions are summarized in the Table1. For instance one can see that up to resolution 5 $N_k = 2^{k-1}$, but at higher resolutions this relations holds only approximately.

k	2	3	4	5	6	7	8	9	10	11	12
N_k	2	4	8	16	32	60	114	212	384	692	1232

TABLE 1. Number of multitrace massless states as function of resolution.

V   CORRELATION FUNCTIONS IN SYM AND DLCQ.

The bound state problem we have studied so far is the traditional one for DLCQ. However this is not the only calculation that can be done using this method. The problem of computing of correlation functions, more traditional for conventional quantum field theory, can also be addressed in the light cone quantization. Unlike the usual methods of QFT the results of DLCQ calculations are valid beyond the perturbation theory and thus they can be used for the testing the duality between the gauge theory and supergravity.

There has been a great deal of excitement during this past year following the realization that certain field theories admit concrete realizations as a string theory on a particular background [60]. By now many examples of this type of correspondence for field theories in various dimensions with various field contents have been reported in the literature (for a comprehensive review and list of references, see [2]). However, attempts to apply these correspondences to study the details of these theories have only met with limited success so far. The problem stems from the fact that our understanding of both sides of the correspondence is limited. On the field theory side, most of what we know comes from perturbation theory where we assume that the coupling is weak. On the string theory side, most of what we know comes from the supergravity approximation where the curvature is small. There are no known situations where both approximations are simultaneously valid. At the present time, comparisons between the dual gauge/string theories have been restricted to either qualitative issues or quantities constrained by symmetry. Any improvement in our understanding of field theories beyond perturbation theory or string theories beyond the supergravity approximation is therefore a welcome development.

We will study the field theory/string theory correspondence motivated by considering the near-horizon decoupling limit of a D1-brane in type IIB string theory [49]. The gauge theory corresponding to this theory is the Yang-Mills theory in two dimensions with 16 supercharges. Its SDLCQ formulation was recently reported in [8]. This is probably the simplest known example of a field theory/string theory correspondence involving a field theory in two dimensions with a concrete Lagrangian formulation.

A convenient quantity that can be computed on both sides of the correspondence is the correlation function of gauge invariant operators [40,82]. We will focus on two point functions of the stress-energy tensor. This turns out to be a very convenient quantity to compute for many reasons that we will explain along the way. Some aspects of this as it pertains to a consideration of black hole entropy was recently discussed in [41]. There are other physical quantities often reported in the literature. In the DLCQ literature, the spectrum of hadrons is often reported. This would be fine for theories in a confining phase. However, we expect the SYM in two dimension to flow to a non-trivial conformal fixed point in the infra-red [49,30]. The spectrum of states will therefore form a continuum and will be cumbersome to handle. On the string theory side, entropy density [23] and the quark anti-quark potential [23,72,61] are frequently reported. The definition of entropy density requires that we place the field theory in a space-like box which seems incommensurate with the discretized light cone. Similarly, a static quark anti-quark configuration does not fit very well inside a discretized light-cone geometry. The correlation function of point-like operators do not suffer from these problems. We should mention that there exists interesting work on computing the QCD string tension [14,15] directly in the field theory. These authors find that the QCD string tension vanishes in the supersymmetric theories which is consistent with the power law quark anti-quark potential found on the supergravity side.

A   Correlation Functions from Supergravity

Let us begin by reviewing the computation of the correlation function of stress energy tensors on the string theory side using the supergravity approximation. The computation is essentially a generalization of [40,82]. The main conclusion on the supergravity side was reported recently in [41] but we will elaborate further on the details. The near horizon geometry of a D1-brane in string frame takes the form

$$ds^2 = \alpha' \left(\frac{U^3}{\sqrt{64\pi^3 g_{YM}^2 N}} dx_{\parallel}^2 + \frac{\sqrt{64\pi^3 g_{YM}^2 N}}{U^3} dU^2 + \sqrt{64\pi^3 g_{YM}^2 N} U d\Omega_{8-p}^2 \right)$$

$$e^\phi = 2\pi g_{YM}^2 \left(\frac{64\pi^3 g_{YM}^2 N}{U^6} \right)^{\frac{1}{2}}. \tag{5.1}$$

In order to compute the two point function, we need to know the action for the diagonal fluctuations around this background to the quadratic order. What we need is an analogue of [53] for this background which unfortunately is not currently available in the literature. Fortunately, some diagonal fluctuating degrees of freedom can be identified by following the early work on black hole absorption cross-subsections [55,39]. In particular, we can show that the fluctuations parameterized according to

$$ds^2 = \Big(1 + f(x^0, U) + g(x^0, U)\Big) g_{00}(dx^0)^2 + \Big(1 + 5f(x^0, U) + g(x^0, U)\Big) g_{11}(dx^1)^2$$

$$+ \Big(1 + f(x^0, U) + g(x^0, U)\Big) g_{UU} dU^2 + \Big(1 + f(x^0, U) - \frac{5}{7}g(x^0, U)\Big) g_{\Omega\Omega} d\Omega_7^2$$

$$e^\phi = \Big(1 + 3f(x^0, U) - g(x^0, U)\Big) e^{\phi_0} \tag{5.2}$$

will satisfy the equations of motion

$$f''(U) + \frac{7}{U} f'(U) - \frac{64\pi^3 g_{YM}^2 N k^2}{U^6} f(U) = 0$$

$$g''(U) + \frac{7}{U} g'(U) - \frac{72}{U^2} g(U) - \frac{64\pi^3 g_{YM}^2 N k^2}{U^6} g(U) = 0 \tag{5.3}$$

by direct substitution into the equations of motion in 10 dimensions. We have assumed without loss of generality that these fluctuation vary only along the x^0 direction of the world volume coordinates like a plane wave e^{ikx^0}. The fields $f(U)$ and $g(U)$ are scalars when the D1-brane is viewed as a black hole in 9 dimensions; in fact there are the minimal and the fixed scalars in this black hole geometry. In 10 dimensions, however, we see that they are really part of the gravitational fluctuation. We expect therefore that they are associated with the stress-energy tensor in the operator field correspondence of [40,82]. In the case of the correspondence between $\mathcal{N} = 4$ SYM and $AdS_5 \times S_5$, superconformal invariance allowed the identification of operators and fields in short multiplets [33]. For the D1-brane, we do not have superconformal invariance and this technique is not applicable. In fact, we expect all fields of the theory consistent with the symmetry of a given operator to mix. The large distance behavior should then be dominated by the contribution with the longest range. The field $f(k^0, U)$ appears to be the one with the longest range since it is the lightest field.

The equation (5.3) for $f(U)$ can be solved explicitly in terms of the Bessel's function

$$f(U) = U^{-3}K_{3/2}(\sqrt{16\pi^3 g_{YM}^2 N U^{-2}}k).$$ (5.4)

By thinking of $f(U)$ in direct analogy with the minimally coupled scalar as was done in [40,82], we can compute the flux factor

$$\mathcal{F} = \lim_{U_0 \to \infty} \frac{1}{2\kappa_{10}^2}\sqrt{g}g^{UU}e^{-2(\phi-\phi_\infty)}\partial_U \log(f(U))\bigg|_{U=U_0} = \frac{NU_0^2 k^2}{2g_{YM}^2} - \frac{N^{3/2}k^3}{4g_{YM}} + \cdots$$ (5.5)

up to a numerical coefficient of order one which we have suppressed. We see that the leading non-analytic (in k^2) contribution is due to the k^3 term, whose Fourier transform scales according to[7]

$$\langle \mathcal{O}(x)\mathcal{O}(0)\rangle = \frac{N^{\frac{3}{2}}}{g_{YM}x^5}.$$ (5.6)

This result passes the following important consistency test. The SYM in 2 dimensions with 16 supercharges have conformal fixed points in both UV and IR with central charges of order N^2 and N, respectively. Therefore, we expect the two point function of stress energy tensors to scale like N^2/x^4 and N/x^4 in the deep UV and IR, respectively. According to the analysis of [49], we expect to deviate from these conformal behavior and cross over to a regime where supergravity calculation can be trusted. The cross over occurs at $x = 1/g_{YM}\sqrt{N}$ and $x = \sqrt{N}/g_{YM}$. At these points, the N scaling of (5.6) and the conformal result match in the sense of the correspondence principle [48].

B   Correlation functions from DLCQ

The challenge then is to attempt to reproduce the scaling relation (5.6), fix the numerical coefficient, and determine the detail of the cross-over behavior using SDLCQ. Ever since the original proposal [31], the question of equivalence between quantizing on a light-cone and on a space-like slice have been discussed extensively. This question is especially critical whenever a massless particle or a zero-mode in the quantization is present. It is generally believed that the massless theories can be described on the light-cone as long as we take $m \to 0$ as a limit. The issue of zero mode have been examined by many authors. Some recent accounts can be found in [44,25,13,9,84]. Generally speaking, supersymmetry seems to save SDLCQ from complicated zero-mode issues. We will not contribute much to these discussions. Instead, we will formulate the computation of the correlation function of stress energy tensor in naive DLCQ. To check that these results are sensible, we will first do the computation for the free fermions. Extension to SYM with 16 supercharges will be essentially straightforward, except for one caveat. In order to actually evaluate the correlation functions, we must resort to numerical analysis at the last

---

[7] It is not difficult to show that for a generic p-brane, $\langle \mathcal{O}(x)\mathcal{O}(0)\rangle = N^{\frac{7-p}{5-p}}g_{YM}^{-\frac{2(3-p)}{5-p}}x^{-\frac{19+2p-p^2}{5-p}}$.

stage of the computation. For the SYM with 16 supercharges, this problem grows too big too fast to be practical on desk top computer where the current calculations were performed. We can only provide an algorithm, which, when executed on an much more powerful computer, should reproduce (5.6). Nonetheless, the fact that we can define a concrete algorithm seems to be a progress in the right direction. One potential pit-fall is the fact that the computation may not show any sign of convergence. If this is the case, or if it converges to a result at odds with (5.6), we must go back and re-examine the issue of equivalence of forms and the issue of zero modes.

The technique of DLCQ is reviewed by many authors [24,29] so we will be brief here. The basic idea of light-cone quantization is to parameterize the space using light cone coordinates x^+ and x^- and to quantize the theory making x^+ play the role of time. In the discrete light cone approach, we require the momentum $p_- = p^+$ along the x^- direction to take on discrete values in units of p^+/k where p^+ is the conserved total momentum of the system and k is an integer commonly referred to as the harmonic resolution. One can think of this discretization as a consequence of compactifying the x^- coordinate on a circle with a period $2L = 2\pi k/p^+$. The advantage of discretizing the light cone is the fact that the dimension of the Hilbert space becomes finite. Therefore, the Hamiltonian is a finite dimensional matrix and its dynamics can be solved explicitly. In SDLCQ one makes the DLCQ approximation to the supercharges and these discrete representations satisfy the supersymmetry algebra. Therefore SDLCQ enjoys the improved renormalization properties of supersymmetric theories. Of course, to recover the continuum result, we must send k to infinity and as luck would have it, we find that SDLCQ usually converges faster than the naive DLCQ. Of course, in the process, the size of the matrices will grow, making the computation harder and harder.

Let us now return to the problem at hand. We would like to compute a general expression of the form

$$F(x^-, x^+) = \langle \mathcal{O}(x^-, x^+)\mathcal{O}(0,0)\rangle \ . \tag{5.7}$$

In DLCQ, where we fix the total momentum in the x^- direction, it is more natural to compute its Fourier transform

$$\tilde{F}(P_-, x^+) = \frac{1}{2L}\langle \mathcal{O}(P_-, x^+)\mathcal{O}(-P_-, 0)\rangle \ . \tag{5.8}$$

This can naturally be expressed in a spectrally decomposed form

$$\tilde{F}(P_-, x^+) = \sum_i \frac{1}{2L}\langle 0|\mathcal{O}(P_-)|i\rangle e^{-iP_+^i x^+}\langle i|\mathcal{O}(-P_-, 0)|0\rangle \ . \tag{5.9}$$

C   Correlator for Free Dirac Fermions

Let us first consider evaluating this expression for the stress-energy tensor in the theory of free Dirac fermions as a simple example. The Lagrangian for this theory is

$$\mathcal{L} = i\bar{\Psi}\partial\!\!\!/\,\Psi - m\bar{\Psi}\Psi \tag{5.10}$$

where for concreteness, we take $\gamma^0 = \sigma^2, \gamma^1 = i\sigma^1$ and we take $\Psi = 2^{-1/4}\binom{\psi}{\chi}$. In terms of the spinor components, the Lagrangian takes the form

$$\mathcal{L} = i\psi^*\partial_+\psi + i\chi^*\partial_-\chi - \frac{im}{\sqrt{2}}(\chi^*\psi - \psi^*\chi) \ . \tag{5.11}$$

Since we treat x^+ as time and since χ does not have any derivatives with respect to x^+ in the Lagrangian, it can be eliminated from the equation of motion, leaving a Lagrangian which depends only on ψ:

$$\mathcal{L} = i\psi^*\partial_+\psi + i\frac{m^2}{2}\psi^*\frac{1}{\partial_-}\psi \ . \tag{5.12}$$

We can therefore express the canonical momentum and energy as

$$P_- = \int dx^- \, i\psi^*\partial_-\psi$$

$$P_+ = \int dx^- \, -\frac{im^2}{2}\psi^*\frac{1}{\partial_-}\psi \ . \tag{5.13}$$

In DLCQ, we compactify x^- to have period $2L$. We can then expand ψ and ψ^* in modes

$$\psi = \frac{1}{\sqrt{2L}}\left(b(n)e^{-\frac{in\pi}{L}x^-} + d(-n)e^{\frac{in\pi}{L}x^-}\right)$$

$$\psi^* = \frac{1}{\sqrt{2L}}\left(b(-n)e^{\frac{in\pi}{L}x^-} + d(n)e^{-\frac{in\pi}{L}x^-}\right) \ . \tag{5.14}$$

Operators $b(n)$ and $d(n)$ with positive and negative n are interpreted as a destruction and creation operators, respectively. In a theory with only fermions, it is customary to take anti-periodic boundary condition in order to avoid zero-mode issues. Therefore, n will take on half-integer values[8]. They satisfy the anticommutation relation

$$\{b(n), b(-m)\} = \{d(n), d(-m)\} = \delta_{n,m} \ . \tag{5.15}$$

Now we are ready to evaluate (5.9) in DLCQ. As a simple and convenient choice, we take

$$\mathcal{O}(-k) = \frac{1}{2}\int dx^- \, (i\psi^*\partial_-\psi - i(\partial_-\psi^*)\psi)\,e^{-\frac{ik\pi}{L}x^-}. \tag{5.16}$$

which is the Fourier transform of the local expression for P_- with the total derivative contribution adjusted to make this operator Hermitian. Therefore, this should be

---

[8] In SDLCQ one must use periodic boundary condition for all the fields to preserve the supersymmetry.

thought of as the T^{++} component of the stress energy tensor. For reasons that will become clear as we go on, this turns out to be one of the simplest things to compute. When acted on the vacuum, this operator creates a state

$$T^{++}(-k)|0\rangle = \frac{\pi}{L}\left(\frac{k}{2} - n\right)b(-k+n)d(-n)|0\rangle .$$ (5.17)

Since the fermions in this theory are free, the plane wave states

$$|n\rangle = b(-k+n)d(-n)|0\rangle$$ (5.18)

constitute an eigenstate. The spectrum can easily be determined by commuting these operators:

$$M_n^2|n\rangle = 2P_-P_+|n\rangle = m^2\left(\frac{k}{n} + \frac{k}{k-n}\right)|n\rangle$$ (5.19)

which is simply the discretized version of the spectrum of a two body continuum. All that we have to do now is calculate eigenstates of the actual theory we are interested in and to assemble these pieces into (5.9), but we can do a little more to make the result more presentable. The point is that since (5.9) is expressed in mixed momentum/position space notation in Minkowski space, the answer is inherently a complex quantity that is cumbersome to display. For the computation of two point function, however, we can go to position space by Fourier transforming with respect to the L variable. After Fourier transforming, it is straight forward to Euclideanize and display the two point function as a purely real function without loosing any information. To see how this works, let us write (5.9) in the form

$$\tilde{F}(P_-, x^+) = \left|\frac{L}{\pi}\langle 0|T^{++}(k)|n\rangle\right|^2 \frac{1}{2L}\frac{\pi^2}{L^2}e^{\frac{-iM_n^2}{2(\frac{k\pi}{L})}x^+} .$$ (5.20)

The quantity inside the absolute value sign is designed to be independent of L. Now, to recover the position space form of the correlation function, we inverse Fourier transform with respect to $P_- = k\pi/L$.

$$F(x^-, x^+) = \left|\frac{L}{\pi}\langle 0|T^{++}(k)|n\rangle\right|^2 \int \frac{d\left(\frac{k\pi}{L}\right)}{2\pi}\frac{1}{2L}\frac{\pi^2}{L^2}e^{-i\frac{M_n^2}{2(\frac{k\pi}{L})}x^+ - i\frac{k\pi}{L}x^-} .$$ (5.21)

The integral over L can be done explicitly and gives

$$F(x^-, x^+) = \left|\frac{L}{\pi}\langle 0|T^{++}(k)|n\rangle\right|^2 \left(\frac{x^+}{x^-}\right)^2 \frac{M_n^4}{8\pi^2 k^3}K_4\left(M_n\sqrt{2x^+x^-}\right)$$ (5.22)

where $K_4(x)$ is the 4-th modified Bessel's function. We can now continue to Euclidean space by taking $r^2 = 2x^+x^-$ to be real and considering the quantity

$$\left(\frac{x^-}{x^+}\right)^2 F(x^-, x^+) = \left|\frac{L}{\pi}\langle 0|T^{++}(k)|n\rangle\right|^2 \frac{M_n^4}{8\pi^2 k^3} K_4(M_n r) . \tag{5.23}$$

This is a fundamental result which we will refer to a number of times in this paper. It has explicit dependence on the harmonic resolution parameter k, but all dependence on unphysical quantities such as the size of the circle in the x^- direction and the momentum along that direction have been canceled. For the free fermion model, (5.23) evaluates to

$$\left(\frac{x^-}{x^+}\right)^2 F(x^-, x^+) = \frac{N}{k}\sum_n \frac{M_n^4}{32\pi^2}\frac{(k-2n)^2}{k^2} K_4(M_n r) \tag{5.24}$$

with M_n^2 given by (5.19). The large k limit can be gotten by replacing $n \to kx$ and $\frac{1}{k}\sum_n \to \int_0^1 dx$. We recover the identical result using Feynman rules. For $r \ll m^{-1}$, this behaves like

$$\left(\frac{x^-}{x^+}\right)^2 F(x^-, x^+) = \frac{N}{k}\sum_n \frac{3(k-2n)^2}{2\pi^2 k^2 r^4} \to \frac{N}{2\pi^2 r^4} . \tag{5.25}$$

D   Correlator for Supersymmetric Yang-Mills Theory with 16 Supercharges.

Finally, let us turn to the problem of computing the two point function of the T^{++} operator for the SYM with 16 supercharges. Adopting light-cone coordinates and choosing the light-cone gauge will eliminate the gauge boson and half of the fermion degrees of freedom. The most significant change comes from the fact that the fields in this theory are in the adjoint rather than the fundamental representations and the theory is supersymmetric. This does not cause any fundamental problem in the DLCQ formulation of these theories. Indeed, the SDLCQ formulation of this [8] as well as many other related models with adjoint fields have been studied in the literature. The main difficulty comes from the fact that in supersymmetric theories low mass states such as $\text{tr}[b(-n_1)b(-n_2)b(-k+n_1+n_2)]|0\rangle$ with an arbitrary number of excited quanta, or "bits," appear in the spectrum. This means that for a given harmonic resolution k, the dimension of the Hilbert space grows like $\exp(\sqrt{k})$, which is roughly the number of ways to partition k into sums of integers.

The fact that the size of the problem grows very fast is somewhat discouraging from a numerical perspective. Nevertheless, it is interesting to note that DLCQ provides a well defined algorithm for computing a physical quantity like the two point function of T^{++} that can be compared with the prediction from supergravity. In the following, we will show that this can be computed for the SYM theory by a straight forward application of (5.23).

As we found in section I the momentum operator P^+ is given by

195

$$P^+ = \int dx^- \text{tr}\left[(\partial_- X_I)^2 + iu_\alpha \partial_- u_\alpha\right].$$ (5.26)

The local Hermitian form of this operator is given by

$$T^{++}(x) = \text{tr}\left[(\partial_- X^I)^2 + \frac{1}{2}\left(iu^\alpha \partial_- u^\alpha - i(\partial_- u^\alpha)u^\alpha\right)\right], \qquad I = 1 \ldots 8, \quad \alpha = 1 \ldots 8$$ (5.27)

where X and u are the physical adjoint scalars and fermions respectively, following the notation of [8]. When discretized, these operators have the mode expansion

$$X^I_{i,j} = \frac{1}{\sqrt{4\pi}} \sum_{n=1}^{\infty} \frac{1}{\sqrt{n}} \left[A^I_{ij}(n)e^{-i\pi nx^-/L} + A^I_{ji}(-n)e^{i\pi nx^-/L}\right]$$

$$u^\alpha_{i,j} = \frac{1}{\sqrt{4L}} \sum_{n=1}^{\infty} \left[B^\alpha_{ij}(n)e^{-i\pi nx^-/L} + B^\alpha_{ji}(-n)e^{i\pi nx^-/L}\right].$$ (5.28)

In terms of these mode operators, we find

$$T^{++}(-k)|0\rangle =$$ (5.29)

$$\frac{\pi}{2L} \sum_{n=1}^{k-1} \left[-\sqrt{n(k-n)}A_{ij}(-k+n)A_{ji}(-n) + \left(\frac{k}{2}-n\right)B_{ij}(-k+n)B_{ji}(-n)\right]|0\rangle.$$

Therefore, $(L/\pi)\langle 0|T^{++}(-k)|n\rangle$ is independent of L and can be substituted directly into (5.23) to give an explicit expression for the two point function.

We see immediately that (5.29) has the correct small r behavior, for in that limit, (5.29) asymptotes to (assuming $n_b = n_f$)

$$\left(\frac{x^-}{x^+}\right)^2 F(x^-, x^+) =$$ (5.30)

$$\frac{N^2}{k} \sum_n \left(\frac{3(k-2n)^2 n_f}{4\pi^2 k^2 r^4} + \frac{3n(k-n)n_b}{\pi^2 k^2 r^4}\right) = \frac{N^2(2n_b + n_f)}{4\pi^2 r^4}\left(1 - \frac{1}{k}\right)$$

which is what we expect for the theory of n_b free bosons and n_f free fermions in the large k limit.

Computing this quantity beyond the small r asymptotics, however, represents a formidable technical challenge. In [8] we constructed the mass matrix explicitly and compute the spectrum for $k = 2$, $k = 3$, and $k = 4$. Even for these modest values of the harmonic resolution, the dimension of the Hilbert space was as big as 256, 1632, and 29056 respectively (the symmetries of the theory can be used to reduce the size of the calculation somewhat). In figure 3, we display the results with the currently available values of k, except for the fact that we display the correlation function multiplied by a factor of $4\pi^2 r^4/N^2(2n_b + n_f)$, so that it now asymptotes to 1 (or 0 in the logarithmic scale) in the $k \to \infty$ limit. In this way any deviation

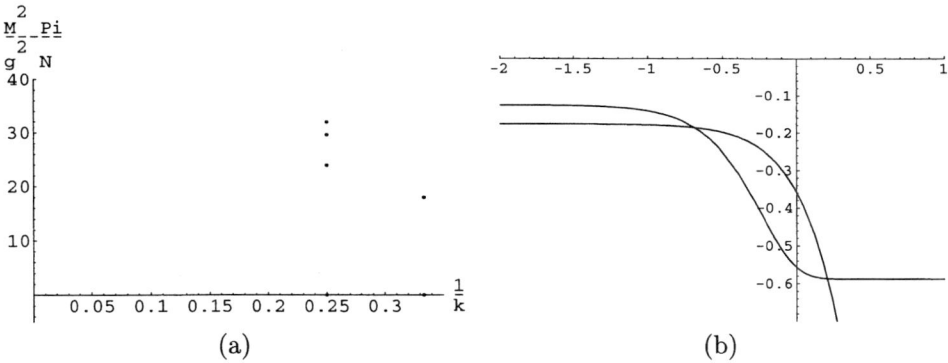

(a)   (b)

FIGURE 3. (a) The spectrum as a function of $1/k$ and (b) the Log-Log plot of the correlation function $\langle T^{++}(x)T^{++}(0)\rangle \left(\frac{x^-}{x^+}\right)^2 \frac{4\pi^2 r^4}{N^2(2n_b+n_f)}$ v.s. r in the units where $g_{YM}^2 N/\pi = 1$ for $k = 3$ and $k = 4$.

from the asymptotic behavior $1/r^4$ is made more transparent. Note that with the values of the harmonic resolution k obtained at present, the spectrum in figure 3.a is far from resembling a dense continuum near $M = 0$. Clearly, we must probe much higher values of k before we can sensibly compare our field theory results with the prediction from supergravity.

E   Supersymmetric Yang-Mills Theories with Less Than 16 Supercharges

The computation of the correlator for the stress energy tensor in the (8,8) model is limited by our inability to carry out the computation for large enough harmonic resolution. It is the (8,8) model which we are ultimately interested in solving in order to compare against the prediction of Maldacena's conjecture in the supergravity limit. Nevertheless, the computation of the correlation function can just as well be applied to models with less supersymmetry. We will conclude by reporting the results of such a computation.

First, let us consider the theory with supercharges (1,1). This theory is argued not to exhibit dynamical supersymmetry breaking in [59,66]. We can also provide a physicist's proof that supersymmetry is not spontaneously broken for this theory by adopting the argument of Witten for the 2+1 dimensional SYM with Chern-Simons interaction [83]. In [83], the index of 2+1 dimensional SYM with gauge group $SU(N)$ and 2 supercharges on $R \times T^2$ was computed and was found to be non-vanishing for Chern-Simons coupling $k_3 > N/2$. If the period L of one of the circles in T^2 is sufficiently small, this theory is approximately the 2-dimensional SYM with (1,1) supersymmetry with gauge coupling $g_2^2 = g_3^2/L$ and BF coupling $k_2 = k_3 L$ [21]. Imagine approaching this theory by taking $L \to 0$ keeping g_2 and k_3 fixed. In this limit, $k_2 \to 0$ in the units of g_2 so the limiting theory is that

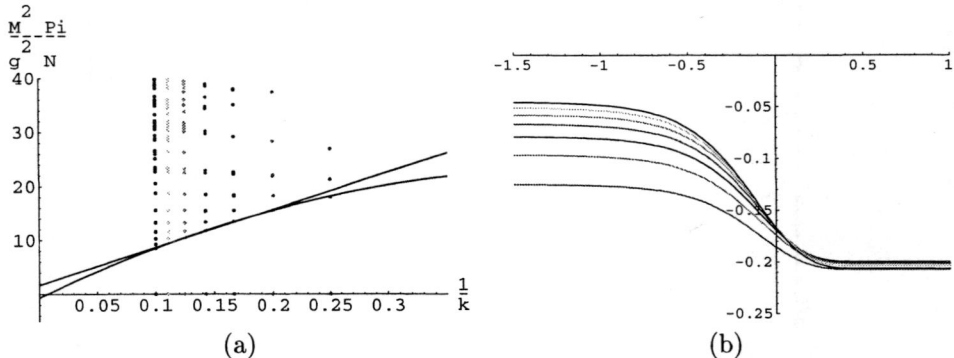

FIGURE 4. (a) The spectrum as a function of $1/k$ and (b) the Log-Log plot of the correlation function $\langle T^{++}(x)T^{++}(0)\rangle \left(\frac{x^-}{x^+}\right)^2 \frac{4\pi^2 r^4}{N^2(2n_b+n_f)}$ v.s. r in the units where $g_{YM}^2 N/\pi = 1$ for $k = 4\ldots 10$.

of pure SYM with (1,1) supersymmetry and a vanishing BF coupling. Choosing different values of k_3 corresponds to a different choice in the path of approach to this limit. If we chose $k_3 > N/2$, we are guaranteed to have a non-zero index for finite L. This means that there will be a state with zero mass in the $L \to 0$ limit also, indicating that supersymmetry is not spontaneously broken in this limit. On the other hand, the index is not a well defined quantity in the $L \to 0$ limit, as a different choice of k_3 will lead to a different value of the index in the $L \to 0$ limit. In fact, the index can be made arbitrarily large by taking k_3 to be also arbitrarily large. This suggests that there are infinitely many states forming a continuum near $m = 0$. The index is therefore an ill defined quantity, akin to counting the number of exactly zero energy states on a periodic box as one takes the volume to infinity.

This theory is also believed not to be confining [14,15] and is therefore expected to exhibit non-trivial infra-red dynamics.

The SDLCQ of the 1+1 dimensional model with (1,1) supersymmetry was solved in [63,5], and we apply these results directly in order to compute (5.23). For simplicity, we work to leading order in the large N expansion. The spectrum of this theory for various values of k, and the subsequent computation of (5.23) is illustrated in figure 4.a.

The spectrum of this theory at finite k, illustrated in figure 4.a, consists of $2k-2$ exactly massless states[9], accompanied by large numbers of massive states separated by a gap. The gap appears to be closing in the limit of large k however. We have tried extrapolating the mass of the lightest massive state as a function of $1/k$ by performing a least square fit to a line and a parabola, giving the extrapolated value of $M^2\pi/g_{YM}^2 N = 1.7$ and $M^2\pi/g_{YM}^2 N = -0.6$, suggesting indeed that at large k, the gap is closed. This is consistent with the expectation that the spectrum is that of a continuum starting at $M = 0$ discussed earlier, although one must be

---

[9] i.e. $k-1$ massless bosons, and their superpartners.

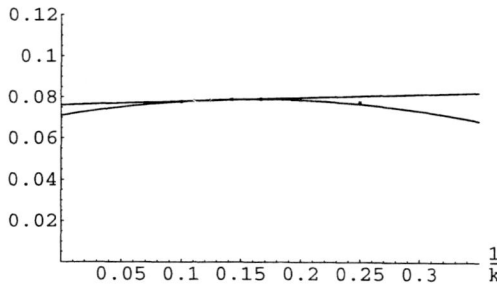

FIGURE 5. $\frac{1}{k^3}\sum_n |\frac{L}{\pi}\langle 0|T^{++}(k)|n\rangle|^2$ v.s. k from states with $M|n\rangle = 0$. This quantity determines the coefficient of the $1/r^4$ asymptotic tail of the correlation function in the large r limit for the (1,1) model.

careful when the order of large N and large k limits are exchanged. At finite N, we expect the degeneracy of $2k - 2$ exactly massless states to be broken, giving rise to precisely a continuum of states starting at $M = 0$ as expected.

In the computation of the correlation function illustrated in figure 4.b, we find a curious feature that it asymptotes to the inverse power law c/r^4 for large r. This behavior comes about due to the coupling $\langle 0|T^{++}|n\rangle$ with exactly massless states $|n\rangle$. The contribution to (5.23) from strictly massless states are given by

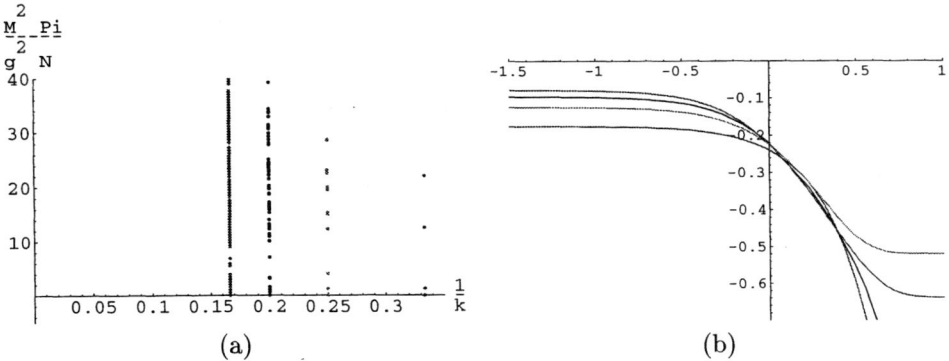

(a)

(b)

FIGURE 6. (a) The spectrum as a function of $1/k$ and (b) the Log-Log plot of the correlation function $\langle T^{++}(x)T^{++}(0)\rangle \left(\frac{x^-}{x^+}\right)^2 \frac{4\pi^2 r^4}{N^2(2n_b+n_f)}$ v.s. r in the units where $g_{YM}^2 N/\pi = 1$ for $k = 3\ldots 6$.

199

$$\left(\frac{x^-}{x^+}\right)^2 F(x^-, x^+) = \left|\frac{L}{\pi}\langle 0|T^{++}(k)|n\rangle\right|^2 \frac{M_n^4}{8\pi^2 k^3} K_4(M_n r)\Bigg|_{M_n=0} = \quad (5.31)$$

$$\left|\frac{L}{\pi}\langle 0|T^{++}(k)|n\rangle\right|^2_{M_n=0} \frac{6}{k^3\pi^2 r^4}.$$

We have computed this quantity as a function of $1/k$ and extrapolated to $1/k \to 0$ by fitting a line and a parabola to the computed values for finite $1/k$. The result of this extrapolation is illustrated in figure 5. The data currently available suggests that the non-zero contribution from these massless states persists in the large k limit.

Let us now turn to the model with (2,2) supersymmetry. The SDLCQ version of this model was solved in [10]. The result of this computation can be applied to (5.23). The result is summarized in figure 6. This model appears to exhibit the onset of a gapless continuum of states more rapidly than the (1,1) model as the harmonic resolution k is increased. Just as we found in the (1,1) model, this theory contains exactly massless states in the spectrum. These massless states appear to couple to $T^{++}|0\rangle$ only for k even, and the overlap appears to be decreasing as k is increased. We believe that this model is likely to exhibit a power law behavior c/r^γ for $\gamma > 4$ for the T^{++} correlator for $r \gg g_{YM}\sqrt{N}$ in the large N limit. Unfortunately, the existing numerical data do not permit the reliable computation of the exponent γ.

VI   BOUND STATES OF THREE DIMENSIONAL SUPERSYMMETRIC THEORY.

Recently, there has been considerable progress in understanding the properties of strongly coupled gauge theories with supersymmetry [74,73]. In particular, there are a number of supersymmetric gauge theories that are believed to be interconnected through a web of strong-weak coupling dualities. Although these dualities provide a deep insight into the dynamics of gauge theory at strong and weak couplings, they do not usually give much information about the spectrum of bound states at intermediate values of coupling constant g. The prominent exception is so called BPS states whose mass is protected by supersymmetry and thus stays the same for all values of g. An interesting new possibility for analytical treatment of bound state problem in SYM_D is based on the duality between SYM and supergravity proposed by Maldacena [60] as discussed in the previous section. This idea was also exploited in [27] to get the glueball spectrum of three dimensional theory and the results agree with lattice calculations.

However it would be interesting to solve the bound state problem for SYM theory directly, starting from the first principles of quantum field theory. As we have seen in the previous sections the solution can be found for the various two dimensional theories by means of applying discrete light cone quantization. Evidently, it would be desirable to extend these DLCQ/SDLCQ algorithms to solve higher dimensional theories. One important difference between two dimensional and higher dimensional theories is the phase diagram induced by variations in the gauge coupling. The spectrum of a $1+1$ dimensional gauge theory scales trivially with respect to the gauge coupling, while a theory in higher dimensions has the potential of exhibiting a complex phase structure, which may include a strong-weak coupling duality. It is therefore interesting to study the phase diagram of gauge theories in $D \geq 3$ dimensions.

Towards this end, we consider three dimensional SU(N) $\mathcal{N} = 1$ super-Yang-Mills compactified on the space-time $\mathbf{R} \times S^1 \times S^1$. In particular, we compactify the light-cone coordinate x^- on a light-like circle via DLCQ, and wrap the remaining transverse coordinate $x^\perp$ on a spatial circle. By retaining only the first few excited modes in the transverse direction, we are able to solve for bound state wave functions and masses numerically by diagonalizing the discretized light-cone supercharge. We show that the supersymmetric formulation of the DLCQ procedure – which was studied in the context of two dimensional theories extends naturally in $2+1$ dimensions, resulting in an exactly supersymmetric spectrum.

A Light-Cone Quantization and SDLCQ

We wish to study the bound states of $\mathcal{N} = 1$ super-Yang-Mills in $2+1$ dimensions. Any numerical approach necessarily involves introducing a momentum lattice – i.e. parton momenta can only take on discretized values. The usual space–time lattice explicitly breaks supersymmetry, so if we wish to discretize our theory *and* preserve supersymmetry, then a judicious choice of lattice is required.

In $1+1$ dimensions, it is well known that the light-cone momentum lattice induced by the DLCQ procedure preserves supersymmetry if the supercharge rather than the Hamiltonian is discretized [63,6]. In $2+1$ dimensions, a supersymmetric prescription is also possible. We begin by introducing light-cone coordinates $x^\pm = (x^0 \pm x^1)/\sqrt{2}$, and compactifying the x^- coordinate on a light-like circle. In this way, the conjugate light-cone momentum k^+ is discretized. To discretize the remaining (transverse) momentum $k^\perp = k^2$, we may compactify $x^\perp = x^2$ on a spatial circle. Of course, there is a significant difference between the discretized light-cone momenta k^+, and discretized transverse momenta $k_\perp$; namely, the light-cone momentum k^+ is always positive[10], while $k_\perp$ may take on positive or negative values. The positivity of k^+ is a key property that is exploited in DLCQ calculations; for any given light-cone compactification, there are only a finite number of choices for k^+ – the total number depending on how finely we discretize the momenta[11]. In the context of two dimensional theories, this implies a finite number of Fock states [67].

In the case we are interested in – in which there is an additional transverse dimension – the number of Fock states is no longer finite, since there are an arbitrarily large number of transverse momentum modes defined on the transverse spatial circle. Thus, an additional truncation of the transverse momentum modes is required to render the total number of Fock states finite, and the problem numerically tractable[12]. In this work, we choose the simplest truncation procedure beyond retaining the zero mode; namely, only partons with transverse momentum $k_\perp = 0, \pm\frac{2\pi}{L}$ will be allowed, where L is the size of the transverse circle.

Let us now apply these ideas in the context of a specific super-Yang-Mills theory. We start with $2 + 1$ dimensional $\mathcal{N} = 1$ super-Yang-Mills theory defined on a space-time with one transverse dimension compactified on a circle:

$$S = \int d^2x \int_0^L dx_\perp \mathrm{tr}(-\frac{1}{4}F^{\mu\nu}F_{\mu\nu} + i\bar{\Psi}\gamma^\mu D_\mu\Psi). \qquad (6.1)$$

After introducing the light–cone coordinates $x^\pm = \frac{1}{\sqrt{2}}(x^0 \pm x^1)$, decomposing the spinor Ψ in terms of chiral projections –

$$\psi = \frac{1+\gamma^5}{2^{1/4}}\Psi, \qquad \chi = \frac{1-\gamma^5}{2^{1/4}}\Psi \qquad (6.2)$$

and choosing the light–cone gauge $A^+ = 0$, the action becomes

$$S = \int dx^+ dx^- \int_0^L dx_\perp \mathrm{tr}\left[\frac{1}{2}(\partial_- A^-)^2 + D_+\phi\partial_-\phi + i\psi D_+\psi+ \right.$$
$$\left. +i\chi\partial_-\chi + \frac{i}{\sqrt{2}}\psi D_\perp\phi + \frac{i}{\sqrt{2}}\phi D_\perp\psi\right]. \qquad (6.3)$$

---

[10] Since we wish to consider the decompactified limit in the end, we omit zero modes. This is a necessary technical constraint in numerical calculations.

[11] The 'resolution' of the discretization is usually characterized by a positive integer K, which is called the 'harmonic resolution' [67,62]; for a given choice of K, the light-cone momenta k^+ are restricted to positive integer multiples of P^+/K, where P^+ is the total light-cone momentum of a state

[12] This truncation procedure, which is characterized by some integer upper bound, is analogous to the truncation of k^+ imposed by the 'harmonic resolution' K.

A simplification of the light–cone gauge is that the non-dynamical fields A^- and χ may be explicitly solved from their Euler-Lagrange equations of motion:

$$A^- = \frac{g}{\partial_-^2} J = \frac{g}{\partial_-^2} \left(i[\phi, \partial_- \phi] + 2\psi\psi \right), \tag{6.4}$$

$$\chi = -\frac{1}{\sqrt{2}\partial_-} D_\perp \psi.$$

These expressions may be used to express any operator in terms of the physical degrees of freedom only. In particular, the light-cone energy, P^-, and momentum operators, $P^+, P^\perp$, corresponding to translation invariance in each of the coordinates $x^\pm$ and $x_\perp$ may be calculated explicitly:

$$P^+ = \int dx^- \int_0^L dx_\perp \text{tr} \left[(\partial_- \phi)^2 + i\psi\partial_- \psi \right], \tag{6.5}$$

$$P^- = \int dx^- \int_0^L dx_\perp \text{tr} \left[-\frac{g^2}{2} J \frac{1}{\partial_-^2} J - \frac{i}{2} D_\perp \psi \frac{1}{\partial_-} D_\perp \psi \right], \tag{6.6}$$

$$P_\perp = \int dx^- \int_0^L dx_\perp \text{tr} \left[\partial_- \phi \partial_\perp \phi + i\psi\partial_\perp \psi \right]. \tag{6.7}$$

The light-cone supercharge in this theory is a two component Majorana spinor, and may be conveniently decomposed in terms of its chiral projections:

$$Q^+ = 2^{1/4} \int dx^- \int_0^L dx_\perp \text{tr} \left[\phi\partial_- \psi - \psi\partial_- \phi \right], \tag{6.8}$$

$$Q^- = 2^{3/4} \int dx^- \int_0^L dx_\perp \text{tr} \left[2\partial_\perp \phi \psi + g \left(i[\phi, \partial_- \phi] + 2\psi\psi \right) \frac{1}{\partial_-} \psi \right]. \tag{6.9}$$

The action (6.3) gives the following canonical (anti)commutation relations for propagating fields at equal x^+:

$$\left[\phi_{ij}(x^-, x_\perp), \partial_- \phi_{kl}(y^-, y_\perp) \right] = \frac{1}{2} i\delta(x^- - y^-)\delta(x_\perp - y_\perp) \left(\delta_{il}\delta_{jk} - \frac{1}{N}\delta_{ij}\delta_{kl} \right), \tag{6.10}$$

$$\left\{ \psi_{ij}(x^-, x_\perp), \psi_{kl}(y^-, y_\perp) \right\} = \frac{1}{2} \delta(x^- - y^-)\delta(x_\perp - y_\perp) \left(\delta_{il}\delta_{jk} - \frac{1}{N}\delta_{ij}\delta_{kl} \right). \tag{6.11}$$

Using these relations one can check the supersymmetry algebra:

$$\{Q^+, Q^+\} = 2\sqrt{2}P^+, \qquad \{Q^-, Q^-\} = 2\sqrt{2}P^-, \qquad \{Q^+, Q^-\} = -4P_\perp. \tag{6.12}$$

We will consider only states which have vanishing transverse momentum, which is possible since the total transverse momentum operator is kinematical[13]. On such states, the light-cone supercharges Q^+ and Q^- anti-commute with each other, and

---

[13] Strictly speaking, on a transverse cylinder, there are separate sectors with total transverse momenta $2\pi n/L$; we consider only one of them, $n = 0$.

the supersymmetry algebra is equivalent to the $\mathcal{N} = (1,1)$ supersymmetry of the dimensionally reduced (i.e. two dimensional) theory [63]. Moreover, in the $P_\perp = 0$ sector, the mass squared operator M^2 is given by $M^2 = 2P^+P^-$.

As we mentioned earlier, in order to render the bound state equations numerically tractable, the transverse momentum of partons must be truncated. First, we introduce the Fourier expansion for the fields ϕ and ψ, where the transverse space-time coordinate $x^\perp$ is periodically identified:

$$\phi_{ij}(0, x^-, x_\perp) =$$
$$\frac{1}{\sqrt{2\pi L}} \sum_{n^\perp = -\infty}^{\infty} \int_0^\infty \frac{dk^+}{\sqrt{2k^+}} \left[a_{ij}(k^+, n^\perp) e^{-ik^+ x^- - i\frac{2\pi n^\perp}{L} x_\perp} + a_{ji}^\dagger(k^+, n^\perp) e^{ik^+ x^- + i\frac{2\pi n^\perp}{L} x_\perp} \right]$$

$$\psi_{ij}(0, x^-, x_\perp) =$$
$$\frac{1}{2\sqrt{\pi L}} \sum_{n^\perp = -\infty}^{\infty} \int_0^\infty dk^+ \left[b_{ij}(k^+, n^\perp) e^{-ik^+ x^- - i\frac{2\pi n^\perp}{L} x_\perp} + b_{ji}^\dagger(k^+, n^\perp) e^{ik^+ x^- + i\frac{2\pi n^\perp}{L} x_\perp} \right]$$

Substituting these into the (anti)commutators (6.11), one finds:

$$\left[a_{ij}(p^+, n_\perp), a_{lk}^\dagger(q^+, m_\perp) \right] = \delta(p^+ - q^+) \delta_{n_\perp, m_\perp} \left(\delta_{il}\delta_{jk} - \frac{1}{N}\delta_{ij}\delta_{lk} \right) \tag{6.13}$$

$$\left\{ b_{ij}(p^+, n_\perp), b_{lk}^\dagger(q^+, m_\perp) \right\} = \delta(p^+ - q^+) \delta_{n_\perp, m_\perp} \left(\delta_{il}\delta_{jk} - \frac{1}{N}\delta_{ij}\delta_{lk} \right). \tag{6.14}$$

The supercharges now take the following form:

$$Q^+ = i2^{1/4} \sum_{n^\perp \in \mathbf{Z}} \int_0^\infty dk \sqrt{k} \left[b_{ij}^\dagger(k, n^\perp) a_{ij}(k, n^\perp) - a_{ij}^\dagger(k, n^\perp) b_{ij}(k, n^\perp) \right], \tag{6.15}$$

$$Q^- = \frac{2^{7/4}\pi i}{L} \sum_{n^\perp \in \mathbf{Z}} \int_0^\infty dk \frac{n^\perp}{\sqrt{k}} \left[a_{ij}^\dagger(k, n^\perp) b_{ij}(k, n^\perp) - b_{ij}^\dagger(k, n^\perp) a_{ij}(k, n^\perp) \right] +$$

$$+\frac{i2^{-1/4}g}{\sqrt{L\pi}} \sum_{n_i^\perp \in \mathbf{Z}} \int_0^\infty dk_1 dk_2 dk_3 \delta(k_1 + k_2 - k_3) \delta_{n_1^\perp + n_2^\perp, n_3^\perp} \Bigg\{$$

$$\frac{1}{2\sqrt{k_1 k_2}} \frac{k_2 - k_1}{k_3} [a_{ik}^\dagger(k_1, n_1^\perp) a_{kj}^\dagger(k_2, n_2^\perp) b_{ij}(k_3, n_3^\perp) - b_{ij}^\dagger(k_3, n_3^\perp) a_{ik}(k_1, n_1^\perp) a_{kj}(k_2, n_2^\perp)]$$

$$\frac{1}{2\sqrt{k_1 k_3}} \frac{k_1 + k_3}{k_2} [a_{ik}^\dagger(k_3, n_3^\perp) a_{kj}(k_1, n_1^\perp) b_{ij}(k_2, n_2^\perp) - a_{ik}^\dagger(k_1, n_1^\perp) b_{kj}^\dagger(k_2, n_2^\perp) a_{ij}(k_3, n_3^\perp)]$$

$$\frac{1}{2\sqrt{k_2 k_3}} \frac{k_2 + k_3}{k_1} [b_{ik}^\dagger(k_1, n_1^\perp) a_{kj}^\dagger(k_2, n_2^\perp) a_{ij}(k_3, n_3^\perp) - a_{ij}^\dagger(k_3, n_3^\perp) b_{ik}(k_1) a_{kj}(k_2, n_2^\perp)]$$

$$(\frac{1}{k_1} + \frac{1}{k_2} - \frac{1}{k_3})[b_{ik}^\dagger(k_1, n_1^\perp) b_{kj}^\dagger(k_2, n_2^\perp) b_{ij}(k_3, n_3^\perp) + b_{ij}^\dagger(k_3, n_3^\perp) b_{ik}(k_1, n_1^\perp) b_{kj}(k_2, n_2^\perp)] \Bigg\}.$$
$$\tag{6.16}$$

We now perform the truncation procedure; namely, in all sums over the transverse momenta $n_i^\perp$ appearing in the above expressions for the supercharges, we restrict summation to the following allowed momentum modes: $n_i^\perp = 0, \pm 1$. More generally, the truncation procedure may be defined by $|n_i^\perp| \leq N_{max}$, where N_{max} is some positive integer. In this work, we consider the simple case $N_{max} = 1$. Note that this prescription is symmetric, in the sense that there are as many positive modes as there are negative ones. In this way we retain parity symmetry in the transverse direction.

How does such a truncation affect the supersymmetry properties of the theory? Note first that an operator relation $[A, B] = C$ in the full theory is not expected to hold in the truncated formulation. However, if A is quadratic in terms of fields (or in terms of creation and annihilation operators), one can show that the relation $[A, B] = C$ implies

$$[A_{tr}, B_{tr}] = C_{tr}$$

for the truncated operators A_{tr}, B_{tr}, and C_{tr}. In our case, Q^+ is quadratic, and so the relations $\{Q_{tr}^+, Q_{tr}^+\} = 2\sqrt{2}P_{tr}^+$ and $\{Q_{tr}^+, Q_{tr}^-\} = 0$ are true in the $P_\perp = 0$ sector of the truncated theory. The $\{Q_{tr}^-, Q_{tr}^-\}$ however is not equal to $2\sqrt{2}P_{tr}^-$. So the diagonalization of $\{Q_{tr}^-, Q_{tr}^-\}$ will yield a different bound state spectrum than the one obtained after diagonalizing $2\sqrt{2}P_{tr}^-$. Of course the two spectra should agree in the limit $N_{max} \to \infty$. At any finite truncation, however, we have the liberty to diagonalize any one of these operators. This choice specifies our regularization scheme.

Choosing to diagonalize the light-cone supercharge, however, has an important advantage: *the spectrum is exactly supersymmetric for any truncation.* In contrast, the spectrum of the Hamiltonian becomes supersymmetric only in the $N_{max} \to \infty$ limit[14].

To summarize, we have introduced a truncation procedure that facilitates a numerical study of the bound state problem, and preserves supersymmetry. The interesting property of the light-cone supercharge Q^- [Eqn(6.16)] is the presence of a gauge coupling constant as an independent variable, which does not appear in the study of two dimensional theories. In the next subsection, we will study how variations in this coupling affects the bound states in the theory.

B Numerical Results: Bound State Solutions

In order to implement the DLCQ formulation of the bound state problem – which is tantamount to imposing periodic boundary conditions $x^- = x^- + 2\pi R$ [62] – we simply restrict the light-cone momentum variables k_i appearing in the expressions for Q^+ and Q^- to the following discretized set of momenta: $\{\frac{1}{K}P^+, \frac{2}{K}P^+, \frac{3}{K}P^+, \dots, \}$. Here, P^+ denotes the total light-cone momentum of a state, and may be thought of as a fixed constant, since it is easy to form a Fock

---

[14] If one chooses anti-periodic boundary conditions in the x^- coordinate for fermions, then there is no choice; one can only diagonalize the light-cone Hamiltonian. See [28] for more details on this approach.

basis that is already diagonal with respect to the operator P^+ [67]. The integer K is called the 'harmonic resolution', and $1/K$ measures the coarseness of our discretization. The continuum limit is then recovered by taking the limit $K \to \infty$. Physically, $1/K$ represents the smallest positive unit of longitudinal momentum-fraction allowed for each parton in a Fock state.

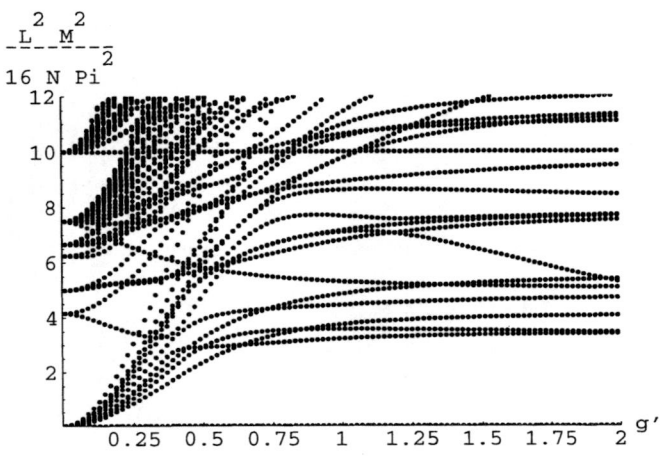

FIGURE 7. Plot of bound state mass squared M^2 in units $16\pi^2 N/L^2$ as a function of the dimensionless coupling $0 \le g' \le 2$, defined by $(g')^2 = g^2 N L/16\pi^3$, at $N = 1000$ and $K = 5$. Boson and fermion masses are identical.

Of course, as soon as we implement the DLCQ procedure, which is specified unambiguously by the harmonic resolution K, and cut-off transverse momentum modes via the constraint $|n_i^\perp| \le N_{max}$, the integrals appearing in the definitions for Q^+ and Q^- are replaced by finite sums, and so the eigen-equation $2P^+P^-|\Psi\rangle = M^2|\Psi\rangle$ is reduced to a finite matrix diagonalization problem. In this last step we use the fact that P^- is proportional to the square of the light-cone supercharge[15] Q^-. In the present work, we are able to perform numerical diagonalizations for $K = 2, 3, 4$ and 5 with the help of Mathematica and a desktop PC. In Figure 7, we plot the bound state mass squared M^2, in units $16\pi^2 N/L^2$, as a function of the dimensionless coupling $g' = g\sqrt{NL}/4\pi^{3/2}$, in the range $0 \le g' \le 2$. We consider the specific case $N = 1000$, although our algorithm can calculate masses for any choice of N, since it enters our calculations as an algebraic variable. Since there is an exact boson-fermion mass degeneracy, one needs only diagonalize the mass

---

[15] Strictly speaking, $P^- = \frac{1}{\sqrt{2}}(Q^-)^2$ is an identity in the continuum theory, and a *definition* in the compactified theory, corresponding to the SDLCQ prescription [63,6].

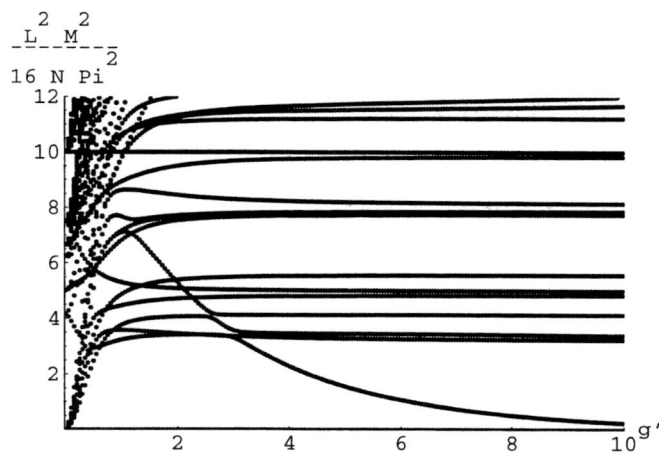

$$\frac{L^2 M^2}{16 N \pi^2}$$

FIGURE 8. Plot of bound state mass squared M^2 in units $16\pi^2 N/L^2$ as a function of the dimensionless coupling $0 \le g' \le 10$, defined by $(g')^2 = g^2 NL/16\pi^3$, at $N = 1000$ and $K = 5$. Note the appearance of a new massless state at strong coupling.

matrix M^2 corresponding to the bosons. For $K = 5$, there are precisely 600 bosons and 600 fermions in the truncated light-cone Fock space, so the mass matrix that needs to be diagonalized has dimensions 600×600. At $K = 4$, there are 92 bosons and 92 fermions, while at $K = 3$, one finds 16 bosons and 16 fermions.

In Figure 8, we plot the bound state spectrum in the range $0 \le g' \le 10$. It is apparent now that new massless states appear in the strong coupling limit $g' \to \infty$.

An interesting property of the spectrum is the presence of exactly massless states that persist for all values of the coupling g'. For $K = 5$, there are 16 such states (8 bosons and 8 fermions). At $K = 4$, one finds 8 states (4 bosons and 4 fermions) that are exactly massless for any coupling, while for $K = 3$, there are 4 states (two bosons and two fermions) with this property. We will have more to say regarding these states in the next subsection, but here we note that the structure of these states become 'string-like' in the strong coupling limit. This is illustrated in Figure 9, where we plot the 'average length' (or average number of partons) of each of these massless states[16]. This quantity is obtained by counting the number of partons in each Fock state that comprises a massless bound state, appropriately weighted by the modulus of the wave function squared. Clearly, at strong coupling, the average

---

[16] The 'noisiness' in this plot for larger values of g' reflects the ambiguity of choosing a basis for the eigen-space, due to the exact mass degeneracy of the massless states.

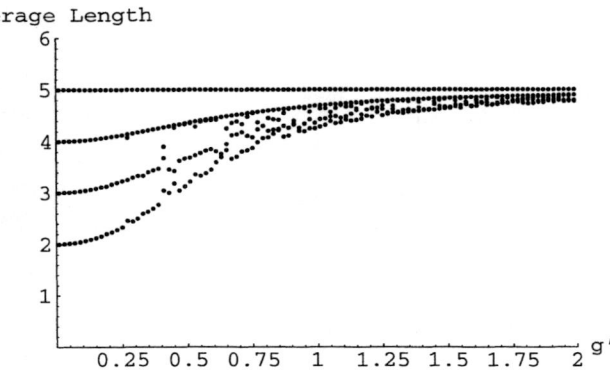

FIGURE 9. Plot of average length for the eight massless bosonic states as a function of the dimensionless coupling g', defined by $(g')^2 = g^2 NL/16\pi^3$, at $N = 1000$ and $K = 5$. Note that the states attain the maximum possible length allowed by the resolution $K = 5$ in the limit of strong coupling.

number of partons saturates the maximum possible value allowed by the resolution – in this case 5 partons. The same behavior is observed at lower resolutions. Thus, in the continuum limit $K \to \infty$, we expect the massless states in this theory to become string-like at strong coupling.

One interesting property of the model studied here is the manifest $\mathcal{N} = (1,1)$ supersymmetry in the $P^\perp = 0$ momentum sector, by virtue of the supersymmetry relations (6.12). Moreover, if we consider retaining only the zero mode $n_i^\perp = 0$, then the light-cone supercharge Q^- for the 2+1 model is identical to the 1+1 dimensional $\mathcal{N} = (1,1)$ supersymmetric Yang-Mills theory studied in [63,4,5], after a rescaling by the factor $1/g'$. (This is equivalent to expressing the mass squared M^2 in units $\tilde{g}^2 N/\pi$, where $\tilde{g} = g/\sqrt{L}$. The quantity $\tilde{g}$ is then identified as the gauge coupling in the 1+1 theory.) We may therefore think of the additional transverse degrees of freedom in the $2 + 1$ model, represented by the modes $n^\perp = \pm 1$, as a modification of the $1 + 1$ model. A natural question that follows from this viewpoint is: How well does the 1+1 spectrum approximate the 2+1 spectrum after performing this rescaling? Before discussing the numerical results summarized in Table 2, let us first attempt to predict what will happen at small coupling g'. In this case, the coefficients of terms in the rescaled Hamiltonian P^- that correspond to summing the transverse momentum squared $|k^\perp|^2$ of partons in a state will be large. So the low energy sector will be dominated by states with $n^\perp = 0$. i.e. those states that appear in the Fock space of the $\mathcal{N} = (1,1)$ model in $1 + 1$ dimensions. This is indeed supported by the results in Table 2.

For large coupling g', however, it is clear that the approximation breaks down. In

Comparison Between $1+1$ and $2+1$ Spectra				
-	$1+1$ Model	Rescaled $2+1$ Model		
K	-	$g' = .01$	$g' = .1$	$g' = 1.0$
$K = 5$	15.63	15.5	15.17	3.7
	18.23	17.6	17.9	3.5
	21.8	21.3	21.7	3.2
$K = 4$	-	-	-	-
	18.0	17.99	17.6	3.56
	21.3	21.3	21.0	3.1
$K = 3$	-	-	-	-
	-	-	-	-
	20.2	20.2	19.8	3.1

TABLE 2. Values for the mass squared M^2, in units $\tilde{g}^2 N/\pi$, with $\tilde{g}^2 = g^2/L$, for bound states in the dimensionally reduced $\mathcal{N} = (1,1)$ model, and the $2+1$ model studied here. The quantity $\tilde{g}$ is identified as the gauge coupling in the $1+1$ model. We set $K = 3, 4$ and 5, and $N = 1000$. Note that the comparison of masses between the $1+1$ model, and the (re-scaled) $2+1$ model is good only at weak coupling g'.

fact, one can show that the tabulated masses in the rescaled $2+1$ model tend to zero in the strong coupling limit, which eliminates any scope for making comparisons between the two and three dimensional models.

Thus, the non-perturbative problem of solving dimensionally reduced models in $1+1$ dimensions can only provide information about bound state masses in the corresponding *weakly coupled* higher dimensional theory.

C   Analytical Results: The Massless Sector

In the previous subsection we presented the results of studying the bound state problem using numerical methods. In performing such a study we conveniently chose the simplest nontrivial truncation of the transverse momentum modes; namely, $n_\perp = 0, \pm 1$. Surprisingly, such a simple scheme provided many interesting insights concerning the massless and massive sector. In particular we see that there are three types of massless states; those that are massless only at $g = 0$ or $g = \infty$ (but not both), and those that are massless for any value of the coupling. In this subsection, we will analyze only the massless sector of the theory, and show that the observed properties of the spectrum with the truncation $n^\perp = 0, \pm 1$ also persists if we include higher modes: $n^\perp = 0, \pm 1, \pm 2, \ldots, \pm N_{max}$. We therefore consider the model with supercharges given by (6.15) and (6.16), and restrict summation of transverse momentum modes via the constraint $|n^\perp| \leq N_{max}$.

For states carrying positive light-cone momentum, P^+ is never zero, and so massless states must satisfy the equation $P^-|\Psi\rangle = 0$, which, using the relation $P^- = \frac{1}{\sqrt{2}}(Q^-)^2$, and hermiticity of Q^-, reduces to

$$Q^-|\Psi\rangle = 0. \tag{6.17}$$

This is the equation we wish to study in detail.

We begin with an analysis of the weak coupling limit of the theory. This limit means that the dimensionless coupling constant is small: i.e. $g\sqrt{L} \ll 1$. We will consider the strong–weak coupling behavior of the theory on a cylinder with fixed circumference L so it is convenient to choose the units in which $L = 1$ for this discussion. The supercharge (6.16) consists of two parts: one is proportional to the coupling and the other is coupling–independent:

$$Q^- = Q_\perp + g\tilde{Q}. \tag{6.18}$$

So at $g = 0$, the equation (6.17) reduces to $Q_\perp|\Psi\rangle = 0$, which means that $|\Psi\rangle$ may be viewed as a state in the Fock space of the two dimensional $\mathcal{N} = (1,1)$ super Yang-Mills theory, which may be obtained by dimensional reduction of the $2+1$ theory. Thus the massless states at $g = 0$ are states with any combination of $a^\dagger(k,0)$ and $b^\dagger(k,0)$ modes, and no partons with nonzero transverse momentum.

What happens with these massless states when one switches on the coupling? To answer this question, we need some information about the behavior of states as functions of the coupling. We assume that wave functions are analytic in terms of g at least in the vicinity of $g = 0$. This means that in this region any massless state $|\Psi\rangle$ may be written in the form:

$$|\Psi\rangle = \sum_{n=0}^{\infty} g^n|n\rangle, \tag{6.19}$$

where states $|n\rangle$ are coupling independent. Then using relation (6.18), the g–dependent equation (6.17) may be written as an infinite system of relations between different $|n\rangle$:

$$Q_\perp|0\rangle = 0, \tag{6.20}$$
$$Q_\perp|n\rangle + \tilde{Q}|n-1\rangle = 0, \qquad n > 0. \tag{6.21}$$

The first of these equations was already used to exclude partons carrying non-zero transverse momentum, which is a property of the massless bound states at zero coupling. The second equation is non-trivial. Let us consider two different subspaces in the theory. The first of these subspaces consists of states with no creation operators for transverse modes which we will label 1. The other is the complement of this space in which the operator $Q_\perp$ is invertible and we label this space 2. Equation (6.21) defines the recurrence relation when $\tilde{Q}|n-1\rangle$ is in subspace 2:

$$|n\rangle = -Q_\perp^{-1}\left(\tilde{Q}|n-1\rangle\big|_2\right), \tag{6.22}$$

The consistency condition is that projection of $\tilde{Q}|n-1\rangle$ in subspace 1 is zero,

$$\tilde{Q}|n-1\rangle\big|_1 = 0. \tag{6.23}$$

This condition implies that not all states of the two dimension theory, $g = 0$, may be extended to such states at arbitrary g using (6.22). Taking $n = 1$, (6.23) implies that $|0\rangle$ is a massless state of the dimensionally reduced theory. The numerical solutions, of course, show this correspondence between the $2+1$ and $1+1$ [63,4,5] massless bound states. Starting from a massless state of the two dimensional theory, and we construct states $|n\rangle$ using (6.22), and for which (6.23) is always satisfied. Then $|\Psi\rangle$ may be found from summing a geometric series:

$$|\Psi\rangle = \sum_{n=0}^{\infty}(-gQ_\perp^{-1}\tilde{Q})^n|0\rangle = \frac{1}{1+gQ_\perp^{-1}\tilde{Q}}|0\rangle. \tag{6.24}$$

So, starting from the massless state of the two dimensional $\mathcal{N} = (1,1)$ model, one can always construct unique massless states in the three dimensional theory at least in the vicinity of $g = 0$.

The state (6.24) turns out to be massless for any value of the coupling:

$$Q^-|\Psi\rangle = Q_\perp(1+gQ_\perp^{-1}\tilde{Q})\frac{1}{1+gQ_\perp^{-1}\tilde{Q}}|0\rangle = Q_\perp|0\rangle = 0, \tag{6.25}$$

though the state itself is dependent on g. Thus, we have shown that massless states of the three dimensional theory, at nonzero coupling, can be constructed from massless states of the corresponding model in two dimensions. All other states containing only two dimensional modes can also be extended to the eigenstates of the full theory. But such eigenstates are massless only at zero coupling. Assuming analyticity, one can then show that their masses grow linearly at g in the vicinity of zero. Such behavior also agrees with our numerical results.

To illustrate the general construction explained above we consider one simple example. Working in DLCQ at resolution $K = 3$ we choose a special two dimensional massless state[17] [63,4,5]:

$$|0\rangle = \text{tr}(a^\dagger(1,0)a^\dagger(2,0))|vac\rangle. \tag{6.26}$$

Then in the SU(N) theory we find:

---

[17] The state $|0\rangle$ denotes a massless state, while $|vac\rangle$ represents the light-cone vacuum.

$$\tilde{Q}|0\rangle = \frac{3}{2\sqrt{2}} \left[\text{tr} \left(a^\dagger(1,0)(b^\dagger(1,-1)a^\dagger(1,1) - a^\dagger(1,1)b^\dagger(1,-1) + \right. \right.$$
$$\left. \left. + b^\dagger(1,1)a^\dagger(1,-1) - a^\dagger(1,-1)b^\dagger(1,1)) \right) \right] |vac\rangle, \tag{6.27}$$

$$|1\rangle = -Q_\perp^{-1}\tilde{Q}|0\rangle = -\frac{\sqrt{L}}{4\pi^{3/2}}\frac{3}{2\sqrt{2}} \left(a^\dagger(1,0)a^\dagger(1,-1)a^\dagger(1,1) - \right.$$
$$\left. - a^\dagger(1,0)a^\dagger(1,1)a^\dagger(1,-1) \right) |vac\rangle \tag{6.28}$$

$$\tilde{Q}|1\rangle = 0. \tag{6.29}$$

The last equation provides the consistency condition (6.23) for $n = 2$, and it also shows that for this special example we have only two states $|0\rangle$ and $|1\rangle$, instead of a general infinite set. The matrix form of the operator $1 + gQ_\perp^{-1}\tilde{Q}$ in the $|0\rangle, |1\rangle$ basis is

$$1 + gQ_\perp^{-1}\tilde{Q} = \begin{pmatrix} 1 & -g \\ 0 & 1 \end{pmatrix} = \begin{pmatrix} 1 & g \\ 0 & 1 \end{pmatrix}^{-1}. \tag{6.30}$$

Then the solution of (6.24) is

$$|\Psi\rangle = |0\rangle + g|1\rangle = \text{tr}(a^\dagger(1,0)a^\dagger(2,0))|vac\rangle + \tag{6.31}$$
$$+ \frac{g\sqrt{L}}{4\pi^{3/2}}\frac{3}{2\sqrt{2}} \left(a^\dagger(1,0)a^\dagger(1,1)a^\dagger(1,-1) - a^\dagger(1,0)a^\dagger(1,-1)a^\dagger(1,1) \right) |vac\rangle.$$

This state was observed numerically, and the dependence of the wave function on the coupling constant is precisely the one given by the last formula.

In principle, we can determine the wave functions of all massless states using this formalism. Our procedure has an important advantage over a direct diagonalization of the three dimensional supercharge. Firstly, in order to find two dimensional massless states, one needs to diagonalize the corresponding supercharge [63]. However, the dimension of the relevant Fock space is much less than the three dimensional theory (at large resolution K, the ratio of these dimensions is of order $(N_{max} + 1)^{\alpha K}$, $\alpha \sim 1/4$). The extension of the two dimensional massless solution into a massless solution of the three dimensional theory requires diagonalizing a matrix which has a smaller dimension than the original problem in three dimensions. Thus, if one is only interested in the massless sector of the three dimensional theory, the most efficient way to proceed in DLCQ calculations is to solve the two dimensional theory, and then to upgrade the massless states to massless solutions in three dimensions.

Finally, we will make some comments on bound states at very strong coupling. Of course, we have states (6.24) which are massless at any coupling, but our numerical calculation show there are additional states which become massless at $g = \infty$ (see Figure 8). To discuss these state it is convenient to consider

$$\bar{Q}^- = \frac{1}{g}Q_\perp + \tilde{Q} \tag{6.32}$$

212

instead of Q^-, and perform the strong coupling expansion. Since we are interested only in massless states, the absolute normalization doesn't matter. We repeat all the arguments used in the weak coupling case: first, we introduce the space 1^* where $\tilde{Q}$ can not be inverted, and its orthogonal complement 2^*. Then any state from 1^* is massless at $g = \infty$, but assuming the expansion

$$|\Psi\rangle = \sum_{n=0}^{\infty} \frac{1}{g^n} |n\rangle^* \tag{6.33}$$

at large enough g, one finds the analogs of (6.22) and (6.23):

$$|n\rangle^* = -\tilde{Q}^{-1} \left(Q_\perp |n-1\rangle^* |_{2^*} \right), \tag{6.34}$$

$$Q_\perp |n-1\rangle^* |_{1^*} = 0. \tag{6.35}$$

As in the small coupling case, there are two possibilities: either one can construct all states $|n\rangle^*$ satisfying the consistency conditions, or at least one of these conditions fails. The former case corresponds to the massless state in the vicinity of $g = \infty$, which can be extended to the massless states at all couplings. The states constructed in this way – and ones given by (6.24) – define the same subspace. In the latter case, the state is massless at $g = \infty$, but it acquires a mass at finite coupling. There is a big difference, however, between the weak and strong coupling cases. While the kernel of $Q_\perp$ consists of "two dimensional" states, the description of the states annihilated by $\tilde{Q}$ is a nontrivial dynamical problem. Since the massless states can be constructed starting from either $g = 0$ or $g = \infty$, we don't have to solve this problem to build them. If, however, one wishes to show that massless states become long in the strong coupling limit (there is numerical evidence for such behavior – see Figure 9), the structure of 1^* space becomes important, and we leave this question for future investigation.

Conclusion.

In these lectures we have reviewed some of the progress in the application of discrete light cone quantization to the supersymmetric systems. Studying such systems is especially interesting because the cancellation between bosonic and fermionic loops make these theories much easier to renormalize than the models without supersymmetry. Although we didn't need this advantage when considering two dimensional systems, it becomes crucial in higher dimensions. From this point of view it is desirable to have exact SUSY in discretized theories to simplify the renormalization in DLCQ.

while we are still far from the point of solving the bound state problem in three and four dimensional theories, we can already make some statements about these theories. For example in section II we described the vacuum structure of SYM_{2+1} on a cylinder. The reason for this is that only zero modes contribute to such structure, thus studying the theory dimensionally reduced to $1+1$ provide all the necessary information. As we saw in section VI, two dimensional models can also be used to determine the behavior of bound states at weak coupling in three dimensions and to count the exact massless states. We performed such counting only for $(1,1)$ theory, the case of $(2,2)$ supersymmetry [10] and even more interesting case of the $(8,8)$ theory [8] which is known to have a mass gap have not been addressed.

Let us now mention the immediate challenges to DLCQ following from our consideration. First of all it is straightforward to extend the numerical results of section V to higher resolution and thus to test the Maldacena's conjecture. The only problem here is the limits in one's computing resources. The better computer power may also help to extend our analysis of three dimensional system to larger values of transverse truncation and it might be possible to extrapolate the results to continuum. The potential problem with such extrapolation is that one has to take two limits: both longitudinal and transverse resolutions should go to infinity and these two limits do not necessarily commute. In any case the simplest transverse truncation we have used does not provide much information about behavior of the spectrum as function of transverse resolution and it was used mainly for illustration of the general concepts. Another serious disadvantage of our approach is that we study the theory on the cylinder and thus the structure of the topological excitations of our system differs from the one of SYM_{2+1} in infinite spacetime. Such excitations become important in the strong coupling limit. However even the spectrum for the theory on the cylinder is of a great interest and we leave the detailed numerical study of this system for future work. Finally solving for bound states of four dimensional theories is still the greatest challenge for DLCQ.

VII ACKNOWLEDGMENTS

This work was supported in part by the US Department of Energy. All of the work reported in these lectures was done in collaboration with Francesco Antonuccio. We would like to acknowledge the other members of the SDLCQ project, S. Tsujimaru, C. Pauli, J. Hiller, Uwe Trittmann, and Igor Filippos.

REFERENCES

1. S. Adler, "Axial vector vertex in spinor electrodynamics," *Phys. Rev.* **177** (1969) 2426.
2. O. Aharony, S. S. Gubser, J. Maldacena, H. Ooguri, and Y. Oz, "Large N field theories, string theory and gravity," `hep-th/9905111`.
3. F. Antonuccio, S.J. Brodsky, S. Dalley, "Light-Cone Wavefunctions at Small x," *Phys.Lett.* **B412** (1997) 104.
4. F. Antonuccio, O. Lunin, S. Pinsky, "Bound States of Dimensionally Reduced SYM_{2+1} at Finite N", *Phys.Lett.* **B429** (1998) 327-335, `hep-th/9803027`.

5. F. Antonuccio, O. Lunin, and S. Pinsky, "Nonperturbative spectrum of two-dimensional (1,1) superYang-Mills at finite and large N," *Phys. Rev.* **D58** (1998) 085009, `hep-th/9803170`.

6. F. Antonuccio, O. Lunin, and S. Pinsky, "On Exact Supersymmetry in DLCQ," *Phys.Lett.* **B442** (1998) 173, `hep-th/9809165`.

7. F. Antonuccio, O. Lunin, and S. Pinsky, "Super Yang-Mills at Weak, Intermediate and Strong Coupling", *Phys.Rev.* D **D59** (1999) 085001, `hep-th/9811083`.

8. F. Antonuccio, O. Lunin, S. Pinsky, H. C. Pauli, and S. Tsujimaru, "The DLCQ spectrum of N=(8,8) superYang-Mills," *Phys. Rev.* **D58** (1998) 105024, `hep-th/9806133`.

9. F. Antonuccio, O. Lunin, S. Pinsky, and S. Tsujimaru, "The Light cone vacuum in (1+1)-dimensional superYang-Mills theory," `hep-th/9811254`.

10. F. Antonuccio, H. C. Pauli, S. Pinsky, and S. Tsujimaru, "DLCQ bound states of N=(2,2) super Yang-Mills at finite and large N," *Phys. Rev.* **D58** (1998) 125006, `hep-th/9808120`.

11. F. Antonuccio, S.S. Pinsky, "Matrix theories from reduced $SU(N)$ Yang–Mills with adjoint fermions," *Phys.Lett* **B397** (1997) 42, `hep-th/9612021`.

12. F. Antonuccio, S. Pinsky, "On the transition from confinement to screening in QCD_{1+1} coupled to adjoint fermions at finite N, " *Phys.Lett.* **B439** (1998) 142, `hep-th/9805188`.

13. F. Antonuccio, S. Pinsky, and S. Tsujimaru, "A Comment on the light cone vacuum in (1+1)-dimensional superYang-Mills theory," `hep-th/9810158`.

14. A. Armoni, Y. Frishman, and J. Sonnenschein, "Screening in supersymmetric gauge theories in two- dimensions," *Phys. Lett.* **B449** (1999) 76, `hep-th/9807022`.

15. A. Armoni, Y. Frishman, and J. Sonnenschein, "The String tension in two-dimensional gauge theories," `hep-th/9903153`.

16. A. Armoni, J. Sonnenschein, "Screening and Confinement in Large N_f QCD_2 and in $N = 1$ SYM_2," `hep-th/9703114`.

17. T. Banks, W. Fischler, S. Shenker, L. Susskind, "M theory as a matrix model: a conjecture," *Phys. Rev.* **D55** (1997) 5112, `hep--th/9610043`.

18. C. Bender, S. Pinsky, B. van de Sande, "Spontaneous symmetry breaking of ϕ^4_{1+1} in light front field theory," *Phys. Rev.* **D48** (1993) 816.

19. G. Bhanot, K. Demeterfi, I.R. Klebanov, "(1+1)–Dimensional Large N QCD Coupled to Adjoint fermions," *Phys. Rev.* **D48** (1993) 4980, `hep--th/9307111`.

20. D. Bigatti, L. Susskind, "Review of Matrix Theory," `hep-th/9712072`.

21. D. Birmingham, M. Blau, M. Rakowski, and G. Thompson, "Topological field theory," *Phys. Rept.* **209** (1991) 129–340.

22. J. Boorstein, D. Kutasov, "Symmetries and mass splittings in QCD_2 coupled to adjoint fermions," *Nucl.Phys.* **B421** (1994) 263, `hep-th/9401044`.

23. A. Brandhuber, N. Itzhaki, J. Sonnenschein, and S. Yankielowicz, "Wilson loops, confinement, and phase transitions in large N gauge theories from supergravity," *JHEP* **06** (1998) 001, `hep-th/9803263`.

24. S. J. Brodsky, H.-C. Pauli, and S. S. Pinsky, "Quantum chromodynamics and other field theories on the light cone," *Phys. Rept.* **301** (1998) 299, `hep-ph/9705477`.

25. M. Burkardt, F. Antonuccio, and S. Tsujimaru, "Decoupling of zero modes and covariance in the light front formulation of supersymmetric theories," *Phys. Rev.*

D58 (1998) 125005, hep-th/9807035.

26. C. G. Callan, N. Coote, and D. J. Gross, "Two-Dimensional Yang-Mills Theory: A Model of Quark Confinement," *Phys. Rev.* **D13** (1976) 1649.

27. C. Csaki, H. Ooguri, Y. Oz, J.Terning, "Glueball mass spectrum from supergravity," *JHEP* **9901** (1999) 017, hep-th/9806021.

28. S. Dalley and I.R. Klebanov, "String Spectrum of 1+1-Dimensional Large N QCD with Adjoint Matter", *Phys. Rev.* **D47** (1993) 2517, hep-th/9209049.

29. K. Demeterfi and I. R. Klebanov, "Matrix models and string theory,". Lectures given at Spring School on String Theory, Gauge Theory and Quantum Gravity, Trieste, Italy, 19-27 Apr 1993.

30. R. Dijkgraaf, E. Verlinde, and H. Verlinde, "Matrix string theory," *Nucl. Phys.* **B500** (1997) 43, hep-th/9703030.

31. P. A. M. Dirac, "Forms of relativistic dynamics," *Rev. Mod. Phys.* **21** (1949) 392.

32. S. Ferrara, "Supersymmetric gauge theories in two dimensions," *Lett. Nuovo. Cimen* **13** (1975) 629.

33. S. Ferrara, C. Fronsdal, and A. Zaffaroni, "On N=8 supergravity on AdS(5) and N=4 superconformal Yang- Mills theory," *Nucl. Phys.* **B532** (1998) 153, hep-th/9802203.

34. A. Giveon, D. Kutasov, "Brane dynamics and gauge theory," hep-th/9802067.

35. M.B. Green, J.H. Schwarz, E. Witten, *Superstring Theory*, Vol.1, CUP (1987).

36. V.N. Gribov, "Quantization of non–abelian gauge theories," *Nucl. Phys.* **B139** (1978) 1.

37. D.J. Gross, I.R. Klebanov, A.V. Matytsin, A.V. Smilga, "Screening vs confinement in $1+1$ dimensions," *Nucl.Phys* **B461** (1996) 109, hep-th/9511104.

38. D.J. Gross, A. Hashimoto and I.R. Klebanov, "The Spectrum of a Large N Gauge Theory Near Transition From Confinement to Screening," hep-th/9710240.

39. S. S. Gubser, A. Hashimoto, I. R. Klebanov, and M. Krasnitz, "Scalar absorption and the breaking of the world volume conformal invariance," *Nucl. Phys.* **B526** (1998) 393, hep-th/9803023.

40. S. S. Gubser, I. R. Klebanov, and A. M. Polyakov, "Gauge theory correlators from noncritical string theory," *Phys. Lett.* **B428** (1998) 105, hep-th/9802109.

41. A. Hashimoto and N. Itzhaki, "A Comment on the Zamolodchikov c function and the black string entropy," hep-th/9903067.

42. A. Hashimoto and I.R. Klebanov, "Non-perturbative solution of matrix models modified by trace–squared terms," *Nucl.Phys* **B434**, (1995) 264, hep-th/9409064.

43. A. Hashimoto, I.R. Klebanov, "Matrix Model Approach to $d > 2$ Non-critical Superstrings", *Mod.Phys.Lett.* **A10** (1995) 2639.

44. S. Hellerman and J. Polchinski, "Compactification in the lightlike limit," *Phys. Rev.* **D59** (1999) 125002, hep-th/9711037.

45. G. 't Hooft, "A Two-Dimensional Model for Mesons," *Nucl. Phys.* **B75** (1974) 461.

46. K. Hornbostel, "The Application of Light Cone Quantization to Quantum Chromodynamics in (1+1)-Dimensions,". PhD Thesis, SLAC-0333.

47. K. Hornbostel, S. J. Brodsky, and H. C. Pauli, "Light Cone Quantized QCD in (1+1)-Dimensions," *Phys. Rev.* **D41** (1990) 3814.

48. G. T. Horowitz and J. Polchinski, "A Correspondence principle for black holes and strings," *Phys. Rev.* **D55** (1997) 6189–6197, hep-th/9612146.

49. N. Itzhaki, J. M. Maldacena, J. Sonnenschein, and S. Yankielowicz, "Supergravity and the large N limit of theories with sixteen supercharges," *Phys. Rev.* **D58** (1998) 046004, hep-th/9802042.

50. A.C. Kalloniatis, "On zero zodes and the vacuum problem – a study of scalar adjoint matter in two-dimensional Yang-Mills theory via light-cone quantisation," *Phys.Rev* **D54** (1996) 2876.

51. A.C. Kalloniatis, H.C. Pauli, S.S. Pinsky, "Dynamical Zero Modes and Pure Glue QCD_{1+1} in Light-Cone Field Theory," *Phys. Rev.* **D50** (1994) 6633.

52. H.C. Pauli, A.C. Kalloniatis, S.S. Pinsky, "Towards solving QCD - the transverse zero modes in light-cone quantization," *Phys. Rev.* **D52** (1995) 1176.

53. H. J. Kim, L. J. Romans, and P. van Nieuwenhuizen, "Mass spectrum of chiral ten-dimensional $N = 2$ supergravity on S^5," *Phys. Rev.* **D32** (1985) 389–399.

54. I. Klebanov and A. Tseytlin, "A Non-Supersymmetric Large N CFT from Type 0 String Theory", hep-th/9901101

55. M. Krasnitz and I. R. Klebanov, "Testing effective string models of black holes with fixed scalars," *Phys. Rev.* **D56** (1997) 2173–2179, hep-th/9703216.

56. D. Kutasov, "Two Dimensional QCD Coupled to Adjoint Matter and String Theory," *Phys. Rev.* **D48** (1993) 4980, hep--th/9306013.

57. F. Lenz, H.W.L. Naus, M. Theis, "QCD in the axial gauge representation," *Ann. Phys.* (N.Y.) **233** (1994) 317.

58. F. Lenz, M. Shifman, M. Thies, "Quantum mechanics of the vacuum state in two-dimensional QCD with adjoint fermions," *Phys. Rev.* **D51** (1995) 7060

59. M. Li, "Large N solution of the 2-d supersymmetric Yang-Mills theory," *Nucl. Phys.* **B446** (1995) 16–34, hep-th/9503033.

60. J. Maldacena, "The Large N limit of superconformal field theories and supergravity," *Adv. Theor. Math. Phys.* **2** (1998) 231, hep-th/9711200.

61. J. Maldacena, "Wilson loops in large N field theories," *Phys. Rev. Lett.* **80** (1998) 4859, hep-th/9803002.

62. T. Maskawa and K. Yamawaki, "The problem of $p^+ = 0$ mode in the null plane field theory and Dirac's method of quantization," *Prog. Theor. Phys.* **56** (1976) 270.

63. Y. Matsumura, N. Sakai, and T. Sakai, "Mass spectra of supersymmetric Yang-Mills theories in (1+1)-dimensions," *Phys. Rev.* **D52** (1995) 2446–2461, hep-th/9504150.

64. G. McCartor, D.G. Robertson, S. Pinsky, "Vacuum structure of two-dimensional gauge theories on the light front," *Phys.Rev* **D56** (1997) 1035, hep-th/9612083.

65. A.S. Mueller, A.C. Kalloniatis, H.C. Pauli, ",Effect of zero modes on the bound-state spectrum in light-cone quantisation" *Phys.Lett* **B435** (1998) 189.

66. H. Oda, N. Sakai, and T. Sakai, "Vacuum structures of supersymmetric Yang-Mills theories in (1+1)-dimensions," *Phys. Rev.* **D55** (1997) 1079–1090, hep-th/9606157.

67. H. C. Pauli and S. J. Brodsky, "Discretized light cone quantization: solution to a field theory in one space one time dimensions," *Phys. Rev.* **D32** (1985) 1993, 2001.

68. S. Pinsky, "The Analog of the t'Hooft Pion with Adjoint Fermions," Invited talk at New Nonperturbative Methods and Quantization of the Light Cone, Les Houches, France, 24 Feb - 7 Mar 1997. hep-th/9705242.

69. S. Pinsky, "(1+1)-dimensional Yang-Mills theory coupled to adjoint fermions on the light front," *Phys.Rev* **D56** (1997) 5040, hep-th/9612073.

70. S. Pinsky, A. Kalloniatis, "Light-front QCD_{1+1} coupled to adjoint scalar matter," *Phys.Lett* **B 365** (1996) 225.

71. S. Pinsky, D. Robertson, "Light-front QCD_{1+1} coupled to chiral adjoint fermions," *Phys.Lett* **B 379** (1996) 169

72. S.-J. Rey and J. Yee, "Macroscopic strings as heavy quarks in large N gauge theory and anti-de Sitter supergravity," hep-th/9803001.

73. N. Seiberg, "Electric-magnetic duality in supersymmetric non-Abelian gauge theories," *Nucl.Phys.* **B435** (1995) 129

74. N. Seiberg, E. Witten, "Monopoles, duality and chiral symmetry breaking in N=2 supersymmetric QCD," *Nucl.Phys.* **B431** (1994) 484.

75. M.J. Strassler, "Manifolds of fixed Points and Duality in Supersymmetric Gauge theories," *Prog.Theor.Phys.Suppl.* **123** (1996) 373, hep-th/960202.

76. L. Susskind, "Another Conjecture About Matrix Theory", hep-th/9704080.

77. W. Taylor, "Lectures on D-Branes, Gauge Theory and M(atrices)," hep-th/9801182.

78. B. van de Sande, "Convergence of discretized light cone quantization in the small mass limit," *Phys.Rev.* **D54** (1996) 6347, hep-ph/9605409.

79. J. Wess, J. Bagger, *Supersymmetry and supergravity*, Princeton University Press(1992).

80. E. Witten, "Theta vacua in two–dimensional quantum chromodynamics," *Nuovo Cim.* **A51** (1979) 325.

81. E. Witten, "Bound states of strings and p–branes," *Nucl.Phys.* **B460**, (1996) 335, hep-th/9510135.

82. E. Witten, "Anti-de Sitter space and holography," *Adv. Theor. Math. Phys.* **2** (1998) 253, hep-th/9802150.

83. E. Witten, "Supersymmetric index of three-dimensional gauge theory," hep-th/9903005.

84. K. Yamawaki, "Zero mode problem on the light front," hep-th/9802037.

Instantons and Spin Physics

Nikolai Kochelev* and Vicente Vento [†]

* Bogoliubov Laboratory of Theoretical Physics,
Joint Institute for Nuclear Research,
Dubna, Moscow region, 141980 Russia
[†] Departament de Física Teòrica and Institut de Física Corpuscular,
Universitat de València-CSIC
E-46100 Burjassot (Valencia), Spain

Abstract. It is shown that a consistent treatment of the axial anomaly leads to an explicit cancellation between the the infrared and ultraviolet contributions to the flavor singlet axial charge. This result is a consequence of the simultaneous crossing of the zero point energy and the ultraviolet cut-off by quark levels of defined chirality from the vacuum in the presence of gluon fields. Based on this consideration we give the arguments in favor of a large violation of the Ellis-Jaffe sum rule by predicting the vanishing of the flavor singlet axial charge g_A^0 in a plausible scenario. From this result the value of the gluon polarization is estimated by using the Kühn-Zakharov value for the matrix element of the axial anomaly.

INTRODUCTION

During the last ten years there have been many attempts, both theoretical and experimental [1–3], to understand how the spin is distributed among the different components of the proton. This investigations have been labeled generically as *the proton spin problem.*

From the experimental point of view there has been much progress since the early EMC experiments [2] and the nucleon spin dependent structure functions are presently determined with high precision [3], except in the small x region where error bars are still large. This lack of precise determination at small x affects in great manner the low moments of the structure functions since they suffer from the ambiguities associated with the extrapolation of the structure functions to the $x \to 0$ region. From our point of view the theoretical knowledge of the low x behavior of spin-dependent structure functions $g_1^{p,n}(x, Q^2)$ is incomplete and therefore the precission of the experimental results is limited by the burden of this dependence.

There are two lines of thought, among the many theoretical developments providing an explanation of the small portion of nucleon spin carried by the quarks,

CP494, *New Directions in Quantum Chromodynamics*, edited by C.-R. Ji and D.-P. Min
© 1999 American Institute of Physics 1-56396-908-4/99/$15.00

which merit our attention here. One of them attributes this fact to the contribution of the polarized gluons via the anomaly to the first moment of $g_1(x, Q^2)$ [4–6], which results from the analysis of the triangle diagram contribution to the singlet axial-vector current. The final result for this gluon term is very sensitive to the factorization procedure used for triangle diagram [1]. For example, if one introduces an off-shell gluon momemtum such that $m_q^2/P^2 \to 0$ one obtains a non-zero value for the gluon contribution to the first moment through axial anomaly. But note that in this case only the large $k_\perp \to \infty$ region of integration over transfer momentum of quark in the triangle diagram is involved, which is associated with the *ultraviolet* description of the axial anomaly.

The other line suggests that a large negative quark polarization, leading to a non-perturbative contribution to the flavor singlet axial charge (FSAC), is responsible for the proton spin. In QCD, this polarization can be generated by the presence of non-perturbative gluon fields called instantons [7], which lead to the famous t' Hooft interaction [8]. Microscopically this interaction arises, because in the presence of the instanton field, quark-antiquark pairs are created by the zero energy quark modes. These pairs cause the large negative polarization needed [9–11]. This approach therefore is connected with the *infrared* treatment of the non conservation of the axial charge.

Our aim here is to proof that one should consider both contributions, the infrared and the ultraviolet, to the singlet axial-vector charge together, to be consistent with the properties of the QCD vacuum. In order to do so we show that these contributions are related to non-perturbative properties of the QCD vacuum, i.e., the behavior of the Dirac level spectrum. Moreover, due to Dirac level number conservation in the vacuum, they are equal in magnitude and of opposite sign cancelling each other. Our message is therefore that the small value of g_A^0 can be understood as a consequence of non-perturbative effects in QCD, namely the level motion of the Dirac spectrum in the vacuum.

TWO FACES OF AXIAL ANOMALY AND THEIR CONTRIBUTION TO FLAVOR SINGLET AXIAL CHARGE

Present wisdom tells us that the key to resolve the *spin crisis* is in the non conservation of the flavor singlet axial-vector current

$$\partial_\mu J_{\mu 5}^0(x) = 2i \sum_i m_i \bar{q} \gamma_5 q + 2N_f \frac{\alpha_s}{8\pi} G_{\mu\nu}^a \widetilde{G_{\mu\nu}^a}, \qquad (1)$$

due to axial anomaly term, which is the last in Eq.(1). A very detailed account of the various ways to get Eq.(1) was given in the review by Shifman [12] and we shall now use them to present our argument.

The anomaly presents itself with two faces [12], the infrared and the ultraviolet one. The reason for this duality being that there are two ways of getting the adequate result: the infrared one, by looking at the motion of the low lying levels of well defined chirality in the Dirac vacuum; the ultraviolet one, by looking to the corresponding high levels after introducing a gauge invariant cut-off. It is now clear that the latter mechanism with opposite flow is the one needed to preserve the axial charge of the vacuum and is the basis of our argument which can be very clearly visualized by looking at Fig. 1 of ref. [12] and which goes as follows.

The number of Dirac levels with definite chirality should be conserved. The crossing of the levels by the zero energy point, which leads to a quark-antiquark pair creation , should be accompanied by a simultaneous crossing by the levels of the ultraviolet boundary. Both contributions are equal in magnitude, but the flow is opposite and therefore their sign opposite, therefore their simultaneous consideration produces a cancellation. In the simplified model of ref. [12] this result is immediate. In terms of the more familiar language in four dimensions, the crossing of the ultraviolet cut-off determines the perturbative contribution of the gluons to the non conservation of the FSAC. The conservation of the number of Dirac level implies therefore, that the two contributions, infrared and *perturbative* cancel each other and that the FSAC is conserved. This conservation implies that a Goldstone boson is not needed in this channel to insure axial-vector current conservation. This statement reflects the solution to the $U_A(1)$ problem and is a realiztion of 't Hooft's consistency condition.

A similar analysis can be done following the description of Mueller [14] for the massive case. The paralelism is complete, since the intermediate non chiral states do not contribute, and the cancellation is explicitly shown. In this case it appears as the cancellation between the charge coming from the mass term, which arises from the infrared region, due to the explicit breaking of chiral symmetry, and the charge coming from the anomaly, i.e., from the ultraviolet region, due to the flow of charge in the fiducial volume. It is necessary to stress that the mass term in Eq. (1) plays a crucial role in the cancellation and therefore it can neither be neglected nor treated as a small perturbation.

The results discussed above has been shown explicitly for QED in 1+1 dimensions. It must be remembered that this theory is confining, but moreover, as discussed in detail by Shifman, confinement plays no role in his description of the properties of the anomaly. The number of level argument holds also in 3+1 dimensions and for QCD, and this is all the necessary requirement for our conclusion. The analysis of Mueller allows moreover a partonic interpretation of the two contributions. The contribution arising from the conversion of the right-handed components of the Dirac sea to the left-handed, by the presence of the explicit chiral symmetry breaking term in the lagrangian, occurs at low tranverse momentum and corresponds to a net chirality being carried by the quark-antiquark partonic components. In the massless case this contribution corresponds to the motion of the levels around the zero point energy, i.e., the gluonic field polarizes the medium. The other contribution to the charge comes from the flow of chirality along the lowest Landau level from $p_\perp = \infty$ and it is the due to the anomaly term. This piece cannot be given a Fock space description in terms of quarks at finite momentum.

It seems more appropriate to associate this chirality directly with the gauge field, i.e., with the gluonic components of the Fock space. This mechanism corresponds in our interpretation to the ultraviolet contribution. Thus in the massless case the two faces of the anomaly conspire to obtain an analogous partonic interpretation. We are dealing with the vacuum, thus no valence quarks have been considered in the argument, only levels in a Dirac sea and therefore the mechanism discussed arises just as a property of the QCD vacuum.

For the proton, the same analysis is valid, but one has to take care of the additional valence quark terms. In the massless case, chirality is a good quantum number and the analysis of number level conservation follows through without problem. In the massive case, there are additional contributions coming from the region where chirality is not a good quantum number. One expects these contributions, which arise from the mass terms, to be small [15]. In the case of the instanton vacuum they have been shown to be zero since, besides the negative sea quark polarization mentioned above, the zero modes flip the spin of the valence quarks [10]. This is the main difference one encounters between QED and QCD, the chiral limit is good in QCD but not in QED. In the former the spin flips, due to the presence of the instantons, and the valence contribution vanishes. In the latter it does not and the valence contribution remains.

The authors of ref. [15] have arrived to similar conclusion, however we must stress, that it is the cancellation between the anomaly (ultraviolet ΔG), and the screaned (due to depolarization) valence contributions which make the FSAC vanish. In particular within instanton model for $N_f = 1$, and only in this case, the two terms vanish independently.

The work of Shore and Veneziano reviewed in ref. [16] relates also the value of g_A^0 to the properties of the QCD vacuum. In their case the analysis leads to a relation between the FSAC and the topological susceptibility. The present arguments imply that in the chiral limit this quantity should vanish and that it should be related to the properties of the chiral quark fields. If chiral symmetry is explicitly broken through mass terms the susceptibility will be small.

Finally we would like to confront our calculation with the detailed analysis done in the Chiral Bag Model [17]. In the latter all possible contributions to the FSAC are explicitly calculated. The authors take into account contributions from the Valence quarks, the Dirac sea, the Gluonic sea and the meson fields. An exciting consequence is that in their calculation a tremendous cancellation occurs between relatively big contributions. This result certainly supports our analysis. Moreover the calculation is basically radius independent, i.e., satisfies approximately the so called Cheshire Cat principle [18], from our point of view it means that the result is fundamental and not model dependent. However the final number is small and not zero. In the chiral limit it should certainly be zero if our argument holds. We understand that the non-vanishing of the FSAC may be related to the various approximations involved in this realistic calculation. However it surprises us the sign of the Gluon Casimir calculation. We would relate it to the ultraviolet flow in our scheme and we would expect it to be negative. If this were the case the cancellation would be almost complete. We encourage the authors to reanalyze that difficult term in detail.

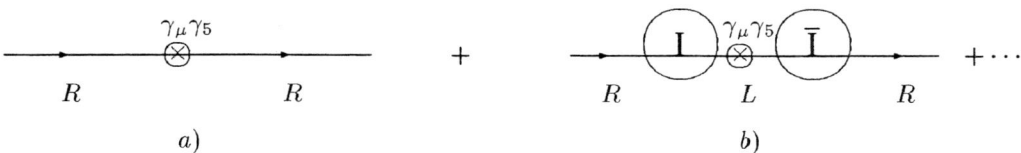

FIGURE 1. *The contribution to the matrix elements between quark states induced by instantons.*

One approach to show explicitly the non-perturbative QCD mechanisms in DIS is the instanton model for the QCD vacuum [19]. In the framework of this model the matrix element of the flavor singlet axial current (see Fig.(1b)) between gluon states with different momenta $Q = p' - p$ is [21]

$$< g', s|J^0_{\mu 5}|g, s >_{inst} \sim (K_2(\rho Q) - \frac{2K_1(\rho Q)}{Q\rho}), \tag{2}$$

where ρ is the instanton size. In Eq.(2) the first term comes from *non-zero modes* in the instanton field and corresponds to the anomaly term in Eq.(1). The second term has its origin in the *zero modes* and is related to the mass term in Eq.(1). Again, as in the perturbative case, at $Q^2 \to 0$ by using expansion of the functions $K_2(Q\rho) \to 2/(Q\rho)^2$, $K_1(Q\rho) \to 1/Q\rho$, we observe an *explicit* cancellation between the two contributions in Eq.(2). This cancellation is the fundamental reason behind the resolution of the $U_A(1)$ problem. At $Q^2 \neq 0$ there are two massless poles in Eq.(2). One of them is the usual Goldstone pole which is related to the mass term in Eq.(1), the other comes from the anomaly. The consequence of the explicit cancellation between these two poles is the appearence of the massive η' meson. The massless pole which is connected with first term in Eq. (2) can be interpreted as the Kogut-Susskind ghost pole [22] which was used to resolve $U_A(1)$ problem [23], [24] [1]. Therefore only the contribution from the ultraviolet region provides the mechanism which makes the η' different from octet Goldstone bosons [2].

---

[1] The connection of ghost pole with proton spin problem was discussed in [25].
[2] It should be mentioned that in instanton model of the QCD vacuum it is sufficient to take into account only the zero mode contributions to obtain the massless pseudoscalar octet bosons in the chiral limit.

The calculation just described is the conventional one in the framework of the instanton model for the QCD vacuum (see for example [10,19]), and as such purely non-perturbative. However, we must point out, that in many papers where this type of calculations were performed only zero modes were taken into account. We have shown the importance of non-zero modes in the calculation of the matrix element of the flavor singlet axial current

Recent data on flavor asymmetry for the sea [20] confirm the presence of both zero and non-zero modes in the calculation. The data show a strong flavor asymmetry at low $x < 0.1$, a characteristic of the zero-mode contribution, while an almost symmetric flavor sea at high $x > 0.1$, a very clear signal of the contribution of the non-zero modes.

One could also calculate the matrix element of divergence of the flavor singlet axial-vector current between proton states in the framework of the same instanton model for the QCD vacuum. It has been done in the particular case of $N_f = 1$ in a paper by Forte and Shuryak [10] by taking into account the contributions which come from the diagrams on Fig. 2. Again the explicit cancellation has been found. This is however a model dependent statement.

ANALYSIS OF THE PROTON SPIN

We proceed to analyze the physical consequences of the vanishing of the matrix element of the divergence of the flavor singlet axial-vector current in the extreme instanton vacuum model scenario. The matrix element between nucleon states of the flavor singlet axial-vector current is given by

$$< p'|J_{\mu 5}^0|p> = g_A^0(Q^2)\bar{P}\gamma_\mu\gamma_5 P + g_P^0(Q^2)q_\mu\bar{P}\gamma_5 P \qquad (3)$$

and by recovering the matrix element of Eq.(1) we immediately arrive to the conclusion that

$$g_A^0 = 0, \qquad (4)$$

which is the main result of this paper. Therefore, we predict very large violation of the Ellis-Jaffe sum rule [29] [3].

The modern experimental data on spin-dependent structure function g_1 [3] give the value of singlet axial-vector current $g_A^0(5 \div 10 GeV^2) \approx 0.2 \div 0.28$. From our point of view, the difference between Eq.(4) and data comes from the inaccessible low x region where the *negative* contribution from quark zero-modes should occur [11]. Naive Regge extrapolation at $x \to 0$, used in experimental papers should be used with utmost precaution.

As it has been mentioned above, one can interpret the matrix element of the anomaly as the negative gluon contribution to g_A^0

---

[3] Recall that for QED the analog of the Ellis-Jaffe sum rule would be satisfied.

$$g_{AG}^0 2 M_P \bar{P} i \gamma_5 P = < P |2 N_f \frac{\alpha_s}{8\pi} G_{\mu\nu}^a \widetilde{G_{\mu\nu}^a}| P > = - \frac{2 N_f \alpha_s}{4\pi} \Delta G 2 M_P \bar{P} i \gamma_5 P \qquad (5)$$

Note that in [30], the quark mass term was not considered and therefore g_{AG}^0 has been identified with g_A^0. This is

not the case here since, as we have shown, the quark mass term contribution is very important.

The matrix element in Eq.(5) for the anomaly has been calculated in a paper by Kühn and Zakharov [30]

$$g_{AG}^0 = - \frac{2 N_f}{3 b_0}, \qquad (6)$$

where $b_0 = 11 N_c / 3 - 2 N_f / 3$. By using the last experimental data for α_s from CCFR collaboration $\alpha_s (3 GeV^2) = 0.278$ [26], and Eqs.(5) and (6) for $N_f = 3$ our estimate for the gluon polarization becomes

$$\Delta G (3 GeV^2) = 1.67 \ . \qquad (7)$$

It is interesting to mention, that if we use the fixed value of $\alpha_s (Q^2)$ at small $Q^2 \to 0$, which has been obtained by Shirkov and Solovtsov [31]

$$\alpha_s (Q^2 = 0) = \frac{4\pi}{b_0}, \qquad (8)$$

the result for gluon polarization is

$$\Delta G (0) = \frac{1}{3}. \qquad (9)$$

The very interesting feature of Eq.(9) is absence any dependence from N_f, N_c and another parameters of theory.

There have been many attempts to estimate the gluon polarization in nucleon by using the different approaches [27]. However, there is no agreement in the final result, not only for absolute value, but even for the sign of the gluon polarization. We have presented a scenario, where this value is uniquely given.

CONCLUSION

We have presented a new point of view which emphasizes, how the non-perturbative behavior of the QCD vacuum, leads to a better understanding of the proton spin problem.

We have shown that there is a cancellation between the contributions from the ultraviolet and the infrared regions to the FSAC and that it appears as a consequence of the general properties of the QCD vacuum. Therefore it should be respected by all models. This cancellation leads to a small matrix element between proton states of the right side of divergence of the singlet axial-vector current, which implies an almost conserved FSAC[4].

Within a simple version of the instanton vacuum model we have shown that, besides the above cancellation, the valence contribution is also zero and therefore the FSAC vanishes for the proton. Thus we are led to a scenario of maximal violation of the Ellis-Jaffe sum rule. This calculation and the validity of the chiral limit lead us to conclude, that nature seems to be deviating strongly from the Ellis-Jaffe sum rule and therefore data in the low x region are crucial to establish the magnitude of the deviation and put stringent conditions on the models of the QCD vacuum.

A main observation of our work is the relevance that the quark mass term in the equation for divergence of singlet axial-vector current plays in the cancellation due to existence of zero quark modes in the QCD vacuum. If one does not include this term one will wrongly conclude that the contribution of the gluons to proton spin is small see [32].

In spite of the fact that these two contributions combined explicitly cancel in the FSAC, their contributions to the polarized structure functions does not, because each of them has a different x dependence. So in the large x region the contribution from gluons should be large while at low x a large contribution from zero modes is expected. These two contributions have also a different structure for the particles in the final state. The ultraviolet contribution leads to quark and antiquark jets with high $k_\perp$ and different helicities for the quark and antiquark. On the contrary, the infrared contribution is associated with the production of a $q\bar{q}$ pairs with small $k_\perp$ and the same helicity for quark and antiquark [33].

It is evident that the cancellation phenomenon is a general statement of the QCD vacuum and can be explained by quotating from Shifman's review [12] : *Both phenomena, though - the ossing of the zero-energy point and the departure (arrival) of the levels via the ultraviolet cut-off - occur simultaneously and represent, actually, two different facets of the one and the same anomaly, which admits both, the infrared and ultraviolet interpretation.*

ACKNOWLEDGEMENTS

N. Kochelev is very grateful to Prof. D.-P. Min and Prof. C.-R. Ji for the warm hospitality in Korea and APCTP for the financial support.

REFERENCES

1. M.Anselmino, A.Efremov, and E.Leader *Phys. Reports* **261** 1 (1995).
2. EMC, J.Ashman et al, *Phys.Lett.* **B206** 345 (1988).

---

[4] The possibility of the conservation of the flavor singlet axial charge was discussed in papers [28] from different point of view.

3. K. Abe et al., E154 collaboration, hep-ph/9705344v2;
   SMC, talk by J.Cranshaw; HERMES, talk by K.Rith; E155, talk by H.Borel.
   Workshop "Deep Inelastic Scattering 1998", Brussel, April, 1998;
   P.J.Mulders and T.Sloan, hep-ph/9806314.
4. A.V.Efremov and O.A.Teryaev, JINR preprint E2-88-287 (1988);
5. G.Altarelli and G.G.Ross, *Phys. Lett.* **B212** 391 (1988).
6. R.D.Carlitz,J.C.Collins and A.H.Muller, *Phys. Lett.* **B214** 229 (1988).
7. A.A. Belavin, A.M. Polyakov, A.S. Swartz, and Yu.S. Tyupkin *Phys. Lett.* **59B** 85 (1975).
8. 't Hooft *Phys. Rev.* **D14** 3432 (1976).
9. S.Forte *Phys. Lett.* **B224** 189 (1989);
   B.L.Ioffe and M.Karliner *Phys. Lett.* **B247** 387 (1990);
   A.E.Dorokhov, N.I.Kochelev, and Yu.A. Zubov *Int. J. Mod. Phys.* **8A** 603 (1993);
   A.E.Dorokhov and N.I.Kochelev *Phys. Lett.* **304B** 167 (1993).
10. S.Forte and E.V.Shuryak *Nucl.Phys.* **B357** 153 (1991).
11. N.I.Kochelev, *Phys. Rev.* **D57** 5539 (1998).
12. M.A. Shifman *Phys. Rep.* **209** 341 (1991).
13. V.N. Gribov, Hungarian Academy of Sciences report KFKI-1981-66 (1981).
14. A.H. Mueller *Phys. Lett.* **B234** 517 (1990).
15. Igor Halperin and Ariel Zhitnitsky, hep-ph/9706251.
16. G.M. Shore, Nucl. Phys. (Proc. Suppl.) **64** 167 (1998).
17. H-J. Lee, D-P. Min, B-Y. Park, M. Rho and V.Vento, hep-ph/9810539 (Nucl. Phys. A to be published); M. Rho and V. Vento, Nucl. Phys. **A662** 413 (1997).
18. M. A. Nowak, M. Rho and I. Zahed in *Chiral Nuclear Dynamics* (World Scientific Pub. Singapore 1996) Chapter 8.
19. T.Schäfer, E.V.Shuryak, Rev.Mod.Phys. **70** 323 (1998).
20. J.C. Peng et al., Phys.Rev. **D58** 092004 (1998).
21. B.V.Geshkenbein and B.L.Ioffe, *Nucl. Phys.* **B166** 340 (1980).
22. J.Kogut and L.Susskind, *Phys.Rev.* **D11** 3594 (1975);
    see also S.Weinberg *Phys.Rev.* **D11** 3583 (1975).
23. G.Veneziano, *Nucl.Phys.* **B159** 213 (1979).
24. S.Witten,*Nucl.Phys.* **B156** 269 (1979).
25. G.Veneziano, *Mod. Phys. Lett.* **A4** 1605 (1989);
    G.M.Shore and G.Veneziano, *Phys. Lett.* **B244** 75 (1990);
    A.V.Efremov, J.Soffer and N.A. Törnqvist, *Phys. Rev.* **D44** 1369 (1991).
26. The CCFR/NuTeV Collaboration, J.Yu et. al, hep-ex/9806031.
27. S.J.Brodsky, M.Burkardt, and I.Schmidt, *Nucl. Phys.* **B441** 197 (1995);
    R.L.Jaffe, *Phys. Lett.* **B365** 359 (1996) ;
    M.Glück, E.Reya, M.Stratmann and W.Vogelsang, *Phys.Rev.* **D53** 4775 (1996);
    G.Altarelli,R.D.Ball,S.Forte, and G.Ridolfi, *Acta Phys. Polon.* **B29** 1145 (1998);
    A.Saalfeld, G.Piller, and L.Mankiewicz, Eur. Phys. J. **C4** 307 (1998);
    E.Leader, A.V.Sidorov and D.B.Stamenov, Phys. Lett. **B445** 232 (1998) , Phys. Rev. **D58** 114028 (1998);
    S. Scopetta, V. Vento and M. Traini, *Phys. Lett.* **B421** 64 (1998) , Phys.Lett. B442 28 (1998) 28; P. Faccioli, M. Traini, V. Vento, Few Body Syst.Suppl. **11** 347 (1999),

hep-ph/9808201.

28. R.J.Crewther, *Riv. Nuovo Cimento* **2** 63 (1979);
    G.A.Christos, *Phys. Rep.* **116** 251 (1984).
29. J.Ellis and R.L.Jaffe, *Phys. Rev.* **D9** 1444 (1974).
30. J.H.Kühn snd V.I.Zakharov, *Phys. Lett.* **252** 615 (1990).
31. D.V.Shirkov and I.L.Solovtsov, *Phys. Rev. Lett.* **79** 1209 (1997).
32. H.Fritzsch, *Phys. Lett.* **B229** 122 (1989).
33. A. Freund and L.M. Sehgal, *Phys. Lett.* **B341** 90 (1994).

WORKSHOP TALKS

LIGHT-CONE FIELD THEORY

Embedded States in Light-Front Dynamics

B.L.G. Bakker

Department of Physics and Astronomy, Free University, Amsterdam, The Netherlands

Abstract The role of embedded states in the calculation of current matrix elements in light-front dynamics is discussed. As an example the current of a scalar boson with spin-1/2 constituents is calculated and numerical results are produced. They demonstrate the importance of the non-valence parts.

INTRODUCTION

Light-front Hamiltonian theory is expected to be a frame work in which realistic field theories, including QCD, can be solved non-perturbatively (1). In simple cases one has been able to calculate the wave functions of bound states using this method. For realistic situations, like QCD in 3+1 dimensions, model wave functions for mesons and baryons are still much used. In either case, a natural way to calculate the response of a composite particle is to consider the triangle diagram, like the one depicted in Fig. 1.

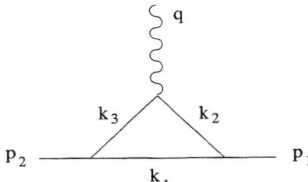

Figure 1. Covariant triangle diagram

First-order perturbation theory is used for the probing interaction, while the vertices connecting the composite particle and its constituents are described in terms of wave functions.

The concept of a wave function is very natural in a time-ordered context, but in the framework of covariant perturbation theory the covariant vertices are more appropriate. If one expands the covariant triangle diagram into time-ordered parts,

CP494, *New Directions in Quantum Chromodynamics*, edited by C.-R. Ji and D.-P. Min
© 1999 American Institute of Physics 1-56396-908-4/99/$15.00

one sees immediately that the covariant vertices give rise to time-ordered amplitudes that can be written entirely in terms of valence wave functions, and some that cannot. The vertices that cannot be expressed in terms of valence wave functions are called *embedded states*. It is important to emphasize that the non-valence contributions are needed to guarantee covariance of the calculated amplitude. Neglect of those contributions leads to all kinds of problems, e.g., violation of current conservation.

MODEL

We consider the electromagnetic vector current matrix element of a composite scalar boson of mass m composed of two charged fermions, masses m_a, m_b and charges e_a and e_b resp. The light-cone wave functions are taken in zero-range approximation corresponding to point-like vertices. To specify the model, we consider the Lagrangian:

$$\mathcal{L} = \overline{\psi_a}(i(\not{\partial} + ie_a \not{A}) - m_a)\psi_a + \overline{\psi_b}(i(\not{\partial} + ie_b \not{A}) - m_b)\psi_b$$
$$+ \frac{1}{2}\partial_\mu \phi \partial^\mu \phi - \frac{1}{2}m^2 \phi\phi + g\phi(\overline{\psi_a}\psi_b + \overline{\psi_b}\psi_a). \tag{1}$$

This admittedly non-realistic model is considered because the effects of the embedded states can be demonstrated explicitly. We consider it both in 1+1 and 3+1 dimensions. In general, there are two contributions: the photon may couple to particle a or particle b. The corresponding currents are denoted by $\mathcal{M}_a^\mu$ and $\mathcal{M}_b^\mu$ respectively. They are related by the interchange $a \leftrightarrow b$ in the formulas written below.

COVARIANT CALCULATION

The covariant amplitude follows from the Feynman rules. It is

$$\mathcal{M}_a^\mu = -e_a g^2 \int \frac{d^4 k}{(2\pi)^4} \frac{\text{Tr}[(\not{k}_1 + m_b)(\not{k}_2 + m_a)\gamma^\mu(\not{k}_3 + m_a)]}{(k_1^2 - m_b^2 + i\epsilon)(k_2^2 - m_a^2 + i\epsilon)(k_3^2 - m_a^2 + i\epsilon)}. \tag{2}$$

The traces can be computed using well-known techniques. We shall not write them down explicitly here. The momenta are chosen as in Fig. 1. By using the Feynman-parameter technique one is left with a two-dimensional integral over the Feynman-parameters and an integral over $d^4 k$ that can be evaluated in closed form. If the external legs with momenta p_1 and p_2 are on shell, Lorentz invariance and current conservation require the matrix element to be of the form

$$\mathcal{M}_a^\mu = -i\, e_a\, F_a\, (p_1^\mu + p_2^\mu). \tag{3}$$

234

The form factor can be written as an integral over the Feynman parameters of two space-time integrals

$$I_0 = \int \frac{d^4k}{(2\pi)^4} \frac{1}{D(k^2; x, y)^3}, \quad I_2 = \int \frac{d^4k}{(2\pi)^4} \frac{k^2}{D(k^2; x, y)^3}. \tag{4}$$

The denominator function $D(k^2; x, y)$ depends on the momenta and the Feynman parameters. It can be found in a straightforward way.

The actual computation of the integrals involves a Wick rotation of $k_M = (k^0, k^1, k^2, k^3)$ to $k_E = (k^1, k^2, k^3, k^4 = ik^0)$. This leads to a denominator

$$D(k^2; x, y)|_E = -(k_E^2 + M^2(x, y; Q^2)), \tag{5}$$

where Q^2 is the negative of the square of q. We limit our calculations to the mass region $m < m_a + m_b$ and $m > |m_a - m_b|$. Then the mass function M^2 is positive definite and the integrals are well defined.

Using the form defined above we find the Wick-rotated integrals $-iI_0^E$ and iI_2^E. The standard rules give in D dimensions

$$I_0^E = \frac{\Gamma(3 - D/2)}{(4\pi)^{D/2}\Gamma(3)} M^{D-6}, \quad I_2^E = \frac{D}{2} \frac{\Gamma(2 - D/2)}{(4\pi)^{D/2}\Gamma(3)} M^{D-4}, \tag{6}$$

exposing the logarithmic singularity of I_2 for $D = 4$. We use BPHZ regularization as detailed in (2). The mass function M^2 is split into a part depending on the external invariants, M_e^2, and a remainder that depends on the masses of the internal lines only, M_i^2:

$$M_e^2 = [(x + y)(x + y - 1)m^2 + xyQ^2], \quad M_i^2 = [(1 - x - y)m_b^2 + (x + y)m_a^2]. \tag{7}$$

The BPHZ step is to write the regularized integral $I_2^{E \, Reg}$ as

$$I_2^{E \, Reg} = \int_0^1 d\lambda \int_E \frac{d^4k}{(2\pi)^2} \frac{\partial}{\partial\lambda} \frac{k^2}{(k^2 + \lambda M_e^2 + M_i^2)^3} = \frac{-1}{(4\pi)^2} \log\left(\frac{M_e^2 + M_i^2}{M_i^2}\right). \tag{8}$$

(This result agrees with the one obtained using dimensional regularization.) The real form factor $F_a = F_{a0} + F_{a2}$ can be expressed in terms of the regulated integrals I_0^E and $I_2^{E \, Reg}$.

LIGHT-FRONT TIME-ORDERED AMPLITUDES

We derive the light-front amplitudes from the covariant amplitude Eq. (2) by integrating over the k^- variable (3). For theories containing scalar particles only, this construction is well defined, but as soon as particles with spin different from

0 are included one needs more care. For spin-1/2 this becomes apparent from the splitting of the Dirac propagator into a propagating part and an instantaneous part

$$\frac{\not{k} + m}{k^2 - m^2} = \frac{\not{k}_{\rm on} + m}{2k^+(k^- - k_{\rm on}^-)} + \frac{\gamma^+}{2k^+}, \quad k_{\rm on} = \left(k^+, \frac{k^{\perp 2} + m^2}{2k^+}, \vec{k}^\perp\right). \tag{9}$$

(We use the metric $k^2 = 2k^+k^- - \vec{k}^{\perp 2}$.) The occurrence of $1/k^+$ in the numerators of Dirac propagators may cause additional divergences to occur. They are absent in the $\mathcal{M}^+$ component of the current (4), so we concentrate here on this "good component" and consider the kinematical situation $p_2^+ - p_1^+ = q^+ \geq 0$.

Straightforward application of the k^- integration gives for the good component of the current three contributions. We summarize the situation graphically in Fig. 2.

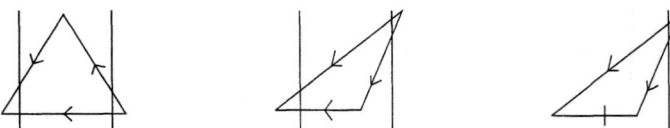

Figure 2. LF-time ordered diagrams (a, v), (a, nv), and (b, nv) for $q^+ > 0$. The vertical lines denote energy denominators. The tag denotes an instantaneous propagator.

We write the three amplitudes in terms of phase-space factors, traces and energy denominators. The integration variable is $k = -k_1$. For $0 < k^+ < p_1^+$ we have only a valence contribution

$$\mathcal{M}_a^\mu({\rm a, v}) = \frac{ie_a g^2}{(2\pi)^3} \int d^2 k_\perp \int_0^{p_1^+} dk^+ \frac{1}{\Phi({\rm v})} \frac{{\rm Tr}({\rm a, v})}{D_2({\rm v}) D_1({\rm v})}. \tag{10}$$

The non-valence contribution corresponds to the domain $p_1^+ < k^+ < p_2^+$.

$$\mathcal{M}_a^\mu({\rm a, nv}) = \frac{ie_a g^2}{(2\pi)^3} \int d^2 k_\perp \int_{p_1^+}^{p_2^+} dk^+ \frac{1}{\Phi({\rm nv})} \frac{{\rm Tr}({\rm a, nv})}{D_2({\rm v}) D_1({\rm nv})}. \tag{11}$$

In the same domain there is a contribution with one instantaneous propagator

$$\mathcal{M}_a^\mu({\rm b, nv}) = \frac{ie_a g^2}{(2\pi)^3} \int d^2 k_\perp \int_{p_1^+}^{p_2^+} dk^+ \frac{1}{\Phi({\rm nv})} \frac{{\rm Tr}({\rm b, nv})}{D_1({\rm nv})}. \tag{12}$$

The phase space factors $\Phi({\rm v})$ and $\Phi({\rm nv})$ are given by

$$\Phi({\rm v}) = 2k^+ 2(p_1^+ - k^+) 2(p_2^+ - k^+), \quad \Phi({\rm nv}) = 2k^+ 2(p_2^+ - k^+) 2(k^+ - p_1^+). \tag{13}$$

The energy denominators are

$$D_1({\rm v}) = p_1^- - \frac{k^{\perp 2} + m_b^2}{2k^+} - \frac{(p_1^\perp - k^\perp)^2 + m_a^2}{2(p_1^+ - k^+)},$$

$$D_2({\rm v}) = p_2^- - \frac{k^{\perp 2} + m_b^2}{2k^+} - \frac{(p_2^\perp - k^\perp)^2 + m_a^2}{2(p_2^+ - k^+)},$$

$$D_1({\rm nv}) = p_2^- - p_1^- - \frac{(k - p_1)^{\perp 2} + m_a^2}{2(k^+ - p_1^+)} - \frac{(p_2^\perp - k^\perp)^2 + m_a^2}{2(p_2^+ - k^+)}. \tag{14}$$

The amplitudes may contain separately longitudinal singularities as analyzed in (5). If we limit ourselves to the "good" component of the current, then these singularities do not occur. The transverse divergencies of the 3+1 dimensional amplitudes are removed using minus-regularization (2), (5).

NUMERICAL RESULTS AND DISCUSSION

It is clear that a composite-particle interpretation of the current matrix element fails in the 3+1 dimensional case, as the divergence of I_2 precludes the normalization of the wave function. Normalization is possible if one includes additional vertex functions that remove the divergencies.

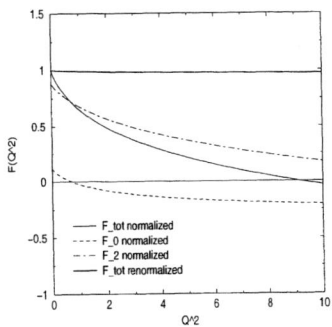 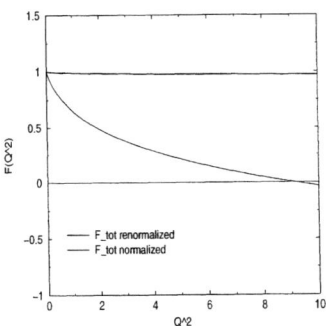

Figure 3. Results of the covariant calculation (left) and light-front calculation (right) in 3+1 dimensions. The boson mass m is 1.99. F_a is plotted only.

We limited ourselves in 3+1 dimensions to the kinematically simple case where $q^+ = 0$. The results are shown in Fig. 3. Our parameters were $e_a = 1$, $g = 1$, $m_a = m_b = 1$, $m = 1.99$. It is useful for a comparison of the two contributions F_{a0} and F_{a2} to plot the form factor divided by its value at $q^2 = 0$. The results are labeled "F normalized". An interpretation of the triangle diagram as a renormalization of the bare $\gamma\phi\phi$-vertex would require the sum of the bare current and the contribution from the triangle diagram at $q^2 = 0$ to correspond to the physical charge. The curves labeled "F renormalized" correspond to this case. Our calculations show that the regulated covariant current component $\mathcal{M}_a^+$ is reproduced by the regulated light-front calculation. The contribution of the triangle diagram appears to be very small in all cases.

In 1+1 dimensions one can use this model for a genuine bound state, the normalization of the form factor to 1 at $q^2 = 0$ being guaranteed by the normalization of the wave function. In Fig. 4 we show the results. The valence contribution dominates only at very small momentum transfer and binding energy ($Q^2 < 1$, $m = 1.99$, lower-right panel). So the embedded states are essential for obtaining a covariant result using light-front methods.

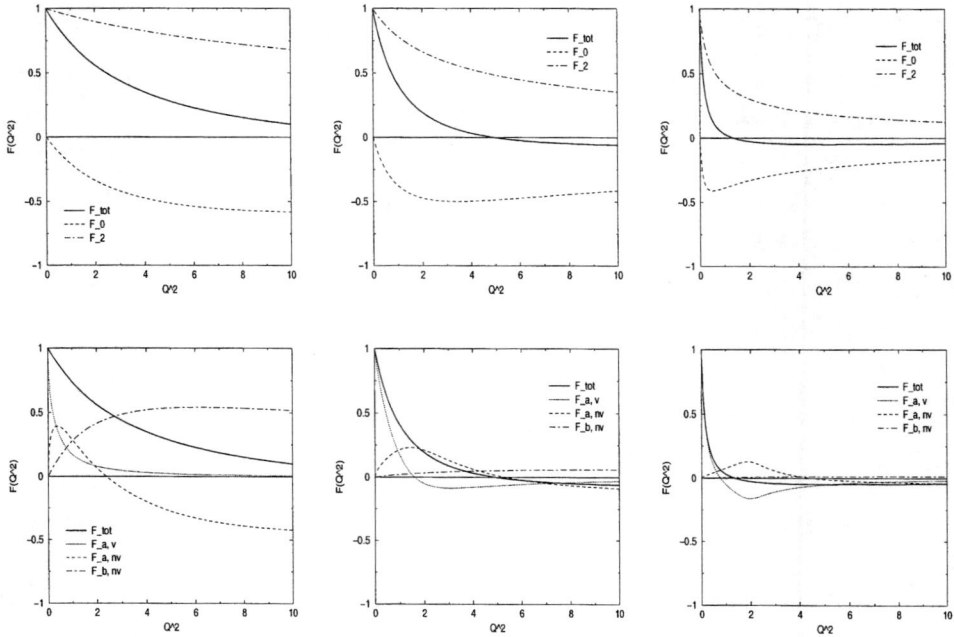

Figure 4. Results of covariant calculations (upper panels) and light-front calculations (lower panels) in 1+1 dimensions. The boson mass is 1.00, 1.90 and 1.99 respectively.

If one would calculate the other components of the current, $\mathcal{M}^{\perp}$ and $\mathcal{M}^{-}$, one might expect singularities to occur that are not present in a covariant calculation. Some of them are associated with $1/k^{+}$ singularities in the interior of the integration interval. They can be removed using the the "blink" mechanism described in (5). Other singularities of this type occur at the endpoints of the k^{+} integration and cannot be removed in the same way. To regularize them one needs other methods. Those are presently being studied by the author.

REFERENCES

1. S.J. Brodsky, H.-C. Pauli, and S.S. Pinsky, *Physics Reports* **301**, 299-486 (1998)

2. N.C.J. Schoonderwoerd and B.L.G. Bakker, *Phys. Rev.* **D57**, 4965-4975 (1998)

3. J.B. Kogut and D.S. Soper, *Phys. Rev.* **D 1**, 2901-2914 (1970)

4. J.P.B.C. de Melo, J.H.O. Sales, T. Frederico, and P.U. Sauer, *Nucl. Phys.* **A631**, 574c-579c (1998)

5. N.E. Ligterink and B.L.G. Bakker, *Phys. Rev.* **D52**, 5954-5979 (1995)

Gauge Field Theories on a ⊥ lattice

Matthias Burkardt

New Mexico State University
Las Cruces, NM 88003, U.S.A.

Abstract. In these notes, the transverse (⊥) lattice approach is presented as a means to control the $k^+ \to 0$ divergences in light-front QCD. Technical difficulties of both the canonical compact formulation as well as the non-compact formulation of the ⊥ lattice motivate the color-dielectric formulation, where the link fields are linearized.

INTRODUCTION

The main subject of these notes are difficulties associated with the formulation of gauge field theories on a transverse (⊥) lattice using light-front (LF) quantization. Because of these difficulties, the reader may wonder about the advantages of this approach — particularly given the successes of Euclidean lattice gauge theory (LGT). The primary motivations for formulating QCD in this framework is that LF quantization is the most physical approach towards a microscopic description of the parton distributions measured in deep-inelastic scattering as well as many other hard processes. [1] It is important to emphasize this fact in this brief introduction since it explains why LF quantization of QCD and the ⊥ lattice should be investigated as a possible alternative to Euclidean and Hamiltonian LGT formulations — despite the difficulties that will be discussed in the remainder of these notes.

Why LF gauge?

Although the choice of quantization hyperplane and the choice of gauge are in principle independent issues, the so-called LF gauge ($A^+ = 0$) turns out to be highly preferable for the canonical formulation of LFQCD. The main reason is that in the kinetic energy term for $\vec{A}_\perp$ (from $-\frac{1}{4}F^{\mu\nu}F_{\mu\nu}$)

$$\mathcal{L}_{kin,A_\perp} = D_+\vec{A}_\perp D_-\vec{A}_\perp = (\partial_+ - igA_+)\,\vec{A}_\perp\,(\partial_- - igA_-)\,\vec{A}_\perp, \qquad (1)$$

the term multiplying the 'time' derivative of $\vec{A}_\perp$ (i.e. $\partial_+\vec{A}_\perp$) contains also $A_- = A^+$. Therefore, the canonical momenta

---

[1] This and other motivations are discussed in more detail in Ref. [1] and in references therein.

CP494, *New Directions in Quantum Chromodynamics*, edited by C.-R. Ji and D.-P. Min
© 1999 American Institute of Physics 1-56396-908-4/99/$15.00

$$\Pi = \frac{\partial \mathcal{L}}{\partial(\partial_+ A_\perp)} = (\partial_- - igA_-) A_\perp, \tag{2}$$

which are the LF analog to $\Pi = \frac{\partial \mathcal{L}}{\partial(\partial_0 A_\perp)}$ in equal time quantization, are "simple" (i.e. linear in the fields) if and only if $A_- = A^+ = 0$. Therefore, in order to avoid having to deal with a system that has to satisfy nonlinear constraints [2] one normally selects $A^+ = 0$ gauge before quantizing in LF coordinates.

However, this choice of gauge is not entirely free of problems. To illustrate this fact, let us start from the Euler-Lagrange equation for A^- in QED [3]

$$-\partial_-^2 A^- = gJ^+ + \partial_- \vec{\partial}_\perp \vec{A}_\perp \equiv g\tilde{J}^+ \tag{3}$$

(the LF analog to the Poisson equation), which is also a constraint equation. It is convenient to eliminate A^-, using the solution to Eq.(3), i.e. $A^-(x^-, \vec{x}_\perp) = -\frac{g}{2} \int_{-\infty}^{\infty} dy^- |x^- - y^-| \tilde{J}^+(y^-, \vec{x}_\perp)$, yielding an instantaneous interaction term

$$V^{inst} = -\frac{g^2}{4} \int_{-\infty}^{\infty} dx^- \int_{-\infty}^{\infty} dy^- \int d^2 x_\perp \tilde{J}^+(x^-, \vec{x}_\perp) |x^- - y^-| \tilde{J}^+(y^-, \vec{x}_\perp). \tag{4}$$

This linearly rising interaction in the LF Hamiltonian causes IR divergences, unless

$$\int_{-\infty}^{\infty} dx^- \tilde{J}^+(x^-, \vec{x}_\perp) = 0 \qquad \forall \vec{x}_\perp. \tag{5}$$

The origin of this problem lies in the fact that setting $A^+ = 0$ does not completely fix the gauge freedom. An x^- independent gauge transformation $A^\mu \longrightarrow \mathcal{U}^\dagger A^\mu \mathcal{U} - \frac{i}{g} \mathcal{U}^\dagger \partial^\mu \mathcal{U}$, with $\mathcal{U} = \mathcal{U}(\vec{x}_\perp)$, leaves $A^+ = 0$ unchanged and Eq. (5) is just the Gauß' law constraint associated with this residual gauge symmetry[4]. As long as $\vec{x}_\perp$ is a continuous variable, Eq. (5) implies an infinite number of constraint on the states (∞ number of $\vec{x}_\perp$!), which is again difficult to deal with. This is one of the motivations for discretizing the $\perp$ space direction in the context of LF quantization.

THE TRANSVERSE LATTICE

The basic idea behind the $\perp$ lattice [2] is to work in two continuous (x^0 and x^3 or x^+ and x^-) space time directions and two discrete [$\vec{x}_\perp \equiv (x_1, x_2)$] space directions, i.e. space-time consists of a 2-dimensional array of 2-dimensional sheets. [5] The motivation for working with such a 'hybrid' formulation is that the discretized $\perp$ directions provide a the possibility to introduce a gauge invariant cutoff, while the

---

[2] Eq. (2) is a constraint equation since it involves no time-derivative.
[3] Since the problem that we are going to discuss occurs already in QED, we will discuss it there because of the simpler algebra.
[4] Fixing the remaining gauge freedom requires dealing with explicit zero-mode degrees of freedom and it is still not completely understood how to do this!
[5] In the closely related Hamiltonian LGT space-time consists of a 3-dim. array of 1-dim. lines.

continuous longitudinal directions allow to maintain manifest longitudinal boost invariance (which is one of the advantages of the LF formulation).

The natural way to introduce gauge fields within this framework seems to be to work with compact link-fields $U_\perp \in SU(N_C)$ in the discretized $\perp$ directions (as is done in conventional LGT) and with non-compact gauge fields $A^\pm$ in the continuous longitudinal directions. It should be emphasized that both the $U_\perp$'s as well as the $A^\pm$ (which are defined on the links and sites of the $\perp$ lattice respectively), are functions of two discrete and two continuous variables, i.e. one can think of the $\perp$ lattice action as consisting of many 1+1 dim. gauge theories coupled together.

The trouble with this formulation is the nonlinear $U_\perp \in SU(N_C)$ constraint on the link fields. The reason that this constraint is more difficult to handle on the $\perp$ lattice than in Euclidean or Hamiltonian LGT is due to the fact that the $U_\perp$s are still two dimensional fields (and not just variables, as in Euclidean LGT, or quantum mechanical rotors, as in Hamiltonian LGT). Despite several attempts in this direction [3], nobody has been able to construct a Fock space basis out of these "nonlinear σ model" degrees of freedom which still allows one to evaluate matrix elements of the LF Hamiltonian.

Two possibilities to avoid the problems associated with the $SU(N)$-constraint have been pursued: The first is to work with non-compact gauge fields also in the $\perp$ direction and the other is to keep compact fields, but to relax the $SU(N)$ constraint and linearize the degrees of freedom.

Non-compact formulation of the $\perp$ lattice

Again, we illustrate the main difficulties in the context of QED. In order to satisfy the $U(1)$ constraint, one starts with the ansatz $U_\perp = \exp(ieA_\perp)$, yielding

$$P^- \sim -\frac{1}{4}\sum_{\vec{n}_\perp} \int dx^- \int dy^- \tilde{j}^+(x^-, \vec{n}_\perp)\left|x^- - y^-\right| \tilde{j}^+(y^-, \vec{n}_\perp) + P^-_{plaq}, \qquad (6)$$

where P^-_{plaq} is the $\perp$ plaquette interaction (xy orientation) and

$$\tilde{j}^+(x^-, \vec{n}_\perp) = j_q^+(x^-, \vec{n}_\perp) + \frac{1}{e}\Delta_\perp \partial_- A_\perp, \qquad (7)$$

where $\Delta_\perp$ is the discrete approximation to the $\perp$ Laplace operator and j_q^+ is the portion of the current carried by the fermions.

As long as one restricts oneself to gauge fields with *local* fluctuations only (as in the Fock expansion!) one finds $A_\perp(+\infty) = A_\perp(-\infty)$ and thus

$$\int_{-\infty}^{\infty} dx^- \tilde{j}^+(x^-, \vec{n}_\perp) = \int_{-\infty}^{\infty} dx^- j_q^+(x^-, \vec{n}_\perp). \qquad (8)$$

Together with Gauß' law this implies that $\int_{-\infty}^{\infty} dx^- j_q^+(x^-, \vec{n}_\perp) = 0$, i.e. charges must add up to zero at each site! Transversely separated charges are only allowed in the presence of "soliton-like" gauge fields with $A_\perp(x^- = +\infty) - A_\perp(x^- = -\infty) = ke$, where k is an integer! The physics behind this result becomes clear by noting that $\exp(ieA_\perp)$ acts like such a soliton operator: from the canonical commutation relations $[A_\perp(x^-), A_\perp(y^-)] = \frac{i}{2}\varepsilon(x^- - y^-)$, together with $[e^A, B] = [A, B]e^A$ if $[A, B]$ is a c-number, one finds

$$\left[A_\perp(x^- = +\infty) - A_\perp(x^- = -\infty), \exp\left(ieA_\perp(y^-)\right)\right] = -e\exp\left(ieA_\perp(y^-)\right). \tag{9}$$

Therefore the Gauß' law constraint is satisfied if transversely separated charges are separated by a string of exponentials — just as one would have expected from gauge invariance — and the good news is that the infrared divergences cancel if states are constructed in a gauge invariant way.

It is instructive to examine in detail how the $k^+ \to 0$ divergences cancel in QED_{2+1}. In 2+1 dimensions, there is only one $\perp$ direction and therefore purely $\perp$ plaquette terms are absent. As a result, the whole dynamics of $A_\perp$ is described by its coupling to A^- and pure gauge, coupled to external sources, becomes exactly solvable in the non-compact formulation.

The rest frame energy of an external source j^+ in QED_{2+1} on a $\perp$ lattice is given by $H_{RF} = v^+ P^- + H_{recoil}$, where v^+ is the velocity of the source and

$$P^- = \frac{e^2 a}{2}\sum_n \int_{-\infty}^{\infty} dq^+ \tilde{j}_n(q^+)\frac{1}{q^{+2}}\tilde{j}_n(-q^+). \tag{10}$$

is the instantaneous interaction arising from eliminating A^- and

$$H_{recoil} = \frac{1}{2v^+}\sum_n \int_0^{\infty} dk^+ a_n^\dagger(k^+)a_n(k^+)k^+ \tag{11}$$

is a recoil term which appears in the LF description of fixed sources [4]. The effective current $\tilde{j}^+$ receives contributions from both $A_\perp$ and the external current j^+ [see also Eq.(7)]. In momentum space, one finds for the current on the nth site

$$\tilde{j}_n^+(q^+) = ej_n^+(q^+) + \frac{iq^+}{a}\left[A_{\perp,n}(q^+) - A_{\perp,n-1}(q^+)\right], \tag{12}$$

where we define the nth site to be the one between the $(n-1)^{th}$ and the nth link. It is instructive to decompose the instantaneous interaction into terms quadratic in $A_\perp$ and j^+ respectively and a mixed term, i.e. $P^- = P_{AA}^- + P_{jj}^- + P_{JA}^-$, where

$$P_{AA}^- = \sum_n \int_0^{\infty}\frac{dk^+}{2a^2 k^+}\left[a_n^\dagger(k^+) - a_{n-1}^\dagger(k^+)\right]\left[a_n(k^+) - a_{n-1}(k^+)\right]$$

$$= \frac{4}{a^2}\int_0^{\infty}\frac{dk^+}{2k^+}\int_{-\pi/a}^{\pi/a}dk_\perp \sin^2\left(\frac{ak_\perp}{2}\right)a^\dagger(k^+, k_\perp)a(k^+, k_\perp). \tag{13}$$

where $a_n(q^+) = \int_{-\pi/a}^{\pi/a}\frac{dq_\perp}{\sqrt{2\pi}}a(q^+, q_\perp)\exp(iq_\perp an)$. Furthermore

$$P_{jj}^- = \frac{e^2 a}{2} \sum_n \int_{-\infty}^{\infty} dq^+ j_n(q^+) \frac{1}{q^{+2}} j_n(-q^+) \tag{14}$$

is the self-energy of the source, and the coupling of the source to $A_\perp$ reads

$$P_{jA}^- = -ie \sum_n \int_{-\infty}^{\infty} \frac{dq^+}{q^+} j_n(q^+) \left[\tilde{A}_n(q^+) - \tilde{A}_{n-1}(q^+) \right] \tag{15}$$

$$= \frac{ie}{\sqrt{2a}} \int_0^{\infty} \frac{dq^+}{q^{+3/2}} \left\{ j_n^+(q^+) \left[a_n(q^+) - a_{n-1}(q^+) \right] - j_n^+(-q^+) \left[a_n^\dagger(q^+) - a_{n-1}^\dagger(q^+) \right] \right\}.$$

In order to calculate the self-energy of an external charge-distribution $j_q(q^+)$ to order e^2, one needs to add the instantaneous self-interaction (10) [which is of $\mathcal{O}(e^2)$ already] in first order to the contribution from the coupling to $A_\perp$ (15) [which is only $\mathcal{O}(e)$] treated in 2nd order perturbation theory. The latter yields

$$\delta E^{(2)} = -\frac{v^{+2}e^2}{a^2} \int_0^{\infty} \frac{dq^+}{q^{+3}} \int_{-\pi/a}^{\pi/a} dq_\perp \frac{2 \sin^2\left(\frac{aq_\perp}{2}\right) \tilde{j}(q) \tilde{j}(-q)}{\frac{q^+}{2v^+} + \frac{2v^+}{a^2 q^+} \sin^2\left(\frac{aq_\perp}{2}\right)}, \tag{16}$$

where we used the shorthand notation $j(q) \equiv j(q^+, q_\perp)$ and where $j_n(q^+) = \int_{-\pi/a}^{\pi/a} \frac{dq_\perp}{\sqrt{2\pi}} j(q^+, q_\perp) \exp(iq_\perp an)$. In general, $\delta E^{(2)}$ behaves for $q^+ \to 0$ like

$$\delta E_{div}^{(2)} = -\frac{v^+ e^2}{2} \int_{-\infty}^{\infty} \frac{dq^+}{q^{+2}} \int_{-\pi/a}^{\pi/a} dq_\perp j(q) j(-q) = -\frac{v^+ e^2 a}{2} \int_{-\infty}^{\infty} \frac{dq^+}{q^{+2}} \sum_n j_n(0) j_n(0), \tag{17}$$

which diverges unless the net charge $j_n(0)$ on each (!) site is zero. However, a similar divergence (with opposite sign) arises from the instantaneous interaction, as can be directly read off from Eq. (14). The sum of the two terms is IR finite as long as the total (i.e. sum of charge on all sites) charge is zero[6]

$$\delta E \equiv v^+ P_{jj}^- + \delta E^{(2)} = \frac{e^2 a}{2} \int_{-\infty}^{\infty} \frac{dq^+}{v^+} \int_{-\pi/a}^{\pi/a} dq_\perp \frac{\left(\frac{a}{2}\right)^2 j(q) j(-q)}{\left(\frac{aq^+}{2v^+}\right)^2 + \sin^2\left(\frac{aq_\perp}{2}\right)}. \tag{18}$$

In particular, for the case of two (oppositely charged) point charges one finds in the limit $a \to 0$ the logarithmic interaction energy, characteristic for an Abelian gauge theory in $2+1$ dimensions

$$\delta E = \frac{e^2}{2\pi} \log \sqrt{R_\perp^2 + R_L^2} + \text{const.}. \tag{19}$$

At first this result (i.e. perfect cancellation of the IR singularity in perturbation theory) seems to contradict the general discussion of the non-compact formulation

---

[6] This is not surprising since QED_{2+1} confines.

of gauge theories above, where it is shown that transversely separated charges need to be connected by a gauge string in order to cancel the IR divergences.

However, this apparent contradiction is resolved by the simple observation the $\perp$ lattice Hamiltonian for non-compact QED_{2+1} coupled to external sources is, technically speaking, just a bunch of coupled shifted harmonic oscillators (the Hamiltonian is Gaussian!). This has two important consequences: First of all, for a shifted harmonic oscillators, the exact ground state energy is obtained already in 2nd order perturbation theory. Therefore, the calculated ground state energy (18) is the exact one. Secondly, in the language of Fock space operators, the eigenstate of a shifted harmonic oscillator are coherent states, i.e. exponentials of raising operators.

Keeping these facts in mind, everything fits together: The coherent states, which *are* the eigenstates of the the $\perp$ lattice in the presence of the external source, accomplish the same effect as the exponentials of link fields, namely the act as soliton like operators which are necessary to cancel the small q^+ divergence of the instantaneous self-interaction. Furthermore, the fact that we are dealing only with shifted harmonic oscillators in QED_{2+1} also guarantees that we observed this cancellation already in perturbation theory — even though we did not construct these coherent states explicitly.

However, the bad news is that it is difficult to construct a Fock space basis containing exponentials of the gauge fields. [7] Furthermore, when one cannot solve the problem exactly, it is very difficult to achieve this cancellation of IR singularities — unless one works with a basis of gauge invariant states, which is very hard in the non-compact formulation. As usual, in QCD, the situation is worse because the gauge fields themselves carry color-charge. Furthermore, if one tries to maintain gauge invariance then the ansatz $U_\perp = \exp(igA_\perp)$ leads to $\frac{1}{g^2}\partial_\mu \exp(igA_\perp)\partial^\mu \exp(-igA_\perp) = \partial_\mu A_\perp \partial^\mu A_\perp +$ "higher orders" and these higher order terms make it again very difficult to quantize the theory. For these reasons, the non-compact formulation of the $\perp$ lattice has been abandoned.

Color-Dielectric Formulation of the $\perp$ Lattice

Naively, one might think that one possibility to introduce $U_\perp \in SU(N)$ fields would be to work with linearized general complex matrix fields and to add a potential $V_{eff}(U_\perp)$ that has a minimum for $U_\perp \in SU(N)$ (e.g. "Mexican hat" type potential). However, in the Fock expansion (an essential ingredient in the LF Hamiltonian formulation), the fields are usually expanded around the origin (or some other fixed value). Expanding around the origin makes no sense for a Mexican hat shaped potential since the origin corresponds to the false vacuum. On the other hand, expanding around any point along the minimum breaks the manifest global gauge symmetry. Therefore, this idea of adding a (Lagrange multiplier) effective self interaction to enforce the constraint must be abandoned as well.

---

[7] Even in QED_{2+1}, the exact solution contains exponentials of $a^\dagger$ and thus leads to non-normalizable states.

A solution out of this dilemma is to work with blocked (smeared, averaged) degrees of freedom $\mathcal{M}$, which are obtained from the original $U_\perp \in SU(N)$ by averaging $U_\perp$s or strings of $U_\perp$s over some finite volume, e.g. by defining $\mathcal{M} = \sum_{av} U$ [5]. The advantage of this procedure is that the smeared $\mathcal{M}$s are no longer subject to the strict $SU(N)$ constraint.

The blocked theory is still equivalent to the original theory, provided the action for the $\mathcal{M}$s contains an effective interaction *defined* by integrating out the $U_\perp$s

$$\exp\left[-V_{eff}(\mathcal{M})\right] = \int \mathcal{D}U_\perp \delta \left(\mathcal{M} - \sum_{av} U\right) \exp\left[-S_{canonical}(U_\perp)\right]. \qquad (20)$$

The catch in the whole procedure is that $V_{eff}(\mathcal{M})$ as defined through Eq. (20) can be infinitely complicated and for its exact determination one would have to perform a path integral. However, for approximate calculations, one can always make an ansatz for $V_{eff}(\mathcal{M})$ and there exist various options to determine the parameters appearing in this ansatz. Note that a direct use of Eq. (20) to calculate $V_{eff}(\mathcal{M})$ within the LF framework does not seem to be possible: in order to evaluate the r.h.s. of Eq. (20) one needs to work with link-fields $U_\perp \in SU(N)$ and the difficulties in doing this were the main motivation to introduce the color-dielectric formulation in the first place. In the Euclidean, calculating V_{eff} seems to be more straightforward, but for using it on the $\perp$ lattice, the issue arises of translating V_{eff} from the Euclidean to the LF. [8] An alternative procedure, based on covariance requirements, appears to be very promising. Due to lack of space, the reader is referred to Refs. [7,5], where the procedure has been discussed in detail.

Acknowledgements: I would like to thank the organizers for the invitation and DFG for financial assistance. It is a pleasure to thank P. Griffin, B. vande Sande and S. Dalley for many enlightening discussions.

REFERENCES

1. Burkardt, M. *Advances Nucl. Phys.* **23**, 1 (1995); Brodsky, S.J., Pauli, H.C., and Pinsky, S.S., *Phys. Rept.* **301**, 299 (1998).
2. Bardeen, W.A., and Pearson, R.B., *Phys. Rev.* **D14**, 547 (1976); Bardeen, W.A., Pearson, R.B., and Rabinovici, E., *Phys. Rev.* **D21**, 1037 (1980).
3. Griffin, P.A., in *Theory of Hadrons and Light Front QCD*, ed. S.D. Glazek, World Scientific, 1995, p. 240, hep-th/9410243.
4. Burkardt, M., in *Theory of Hadrons and Light Front QCD*, ed. S.D. Glazek, World Scientific, 1995, p. 233, hep-ph/9410219; Blunden, P.G. et al., hep-ph/9908067.
5. Dalley, S., and vande Sande, B., *Phys. Rev.* **D59**, 065008 (1999); *Phys. Rev. Lett.* **82**, 1088 (1999). Klindworth, B., and Burkardt, M., talk@ConfinementIII, hep-ph/9809283; Burkardt, M., and Klindworth, B., *Phys. Rev.* **D55**, 1001 (1997).
6. Burkardt, M., *Phys. Rev.* **D47**, 4628 (1993).
7. Dalley, S., *these proceedings*.

---

[8] See Ref. [6] for a discussion of this issue in the context of scalar field theories.

Causality in Light-Cone Field Theory

Thomas Heinzl

Theoretisch-Physikalisches Institut, Friedrich-Schiller-Universität Jena
Max-Wien-Platz 1, D-07743 Jena, Germany

Abstract. We show that the method of "discretized light-cone quantization" (DLCQ) is in conflict with the principle of microcausality.

INTRODUCTION

In this contribution I will discuss the relation between requirements of causality and the choice of boundary conditions for different quantization schemes. The presented results are based on a collaboration with N. Scheu and H. Kröger [1]. Let me begin with recalling the principle of microcausality. It states that the commutator of two observables $\mathcal{O}_1(x)$ and $\mathcal{O}_2(y)$ must vanish whenever their separation $x - y$ is space-like. This implies that measurements of the observable $\mathcal{O}$ performed at x and y do not interfere. Some consequences of this principle are the spin-statistics theorem, analyticity properties of Green functions which lead to dispersion relations etc. [2].

For simplicity let me specialize to a free massive scalar field ϕ in $d = 2$ dimensions. The commutator of two such fields is of course known. It is the Pauli-Jordan [3] or Schwinger [4] function,

$$\Delta(x) = -\frac{1}{2}\mathrm{sgn}(x^0)\theta(x^2)J_0(m\sqrt{x^2}) \ . \tag{1}$$

It oscillates inside the light-cone (LC) and vanishes outside (see Fig. 1).

The logical connection between the covariant commutator function (1) and canonical (Hamiltonian) quantization can be made as follows. In a first step, one choses a 'time arrow', i.e. an evolution parameter in which the system under consideration develops. In a second step, one imposes canonical commutators (at equal 'time') such that they uniquely determine the Pauli-Jordan function Δ via the equations of motion. As Dirac has shown, the choice of 'time' is not unique [5]. In what follows, I will discuss two possibilities.

The standard choice of time, $t = x^0$, corresponds to Dirac's *instant form* (IF) of dynamics. Alternatively, one can chose the LC time variable, $x^+ = x^0 + x^1$, corresponding to Dirac's *front form* (FF) of dynamics. LC quantization (for a

CP494, *New Directions in Quantum Chromodynamics*, edited by C.-R. Ji and D.-P. Min
© 1999 American Institute of Physics 1-56396-908-4/99/$15.00

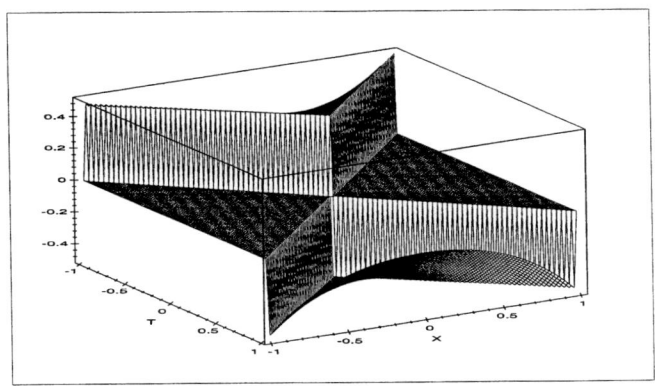

FIGURE 1. The Pauli-Jordan function as a function of $T = mx^0/2$ and $X = mx^1/2$. It vanishes outside the LC.

recent review, see [6]) amounts to prescribing commutators for equal LC time x^+, namely,

$$[\phi(x), \phi(0)]_{x^+=0} = i\Delta(x^+ = 0, x^-) = -\frac{i}{4}\text{sgn}(x^-) . \tag{2}$$

At variance with the IF, the (LC) time derivative of this expression does not constitute independent information as the normal to a null-plane (like $x^+ = 0$) lies within the null-plane [7]. It thus seems as if we need only half the data as compared to the IF. This, however, is not true. It turns out that data on a single null-plane (or light-front) are not sufficient because they do not uniquely determine the Pauli-Jordan function [7].

A way out of this dilemma has been presented in [8]. One encloses the system in a finite volume, $-L \le x^- \le L$ and imposes periodic boundary conditions (pBC). Accordingly, the longitudinal momentum k^+ conjugate to x^- takes on discrete values, hence the name "discretized light-cone quantization" (DLCQ) [9,10]. The associated torus geometry leads to a well-posed initial-boundary-value problem [8]. However, it has been taken for granted that causality is maintained within this approach. This is illegitimate, as has first been observed in [11]. Let me discuss what is involved.

I CAUSALITY ON THE TORUS

Our starting point are the Fourier representations,

$$\text{IF:} \quad \Delta(x) = -\int \frac{dk^1}{2\pi\omega_k}\sin(k \cdot x) \equiv \int dk^1\, I(k^1) , \tag{3}$$

$$\text{FF:} \quad \Delta(x) = -\int_0^\infty \frac{dk^+}{2\pi k^+}\sin(k \cdot x) \equiv \int dk^+\, I(k^+) . \tag{4}$$

The scalar products are $k \cdot x = \omega_k x^0 - k^1 x^1 = (\kappa^- x^+ + k^+ x^-)/2$ with the on-shell energies $\omega_k = (k_1^2 + m^2)^{1/2}$ and $\kappa^- = m^2/k^+$, respectively. Both integrals yield the same result (1) for the Pauli-Jordan function. Note, however, that $I(k^+)$ is exploding and rapidly oscillating for $k^+ \to 0$ so that the result is due to sizable cancellations that occur upon integration.

To obtain the torus representations, one restricts the spatial coordinates, $-L \leq x^1, x^- \leq L$ and impose pBC for the field ϕ. The conjugate momenta become discrete, $k_n^1 \equiv \pi n/L$, and $k_n^+ \equiv 2\pi n/L$. The finite-volume representations are *defined* by replacing the integrals (3) and (4) by the discrete sums,

$$\Delta_{IF}(x) \equiv - \sum_{n=-N}^{N} \frac{1}{2\omega_n L} \sin(k_n \cdot x) , \tag{5}$$

$$\Delta_{FF}(x) \equiv - \sum_{n=1}^{N} \frac{1}{2\pi n} \sin(k_n \cdot x) . \tag{6}$$

The on-shell energies for discrete momenta are defined as $\omega_n = (n^2 \pi^2/L^2 + m^2)^{1/2}$ and $\kappa_n^- = m^2 L/2\pi n$. For both functions, Δ_{IF} and Δ_{FF}, the periodicity in x^1 and x^-, respectively, with periodicity length $2L$, is obvious. The limit $N \to \infty$ is understood unless we perform numerical calculations where N is kept finite.

The evaluation of the sums (5) and (6) is not straightforward. To gain some intuition, we evaluate them numerically beginning with the IF expression (5). The resulting Δ_{IF} is plotted in Fig. 2.

Upon inspection, one notes the following: Up to small oscillations stemming from the (unavoidable) Gibbs phenomenon, Δ_{IF} vanishes outside the LC ($|x^0| < |x^1| < L$), and thus is causal even in finite volume. If we let the summation cutoff N go to infinity, Δ_{IF} approaches the continuum Pauli-Jordan function Δ (for $-L < x^0, x^1 < L$). There is a clear physical picture behind these observations. One can imagine a periodic array of sources located at the quantization hypersurface $x^0 = 0$ at points $x^1 = 2Ln$. These sources 'emit' spherical 'waves' into their own future LCs which start to overlap after time $x^0 > L$. At this point the 'waves'

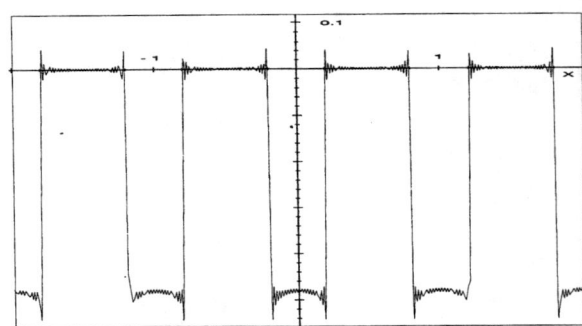

FIGURE 2. $\Delta_{IF}(X, T)$ as a function of $X = x^1/2L$. $T = X^0/2L = 0.2$, $mL = 1$, $N = 50$.

emanating from the sources begin to interfere. Thus, the influence of the BC is felt only after a long time (as large as the spatial extension L of the system). This picture can be confirmed analytically. An application of the Poisson resummation formula yields $\Delta_{IF}(x) = \sum_n \Delta(x^0, x^1 + 2Ln)$, i.e. a periodic array of Δ's which are non-overlapping as long as $x^0 < L$.

For the front form, the situation turns out to be more complicated. Using Poisson resummation one can derive the finite-volume version of the canonical LC commutator (2) at $x^+ = 0$, which is a periodic sign function [1]. For $x^+ \neq 0$, I have evaluated Δ_{FF} numerically. The result is shown in Fig. 3 as a function of the dimensionless variables $v \equiv x^-/2L$, $w = m^2 L x^+/2$. For large values of w, Δ_{FF} attains a very irregular shape, though numerically the representation (6) converges to a periodic function. The most important observation, however, is that Δ_{FF} does not vanish outside the LC, i.e. for $x^- < 0$, if $x^+ > 0$ as in Fig. 3. This a clear *violation of microcausality* [11].

As already stated, it is not straightforward to confirm these findings analytically. Poisson resummation does not work, first, because of the weak localization properties of Δ in x^- (asymptotically, is goes like $(x^-)^{-1/4}$); second, and even worse, because the zero mode $I(k^+ = 0)$ does not exist for $x^+ \neq 0$. Nevertheless, an independent confirmation of causality violation *can* be obtained from resumming (6) in terms of Bernoulli polynomials, thereby replacing the Fourier series by a (rapidly converging) power series in w. The result is shown in Fig. 4 for $w = 5$. There is nice agreement with the Fourier representation (6) (and no Gibbs phenomenon, as expected). Again, causality violation is obvious.

II   RESOLUTION

From the discussion above a natural questions arises: is there a remedy for the causality violation? The answer is positive. We have found two ways around the

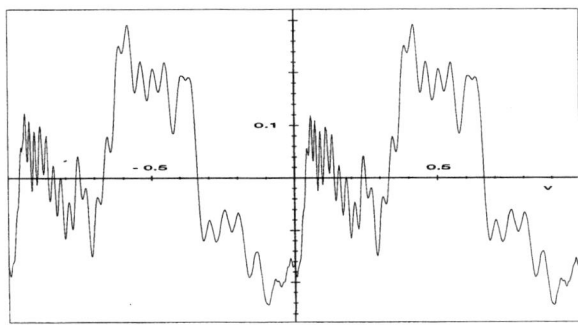

FIGURE 3. $\Delta_{FF}(v, w)$ as a function of $v = x^-/2L$. $w = 10000$, $N = 70$. It does not vanish outside the LC, $-1 < v < 0$.

249

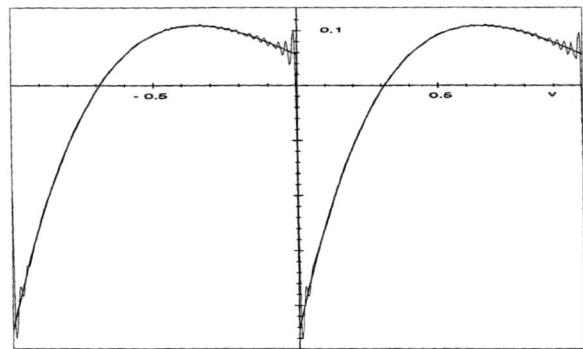

FIGURE 4. Comparison of the Fourier representation (6) with the result of Bernoulli resummation (smooth, heavy line).

problem, both, however, with shortcomings of their own. The first way is to regularize the integral (4), replacing Δ by Δ_ϵ in such a way that the associated integrand satisfies $I_\epsilon(k^+ = 0) = 0$. One can chose e.g. a principal value regularization [12] or a more sophisticated prescription [13]. In this way, one suppresses the oscillations and the divergence at $k^+ = 0$ at the price of introducing a small causality violation of order ϵ. But now Δ_ϵ *can* be approximated by a discrete sum if the momentum grid is sufficiently fine, $\Delta k^+ \ll \epsilon$. The order of limits, however, becomes important. First, one has to perform the continuum limit, $\Delta k^+ \to 0$, $L \to \infty$, and only then the limit $\epsilon \to 0$. For this method, Poisson resummation should work [14]. However, it seems somewhat absurd and not very economic to perform *two* regularizations (finite L *and* ϵ).

An alternative way of resolving the problem is the following: instead of an equally spaced grid à la DLCQ (i.e. $\Delta k_n^+ = const$) one can chose an adapted momentum grid with spacing $\Delta k_n^+ \sim 1/n$ for small n. In this way, one is sampling the small-k^+-region of $I(k^+)$ in a more reasonable way. Practically, the method amounts to viewing Δ_{IF} as the correct finite-volume expression and replacing x^0 and x^1 by $(x^+ \pm x^-)/2$, respectively. This is equivalent to introducing *new* discrete momenta, $k_n^\pm \equiv \omega_n \pm k_n^1$.

As a result, the point $k^+ = 0$ becomes an accumulation point of the momentum grid which leads to a causal finite-volume representation $\Delta_c(x^+, x^-)$ of Δ (see Fig. 5). This function, however, is no longer periodic in x^-. We thus find that, with a light-like direction being compactified, one cannot have both, periodicity *and* causality. It is an open question, however, whether the 'anharmonic' momentum grid will work for arbitrary causal Green functions and not just the commutator.

In summary, we have seen that one can find a causal finite-volume expression for the commutator of two scalar fields, $\Delta_{IF}(x^0, x^1) \equiv \Delta_c(x^+, x^-)$, both within the instant and front form of dynamics. The DLCQ representation Δ_{FF} is neither an approximation to the continuum Pauli-Jordan function Δ, nor does it satisfy the

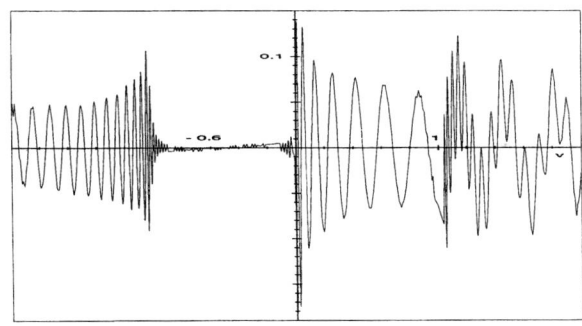

FIGURE 5. The causal commutator Δ_c as a function of v. $x^+/2L = 0.2$, $mL = 50$, $N = 50$.

requirement of causality: it does not vanish outside the LC. There is numerical evidence that the same is true in dimensions $d > 2$.

ACKNOWLEDGEMENTS

This work was supported by DFG under grant WI 777/3-2. The author thanks H. Kröger and N. Scheu for a fruitful collaboration. He gratefully acknowledges valuable discussions with B. Bakker, F. Lenz, T. Tok, and A. Wipf. It is a great pleasure to thank the organizers, D.-P. Min, C.-R. Ji, and Y. Oh, for their efforts to create such a stimulating workshop atmosphere.

REFERENCES

1. Heinzl, T., Kröger, H., and Scheu, N., "Loss of Causality in Discretized Light-Cone Quantisation", to appear.
2. Streater, R.F., and Wightman, A.S., *PCT, Spin & Statistics, and All That*, New York: Benjamin, 1963.
3. Jordan, P., and Pauli, W., *Z. Phys.* **47**, 151 (1928).
4. Schwinger, J., *Phys. Rev.* **74**, 1439 (1948), *ibid.* **75**, 651 (1949).
5. Dirac, P.A.M., *Rev. Mod. Phys.* **21**, 392 (1949).
6. Heinzl, T., "Light-Cone Dynamics of Particles and Fields", hep-th/9812190
7. Neville, R., and Rohrlich, F., *Nuovo Cimento* **A1**, 625 (1971).
8. Heinzl, T., and Werner, E., *Z. Phys.* **C62**, 521 (1994), hep-th/9311108.
9. Maskawa, T., and Yamawaki, K., *Progr. Theor. Phys.* **56**, 270 (1976).
10. Pauli, H.C., and Brodsky, S.J., *Phys. Rev.* **D32** (1985) 1993; 2001.
11. Scheu, N., Ph.D. Thesis, Université Laval (1997), hep-th/9804190.
12. Hornbostel, K., Brodsky, S.J., and Pauli, H.-C., *Phys. Rev.* **D37**, 2363 (1988).
13. Nakanishi, N., and Yamawaki, K., *Nucl. Phys.* **B122**, 15 (1977).
14. Grange, P., Salmons, S., and Werner, E., "Field Dynamics on the Light-Cone: Compact vs. Continuum Quantization", hep-th/9903101.

Pauli–Villars regularization in DLCQ

John R. Hiller

Department of Physics
University of Minnesota Duluth
Duluth Minnesota 55812

Abstract. Calculations in a (3+1)-dimensional model indicate that Pauli-Villars regularization can be combined with discrete light-cone quantization (DLCQ) to solve at least some field theories nonperturbatively. Discrete momentum states of Pauli-Villars particles are included in the Fock basis to automatically generate needed counterterms; the resultant increase in basis size is found acceptable. The Lanczos algorithm is used to extract the lowest massive eigenstate and eigenvalue of the light-cone Hamiltonian, with basis sizes ranging up to 10.5 million. Each Fock-sector wave function is computed in this way, and from these one can obtain values for various quantities, such as average multiplicities and average momenta of constituents, structure functions, and a form-factor slope.

INTRODUCTION

Field-theoretic calculations of bound-state properties, such as those one would like to do for quantum chromodynamics (QCD), require regularization of infinities and renormalization of parameters. As a way of providing a systematic regularization of ultraviolet infinities we study [1,2] the use of Pauli–Villars (PV) regularization [3] in the context of discrete light-cone quantization (DLCQ) [4,5]. Renormalization is accomplished by adjusting bare parameters to fit selected state properties with "data." The problem to be solved is then a bound-state eigenvalue problem, which includes PV constituents, combined with renormalization conditions. The couplings of the PV constituents are chosen to produce desired cancellations in perturbation theory.

We have tested these ideas for two related (3+1)-dimensional Hamiltonians [1,2]. The first [1] was constructed to have an analytic solution, in analogy with the equal-time model of Greenberg and Schweber [6]. The second Hamiltonian [2] is a generalization of the first which assigns proper light-cone energies to all particles, but does not have an analytic solution. Both Hamiltonians are distantly related to Yukawa theory, in that a fermion field acts as a source and sink for bosons. The second Hamiltonian and some of the results obtained will be described here. Work on direct application to Yukawa theory is in progress.

CP494, *New Directions in Quantum Chromodynamics,* edited by C.-R. Ji and D.-P. Min

The choice of light-cone coordinates ($x^\pm = t \pm z$, $\mathbf{x}_\perp = (\mathbf{x}, \mathbf{y})$) [7,5] is driven by important advantages, which include kinematical boosts, a simple vacuum, and well-defined Fock-state expansions with no disconnected pieces. The latter two derive from the positivity of the longitudinal light-cone momentum $p^+ = E + p_z$. When momenta are discretized [4] this positivity brings the additional advantage of a finite limit on the number of constituents.

MODEL EIGENVALUE PROBLEM

The bound-state eigenvalue problem is $H_{LC}\Phi_\sigma = M^2\Phi_\sigma$, where $H_{LC} = P^+P^-$ is known as the light-cone Hamiltonian, P^+ is the longitudinal momentum operator, and P^- is the generator for evolution in light-cone time. We work in the frame with no net transverse momentum and in a basis diagonal in P^+.

The Hamiltonian that we consider is

$$
\begin{aligned}
H_{LC} = \int & \frac{dp^+ d^2 p_\perp}{16\pi^3 p^+} \left(\frac{M^2 + p_\perp^2}{p^+/P^+} + M_0' p^+/P^+ \right) \sum_\sigma b_{\underline{p}\sigma}^\dagger b_{\underline{p}\sigma} \\
+ & \int \frac{dq^+ d^2 q_\perp}{16\pi^3 q^+} \left[\frac{\mu^2 + q_\perp^2}{q^+/P^+} a_{\underline{q}}^\dagger a_{\underline{q}} + \frac{\mu_1^2 + q_\perp^2}{q^+/P^+} a_{1\underline{q}}^\dagger a_{1\underline{q}} \right] \\
+ g & \int \frac{dp_1^+ d^2 p_{\perp 1}}{\sqrt{16\pi^3 p_1^+}} \int \frac{dp_2^+ d^2 p_{\perp 2}}{\sqrt{16\pi^3 p_2^+}} \int \frac{dq^+ d^2 q_\perp}{16\pi^3 q^+} \sum_\sigma b_{\underline{p}_1\sigma}^\dagger b_{\underline{p}_2\sigma} \\
\times & \left[a_{\underline{q}}^\dagger \delta(\underline{p}_1 - \underline{p}_2 + \underline{q}) + a_{\underline{q}} \delta(\underline{p}_1 - \underline{p}_2 - \underline{q}) \right. \\
& \left. + i a_{1\underline{q}}^\dagger \delta(\underline{p}_1 - \underline{p}_2 + \underline{q}) + i a_{1\underline{q}} \delta(\underline{p}_1 - \underline{p}_2 - \underline{q}) \right] .
\end{aligned}
\tag{1}
$$

The creation operators $b_{\underline{p}\sigma}^\dagger$, $a_{\underline{q}}^\dagger$, and $a_{1\underline{q}}^\dagger$ are associated with fermion, boson, and PV boson fields, respectively. The corresponding masses are M, μ, and μ_1. Each operator depends on a light-cone three-momentum such as $\underline{p} \equiv (p^+, p_x, p_y)$. The nonzero commutation relations are

$$
\left\{ b_{\underline{p}\sigma}, b_{\underline{p}'\sigma'}^\dagger \right\} = 16\pi^3 p^+ \delta(\underline{p} - \underline{p}')\delta_{\sigma\sigma'} ,
\tag{2}
$$

$$
\left[a_{\underline{q}}, a_{\underline{q}'}^\dagger \right] = 16\pi^3 q^+ \delta(\underline{q} - \underline{q}') , \quad \left[a_{1\underline{q}}, a_{1\underline{q}'}^\dagger \right] = 16\pi^3 q^+ \delta(\underline{q} - \underline{q}') .
$$

The structure of the Hamiltonian provides for emission and absorption of bosons by the fermion, but no change in fermion number. We explore only the one-fermion sector. The particular form of the interaction causes the fermion mass counterterm to have an unusual momentum dependence. The coefficient of this counterterm is finite because of cancellations arranged by assigning an imaginary coupling to the PV boson.

The state vector Φ_σ describes a dressed fermion with spin σ. Its Fock-state expansion is given by

$$\Phi_\sigma = \sqrt{16\pi^3 P^+} \sum_{n,n_1} \int \frac{dp^+ d^2 p_\perp}{\sqrt{16\pi^3 p^+}} \prod_{i=1}^n \int \frac{dq_i^+ d^2 q_{\perp i}}{\sqrt{16\pi^3 q_i^+}} \prod_{j=1}^{n_1} \int \frac{dr_j^+ d^2 r_{\perp j}}{\sqrt{16\pi^3 r_j^+}} \tag{3}$$

$$\times \delta(\underline{P} - \underline{p} - \sum_i^n \underline{q}_i - \sum_j^{n_1} \underline{r}_j) \phi^{(n,n_1)}(\underline{q}_i, \underline{r}_j; \underline{p}) \frac{1}{\sqrt{n! n_1!}} b_{p\sigma}^\dagger \prod_i^n a_{q_i}^\dagger \prod_j^{n_1} a_{1 r_j}^\dagger |0\rangle \, ,$$

with normalization $\Phi_\sigma'^\dagger \cdot \Phi_\sigma = 16\pi^3 P^+ \delta(\underline{P}' - \underline{P})$, which implies

$$1 = \sum_{n,n_1} \prod_i^n \int dq_i^+ d^2 q_{\perp i} \prod_j^{n_1} \int dr_j^+ d^2 r_{\perp j} \left| \phi^{(n,n_1)}(\underline{q}_i, \underline{r}_j; \underline{P} - \sum_i \underline{q}_i - \sum_j \underline{r}_j) \right|^2 . \tag{4}$$

To satisfy the eigenvalue condition, the Fock-sector wave functions $\phi^{(n,n_1)}$ must solve the following coupled system:

$$\left[M^2 - \frac{M^2 + p_\perp^2}{p^+/P^+} - M_0' p^+/P^+ - \sum_i \frac{\mu^2 + q_{\perp i}^2}{q_i^+/P^+} - \sum_j \frac{\mu_1^2 + r_{\perp j}^2}{r_j/P^+} \right] \phi^{(n,n_1)}(\underline{q}_i, \underline{r}_j, \underline{p})$$

$$= g \left\{ \sqrt{n+1} \int \frac{dq^+ d^2 q_\perp}{\sqrt{16\pi^3 q^+}} \phi^{(n+1,n_1)}(\underline{q}_i, \underline{q}, \underline{r}_j, \underline{p} - \underline{q}) \right. \tag{5}$$

$$+ \frac{1}{\sqrt{n}} \sum_i \frac{1}{\sqrt{16\pi^3 q_i^+}} \phi^{(n-1,n_1)}(\underline{q}_1, \ldots, \underline{q}_{i-1}, \underline{q}_{i+1}, \ldots, \underline{q}_n, \underline{r}_j, \underline{p} + \underline{q}_i)$$

$$+ i\sqrt{n_1+1} \int \frac{dr^+ d^2 r_\perp}{\sqrt{16\pi^3 r^+}} \phi^{(n,n_1+1)}(\underline{q}_i, \underline{r}_j, \underline{r}, \underline{p} - \underline{r})$$

$$\left. + \frac{i}{\sqrt{n_1}} \sum_j \frac{1}{\sqrt{16\pi^3 r_j^+}} \phi^{(n,n_1-1)}(\underline{q}_i, \underline{r}_1, \ldots, \underline{r}_{j-1}, \underline{r}_{j+1}, \ldots, \underline{r}_{n_1}, \underline{p} + \underline{r}_j) \right\} .$$

The bare parameters M_0' and g are determined by fitting $\langle :\phi^2(0): \rangle \equiv \Phi_\sigma^\dagger :\phi^2(0): \Phi_\sigma$ and M^2 to chosen values. The quantity $\langle :\phi^2(0): \rangle$ was selected for ease of computation; it can be computed from a form similar to the normalization sum (4):

$$\langle :\phi^2(0): \rangle = \sum_{n=1, n_1=0}^{\infty} \prod_i^n \int dq_i^+ d^2 q_{\perp i} \prod_j^{n_1} \int dr_j^+ d^2 r_{\perp j} \tag{6}$$

$$\times \left(\sum_{k=1}^n \frac{2}{q_k^+/P^+} \right) \left| \phi^{(n,n_1)}(\underline{q}_i, \underline{r}_j; \underline{P} - \sum_i \underline{q}_i - \sum_j \underline{r}_j) \right|^2 .$$

The renormalization conditions that determine M_0' and g are then solved simultaneously with the eigenvalue problem (5). In practice this is done by rearranging (5) into an eigenvalue problem for $1/g$ and simultaneously solving for M_0' in a single nonlinear equation where $\langle :\phi^2(0): \rangle$ is equal to a fixed value. The simultaneous solution is done by iterative means.

Once the wave functions $\phi^{(n,n_1)}$ have been obtained, they can be used to compute various quantities, such as the boson structure function

$$f_B(y) \equiv \sum_{n,n_1} \prod_i^n \int dq_i^+ d^2 q_{\perp i} \prod_j^{n_1} \int dr_j^+ d^2 r_{\perp j} \sum_{i=1}^n \delta(y - q_i^+/P^+) \tag{7}$$

$$\times \left| \phi^{(n,n_1)}(\underline{q}_i, \underline{r}_j; \underline{P} - \sum_i \underline{q}_i - \sum_j \underline{r}_j) \right|^2$$

and the average boson multiplicity and momentum

$$\langle n_B \rangle = \int_0^1 f_B(y) dy, \quad \langle y \rangle = \int_0^1 y f_B(y) dy. \tag{8}$$

We also compute the slope of the fermion "charge" form factor $F'(0)$, from an expression derived in Ref. [1].

NUMERICAL METHODS AND RESULTS

The coupled equations (5) are converted to a finite matrix eigenvalue problem by applying the DLCQ procedure [4]. Integrals are approximated by sums over discrete momentum values $(n\pi/L, n_x \pi/L_\perp, n_y \pi/L_\perp)$, and the transverse range is limited by a cutoff Λ^2 such that $(m_i^2 + p_{\perp i}) \le \Lambda^2 p_i^+/P^+$ for each particle, with m_i being its mass. The length scales L and $L_\perp$ are associated with a light-cone coordinate box. Bosons are assigned periodic boundary conditions and the fermion is assigned an antiperiodic boundary condition in the longitudinal direction. The momentum integer n is then even for bosons and odd for the fermion.

The total longitudinal momentum P^+ of the dressed fermion defines an odd integer $K = P^+ L/\pi$, called the harmonic resolution [4]. A longitudinal momentum fraction $x = p^+/P^+$ then reduces to a rational number n/K. The positivity of longitudinal momentum implies that $0 < n \le K$ and that the maximum number of constituents is of order $K/2$. The integers n_x and n_y range between $-N_\perp$ and $N_\perp$, with $N_\perp$ set to reach the limit imposed by the transverse cutoff for the one-boson physical state.

Thus the discretization is determined by three parameters: K, $N_\perp$, and Λ^2. The transverse scale $L_\perp$ is computed from these as $L_\perp = \pi N_\perp \sqrt{2/(\Lambda^2 - M^2 - \mu^2)}$. The longitudinal scale does not appear [4], but the limit $L \to \infty$ is equivalent to $K \to \infty$. We therefore study the limit where K, $N_\perp$, and Λ^2 all become large. This recovers the continuum form of the theory, which is regulated by the PV mass μ_1, not by Λ. We must then also study the large μ_1 limit.

Typical discretizations, such as $K = 15$, $N_\perp = 6$, and $\Lambda^2 = 50\mu^2$, with $\mu_1^2 = 10\mu^2$ produce matrices with ranks on the order of 5 million. The largest calculations carried out were of rank 10.5 million, which required approximately 2 hours of cpu time on a single 4-processor node of an IBM SP. The diagonalization method used is the Lanczos algorithm for complex symmetric matrices [8].

Although the automatic truncation of particle number imposed by DLCQ can be sufficient, further truncation can be made when the coupling is weak. Such truncation, typically to 4 bosons, was used to permit increased resolution within fixed memory limits. The validity of such an approximation was checked by computing the contribution of individual Fock sectors to the total norm.

Most quantities are remarkably insensitive to numerical resolution. This can be seen in Fig. 1 where we display the boson structure function f_B for various mass ratios M/μ. Different values of K and $N_\perp$ generally do not yield significantly different results. Notice that a smaller mass ratio is associated with a state where the boson constituents carry more of the momentum.

Values for a set of bound-state observables are given in Table 1. The quantity $|\phi_0|^2$ represents the probability for the bare fermion state. Each entry has been extrapolated from the numerical results by fits to the form $\alpha + \beta/K^2 + \gamma/N_\perp^2$. To obtain the behavior of this form the use of weighting factors [1] in DLCQ sums is important. The numerical resolutions used ranged from 9 to 19 for K and 5 to as much as 10 for $N_\perp$; the larger values of $N_\perp$ were available only for smaller K. Most observables converge quickly with respect to the PV regulator mass μ_1. Only M_0' is strongly dependent on μ_1. The form factor slope is sensitive to the transverse resolution and range.

SUMMARY

This work shows that PV regularization is feasible for DLCQ calculations. The matrix size does increase but not beyond the capacity of present-day machines for

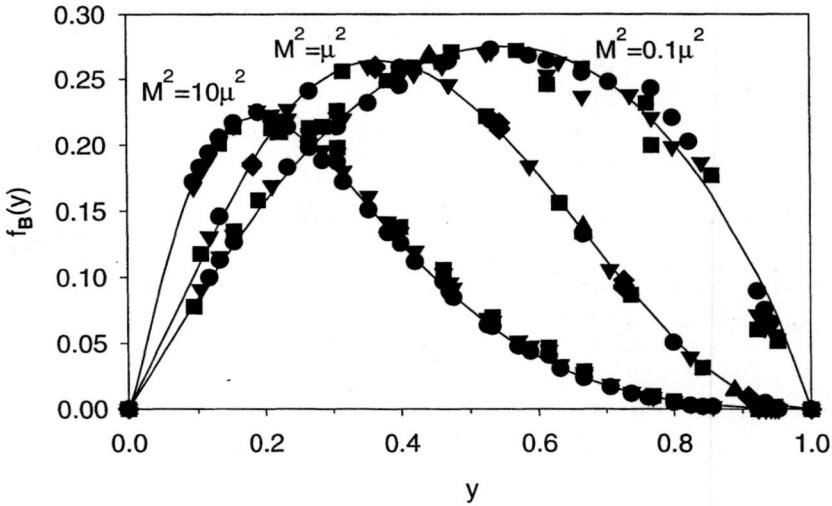

FIGURE 1. The boson structure function f_B at various numerical resolutions and mass values, with $\langle :\phi^2(0): \rangle = 1$, $\Lambda^2 = 50\mu^2$, and $\mu_1^2 = 10\mu^2$. The solid lines are parameterized fits of the form $Ay^a(1-y)^b e^{-cy}$.

TABLE 1. Extrapolated bare parameters and observables, with $\langle :\phi^2(0): \rangle = 1$.

$(M/\mu)^2$	1	1	1	1	1	1	1	0.1	5	10
$(\mu_1/\mu)^2$	5	5	5	10	10	20	20	10	10	10
$(\Lambda/\mu)^2$	12.5	25	50	25	50	50	100	50	100	100
g/μ	21.4	17.7	16.3	17.8	16.0	16.0	15.5	15.1	18.1	19.0
M_0'/μ^2	1.26	1.10	1.10	1.48	1.4	1.8	1.9	1.39	1.66	1.60
$|\phi_0|^2$	0.82	0.83	0.84	0.85	0.86	0.87	0.87	0.83	0.89	0.90
$-100\mu^2 F'(0)$	1.04	0.78	0.66	0.72	0.59	0.59	0.51	2.0	0.14	0.07
$\langle n_B \rangle$	0.18	0.15	0.14	0.15	0.14	0.13	0.13	0.16	0.10	0.09
$\langle y \rangle$	0.077	0.062	0.057	0.062	0.056	0.056	0.053	0.073	0.032	0.024

the models considered. More complicated theories will require multiple computing nodes and message-passing technology. Use of multiple nodes is facilitated by a natural block structure that arises in the matrix due to limited coupling between Fock sectors.

Work on Yukawa theory in a single-fermion truncation is now in progress. The complications include additional PV boson flavors and nontrivial spin dependence. Quantum electrodynamics is perhaps the next logical step. QCD could also be considered in a broken supersymmetric form that contains heavy particles analogous to the Abelian PV particles introduced here.

ACKNOWLEDGMENTS

This work was done in collaboration with S.J. Brodsky and G. McCartor and was supported in part by the Minnesota Supercomputing Institute through grants of computing time and by the Department of Energy contract DE-FG02-98ER41087.

REFERENCES

1. Brodsky, S.J., Hiller, J.R., and McCartor, G., *Phys. Rev. D* **58**, 025005 (1998).
2. Brodsky, S.J., Hiller, J.R., and McCartor, G., *Phys. Rev. D*, **60**, 054506 (1999).
3. Pauli, W., and Villars, F., *Rev. Mod. Phys.* **21**, 4334 (1949).
4. Pauli, H.-C., and Brodsky, S.J., *Phys. Rev. D* **32**, 1993 (1985); **32**, 2001 (1985).
5. For a review, see Brodsky, S.J., Pauli, H.-C., and Pinsky, S.S., *Phys. Rep.* **301**, 299 (1997).
6. Greenberg, O.W., and Schweber, S.S., *Nuovo Cim.* **8**, 378 (1958); Schweber, S.S., *An introduction to relativistic quantum field theory*, Evanston, IL: Row, Peterson, 1961, p. 339. See also Głazek, St.D., and Perry, R.J., *Phys. Rev. D* **45**, 3734 (1992).
7. Dirac, P.A.M., *Rev. Mod. Phys.* **21**, 392 (1949).
8. Lanczos, C., *J. Res. Nat. Bur. Stand.* **45**, 255 (1950); Cullum, J., and Willoughby, R.A., in *Large-Scale Eigenvalue Problems*, eds. Cullum, J., and Willoughby, R.A., Amsterdam: Elsevier, 1986, *Math. Stud.* **127**, p. 193.

Hadronic Electroweak Transition Matrix Elements in Light-Cone Field Theory

Dae Sung Hwang

Department of Physics, Sejong University, Seoul 143–747, Korea

Abstract. In general, each semileptonic exclusive decay amplitude receives two contributions, a diagonal $\Delta n = 0$ parton-number-conserving amplitude and a $\Delta n = -2$ contribution in which a quark and an antiquark from the initial hadron Fock state annihilate to the leptonic current. The general formalism can be used as a basis for systematic approximations to heavy hadron decay amplitudes such as hard perturbative QCD contributions. We illustrate the general formalism using a simple perturbative model of composite hadrons. Our analysis demonstrates the occurrence of "zero-mode" endpoint contributions to matrix elements of the "bad" j^- currents in the Drell-Yan frame when $q^+ \to 0$.

INTRODUCTION

In this proceeding we shall give formulas for the current matrix elements $\langle A|J^\mu|B \rangle$ describing general transition between hadrons B and A. The formulas are in principle exact, given the light-cone wavefunctions of hadrons. Our results generalize the expressions for the elastic form factors obtained by Drell and Yan [1,2] and West [3]. The underlying formalism is the light-cone Hamiltonian Fock expansion in which hadron wavefunctions are decomposed on the free Fock basis of QCD [4]. In this formalism, the full Heisenberg current J^μ can be equated to the current j^μ of the non-interacting theory which in turn has simple matrix elements on the free Fock basis. The entire electroweak current matrix element is in general given by the sum of the diagonal $n \to n$ and off-diagonal $n + 1 \to n - 1$ transitions. As we shall see, an important feature of a general analysis is the emergence of singular $\delta(x)$ "zero-mode" contributions in the "bad" $-$ current $j^- = j^0 - j^z$ from the off-diagonal matrix elements if the choice of frame dictates $q^+ = 0$. The contents of this proceeding are from the work done in collaboration with Brodsky and details can be found in Ref. [5].

CP494, *New Directions in Quantum Chromodynamics*, edited by C.-R. Ji and D.-P. Min
© 1999 American Institute of Physics 1-56396-908-4/99/$15.00

MATRIX ELEMENTS OF ELECTROWEAK CURRENTS

The light-cone Fock expansion is defined as the projection of an exact eigen-solution of the full light-cone quantized Hamiltonian on the solutions of the free Hamiltonian with the same global quantum numbers. The coefficients of the Fock expansion are the complete set of n-particle light-cone wavefunctions, $\{\psi_n(x_i, k_{\perp i}, \lambda_i)\}$. The coordinates $x_i, k_{\perp i}$ are internal relative coordinates, independent of the total momentum of the bound state, and satisfy $0 < x_i < 1$, $\sum_i^n x_i = 1$ and $\sum_i^n k_{\perp i} = 0_\perp$. Here $x = \frac{k^+}{P^+} = \frac{k^0 + k^3}{P^0 + P^3}$ and we use the metric convention $a \cdot b = \frac{1}{2}(a^+ b^- + a^- b^+) - \vec{a}_\perp \cdot \vec{b}_\perp$. The evaluation of the semileptonic decay amplitude $B \to A \ell \bar{\nu}$ requires the matrix element of the weak current between hadron states $\langle A|j^\mu(0)|B\rangle$. The outgoing leptonic current carries $q^\mu = (q^+, q_\perp, q^-) = \left(\Delta P^+, q_\perp, \frac{q^2 + q_\perp^2}{\Delta P^+}\right)$, and the value of $\Delta = q^+/P^+$ is determined from four-momentum conservation $\frac{q^2 + q_\perp^2}{\Delta} + \frac{m_A^2 + q_\perp^2}{1 - \Delta} = m_B^2$.

For the $n \to n$ diagonal term ($\Delta n = 0$), the final-state hadron wavefunction has arguments $\frac{x_1 - \Delta}{1 - \Delta}$, $\vec{k}_{\perp 1} - \frac{1 - x_1}{1 - \Delta}\vec{q}_\perp$ for the struck quark and $\frac{x_i}{1 - \Delta}$, $\vec{k}_{\perp i} + \frac{x_i}{1 - \Delta}\vec{q}_\perp$ for the $n-1$ spectators. We thus have a formula for the diagonal (parton-number-conserving) matrix element of the form [5]:

$$\langle A|J^\mu|B\rangle_{\Delta n=0} = \sum_{n,\,\lambda} \prod_{i=1}^n \int_\Delta^1 dx_1 \int_0^1 dx_{i(i\neq 1)} \int \frac{d^2 \vec{k}_{\perp i}}{2(2\pi)^3}\, \delta\left(1 - \sum_{j=1}^n x_j\right)$$

$$\times \delta^{(2)}\left(\sum_{j=1}^n \vec{k}_{\perp j}\right) \psi_{A(n)}^\dagger(x_i', \vec{k}_{\perp i}', \lambda_i)\, j^\mu\, \psi_{B(n)}(x_i, \vec{k}_{\perp i}, \lambda_i), \qquad (1)$$

where

$$\begin{cases} x_1' = \frac{x_1 - \Delta}{1 - \Delta}, & \vec{k}_{\perp 1}' = \vec{k}_{\perp 1} - \frac{1 - x_1}{1 - \Delta}\vec{q}_\perp & \text{for the struck quark} \\ x_i' = \frac{x_i}{1 - \Delta}, & \vec{k}_{\perp i}' = \vec{k}_{\perp i} + \frac{x_i}{1 - \Delta}\vec{q}_\perp & \text{for the } (n-1) \text{ spectators.} \end{cases} \qquad (2)$$

A sum over all possible helicities λ_i is understood. If quark masses are neglected the vector and axial currents conserve helicity. We also can check that $\sum_i^n x_i' = 1$, $\sum_i^n \vec{k}_{\perp i}' = \vec{0}_\perp$.

For the $n+1 \to n-1$ off-diagonal term ($\Delta n = -2$), let us consider the case where partons 1 and $n+1$ of the initial wavefunction annihilate into the leptonic current leaving $n-1$ spectators. Then $x_{n+1} = \Delta - x_1$, $\vec{k}_{\perp n+1} = \vec{q}_\perp - \vec{k}_{\perp 1}$. The remaining $n-1$ partons have total momentum $((1-\Delta)P^+, -\vec{q}_\perp)$. The final wavefunction then has arguments $x_i' = \frac{x_i}{(1-\Delta)}$ and $\vec{k}_{\perp i}' = \vec{k}_{\perp i} + \frac{x_i}{1-\Delta}\vec{q}_\perp$. We thus obtain the formula for the off-diagonal matrix element [5]:

$$\langle A|J^\mu|B\rangle_{\Delta n=-2} = \sum_{n,\,\lambda} \int_0^\Delta dx_1 \int_0^1 dx_{n+1} \int \frac{d^2 \vec{k}_{\perp 1}}{2(2\pi)^3} \int \frac{d^2 \vec{k}_{\perp n+1}}{2(2\pi)^3}$$

$$\times \prod_{i=2}^{n} \int_0^1 dx_i \int \frac{d^2 \vec{k}_{\perp i}}{2(2\pi)^3} \delta \left(1 - \sum_{j=1}^{n+1} x_j\right) \delta^{(2)} \left(\sum_{j=1}^{n+1} \vec{k}_{\perp j}\right)$$

$$\times \psi_{A(n-1)}^{\dagger}(x_i', \vec{k}_{\perp i}', \lambda_i) \; j^{\mu} \; \psi_{B(n+1)}(\{x_1, x_i, x_{n+1} = \Delta - x_1\},$$

$$\{\vec{k}_{\perp 1}, \vec{k}_{\perp i}, \vec{k}_{\perp n+1} = \vec{q}_{\perp} - \vec{k}_{\perp 1}\}, \{\lambda_1, \lambda_i, \lambda_{n+1} = -\lambda_1\}). \tag{3}$$

Here $i = 2, 3, \cdots, n$ with

$$x_i' = \frac{x_i}{1 - \Delta}, \qquad \vec{k}_{\perp i}' = \vec{k}_{\perp i} + \frac{x_i}{1 - \Delta} \vec{q}_{\perp} \tag{4}$$

label the $n - 1$ spectator partons which appear in the final-state hadron wavefunction. We can again check that the arguments of the final-state wavefunction satisfy $\sum_{i=2}^{n} x_i' = 1$, $\sum_{i=2}^{n} \vec{k}_{\perp i}' = \vec{0}_{\perp}$.

EXAMPLE—ϕ^3 PERTURBATION THEORY

As an explicit example and check on the above formalism, we shall consider the electromagnetic vector current matrix element of a neutral composite system composed of two charged scalars ($e_a + e_b = 0$) where the light-cone wavefunctions are known explicitly from perturbation theory. To construct the model, we consider a 3+1 dimensional system represented by the Lagrangian:

$$\mathcal{L} = (\partial_\mu \phi_a + ie_a A_\mu \phi_a)^{\dagger}(\partial^\mu \phi_a + ie_a A^\mu \phi_a) - m_a^2 \phi_a^{\dagger} \phi_a \tag{5}$$

$$+ (\partial_\mu \phi_b - ie_b A_\mu \phi_b)^{\dagger}(\partial^\mu \phi_b - ie_b A^\mu \phi_b) - m_b^2 \phi_b^{\dagger} \phi_b$$

$$+ \frac{1}{2} \partial_\mu \Phi \partial^\mu \Phi - \frac{1}{2} M^2 \Phi \Phi + g\Phi(\phi_a^{\dagger} \phi_b + \phi_b^{\dagger} \phi_a).$$

We can derive the light-cone amplitudes from the covariant amplitude by integrating over the k^- variable. The amplitude of the transition from Φ to Φ in one loop is given as follows from the Feynman rules:

$$\mathcal{M}^{\mu} = ie_a g^2 \int \frac{d^4 k}{(2\pi)^4} \frac{(2k - q)^{\mu}}{(k^2 - m_a^2 + i\epsilon) \left((k - q)^2 - m_a^2 + i\epsilon\right)((k - P)^2 - m_b^2 + i\epsilon)}$$

$$+ \left(a \leftrightarrow b\right). \tag{6}$$

When we perform the integration over k^-, the integral does not vanish only for $0 \le x \le 1$ where $k^+ = xP^+$.

After the k^- integration and some arrangements, we obtain [5]

$$\mathcal{M}^\mu = (2P - q)^\mu F(q^2) \tag{7}$$

$$= \int_\Delta^1 dx \int \frac{d^2\vec{k}_\perp}{2(2\pi)^3} \, \psi_{(2)}(x'_a, x'_b, \vec{k}'_{a\perp}, \vec{k}'_{b\perp}; M, m_a, m_b) j^\mu_{(2\to2)a}$$

$$\times \psi_{(2)}(x_a, x_b, \vec{k}_{a\perp}, \vec{k}_{b\perp}; M, m_a, m_b)$$

$$+ \int_0^\Delta dx \int \frac{d^2\vec{k}_\perp}{2(2\pi)^3} \int_0^1 dy \int \frac{d^2\vec{k}_{y\perp}}{2(2\pi)^3} \, \psi_{(1)}\left(\frac{y}{1-\Delta}, \vec{k}_{y\perp} - \vec{P}'_\perp\right)$$

$$\times j^\mu_{(3-1)a} \psi_{(3)}(x, y, \Delta - x, \vec{k}_\perp, \vec{k}_{y\perp}, \vec{q}_\perp - \vec{k}_\perp; M, m_a, m_b)$$

$$+ \left(a \leftrightarrow b\right),$$

where

$$j_{(2\to2)a} = e_a \frac{P^+}{\sqrt{x(x-\Delta)}} \tag{8}$$

$$\times \left((2x - \Delta) \,,\, \frac{1}{P^{+2}} \left(2\left(M^2 - \frac{(m_b^2 + \vec{k}_\perp^2)}{1-x} \right) - \frac{(q^2 + \vec{q}_\perp^2)}{\Delta} \right) \,,\, (2\vec{k}_\perp - \vec{q}_\perp) \right),$$

$$\psi_{(2)}(x_a, x_b, \vec{k}_{a\perp}, \vec{k}_{b\perp}; M, m_a, m_b) = g \frac{1}{\sqrt{x_a x_b}} \frac{1}{\left(M^2 - \frac{m_a^2 + \vec{k}_{a\perp}^2}{x_a} - \frac{m_b^2 + \vec{k}_{b\perp}^2}{x_b} \right)},$$

$$j^\mu_{(3\to1)a} = e_a \frac{P^+}{\sqrt{x(\Delta - x)}}$$

$$\times \left((2x - \Delta) \,,\, \frac{1}{P^{+2}} \left(2\frac{(m_a^2 + \vec{k}_\perp^2)}{x} - \frac{(q^2 + \vec{q}_\perp^2)}{\Delta} \right) \,,\, (2\vec{k}_\perp - \vec{q}_\perp) \right),$$

$$\psi_{(3)}(x, y, \Delta - x, \vec{k}_\perp, \vec{k}_{y\perp}, \vec{q}_\perp - \vec{k}_\perp; M, m_a, m_b) = g^2 \frac{1}{\sqrt{x(1-x)^2(\Delta - x)y}}$$

$$\times \frac{1}{\left(M^2 - \frac{m_a^2 + \vec{k}_\perp^2}{x} - \frac{m_b^2 + (-\vec{k}_\perp)^2}{1-x} \right) \left(M^2 - \frac{m_a^2 + \vec{k}_\perp^2}{x} - \frac{m_a^2 + (\vec{q}_\perp - \vec{k}_\perp)^2}{\Delta - x} - \frac{M^2 + \vec{k}_{y\perp}^2}{y} \right)},$$

$$\psi_{(1)}(y, \vec{k}_{y\perp}) = \frac{2(2\pi)^3}{\sqrt{y}} \delta(y - 1)\delta^2(\vec{k}_{y\perp}),$$

$$x_a = x, \quad x_b = 1 - x, \quad \vec{k}_{a\perp} = \vec{k}_\perp, \quad \vec{k}_{b\perp} = -\vec{k}_\perp,$$

$$x'_a = \frac{x_a - \Delta}{1 - \Delta}, \quad x'_b = \frac{x_b}{1 - \Delta}, \quad \vec{k}'_{a\perp} = \vec{k}_{a\perp} + (1 - x'_a)\vec{P}'_\perp,$$

$$\vec{k}'_{b\perp} = \vec{k}_{b\perp} - x'_b \vec{P}'_\perp, \quad \vec{P}'_\perp = -\vec{q}_\perp.$$

In (7) $\mathcal{M}^\mu = (2P - q)^\mu F(q^2)$ follows from $q_\mu \mathcal{M}^\mu = 0$.

For $q^2 \to 0$, $\vec{q}_\perp \to \vec{0}_\perp$ and $\Delta \to 0$, + component of (7) gives

$$F(0) = e_a \int \frac{d^2\vec{k}_\perp}{(2\pi)^3} \int_0^1 dx \, |\psi_{(2)}(x, 1 - x, \vec{k}_\perp, -\vec{k}_\perp; M, m_a, m_b)|^2 \tag{9}$$

$$+ e_b \int \frac{d^2\vec{k}_\perp}{(2\pi)^3} \int_0^1 dx \, |\psi_{(2)}(x, 1 - x, \vec{k}_\perp, -\vec{k}_\perp; M, m_b, m_a)|^2 = 0,$$

where $\psi_{(2)}$ is given in (8). Each term can be normalized to unit charge, thus providing wavefunction renormalization in the model. Alternatively we can evaluate the $-$ component of (7) to obtain

$$F(0) = e_a \int \frac{d^2\vec{k}_\perp}{(2\pi)^3} \frac{1}{M^2} \left(\int_0^1 dx \, \frac{1}{x} \left(M^2 - \frac{m_b^2 + \vec{k}_\perp^2}{1 - x} \right) \right. \tag{10}$$

$$\times |\psi_{(2)}(x, 1 - x, \vec{k}_\perp, -\vec{k}_\perp; M, m_a, m_b)|^2 \quad + \quad \left. \frac{1}{m_a^2 + \vec{k}_\perp^2} g^2 \right)$$

$$+ e_b \int \frac{d^2\vec{k}_\perp}{(2\pi)^3} \frac{1}{M^2} \left(\int_0^1 dx \, \frac{1}{x} \left(M^2 - \frac{m_a^2 + \vec{k}_\perp^2}{1 - x} \right) \right.$$

$$\times |\psi_{(2)}(x, 1 - x, \vec{k}_\perp, -\vec{k}_\perp; M, m_b, m_a)|^2 \quad + \quad \left. \frac{1}{m_b^2 + \vec{k}_\perp^2} g^2 \right),$$

where the $\frac{e_a}{m_a^2 + \vec{k}_\perp^2} g^2$ and $\frac{e_b}{m_b^2 + \vec{k}_\perp^2} g^2$ terms come from the singular contributions of the $\int_0^\Delta dx \, \psi_{(1)} \, j_{(3-1)}^- \, \psi_{(3)}$ terms in (7) when we take the limit $\Delta \to 0$. The $\perp$ components of (7) do not give more information since $(2\vec{P} - \vec{q})_\perp \to \vec{0}_\perp$ in the left hand side and the integrand of the right hand side is odd about $\vec{k}_\perp$.

The above analysis provides an explicit realization of the general formulas (1) and (3). In this simple model two transition matrix elements appear: $2 \to 2$ and $3 \to 1$. The equality of the formulas for (9) and (10) is a general condition which follows from gauge invariance and consistency of the light-cone formalism. We have verified the equality for the perturbative model by direct evaluation of the integrals. Note that the singular contributions in (10) includes the zero mode $\delta(x)$ contributions from the $n + 1 \to n - 1$ off-diagonal matrix element.

CONCLUSION

Glazek and Sawicki [6] have studied the relativistic two-body bound-state form factor in the 1+1 dimensional scalar field theory by considering both the spectator diagram and the Z diagram in the light-cone formalism. The necessity for this zero mode $\delta(x)$ terms were first noted in the pioneering work of Chang, Root and Yan [7], and Burkardt analyzed it in his studies of higher-twist parton distributions [8].

Here we see that the presence of such terms is a general feature of local operator matrix elements when one selects the simplified $q^+ = 0$ frame.

The off-diagonal $n + 1 \rightarrow n - 1$ contributions provide a new perspective on the physics of B-decays. A semileptonic decay involves not only matrix element where a quark changes flavor, but also a contribution where the leptonic pair is created from the annihilation of a $q\bar{q}'$ pair within the Fock states of the initial B wavefunction. The semileptonic decay thus can occur from the annihilation of a nonvalence quark and a valence antiquark in the initial hadron. This feature will carry over to exclusive hadronic B-decays, such as $B^0 \rightarrow \pi^- D^+$. In this case the pion can be produced from the coalescence of a $d\bar{u}$ pair emerging from the initial higher particle number Fock wavefunction of the B. The D meson is then formed from the remaining quarks after the internal exchange of a W boson.

ACKNOWLEDGMENTS

This work was done in collaboration with Stanley J. Brodsky. This work was supported in part by the Basic Science Research Institute Program, Ministry of Education, Project No. BSRI-97-2414, and in part by Non-Directed-Research-Fund, Korea Research Foundation 1997.

REFERENCES

1. S.D. Drell and T.M. Yan, *Phys. Rev. Lett.* **24**, 181 (1970).
2. S.J. Brodsky and S.D. Drell, *Phys. Rev.* D **22**, 2236 (1980).
3. G.B. West, *Phys. Rev. Lett.* **24**, 1206 (1970).
4. G.P. Lepage and S.J. Brodsky, *Phys. Rev.* D **22**, 2157 (1980); *Phys. Lett.* B **87**, 359 (1979); *Phys. Rev. Lett.* **43**, 545, 1625(E) (1979).
5. S.J. Brodsky and D.S. Hwang, *Nucl. Phys.* B **543**, 239 (1999).
6. St. Glazek and M. Sawicki, *Phys. Rev.* D **41**, 2563 (1990).
7. S.J. Chang, R.G. Root and T.M. Yan, *Phys. Rev.* D **7**, 1133 (1973).
8. M. Burkardt, *Nucl. Phys.* B **373**, 613 (1992).

On DLCQ M Theory

Seungjoon Hyun

School of Physics, Korea Institute for Advanced Study, Seoul 130-012, Korea

Abstract. I discuss some basic aspects of DLCQ of M theory. Among others I explain the role of gravitational effect to resolve the zero mode divergence problem existing in the light-cone field theory on flat Minkowski background and to lift ten-dimensional theory to eleven-dimensional one. Talk presented at the Eleventh International Light-Cone School and Workshop, June 1999.

INTRODUCTION

Ever since it was realized that superstring theories have 11-dimensional origin [1], one of the central issue has been the formulation of this 11-dimensional theory, named M theory. One noble approach toward the formulation of the M theory is matrix model, proposed by Banks, Fishler, Shenker and Susskind [2]. In its most naive terms, the basic idea is that if we put the whole system in an infinite momentum frame, most of the degrees of freedom get infinite energy and the only relevant degrees of freedom remained would be those of gravitational waves moving in the same direction, which is nothing but the D0 branes in 10 dimensions. Since the whole theory must be Lorentz invariant, by taking this infinite momentum frame, it is conjectured that the M theory is described by D0 matrix quantum mechanics.

This matrix model can be understood as the large N limit of the discrete light-cone quantization (DLCQ) of M theory as suggested by Susskind [3] and further developed by Seiberg and Sen [4]. In this approach we consider a light-like compactification which results in the discretized light-cone momentum $p_- = \frac{N}{R}$. Since p_- is conserved, we consider the fixed N sector and the decompactification limit $R \to \infty$ may be achieved by taking $N \to \infty$ simultaneously.

Apparently there seem to be some problems in the above arguments. First of all it is implausible to get (spatial) decompactification limit as the proper length of R is null. Another related issue is the zero mode divergence problem in the light-cone field theory [5,6]. The resolution to these problems is to take into account the gravitational effect. In the M theory, which contains gravity, self gravitational effects of D0 branes turn the light-like coordinate into the space-like one almost everywhere and resolve both issues [7]. See also [8,9]. In this talk I explain some

CP494, *New Directions in Quantum Chromodynamics*, edited by C.-R. Ji and D.-P. Min
© 1999 American Institute of Physics 1-56396-908-4/99/$15.00

basic aspects of the DLCQ M theory. First I show the emergence of supergraviton in the infrared limit of matrix model as a motivation for the matrix model as a candidate of M theory. Then I briefly describe the DLCQ M theory and explain the role of the gravitational effects in this formalism. For an illuminating review on this subject, see [10].

ELEVEN-DIMENSIONAL SUPERGRAVITON IN THE LIGHT-CONE GAUGE

Consider the action of eleven-dimensional supergraviton,

$$S = \int d\lambda \mathcal{L} = \frac{1}{2} \int d\lambda e^{-1} (\dot{x}^\mu + i\bar{\theta}\Gamma^\mu \dot{\theta})^2 , \qquad (1)$$

where θ and $\bar{\theta} = \theta^T \Gamma^0$ are 32 component real spinors.

The equations of motion of this action are those of linearized eleven-dimensional supergravity. This action has world-line reparametrization invariance under which x^μ and θ transform as world-line scalar. If we take the light-cone coordinate x^- periodic, $x^- \equiv x^- + R$, it is natural to identify $\frac{x^+}{2}$ as time coordinate. The natural gauge for the world-line diffeomorphism in this DLCQ formulation is the static gauge

$$\dot{x}^+ = 2\dot{x}^\tau = 2. \qquad (2)$$

The spacetime supersymmetry transformation laws with parameter ξ are given by

$$\delta\theta = \xi, \quad \delta x^\mu = -i\bar{\xi}\Gamma^\mu \theta ,$$
$$\delta\bar{\theta} = \bar{\xi}, \quad \delta e = 0 . \qquad (3)$$

In addition, the action has local fermionic symmetry with parameter $\kappa(\lambda)$,

$$\delta\theta = ie\Gamma \cdot p\kappa , \quad \delta x^\mu = -i\bar{\theta}\Gamma^\mu \delta\theta , \quad \delta e = 4e\dot{\bar{\theta}}\kappa , \qquad (4)$$

which, in effect, reduces the degrees of freedom of θ by half. Here $p_\mu = e^{-1}\eta_{\mu\nu}(\dot{x}^\nu + i\bar{\theta}\Gamma^\nu \dot{\theta})$ denotes the conjugate momentum of x^μ. We can fix this kappa symmetry by choosing

$$\Gamma^+\theta = 0 , \quad \Gamma^\pm = (\Gamma^{10} \pm \Gamma^0) . \qquad (5)$$

Note that with this gauge fixing the conjugate $\bar{\theta} \equiv \theta^T \Gamma^0$ becomes

$$\bar{\theta} = \theta\frac{1}{2}(\Gamma^+ - \Gamma^-) = \theta\Gamma^\tau ,$$

where $\Gamma^\tau = \Gamma^+/2$. This is in accord with the gauge fixing (2).

The light-cone momentum p_- is quantized and given by $p_- = \frac{N}{R}$. From the action one can see the coordinate x^- is cyclic and thus the conjugate momentum p_- is conserved. Henceforth we consider the fixed N-sector of the theory and the appropriate effective action is given by Routhian [11]

$$\mathcal{L}_{eff} = \mathcal{L} - p_- \dot{x}^- (p_-) = -p_- \dot{x}^- \ . \tag{6}$$

The last relation in the above comes from the constraint equation

$$\eta_{\mu\nu} p^\mu p^\nu = 0 \ , \tag{7}$$

from which we can also solve for $\dot{x}^-$.

The effective action after the gauge fixing (2) and (5) becomes

$$\mathcal{L}_{eff} = p_- \left(\frac{(v^i)^2}{2} + 4i\theta_{(16)} \dot{\theta}_{(16)} \right) \ , \tag{8}$$

where $v^i = \dot{x}^i$. This action corresponds to the $(0+1)$-dimensional $U(1)$ supersymmetric Yang-Mills theory with adjoint fermions $\psi \equiv 2\sqrt{2}\theta_{(16)}$.

Now let us examine the supersymmetry transformation law of the action (8). Since we fixed the gauge for the worldline diffeomorphism and kappa symmetry, the original supersymmetry transformation laws (3) are modified to preserve those gauge fixing (2) and (5). Half of the original supersymmetries with $\Gamma^+ \xi = 0$ remain to be the symmetry in which

$$\delta\theta = \xi \ , \quad \delta x^i = 0,$$

even after gauge fixing. This is just constant shift in spinors which also appears as trivial kinematic supersymmetry transformations in the original matrix model [2].

On the other hand, other half of the form $\xi = \begin{pmatrix} \xi_{(16)} \\ \xi_{(16)} \end{pmatrix}$, with $\Gamma^- \xi = 0$, do not preserve the gauge fixing conditions and thus are modified by taking appropriate κ-transformations (4). From the condition

$$\Gamma^+ (\delta\theta) = \Gamma^+ (\xi + i\Gamma \cdot p\kappa) = 0 \ , \tag{9}$$

the κ-transformations of the form,

$$\kappa = \begin{pmatrix} \kappa_{(16)} \\ -\kappa_{(16)} \end{pmatrix}, \tag{10}$$

should satisfy

$$\kappa_{(16)} = \frac{i}{2} \xi_{(16)}. \tag{11}$$

The total supersymmetry transformations preserve the static gauge (2) and are given by

$$\delta\psi = v^i \gamma^i \epsilon \ , \quad \delta x^i = -i\epsilon \gamma^i \psi \tag{12}$$

with $\epsilon \equiv \sqrt{2}\xi_{(16)}$. These are the supersymmetry transformation laws of $(0+1)$-dimensional matrix model in the infrared limit.

266

DLCQ M THEORY

In this section we describe briefly the DLCQ of M theory [3] following [4]. Consider a nearly light-like compactification of M theory with the radius R. After the compactification along x^-, the theory becomes the ten-dimensional theory with hidden eleventh dimension with radius R and thus the string coupling g_s, and string length scale are given by

$$g_s = \left(\frac{R}{l_p}\right)^{3/2} , \quad l_s^2 = \frac{l_p^3}{R}$$

where l_p is eleven-dimensional Planck scale. This DLCQ M theory can be understood as a limit of infinite boosts of M theory on a vanishingly small spatial circle of radius R_s. In order to set up the energy scale probed at the same order of magnitude before and after the infinite boosts, we readjust the Planck scale requiring

$$\frac{R_s}{\tilde{l}_p^2} = \frac{R}{l_p^2} , \tag{13}$$

kept fixed. M theory compactified on the spatial circle in the limit $R_s \to 0$ with (13), denoted as $\tilde{M}$ theory, has vanishingly small string length scale $\tilde{l}_s \sim R_s^{1/4} \to 0$ and string coupling $\tilde{g}_s \sim R_s^{3/4} \to 0$ and is described by D0 quantum mechanics. Then DLCQ M theory obtained by the infinite boosts of $\tilde{M}$ theory with energy rescaling in such a way to satisfy (13) is also described by D0 quantum mechanics. Eleven-dimensional M theory may be recovered by the large N limit, while keeping p_- fixed.

GRAVITATIONAL EFFECT

Since M theory contains gravity as a low energy effective theory, we need to take into account the gravitational effect. In this section we discuss the background geometry appropriate for the DLCQ M theory.

As a solution of $\tilde{M}$ theory, the eleven-dimensionally lifted geometry of the D0 solution is given by [12]

$$ds_{11}^2 = -\frac{1}{f}dt^2 + f(dx_{11} - (1 - \frac{1}{f})dt)^2 + dx_1^2 + \cdots + dx_9^2 , \tag{14}$$

where

$$f = 1 + \frac{N\tilde{l}_p^9}{R_s^2}\frac{1}{r^7}$$

is a nine-dimensional harmonic function and $r^2 = x_1^2 + \cdots + x_9^2$. In terms of asymptotic light-cone coordinates $x^\pm = x_{11} \pm t$ and the metric Eq.(14) becomes

$$ds_{11}^2 = dx^+ dx^- + h_s dx^- dx^- + dx_1^2 + \cdots + dx_9^2 \qquad (15)$$

where $h_s = f - 1$.

Under a Lorentz boost along the M-theory circle x_{11} given by the boost parameter

$$\beta = \frac{R}{\sqrt{R^2 + 4R_s^2}},$$

the original spatial circle with the period R_s approaches the light-cone circle with the period R, as we take the limit $\beta \to 1$ and $R_s \to 0$ while keeping R fixed. To see this, we note that the unboosted identification under the lattice translation

$$x^+ \simeq x^+ + R_s \quad , \quad x^- \simeq x^- + R_s$$

changes into

$$x^+ \simeq x^+ + \frac{2}{1 + \sqrt{1 + 4R_s^2/R^2}} \frac{R_s^2}{R} \quad , \quad x^- \simeq x^- + \frac{1 + \sqrt{1 + 4R_s^2/R^2}}{2} R$$

under the boost. This shows that, under the infinite boosts, we have a discrete light cone where we identify $x^- \simeq x^- + R$. Therefore the asymptotically light-like coordinate x^- plays the role as spatial coordinate and $x^+/2$ becomes the time-like coordinate. The effect of the Lorentz boost on the metric corresponds to the substitution

$$h_s \to h = \frac{4}{(1 + \sqrt{1 + 4R_s^2/R^2})^2} \frac{R_s^2}{R^2} h_s \qquad (16)$$

in Eq. (15). After rewritten in terms of the parameters of M theory, the final form of the solution is given by [7,8]

$$ds_{11}^2 = dx^+ dx^- + h dx_-^2 + dx_1^2 + \cdots + dx_9^2 , \qquad (17)$$

where

$$h = \frac{N l_p^9}{R^2 r^7} .$$

Note that in order to deal with the light-cone quantization of field theories in the flat Minkowski spacetime, we need to introduce fictitious cutoff parameter to turn the light-like coordinate into the space-like one and circumvent the infamous zero mode divergence problem [5,6]. In this case, on the contrary, x^-, being only asymptotically null due to gravitational effect, can be considered as true eleventh spatial coordinate with the compactification radius R. Furthermore the effective radius $R(r)$ of compact dimension at the radial position r in units of Planck scale is given by

$$\frac{R(r)}{l_p} = \left(\frac{N l_p^7}{r^7} \right)^{1/2} ,$$

and therefore it becomes indeed the eleven-dimensional M theory in the region $r < N^{1/7}l_p$, justifying the DLCQ description of M theory.

Acknowledgements. I would like to thank the organizers of the workshop, in particular Yoonbai Kim for invitation. I am grateful to Youngjai Kiem and Hyeonjoon Shin for useful discussions and collaborations on the various aspects of the DLCQ M theory.

REFERENCES

1. P. K. Townsend, Phys. Lett. **B350**, 184 (1995);
   E. Witten, Nucl. Phys. **B443**, 85 (1995).
2. T. Banks, W. Fischler, S. Shenker, L. Susskind, Phys. Rev. **D55**, 5112 (1997).
3. L. Susskind, hep-th/9704080.
4. N. Seiberg, Phys. Rev. Lett. **79**, 3577 (1997);
   A. Sen, Adv. Theor. Math. Phys. **2**, 51 (1998).
5. H. W. L. Naus, H. J. Pirner, T. J. Fields and J. P. Vary, Phys. Rev. **D56**, 8062 (1997).
6. S. Hellerman and J. Polchinski, Phys. Rev. **D59**, 125002 (1999).
7. S. Hyun, Y. Kiem and H. Shin, Phys. Rev. **D57**, 4856 (1998);
   S. Hyun, Phys. Lett. **B441**, 116 (1998);
   S. Hyun and Y. Kiem, Phys. Rev. **D59**, 026003 (1999).
8. V. Balasubramanian, R. Gopakumar and F. Larsen, Nucl. Phys. **B526**, 415 (1998).
9. N. Itzhaki, J. M. Maldacena, J. Sonnenschein and S. Yankielowicz, Phys. Rev. **D58**, 046004 (1998).
10. J. Polchinski, hep-th/9903165.
11. K. Becker, M. Becker, J. Polchinski and A. Tseytlin, Phys. Rev. **D56**, 3174 (1997).
12. G. T. Horowitz and A. Strominger, Nucl. Phys. **B360**, 197 (1991).

1+1 Gauge Theories in the Light-Cone Representation

Gary McCartor* and Yuji Nakawaki†

*Department of Physics, SMU, Dallas, Texas 75275
†Division of Physics and Mathematics, Setsunan University, Osaka 572-8508

Abstract. We present a representation independent solution to the continuum Schwinger model in light-cone ($A^+ = 0$) gauge. We then discuss the problem of finding that solution using various quantization schemes. In particular we shall consider equal-time quantization and quantization on either characteristic surface, $x^+ = 0$ or $x^- = 0$.

INTRODUCTION

We shall give a solution to the light-cone gauge Schwinger model in the continuum [1]. In early light-cone meetings many people gave talks on the Schwinger model and a number of papers were published. All of these solutions made use of periodicy conditions to regulate the infrared singularities of the Schwinger model. Since a solution with nice periodicity conditions on a space-like surface does not posess nice periodicity conditions on a light-like surface (or even another space-like surface) the comparisons of quantization on a light-like surface with quantization on a space-like surface were never direct. Since the continuum is the continuum on any surface, in the present paper we shall be able to compare quantizaton proceedures which must arrive at a common solution.

Some of the points we will make about the various cases are as follows:
For equal-time quantization we find

> EASY FORMULATION
> DIFFICULT SOLUTION
> COMPLEX, DYNAMICALLY DETERMINED VACUUM

For light-cone ($x^+ = 0$) quantization we find

> MORE DIFFICULT FORMULATION
> EASIER SOLUTION
> VACUUM FIXED BY KINEMATICS AND GAUGE INVARIANCE

CP494, *New Directions in Quantum Chromodynamics,* edited by C.-R. Ji and D.-P. Min

For light-cone ($x^- = 0$) quantization (note that in the continuum this is precisely the same as quantizing on $x^+ = 0$ but in the gauge $A^- = 0$ — the anti-light-cone gauge) we find

UNFAMILIAR FORMULATION
UNFAMILIAR DEGREES OF FREEDOM
EASIER SOLUTION
VACUUM FIXED BY KINEMATICS AND GAUGE INVARIANCE

An important point is that even though the vacuum is fixed by kinematics in both the light-cone representations, physical quantities such as the chiral condensate are dynamical just as they are in the equal-time representation.

The solution contains fields which are functions of x^+; they therefore appear as dynamical fields when quantized on $t = 0$, zero-mode fields when quantized on $x^+ = 0$ and static fields when quantized on $x^- = 0$. We shall emphasize the essential nature of these fields to the solution and shall argue that it is natural and not difficult to include them in any of the quantization schemes.

LIGHT-CONE GAUGE SCHWINGER MODEL

The Lagrangian is

$$\mathcal{L} = i\bar{\psi}\gamma^\mu \partial_\mu \psi - \frac{1}{4}F^{\mu\nu}F_{\mu\nu} - A^\mu J_\mu - \lambda A^+$$

Where λ is a Lagrange multiplier field.

The solution is given by

$$\Psi_+ = Z_+ e^{\Lambda_+^{(-)}} \sigma_+ e^{\Lambda_+^{(+)}}$$

$$\Lambda_+ = -i2\sqrt{\pi}(\eta(x^+) + \tilde{\Sigma}(x^+, x^-))$$

$$Z_+^2 = \frac{m^2 e^\gamma}{8\pi\kappa}$$

$$\Psi_- = \psi_- = Z_- e^{\Lambda_-^{(-)}} \sigma_- e^{\Lambda_-^{(+)}}$$

$$Z_-^2 = \frac{\kappa e^\gamma}{2\pi}$$

$$\Lambda_- = -i2\sqrt{\pi}\phi(x^+)$$

$$\lambda = m\partial_+(\eta - \phi)$$

$$A_+ = \frac{2}{m}\partial_+(\eta + \tilde{\Sigma})$$

271

In these equations, $\tilde{\Sigma}$ is a massive ($\frac{e}{\sqrt{\pi}}$) pseudoscalar; ϕ is the x^+-dependent piece of a massless scalar and η is the x^+-dependent piece of a massless ghost. These last two fields are regulated with the Klaiber procedure; for example

$$\phi^{(+)}(x^+) = i(4\pi)^{-\frac{1}{2}} \int_0^\infty dk_+ k_+^{-1} d(k_+) \left(e^{-ik_+ x^+} - \theta(\kappa - k_+) \right)$$

We again wish to emphasize that the above soloution is representation independent. It is not light-cone quantized or equal-time quantized but is simply the answer. Any quantization scheme is an attempt to find that answer.

ROLE OF THE ZERO-MODE FIELDS

We shall refer to the x^+-dependent fields as zero-mode fields, although they are such only if quantizing on $x^+ = 0$. The only operators in the above solution which carry a charge are the spurions, σ_+ and σ_-

$$\sigma_+ = \exp\left[i\sqrt{\pi}\{Q_5 + Q\}(4m)^{-1} + \int_0^\kappa dk_1 k_1^{-1}\{\eta(k_1) - \eta^*(k_1)\} \right]$$

$$\sigma_- = \exp\left[i\sqrt{\pi}\{Q_5 - Q\}(4m)^{-1} + \int_0^\kappa dk_1 k_0^{-1}\{d(k_1) - d^*(k_1)\} \right]$$

The charges themselves are made up entirely from operators from the zero-mode fields. We have

$$Q = \int_{-\infty}^\infty m\partial_+(\phi - \eta)dx^+$$

$$Q_5 = \int_{-\infty}^\infty m\partial_+(\phi + \eta)dx^+$$

$$\sigma_\pm^* = \sigma_\pm^{-1}$$

$$[Q, \sigma_+^*\sigma_-] = [Q, \sigma_-^*\sigma_+] = 0$$

$$[Q_5, \sigma_+^*\sigma_-] = -2\sigma_+^*\sigma_- \; ; \; [Q_5, \sigma_-^*\sigma_+] = 2\sigma_-^*\sigma_+$$

Without the zero-mode fields the model would be electrodynamics without charges.

The spurions (made from zero-mode operators) are the generators of large gauge transformations and are necessary to create a gauge invariant vacuum and therefore, to give a gauge invariant solution. We have

$$(\sigma_+^*\sigma_-) \, A_+ \, (\sigma_+\sigma_-^*) = A_+ + \frac{\sin \kappa x^+}{\kappa}$$

So the vacuum must be an eigenstate of $\sigma_+^*\sigma_-$. In fact, the physical vacuum is given by

272

$$|\Omega(\theta)\rangle \equiv \sum_{M=-\infty}^{\infty} e^{iM\theta}|\Omega(M)\rangle \quad ; \quad |\Omega(M)\rangle = (-\sigma_+^*\sigma_-)^M|0\rangle$$

This family of vectors are the only gauge invariant vectors which satisfy

$$\tilde{\Sigma}^{(+)}|0\rangle = \phi^{(+)}|0\rangle = \eta^{(+)}|0\rangle = Q_5|0\rangle = 0 \tag{1}$$

It is particularly easy to calculate the chiral condensate in light-cone gauge and obtain

$$\langle\Omega(\theta)|\bar{\Psi}\Psi|\Omega(\theta)\rangle = -\frac{m}{2\pi}e^{\gamma}\cos\theta$$

The zero-mode fields also play an essential role in regulating the operator products in the theory. To illustrate this fact consider the product of Ψ_+ with its conjugate

$$\langle\Psi_+^*(x+\epsilon)\Psi_+(x)\rangle \sim \frac{1}{2\pi\epsilon^-}$$

This behavior allows us to define a gauge invariant Fermi product. But it comes about through the combination

$$\langle : e^{i2\sqrt{\pi}\tilde{\Sigma}(x+\epsilon)} :: e^{-i2\sqrt{\pi}\tilde{\Sigma}(x)} : \rangle \sim e^{-2\gamma}\frac{4}{m^2}\frac{1}{\epsilon^+\epsilon^-}$$

with

$$\langle e^{i2\sqrt{\pi}\eta^{(-)}(x+\epsilon)}\sigma_+^* e^{i2\sqrt{\pi}\eta^{(+)}(x+\epsilon)}e^{-i2\sqrt{\pi}\eta^{(-)}(x)}\sigma_+ e^{-i2\sqrt{\pi}\eta^{(+)}(x)}\rangle \sim e^{\gamma}\kappa\epsilon^+$$

Without the zero-mode field the operator product would be too singular to treat. With it we can define

$$:\psi^{\dagger}\psi: = \lim_{\substack{\epsilon\to 0 \\ \epsilon^2<0}}\left\{e^{-ie\int_x^{x+\epsilon}A_\nu^{(-)}dx^\nu}\psi^{\dagger}(x+\epsilon)\psi(x)e^{-ie\int_x^{x+\epsilon}A_\nu^{(+)}dx^\nu} - \text{V.E.V.}\right\}$$

Which allows the existence of the operator solution. The zero-mode fields enter the solution in other, essential ways, but we shall now proceed to finding the solution by formulating the problem in the various quantization schemes.

QUANTIZING THE MODEL

Looking at the solution we see that it is straightforward to quantize the model on $t = 0$. The degrees of freedom are

$$\Psi_+ \; ; \; A_+ \; ; \; \Pi_A \; ; \; \Psi_-$$

The quantization follows completely standard lines. We initialize ψ_+ and write it in the Bosonized basis.

$$\Psi_+ = Z_+ e^{\Lambda_+^{(-)}}\sigma_+ e^{\Lambda_+^{(+)}}$$

273

$$\Lambda^{(+)}(0, x^1) = i(4\pi)^{-\frac{1}{2}} \int_0^\infty dk_1 k_1^{-1} c(k_1) \left(e^{-ik_1 x^1} - \theta(\kappa - k_1) \right)$$

Looking at the solution we see that, in terms of the modes which diagonalize the Poincaré generators we have

$$c(k^1) = \eta^*(-k^1) + \sqrt{\frac{|k_1|}{\omega}} \left(\Sigma(k_1) + \Sigma^*(-k_1) \right)$$

From (1) we see that

$$c(k^1)|0_{ET}\rangle = A_+^{(+)}|0_{ET}\rangle = \Pi_A^{(+)}|0_{ET}\rangle = 0 \; ; \;\; |0_{ET}\rangle \neq |0\rangle$$

Not only is the physical vacuum not the perturbative vacuum in the equal-time basis, it is extremely complicated, projecting on to every basis vector allowed by kinematics and the charge super selection rule. In addition, the problem of finding the eigenstates of the Hamiltonian is a complicated dynamical problem.

We now consider the problem of quantizing the model on the surface $x^+ = 0$. The degrees of freedom are

$$\Psi_+ \; ; \; \Psi_- \; ; \; \lambda$$

The main subtlety encountered in this quantization scheme is not the existence of the zero-mode fields. If these fields are left out a solution cannot be found; but the fields are right there in the Lagrangian, the fact that they are zero-mode fields is easily learned from the equations of motion and the algebra of the zero mode fields can be determined by fairly straightforward procedures. The most serious problem is that the Fermi product cannot be regulated by splitting in the x^- direction. There are various ways to get around the difficulty. Here we will describe one which involves over regulating the theory, solving the dynamics then removing the over-regulation.

We first use the equations of motion to discover the dependence of Ψ_+ on η:

$$2\partial_+\partial_- A_+ = \lambda + m\partial_+\phi(x) - \frac{m^2}{2} A_+(x)$$

$$\Longrightarrow A_+ \supset m\partial_+\eta$$

$$i\partial_+\Psi_+ = e\Psi_+ A_+$$

$$\Longrightarrow \Psi_+ = e^{-i2\sqrt{\pi}(\eta(x^+))}\Psi_R$$

Where we have defined Ψ_R simply to be the rest of Ψ_+. We find that Ψ_R satisfies the canonical commutation relations on $x^+ = 0$ so we initialize it as

$$\{\Psi_R(x^+, x^-), \Psi_R^*(x^+, y^-)\} = \delta(x^- - y^-)$$

$$\Psi_R(0, x^-) = Z\exp[-2i\sqrt{\pi}\tilde{\phi}^{(-)}(0, x^-)]\tilde{\sigma}_+\exp[-2i\sqrt{\pi}\tilde{\phi}^{(+)}(0, x^-)]$$

$$\tilde{\sigma}_+ = exp\left[i\sqrt{\pi}\{\tilde{Q}_5 + \tilde{Q}\}(4m)^{-1} + \int_0^{\tilde{\kappa}} dk_1 k_0^{-1}\{\tilde{c}(k_1) - \tilde{c}^*(k_1)\}\right]$$

$$\tilde{\phi}^{(+)}(x^-) = i(4\pi)^{-\frac{1}{2}}\int_{-\infty}^0 dk_1 k_0^{-1}\tilde{c}(k_1)\left(e^{-ik\cdot x} - \theta(\tilde{\kappa} - k_0)\right)$$

We now calculate P_+

$$P_+ = \frac{m^2}{4}\int_{-\infty}^{\infty}\tilde{\phi}^2 dx^- + \int_{-\infty}^{\infty}\{(\partial_+\phi)^2 - (\partial_+\eta)^2\}dx^+$$

Since this operator is already diagonal, we now know the space-time dependence of Ψ_+. As it stands the construction is not translationally invariant. But we can now define Fermi products by splitting in a space-like direction and recover covariance by taking the limit

$$\lim \tilde{\kappa} \to 0 \implies \Psi_+ = Z_+ e^{\Lambda_+^{(-)}}\sigma_+ e^{\Lambda_+^{(+)}}$$

We note that here, $|0\rangle$ is the perturbative vacuum so that the physical vacuum, $|\Omega\rangle$, is set by kinematics and gauge invariance. In the same way we can discover that $\langle\theta|\bar{\Psi}\Psi|\theta\rangle \neq 0$. But we cannot calculate the value of the condensate from kinematics and gauge invariance: the value depends on the Ψ_+ wavefunction renormalization constant and that quantity is determined by maintaining the canonical commutation relation for Ψ_+ with the product defined by space-like splittings — which requires a dynamical calculation.

We now turn to the problem of quantization on the surface $x^- = 0$ — the anti-light-cone gauge. The degrees of freedom are

$$\Psi_- \; ; \; A_+ \; ; \; \partial_- A_+$$

A very unusual feature here is that not only is Ψ_+ not a degree of freedom, it is not a constraint which must be resolved prior to solving for the dynamics of the degrees of freedom. Indeed, Ψ_+ must be solved for at the very end of the calculation, as we shall discuss below. A very important feature of this formulation is that the Ψ_- products can be defined by splitting in the initial value surface. We can thus calculate

$$J_+(x) = \lim_{y^+ \to x^+}\frac{e}{2}\{\psi_-^*(x^+)\psi_-(y^+)\exp[-ie\int_{y^+}^{x^+} A_+(z^+, x^-)dz^+] + h.c.\}$$

$$J_+(x) = m\partial_+\phi(x) - \frac{m^2}{2}A_+(x)$$

$$\lim_{y^+ \to x^+}\{\frac{i}{2}\psi_-^*(x^+)\partial_+\psi_-(y^+)\exp[-ie\int_{y^+}^{x^+} A_+(z, x^-)dz] + h.c. - \frac{1}{2\pi}\frac{1}{(x^+ - y^+)^2}\}$$

$$i\Psi_-^*\partial_+\Psi_- = (\partial_+\phi)^2 - \frac{m^2}{4}(A_+)^2$$

Perhaps the main subtlety of the current formulation is not to be too quick to think that A_+ and $\partial_- A_+$ are canonical conjugates. Indeed, although

$$[A_+(x^+, x^-), A_+(y^+, x^-)] = 0$$

and

$$[\partial_- A_+(x^+, x^-), A_+(y^+, x^-)] = -\frac{i}{2}\delta(x^+ - y^+)$$

we have

$$[\partial_- A_+(x^+, x^-), \partial_- A_+(y^+, x^-)] = -i\frac{m^2}{16}\epsilon(x^+ - y^+)$$

This problem is mearly a matter of being careful; the correct algebra can be obtained either by forcing agreement between the Heisenberg equations and the equations of motion or by the Dirac procedure.

Combining the last equation with

$$P_- = \int_{-\infty}^{\infty} (\partial_- A_+)^2 dx^+$$

gives

$$\partial_- A_+(x) = -\frac{m}{2}\tilde{\Sigma}$$

Again the dynamical operator is diagonal in the degrees of freedom. From

$$-2\partial_+\partial_- A_= = \lambda + m\partial_+\phi(x) - \frac{m^2}{2}A_+$$

we get

$$A_+(x) = \frac{2}{m^2}\{\lambda + m\partial_+(\phi + \tilde{\Sigma})\}$$

or

$$A_+ = \frac{2}{m}\partial_+(\tilde{\Sigma} + \eta)$$

We now turn to the problem of constructing Ψ_+. We do that through the equation

$$i\partial_+\Psi_+ = e\Psi_+ A_+$$

Formally, this equation is easy to solve. In doing so we encounter some singular objects; but if we stick to the plan of regulating ultraviolet singularities by spacelike point splitting, infrared singularities by a Klaiber subtraction and defining exponentials by Wick ordering, we recover the correct solution.

REFERENCES

1. For details and a complete list of references see: Nakawaki Y., and McCartor G. **hep-th/9903017**.

Gauge Theories in the Light-Cone Representation

Yuji Nakawaki* and Gary McCartor†

*Division of Physics and Mathematics, Faculty of Engineering, Setsunan University, Osaka
572-8508, Japan
†Department of Physics, SMU, Dallas, Texas 75275

Abstract. We attempt in McCartor and Robertson's framework to formulate a perturbation theory of light-cone axial gauge QED in which zero-mode fields play roles as regulator fields yielding well-defined Mandelstam-Leibbrandt form of gauge field propagator. We find that zero-mode fields make up for degrees of freedom of A_+ and its canonical conjugate in the light-cone temporal gauge formulation and that they are retained in the interaction term $j^+ A_+$ through A_+, if and only if the integral $\int_{-\infty}^{\infty} dx^- j_-$ does not vanish. It is pointed out that from the boundary surface contributions $T_{++}(x^- = \pm\infty)$, which are added to obtain P_+ identical to those in ordinary coordinates, an infinite number of noncovariant interaction terms might be obtained so as to cancel corresponding infinite number of noncovariant diagrams yielded by the contact term of the Fermion propagator.

INTRODUCTION

Recently the search for nonperturbative solutions of QCD has led to extensive studies of light-front field theory, in which the infinite-momentum limit is incorporated by the change of variables [1] $x^+ = \frac{x^0 + x^3}{\sqrt{2}}$, $x^- = \frac{x^0 - x^3}{\sqrt{2}}$, and a change of the representation space so that one is able to have vacuum state composed only of particles with nonnegative longitudinal momentum and also to have relativistic bound-state equations of the Schrödinger-type [2].

Quantization has traditionally been carried out in parallel with the axial gauge formulation of QED in ordinary space-time coordinates. Thus x^+ and $A_-^a = \frac{A_0^a - A_3^a}{\sqrt{2}} = 0$ have been chosen respectively to be the evolution parameter and the gauge fixing condition, and minus component of the gauge field equations, on which the gauge fixing condition is imposed, has been solved as the constraint by using the operator $(\partial_-)^{-1}$ to express the Hamiltonian in terms of the physical degrees of freedom [3]. In what follows we call this formulation light-cone axial gauge formulation. It turns out however that the formulation encounters inherent difficulties. The quantity $((\partial_-)^{-1})^2$, which is needed for this construction, turns out to be ill de-

CP494, *New Directions in Quantum Chromodynamics*, edited by C.-R. Ji and D.-P. Min
© 1999 American Institute of Physics 1-56396-908-4/99/$15.00

fined in a positive definite Hilbert space [4]. One result of this is that the light-cone gauge formulations are not ghost free, contrary to what was originally expected and is still sometimes claimed. It was found first that in order to bring perturbative calculations done in light-cone gauge into agreement with calculations done in co-variant gauges, the spurious singurarity of the free gauge field propagator, which arises from applying $(\partial_-)^{-1}$ to the physical operators, has to be regularized not as a principal value but according to Mandelstam-Leibbrandt (ML) prescription in such a way that causality is preserved [5]. Shortly afterwards Bassetto et al. found that the ML form of the propagator is realized in a light-cone gauge canonical operator formalism in ordinary space- time coordinates, if one introduces a Lagrange mul-tiplier field and its conjugate as residual gauge's degrees of freedom [6]. Moreover, Morara and Soldati found just recently [7] that if one takes x^+ as the evolution parameter and $A_+ = \frac{A_0 + A_-}{\sqrt{2}} = 0$ instead of $A_- = 0$ as the gauge fixing condition and if one introduces the Lagrange multiplier field and its conjugate, then one can obtain less constraints among the canonical variables so that one can construct a temporal gauge canonical operator formalism in light-cone coordinates (light-cone temporal gauge formulation), which realizes the ML form of the free gauge field propagator.

We may therefore think that restoring the residual gauge degrees of freedom overcomes the difficulties inherent to the light-cone axial gauge formulation. In doing so we can expect that the residual gauge degrees of freedom make up for degrees of freedom of A_+ and its canonical conjugate in the light-cone temporal gauge formulation. Those operators, because they become zero-mode operators in the light-cone axial gauge formulation, make the quantization somewhat nonstan-dard so that Dirac's canonical quantization procedure cannot be used to specify the algebra. Recently McCartor and Robertson [8] found that the zero-mode operators are introduced as nondynamical integration constants (which are independent of x^- by definition but are, however, functions of x^+ and $\boldsymbol{x}_\perp$.) They also found that since, with the introduction of the integration constants, we can assume that other operators vanish in the limit $x^- \to \pm\infty$, Hamiltonian formalism can be extended in such a way that the x^+-dependence of the dynamical operators can be solved for in the presence of the x^+-dependent integration constants.

These motivate us to consider further formulating a perturbation theory of light-cone axial gauge QED in McCartor and Robertson's framework. In a preliminary investigation [9] we have found that the light-cone gauge solution of free electro-magnetic field equations is described independently of representations as

$$A_- = 0, \quad A_i = T_i - \partial_i\phi, \quad A_+ = \frac{1}{\partial_-}\partial_i T_i - \frac{1}{\partial_\perp^2}\lambda - \partial_+\phi$$

where $T_i(i = 1, 2)$ is the physical field and λ and ϕ are Lagrange multiplier field and its conjugate, respectively. We use latin indices $(i, j, \cdots)$ throughout this paper to denote transverse components. These zero-mode fields regularize infrared divergences given rise to by T_i. Consequently with a suitable choice of vacuum, the Mandelstam-Leibbrant form of the propagator has been derived as the vacuum expectation value of x^+-ordered product of the free electromagnetic field:

$$\langle 0 | \{ \theta(x^+ - y^+) A_\mu(x) A_\nu(y) + \theta(y^+ - x^+) A_\nu(y) A_\mu(x) \} | 0 \rangle$$
$$= \tfrac{1}{(2\pi)^4} \int d^4 k D_{\mu\nu}(k) e^{-ik\cdot(x-y)}$$

where

$$D_{\mu\nu}(k) = \frac{i}{k^2 + i\epsilon} \left(-g_{\mu\nu} + \frac{n_\mu k_\nu + n_\nu k_\mu}{k_- + i\epsilon \operatorname{sgn}(k_+)} \right) - \frac{i}{2} \delta_{\mu+} \delta_{\nu+} \left\{ \frac{1}{(k_- + i\epsilon)^2} + \frac{1}{(k_- - i\epsilon)^2} \right\}.$$

In this contribution we consider retaining λ and ϕ in the light-cone axial gauge Hamiltonian of QED. It is found that they are retained in the interaction term $j^+ A_+$ through A_+, if and only if the integral $\int_{-\infty}^{\infty} dx^- j_-$ does not vanish. It is also pointed out that the bondary surface contributions $T_{++}(x^- = \pm\infty)$ are indispensable to obtain x^+ independent translational generators and also to give rise to an infinite number of interaction terms which cancel corresponding noncovariant diagrams yielded by the contact term of the Fermion propagator.

We will use the following notation:

$$x^+ = \tfrac{x^0 + x^3}{\sqrt{2}}, \quad x^- = \tfrac{x^0 - x^3}{\sqrt{2}}, \quad \partial_+ = \tfrac{\partial_0 + \partial_3}{\sqrt{2}}, \quad \partial_- = \tfrac{\partial_0 - \partial_3}{\sqrt{2}},$$
$$\gamma^+ = \tfrac{\gamma^0 + \gamma^3}{\sqrt{2}}, \quad \gamma^- = \tfrac{\gamma^0 - \gamma^3}{\sqrt{2}}, \quad \psi_+ = \tfrac{1}{\sqrt{2}} \gamma^0 \gamma^+ \psi, \quad \psi_- = \tfrac{1}{\sqrt{2}} \gamma^0 \gamma^- \psi,$$
$$\boldsymbol{x} = (x^1, x^2, x^3), \quad \boldsymbol{x}^\pm = (x^\pm, x^1, x^2), \quad \boldsymbol{x}_\perp = (x^1, x^2),$$
$$d^2 x = dx^1 dx^2, \quad d^3 x = dx^1 dx^2 dx^3, \quad d^3 x^\pm = dx^1 dx^2 dx^\pm.$$

LIGHT-CONE AXIAL GAUGE QED

We begin by pointing out that zero-mode fields can be introduced as integration constants in the light-cone axial gauge formulation. For details refer to [8]. Light-cone gauge QED is defined by the Lagrangian

$$L = -\frac{1}{4} F_{\mu\nu} F^{\mu\nu} - \lambda n \cdot A + \bar{\psi} \{ i\gamma^\mu (\partial_\mu + ieA_\mu) - m \} \psi$$

where $n_\mu = (n_+, n_-, n_1, n_2) = (1, 0, 0, 0)$. As Euler-Lagrange equations, field equations and gauge fixing condition are derived as

$$\partial^\mu F_{\mu\nu} = n_\nu \lambda + j_\nu, \quad j_\nu = e\bar{\psi}\gamma_\nu\psi$$
$$\{ i\gamma^\mu (\partial_\mu + ieA_\mu) - m \} \psi = 0$$
$$A^+ = A_- = 0.$$

The field equation of Lagrange multiplier field, $\partial_- \lambda = 0$, is obtained by operating on the gauge field equations with ∂_ν and by making use of current conservation. Because $A_- = 0$, minus component of the gauge field equation becomes the following constraint in the light-cone axial gauge formulation:

$$-\partial_-(\partial_- A_+ - \partial_i A_i) = j_-. \tag{1}$$

Integrating this in terms of x^- enables us to introduce ϕ as an integration constant as in the following form

$$\partial_- A_+ - \partial_i A_i = -\frac{1}{\partial_-} j_- + \partial_\perp^2 \phi \tag{2}$$

where $\partial_\perp^2 = \partial_1^2 + \partial_2^2$ and the operator $\frac{1}{\partial_-}$ is defined by

$$\frac{1}{\partial_-} j_-(x) = \frac{1}{2} \int_{-\infty}^{\infty} dy^- \epsilon(x^- - y^-) j_-(x^+, y^-, \mathbf{x}_\perp). \tag{3}$$

From (2) we see that because $\partial_- A_+$ and j_- are gauge invariant quantities,

$$T_i \equiv A_i + \partial_i \phi, \quad (i = 1, 2) \tag{4}$$

must be also gauge invariant. We therefore regard T_i as physical operators in the light-cone axial gauge formulation. We also see that in the limit $x^- \to \pm\infty, \partial_- A_+$ acquires nonvanishing boundary surface $(x^- \to \pm\infty)$ contributions

$$\lim_{x^- \to \pm\infty} \partial_- A_+(x) = \mp \frac{1}{2} \int_{-\infty}^{\infty} dy^- j_-(x^+, y^-, \mathbf{x}_\perp). \tag{5}$$

Here we have assumed that T_i tends to 0 in the limit $x^- \to \pm\infty$. Integrating the constraint in (2) once more in terms of x^- and taking consistency among the gauge field equations into account, we obtain A_μ described in terms of T_i, λ, ϕ and j_-:

$$A_- = 0, \quad A_i = T_i - \partial_i \phi,$$
$$A_+ = \frac{1}{\partial_-} \partial_i T_i - (\frac{1}{\partial_-})^2 j_- - \frac{1}{\partial_\perp^2} \lambda - \partial_+ \phi. \tag{6}$$

Furthermore field equation of T_i turns out to be

$$(2\partial_+ \partial_- - \partial_\perp^2) T_i = j_i - \frac{\partial_i}{\partial_-} j_-. \tag{7}$$

When we substitute the obtained A_μ into the translational generator $P_+ = \int d^3 x^- T_{+-}$, where T_{+-} is the canonical energy-momentum tensor given by

$$T_{+-} = \partial_- A_i \partial_i A_+ - \frac{1}{2}(\partial_- A_+)^2 + \frac{1}{2}(F_{12})^2 + i\bar{\psi}\gamma^+ \partial_+ \psi, \tag{8}$$

we find that the zero-mode fields are retained in the interaction term $j^+ A_+$ through A_+, if and only if $\lim_{x^- \to \pm\infty} \partial_- A_+ \neq 0$. Actually by the help of the field equation of ψ_+ and (1) we can rewrite P_+ as

$$P_+ = \int d^3 x^- [-\partial_-^2 A_+ \cdot A_+ - \frac{1}{2}(\partial_- A_+)^2 + \frac{1}{2} F_{12}^2 + \psi_+^* \gamma^0 \{m - i\gamma^i(\partial_i + ieA_i)\}\psi_-]. \tag{9}$$

Because A_+ retains the zero-mode operator $\frac{1}{\partial_\perp^2}\lambda + \partial_+\phi$, which is independent of x^-, we obtain

$$-\int dx^- \partial_-^2 A_+ \cdot A_+|_{zero} = \{\lim_{x^-\to\infty}\partial_- A_+ - \lim_{x^-\to-\infty}\partial_- A_+\}(\frac{1}{\partial_\perp^2}\lambda + \partial_+\phi). \quad (10)$$

Therefore in case that the integral in (5) vanishes, the first term is rewritten as $(\partial_- A_+)^2$ so that the first and second terms are combined into the standard expression $\frac{1}{2}(\partial_- A_+)^2$. Because gauge invariant operator $\partial_- A_+$ does not possess the zero-mode fields, which is seen from (2) and (4), we can further carry out a residual gauge transformation in such a way that the zero-mode fields disappear from P_+ at all. Not to obtain this undesirable result, we cannot require that the integral in (5) vanishes.

We express P_+ in (8) expicitly in terms of T_i, λ and ϕ as follows:

$$P_+ = \int d^3x \, [\frac{1}{2}(\partial_i T_i)^2 + \frac{1}{2}(F_{12})^2 - \partial_i T_i \frac{1}{\partial_-} j_- + \frac{1}{2}(\frac{1}{\partial_-} j_-)^2$$
$$- j_-(\frac{1}{\partial_\perp^2}\lambda + \partial_+\phi) + \psi_+^* \gamma^0 \{m - i\gamma^i(\partial_i + ieA_i)\}\psi_-]. \quad (11)$$

Then it is straightforward to examine that the field equations of T_i and ψ_+ are derived as Heisenberg equations. In fact the light-cone axial gauge quantization conditions of T_i

$$[T_i(x), T_j(y)]|_{x^+=y^+} = -\delta_{ij}\frac{i}{2\partial_-}\delta^{(3)}(\boldsymbol{x}^- - \boldsymbol{y}^-),$$
$$[T_i(x), \lambda(y)] = [T_i(x), \phi(y)] = [T_i(x), \psi_+(y)]|_{x^+=y^+} = 0, \quad (12)$$

yield Eq.(7) and those of ψ_+

$$\{\psi_+(x), \psi_+^*(y)\}_+|_{x^+=y^+} = \frac{\gamma^0\gamma^+}{2}\delta^{(3)}(\boldsymbol{x}^- - \boldsymbol{y}^-),$$
$$\{\psi_+(x), \psi_+(y)\}_+|_{x^+=y^+} = [\psi_+(x), \lambda(y)] = [\psi_+(x), \phi(y)] = 0, \quad (13)$$

provides us with the equation ψ_+ satisfies.

It is evident that traditional Heisenberg formalism cannot give rise to the equation of the zero-mode fields λ and ϕ as Heisenberg equation. As a matter of fact P_+ in (11) does not possess kinetic term of them. We follow McCartor and Robertson [8] to overcome this difficulty. We add to P_+ the operators necessary to properly translate the zero-mode fields by defining the translational generators in the light-cone coordinates by requiring that they are identical to those in ordinary coordinates. From the divergence equation

$$\partial^\nu T_{\mu\nu} = 0$$

we obtain

$$\oint T_{\mu\nu} d\sigma^\nu = 0.$$

If we perform the integral over the closed surface shown in Fig.1, we obtain

281

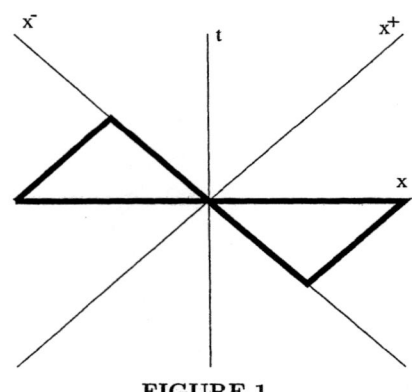

FIGURE 1.

$$P_+ = \int d^3x\, T_{+0}(x^0, \boldsymbol{x}) = \int d^3x^-\, T_{+-}(x^+ = \tfrac{x^0}{\sqrt{2}}, \boldsymbol{x}^-)$$

$$+ \int d^2x \left(\int_{-\infty}^{\frac{x^0}{\sqrt{2}}} dx^+\, T_{++}(x^+, x^- = \infty, \boldsymbol{x}_\perp) + \int_{\frac{x^0}{\sqrt{2}}}^{\infty} dx^+\, T_{++}(x^+, x^- = -\infty, \boldsymbol{x}_\perp) \right). \quad (14)$$

By substituting A_μ in (6) into $T_{++} = \partial_- A_+ \partial_+ A_+ + F_{+i}\partial_+ A_i + i\bar{\psi}\gamma^-\partial_+\psi$ we obtain the desired kinetic term $\int d^3x^+\, \lambda\partial_+\phi$ from the zero-mode part of $F_{+i}\partial_+ A_i$. Here it should be noted that owing to the terms on the second line of (14) it can be verified formally that P_+ is independent of $x^0 (= \sqrt{2}x^+)$. Actually the first term on the second line of (11) is dependent on x^+. This x^+ dependence is partly canceled by the term

$$\int d^3x^-\, \frac{1}{\partial_+} \left(j_- \,\partial_+(\frac{1}{\partial_\perp^2}\lambda + \partial_+\phi) \right) \quad (15)$$

which results from the term $\frac{1}{\partial_-}j_- \cdot \partial_+(\frac{1}{\partial_\perp^2}\lambda + \partial_+\phi)$ retained in the term $\partial_- A_+ \partial_+ A_+$. This implies that there have to exist other terms to cancel the remaining x^+ dependence completely and also to cancel extra terms, which result from commutation relations of P_+ in (11) with λ and ϕ. It seems that necessary terms can be found, if we can succeed in solving the field equation of ψ_- and if we succeed in calculating $\lim_{x^-\to\pm\infty}\psi_-$. Therefore we do not consider them here. Note furthermore that we encounter a new difficulty, if we add (15) to P_+. Owing to the operator $\frac{1}{\partial_+}$ we can not make use of equal x^+-time anticommutation relations of ψ_+. Therefore we assume that the term in (15) and any other terms, which result from the second terms in (14), do not give rise to extra terms to Heisenberg equation of ψ_+.

With this understanding we employ the following as Hamiltonian

$$H = \int d^3x^-\, T_{+-}(x) + \int d^3x^+\, \lambda\partial_+\phi. \quad (16)$$

282

It is straightforward to rewrite the Hamiltonian in the interaction representation as in the form:
$$H = H_0 + H_I$$
where

$$H_0 = \int d^3x^- \{\tfrac{1}{2}(\partial_i T_i)^2 + \tfrac{1}{2}(F_{12})^2 + i\bar{\psi}\gamma^-\partial_-\psi\} + \int d^3x^+ \ \lambda\partial_+\phi,$$
$$H_I = \int d^3x^- \{\tfrac{1}{2}(j^\mu, A_\mu)_+ + \tfrac{1}{2}(\tfrac{1}{\partial_-}j_-)^2 + e^2\bar{\psi}\gamma^\mu A_\mu \tfrac{\gamma^+}{2i\partial_-}\gamma^\nu A_\nu\psi\}. \tag{17}$$

We have obtained two noncovariant interaction terms. Coulomb like term cancels the contact term of the gauge field propagator and the sea-gull term cancels the contact term of the Fermion propagator in perturbation calculation of S matrices. It is evident however that the sea-gull term cannot cancel all noncovariant diagrams which the contact term of the Fermion propagator gives rise to. In fact in the light-cone temporal gauge formulation and in covariant gauge formulations with nonvanishing A_-, interaction Hamiltonians possess an infinite number of noncovariant terms in addition to the sea-gull term in order to cancel an infinite number of noncovariant diagrams, which the contact term of the Fermion propagator gives rise to. We can expect to derive those interaction terms from the second terms in (14). We leave this task for subsequent studies.

REFERENCES

1. L. Susskind, Phys. Rev.**165**, 1535, (1968).
   K. Bardakci and M. B. Halpern, Phys. Rev.**176**, 1686, (1968).
2. *Theory of Hadrons and Light-Front QCD*, S. D. Glazek, Ed.,(World Scientific, Singapore, 1995).
   S. J. Brodsky, H. C. Pauli, and S. S. Pinsky, Phys. Reports, **301**, 299, (1998).
3. J. B. Kogut and D. E. Soper, Phys. Rev. **D1**, 2901, (1970).
   R.A. Neville and F. Rohrlich, Phys. Rev.**D3**, 1692, (1971).
4. N. Nakanishi, Phys. Lett. **131B**, 381, (1983).
   N. Nakanishi, *Quantum Electrodynamics*, ed.T. Kinoshita (World Scientific, Singapore, 1990), p. 36.
5. S. Mandelstam, Nucl. Phys. **B213** (1983), 149.
   G. Leibbrandt, Phys. Rev. **D29**, 1699, (1984).
6. A. Bassetto, M. Dalbosco, I. Lazziera and R. Soldati, Phys. Rev. **D31**, 2012, (1985).
   A. Bassetto, G. Nardelli and R. Soldati, *Yang-Mills Theories in Algebraic Non-Covariant Gauges*, (World Scientific, Singapore,1991).
7. M. Morara and R. Soldati, Phys. Rev. **D58**, 105011, (1998). This gauge was discussed in an unpublished talk by K. Hornbostel,1992, Dallas, Texas.
8. G. McCartor and D.G. Robertson, Z. Phys. **C62**, 349, (1994); **C68**, 345, (1995).
9. Y. Nakawaki and G. McCartor, Prog. Theor. Phys. **102**, 149, (1999).

Consistent Perturbative Light Front Formulation of Yang-Mills Theories

M. Morara*, R. Soldati* and G. McCartor†

Dipartimento di Fisica "A. Righi", Università di Bologna
† *Department of Physics, SMU, Dallas TX*

Abstract. It is shown how to obtain the consistent light front form quantization of a non-Abelian pure Yang-Mills theory (gluondynamics) in the framework of the standard perturbative approach. After a short review of the previous attempts in the light cone gauge $A_- = 0$, it is explained how the difficulties can be overcome after turning to the anti light cone gauge $A_+ = 0$. In particular, the generating functional of the renormalized Green's functions turns out to be the same as in the conventional instant form approach, leading to the Mandelstam-Leibbrandt prescription for the free gluon propagator.

LIGHT FRONT FORM FOR FREE FIELDS

The light front form (**LFF**) formulation[1] of field theories [1] is given in terms of the evolution parameter x^+, the light front "time", and the "volume" coordinates $\mathbf{x} = (x^-, x^\perp)$, which label the light front "space". The LFF formulation of gauge theories in the light cone gauge $A_- = n^\mu A_\mu = 0$, is known since almost thirty years [2]. This original formulation involves only physical degrees of freedom and, at the perturbative level, it unavoidably leads to the Cauchy's principal value (CPV) prescription for the non-covariant spurious singularity of the free gauge boson propagator: namely,

$$D^+_{\mu\nu}(x) = i \int \frac{d^4 k}{(2\pi)^4} \frac{e^{ikx}}{k^2 + i\epsilon} \left\{ -g_{\mu\nu} + \frac{n_\mu k_\nu + n_\nu k_\mu}{[nk]_{\text{CPV}}} \right\},$$

where

$$\frac{1}{[nk]^2_{\text{CPV}}} \equiv \mathcal{S}' - \lim_{\varepsilon \to 0} \frac{nk}{nk^2 + \varepsilon^2},$$

the limit being understood in the tempered distributions' topology.

---

1) *Notations*:

$$v^\mu \equiv (v^+, v^-, v^\perp), \ v^\pm = \frac{v^0 \pm v^3}{\sqrt{2}}, \ v^\perp = (v^1, v^2) = v^\alpha.$$

CP494, *New Directions in Quantum Chromodynamics*, edited by C.-R. Ji and D.-P. Min
© 1999 American Institute of Physics 1-56396-908-4/99/$15.00

Unfortunately the CPV prescription so roughly violates power counting and causality, that it even fails to reproduce the correct one loop beta function in the non-Abelian case [3]. It follows therefrom that the original light front form perturbation theory in the light cone gauge is inconsistent.

On the other hand, it turns out that the usual instant form (**IF**) formulation, which is in terms of the evolution parameter $x^0 = ct$, the ordinary time, and volume coordinates $\vec{x}$ for the ordinary space, is fully consistent in perturbative approaches to gauge theories [4], both in the light-cone gauge $A_0 - A_3 = 0$ as well as in the anti light cone gauge $A_0 + A_3 = 0$.

This **IF** formulation necessarily involves extra unphysical degrees of freedom (ghosts) and canonical quantization necessarily leads to the Mandelstam-Leibbrandt (**ML**) tempered distribution [5] to regulate the non-covariant singularity in the free gauge boson propagator, i.e.,

$$\frac{1}{[k_0 \mp k_3]_{\text{ML}}} \equiv \mathcal{S}' - \lim_{\varepsilon \to 0} \frac{k_0 \pm k_3}{(k_0 \pm k_3)(k_0 \mp k_3) + i\varepsilon}.$$

The **ML** prescription guarantees power counting and causality and in so doing the **IF** perturbative formulation does fulfill renormalizability, unitarity and covariance of the formal S-matrix elements. The natural question arises: is it possible to find some **LFF** formulation which reproduces those remarkable results? For Quantum Electrodynamics (the Abelian theory) the answer is yes [6], provided we set up canonical light front form quantization in the Weyl's gauge $A_+ = 0$. In order to obtain some consistent **LFF** perturbation theory we have to find some canonical **LFF** framework leading to the **ML** form of the free propagator. The first attempt towards this task has been pionereed, in the light-cone gauge, by G. McCartor and D. G. Robertson [7].

Their starting point is the free Lagrange density

$$\mathcal{L}_{\text{rad}} = -\frac{1}{4} F_{\mu\nu} F^{\mu\nu} - \Lambda n^\mu A_\mu,$$

in which $n^\mu = (n^+, n^-, n^\perp) = (0, 1, 0, 0)$. After the introduction of the new field variables $(A_\alpha, A_+, \Lambda) \longmapsto (T_\alpha, \varphi, \lambda)$, i.e.,

$$A_\alpha = T_\alpha - \partial_\alpha \varphi, \quad \Lambda = \partial_\perp^2 \lambda, \tag{1}$$
$$A_+ = \partial_\alpha \partial_-^{-1} * T_\alpha - \partial_+ \varphi - \lambda, \tag{2}$$

the equations of motion become

$$2\partial_+ \partial_- T_\alpha = \partial_\perp^2 T_\alpha, \quad \partial_- \varphi = \partial_- \lambda = 0.$$

We remark that φ and λ fulfill constraint equations and thereby φ and λ can not be quantized on the null hyperplanes at constant x^+. Owing to this feature, in [7] some new **LFF** quantization procedure was suggested involving two characteristic

285

surfaces, i.e., transverse fields T_α are quantized on null hyperplanes at equal x^+, whereas longitudinal fields φ and λ are quantized on null hyperplanes at equal x^-. The above quantization procedure leads to the light front form canonical commutation relations listed below

$$[T_\alpha(x), T_\beta(y)]_{x^+=y^+} = \frac{\delta_{\alpha\beta}}{2i}\mathrm{sgn}(x^- - y^-)\delta^{(2)}(x^\perp - y^\perp), \qquad (3)$$

$$[\varphi(x), \lambda(y)]_{x^-=y^-} = i\delta(x^+ - y^+)\partial_\perp^{-2} * \delta^{(2)}(x^\perp - y^\perp), \qquad (4)$$

$$[T_\alpha(x), \varphi(y)] = [T_\alpha(x), \lambda(y)] = [\varphi(x), \varphi(y)] = [\lambda(x), \lambda(y)] = 0. \qquad (5)$$

However, when we compute the **LFF** ordered product

$$D_{\mu\nu}^+(x - y) \equiv \theta(x^+ - y^+)\langle 0|A_\mu(x)A_\nu(y)|0\rangle + \theta(y^+ - x^+)\langle 0|A_\nu(y)A_\mu(x)|0\rangle, \quad (6)$$

the **LFF** quantization scheme of Eqs. (3), (4) and (5) drives to ill-defined convolution products and not to the **ML** form of the vector boson propagator. As a consequence, it turns out that the **LFF** quantization of gauge theories in the light cone gauge $A_- = 0$ is indeed troublesome, when choosing x^+ as the evolution parameter. The simplest way to overcome the above barring is the transition to the anti light cone gauge, or light front form Weyl's gauge, $n^{*\mu}A_\mu = A_+ = 0$, $n^{*\mu} \equiv (1,0,0,0)$. Consider therefore the new Lagrange density for the free Maxwell's radiation field

$$\mathcal{L}_{\mathrm{rad}} = -\frac{1}{4}F_{\mu\nu}F^{\mu\nu} - \Lambda n^{*\mu}A_\mu;$$

The best way to set up the light front form quantization of the above constrained system is to follow Dirac's method [8] of canonical quantization. The free (unconstrained) canonical momentum is $\pi^- = F_{+-}$, and we have the second class primary constraints $\pi^\alpha - F_{-\alpha} = 0$, as well as the first class primary constraints $\pi^+ = \pi^\Lambda = 0$. The canonical Hamilton density is

$$\mathcal{H}_{\mathrm{rad}} = \frac{1}{2}\left(\pi^-\right)^2 + \frac{1}{4}F_{\alpha\beta}F_{\alpha\beta} - A_+\left(\partial_\alpha\pi^\alpha + \partial_-\pi^- - \Lambda\right),$$

whence we derive the secondary constraints $A_+ = 0$, $\partial_\alpha\pi^\alpha + \partial_-\pi^- = \Lambda$. The full set of constraints is now second class and thereby we can compute equal x^+ Dirac's brackets, whose explicit form can be found in [6].

After introduction of the new set of variables

$$A_\alpha = T_\alpha - \partial_\alpha\varphi, \quad \pi^- = \partial_\alpha T_\alpha, \qquad (7)$$

$$A_- = 2\partial_-\partial_\alpha\partial_\perp^{-2} * T_\alpha - \partial_-\varphi - \lambda, \qquad (8)$$

we obtain the genuine equations of motion for all the fields, i.e.,

$$2\partial_-\partial_+T_\alpha = \partial_\perp^2 T_\alpha, \quad \partial_+\varphi = \partial_+\lambda = 0.$$

286

The transition to the quantum theory is achieved after replacement of the light front form Dirac's brackets with the corresponding light front form canonical commutation relations, which now read the same as in Eq.s (3), (4) and (5), but for the crucial Eq. (4) which is replaced by

$$[\varphi(x), \lambda(y)]_{x^+ = y^+} = i\delta(x^- - y^-)\partial_\perp^{-2} * \delta^{(2)}(x^\perp - y^\perp).$$

Actually, the quantization characteristic surface is the very same for all the field variables, at variance with Eq.s (3), (4) and (5). It is convenient to introduce the longitudinal (unphysical) components of the gauge potential $\Gamma_\mu = -\left(\partial_\mu\varphi + n_\mu^*\lambda\right)$,

$$\Gamma_\mu(x) = \int \frac{d^2k_\perp dk_-}{(2\pi)^{3/2}} \frac{\theta(k_-)}{\sqrt{|k_\perp|}} \left\{ \left[-\frac{k_\mu}{|k_\perp|} f(k_\perp, k_-) \right. \right. \tag{9}$$

$$\left. \left. + n_\mu^* g(k_\perp, k_-) \right] e^{-ikx} + \text{h. c.} \right\}_{k^+=0}, \tag{10}$$

whilst the transversal (physical) components become $T_\mu(x) \equiv A_\mu(x) - \Gamma_\mu(x)$,

$$T_\mu(x) = \int \frac{d^2k_\perp dk_-}{(2\pi)^{3/2}} \frac{\theta(k_-)}{\sqrt{2k_-}} \varepsilon_\mu^{(\alpha)}(k_\perp, k_-) \tag{11}$$

$$\times \left\{ a_\alpha(k_\perp, k_-) e^{-ikx} + a_\alpha^\dagger(k_\perp, k_-) e^{ikx} \right\}_{k^+ = k_\perp^2/2k_-}, \tag{12}$$

the real polarization vectors being given, e.g., in [6].

It is very easy to verify that the canonical light front form algebra entails

$$\left[a_\alpha(k_\perp, k_-), a_\beta^\dagger(p_\perp, p_-) \right] = \delta_{\alpha\beta}\delta^{(2)}(k_\perp - p_\perp)\delta(k_- - p_-),$$

$$\left[f(k_\perp, k_-), g^\dagger(p_\perp, p_-) \right] = \delta^{(2)}(k_\perp - p_\perp)\delta(k_- - p_-), \tag{13}$$

all the other commutators vanishing. Owing to the **LFF** canonical commutation relations (13), it is clear that the theory involves an indefinite metric space of states [4], [9] and the physical Hilbert's subspace $\mathcal{V}_{\text{phys}}$ is defined through $g(k_\perp, k_-)|v\rangle = 0, \forall |v\rangle \in \mathcal{V}_{\text{phys}}$. Now, taking Eq.s (10-13) into account, Eq. (6) precisely yields [6] the standard Mandelstam-Leibbrandt form of the light front form propagator in the anti light cone gauge

$$\tilde{D}_{\mu\nu}^+(k) = \frac{i}{k^2 + i\epsilon} \left\{ -g_{\mu\nu} + \frac{n_\mu^* k_\nu + n_\nu^* k_\mu}{[n^* k]_{\text{ML}}} \right\}.$$

LIGHT-FRONT PURE YANG-MILLS THEORY

Let us start now from the $SU(N)$-YM Lagrange density

$$\mathcal{L}_{\text{YM}} = -\frac{1}{4}\langle F_{\mu\nu}, F^{\mu\nu}\rangle - n^{*\mu}\langle\Lambda, A_\mu\rangle,$$

287

in which we understand gauge potentials as well as non-Abelian field strengths to be $su(N)$-Lie algebra valued fields, $\langle\,,\,\rangle$ being the inner product. In order to quantize the system we shall follow, as it is somewhat customary in the non-Abelian case, the Hamiltonian path-integral quantization [10]. The free (unconstrained) canonical momentum is $\pi^- = F_{+-}$, and we have primary second class constraints $\phi^\alpha \equiv \pi^\alpha - F_{-\alpha} = 0$ and primary first class constraints $\pi^+ = \pi^\Lambda = 0$.

The canonical Hamilton density is

$$\mathcal{H}_{\mathrm{YM}} \equiv \left\langle \pi^-, \partial_+ A_- \right\rangle + \left\langle \pi^\alpha, \partial_+ A_\alpha \right\rangle - \mathcal{L}_{\mathrm{YM}} \tag{14}$$

$$= \frac{1}{2}\left\langle \pi^-, \pi^- \right\rangle + \frac{1}{4}\left\langle F_{\alpha\beta}, F^{\alpha\beta} \right\rangle - \left\langle A_+, D_\alpha \pi^\alpha + D_- \pi^- - \Lambda \right\rangle, \tag{15}$$

with $D_\mu \equiv \mathbf{1}\partial_\mu + ig[A_\mu,\ ]$. Consequently, we derive the secondary constraints $A_+ = 0$, $D_\alpha \pi^\alpha + D_- \pi^- = \Lambda$, and since we have primary second class constraints $\phi^\alpha \equiv \pi^\alpha - D_- A_\alpha + D_\alpha A_- = 0$, satisfying

$$\left\{ \phi^\alpha(x), \phi^\beta(y) \right\}\Big|_{x^+=y^+} = 2\delta^{\alpha\beta} D_-(x-y),$$

with

$$D_-(x-y) \equiv \left\{ \mathbf{1}\frac{\partial}{\partial x^-} + ig[A_-(x),\ ] \right\} \delta^{(3)}(\mathbf{x}-\mathbf{y}),$$

then the Hamiltonian generating functional takes the form

$$\mathcal{Z}[J^\mu] = \mathcal{N}^{-1} \int \mathcal{D}A_\mu \mathcal{D}\Lambda \mathcal{D}\pi^- \mathcal{D}\pi^\perp \, \delta\left(\pi^\perp - F_{-\perp}\right) \det\|D_-\| \tag{16}$$

$$\exp i \int d^4x \left\{ \left\langle \pi^-, \partial_+ A_- \right\rangle + \left\langle \pi^\alpha, \partial_+ A_\alpha \right\rangle - \mathcal{H}_{\mathrm{YM}} + \left\langle A_\mu, J^\mu \right\rangle \right\}. \tag{17}$$

Now the key point: it is well known [11] that within dimensional regularization it turns out that

$$\det\|D_-\|\big|_{\mathrm{dim\ reg}} = \det\|\partial_-\|,$$

and after integration over A_+, π^α, π^- and Λ one gets

$$\mathcal{Z}[J^\perp, J^-] = \mathcal{N}'^{-1}\det\|\partial_-\| \int \mathcal{D}A_- \mathcal{D}A_\perp \, \exp\, i \int d^4x$$

$$\left\{ \frac{1}{2}\langle \partial_+ A_-, \partial_+ A_- \rangle - \frac{1}{4}\left\langle f_{\alpha\beta}, f^{\alpha\beta} \right\rangle + \langle f_{-\alpha}, \partial_+ A_\alpha \rangle \right\}$$

$$\times \exp\, i \int d^4x \left\{ \mathcal{L}_{\mathrm{Int}} + \langle A_\alpha, J^\alpha \rangle + \left\langle A_-, J^- \right\rangle \right\},$$

where it is convenient to separate the Abelian part of the gauge field strengths $f_{\mu\nu} \equiv \partial_\mu A_\nu - \partial_\nu A_\mu$, whereas the interaction Lagrange density

$$\mathcal{L}_{\mathrm{Int}}(A_-, A_\perp) = \frac{i}{2}g\left\langle [A_\alpha, A_\beta], F_{\alpha\beta} \right\rangle - ig\left\langle [A_-, A_\alpha], \partial_+ A_\alpha \right\rangle,$$

288

leads to the conventional Feynman's rules for the non-Abelian three- and four-gluon vertices. It follows therefrom that the perturbation theory generating functional takes the form

$$\mathcal{Z} = \exp\left\{ i \int d^4y \, \mathcal{L}_{\text{Int}} \left\langle \frac{\delta}{i\delta J^-(y)}, \frac{\delta}{i\delta J^\perp(y)} \right\rangle \right\} \mathcal{Z}_0[J^-, J^\perp],$$

in which the free gaussian Abelian generating functional reads

$$\mathcal{Z}_0[J^\perp, J^-] = \mathcal{N}_0^{-1} \int \mathcal{D}A_- \mathcal{D}A_\perp \mathcal{D}\pi^- \exp i \int d^4x$$

$$\left\{ -\frac{1}{2} \langle \pi^-, \pi^- \rangle - \frac{1}{4} \langle f_{\alpha\beta}, f^{\alpha\beta} \rangle + \langle f_{-\alpha}, \partial_+ A_\alpha \rangle \right\}$$

$$\times \exp i \int d^4x \left\{ \langle \pi^-, \partial_+ A_- \rangle + \langle A_\alpha, J^\alpha \rangle + (A_-, J^-) \right\}. \tag{18}$$

We have now to show that the above expression (18) for the free generating functional actually gives rise to the **ML** form of the free gluon propagator. To this aim, let us perform the change of variables of Eq. (8) in the functional integral, the corresponding Jacobian being $J = \|\partial_\perp^2\|$, together with the sources redefinition

$$j^\alpha = J^\alpha + 2\partial_\alpha \partial_- \partial_\perp^{-2} * J^-, \quad \eta = \partial_\alpha J^\alpha + \partial_- J^-.$$

Then the free generating functional exactly becomes

$$\mathcal{Z}_0[\eta, j^\perp, J^-] = \mathcal{N}_0^{-1} \text{det} \|\partial_\perp^2\| \int \mathcal{D}T_\perp \mathcal{D}\varphi \mathcal{D}\lambda \exp i \int d^4x$$

$$\left\{ \langle \partial_+ T_\alpha, \partial_- T_\alpha \rangle - \frac{1}{2} \langle \partial_\alpha T_\beta, \partial_\alpha T_\beta \rangle + \langle \varphi, \partial_+ \partial_\perp^2 \lambda \rangle \right\}$$

$$\times \exp i \int d^4x \left\{ \langle T_\alpha, j^\alpha \rangle + \langle \varphi, \eta \rangle - \langle \lambda, J^- \rangle \right\},$$

and after choosing standard causal asymptotic conditions at $x^+ \to \pm\infty$ for all the integration field variables [12], it is not difficult to prove that

$$\mathcal{Z}_0[J^\mu] = \exp \frac{1}{2} \int d^4x \int d^4y \, \langle J^\mu(x), D^+_{\mu\nu}(x-y) J^\nu(y) \rangle,$$

where

$$D^+_{\mu\nu}(x) = i \int \frac{d^4k}{(2\pi)^4} \frac{e^{ikx}}{k^2 + i\epsilon} \left\{ -g_{\mu\nu} + \frac{n^*_\mu k_\nu + n^*_\nu k_\mu}{[n^*k]_{\text{ML}}} \right\},$$

In conclusion, we can say that the light front form perturbative approach in the light cone gauge $A_- = 0$ is still unclear, whereas the **LFF** perturbative approach in the anti light cone gauge $A_+ = 0$ is fully consistent. In particular, since we have seen that the light front form and the instant form of the generating functional actually coincide - they are nothing but the same formal expression, although written using different coordinates systems and after a suitable rearrangement of the external sources - it immediately follows that the structure of the counterterms is the very same [4]. It has been quite recently noticed that quark field can also be included into the same approach [6], [13]. Therefore, after fifty years since the original Dirac's attempt [1], we have at our disposal a light front form perturbative approach for gauge theories on equal footing as the standard covariant one, what is a highly non-trivial achievement. Once a completely consistent light front form formulation has been reached, we can turn now to attack non-perturbative **LFF** open issues such as, e.g., the light front vacuum structure of gauge theories and the discretized light front quantization.

REFERENCES

1. Dirac, P. A. M., *Rev. Mod. Phys.* **21**, 392 (1949).
2. Kogut, J. B., Soper, D. E., *Phys. Rev. D* **1**, 2901 (1970); Röhrlich, F., *Acta Phys. Austriaca*, Suppl. **8**, 2777 (1971); Neville, R. A., Röhrlich, F., *Phys. Rev. D* **3**, 1692 (1971); Cornwall, J. M., Jackiw, R., *ibid.* **4**, 367 (1971); Tomboulis, E., *ibid.* **8**, 2736 (1973); Ten Eyck, J. H., Röhrlich, F., *ibid.* **9**, 436 (1974).
3. Capper, D. M., Dulvich, J. J., Litvak, M. J., *Nucl. Phys. B* **241**, 463 (1984).
4. Bassetto, A., Dalbosco, M., Lazzizzera I., Soldati, R., *Phys. Rev. D* **31**, 2012 (1985); Bassetto, A., Dalbosco, M., Soldati R., *ibid.* **36**, 3138 (1987); Becchi, C., *Physical and Nonstandard Gauges*, Berlin: Springer-Verlag, 1990, ch. III, pp. 177-184.
5. Mandelstam, S., *Nucl. Phys. B* **213**, 149 (1983); Leibbrandt, G., *Phys. Rev. D* **29**, 1699 (1984).
6. Morara, M., Soldati, R., *Phys. Rev. D* **58**, 105011 (1998)
7. McCartor, G., Robertson, D. G., *Z. Phys. C* **62**, 349 (1994); *ibid.* **68**, 345 (1995).
8. Dirac, P. A. M., *Lectures on Quantum Mechanics*, Belfer Graduate School of Science, Yeshiva University, New York: Academic Press (1964).
9. Nakawaki, Y., McCartor, G., `hep-th/9903018`.
10. Weinberg, S., *The Quantum Theory of Fields, Volume II Modern Applications*, Cambridge: Cambridge University Press, 1998, ch. 15, pp. 14-27.
11. Bassetto, A., Nardelli, G., Soldati, R., *Yang-Mills Theories in Algebraic Noncovariant Gauges*, Singapore: World Scientific, ch. 3, pp. 64-67.
12. Faddeev, L. D., Slavnov, A. A., *Gauge Fields: Introduction to Quantum Theory*, Reading: Benjamin/Cummings, ch. 2, pp. 35-59.
13. Przeszowski, J. A., `hep-th/9906037`.

Variational Calculation
of the Effective Action

Takanori Sugihara

Department of Physics, Nagoya University
Nagoya 464-8602, Japan

Abstract. An indication of spontaneous symmetry breaking is found in the two-dimensional $\lambda\phi^4$ model, where attention is paid to the functional form of an effective action. An effective energy, which is an effective action for a static field, is obtained as a functional of the classical field from the ground state of the Hamiltonian $H[J]$ interacting with a constant external field. The energy and wavefunction of the ground state are calculated in terms of DLCQ (Discretized Light-Cone Quantization) under antiperiodic boundary conditions. A field configuration that is physically meaningful is found as a solution of the quantum mechanical Euler-Lagrange equation in the $J \to 0$ limit. It is shown that there exists a nonzero field configuration in the broken phase of Z_2 symmetry because of a boundary effect.

INTRODUCTION

Light-front field theory has been considered to be useful technique for bound state problems such as hadron physics due to vacuum triviality. However, this advantage is also a drawback to the completeness of field theory since one cannot extract information of spontaneous symmetry breaking from the trivial light-front vacuum. It is the common understanding that the entire effect of spontaneous symmetry breaking in the light-front field theory comes from the constrained zero mode, which is a dependent variable and then should be represented as a superposition of other oscillation modes [1]. It has been numerically confirmed, with a rough approximation, that the zero mode gives rise to a non-zero vacuum expectation value [2]. However, it does not work in practice solving the constraint and calculating the vacuum energy including the zero-mode effect, since the constraint equation is highly complicated and it is difficult to find some reasonable technique to solve it accurately. It is worthwhile to discuss the problem avoiding the zero-mode problem.

In this work, we look for an indication of spontaneous symmetry breaking in the two-dimensional $\lambda\phi^4$ model by paying attention to a functional form of the effective energy, where the system is defined and solved using DLCQ [1,3]. The

CP494, *New Directions in Quantum Chromodynamics,* edited by C.-R. Ji and D.-P. Min
© 1999 American Institute of Physics 1-56396-908-4/99/$15.00

effective energy is obtained as a functional of the classical field (expectation value of the field) and space-dependent non-uniform solution to the quantum mechanical Euler-Lagrange equation is found [4].

EFFECTIVE ENERGY

Let us consider a Hamiltonian that interacts with an external field $J(\mathbf{x})$

$$H[J] = H - \int d^{n-1}x J(\mathbf{x})\phi(\mathbf{x}), \tag{1}$$

where $\mathbf{x}$ stands for spatial coordinate and $\phi(\mathbf{x})$ is a field operator in the Schrödinger picture. The effective energy is defined as a Legendre transform of the ground-state energy $w[J]$ of the Hamiltonian (1) [5,6],

$$\mathcal{E}[\varphi] \equiv \langle 0_J|H|0_J\rangle = w[J] + \int d^{n-1}x J(\mathbf{x})\varphi(\mathbf{x}), \tag{2}$$

where $H[J]|0_J\rangle = w[J]|0_J\rangle$ and $\varphi(\mathbf{x}) = \langle 0_J|\phi(\mathbf{x})|0_J\rangle$. The ground state $|0_J\rangle$ is normalized as $\langle 0_J|0_J\rangle = 1$. An actual expectation value $\varphi(\mathbf{x})$ should be given as a solution of the following generalized Euler-Lagrange equation in the vanishing J limit,

$$\frac{\delta\mathcal{E}[\varphi]}{\delta\varphi(\mathbf{x})} = J(\mathbf{x}), \quad J(\mathbf{x}) \to 0. \tag{3}$$

In order to obtain the solution $\varphi(\mathbf{x})$, we have to make three steps: i) solve the eigenvalue problem $H[J]|\Psi\rangle = E|\Psi\rangle$, ii) evaluate the energy $\mathcal{E}[\varphi]$, and iii) find the stationary point of $\mathcal{E}[\varphi]$. It is difficult to clear the first step if the field is quantized in the ordinary equal-time coordinate, because vacuum fluctuations dominate and higher Fock states seem to be needed to represent the ground state of the Hamiltonian. In order to solve the eigenvalue problem we use DLCQ with antiperiodic boundary conditions.

NUMERICAL RESULT

In Fig. 1, the effective energy $\mathcal{E}[\varphi]$ for the two-dimensional $\lambda\phi^4$ model is plotted as a function of $\varphi_{\max}$, which is a maximum value of the classical field $\varphi(x^-)$, instead of as a functional of $\varphi(x^-)$. In the symmetric phase ($\lambda = 0.1\mu^2$), we can see that the energy $\mathcal{E}[\varphi]$ has a minimum at the origin, where the state is composed only of a zero-momentum state $|0\rangle$ (Fock vacuum). The ground state of the Hamiltonian $H[J]$ goes to $|0\rangle$ and gives zero energy $\mathcal{E}[\varphi] = 0$ in the $J \to 0$ limit. In the broken phase ($\lambda = 50\mu^2$), however, a situation is completely different from the symmetric one. The effective energy has a flat bottom. There are infinite number of states that have the same energy. The states on the flat region have wavefunction where

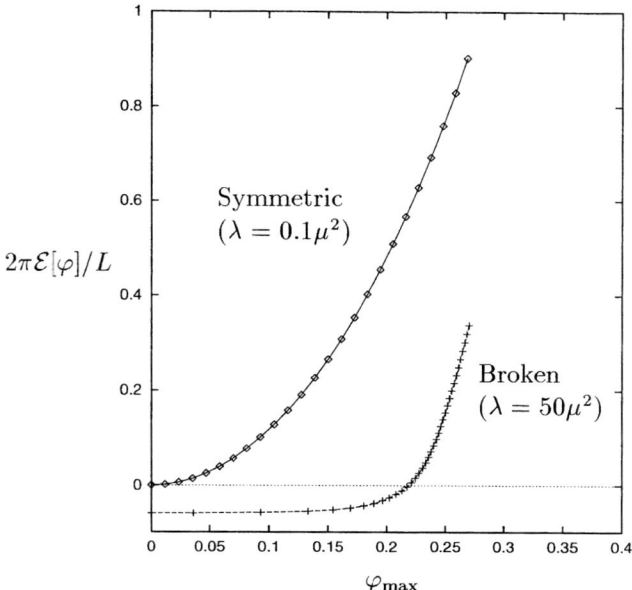

FIGURE 1. The effective energy $2\pi\mathcal{E}[\varphi]/L$ is plotted as a function of $\varphi_{max} \equiv \max\{\varphi(x^-)\}$ instead of as a functional of $\varphi(x^-)$, where Fock space is truncated with $K_{cut} = 31/2$ and $N_{TD} = 7$. Diamonds and pluses correspond to symmetric ($\lambda = 0.1\mu^2$) and broken ($\lambda = 50\mu^2$) phases, respectively. In the symmetric phase, a physically meaningful configuration is at the origin, where the state is composed only of the Fock vacuum $|0\rangle$. In the broken phase, there seems to exist a nonzero field configuration as a solution of the extended Euler-Lagrange equation in the $J \to 0$ limit, since the bottom of the effective energy is flat. This says the existence of an infinite number of configurations $\varphi(x^-)$ which are energetically equivalent.

small K components are dominant and do not go to trivial Fock vacuum in the $J \to 0$ limit . In the thermodynamic limit, the small K components accumulate to zero momentum, $p^+ \sim 0$. This fact supports the existence of an infinite number of configurations $\varphi(x^-)$, which are energetically equivalent, and a nonzero field configuration as a kink solution of the Euler-Lagrange equation (3) in the broken phase.

SUMMARY AND DISCUSSIONS

In the symmetric phase, the effective energy has a minimum at the origin, which is composed only of the trivial Fock vacuum. In the broken phase, Z_2 symmetry spontaneously breaks, which has been confirmed by seeing that the bottom of the effective energy is flat. In the vanishing J limit, a nonzero expectation value of the classical field seems to remain. A field configuration that has a twist can be a solution to the quantum mechanically extended Euler-Lagrange equation.

We have discussed spontaneous symmetry breaking by searching for a state that minimizes the effective energy. We have succeeded in finding an indication of spontaneous symmetry breaking, which is just contained in the Hamiltonian with antiperiodic boundary conditions. This suggests that the Hamiltonian knows the existence of symmetry breaking in spite of an absence of the constrained light-front zero mode. We have considered how to extract information about symmetry breaking from the effective energy rather than the effective potential though it is common to use the effective potential for such investigation. Since no sophisticated non-perturbative technique has not been found for solving the zero-mode constraint equation, the effective energy calculation seems to be much suitable for such a consideration at the present time.

REFERENCES

1. T. Maskawa and K. Yamawaki, Prog. Theo. Phys. **56**, 270(1976).
2. C. Bender, S. Pinsky and B. Sande, Phys. Rev. D **48**, 816(1993); S. Pinsky and B. Sande, *ibid.* **49**, 2001(1994); S. Pinsky, B. Sande and J. Hiller, *ibid.* **51**, 726(1995).
3. H. C. Pauli, and S. J. Brodsky, Phys. Rev. D **32**, 1993(1985); H. C. Pauli, and S. J. Brodsky, *ibid.* **32**, 2001(1985).
4. T. Sugihara, Phys. Rev. D **57**, 7373(1998).
5. S. Coleman, *Aspects of Symmetry* (Cambridge Univ. Press, 1985).
6. T. Kugo, *Quantum Theory of Gauge Fields I* (Baifukan, 1986), in Japanese.

Physical Role of the Light-Front Zero Mode

Masa-aki Taniguchi

Department of Physics, Nagoya University, Nagoya 464-8602, Japan
E-mail: mass@eken.phys.nagoya-u.ac.jp

Abstract. We study $(1+1)$-dimensional ϕ^4 theory quantized on the light cone. We pay much attention to the light-front zero mode, and find the second-order phase transition in the strong coupling region by solving the zero-mode constraint with a new method. We also obtain the critical coupling constant, $\lambda_c = 24\mu_R^2$, which is consistent with the equal-time calculations, and the critical exponent, $\beta = 1/2$.

The light-front Tamm-Dancoff (LFTD) approximation [1] is a new promising numerical approach to solve the nonperturbative problems. It has been successfully applied to two-dimensional models, [2–4] and there is a possibility to make new progress in the study of the relativistic bound-state problems once it is applied to QCD. [5] It is important to note that this new approach not only reproduces known results correctly, but also brings us new results. [1–5]

The success of the LFTD approximation is based on the vacuum simplicity of the light-front field theory (LFFT). In LFFT the vacuum is almost trivial and therefore the particle picture matches with LFFT very well. However there is a problem in LFFT how we can derive the spontaneous symmetry breaking, such as the chiral symmetry breaking with this simple vacuum. It is widely believed [6] that the light-front zero mode plays an important role for the spontaneous symmetry breaking. In order to treat it properly, it is convenient to put the system into a finite box $(x^- \in [-L, L])$ and impose the periodic boundary condition. [6] We can treat the zero mode separately from the other modes under this condition.

In this article, we study $(1 + 1)$-dimensional ϕ^4 theory to clarify that we can derive the spontaneous symmetry breaking by treating the light-front zero mode properly. There are several papers on this subject. [7,8] They solve the zero-mode constraint by applying the approximations which are based on the truncation of the particle number. However, it is technically difficult to extend their methods to higher orders. We investigate the zero mode more precisely with a new method, which has been never used in the previous studies. We show that there is the second-order phase transition in the strong coupling region and obtain the critical coupling constant and critical exponent in ϕ_{1+1}^4 theory.

CP494, *New Directions in Quantum Chromodynamics*, edited by C.-R. Ji and D.-P. Min
© 1999 American Institute of Physics 1-56396-908-4/99/$15.00

The Lagrangian density of ϕ_{1+1}^4 theory is given by $\mathcal{L} = \partial_+ \phi \partial_- \phi - \mu^2 \phi^2 / 2 - \lambda \phi^4 / 4!$. We put the system into a finite light-like box $(x^- \in [-L, L])$ and impose the periodic boundary condition, $\phi(x^- = L) = \phi(x^- = -L)$, to separate the zero mode from the non-zero mode,

$$\phi(x) = \phi_0(x^+) + \tilde{\phi}(x),$$
$$= \phi_0(x^+) + \sum_{n>0} [\phi_n(x^+) e^{-ip_n^+ x^-} + \phi_{-n}(x^+) e^{ip_n^+ x^-}], \quad (p_n^+ \equiv \frac{n\pi}{L}) \qquad (1)$$

where the zero mode is defined as $\phi_0(x^+) \equiv (1/2L) \int_{-L}^{L} dx^- \phi(x)$, and ϕ_n's are the Fourier components of the non-zero mode $\tilde{\phi}$.

This system has constraints because the Lagrangian density is the first order in the light-cone time derivative. Non-trivial one is, so called, the zero-mode constraint,

$$\Phi \equiv \mu^2 \phi_0 + \frac{\lambda}{3!} \phi_0{}^3 + \frac{\lambda}{3!} ([\tilde{\phi}^2]_0 \phi_0 + [\tilde{\phi}\phi_0\tilde{\phi}]_0 + \phi_0[\tilde{\phi}^2]_0) + \frac{\lambda}{3!} [\tilde{\phi}^3]_0 \approx 0, \qquad (2)$$

where $[O]_0$ is the zero mode of an (composite) operator O and defined by $[O]_0 \equiv \frac{1}{2L} \int_{-L}^{L} dx^- O(x)$. The composite operators in Eq. (2) have been made to symmetric ordering, since it is natural in quantizing the theory.

The equal-x^+ commutation relations are as follows:

$$[\phi_m, \phi_n] = \frac{1}{4\pi m} \delta_{m+n,0}, \quad [\phi_0, \phi_0] = 0, \qquad (3)$$

$$x_n \equiv [\phi_0, \phi_n] = \frac{\lambda}{8\pi n} (\phi_0 \phi_n + \phi_n \phi_0 + \sum_{m \neq 0,n} \phi_m \phi_{n-m}) \alpha_0^{-1}, \qquad (4)$$

where $\alpha_0 \equiv \mu^2 + \lambda/2(\phi_0{}^2 + [\tilde{\phi}^2]_0)$. Eq. (3) are the same as the canonical ones but Eq. (4) is not. Although the right-hand side of Eq. (4) is q-number, we have just written it down formally here, neglecting the ordering of the operators. In this work, we will show that the operator ordering in Eq. (2) and (4) is essential to lead to the spontaneous symmetry breakdown in this system.

The zero mode is not a dynamical degree of freedom in this system. It is clear that the zero mode should be written as a function of the non-zero modes according to Eq. (2). The Fock space should be constructed only with the non-zero modes, and therefore the vacuum is trivial, which is defined by $\phi_n|0\rangle = 0$ $(n > 0)$, where the ϕ_n $(n > 0)$ is the annihilation operator as usual. The creation operator is defined as $\phi_n^\dagger \equiv \phi_{-n}$ $(n > 0)$ which creates a particle with the momentum $n\pi/L$. The order parameter of Z_2-symmetry is $\langle \phi \rangle = \langle \phi_0 \rangle = v$ in this model. From now on, we denote the zero mode as $\phi_0 \equiv v + w$, where w is the normal-ordered operator part of the zero mode.

We are now going to obtain the vacuum expectation value v by solving the zero-mode constraint (2). It is clear that the vacuum expectation value of Eq. (2) is written as follows:

$$\mu^2 v + \frac{\lambda}{3!} v^3 + \frac{\lambda}{2!} \langle [\tilde{\phi}^2]_0 \rangle v + \frac{\lambda}{3!} \sum_m \langle \phi_{-m} w \phi_m \rangle = 0 \tag{5}$$

where a relation, $\langle \phi_0^3 \rangle = \langle \phi_0 \rangle^3 = v^3$, has been used. It holds because there is no independent state with zero momentum except the vacuum state in this system. We also used $\langle [\tilde{\phi}^3]_0 \rangle = 0$ according to the particle number conservation. Here, we define the renormalized mass as $\mu_R^2 \equiv \mu^2 + \frac{\lambda}{2!} \langle [\tilde{\phi}^2]_0 \rangle$, because the second term corresponds to the one loop self-energy diagram in the two point function which is the only divergent diagram in ϕ_{1+1}^4 theory.

The last term in Eq. (5) can be evaluated easily by using the commutation relation (4). However, Eq. (4) is ambiguous as to the operator ordering. It was calculated through Dirac bracket, and does not contain the information of the operator ordering. Therefore we employ the other expression for x_n,

$$\mu^2 x_n + \frac{\lambda}{3!} (\phi_0^2 x_n + \phi_0 x_n \phi_0 + x_n \phi_0^2) + \frac{\lambda}{3!} \sum_m (\phi_{-m} \phi_m x_n + \phi_{-m} x_n \phi_m + x_n \phi_{-m} \phi_m)$$

$$= \frac{\lambda}{8\pi n} (\phi_0 \phi_n + \phi_n \phi_0 + \sum_m \phi_m \phi_{n-m}), \tag{6}$$

which is derived from the commutation relation between the zero-mode constraint and the non-zero mode, $[\Phi, \phi_n] = 0$. In this expression, the operator ordering is obvious. Moreover, we can derive the natural relation $[P^+, \phi_0] = 0$ only when using Eq. (6). This equation is equivalent to Eq. (4) if we neglect the operator ordering. We can obtain x_n by using the semi-classical approximation as follows:

$$[\phi_m, \phi_n] = \frac{\hbar}{4\pi m} \delta_{m+n,0}, \quad [\phi_0, \phi_n] \equiv x_n = \hbar x_n^{(1)} + \hbar^2 x_n^{(2)} + \cdots. \tag{7}$$

Combining Eq. (6) and (7), we obtain

$$\langle \phi_{-n} x_n^{(1)} \rangle = -\frac{\lambda v \hbar}{16\pi^2 n^2 (\mu_R^2 + \lambda v^2/2)} \tag{8}$$

in the lowest order calculation. Using this equation to Eq. (5) we obtain the equation for v,

$$(\lambda^2 v^4 + 8\lambda \mu_R^2 v^2 + 12\mu_R^4 - \frac{\lambda^2}{48}) v = 0, \tag{9}$$

and the solutions are $v = 0$ or $v^2 = -4\frac{\mu_R^2}{\lambda} \pm \sqrt{4\frac{\mu_R^4}{\lambda^2} + \frac{1}{48}}$. For the small coupling constant, $\lambda < \lambda_c = 24\mu_R^2$, we find that the only solution for Eq. (9) is $v = 0$, and therefore this phase is the symmetric phase. When the coupling constant is strong enough, $\lambda > \lambda_c$, there are two solutions other than $v = 0$. These solutions behave as $v \propto \pm(\lambda - \lambda_c)^\beta$ and give a same energy smaller than that of $v = 0$. Therefore, this phase is called broken and this phase transition is the second-order, because the order parameter, $v = \langle \phi \rangle$, is continuous to $v = 0$ at the critical coupling constant λ_c. We also obtain the critical exponent β and its value is $1/2$, which is known as one in the mean field approximation.

It is well-established that there exists the second-order phase transition [9] between weak and strong coupling regions in ϕ^4_{1+1} theory, and the critical coupling is given as $22\mu_R^2 < \lambda_c < 55\mu_R^2$ in the usual equal-time calculations. [10] Therefore our result is consistent with the equal-time one. However, our critical exponent β is the same as the mean field value. The correct value is known as $\beta = 1/8$. [11] We hope to obtain the correct critical exponent in the higher order calculations.

In this work, we study $(1+1)$-dimensional ϕ^4 theory quantized on the light cone. We pay much attention to the light-front zero mode, and find the second-order phase transition in the strong coupling region by solving the zero-mode constraint with a new method which is based on the semi-classical approximation. We also obtain the critical coupling constant, $\lambda_c = 24\mu_R^2$, which is consistent with the equal-time calculations, and the critical exponent, $\beta = 1/2$, which corresponds to the mean field value. We hope our method is also applicable to higher dimensional models and models with continuous symmetry.

The author is grateful for T. Sugihara, M. Yahiro and K. Yamawaki for fruitful discussions. I also thank K. Harada for comments. This work has been partially supported by a Grant-in-Aid for Scientific Research from the Ministry of Education, Science and Culture of Japan (No.90001951).

REFERENCES

1. R. J. Perry, A. Harindranath, and K. G. Wilson, Phys. Rev. Lett. **65**, 2959 (1990); S. J. Brodsky, G. McCartor, H. C. Pauli and S. S. Pinsky, Particle World **3**, 109 (1993); K. G. Wilson, T. S. Walhout, A. Harindranath, W. M. Zhang, R. J. Perry, and S .D. Glazek, Phys. Rev **D49**, 6720 (1994).
2. H. Bergknoff, Nucl.Phys. **B122**, 215 (1977); Y. Mo and R. J. Perry, J. Comp. Phys. **108**, 159 (1993); K. Harada, M. Sugihara, M. Taniguchi, M. Yahiro, Phys. Rev. **D49**, 4226 (1994).
3. T. Sugihara, M.Matsuzaki, and M. Yahiro, Phys. Rev. **D50**, 5274 (1994).
4. K. Harada, A. Okazaki, and M. Taniguchi, Phys. Rev. **D55**, 4910 (1997).
5. M. Brisudova, R.J. Perry, K.G. Wilson, Phys.Rev.Lett.**78**, 1227 (1997).
6. T. Maskawa and K. Yamawaki, Prog. Theo. Phys. **56**, 270 (1976).
7. S. S. Pinsky, and B. van de Sande, Phys. Rev. **D49**, 2001, (1994)
8. T.Heinzl, C.Stern, E.Werner, and B.Zellermann, Z.Phys.**C72**, 353 (1996)
9. S. Chang, Phys. Rev. **D13**, 2778 (1976).
10. M. Funke, U. Kaulfuss, and H. Kummel, Phys. Rev. **D35**, 621 (1987); H. Kroger, R.Girard, and G. Dufour, Phys. Rev. **D35**, 3944 (1987).
11. G.A. Baker Jr., B.G. Nickel, and D.I. Meiron, Phys.Rev.**B17**, 1365, (1978); J.C. Le Guillou, and J. Zinn-Justin, Phys.Rev.**B21**, 3976 (1980).

NEW METHODS IN
NONPERTURBATIVE QCD

Solving the QCD Hamiltonian for bound states

Elena Gubankova, Chueng-Ryong Ji

Department of Physics, North Carolina State University, Raleigh, NC 27696-8202

Abstract. The systematic approach to study bound states in quantum chromodynamics is presented. The method utilizes nonperturbative flow equations in the confining background, that makes possible to perform perturbative renormalization and to bring the QCD Hamiltonian to a block-diagonal form with the number of quasiparticles conserving in each block. The effective block-diagonal Hamiltonian provides constituent description for hadron observables. The effective Hamiltonian of gluodynamics in the Coulomb gauge, renormalized to the second order, is obtained at low energies. The masses for scalar and pseudoscalar glueballs are predicted.

I   INTRODUCTION

One of the most difficult and less understood problems in quantum chromodynamics is the treatment of the bound state systems. There are different sources of difficulties. For example, it is quite a common observation known in the spectroscopy, that the splitting between the vector mesons does not depend on flavor, say

$$m(\rho') - m(\rho) \sim m(\psi') - m(J/\psi) \tag{1}$$

which is true experimentally. This fact can not be explained in terms of the canonical QCD interaction, which is given essentially by the strong coupling constant. Indeed, if expressed in terms of an invariant mass s, Eq.(1) implies that the J/ψ is dual to a much larger interval of s than the ρ because the c-quark is heavy. However, the coupling constant runs as a function of an invariant mass, $\alpha(s)$, and is flavor blind, thus the canonical QCD interaction should be much weaker for the J/ψ than for the ρ. This suggests that something is missing when described only in terms of the perturbation theory. The strong coupling constant alone does not provide for strong interactions being *strong*.

Consider the scaling of Quantum Chromodynamics from high to low energies. In the ultraviolet region (at the bare cutoff scale $\Lambda \to \infty$) the strong interactions are given by canonical QCD, which is conformally invariant, in particular this means scaling invariance (there is no scale in the theory) in the chiral limit. Moreover

CP494, *New Directions in Quantum Chromodynamics*, edited by C.-R. Ji and D.-P. Min
© 1999 American Institute of Physics 1-56396-908-4/99/$15.00

a perturbative treatment is possible due to asymptotic freedom. In asymptotic free theories (QCD) the coupling constant grows at low-energies and gets strong, that stops the asymptotic freedom at some moderate scale $\Lambda_0 \sim \Lambda_{QCD}$. This scale appears in the theory when the perturbative renormalization of the coupling constant is performed, and the Λ_0 is the Landau pole in the effective running coupling constant provided the renormalization group invariance. This is called dimensional transmutation, when scaling invariance breaks through the renormalization. Then experiment tells us that this is not the only scale in the theory. There are at least two more characteristic scales in the hadron physics. The first scale is the mass gap of the hadron bound state, given by the square root of the string tension. At this scale the nonperturbative phenomenon of confinement takes place and the bound states of quarks and gluons form. The second scale is the scale of chiral symmetry breaking, which is set by the scalar π-meson. These scales are displayed as $m(\pi) \ll \sqrt{\sigma} \ll \Lambda_{QCD}$.

To summarize: Knowing perturbation theory alone is not enough to describe bound states in hadronic physics.

Other difficulties in the QCD bound state problem are of a more general kind - relativistic nature of quantum field theory. First, the number of particles in any state is not fixed because of particle creation and annihilation in vacuum. Second, there are states with (infinitely) large energies. An attempt to treat both problems was done in [3] by using the method of flow equations. The idea is to find a unitary transformation that transforms the Hamiltonian operator to a block-diagonal form, where each block conserves the number of particles. The Hamiltonian matrix can be represented in the particle number space as

$$H = \begin{pmatrix} PHP & PHQ \\ QHP & QHQ \end{pmatrix} \tag{2}$$

where P and Q are projection operators on the subspaces with different particle number content. The flow equations [1] bring the Hamiltonian matrix Eq.(2) to the form

$$H_{\text{eff}} = \begin{pmatrix} PH_{\text{eff}}P & 0 \\ 0 & QH_{\text{eff}}Q \end{pmatrix} \tag{3}$$

where the two blocks of the effective Hamiltonian decouple from each other. It may be simpler then to solve for bound states within one block, say $PH_{eff}P$, than to diagonalize the complete Hamiltonian, Eq.(2), of the original problem. Since, generally, the number of particles in P and Q spaces is arbitrary, one can reduce in this way the bound state problem with many particles to a few-body problem.

It turns out that the second question on possible presence of the ultraviolet divergences is solved also by the method of flow equations. Flow equations perform a set of (infinitesimal small) unitary transformations, where the flow parameter, which controls the transformation, has the dimension of the inverse of the energy square,

$l \sim 1/E^2$. Therefore by using flow equations to block-diagonalize the Hamiltonian one eliminates the particle number changing contributions not in one step but rather continuous for the different energy differences in sequence. This procedure enables one to separate the ultraviolet divergent contributions and to find the counterterms associated with these divergences. This covers the UV-renormalization for Hamiltonians [2].

To summarize: More generally, flow equations perform Hamiltonian renormalization in the "particle number" and in the "energy" spaces in the sense that the effects of the high Fock states and the effects of the large energies, respectively, are encoded in the effective low-energy Hamiltonian, which operates in the space of the few low Fock components.

This program was applied quite successfully to QED to calculate the positronium spectrum [3]. The key is the validity of the perturbation theory in the bare coupling constant for the characteristic energy scale of positronium bound state. Obviously, it is a bad idea to apply naively the same scheme for QCD. It is not possible to find the fixed number representation for the Hamiltonian in the case of *strong* interactions, where one does not have any control over the process of the creation and the annihilation in vacuum of bare quarks and gluons with *small* current masses. In the language of flow equations convergence can not be achieved when calculated in terms of bare parameters.

The way out is suggested by nature itself. One should consider "*confined* QCD". By using flow equations one constructs then the effective QCD Hamiltonian $H_{eff}(q,g)$, where current quarks and gluons acquire masses of the order of $m_{constituent} \sim 1 GeV$ and become constituent degrees of freedom (see below). The value of the constituent mass plays the role of the energy gap between the sectors in H_{eff}. Schematically, the block-diagonal effective Hamiltonian has the form

$$H_{\text{eff}} = \begin{pmatrix} q\bar{q} & & & \\ \hline & q\bar{q}g & & \\ \hline & & q\bar{q}q\bar{q} & \\ & & gg & \end{pmatrix} \qquad (4)$$

where q and g are constituent quarks and gluons, respectively; empty cells denote zero, to the order calculations are done, matrix elements. To this order the different sectors of the effective Hamiltonian, Eq.(4), describe approximately, when going down, the bound states of mesons, hybrids, glueballs. Actually, such a description with the fixed number of constituents is quasiclassical and nonrelativistic, and is known from the constituent quark model. Physically, the picture is the following. The strong confining interaction, acting inside each "diagonal" (particle number conserving) sector, produces heavy gluon (quark) only in the small volume – in the "bag". No free propagating heavy gluons (quarks) are produced. Therefore the "bags" do not interact with each other, decoupling in H_{eff} and approximating the hadron bound states.

The matrix elements of the "off-diagonal" (particle number changing) sectors are governed by the canonical interaction – typically by the Coulomb term of the strength equal to the inverse of Bohr radius or the current quark mass ($m_{current}$ is of the order of several MeV). In the presence of the strong confining interaction in the "diagonal" sectors the mixing between the sectors is strongly suppressed. One can introduce a small parameter, say,

$$\alpha_s \frac{V_{12}}{E_1 - E_2} \sim 1 \cdot \frac{m_{current}}{m_{constituent}} \sim 0.1 - 0.01, \tag{5}$$

where V_{12} is the Coulomb interaction, and $E_1 - E_2$ is the energy difference between the first and the second "diagonal" sectors. Perturbation theory with respect to the small parameter, Eq.(5), holds between the sectors (but not inside the sector where the confining interaction is strong). By applying flow equations to block-diagonalize the Hamiltonian one gets to leading order a closed chain of decoupled equations, which can be solved analytically. The whole is true provided there is a strong confining interaction in the "diagonal" sectors.

To summarize: In the theory of strong interactions confinement is important to provide the bound states. In the present approach confinement makes it possible to bring, by flow equations, the QCD Hamiltonian to a block-diagonal form with a fixed number of quasiparticles in each sector. The elementary degrees of freedom (quasiparticles) become constituent quarks and gluons, which acquire masses of order $1GeV$. The block-diagonal effective Hamiltonian approximately describes then the different hadronic bound states.

The main idea of the approach is to find the representation for QCD Hamiltonian with the fixed number of quasiparticles, where the sectors with different particle number content decouple from each other. There can be some special cases when one should take into account the mixing between the sectors. In other words the physical state is not given by the pure component of the composite system. The mixing between the high excited state from the previous sector and the ground state from the next sector of the effective Hamiltonian may be possible (for example, the mixing between some excited meson and the low lying hybrid state). In systems with light quarks the influence of coupled channels can be essential. In the strongly coupled effective meson models one includes the effects from the coupled channels directly by mixing the scalar and pseudoscalar channels ($q\bar{q}$ and $q\bar{q}q\bar{q}$). The effect is about 50 percent.

Special consideration is required in the case of the light quarks, where chiral symmetry breaking (CSB) is important. The present approach includes confinement and is like the "bag model" or the "constituent quark model", but it does not include CSB. By implementing CSB in this picture, the scalar π-meson can be viewed as a bound state of the two constituent quarks and simultaneously manifests the Goldstone nature.

Motivation: In order to disentangle the both problems of confinement and CSB we consider the pure gluodynamics (see the next section) [5]. The motivation for this study is to set up a kind of a constituent gluon model, with the confining interaction imposed, to describe glueball bound states.

The specific difficulty of QCD is that the canonical QCD Lagrangian does not manifest explicitly confinement. As far as the mechanism of confinement is concerned one can proceed along several ways. To reveal confinement one uses the suitable formulation of QCD: lattice form, or the special choice of the gauge fixing (for example, maximal Abelian projection). Another option is to study other than QCD theories, but that have the same infrared behavior as QCD and are confined: Super Yang Mills theory, some toy gauge models (for example, Abelian Higgs Model). There may appear some unphysical degrees of freedom in these theories. If one is not interested in the mechanism of confinement, one includes the latter explicitly into QCD. The simplest way is to use the potential model, successfully tested in phenomenology, where the potential between the color charges is given by a sum of Coulomb and confining interactions. This suggests the definite choice of the gauge for the Hamiltonian. We work in the Coulomb gauge, where the Coulomb interaction arises from the gauge fixing procedure. We add then confinement to be able to block-diagonalize to the effective Hamiltonian, which describes the bound states.

The Coulomb gauge is the natural gauge to get the constituent hadron picture. The Coulomb interaction appears there not as a perturbative (propagating in time) one gluon exchange, but rather as a solution of the gauge fixing constrains. Therefore the Coulomb term describes an instantaneous interaction, which is consistent with the nonpropagating massive gluon arising in our approach. Note that the massive gluon mode arises only in the presence of confinement and confinement sets the scale for the gluon mass.

To summarize: The Coulomb gauge has an appealing property of the simple extension of the model to the confining case and is consistent with the constituent picture for hadrons.

II   LOW-ENERGY GLUODYNAMICS IN THE COULOMB GAUGE

As noted above we applied the method to pure gluodynamics [5]. In this section we outline the strategy of this study.

Step 1. *QCD* H_{can}.
The starting point is the canonical QCD Hamiltonian (pure gluodynamics) in the Coulomb gauge $\nabla \cdot \mathbf{A} = 0$: $H_{can}(\mathbf{A}, \mathbf{\Pi})$, where physical degrees of freedom are the transverse gauge fields $\mathbf{A}$ and their conjugate transverse momenta $\mathbf{\Pi}$.

Step 2. $H_{can} = H(g = 0) + O(g) + O(g^2)$.
We expand the canonical QCD Hamiltonian in the Coulomb gauge perturbatively to the second order in the bare coupling constant. Then to the leading order

the Faddeev-Popov determinant can be approximated by unity, that reduces the instantaneous term, arising from the gauge fixing, to the pure Coulomb interaction.

Step 3. *Current (perturbative) basis and the trivial vacuum $|0\rangle$.*
We choose the trivial (perturbative) vacuum $|0\rangle$ and construct the perturbative basis of free (*current*) particles: $a^\dagger(\mathbf{k})|0\rangle$ creates one (perturbative) gluon with zero mass, i.e. the gluon energy $\omega_\mathbf{k} = |\mathbf{k}|$, etc., and the vacuum is defined as $a|0\rangle = 0$.

We express the canonical QCD Hamiltonian (step 2) in this perturbative Fock space, and normal order the result with respect to the trivial vacuum state $|0\rangle$. Denote the normal ordered canonical QCD Hamiltonian as $:H_{can}:$.

Step 4. *Regularization and perturbative renormalization (scheme).*
The normal ordered Hamiltonian $:H_{can}:$ contains ultraviolet (UV) divergent terms (UV-divergent loop integrals). We regulate UV-divergences by the cutoff function $f(q, \Lambda)$ (the explicit form of the cutoff function is specified further). This is the first time when we have introduced an energy scale in the theory – the bare cutoff $\Lambda \to \infty$. To remove the cutoff sensitivity we renormalize the Hamiltonian by adding the counterterms associated with these divergences. Schematically, the renormalized Hamiltonian is written as $H_{ren}(\Lambda) = H_{can} + \delta X_{CT}(\Lambda)$, where $\delta X_{CT}(\Lambda)$ is a set of (unknown) counterterm operators, which we define further [1].

Step 5. *Flow equations perturbatively.*
To find the explicit form of the counterterms and to scale down the Hamiltonian we run flow equations perturbatively. Also the form of the cutoff (regulating) function is specified by flow equations. Generally, flow equations define the prescription of regularization and make possible to perform the perturbative renormalization [2]. Technically, since the Hamiltonian depends on the cutoff scale through the flow parameter, one finds in the given order of perturbation theory (PT) the divergent part of the difference between the Hamiltonian operators given at the two scales, say $(H(\Lambda_2) - H(\Lambda_1))$ with $\Lambda_{QCD} \ll \Lambda_2 \le \Lambda_1 \le \Lambda$. One absorbs then these divergences in the counterterms – local operators with the symmetries of the canonical Hamiltonian, to provide the renormalization group invariance (called in the context of Hamiltonian renormalization "coupling coherence" [2]). This completes the procedure of renormalization, performed by flow equations, to this order. One can proceed in this way order by order in PT to find (all) the counterterms systematically. Note that it is enough to find the gradient of the Hamiltonian in the energy space to define the counterterms. Renormalization group invariance (RGI) insures, that the renormalized Hamiltonian preserves the form of the (original) canonical Hamiltonian, but only the coupling constants and the mass operators (that are usually classified as relevant and marginal operators in renormalization group sense) start to run with the cutoff scale. (We do not consider here, at least

---

[1] In the given (perturbative) basis this equation reads $:H_{ren}(\Lambda): = :H_{can}: + :\delta X_{CT}(\Lambda):$, where ":" stands for normal ordering in the (perturbative) vacuum

[2] Flow equations perform a set of unitary transformations to block-diagonalize the Hamiltonian $H(l, l_0) = U^{-1}(l, l_0)H(l_0)U(l, l_0)$, where l is the flow parameter with the connection to the energy scale $l = 1/\lambda^2$, l_0 is the initial value corresponding to the bare cutoff Λ introduced by the regularization before (section 4).

in the few lowest orders of PT, possible irrelevant operators, that may cause new type of divergences than are carried by coupling constants and masses).

Using flow equations, we run the effective Hamiltonian downwards from the bare cutoff Λ to some intermediate scale $\Lambda_0 \sim \Lambda_{QCD}$, where perturbation theory breaks down. Due to the RGI the "physical gluon" stays massless through this perturbative scaling. We can not proceed with flow equations perturbatively further. The result of this stage is the renormalized (to the second order of PT [5]), effective Hamiltonian, defined at some compositness scale Λ_0: $:H_{ren}(\Lambda, \Lambda_0):$ with $\Lambda_0 \sim \Lambda_{QCD}$ and bare cutoff $\Lambda \to \infty$, and semicolon means normal-ordering in the trivial vacuum $|0\rangle$. Though the renormalized Hamiltonian is obtained in the perturbative frame, it can be represented (regardless of the Fock basis) in terms of the fields $\mathbf{A}$ and $\mathbf{\Pi}$ [5]. (It is a consequence of the RGI). We denote the resulting renormalized Hamiltonian at the scale Λ_0 as $H_{ren}(\Lambda, \Lambda_0)$.

Step 6. *Confinement.*

We introduce confinement as a linear rising potential, that enables to run flow equations "nonperturbatively" (see introduction) until complete diagonalization of the Hamiltonian. In the renormalization group sense, this "spoils" the theory: there arises the massive gluon mode. But the presence of confinement is necessary to find the representation with a fixed number of quasiparticles (constituent massive gluons) for the effective Hamiltonian, which provides the constituent picture (see introduction). Confinement (string tension) sets the (hadron) scale for the gluon mass.

The instantaneous interaction contains two pieces, the sum of the Coulomb and confining potentials. Denote the renormalized effective Hamiltonian with confinement embedded as $H_{eff}(\Lambda, \Lambda_0)$.

Step 7. *Constituent (nonperturbative) basis and the QCD vacuum $|\Omega\rangle$.*

As far as confinement is introduced the trivial vacuum $|0\rangle$ and the perturbative basis of free (current) particles, $\omega_{\mathbf{k}} = |\mathbf{k}|$, define no longer the minimum ground state. Therefore, we introduce the (arbitrary) basis, where the gluon energy $\omega_{\mathbf{k}}$ is kept unknown, and is defined further variationally. Correspondingly, the (nontrivial) QCD vacuum $|\Omega\rangle$ is defined as $\alpha|\Omega\rangle = 0$, and the Fock space of *constituent* particles is given: $\alpha^\dagger|\Omega\rangle$ creates the quasiparticle with the energy $\omega_{\mathbf{k}}$, etc. [3]. The renormalized effective Hamiltonian $H_{eff}(\Lambda, \Lambda_0)$ at the scale Λ_0 (step 6), written through the physical fields $\mathbf{A}$ and $\mathbf{\Pi}$ and having confinement, is decomposed in the trial (constituent) basis and normal-ordered with respect to QCD vacuum $|\Omega\rangle$. The unknown gluon energy is variational parameter in the calculations. We combine the terms in the effective Hamiltonian in each particle number sector according to the

---

[3] The change of basis from the (perturbative) current, $\omega_{\mathbf{k}} = |\mathbf{k}|$, to the (nonperturbative) constituent, with some $\omega_{\mathbf{k}}$, can be written as Bogoliubov-Valatin (BV) transformation from the "old", $a, a^\dagger$, to the "new", $\alpha, \alpha^\dagger$, operators: $a_{\mathbf{k}} = \mathrm{ch}\phi_{\mathbf{k}}\alpha_{\mathbf{k}} + \mathrm{sh}\phi_{\mathbf{k}}\alpha^\dagger_{-\mathbf{k}}$ with BV angle $\phi_{\mathbf{k}}$ given by $\mathrm{ch}\phi_{\mathbf{k}} = 1/2(\sqrt{k/\omega_{\mathbf{k}}} + \sqrt{\omega_{\mathbf{k}}/k})$. The connection between the "old", $|0\rangle$, and the "new", $|\Omega\rangle$, vacuum states is given $|\Omega\rangle = \exp\left(\frac{1}{2}\sum_k \mathrm{th}\phi_{\mathbf{k}} a^\dagger_{\mathbf{k}} a^\dagger_{-\mathbf{k}}\right)|0\rangle$. It was used in the work [4] to transform the QCD Hamiltonian into the constituent basis.

power of coupling constant $O(g^n)$ $(n = 0, 1, 2)$ [4]. In the absence of confinement the effective Hamiltonian preserves the form of canonical Hamiltonian due to RGI, with the proper change $|\mathbf{k}| \rightarrow \omega_{\mathbf{k}}$. In the presence of confinement the canonical form is violated by the second order terms in the effective Hamiltonian, which contribute higher orders $O(g^3)$, etc. in flow equations.

We aim to find the effective Hamiltonian after the scaling downwards from Λ_0 to a hadron scale, say $\sqrt{\sigma}$. Since the effective Hamiltonian preserves the canonical form at least to the second order, the "perturbative" terms obtained by flow equations in step 5 match the "nonperturbative" terms arising when the flow equations are applied to $H_{eff}(\Lambda, \Lambda_0)$.

We denote the effective Hamiltonian in constituent basis as $::H_{eff}(\Lambda, \Lambda_0)::$, where $"::"$ stands for normal-ordering in the QCD vacuum.

Step 8. *Flow equations in the confining background.*

We run flow equations in the confining background to block-diagonalize the effective Hamiltonian $::H_{eff}(\Lambda, \Lambda_0)::$ in the nonperturbative basis and to find consistently all the terms to the second order. Free Hamiltonian and confining interaction are included in "diagonal" sector, the triple-gluon vertex forms "nondiagonal" sectors, that should be eliminated. We bring the Hamiltonian $::H_{eff}(\Lambda, \Lambda_0)::$ to a block-diagonal form, where diagonal blocks decouple from each other including the second order. The leading UV-behavior of the arising to the second order terms is cancelled by the mass counterterm. Generally, this approach allows to include perturbative QCD corrections into nonperturbative calculations of many-body techniques. The resulting block-diagonal effective, renormalized Hamiltonian $::H_{eff}(\Lambda, \sqrt{\sigma})::$ is given at hadronic scale. For simplicity we denote it as H_{eff}.

Step 9. *Gap equation (variational calculations).*

The requirement of block-diagonal form does not fix the effective Hamiltonian completely. The remaining freedom to unitary transform inside of each block is fixed by minimizing the ground state (the vacuum expectation value of the effective Hamiltonian with respect to QCD vacuum)

$$d\langle \Omega | H_{eff} | \Omega \rangle / d\omega = 0 \tag{6}$$

to find the trial gluon energy $\omega(\mathbf{k})$. The variational function $\omega(\mathbf{k})$ is defined. As a result the gluon acquires a nonzero mass, $m \sim 0.5 GeV$, and an iterative procedure of flow equations is performed with respect to the small parameter $1/m$. The next following sector in the effective Hamiltonian is suppressed by this factor, that provides the convergence for the flow equations (see introduction).

Step 10. *Solving for H_{eff}.*

There are two parameters in the method, the two scales: Λ_0, which defines the counterterms and regulates the perturbative radiative corrections to the effective Hamiltonian H_{eff}, and $\sqrt{\sigma}$, where σ is the string tension defining the nonperturbative confining potential.

---

[4] The higher order terms in the effective Hamiltonian are suppressed by the inverse powers of (heavy) gluon mass, which is of order of hadron scale (see introduction).

Since the effective Hamiltonian is block-diagonal, one can solve for the bound states in any interesting sector (actually in the few lowest sectors). We solve for the glueball bound state in the two-body sector. The result is the glueball spectrum.

III SUMMARY

1. *Renormalization*

Renormalization was performed to the second order. We combined the individual counterterms in one- and zero-body sectors to the resulting mass counterterm, written in the field representation (independent of the basis) $\delta X_{CT}(\Lambda) = m^2 \text{Tr} \int d\mathbf{x} \mathbf{A}^2(\mathbf{x})$ with $m^2 = -\frac{\alpha_s}{\pi} N_c \frac{11}{12} \Lambda^2$ Remarkably, when the quark sector is added in the same fashion, the algebraic coefficient in the propagator correction reproduces the QCD β-function. This particular feature of the Coulomb gauge supports our regularization prescription, which follows from flow equations.

2. *Glueball*

We specify the two parameters: the string tension is defined by the lattice calculations $\sigma = 0.2 GeV^2$; the cutoff Λ_0 is found from the gluon condensate to agree with the result of the sum rules, the condensate term is obtained $\langle G^2 \rangle \sim 1.3 \cdot 10^{-2} GeV^4$ with $\Lambda_0 = 4 GeV$.

The solution of the gap equation can be parameterized as $\omega(\mathbf{k}) = k + m(0)\exp(-k/\kappa)$, where the effective gluon mass is obtained $m(0) = 0.90 GeV$ and $\kappa = 0.95 GeV$.

The glueball mass spectrum in scalar and pseudoscalar channels is given in Tamm-Dancoff approximation in [5]. Roughly, the mass of the lowest scalar glueball 0^{++}, $1760 MeV$, is twice of the effective gluon mass $m(0)$.

Acknowledgments. One of the authors (E.G.) is thankful to Karen Avetovich Ter-Martirosyan for useful discussions and introducing the idea. This work was supported by DOE grants DE-FG02-96ER40944, DE-FG02-97ER41048 and DE-FG02-96ER40947.

REFERENCES

1. F. Wegner, Ann. Phys. **3**, 77 (1994).
2. St. D. Glazek and K. G. Wilson, Phys. Rev. **D 48**, 5863 (1993); *ibid.* **49**, 4214 (1994).
3. E. Gubankova and F. Wegner, Phys. Rev. **D 58**, 025012 (1998), hep-th/9710233.
4. A. Szczepaniak, E. S. Swanson, C.-R.Ji, and S. R. Cotanch, Phys. Rev. Lett. **76**, 2011 (1996).
5. E. Gubankova, et.al, hep-ph/9905527.

Fixed Point Structure of 3D Abelian Gauge Theories

Taichi Itoh and Yoonbai Kim

Department of Physics and Institute of Basic Science, Sungkyunkwan University
Suwon 440-746, Korea
taichi@newton.skku.ac.kr, yoonbai@skku.ac.kr

Abstract. We study renormalization group properties of massive quantum electro-dynamics in three dimensions (QED3) and a chiral conformal fixed point (χCFP) is identified in large N limit. The χCFP is located at the intersection point of two renormalization group flow lines of massless QED3 and Thirring model, which are the only two paths to reach the fixed point. The effect of finite N correction is briefly discussed, and a plausible argument is suggested on a phenomenon that both the massless QED3 and the Thirring model result in the same critical number $N_{\mathrm{cr}} = 128/3\pi^2$.

It seems a characteristic feature of three dimensional (3D) quantum field theories that there arise conformal fixed points naturally through the analytic continuation of dimensionality from either the upper or the lower critical dimensions. 4D marginal interactions become relevant in $4 - \epsilon$ dimensions, while 2D marginal ones are perturbatively irrelevant in $2 + \epsilon$ dimensions but turn out to be marginal through some nonperturbative resummation like the $1/N$ expansion. Thus the marginal interactions in both of two critical dimensions can be merged into the 3D massive superrenormalizable theories characterized by two renormalization group (RG) invariant mass scales. The conformal fixed point (CFP) appears in the criticality achieved in the fine-tuning limit where one (M_{UV}) of the mass scales vanishes and the other (M_{IR}) diverges. Representative known examples are U(N) Yukawa model [1,2], O(N) $(\vec{\phi}^2)^2$ theory [2]. In this note we shall show by using the RG method that the massive QED3 shares the same RG and FP structure with them as given in Table 1 below :

TABLE 1. Structure of CFPs in 3D superrenormalizable field theories

Model	Fixed Points	$M_{\mathrm{UV}} \to 0$	$M_{\mathrm{IR}} \to \infty$
U(N) Yukawa model	GFP, χCFP	massless Yukawa model	Gross-Neveu model
O(N) $(\vec{\phi}^2)^2$ theory	GFP, CFP	critical $(\vec{\phi}^2)^2$ theory	nonlinear σ model
massive QED3	GFP, χCFP	massless QED3	Thirring model

CP494, *New Directions in Quantum Chromodynamics*, edited by C.-R. Ji and D.-P. Min
© 1999 American Institute of Physics 1-56396-908-4/99/$15.00

When an Abelian gauge field couples minimally to O(2) nonlinear σ model (or equivalently the Higgs in the limit of infinitely heavy Higgs mass), the Lagrangian of the massive QED3 is given by

$$\mathcal{L} = \bar{\psi}_a \, i\gamma^\mu \left(\partial_\mu - iA_\mu \right) \psi_a - \frac{1}{4e^2} F_{\mu\nu} F^{\mu\nu} + \frac{v^2}{2} \left(A_\mu - \partial_\mu \phi \right)^2 , \tag{1}$$

where ψ_a $(a = 1, \cdots, N)$ denote N four-component fermions and the large-N limit is taken with both $\alpha = Ne^2$ and $M = v^2/N$ held constant. The scalar field ϕ is the Nambu-Goldstone mode for the spontaneously broken Abelian gauge symmetry (the hidden local U(1) symmetry). Then the model of our interest (1) is invariant under the gauge transformation : $\psi' = e^{i\theta}\psi$, $A'_\mu = A_\mu - \partial_\mu\theta$, $\phi' = \phi + \theta$, of which extension to the BRS symmetry is straightforward and the Faddeev-Popov ghost fields decouple due to the Abelian nature of the gauge symmetry [3].

At the tree level, one can get the massless QED3 Lagrangian by taking the limit of $M \to 0$ with α held constant in Eq. (1). To obtain the Thirring model, one only has to take the limit of $\alpha \to \infty$ with M held constant and then to perform the gauge transformation $\Psi = e^{-i\phi}\psi$, $V_\mu = A_\mu - \partial_\mu\phi$. The gauge sector in Eq. (1) shows the Hodge self-duality in the scheme of path integral that is the invariance of the gauge sector under the interchange between the gauge field A_μ and its Hodge dual field $H_\mu = \sqrt{\alpha M}(1/\partial^2)\epsilon_{\mu\nu\rho}\partial^\nu A^\rho$.

Now let us derive the large-N effective gauge action and figure out the FP structure in the massive QED3. The large-N contributions come from all of the fermion one-loop diagrams which consist of the vacuum polarization $\Pi^{\mu\nu} \equiv \Pi(\partial^2)\left(\partial^2 g^{\mu\nu} - \partial^\mu\partial^\nu\right)$, regularized in the gauge invariant manner and the nonlocal gauge vertices. Although the linear divergence is forbidden to appear in the vacuum polarization due to the gauge symmetry, it provides the correction to the scaling of gauge coupling through the UV cutoff Λ. Thus the effective gauge action including quadratic terms is given by

$$\Gamma = \frac{N}{2} \int d^3x \, \hat{A}_\mu \left\{ \frac{\partial^2}{\alpha} + \partial^2 \, \Pi(\partial^2) + M \right\} \hat{A}^\mu, \tag{2}$$

where transverse gauge field is defined by $\hat{A}_\mu \equiv \left(g_{\mu\nu} - \frac{\partial_\mu\partial_\nu}{\partial^2} \right) A^\nu \equiv A_\mu - \partial_\mu\phi$, and the vacuum polarization part is computed as $\partial^2 \, \Pi(\partial^2) = \frac{1}{8}\sqrt{\partial^2} - \frac{1}{3\pi^2}\frac{\partial^2}{\Lambda}$, of which second term in the right-hand side improves the scaling of the gauge coupling. Introducing dimensionless couplings $u = \alpha\Lambda^{-1}$ and $r = \alpha M\Lambda^{-2}$, the effective action is simplified as

$$\Gamma = \frac{N}{2} \int d^3x \, \hat{A}_\mu \left\{ \left(\frac{1}{u} - \frac{1}{u_*} \right) \frac{\partial^2}{\Lambda} + \frac{\sqrt{\partial^2}}{8} + \left(\frac{r}{u} - \frac{r_*}{u_*} \right) \Lambda \right\} \hat{A}^\mu, \tag{3}$$

where one can read $u_* = 3\pi^2$ and $r_* = 0$ from the vacuum polarization.

To achieve renormalization in the large N limit, one only has to impose the cutoff dependence on both of u and r so as to make two mass scales

$$\tilde{\alpha} = \frac{1}{8}\left(\frac{1}{u} - \frac{1}{u_*}\right)^{-1}\Lambda, \quad \tilde{M} = 8\left(\frac{r}{u} - \frac{r_*}{u_*}\right)\Lambda, \tag{4}$$

cutoff independent. Since both $\tilde{\alpha}$ and $\tilde{M}$ should be invariant under the rescaling of the cutoff Λ, they have to obey RG equations $d\tilde{\alpha}/d\Lambda = 0$ and $d\tilde{M}/d\Lambda = 0$, respectively. The Callan-Symanzik β-functions $\beta_u := du/d\ln\Lambda$ and $\beta_r := dr/d\ln\Lambda$ are therefore determined as

$$\beta_u = -u\left(1 - \frac{u}{u_*}\right), \tag{5}$$

$$\beta_r = -r\left(2 - \frac{u}{u_*}\right) + \frac{r_*}{u_*}u. \tag{6}$$

both of which become zero at a conformal fixed point $(u, r) = (u_*, r_*)$. We notice that there are two specific RG flows, $u = u_*$ and $r = (r_*/u_*)u \equiv 0$, where either $\tilde{M}$ or $1/\tilde{\alpha}$ reduces to zero. The CFP is located at the intersection of the two RG flows.

Along $u = u_*$ the kinetic term in the large N gauge sector (3) vanishes so that the gauge boson is converted to the composite excitation of fermions and anti-fermions and it is nothing but the large-N Thirring model. The mapping from the massive QED3 to the Thirring model is given by $(u, r) = (u_*, u_*/g)$ where g/Λ denotes the coupling constant for the Thirring term. The β-functions read $\beta_u \equiv 0$ and $\beta_r = r_* - r$ on $u = u_*$ so that the CFP can be identified with the nontrivial UV FP in the large-N Thirring model [4] though $g_* \equiv 1/r_* = \infty$ in our RG scheme. On the other hand, the mass term in the large-N gauge sector (3) vanishes along $r = (r_*/u_*)u \equiv 0$ where the theory arises as the large-N QED3 of which β-function is given by Eq. (5). The CFP therefore corresponds to the nontrivial IR FP in QED3 though our RG scheme is different from that in Ref. [5].

The FP structure in the massive QED3 is summarized in Fig. 1-(a). The RG flows around the CFP can be classified as either the flows which avoid the CFP or the horizontal or vertical flows which pass through the CFP (the RG flows of QED3 and the Thirring model). Let us focus on the theory located on the flow inside the physical region of $0 \le u \le u_*$ and $r \ge 0$. It converges at the Gaussian fixed point (GFP) in the UV limit (the GFP of QED3), while it goes to $(u, r) = (u_*, \infty)$ in the IR limit (the GFP of the Thirring model where the coupling g reduces to zero).

The radiative correction to the fermion self-energy is given by the gauge boson exchange which appears from the next to leading order in $1/N$ expansion. Throughout the study of the chiral phase transition (χPT) of spontaneous symmetry breaking U(2N)/U(N) $\otimes$ U(N) by use of the Schwinger-Dyson equation both in QED3 and in the Thirring model, the dominant contribution of gauge boson exchange comes from the CFP which corresponds to the low and high momentum regime in QED3 and in the Thirring model, respectively [6,3]. For the theory with small fermion flavors ($N < N_c = 128/3\pi^2$), the effect of finite N appears and thereby the CFP is lifted as the UV FP of the χPT along the line of the Thirring model as shown in Fig. 1-(b). The critical line of χPT is identified with a line which connects the

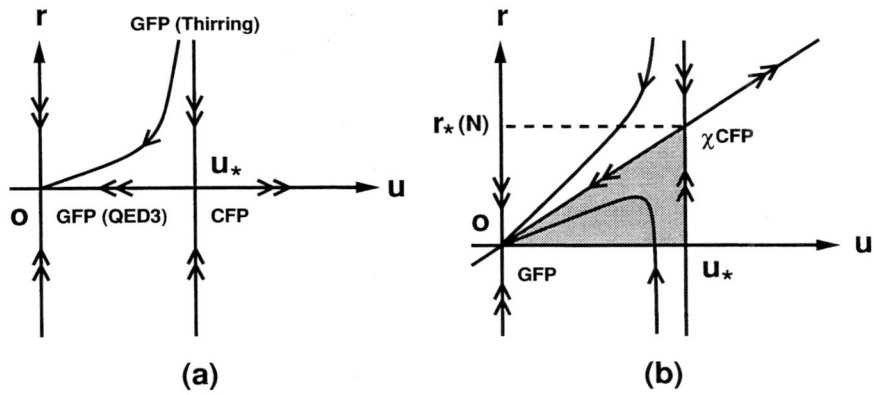

FIGURE 1. (a) The RG flow lines at zero temperature in the massive QED3i in the large N limit. A CFP appears at $(u_*, r_*) = (3\pi^2, 0)$. (b) The RG flow lines for a finite N.

GFP and the χCFP. The shaded area between the above critical line and the line of QED3 corresponds to the broken phase with the dynamically generated fermion mass. We notice that the RG flow line starting from the point in $u < u^*$ finally goes to the GFP so that the direction of dimensionless gauge coupling u is irrelevant to the chiral symmetry breaking. To reach the UV FP on the critical line, we have to start from the original CFP. This situation provides us the reason why the χPT in QED3 is governed by the infrared dynamics of gauge boson. It is now identified with the dynamics of the 3D CFP Abelian gauge theory. To figure out the χPT in QED3 in the context of RG, we necessarily need the relevant direction of r associated with the Thirring coupling.

Acknowledgement. We are indebted to T. Appelquist, Youngjai Kiem, and Chanju Kim for helpful discussions. This work was supported in part by the Korea Research Foundation (1998-015-D00075).

REFERENCES

1. G. Gat, A. Kovner and B. Rosenstein, Nucl. Phys. B 385 (1992) 76.
2. T. Itoh and Y. Kim, in this proceeding.
3. T. Itoh, Y. Kim, M. Sugiura and K. Yamawaki, Prog. Theor. Phys. 93 (1995) 417; K.-I. Kondo, Nucl. Phys. B 450 (1995) 251; M. Sugiura, Prog. Theor. Phys. 97, (1997) 311.
4. S. Hikami and T. Muta, Prog. Theor. Phys. 57 (1977) 785.
5. T.W. Appelquist, M. Bowick, D. Karabali and L.C.R. Wijewardhana Phys. Rev. D 33 (1986) 3704.
6. T. Appelquist, D. Nash and L.C.R. Wijewardhana, Phys. Rev. Lett. 60 (1988) 2575.

Color Confinement in QCD due to Topological Defects

Kei-Ichi Kondo*

*Department of Physics, Faculty of Science, Chiba University, Chiba 263-8522, Japan [1]

Abstract. We outline a derivation of area law of the Wilson loop in SU(N) Yang-Mills theory in the maximal Abelian gauge. This is based on a new version of non-Abelian Stokes theorem and the novel reformulation of the Yang-Mills theory. Abelian dominance and monopole dominance of the string tension in SU(N) QCD are immediate consequences of this derivation.

INTRODUCTION

In the dual superconductor picture [1] of quark confinement, the magnetic monopoles give the dominant contribution to the area law of the Wilson loop or the string tension. Based on 't Hooft argument [2], the partial gauge fixing $G \to H$ realizes the magnetic monopole in Yang-Mills gauge theory in the absence of elementary scalar field. In the conventional approach based on the maximal Abelian (MA) gauge, the residual gauge group was chosen to be the maximal torus subgroup $H = U(1)^{N-1}$ for $G = SU(N)$. This choice immediately determines the type of magnetic monopoles. We re-examine this issue. We learn that the magnetic monopole which is responsible for area law of the Wilson loop is determined by the maximal stability group $\tilde{H}$ rather than the residual gauge group H. This is a new feature appeared in $SU(N), N \geq 3$, which was overlooked so far in the lattice community as far as I know. Indeed, this situation occurs only for $SU(N), N \geq 3$. Therefore, we must distinguish the maximal stability group $\tilde{H}$ from the residual gauge group H. In general, the maximal stability group $\tilde{H}$ is larger than the maximal torus subgroup, $H = U(1)^{N-1} \subset \tilde{H}$. So, the coset $G/\tilde{H} \subset G/H$. We derive area law of the Wilson loop in $SU(N)$ Yang-Mills theory in the MA gauge. This is performed based on the results of a series of works [3–8].

---

[1] This work is supported in part by the Grant-in-Aid for Scientific Research from the Ministry of Education, Science and Culture (No.10640249).

COHERENT STATE

First of all, we construct the coherent state $|\xi, \Lambda\rangle$ corresponding to the coset representatives $\xi \in G/\tilde{H}$. We define the maximal stability subgroup (isotropy subgroup) $\tilde{H}$ as a subgroup of G that consists of all the group elements h that leave the highest-weight state $|\Lambda\rangle$ invariant up to a phase factor, i.e., $h|\Lambda\rangle = |\Lambda\rangle e^{i\phi(h)}, h \in \tilde{H}$. The phase factor is unimportant in the following discussion because we consider the expectation value of any operators in the coherent state. The maximal stability subgroup $\tilde{H}$ includes the Cartan subgroup $H = U(1)^r$, i.e., $H \subset \tilde{H}$.

Let H be the Cartan subgroup of G and $\mathcal{H}$ be the Cartan subalgebra in $\mathcal{G}$. For every element $g \in G$, there is a unique decomposition of g into a product of two group elements, $g = \xi h, \xi \in G/\tilde{H}, h \in \tilde{H}$, for $g \in G$. We can obtain a unique coset space $G/\tilde{H}$ for a given $|\Lambda\rangle$. The action of arbitrary group element $g \in G$ on $|\Lambda\rangle$ is given by $g|\Lambda\rangle = \xi h|\Lambda\rangle = \xi|\Lambda\rangle e^{i\phi(h)}$.

The coherent state is constructed as $|\xi, \Lambda\rangle = \xi|\Lambda\rangle$. This definition of the coherent state is in one-to-one correspondence with the coset space $G/\tilde{H}$ and the coherent states preserve all the algebraic and topological properties of the coset space $G/\tilde{H}$. If $\Gamma^\Lambda(\mathcal{G})$ is Hermitian, then $H_i^\dagger = H_i$, and $E_\alpha^\dagger = E_{-\alpha}$. Every group element $g \in G$ can be written as the exponential of a complex linear combination of diagonal operators H_i and off-diagonal shift operators E_α. Thus the coherent state is given by

$$|\xi, \Lambda\rangle = \xi|\Lambda\rangle = \exp\left[\sum_{\beta'}(\eta_\beta E_\beta - \bar{\eta}_\beta E_{-\beta})\right]|\Lambda\rangle, \quad \eta_\beta \in \mathbf{C}, \tag{1}$$

where $|\Lambda\rangle$ is the highest-weight state $(H_j|\Lambda\rangle = \Lambda_j|\Lambda\rangle)$, $E_\alpha|\Lambda\rangle = 0$ for $\alpha \in R_+$, $R_+(R_-)$ is a subsystem of positive (negative) roots.) such that

(i) $|\Lambda\rangle$ is annihilated by all the (off-diagonal) shift-up operators E_α with $\alpha > 0$, $E_\alpha|\Lambda\rangle = 0(\alpha > 0)$;

(ii) $|\Lambda\rangle$ is mapped into itself by all diagonal operators H_i, $H_i|\Lambda\rangle = \Lambda_i|\Lambda\rangle$;

(iii) $|\Lambda\rangle$ is annihilated by some shift-down operators E_α with $\alpha < 0$, not by other E_β with $\beta < 0$: $E_\alpha|\Lambda\rangle = 0(\text{some } \alpha < 0)$; $E_\beta|\Lambda\rangle = |\Lambda + \beta\rangle(\text{some } \beta < 0)$; and the sum $\sum_{\beta'}$ is restricted to those shift operators E_β which obey (iii).

The coherent states are normalized to unity, $\langle\xi, \Lambda|\xi, \Lambda\rangle = 1$, but, are non-orthogonal, $\langle\xi', \Lambda|\xi, \Lambda\rangle \neq 0$. The coherent state spans the entire space V^Λ. By making use of the the group-invariant measure $d\mu(\xi)$ of G which is appropriately normalized, we obtain $\int|\xi, \Lambda\rangle d\mu(\xi)\langle\xi, \Lambda| = I$, which shows that the coherent states are complete, but overcomplete. This resolution of identity is very important to obtain the path integral formula given below. The coherent states $|\xi, \Lambda\rangle$ are in one-to-one correspondence with the coset representatives $\xi \in G/\tilde{H}, |\xi, \Lambda\rangle \leftrightarrow G/\tilde{H}$. In other words, $|\xi, \Lambda\rangle$ and $\xi \in G/\tilde{H}$ are topologically equivalent.

For concreteness, we first focus on the SU(3) case. The highest weight Λ of the representation specified by the Dynkin index $[m,n]$ (m,n: integers) can be written as $\vec{\Lambda} = m\vec{h}_1 + n\vec{h}_2$ (m,n are non-negative integers for the highest weight) where h_1, h_2 are highest weights of two fundamental representations of SU(3) corresponding to $[1,0],[0,1]$ respectively, i.e., $\vec{h}_1 = \left(\frac{1}{2}, \frac{1}{2\sqrt{3}}\right)$, $\vec{h}_2 = \left(\frac{1}{2}, \frac{-1}{2\sqrt{3}}\right)$. Therefore, we obtain $\vec{\Lambda} = \left(\frac{m+n}{2}, \frac{m-n}{2\sqrt{3}}\right)$. If $mn = 0$, i.e., $m = 0$ or $n = 0$, the maximal stability group $\tilde{H}$ is given by $\tilde{H} = U(2)$ (case (I)). Such a degenerate case occurs when the highest-weight vector $\vec{\Lambda}$ is orthogonal to some root vectors. If $mn \neq 0$, i.e., $m \neq 0$ and $n \neq 0$, H is the maximal torus group $\tilde{H} = U(1) \times U(1)$ (case (II)). This is a non-degenerate case. Therefore, for the highest weight Λ in the case (I), the coset $G/\tilde{H}$ is given by

$$SU(3)/U(2) = SU(3)/(SU(2) \times U(1)) = CP^2, \tag{2}$$

whereas in the case (II)

$$SU(3)/(U(1) \times U(1)) = F_2. \tag{3}$$

Here, CP^n is the complex projective space and F_n is the flag space. Therefore, the two fundamental representations belong to the case (I), so the maximal stability group is $U(2)$, rather than the maximal torus group $U(1) \times U(1)$.

NON-ABELIAN STOKES THEOREM

We have derived a new version of non-Abelian Stokes theorem (NAST) [9,10]. For the non-Abelian Wilson loop defined by the trace of the path-ordered exponent along the closed loop C,

$$W^C[\mathcal{A}] := \text{tr}\left[\mathcal{P}\exp\left(ig\oint_C \mathcal{A}\right)\right], \tag{4}$$

with $\mathcal{A}$ being the connection one-form, $\mathcal{A}(x) = \mathcal{A}_\mu^A(x)T^A dx^\mu = \mathcal{A}^A(x)T^A$, the NAST for $SU(N)$ is given by

$$W^C[\mathcal{A}] = \int [d\mu(\xi)]_C \exp\left(ig\oint_C \left[n^A\mathcal{A}^A + \frac{1}{g}\omega\right]\right)$$

$$= \int [d\mu(\xi)]_C \exp\left(ig\int_{S:\partial S=C} \left[d(n^A\mathcal{A}^A) + \frac{1}{g}\Omega_K\right]\right), \tag{5}$$

where we have defined

$$n^A(x) := \langle\Lambda|V(x)T^A V^\dagger(x)|\Lambda\rangle, \tag{6}$$

$$\omega(x) := \langle\Lambda|iV(x)dV^\dagger(x)|\Lambda\rangle = \langle\Lambda|i\xi^\dagger(x)d\xi(x)|\Lambda\rangle, \tag{7}$$

and Ω_K is the Kähler two-form given by

$$\Omega_K := d\omega. \tag{8}$$

The NAST (5) implies that the Wilson loop is rewritten into

$$W^C[\mathcal{A}] = \int [d\mu(\xi)]_C \exp\left(ig \oint_C a\right) = \int [d\mu(\xi)]_C \exp\left(ig \int_{S:C=\partial S} f\right). \tag{9}$$

First, a is the connection one-form,

$$a := n^A \mathcal{A}^A + \frac{1}{g}\omega = \langle \Lambda | \mathcal{A}^V | \Lambda \rangle, \tag{10}$$

where $\mathcal{A}^V := V\mathcal{A}V^\dagger + \frac{i}{g}V dV^\dagger$ is the gauge transformation of $\mathcal{A}$ by $V \in F_{N-1}$. For quark in the fundamental representation,

$$a = \mathrm{tr}(\mathcal{H}\mathcal{A}^V). \tag{11}$$

Therefore, the one-form a is equal to the diagonal piece of the gauge-transformed potential $\mathcal{A}^V$. Next, f is the curvature two-form,

$$f := da = dC + \frac{1}{g}d\omega = dC + \frac{1}{g}\Omega_K, \tag{12}$$

where we defined the one-form, $C := n^A \mathcal{A}^A$. The first piece dC in f does not contribute to the magnetic current, due to the Bianchi identity. On the other hand, the second term Ω_K in f leads to the non-vanishing magnetic current,

$$k_\mu := \partial_\nu{}^* f_{\mu\nu} \neq 0, \tag{13}$$

where $^* f_{\mu\nu}$ is the Hodge dual of $f_{\mu\nu}$ in four dimensions, $^* f_{\mu\nu} := \frac{1}{2}\epsilon_{\mu\nu\rho\sigma}f_{\rho\sigma}$. In general, the (curvature) two-form $f = d(n^A\mathcal{A}^A) + \Omega_K$ in the NAST is the Abelian field strength (which is invariant even under the non-Abelian gauge transformation of $G = SU(N)$), i.e., the generalized 't Hooft-Polyakov tensor for $SU(N)$,

$$f_{\mu\nu}(x) := \partial_\mu(n^A(x)\mathcal{A}_\nu^A(x)) - \partial_\nu(n^A(x)\mathcal{A}_\mu^A(x)) + \frac{i}{g}\mathbf{n}(x) \cdot [\partial_\mu\mathbf{n}(x), \partial_\nu\mathbf{n}(x)]. \tag{14}$$

The invariance of f is obvious from the NAST (9), since the L.H.S. of (9), i.e., $W^C[\mathcal{A}]$ is gauge invariant and the measure $[d\mu(\xi)]_C$ in the R.H.S. is also invariant under the G gauge transformation. Otherwise, the R.H.S. is zero. In the case of fundamental representation, the invariance is easily seen, because it is possible to rewrite (14) into the manifestly gauge invariant form: $f_{\mu\nu}(x) := \mathrm{tr}\left(\mathbf{n}(x)\mathcal{F}_{\mu\nu}(x) + \frac{i}{g}\mathbf{n}(x) \cdot [D_\mu\mathbf{n}(x), D_\nu\mathbf{n}(x)]\right)$, where $\mathcal{F}_{\mu\nu}(x) := \partial_\mu\mathcal{A}_\nu(x) - \partial_\nu\mathcal{A}_\mu(x) - ig[\mathcal{A}_\mu(x), \mathcal{A}_\nu(x)]$, and $D_\mu\mathbf{n}(x) := \partial_\mu\mathbf{n}(x) - ig[\mathcal{A}_\mu(x), \mathbf{n}(x)]$.

ABELIAN DOMINANCE

In our framework, the Abelian dominance and the monopole dominance are understood as implying the first and the second equality respectively,

$$\left\langle W^C[\mathcal{A}] \right\rangle_{YM} \cong \left\langle \exp\left(ig \oint_C a\right) \right\rangle_{APEGT} \cong \left\langle \exp\left(i \oint_C \omega\right) \right\rangle_{APEGT}, \tag{15}$$

where APEGT denotes the Abelian-projected effective gauge theory [3]. Numerical simulations show that the monopole part exhibits the area law and σ_{Abel} exhausts the full string tension obtained from the non-Abelian Wilson loop (i.e., monopole dominance in the string tension or area law),

$$\left\langle \exp\left(i \oint_C \omega\right) \right\rangle_{APEGT} \sim \exp(-\sigma_{Abel}|S|), \tag{16}$$

while $\left\langle \exp\left(ig \oint_C a - i \oint_C \omega\right) \right\rangle_{APEGT}$ does not exhibits the area law. This result implies that the area law of the original non-Abelian Wilson loop,

$$\left\langle W^C[\mathcal{A}] \right\rangle_{YM} \sim \exp(-\sigma|S|), \quad \sigma \cong \sigma_{Abel}. \tag{17}$$

The monopole dominance in this sense was derived for $SU(2)$ in [6] by showing that the dominant contribution to the area law comes from the monopole piece alone, $\Omega_K = d\omega$. In [7], the monopole dominance and the area law of the Wilson loop have been shown based on the APEGT for $G = SU(2)$. Now this scenario can be extended into $G = SU(N)$.

MAGNETIC MONOPOLES IN YANG-MILLS THEORY

The existence of magnetic monopole is suggested from the non-trivial Homotopy groups $\pi_2(G/H)$. In the case (II), $\pi_2(F_2) = \pi_2(SU(3)/(U(1) \times U(1))) = \pi_1(U(1) \times U(1)) = \mathbf{Z} + \mathbf{Z}$. On the other hand, in the case (I), i.e., [m,0] or [0,n] $\pi_2(CP^2) = \pi_2(SU(3)/U(2)) = \pi_1(U(2)) = \pi_1(SU(2) \times U(1)) = \pi_1(U(1)) = \mathbf{Z}$. Note that CP^n NLSM has only the local $U(1)$ invariance for any n. It is this U(1) invariance that corresponds to a kind of Abelian magnetic monopole in the case (I). This magnetic monopole may be related to the non-Abelian magnetic monopole proposed by E. Weinberg et al.

This situation should be compared with the $SU(2)$ case where the maximal stability group is always given by the maximal torus $H = U(1)$ irrespective of the representation. Therefore, the coset is given by $G/H = SU(2)/U(1) = F_1 = CP^1 \cong S^2 \cong SO(3)$ and $\pi_2(SU(2)/U(1)) = \pi_2(F_1) = \pi_2(CP^1) = \mathbf{Z}$, for *arbitrary* representation . Actually, the NAST derived in this paper shows that the fundamental quark feels only the $U(1)$ embedded in the maximal stability group $U(2)$ as a magnetic monopole (This is a component along the highest-weight).

AREA LAW OF THE WILSON LOOP

The (full) string tension σ is defined by

$$\sigma := - \lim_{A(C) \to \infty} \frac{1}{A(C)} \ln \langle W^C[\mathcal{A}] \rangle_{YM_4}, \qquad (18)$$

where $A(C)$ is the minimal area spanned by the Wilson loop C.

We estimate the Wilson loop in the reformulation of the Yang-Mills theory which was proposed by the author and was called the perturbative deformation of a topological quantum field theory [4]. The Wilson loop is written as

$$\langle W^C[\mathcal{A}] \rangle_{YM_4} = \left\langle\!\!\left\langle \exp\left[ig \oint_C dx^\mu n^A(x) \mathcal{V}_\mu^A(x)\right] \right\rangle\!\!\right\rangle_{pYM} \exp\left[i \oint_C \omega\right] \right\rangle_{TQFT_4}. \qquad (19)$$

Here the expectation value $\langle\!\langle(\cdots)\rangle\!\rangle_{pYM}^{\mathcal{V}}$ for the field $\mathcal{V}$ is calculated in the perturbation theory in the coupling constant g. On the other hand, the expectation value $\langle\!\langle(\cdots)\rangle\!\rangle_{TQFT}^{U}$ should be calculated in a non-perturbative way to incorporate the topological contribution where U is a compact gauge variable,

$$S_{TQFT}[\Omega, C, \bar{C}, B] := \int_{\mathbf{R}^4} d^4x \; i\delta_B \bar{\delta}_B \mathrm{tr}_{G/H} \left[\frac{1}{2}\Omega_\mu(x)\Omega_\mu(x) + iC(x)\bar{C}(x)\right], \qquad (20)$$

where $\Omega_\mu(x) := \frac{i}{g} U(x)\partial_\mu U^\dagger(x)$. This reformulation leads to the result:

$$\langle W^C[\mathcal{A}] \rangle_{YM_4} = \left\langle \exp\left[i \oint_C \omega\right] \right\rangle_{TQFT_4} \left[1 + O(g^2)\right]. \qquad (21)$$

This implies the magnetic monopole dominance in the area law of the Wilson loop.

For the planar Wilson loop C, the Parisi-Soulous dimensional reduction [4] leads to

$$\left\langle \exp\left[i \oint_C \omega\right] \right\rangle_{TQFT_4} = \left\langle \exp\left[i \oint_C \omega\right] \right\rangle_{NLSM_2}, \qquad (22)$$

where the two-dimensional nonlinear sigma model (NLSM) has the action,

$$S_{NLSM}[U, C, \bar{C}] := 2\pi \int_{\mathbf{R}^2} d^2x \; \mathrm{tr}_{G/H} \left[\frac{1}{2}\Omega_\mu(x)\Omega_\mu(x) + iC(x)\bar{C}(x)\right]. \qquad (23)$$

By making use of the complex coordinates of the flag space G/H, the action is rewritten as

$$S_{NLSM} = \frac{\pi}{g^2} \int_{\mathbf{R}^2} d^2x \, g_{\alpha\bar{\beta}} \frac{\partial w^\alpha}{\partial x_a} \frac{\partial \bar{w}^\beta}{\partial x_a} \quad (a = 1, 2) \qquad (24)$$

$$= \frac{\pi}{g^2} \int_{\mathbf{C}} dz d\bar{z} \; g_{\alpha\bar{\beta}} \left(\frac{\partial w^\alpha}{\partial z} \frac{\partial \bar{w}^\beta}{\partial \bar{z}} + \frac{\partial w^\alpha}{\partial \bar{z}} \frac{\partial \bar{w}^\beta}{\partial z}\right), \qquad (25)$$

319

where $z = x + iy = x_1 + ix_2 \in \mathbf{C} \cong \mathbf{R}^2$, and $dxdy = dx_1dx_2 = \frac{i}{2}dzd\bar{z}$, and we have omitted the ghost term, $C(x)\bar{C}(x)$. Here $g(\mu)$ is the running Yang-Mills coupling constant whose running is given by the perturbative deformation in four-dimensional Yang-Mills theory. For the quark in the fundamental representation of $SU(N)$, the relevant NLSM is given by CP^{N-1} model. We can show the area law,

$$\left\langle \exp\left[i \oint_C \omega\right] \right\rangle_{CP_2^{N-1}} \sim \exp(-\sigma_0 TR), \tag{26}$$

by the instanton calculus (dilute instanton-gas approximation) or by the large N expansion in the two-dimensional CP^{N-1} model, see [10].

In summary, we have given a new derivation of non-Abelian Stokes theorem for $G = SU(N)$ for $N \geq 2$ which reduces to the previous result for $SU(2)$. This version of non-Abelian Wilson loop is very helpful to see the role played by the magnetic monopole in the calculation of the non-Abelian Wilson loop. Combining this non-Abelian Stokes theorem with the Abelian-projected effective gauge theory for $SU(N)$, we have explained the Abelian dominance for the Wilson loop in $SU(N)$ Yang-Mills gauge theory. For $SU(N)$ with N greater than two, we must distinguish the maximal stability group $\tilde{H}$ and the residual gauge group $H = U(1)^{N-1}$.

By making use of the non-Abelian Stokes theorem in a novel reformulation of the Yang-Mills theory proposed by one of the authors [4], the derivation of the area law of the non-Abelian Wilson loop in four-dimensional Yang-Mills theory has been reduced to the two-dimensional problem of calculating the expectation value of the Abelian Wilson loop in the coset G/H non-linear sigma model. Especially, in order to show confinement of the fundamental quark in four-dimensional $SU(N)$ Yang-Mills theory in the MA gauge, we have only to consider the two-dimensional CP^{N-1} model. The details will be given in [9,10]. A Monte Carlo simulation on a lattice will be efficient to confirm the above picture of quark confinement.

REFERENCES

1. Nambu, Y., *Phys. Rev. D* **10**, 4262 (1974).
   't Hooft, G., in: High Energy Physics, edited by A. Zichichi (Editorice Compositori, Bologna, 1975).
   Mandelstam, S., *Phys. Report* **23**, 245 (1976).
2. 't Hooft, G., *Nucl. Phys. B* **190** [FS3], 455 (1981).
3. Kondo, K.-I., hep-th/9709109, *Phys. Rev. D* **57**, 7467 (1998).
4. Kondo, K.-I., hep-th/9801024, *Phys. Rev. D* **58**, 105019 (1998).
5. Kondo, K.-I., hep-th/9803133, *Phys. Rev. D* **58**, 085013 (1998).
6. Kondo, K.-I., hep-th/9805153, *Phys. Rev. D* **58**, 105016 (1998).
7. Kondo, K.-I., hep-th/9810167, *Phys. Lett. B* **455**, 251 (1999).
8. Kondo, K.-I., hep-th/9904045.
9. Kondo, K.-I., and Taira, Y., hep-th/9906129.
10. Kondo, K.-I., and Taira, Y., in preparation.

Glueball Mass Spectrum from Supergravity

Csaba Csáki[1] and John Terning[2]

Theoretical Physics Group
Ernest Orlando Lawrence Berkeley National Laboratory
University of California, Berkeley, CA 94720
and
Department of Physics
University of California, Berkeley, CA 94720

Abstract. We review the calculation of the spectrum of glueball masses in non-supersymmetric Yang-Mills theory using the conjectured duality between supergravity and large N gauge theories. The glueball masses are obtained by solving the supergravity wave equations in a black hole geometry. The glueball masses found this way are in unexpected agreement with the available lattice data. We also show how to use a modified version of the duality based on rotating branes to calculate the glueball mass spectrum with some of the Kaluza-Klein states of the supergravity theory decoupled from the spectrum.

I INTRODUCTION

Maldacena's conjecture [1] relates $\mathcal{N} = 4$ supersymmetric $SU(N)$ gauge theories in the large N limit to Type IIB string theory on an $\text{AdS}_5 \times \mathbf{S}^5$ background, where AdS_5 is a five dimensional anti-de Sitter space. The metric of this space is given by

$$\frac{ds^2}{l_s^2\sqrt{4\pi g_s N}} = \rho^{-2}d\rho^2 + \rho^2 \sum_{i=1}^{4} dx_i^2 + d\Omega_5^2 \tag{1}$$

where l_s is the string length related to the superstring tension, g_s is the string coupling constant and $d\Omega_5$ is the line element on $\mathbf{S}^5$. The $x_{1,2,3,4}$ directions in AdS_5 correspond to $\mathbf{R}^4$ where the gauge theory lives. The gauge coupling constant g_4 of the 4D theory is related to the string coupling constant g_s by $g_4^2 = g_s$. In the 't Hooft limit ($N \to \infty$ with $g_4^2 N = g_s N$ fixed), the string coupling constant vanishes $g_s \to 0$. Therefore we can study the 4D theory using the first quantized string theory in the

---

[1] Research fellow, Miller Institute for Basic Research in Science.
[2] Talk presented by J. Terning.

AdS space (1). Moreover if $g_s N \gg 1$, the curvature of the AdS space is small and the string theory is approximated by classical supergravity. Witten extended this proposal to non-supersymmetric theories [2]. In his setup supersymmetry is broken by heating up the $\mathcal{N} = 4$ theory, which corresponds to putting the four dimensional theory on a circle and assigning anti-periodic boundary conditions to the fermions. In this case the fermions will get a supersymmetry breaking mass term of the order $T = 1/2\pi R$, where R is the radius of the compact coordinate and T is the corresponding temperature, while the scalars (not protected by supersymmetry anymore) will get masses from loop corrections. Thus in the $T \to \infty$ limit this should reproduce a pure (3 dimensional) $SU(N)$ theory in the large N limit, which we will refer to as QCD$_3$. On the string theory side this corresponds to replacing the anti-de Sitter metric by a Schwarzschild metric describing a black hole in the anti-de Sitter space. This metric is given by

$$\frac{ds^2}{l_s^2 \sqrt{4\pi g_s N}} = \left(\rho^2 - \frac{b^4}{\rho^2} \right)^{-1} d\rho^2 + \left(\rho^2 - \frac{b^4}{\rho^2} \right) d\tau^2 + \rho^2 \sum_{i=1}^{3} dx_i^2 + d\Omega_5^2, \qquad (2)$$

where τ parameterizes the compactifying circle and the $x_{1,2,3}$ direction corresponding to the $\mathbf{R}^3$ where QCD$_3$ lives. The horizon of this geometry is located at $\rho = b$ with

$$b = \frac{1}{2R} = \pi T. \qquad (3)$$

The supergravity approximation is valid for this theory when the curvature of the space is small, thus when $g_s N \to \infty$. However, in order to obtain the pure gauge theory we have to take the temperature to infinity. In order to keep the intrinsic scale $g_3^2 N = g_4^2 N/R$ of the resulting theory at the scale of QCD, we simultaneously would need to take $g_4^2 N = g_s N \to 0$. Here g_3 is the dimensionful gauge coupling of QCD$_3$. This is exactly the opposite limit in which the supergravity approximation is applicable! Thus as expected for any strong-weak duality, the weakly coupled classical supergravity theory and the QCD$_3$ theory are valid in different limits of the 't Hooft coupling $g_4^2 N$.

From the point of view of QCD$_3$, the radius R of the compactifying circle provides the ultraviolet cutoff scale. Therefore, with the currently available techniques, the Maldacena-Witten conjecture can only be used to study large N QCD with a fixed ultraviolet cutoff R^{-1} in the strong ultraviolet coupling regime, and hope that the results one obtains this way are not very sensitive to removing the cutoff, that is on going from one limit to the other. Since the theory is non-supersymmetric, there is a priori no reason to believe that these two limits have anything to do with each other, since for example there might very well be a phase transition when the 't Hooft coupling is decreased from the very large values where the supergravity description is valid to the small values where the theory should describe QCD$_3$. Nevertheless, Witten showed that the supergravity theory correctly reproduces several of the qualitative features of a confining 3 dimensional pure gauge theory correctly [2].

In particular, he showed that there is an area law in the Wilson loop and that there is a mass gap in the spectrum, both of which are expected features of a confining gauge theory. Here we will address the question of whether any of the quantitative features of the gauge theories are reproduced as well. In particular, we will calculate the glueball mass spectrum of the theory, and find, that it is in reasonable agreement with recent lattice simulations [3].

II THE GLUEBALL SPECTRUM IN 3 DIMENSIONS

In this section we will show how to calculate the glueball spectrum of some of the glueballs in the supergravity approximation in the 3 dimensional case. In the following we will use the notation J^{PC} for the glueballs, where J is the glueball spin, and P, C refer to the parity and charge conjugation quantum numbers respectively. In the field theory, one can find operators that have the quantum numbers corresponding to the given glueball states. For example, an operator with quantum numbers 0^{++} is given by $\mathcal{O}_4 = \text{Tr} F^2$. According to the refinement of the Maldacena conjecture given in [4], one should find a supergravity state corresponding to the chiral primary operators of the original $\mathcal{N} = 4$ conformal theory, which will couple to the supergravity states on the boundary of the AdS space. Assuming this coupling is maintained while heating the system, we would like to calculate the actual glueball mass spectrum corresponding to operators like $\mathcal{O}_4$. To calculate the masses of these states in field theory one would need to evaluate the correlators $\langle \mathcal{O}_4(x)\mathcal{O}_4(y)\rangle = \sum_i c_i e^{-m_i|x-y|}$, where the m_i's are the glueball masses. According to the refinement of the Maldacena conjecture [4], this just amounts to solving the supergravity wave equations for the fields that couple to these operators on the boundary. In the case of the 0^{++} glueballs, we need to find the solutions of the dilaton equations of motion of the form $\Phi = f(\rho)e^{ikx}$. This is because in the supergravity theory on $\text{AdS}_5 \times \mathbf{S}^5$, the Kaluza-Klein modes on the $\mathbf{S}^5$ can be classified according to the spherical harmonics of the $\mathbf{S}^5$, which form representations of the isometry group $SO(6)$ (which is the R-symmetry group of the $\mathcal{N} = 4$ theory). When we put the theory at finite temperature, the states carrying non-trivial $SO(6)$ quantum numbers should eventually decouple from the spectrum, thus the glueballs should be identified with the $SO(6)$ singlet states, which implies a solution of the form $\Phi = f(\rho)e^{ikx}$ for the dilaton. Thus we will look for normalizable regular solutions to the dilaton equation of motion which will give a discrete spectrum with the glueball masses determined as $k_i^2 = -M_i^2$.

In the supergravity description we have to solve the classical equation of motion of the massless dilaton,

$$\partial_\mu \left[\sqrt{g}\partial_\nu \Phi g^{\mu\nu}\right] = 0 \,, \tag{4}$$

on the AdS_5 black hole background (2). Plugging the ansatz $\Phi = f(\rho)e^{ikx}$ into this equation and using the metric of (2) one obtains the following differential equation for f:

$$\rho^{-1} \frac{d}{d\rho}\left(\left(\rho^4 - b^4\right)\rho\frac{df}{d\rho}\right) - k^2 f = 0 \tag{5}$$

Since the glueball mass M^2 is equal to $-k^2$, the task is to solve this equation as an eigenvalue problem for k^2. In the following we set $b = 1$, so the masses are computed in units of b. We need to find normalizable solutions to this equations which are also regular at the horizon. For large ρ, the black hole metric (2) asymptotically approaches the AdS metric, and the behavior of the solution for a p-form for large ρ takes the form ρ^λ, where λ is determined from the mass m of the supergravity field:

$$m^2 = \lambda(\lambda + 4 - 2p) . \tag{6}$$

Indeed both (5) and (6) give the asymptotic forms $f \sim 1, \rho^{-4}$, and only the later is a normalizable solution [2]. Since the black hole geometry is regular at the horizon $\rho = 1$, k^2 has to be adjusted so that f is also regular at $\rho = 1$ [2]. This can be done numerically in a simple fashion using a "shooting" technique as follows. For a given value of k^2 the equation is numerically integrated from some sufficiently large value of ρ ($\rho \gg k^2$) by matching $f(\rho)$ with the asymptotic solution. The glueball mass M is related to the eigenvalues of k^2 by $M^2 = -k^2$ in units of b^2. The results obtained this way, together with the results of the lattice simulations [5] are displayed in Table 1. Since the lattice results are in units of string tension, we normalize the supergravity results so that the lightest 0^{++} state agrees with the lattice result. One should also expect a systematic error in addition to the statistical error denoted in Table 1 for the lattice computations. Similar numerical results have been obtained in [6], while a WKB approximation for the eigenvalues of (5) has been obtained in [7].

TABLE 1. 0^{++} glueball masses in QCD$_3$ coupled to tr $F_{\mu\nu}F^{\mu\nu}$. The lattice results are in units of the square root of the string tension. The denoted error in the lattice results is only the statistical one.

state	lattice, $N = 3$	lattice, $N \to \infty$	supergravity
0^{++}	4.329 ± 0.041	4.065 ± 0.055	4.07 (input)
0^{++*}	6.52 ± 0.09	6.18 ± 0.13	7.02
0^{++**}	8.23 ± 0.17	7.99 ± 0.22	9.92
0^{++***}	-	-	12.80

The 0^{--} glueballs can be dealt with similarly by considering the two-form of the supergravity theory, which couples to the operator $\mathcal{O}_6$. The supergravity equation of motion for the s-wave component of this field is given by

$$\frac{3}{\sqrt{g}}\partial_\mu\left[\sqrt{g}\,\partial_{[\mu'}A_{\mu'_1\mu'_2]}\,g^{\mu'\mu}g^{\mu'_1\mu_1}g^{\mu'_2\mu_2}\right] - 16g^{\mu'_1\mu_1}g^{\mu'_2\mu_2}A_{\mu'_1\mu'_2} = 0, \tag{7}$$

where [ ] denotes antisymmetrization with strength one. For the pseudoscalar component of A_{ij} this can be solved similarly as for the case of the 0^{++} glueballs, and the results are displayed in Table II. Since the supergravity method and the lattice gauge theory compute the glueball masses in different units, one cannot compare the absolute values of the lowest glueball mass obtained using these methods. However it makes sense to compare the lowest glueball masses of different quantum numbers. Using Tables 1 and II, we find that the supergravity results are in good agreement with the lattice gauge theory computation [5]:

$$\left(\frac{M_{0^{--}}}{M_{0^{++}}}\right)_{\text{supergravity}} = 1.50$$

$$\left(\frac{M_{0^{--}}}{M_{0^{++}}}\right)_{\text{lattice}} = 1.45 \pm 0.08 \tag{8}$$

TABLE 2. 0^{--} glueball masses in QCD_3 coupled to $\mathcal{O}_6$. The normalization of the supergravity results is the same as in Table 1.

state	lattice, $N = 3$	lattice, $N \to \infty$	supergravity
0^{--}	6.48 ± 0.09	5.91 ± 0.25	6.10
0^{--*}	8.15 ± 0.16	7.63 ± 0.37	9.34
0^{--**}	9.81 ± 0.26	8.96 ± 0.65	12.37
0^{--***}	-	-	15.33

One can see, that the glueball mass ratios obtained from the supergravity calculation are in reasonable agreement with the lattice results, even though as explained in the introduction these two calculations are in the opposite limits for the 't Hooft coupling. Therefore, it is important to see, how the ratios are modified once corrections due to string theory are taken into account. The leading string theory corrections can be calculated by using the results of [8], who calculated the first α' corrections to the AdS black-hole metric (2). The details of the calculation can be found in [3], where it was shown that the string theory corrections are somewhat uniform for the different excited states of the 0^{++} glueball, and therefore one could hope that these corrections to the ratios of the glueball masses are small. However, it can be seen that this is probably a too optimistic assumption, by considering the Kaluza-Klein partners of the glueball states. The Kaluza-Klein modes do not correspond to any state in the QCD theory, but rather they should decouple in the $R \to 0, g_4^2 N \to 0$ limit from the spectrum. However, in the supergravity limit of finite R, $g_4^2 N \to \infty$ these states have masses comparable to the light glueballs [9]. This is simply a consequence of the fact, that the masses of the fermions and scalars is of the order of the temperature T, thus their bound states are expected to also have masses of the order of the temperature. However, the temperature is the only scale in the theory, so this will also be the cutoff scale of the QCD theory, and thus the mass scale for the glueballs. This situation is clearly unsatisfactory, therefore one may try to improve on it by introducing a different supergravity background, where some of these KK modes are automatically decoupled. We will consider this possibility in the next section where we discuss the construction based on rotating branes [10–12].

III THE GLUEBALL SPECTRUM IN 4 DIMENSIONS AND THE CONSTRUCTION BASED ON ROTATING BRANES

Results similar to the the ones presented in the previous section can be obtained for the glueball mass spectrum in QCD$_4$ by starting from a slightly different construction where the M-theory 5-brane is wrapped on two circles [2]. The details of these results can be found in [3,13]. Here we will review only the generalized construction based on the rotating M5 brane with one angular momentum, first constructed in [10], and explored in [11]. The metric for this background is given by

$$ds_{\text{IIA}}^2 = \frac{2\pi\lambda A}{3u_0} u \Delta^{1/2} \Big[4u^2(-dx_0^2 + dx_1^2 + dx_2^2 + dx_3^2) + \frac{4A^2}{9u_0^2} u^2 \,(1 - \frac{u_0^6}{u^6\Delta})d\theta_2^2 \tag{9}$$

$$+ \frac{4\,du^2}{u^2(1 - \frac{a^4}{u^4} - \frac{u_0^6}{u^6})} + d\theta^2 + \frac{\tilde{\Delta}}{\Delta}\sin^2\theta d\varphi^2 + \frac{1}{\Delta}\cos^2\theta d\Omega_2^2 - \frac{4a^2 A u_0^2}{3u^4\Delta}\sin^2\theta d\theta_2 d\varphi \Big],$$

where $x_{0,1,2,3}$ are the coordinates along the brane where the gauge theory lives, u is the "radial" coordinate of the AdS space, while the remaining four coordinates parameterize the angular variables of S^4, a is the angular momentum parameter, and we have introduced

$$\Delta = 1 - \frac{a^4 \cos^2\theta}{u^4} \,, \qquad \tilde{\Delta} = 1 - \frac{a^4}{u^4} \,, \qquad A \equiv \frac{u_0^4}{u_H^4 - \frac{1}{3}a^4} \,, \qquad u_H^6 - a^4 u_H^2 - u_0^6 = 0 \,.$$
$$\tag{10}$$

u_H is the location of the horizon, and the dilaton background and the temperature of the field theory are given by

$$e^{2\phi} = \frac{8\pi}{27} \frac{A^3 \lambda^3 u^3 \Delta^{1/2}}{u_0^3} \frac{1}{N^2} \,, \qquad R = (2\pi T_H)^{-1} = \frac{A}{3u_0} \,. \tag{11}$$

Note, that in the limit when $a/u_0 \gg 1$, the radius of compactification R shrinks to zero, thus the KK modes on this compact direction are expected to decouple in this theory when we increase the angular momentum a. In order to find the mass spectrum of the 0^{++} glueballs, we need to again solve the dilaton equations of motion as a function of a. This can be done by plugging the background (9) into the dilaton equation of motion

$$\partial_\mu \Big[\sqrt{g} e^{-2\Phi} g^{\mu\nu} \partial_\nu \Phi \Big] = 0. \tag{12}$$

This can be solved the same way as explained in the previous section, where the eigenvalues are now a function of the angular momentum parameter a. The results of this are summarized in Table 3. Note, that while some of the KK modes decouple in the $a \to \infty$ limit, the 0^{++} glueball mass ratios change only very slightly, showing that the supergravity predictions are robust for these ratios against the change of the angular momentum parameter.

TABLE 3. Masses of the first few 0^{++} glueballs in QCD$_4$, in GeV. The change from $a = 0$ to $a = \infty$ is tiny.

state	lattice, $N = 3$	supergravity $a = 0$	supergravity $a \to \infty$
0^{++}	1.61 ± 0.15	1.61 (input)	1.61 (input)
0^{++*}	2.48 ± 0.23	2.55	2.56
0^{++**}	-	3.46	3.48

One can similarly calculate the mass ratios for the 0^{-+} glueballs, by considering the equations of motion of the RR 1-form in the background (9), since on the D4 brane worldvolume this couples to the operator $\mathrm{Tr} F \tilde{F}$. To find the glueball spectrum we have to solve the supergravity equation of motion of the RR 1-form

$$\partial_\nu \left[\sqrt{g} g^{\mu\rho} g^{\nu\sigma} (\partial_\rho A_\sigma - \partial_\sigma A_\rho) \right] = 0 \tag{13}$$

in the background (9). The results are summarized in Table 4. Note, that the change in the 0^{-+} glueball mass is sizeable when going from $a = 0$ to $a \to \infty$, and is in the right direction as suggested by lattice results [14,15].

TABLE 4. Masses of the first few 0^{-+} glueballs in QCD$_4$, in GeV. Note that the change from $a = 0$ to $a = \infty$ in the supergravity predictions is of the order $\sim 25\%$.

state	lattice, $N = 3$	supergravity $a = 0$	supergravity $a \to \infty$
0^{-+}	2.59 ± 0.13	2.00	2.56
0^{-+*}	3.64 ± 0.18	2.98	3.49
0^{-+**}	-	3.91	4.40

One can also calculate the masses of the different Kaluza-Klein modes in the background of (9). One finds, that as expected from the fact that for $a \to \infty$ the compact circle shrinks to zero, the KK modes on this compact circle decouple from the spectrum, leading to a 4 dimensional field theory in this limit. However, the KK modes of the sphere S^4 do not decouple from the spectrum even in the $a \to \infty$ limit. These conclusions remain unchanged even in the case when one considers the theory with the maximal number of angular momenta (which is two for the case of QCD$_4$) [12,16]. In the limit when the angular momentum becomes large, $a/u_0 \gg 1$, the theory approaches a supersymmetric limit [10,12] since the supersymmetry breaking fermion masses get smaller with increasing angular momentum [17]. Therefore, the limit of increasing angular momentum on one hand does decouple some of the KK modes which makes the theory four dimensional, but at the same time reintroduces the light fermions into the spectrum [17].

IV CONCLUSIONS

We have seen how the Witten extension of Maldacena's conjecture can be used to study pure Yang-Mills theories in the large N limit. These theories reproduce several of the qualitative features of QCD, and one can also study the predictions for the glueball mass spectra. One finds, that the supergravity calculations are in a reasonable agreement with the lattice results, even though they are obtained in the opposite limit of the 't Hooft coupling. It would be very important to understand, whether this unexpected agreement is purely a numerical coincidence or whether there is any deeper reason behind it.

ACKNOWLEDGEMENTS

We thank H. Ooguri, Y. Oz, J. Russo and K. Sfetsos for several collaborations, based on which this paper has been written. C. C. is a research fellow of the Miller Institute for Basic Research in Science. This work was supported in part the U.S. Department of Energy under Contract DE-AC03-76SF00098, and in part by the National Science Foundation under grant PHY-95-14797.

REFERENCES

1. J. M. Maldacena, *Adv. Theor. Math. Phys.* **2** 231 (1998).
2. E. Witten, *Adv. Theor. Math. Phys.* **2** 505 (1998).
3. C. Csáki, H. Ooguri, Y. Oz and J. Terning, *JHEP* **9901** 017 (1999).
4. S. Gubser, I. Klebanov and A. Polyakov, *Phys. Lett.* **B428** 105 (1998); E. Witten, *Adv. Theor. Math. Phys.* **2** 253 (1998).
5. M. J. Teper, hep-lat/9711011.
6. R. de Mello Koch, A. Jevicki, M. Mihailescu and J. Nunes, *Phys. Rev.* **D58** 105009 (1998); M. Zyskin, *Phys. Lett.* **B439** 373 (1998).
7. J. Minahan, hep-th/9811156.
8. S. Gubser, I. Klebanov and A. Tseytlin, *Nucl. Phys.* **B534** 202 (1998).
9. H. Ooguri, H. Robins and J. Tannenhauser, *Phys. Lett.* **B437** 77-81 (1998).
10. J. Russo, hep-th/9808117.
11. C. Csáki, Y. Oz, J. Russo and J. Terning, *Phys. Rev.* **D59** 065008 (1999).
12. C. Csáki, J. Russo, K. Sfetsos and J. Terning, hep-th/9902067.
13. A. Hashimoto and Y. Oz, hep-th/9809106.
14. C. Morningstar and M. Peardon, *Phys. Rev.* **D56** 4043 (1997).
15. C. Morningstar and M. Peardon, hep-lat/9901004.
16. J. Russo and K. Sfetsos, hep-th/9901056.
17. M. Cvetič and S. Gubser, hep-th/9903132.

Chiral Phase Transition for $SU(N)$ Gauge Theories

Francesco Sannino

Department of Physics, Yale University, New Haven, CT 06520-8120, USA

Abstract. We describe [1] the chiral phase transition for vector-like $SU(N)$ gauge theories as a function of the number of quark flavors N_f by making use of an anomaly-induced effective potential. The potential depends explicitly on the full β-function and the anomalous dimension γ of the quark mass operator. By using this potential we argue that chiral symmetry is restored for $\gamma < 1$. A perturbative computation of γ then leads to an estimate of the critical value N_f^c for the transition.

INTRODUCTION

The phase structure of strongly coupled gauge field theories as a function of the number of matter fields N_f is a problem of general interest. Much has been learned about the phases of supersymmetric theories in recent years [2–6]. An equally interesting problem is the phase structure of a non-supersymmetric $SU(N)$ theory as a function of the number of fermion fields N_f. At low enough values of N_f, the chiral symmetry $SU(N_f)_L \times SU(N_f)_R$ is expected to break to the diagonal subgroup. At some value of N_f less than $11N/2$ (where asymptotic freedom is lost), there will be a phase transition to a chirally symmetric phase. Whether the transition takes place at a relatively small value of N_f [7] or a larger value remains unknown. The larger value ($N_f/N \approx 4$) is suggested by studies of the renormalization group improved gap equation [8] and is associated with the existence of an infrared fixed point. A recent analysis [9] indicates that instanton effects could also trigger chiral symmetry breaking at comparably large value of N_f/N. Besides being of theoretical interest, the physics of a chiral transition could have consequences for electroweak symmetry breaking [10], since near-critical gauge theories provide a natural framework for walking technicolor theories [11].

If a phase transition is second order, a useful approach is to find a tractable model in the same universality class. For chiral symmetry, a natural order parameter is the $N_f \times N_f$ complex matrix field M describing mesonic degrees of freedom. If the meson degrees of freedom are the only ones that develop large correlation lengths at the phase transition, then the transition may be studied using an effective Landau-Ginzburg theory.

CP494, *New Directions in Quantum Chromodynamics*, edited by C.-R. Ji and D.-P. Min

For the zero-temperature transition as a function of N_f, a similar approach might also be tried. It was suggested in Ref. [8], however, that while the order parameter vanishes continuously as $N_f \to N_f^c$, the transition is not second order. With the gap equation dominated by an infrared fixed point of the gauge theory, the transition was argued to be continuous but infinite order. It has also been noted [12] that because of the associated long range conformal symmetry, the masses of all the physical states, not just the scalar mesons are expected to scale to zero with the order parameter.

We nevertheless suggest that an effective potential using only the low lying mesonic degrees of freedom might be employed to model at least some aspects of the zero-temperature chiral phase transition. The key ingredient is the presence of a new non-analytic potential term that emerges naturally once the anomaly structure of the theory is considered. The anomalies also provide a link between this effective potential term and the underlying gauge theory.

To deduce the anomaly induced effective potential we modify an effective potential [13–15] developed for $N_f < N$, and apply it to the range $N_f > N$.

We use this potential to discuss the zero-temperature phases of an $SU(N)$ gauge theory as a function of N_f. Assuming that the transition is governed by an infrared fixed point of the theory, we deduce that chiral symmetry is restored, together with long-range conformal symmetry, when $\gamma < 1$, where γ is the anomalous dimension of the mass operator. Finally we note that by using the perturbative expansion of γ, chiral symmetry is predicted to be restored above $N_f^c \approx 4N$, in agreement with a gap equation analysis.

THE EFFECTIVE POTENTIAL

In this section we construct an effective potential valid to all orders in the loop expansion and appropriate for the range $N_f > N$. The new ingredients are:

i) Using the full, rather than the one loop, beta function in the trace anomaly saturation.

ii) Taking account of the anomalous dimension of the fermion mass operator.

This anomaly-induced effective potential is based on the QCD trace and $U_A(1)$ anomalies (see [1] for more details).

We build the potential out of the $N_f \times N_f$ complex meson matrix M_i^j transforming as the operator $q_i \bar{q}^j$. So we assign naive mass dimension 3 to M_i^j. The operator $q\bar{q}$ acquires an anomalous dimension γ when quantum corrections are considered and the full dynamical dimension is thus $3 - \gamma$. To make our effective potential capture the low-energy quantum dynamics of the underlying theory, we take $3 - \gamma$ to be the scaling dimension of M_i^j. The anomalous dimension γ is of course a function of the coupling g, which in turn depends on the relevant scale.

To build the final meson potential we only use the chirally invarian term $\det M$. This is plausible (see section VII of Ref. [16]) and would correspond to the "holonomic" structure which emerges if the potential is considered to arise ([13–15]) from broken super QCD. In the same spirit we take $\det M$ to have the scaling dimension $(3 - \gamma)N_f$.

The potential term we find is [1]

$$V = -C\Lambda^4 \left[\frac{\Lambda^{3N_f}}{\det M} \right]^{\frac{4}{f(g)}} + \text{h.c.} , \tag{1}$$

where C is related to A via:

$$C = \frac{f(g)}{4e} \exp \left[\frac{4A}{f(g)} \right] , \tag{2}$$

and

$$f(g) = -\frac{\beta(g)}{g^3} 16\pi^2 - (3 - \gamma)N_f . \tag{3}$$

Finally we integrate out the η' field, which can be isolated by setting

$$\det M = |\det M| e^{i\phi} , \tag{4}$$

where $\phi \propto \eta'$. This is done anticipating that the η' will be heavy with respect to the intrinsic scale of the theory and the other mesonic degrees of freedom. Now using Eq. (4) we derive the field equation $\phi = 0$ which leads to the final potential

$$V = -2C\Lambda^4 \left[\frac{\Lambda^{3N_f}}{|\det M|} \right]^{\frac{4}{f(g)}} . \tag{5}$$

The shape of this potential is determined by the function $f(g)$ Eq. (3).

Our interest here is in the range $N < N_f < (11/2)N$ where the chiral phase transition is expected to occur. For N_f close to $(11/2)N$, a weak infrared fixed point will occur. The β function will be negative and small at all scales and γ will also be small. Thus $f(g)$ will be negative. As N_f is reduced, the fixed point coupling increases as does γ. However in the range of interest $f(g)$ will remain negative $((3 - \gamma)N_f > -(\beta(g)/g^3)16\pi^2)$. The potential in Eq. (5) may then be written as

$$V = +2|C|\Lambda^4 \left[\frac{|\det M|}{\Lambda^{3N_f}} \right]^{\frac{4}{\frac{\beta(g)}{g^3}16\pi^2 + (3-\gamma)N_f}} . \tag{6}$$

It is positive definite and vanishes with the field $|\det M|$.

The Chiral Phase Transition

To study the chiral phase transition, we need the combined effective potential

$$V_{tot} = V + V_I \tag{7}$$

where V_I is a generic potential term not associated with the anomalies. It is instructive, however, to investigate first the extremum properties of the anomaly term (Eq. (6)). Assuming the standard pattern for chiral symmetry breaking $SU_R(N_f) \times SU_L(N_f) \to SU_V(N_f)$, M_j^i may be taken to be the order parameter for the transition. For purposes of this discussion, we restrict attention to the vacuum value of M_j^i, which can be rotated into the form $M_j^i = \delta_j^i \rho$, where $\rho \geq 0$ is the modulus. Substituting the previous expression in the anomaly induced effective potential gives the following expression:

$$V = +2|C|\Lambda^4 \left[\frac{\rho}{\Lambda^3}\right]^{\frac{4N_f}{\frac{\beta(g)}{g^3}16\pi^2+(3-\gamma)N_f}} . \tag{8}$$

Recall that $((3 - \gamma)N_f > -(\beta(g)/g^3)16\pi^2)$ in the range of interest. The first derivative $\partial V/\partial \rho$ vanishes at $\rho = 0$ provided that $\frac{4N_f}{\frac{\beta(g)}{g^3}16\pi^2+(3-\gamma)N_f} > 1$, a condition that is clearly satisfied. The second derivative,

$$\frac{\partial^2 V}{\partial \rho^2} \propto \rho^{\left[\frac{4N_f}{\frac{\beta(g)}{g^3}16\pi^2+(3-\gamma)N_f}-2\right]} , \tag{9}$$

also vanishes at $\rho = 0$ if the exponent in Eq. (9) is positive. The second derivative at $\rho = 0$ is a positive constant when the exponent vanishes, and it is $+\infty$ for

$$\frac{4N_f}{\frac{\beta(g)}{g^3}16\pi^2 + (3 - \gamma)N_f} - 2 < 0 . \tag{10}$$

The curvature of V_{tot} at the origin is given by the sum of the two terms $\frac{\partial^2 V}{\partial \rho^2}$ and $\frac{\partial^2 V_I}{\partial \rho^2}$, evaluated at $\rho = 0$.

To proceed further, we assume that the phase transition is governed by an infrared stable fixed point of the gauge theory. We thus set $\beta(g) = 0$. The curvature of V at the origin is then 0 for $\gamma > 1$, finite and positive for $\gamma = 1$, and $+\infty$ for $\gamma < 1$. The value of γ depends on the fixed point coupling, which in turn depends on N_f. As N_f is reduced from $(11/2)N$, the fixed point coupling increases from 0, as does γ. Assuming that γ remains monotonic in N_f, growing to 1 and beyond as N_f decreases, there will be some critical value N_f^c below which $\frac{\partial^2 V}{\partial \rho^2}$ vanishes at the origin. The curvature of V_{tot} will then be dominated by the curvature of V_I at the origin. For $N_f = N_f^c$, there will be a finite positive contribution to the curvature from the anomaly-induced potential. For $N_f > N_f^c$ ($\gamma < 1$), V possesses an infinite positive curvature at the origin, suggesting that chiral symmetry is necessarily restored. We will here take the condition $\gamma = 1$ to mark the boundary between the broken and symmetric phases, and explore its consequences. This condition was suggested in Ref. [17], based on other considerations.

We next investigate the behavior of the theory near the transition by combining the above behavior with a simple model of the additional, non-anomalous potential V_I. We continue to focus only on the modulus ρ and take the potential to be a traditional Ginzburg-Landau mass term, with the squared mass changing from positive to negative as $\gamma - 1$ goes from negative to positive: $(1 - \gamma) \Lambda^{-2} \rho^2$. Additional, stabilizing terms, such as a ρ^4 term, could be added but will not affect the qualitative conclusions. The full potential is then

$$V_{tot} = 2|C|\Lambda^4 (\frac{\rho}{\Lambda^3})^{\frac{4}{3-\gamma}} - (\gamma - 1)\,\Lambda^{-2}\rho^2 \ . \tag{11}$$

For $\gamma > 1$ (but < 3), the first term stabilizes the potential for large ρ, and the potential is minimized at

$$< \rho > = \Lambda^3 \left[\frac{\gamma - 1}{2|C|}\right]^{\frac{1}{\gamma - 1}} \ , \tag{12}$$

in the limit $\gamma \to 1$. It describes an infinite order phase transition as $\gamma \to 1$, in qualitative agreement with the gap equation studies. This behavior would not be changed by the addition of higher power terms $(\rho^4, \rho^6, ...)$ to the potential.

The curvature of the potential Eq. (11) at the minimum describes a mass associated with the field ρ. To interpret this mass physically, one should construct the kinetic energy term associated with this field (at least to determine its behavior as a function of $\gamma - 1$). We hence rescale ρ to a field σ via $\rho = \sigma^{3-\gamma}\Lambda^\gamma$ with σ possessing a conventional kinetic term $-\frac{1}{2}(\partial^\mu \sigma)^2$. This then leads to the following result for the physical mass M_σ and $< \sigma >$

$$< \sigma > \simeq \left[\frac{\gamma - 1}{2|C|}\right]^{\frac{1}{2(\gamma - 1)}} \Lambda \ , \qquad M_\sigma \simeq 2\sqrt{6}|C| \left[\frac{\gamma - 1}{2|C|}\right]^{\frac{1}{2(\gamma - 1)}} \Lambda \ . \tag{13}$$

Likewise, in the presence of the quark mass term we have (see [1] for details)

$$[< \sigma >]_{\gamma=1} \simeq \left[\frac{mN_f\Lambda}{2|C|}\right]^{\frac{1}{2}} \ , \qquad [M_\sigma]_{\gamma=1} \simeq 2\,[2mN_f\Lambda]^{\frac{1}{2}} \ . \tag{14}$$

Thus the order parameter σ for $\gamma = 1$ vanishes according to the power $1/2$ with the quark mass in contrast with an ordinary second order phase transition where the order parameter is expected to vanish according to the power $1/3$.

Finally we note an important distinction between our effective potential describing an infinite order transition and the Ginzburg-Landau potential describing a second order transition. The latter may be used in both the symmetric and broken phases, describing light scalar degrees of freedom as the transition is approached from either side. Our potential develops infinite curvature at the origin in the symmetric phase, indicating that no light scalar degrees of freedom are formed as the transition is approached from that side. This is in agreement with the conclusions of Ref. [18], indicating that as one crosses to the symmetric phase, mesons melt into quarks and gluons and hence the physics is described via only the underlying degrees of freedom. The present effective Lagrangian formalism for describing the chiral/conformal phase transition is close in spirit to the one developed in Ref. [19].

By perturbatively (see [1]) saturating at two loops the condition $\gamma < 1$ at the fixed point value of the coupling constant leads to the conclusion that chiral symmetry is restored for $N_f > N_f^c \simeq 3.9N$.

CONCLUSIONS

We have explored the chiral phase transition for vector-like $SU(N)$ gauge theories as a function of the number of flavors N_f via an anomaly induced effective potential. The effective potential was constructed by saturating the trace and axial anomalies. It depends on the full beta function and anomalous dimension of the quark-mass operator. We showed that the anomaly induced effective potential for $N_f > N$ is positive definite and vanishes with the field M_i^j. We then investigated the stability of the potential at the origin, and discovered that the second derivative is positive and divergent when the underlying β function and the anomalous dimension of the quark-mass operator satisfy the relation of Eq. (10). We took this to be the signal for chiral restoration. With conformal symmetry being restored along with chiral symmetry (due to the β function vanishing at an infrared fixed point), the criticality relation becomes a constraint on the anomalous dimension of the quark-mass operator:

$$\gamma < 1 \ . \tag{15}$$

To convert this inequality into a condition for a critical number of flavors, we used the perturbative expansion of the anomalous dimension evaluated at the fixed point, deducing that chiral symmetry is restored for $N_f \simeq 4N$, in agreement with gap equation studies.

The core of this talk is the proposal that the chiral/conformal phase transition, suggested by gap equation studies to be continuous and infinite order, may be described by an effective potential whose form is dictated by the trace and axial anomalies of the underlying $SU(N)$ gauge theory.

ACKNOWLEDGMENTS

I am very happy to thank Joseph Schechter for sharing the work on which this talk is based and Thomas Appelquist for enlightening discussions. The work has been partially supported by the US DOE under contract DE-FG-02-92ER-40704.

REFERENCES

1. F. Sannino and J. Schechter, Report Numbers: YCTP-P04-99, SU-4240-694, hep-ph/9903359. To appear in Phys. Rev. D. See here for more details and a complete list of relevant references.
2. N. Seiberg, Phys. Rev. D **49**, 6857 (1994). N. Seiberg, Nucl. Phys. **B435**, 129 (1995).
3. N. Seiberg and E. Witten, Nucl. Phys. **B426**, 19 (1994), E **430**, 495(1994); **B431**, 484 (1994).
4. K. Intriligator and N. Seiberg, Nucl. Phys. Proc. Suppl. **45BC**, 1, 1996.
5. M.E. Peskin,(TASI 96): Fields, Strings, and Duality, Boulder, CO, 2-28 Jun 1996.
6. P. Di Vecchia, Surveys High Energy Phys. **10**, 119 (1997).
7. R.D. Mawhinney, Report Numbers: CU-TP-839, CU-TP-802, hep-lat/9705030; D. Chen, R.D. Mawhinney, Nucl. Phys. Proc. Suppl. **53**, 216, 1997.
8. T. Appelquist, J. Terning and L.C.R. Wijewardhana, Phys. Rev. Lett. **77**, 1214 (1996).
9. T. Appelquist and S. Selipsky, Phys. Lett. B**400**, 364 (1997). M. Velkosky and E. Shuryak, hep-ph/9703345.
10. T. Appelquist and F. Sannino, Phys. Rev. D**59**, 067702 (1999). T. Appelquist, P.S. Rodrigues da Silva and F. Sannino, Report Numbers: YCTP-P14-99, hep-ph/9906555.
11. T. Appelquist, J. Terning and L.C.R. Wijewardhana, Phys. Rev. Lett. **79**, 2767 (1997).
12. S. Chivukula, Phys. Rev. D**55**, 5238 (1997).
13. F. Sannino and J. Schechter, Phys. Rev. D.**57**, 170 (1998).
14. S.D. Hsu, F. Sannino and J. Schechter, Phys. Lett. B**427** , 300 (1998).
15. F. Sannino, Proceedings for the MRST-98 conference: Toward the Theory of Everything, Montreal, Canada, 13-15 May 1998.
16. H. Gomm, P. Jain, R. Johnson and J. Schechter, Phys. Rev. **D33**, 801 (1986).
17. A. Cohen and H. Georgi , Nucl. Phys. **B314** (1989), 7.
18. T. Appelquist, A. Ratnaweera, J. Terning and L.C.R. Wijewardhana, Phys. Rev. D **58** (1998), 105017.
19. V.A. Miransky and K. Yamawaki, Phys. Rev. D**55**, 5051 (1997); erratum ibid. D**56**, 3768 (1997).

PHENOMENOLOGY
CONFRONTING QCD

Semiexclusive Processes: A different way to probe hadron structure

Carl E. Carlson

Nuclear and Particle Theory Group, Physics Department,
College of William and Mary, Williamsburg, VA 23187-8795

Abstract. Hard semiexclusive processes provide an opportunity to design effective currents to probe specific parton distributions, and to probe in leading order parton distributions that fully inclusive reactions probe only via higher order corrections. High transverse momentum pion photoproduction is an example of a such a process. We discuss the perturbative and soft processes that contribute, and show how regions where perturbative processes dominate can give us the parton structure information. Polarized initial states are needed to get information on polarization distributions. Current polarization asymmetry data is mostly in the soft region. However, with somewhat higher energy, determining the polarized gluon distribution using hard pion photoproduction appears quite feasible.

SEMI-EXCLUSIVE PROCESSES AS PROBES OF HADRON STRUCTURE

This talk will discuss hard semiexclusive processes, namely processes of the form $B + A \rightarrow C + X$, where the momentum transfer $t = (p_B - p_C)^2$ is large [1–8]. Both the cases where the hadron C is part of a jet and where it is kinematically isolated are interesting. Semiexclusive processes provide the capability of designing "effective currents" [7] that probe specific parton distributions and for probing in leading order target distributions that are not probed at all in leading order in inclusive reactions.

Particle B can be a hadron or a real or virtual photon. We will here limit ourselves to the latter. The process we will discuss is

$$\gamma + A \rightarrow M + X, \tag{1}$$

where A is the target and M is a meson, for definiteness the pion. The process is perturbative because of the high transverse momentum of the pion, not because of the high Q^2 of the photon. Soft processes are from the present viewpoint an annoyance, but one we need to discuss and we will estimate their size farther below.

CP494, *New Directions in Quantum Chromodynamics*, edited by C.-R. Ji and D.-P. Min
© 1999 American Institute of Physics 1-56396-908-4/99/$15.00

Our considerations also apply to electroproduction,

$$e + A \rightarrow M + X \tag{2}$$

when the final electron is not seen. We use the Weizäcker-Williams equivalent photon approximation [9] to relate the electron and photon cross sections,

$$d\sigma(eA \rightarrow MX) = \int dE_\gamma \, N(E_\gamma) d\sigma(\gamma A \rightarrow MX), \tag{3}$$

where the number distribution of photons accompanying the electron is a well known function.

In the following section, we will describe the subprocesses that contribute to hard pion production and show how the cross sections are dependent upon the parton densities and distribution amplitudes that we wish to probe, and in the subsequent section display some results. There will be a short summary at the end.

THE SUBPROCESSES

At the Highest k_T

At the highest possible transverse momenta, observed pions are directly produced at short range via a perturbative QCD (pQCD) calculable process [3–6]. Two out of four lowest order diagrams are shown Fig. 1. The pion produced this way is kinematically isolated rather than part of a jet, and may be seen either by making an isolated pion cut or by having some faith in the calculation and going to a kinematic region where this process dominates the others. Although this process is higher twist, at the highest transverse momenta its cross section falls less quickly than that of the competition, and we will show plots indicating the kinematics where it can be observed.

The subprocess cross section for direct or short-distance pion production is

$$\frac{d\hat{\sigma}}{dt}(\gamma q \rightarrow \pi^\pm q') = \frac{128\pi^2\alpha\alpha_s^2}{27(-t)\hat{s}^2} I_\pi^2 \left(\frac{e_q}{\hat{s}} + \frac{e_q'}{\hat{u}} \right) \left[\hat{s}^2 + \hat{u}^2 + \lambda h(\hat{s}^2 - \hat{u}^2) \right], \tag{4}$$

where $\hat{s}$, $\hat{t} = t$, and $\hat{u}$ are the subprocess Mandlestam variables; λ and h are the helicities of the photon and target quark, respectively; and I_π is the integral

$$I_\pi = \int \frac{dy_1}{y_1} \phi_\pi(y_1, \mu^2). \tag{5}$$

In the last equation, ϕ_π is the distribution amplitude of the pion, and describes the quark-antiquark part of the pion as a parallel moving pair with momentum fractions y_i. It is normalized through the rate for $\pi^\pm \rightarrow \mu\nu$, and for example,

340

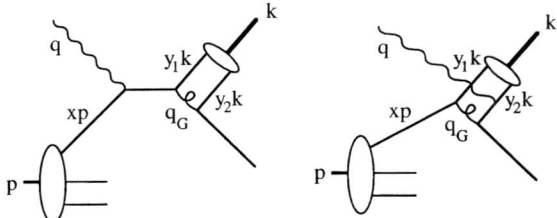

FIGURE 1. Direct pion production. The pion is produced in a short distance perturbatively calculated process, not by fragmentation of an outgoing parton. Thus this pion is kinematically isolated, and not part of a jet. Direct pion production gets important at very high transverse momentum because the pion does not have to share the available momentum with any other particles going in its direction.

$$\phi_\pi = \frac{f_\pi}{2\sqrt{3}} 6y_1(1 - y_1) \tag{6}$$

for the distribution amplitude called "asymptotic" and for $f_\pi \approx 93$ MeV. Overall,

$$\frac{d\sigma}{dx\,dt}(\gamma A \to \pi X) = \sum_q G_{q/A}(x, \mu^2)\frac{d\hat{\sigma}}{dt}(\gamma q \to \pi^\pm q'), \tag{7}$$

where $G_{q/A}(x, \mu^2)$ is the number distribution for quarks of flavor q in target A with momentum fraction x at renormalization scale μ.

There are a number of interesting features about direct pion production.

• For photoproduction, the struck quark's momentum fraction is fixed by experimental observables. This is like deep inelastic scattering, where the experimenter can measure $x \equiv Q^2/2m_N\nu$ and this x is also the momentum fraction of the struck quark (for high Q and ν). For the present case, momenta are defined in Fig. 1 and the Mandlestam variables are

$$s = (p + q)^2; \quad t = (q - k)^2; \quad \text{and} \quad u = (p - k)^2. \tag{8}$$

The Mandlestam variables are all observables, and the ratio

$$x = \frac{-t}{s + u} \tag{9}$$

is the momentum fraction of the struck quark. We will let the reader prove this.

• The gluon involved in direct pion production is well off shell [3–5].

• Without polarization, we can measure I_π, given trust in the other parts of the calculation. This I_π is precisely the same as the I_π in both $\gamma^*\gamma \to \pi^0$ and $e\pi^\pm \to e\pi^\pm$.

• We have polarization sensitivity. For π^+ production at high x,

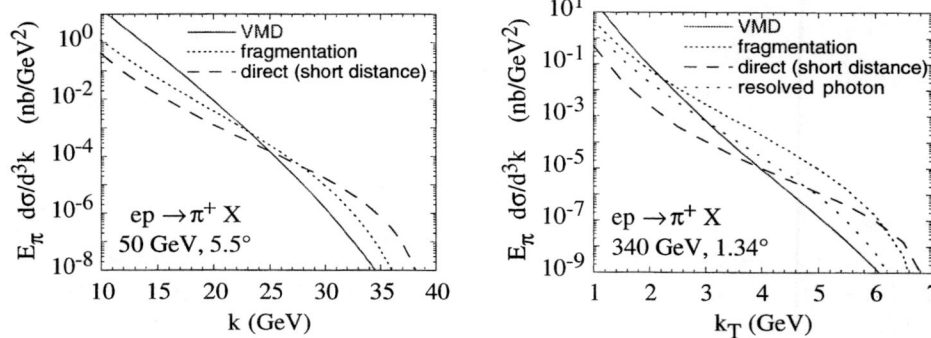

FIGURE 2. Calculated contributions to the cross section for $ep \to \pi^+ X$, with electron beam energies and pions emerging at angles in the target rest frame as indicated; k is the pion momentum. The 50 GeV plot shows a significant region where direct pion production dominates, once there is enough momentum that the soft ("VMD") processes are not large. The 340 GeV plot shows a long window where the fragmentation process dominates.

$$A_{LL} \equiv \frac{\sigma_{R+} - \sigma_{L+}}{\sigma_{R+} + \sigma_{L+}} = \frac{s^2 - u^2}{s^2 + u^2} \cdot \frac{\Delta u(x)}{u(x)} \qquad (10)$$

where R and L refer to the polarization of the photon, and $+$ refers to the target, say a proton, polarization. Also, inside a $+$ helicity proton the quarks could have either helicity, and

$$\Delta u(x) \equiv u_+(x) - u_-(x). \qquad (11)$$

The large x behavior of both $d(x)/u(x)$ and $\Delta d(x)/\Delta u(x)$ are of current interest. Most fits to the data have the down quarks disappearing relative to the up quarks at high x, in contrast to pQCD which has definite non-zero predictions for both of the ratios in the previous sentence. Recent improved work on extracting neutron data from deuteron targets, has tended to support the pQCD predictions [10].

Experimentally, direct or short-range pion production can be seen. To show this, Fig. 2(left) plots the differential cross section for high transverse momentum π^+ electroproduction for a SLAC energy. Specifically, we have 50 GeV incoming electrons, with the pion emerging at 5.5° in the lab. It shows that above about 27 GeV total pion momentum or 2.6 GeV transverse momentum, direct (short distance, isolated) pion production exceeds its competition. Also shown in Fig. 2 is a situation where there is a long region where the fragmentation process—next up for discussion—dominates. Incidentally, the 340 GeV energy for the electron beam on stationary protons was chosen to match recent very preliminary discussions of an Electron Polarized Ion Collider (EPIC) with 4 GeV electrons and 40 GeV protons, and the 1.34° angle in the target rest frame matches 90° in the lab for such a collider.

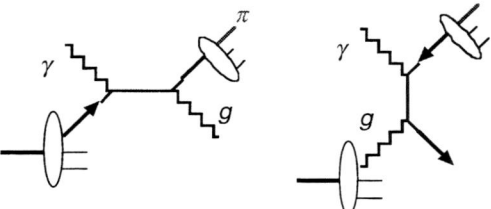

FIGURE 3. The fragmentation process. Pions are produced by fragmentation of partons at long distances from the primary interaction region. The Compton process is on the left; the pion could come from either quark or gluon fragmentation. Quark-gluon fusion is on the right.

Moderate k_T

At moderate transverse momentum, the generally dominant process is still a direct interaction in the sense that the photon interacts directly with constituents of the target, but the pion is not produced directly at short range but rather at long distances by fragmentation of some parton [1,2,5]. Many authors refer to this as the direct process; others of us are in the habit of calling it the fragmentation process. The main subprocesses are called the Compton process and photon-gluon fusion, and one example of each is shown in Fig. 3.

Photon gluon fusion often gives 30–50% of the cross section for the fragmentation process, and the polarization asymmetry is as large as can be in magnitude,

$$\hat{A}_{LL}(\gamma g \to q\bar{q}) = -100\%. \tag{12}$$

Typically for the Compton process, $\hat{A}_{LL}(\gamma q \to gq) \approx 1/2$. We shall show some A_{LL} plots for the overall process after we discuss the soft processes.

We should note that the NLO calculations for the fragmentation process have been done also for the polarized case, though our plots are based on LO. For direct pion production, NLO calculations are not presently completed.

We should also remark that the photon may split into hadronic matter before interacting with the target. If splits into a quark anti-quark pair that are close together, the splitting can be modeled perturbatively or quasi-perturbatively, and we call it a "resolved photon process." A typical diagram is shown in the left hand part of Fig. 4. Resolved photon processes are crucial at HERA energies, but not at energies under discussion here, and we say no more about them.

Soft Processes

This is the totally non-perturbative part of the calculation, whose size can be estimated by connecting it to hadronic cross sections. The photon may turn into hadronic matter, such as $\gamma \to q\bar{q} + \ldots$ with a wide spatial separation. It can be represented as photons turning into vector mesons. See Fig. 4(right).

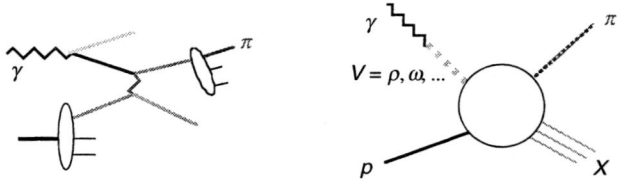

FIGURE 4. Resolved photon process (left) and vector meson dominated process (right).

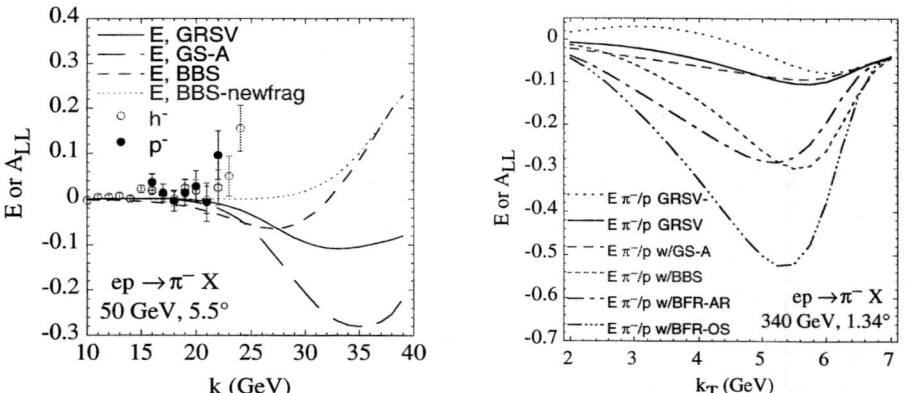

FIGURE 5. Longitudinal polarization asymmetries for π^- production for two energies and target rest frame angles as labeled. A description of the curves is given in the text.

We want a reliable approximation to the non-perturbative cross section so we can say where perturbative contributions dominate and where they do not. To get such an approximation one can start with the cross section given as

$$ d\sigma(\gamma A \to \pi X) = \sum_V \frac{\alpha}{\alpha_V} d\sigma(V + A \to \pi X) + \Big(\text{non} - \text{VMD} \Big), \qquad (13) $$

where the sum is over vector mesons V, $\alpha = e^2/4\pi$, and $\alpha_V = f_V^2/4\pi$. We can get, for example, f_ρ from the decay $\rho \to e^+ e^-$. Then "all" one needs is a parameterization of the hadronic process, based on data. Details of our implementation of this program are given in [12].

We took the soft processes to be polarization insensitive. This agrees with a recent Regge analysis of Manayenkov [13].

RESULTS

Since results for the unpolarized cross section have already been displayed in Fig. 2, we focus on results for A_{LL}, which is also called E by some authors [14]. Fig. 5 shows two plots, both for $\pi-$ production.

The 50 GeV plot, Fig. 5(left), is dominated by direct pion production above the soft region, and is sensitive mainly to the differing polarized quark distributions of the different models. Three different parton distribution models are shown [17–19]. Although the fragmentation process is not the crucial one here, we should mention that mostly we used our own fragmentation functions [3], and that the results using a better known set [15] are not very different. Neither set of fragmentation functions agrees well with the most recent HERMES data [16] for unfavored vs. favored fragmentation functions, and the one curve labeled "newfrag" is calculated with fragmentation functions that agree better with that data.

Below about 20 GeV total pion momentum where the soft process dominates, the data is well described by supposing the soft processes are polarization independent. Above that, with asymmetry due to perturbative processes, the difference among the results for the different sets of parton distributions is quite large for the π^-.

The data is from Anthony *et al.* [11]. Presently most of the data is in the region where the soft processes dominate. The data is already interesting. Further data at even higher pion momenta would be even more interesting. Regarding the differences among the quark distributions, recall that large momentum corresponds to $x \to 1$ for the struck quark, and pQCD predicts that the quarks are 100% polarized in this limit. Only the parton distributions labeled BBS [17] are in tune with the pQCD prediction, and they for large momentum predict even a different sign for A_{LL} for the π^-. Calculated results plotted with the data for the π^+ and for deuteron targets may be examined in [12].

The other plot in Fig. 5 is for 340 GeV electron beam energy, an energy where there is a long region where the fragmentation process dominates. We would like to know how sensitive the possible measurements of A_{LL} are to the different models for Δg. To find out, Fig. 5 (right) presents calculated results for A_{LL} for one set of quark distributions and 5 different distributions for Δg [17–20]. The quark distributions and unpolarized gluon distribution in each case are those of GRSV [18]. There are 6 curves on each figure. One of them (labeled GRSV–) is a benchmark, which was calculated with Δg set to zero. The other curves use the Δg from the indicated distribution. There is a fair spread in the results, especially for the π^- where photon-gluon fusion gives a larger fraction of the cross section. Thus, one could adjudicate among the polarized gluon distribution models.

SUMMARY

Several perturbative processes contribute to hard pion photoproduction. All are calculable. They give us new ways to measure aspects of the pion wave function, and quark and gluon distributions, especially Δq and Δg. The soft processes can be estimated and avoided if the transverse momentum is greater than about 2 GeV. SLAC or HERMES energies would be excellent for finding direct pion production, which is sensitive to Δu and Δd, and higher energies would give a region where the fragmentation process dominates and be excellent for measuring Δg.

ACKNOWLEDGMENTS

My work on this subject has been done with Andrei Afanasev, Chris Wahlquist, and A. B. Wakely and I thank them for pleasant collaborations. I have also benefited from talking to and reading the work of many authors and apologize to those I have not explicitly cited. I thank the NSF for support under grant PHY-9900657.

REFERENCES

1. De Florian, D., and Vogelsang, W., *Phys. Rev.* D **57**, 4376 (1998); Kniehl, B. A., Talk at Ringberg Workshop, hep-ph/9709261; Stratmann, M., and Vogelsang, W., Talk at Ringberg Workshop, hep-ph/9708243.
2. Peralta, J. J., Contogouris, A. P., Kamal, B., and Lebessis, F., *Phys. Rev.* D **49**, 3148 (1994).
3. Carlson, C. E., and Wakely, A. B., *Phys. Rev.* D **48**, 2000 (1993).
4. Afanasev, A., Carlson, C. E., and Wahlquist, C., *Phys. Lett.* B **398**, 393 (1997).
5. Afanasev, A., Carlson, C. E., and Wahlquist, C., *Phys. Rev.* D **58**, 054007 (1998).
6. Brodsky, S. J., Diehl, M., Hoyer, P., and Peigne, S., *Phys. Lett.* B **449**, 306 (1999).
7. Brodsky, S. J., Talk presented at the EPIC'99 Workshop, Indiana University Cyclotron Facility, Bloomington, Indiana, April 8–11, 1999, hep-ph/9907346.
8. Carlson, C. E., Talk presented at the EPIC'99 Workshop, Indiana University Cyclotron Facility, Bloomington, Indiana, April 8–11, 1999, hep-ph/9905492.
9. See for example the Appendix to Brodsky, S. J., Kinoshita, T., and Terazawa, H., *Phys. Rev.* D **4**, 1532 (1971).
10. Melnitchouk, W., Speth, J., and Thomas, A. W., PLB **435**, 420 (1998); Melnitchouk, W., and Peng, J.C., *Phys. Lett.* B **400**, 220 (1997); Melnitchouk, W., and Thomas, A. W., *Phys. Lett.* B **377**, 11 (1996); Yang, U. K., and Bodek, A., *Phys. Rev. Lett.* **82**, 2467 (1999).
11. Anthony, P. L. *et al.*, *Phys. Lett.* B **458**, 536 (1999).
12. Afanasev, A., Carlson, C. E., and Wahlquist, C., hep-th/9903493.
13. Manayenkov, S. I., report DESY 99-016, hep-ph/9903405.
14. Barker, I. S., Donnachie, A., and Storrow, J. K., *Nucl. Phys.* B **95**, 347 (1975).
15. Binneweis, J., Kniehl, B. A., and Kramer, G., *Z. Phys.* C **65**, 471 (1995) and *Phys. Rev.* D **52**, 4947 (1995).
16. Makins, N., Proceedings of CEBAF Workshop on Physics and Instrumentation with 6-12 GeV Beams, ed. S. Dytman, H. Fenker, and P. Roos, JLab, Newport News, June 1998, p.97; Geiger, Ph., *Measurement of Fragmentation Functions at HERMES*, Ph. D. Thesis, Ruprechet-Karls-Universität, Heidelberg, 1998.
17. Brodsky, S. J., Burkardt, M., and Schmidt, I., *Nucl. Phys.* B **441**, 197 (1995).
18. Glück, M., Reya, E., Stratmann, M., and W. Vogelsang, *Phys. Rev.* D **53**, 4775 (1996).
19. Gehrmann, T., and Stirling, W. J., *Phys. Rev.* D **53**, 6100 (1996).
20. Ball, R. D., Forte, S., and Ridolfi, G., *Phys. Lett.* B **378**, 255 (1996).

Exploring the Timelike Region of QCD Exclusive Processes in the Relativistic Quark Model

Ho-Meoyng Choi and Chueng-Ryong Ji

Department of Physics, North Carolina State University, Raleigh, NC 27695-8202

Abstract. We investigate the form factors and decay rates of exclusive $0^- \to 0^-$ semileptonic meson decays using the constituent quark model based on the light-front quantization. Our model is constrained by the variational principle for the linear plus Coulomb interaction motivated by QCD. Our numerical results are in a good agreement with the available experimental data.

One of the distinctive advantages in the light-front approach is the well-established formulation of various form factor calculations using the well-known Drell-Yan-West ($q^+=0$) frame [1]. In $q^+=0$ frame, only parton-number-conserving Fock state (valence) contribution is needed when the "good" components of the current, J^+ and $J_\perp=(J_x, J_y)$, are used [2]. For example, only the valence diagram shown in Fig. 1(a) is used in the light-front quark model (LFQM) analysis of spacelike meson form factors. Successful LFQM description of various hadron form factors can be found in the literatures [3-5].

However, the timelike ($q^2 > 0$) form factor analysis in the LFQM has been hindered by the fact that $q^+=0$ frame is defined only in the spacelike region ($q^2=q^+q^- - q_\perp^2 < 0$). While the $q^+\neq0$ frame can be used in principle to compute the timelike form factors, it is inevitable (if $q^+\neq0$) to encounter the nonvalence diagram arising from the quark-antiquark pair creation (so called "Z-graph"). For example, the nonvalence diagram in the case of semileptonic meson decays is shown in Fig. 1(b). The main source of the difficulty, however, in calculating the nonvalence diagram(see Fig. 1(b)) is the lack of information on the black blob which should contrast with the white blob representing the usual light-front valence wave function. In fact, we noticed [2] that the omission of nonvalence contribution leads to a large deviation from the full results.

In this paper, we circumvent this problem by calculating the semileptonic processes in $q^+=0$ frame and then analytically continuing to the timelike region. The $q^+=0$ frame is useful because only valence contributions are needed. However, one needs to calculate the component of the current other than J^+ to obtain the form

CP494, *New Directions in Quantum Chromodynamics*, edited by C.-R. Ji and D.-P. Min
© 1999 American Institute of Physics 1-56396-908-4/99/$15.00

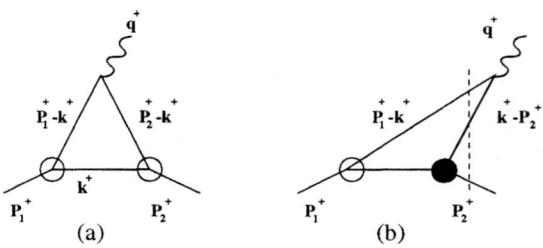

FIGURE 1. The LFQM description of a electroweak meson form factor: (a) the usual light-front valence diagram and (b) the nonvalence(pair-creation) diagram. The vertical dashed line in (b) indicates the energy-denominator for the nonvalence contributions. While the white blob represents the usual light-front valence wave function, the modeling of black blob has not yet been made.

factor $f_-(q^2)$. Since J^- is not free from the zero-mode contributions even in $q^+=0$ frame [6,7], we use $J_\perp$ instead of J^- to obtain f_-.

The key idea in our LFQM [5] for mesons is to treat the radial wave function as a trial function for the variational principle to the QCD-motivated Hamiltonian saturating the Fock state expansion by the constituent quark and antiquark. The spin-orbit wave function is uniquely determined by the Melosh transformation. We take the QCD-motivated effective Hamiltonian as the well-known linear plus Coulomb interaction given by

$$H_{q\bar{q}} = H_0 + V_{q\bar{q}} = \sqrt{m_q^2 + k^2} + \sqrt{m_{\bar{q}}^2 + k^2} + V_{q\bar{q}}, \tag{1}$$

where

$$V_{q\bar{q}} = V_0 + V_{\text{hyp}} = a + br - \frac{4\kappa}{3r} + \frac{2\vec{S}_q \cdot \vec{S}_{\bar{q}}}{3m_q m_{\bar{q}}} \nabla^2 V_{\text{Coul}}. \tag{2}$$

We take the Gaussian radial wave function $\phi(k^2) = N \exp(-k^2/2\beta^2)$ as our trial wave function to minimize the central Hamiltonian [5]. Since the string tension $b=0.18$ GeV^2 and the constituent u and d quark masses $m_u=m_d=0.22$ GeV are rather well known from other quark model analyses commensurate with Regge phenomenology [8], we take them as our input parameters. The model parameters of a, κ, and $\beta_{u\bar{d}}$ are determined by the variational principle using the masses of ρ and π [5,9]. It is very important to note that all other model parameters such as m_c, m_b, β_{uc}, β_{ub}, etc. are then uniquely determined by our variational principle as shown in [9]. More detailed procedure of determining the model parameters and ground state meson mass spectra can be found in [5,9].

Our predictions of the ground state meson mass spectra [9] are in a good agreement with the available experimental data. Furthermore, our model predicts the two unmeasured mass spectra of $^1S_0(b\bar{b})$ and $^3S_1(b\bar{s})$ systems as $M_{b\bar{b}}=9657$ MeV and $M_{b\bar{s}}=5424$ MeV, respectively. Our values of the decay constants [9] are also in a good agreement with the results of lattice QCD [10] anticipating future accurate experimental data.

The matrix element of the current $j^\mu = \bar{q}_2\gamma^\mu Q_1$ for $0^-(Q_1\bar{q}) \to 0^-(q_2\bar{q})$ decays can be parametrized in terms of two hadronic form factors as follows

$$\langle P_1|\bar{q}_2\gamma^\mu Q_1|P_1\rangle = f_+(q^2)(P_1 + P_2)^\mu + f_-(q^2)(P_1 - P_2)^\mu,$$

$$= f_+(q^2)\left[(P_1 + P_2)^\mu - \frac{M_1^2 - M_2^2}{q^2}q^\mu\right] + f_0(q^2)\frac{M_1^2 - M_2^2}{q^2}q^\mu, \quad (3)$$

where $q^\mu = (P_1 - P_2)^\mu$ is the four-momentum transfer to the leptons and $m_l^2 \leq q^2 \leq (M_1 - M_2)^2$. The form factors f_+ and f_0 are related to the exchange of 1^- and 0^+, respectively, and satisfy the following relations:

$$f_+(0) = f_0(0), \quad f_0(q^2) = f_+(q^2) + \frac{q^2}{M_1^2 - M_2^2}f_-(q^2). \quad (4)$$

In the LFQM calculations presented in Ref. [11], the $q^+\neq 0$ frame has been used to calculate the semileptonic decays in the timelike region. However, when the $q^+\neq 0$ frame is used, the inclusion of the nonvalence contributions arising from quark-antiquark pair creation (see Fig. 1(b)) is inevitable and this inclusion may be very important for light-to-light and heavy-to-light decays. Nevertheless, the previous analyses [11] considered only valence contributions in $q^+\neq 0$ frame neglecting non-valence contributions. In this work, we circumvent this problem by calculating the processes in $q^+=0$ frame and analytically continuing to the timelike region. The $q^+=0$ frame is useful because only valence contributions are needed. However, one needs to calculate the component of the current other than J^+ to obtain the form factor $f_-(q^2)$. Since J^- is not free from the zero-mode contributions even in $q^+=0$ frame [6], we use $J_\perp$ instead of J^- to obtain f_-. In the $q^+=0$ frame, we obtain the form factors $f_+(q^2)$ and $f_-(q^2)$ using the matrix element of the "+" and "⊥"-components of the current, J^μ, respectively, and then analytically continue to the timelike $q^2 > 0$ region by changing $q_\perp$ to $iq_\perp$ in the form factors.

Light-to-light decays: For K_{l3} decays, the three form factor parameters, i.e., λ_+, λ_0 and ξ_A, have been measured using the following linear parametrization [12]:

$$f_\pm(q^2) = f_\pm(q^2 = m_l^2)\left(1 + \lambda_\pm\frac{q^2}{M_{\pi^+}^2}\right), \quad (5)$$

where $\lambda_{\pm,0}$ is the slope of $f_{\pm,0}$ evaluated at $q^2 = m_l^2$ and $\xi_A = f_-/f_+|_{q^2=m_l^2}$. Our predictions of the parameters for K_{l3} decays in $q^+=0$ frame, i.e., $f_+(0)$, λ_+, λ_0, $\langle r^2 \rangle_{K\pi} = 6f'_+(0)/f_+(0) = 6\lambda_+/M_{\pi^+}^2$, and $\xi_A = f_-/f_+|_{q^2=m_l^2}$, are summarized in Table 1. Our result for the form factor f_+ at zero momentum transfer, $f_+(0) = 0.962$, is consistent with the Ademollo-Gatto theorem [13] and also in a good agreement with the result of chiral perturbation theory [14], $f_+(0) = 0.961\pm0.008$. Our results for other observables such as λ_+, ξ_A, and $\Gamma(K_{l3})$ are overall in a good agreement

TABLE 1. Model predictions for the parameters of K_{l3} decay form factors obtained from $q^+=0$ frame. The charge radius $r_{\pi K}$ is obtained by $\langle r^2 \rangle_{\pi K} = 6f'_+(q^2 = 0)/f_+(0)$. For comparison, we include the results (in square brackets) of the valence contribution in $q^+ \neq 0$ frame. The CKM matrix used in the calculation of the decay width (in units of 10^6 s^{-1}) is $|V_{us}| = 0.2205 \pm 0.0018$ [12].

Observables	Our model	Other models	Experiment
$f_+(0)$	0.962[0.962]	0.961 ± 0.008 [14], 0.952 [15], 0.93 [16]	
λ_+	0.026[0.083]	0.028 [15], 0.019 [16]	$0.0286 \pm 0.0022[K_{e3}^+]$
			$0.0300 \pm 0.0016[K_{e3}^0]$
λ_0	$-0.009[-0.017]$	0.0026 [15], -0.005 [16]	$0.004 \pm 0.007[K_{\mu3}^+]$
			$0.025 \pm 0.006[K_{\mu3}^0]$
ξ_A	$-0.41[-1.10]$	-0.28 [15], -0.28 [15]	$-0.35 \pm 0.15[K_{\mu3}^+]$
			$-0.11 \pm 0.09[K_{\mu3}^0]$
$\langle r \rangle_{\pi K}$ (fm)	0.56	0.57 [15], 0.48 [16]	
$\Gamma(K_{e3}^0)$	7.36 ± 0.12		$7.7 \pm 0.5[K_{e3}^0]$

with the experimental data [12]. For comparison, we also include the results (in square brackets of the second column of Table 1) of $f_+(0)$, λ_+, λ_0, and ξ_A obtained from the valence contribution in $q^+ \neq 0$ frame. Even though the form factor $f_+(0)$ in $q^+ \neq 0$ frame is free from the nonvalence contributions, its derivative at $q^2=0$, i.e., λ_+, receives the nonvalence contributions. Moreover, the form factor $f_-(q^2)$ in $q^+ \neq 0$ frame is not immune to the nonvalence contributions even at $q^2=0$ [6]. Unless one includes the nonvalence contributions in the $q^+ \neq 0$ frame, one cannot really obtain reliable predictions for the observables such as λ_+, λ_0 and ξ_A for K_{l3} decays.

In Fig. 2, we show the form factors f_+ obtained from both $q^+=0$ and $q^+ \neq 0$ frames for $0 \leq q^2 \leq (M_K - M_\pi)^2$ region. As one can see in Fig. 2, the form factor $f_+(q^2)$ obtained from $q^+=0$ frame (solid lines) [2] appears to be linear functions of q^2 justifying Eq. (5) usually employed in the analysis of experimental data [12]. Note, however, that the $f_+(q^2)$ obtained from only valence contribution in $q^+ \neq 0$ frame (dotted lines) does not exhibit the same behavior.

Heavy-to-light(heavy) decays: Our predicted decay rates for $D \to K$ and $D \to \pi$ are $\Gamma(D^0 \to K^- e^+ \nu_e) = 8.36|V_{cs}|^2 \times 10^{-2}$ ps^{-1} and $\Gamma(D^0 \to \pi^- e^+ \nu_e) = 0.113|V_{cd}|^2$ ps^{-1}, respectively. Using $|V_{cs}| = 1.04 \pm 0.16$ and $|V_{cd}| = 0.224 \pm 0.016$ [12], we obtain the branching ratio of $Br(D \to K) = (3.75 \pm 1.16)\%$ and $Br(D \to \pi) = (2.36 \pm 0.34) \times 10^{-3}$, while the experimental data are $(3.66 \pm 0.18)\%$ for $D \to K$ and $(3.9^{+2.3}_{-1.1} \pm 0.4) \times 10^{-3}$. Also, our predicted decay rates for $B \to \pi$ and $B \to D$ are $\Gamma(B^0 \to \pi^- \ell^+ \nu_\ell) = 8.16|V_{ub}|^2$ ps^{-1}, $\Gamma(B^0 \to D^- \ell^+ \nu_\ell) = 9.39|V_{cb}|^2$, respectively. Using $|V_{ub}| = (3.3 \pm 0.4 \pm 0.7) \times 10^{-3}$ and $|V_{bc}| = 0.0395 \pm 0.003$ [12], we obtain $Br(B \to \pi) = (1.40 \pm 0.34) \times 10^{-4}$ and $Br(B \to D) = (2.28 \pm 0.20)\%$. Our results are quite comparable with the recent experimental data [12], $(1.8 \pm 0.6) \times 10^{-4}$ for $B \to \pi$ and $(2.00 \pm 0.25)\%$ for $B \to D$, within the given error range.

In the heavy quark limit $M_{1(2)} \to \infty$, the form factor $f_+(q^2)$ is reduced to the

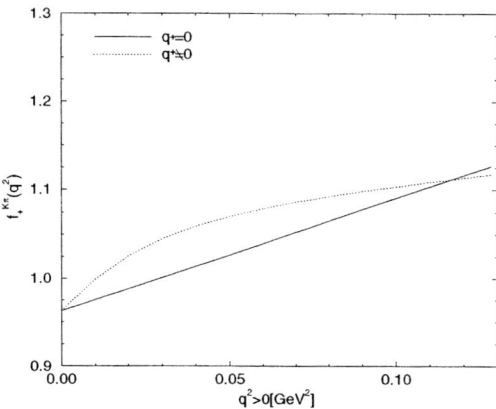

FIGURE 2. The form factors $f_+(q^2)$ for the $K \to \pi$ transition in timelike momentum transfer $q^2 > 0$. The solid and dotted lines are the results from the $q^+=0$ and $q^+\neq0$ frames[2], respectively. The differences of the results between the two frames are the measure of the nonvalence contributions from $q^+\neq0$ frame.

universal Isgur-Wise (IW) function, $\xi(v_1 \cdot v_2) = [2\sqrt{M_1 M_2}/(M_1 + M_2)]f_+(q^2)$, where $v_{1(2)} = P_{1(2)}/M_{1(2)}$. Our prediction of the slope $\rho^2=0.8$ of the IW function at the zero-recoil point defined as $\xi(v_1 \cdot v_2) = 1 - \rho^2(v_1 \cdot v_2 - 1)$ is quite comparable with the current world average $\rho_{\text{avg.}}=0.66\pm0.19$ [12] extracted from exclusive semileptonic $\bar{B} \to D\ell\bar{\nu}$ decay.

In conclusion, we analyzed the exclusive $0^- \to 0^-$ semileptonic meson decays using the LFQM constrained by the variational principle for the QCD-motivated effective Hamiltonian with the well-known linear plus Coulomb interaction [5,9]. The form factors $f_\pm$ are obtained in $q^+=0$ frame and then analytically continued to the timelike region by changing $q_\perp$ to $iq_\perp$ in the form factors. The matrix element of the "$\perp$" component of the current J^μ is used to obtain the form factor f_-. Our model provided overall a good agreement with the available experimental data and the lattice QCD results for the transition form factors and branching ratios of the $0^- \to 0^-$ semileptonic meson decays. Also it rendered a large number of predictions to the heavy meson mass spectra and decay constants [9]. We think that the success of our model hinges on the advantage of light-front quantization realized by the rational energy-momentum dispersion relation. It is crucial to calculate the "good" components of the current in the reference frame which deletes the complication from the nonvalence Z-graph contribution. The present work broadens the utility of the standard light-front frame à la Drell-Yan-West to the timelike form factor calculation. We anticipate further stringent tests of our model with more accurate data from future experiments and lattice QCD calculations.

ACKNOWLEDGEMENT

We thank Professor Dong-Pil Min for his hospitality during our stay at the Center for Theoretical Physics at Seoul National University. This work was supported by the U.S. Department of Energy(DE-FG-02-96ER4-0947). The North Carolina Supercomputing Center and the National Energy Research Scientific Computing Center are also acknowledged for the grant of computing time allocation.

REFERENCES

1. S. D. Drell and T. M. Yan, *Phys. Rev. Lett.* **24**, 181 (1970); G. West, *Phys. Rev. Lett.* **24**, 1206 (1970); G. P. Lepage and S. J. Brodsky, *Phys. Rev.* D **22**, 2157 (1980).
2. H.-M. Choi and C.-R. Ji, *Phys. Rev.* D **59**, 034001 (1999).
3. P. L. Chung, F. Coester, and W. N. Polyzou, *Phys. Lett.* B **205**, 545 (1988); W. Jaus, *Phys. Rev.* D **44**, 2851 (1991); F. Cardarelli et al., *Phys. Lett.* B **332**, 1 (1994); *Phys. Rev.* D **53**, 6682 (1996).
4. H.-M. Choi and C.-R. Ji, *Nucl. Phys.* A **618**, 291 (1997); *ibid.* **56**, 6010 (1997).
5. H.-M. Choi and C.-R. Ji, *Phys. Rev.* D **59**, 074015 (1999).
6. H.-M. Choi and C.-R. Ji, *Phys. Rev.* D **58**, 071901 (1998).
7. S. J. Brodsky and D. S. Hwang, NPB **543**, 239 (1998).
8. S. Godfrey and N. Isgur, *Phys. Rev.* D **32**, 189 (1985); N. Isgur, D. Scora, B. Grinstein, and M. B. Wise, *Phys. Rev.* D **39**, 799 (1989); D. Scora and N. Isgur, *Phys. Rev.* D **52**, 2783 (1992).
9. H.-M. Choi and C.-R. Ji, "Light-Front Quark Model Analysis of Exclusive $0^- \rightarrow 0^-$ Semileptonic Heavy Meosn Decays", to be published in *Phys. Lett.* B [hep-ph/9903496].
10. J. M. Flynn and C. T. Sachrajda, "Heavy Quark Physics From Lattice QCD", to appear in Heavy Flavor (2nd edition) edited by A. J. Buras and M. Lindner (World Scientific, Singapore), hep-lat/9710057.
11. N.B. Demchuk, I. L. Grach, I. M. Narodetskii, and S. Simula, *Phys. At. Nuclei* **59**, 2152 (1996); H.-Y. Cheng, C.-Y. Cheng, and C.-W. Hwang, *Phys. Rev.* D **55**, 1559 (1997).
12. Particle Data Group, C. Caso et al., *Eur. Phys. J.* C **3**, 1 (1998).
13. M. Ademollo and R. Gatto, *Phys. Rev. Lett.* **13**, 264 (1964).
14. H. Leutwyler and M. Roos, *Z. Phys.* C **25**, 91 (1984).
15. A. Afanasev and W. W. Buck, *Phys. Rev.* D **55**, 4380 (1997).
16. D. Scora and N. Isgur; *Phys. Rev.* D **52**, 2783 (1995).

Flux-Tube Model for Gluonic Structures

Jong Bum Choi

Department of Physics Education, Chonbuk National University,
Chonju, Seoul 561-756, Korea

Ho Young Choi

Department of Physics, Chonbuk National University,
Chonju, Seoul 561-756, Korea

Abstract. The gluonic structures in hadrons are usually described by gluon distribution functions parametrized in momentum space with unknown small-x behavior. In order to figure out the non-perturbative structures, we introduce an axiomatic approach in coordinate space with two axioms which can be used to estimate the gluonic densities in hadrons. In momentum space, we can apply the same formalism to account for the fragmentation processes in high energy scatterings resulting in new method to describe jet structures.

I. INTRODUCTION

Gluons are the mediators of strong forces. As is well-known, these gauge bosons are formulated into the standard model via the quantum chromodynamics(QCD). The success of the standard model rests partly on the predictive power of QCD, which can be shown by expansions in α_s and by using appropriate propagators to calculate diagrams related to scattering or bound states. These successful features of perturbative QCD deserve to be taken as the solid ground for the standard model, however, there exist loose ends of the story, i.e., the non-perturbative features in QCD.

QCD becomes non-perturbative in low energy or long distance processes such as small-x physics[1], fragmentations into hadrons[2], or the confining aspects in bound states[3]. In order to circumvent these un-solvable situations, many models were introduced to account for gluon distributions, hadronization procedures resulting in jets, and the confining potentials. These models are quite independent from each other, and it is difficult to have a unified picture of gluonic interactions by considering these models.

CP494, *New Directions in Quantum Chromodynamics*, edited by C.-R. Ji and D.-P. Min
© 1999 American Institute of Physics 1-56396-908-4/99/$15.00

In this talk, we will try to formulate a systematic flux-tube picture which can be applied to the descriptions of non-perturbative aspects of gluonic interactions. In section II, we give general formalisms, and in section III, we apply the general formalisms to deduce the gluonic structures in hadrons. Momentum space formulation is given in section IV, where 3-jet physics are discussed as an example. The final section is devoted to summary and outlook.

II. FLUX-TUBE FORMALISM

The description of gluonic flux-tube was firstly attempted in order to account for baryon mass spectra. This simple string picture was changed into a more quantitative one with the introduction of flux-tube overlap functions which were used to predict hadronic strong decays[4]. In quark pair creation model, the position of the created quark pair affects the decay amplitude, and therefore, we need to specify the functional form of the probability amplitude to have one quark pair. Since the quark pair creation is closely related to the gluon densities, the specification of the functional form can be taken to represent non-perturbative gluonic effects. However, there was no consensus about the form of overlap functions, which became a direct motivation to formulate a new approach to gluonic flux-tubes[5].

Two goals of the new approach are as follows; one is how to treat systematically the changes occurring on flux-tubes, and the other is how to define measures on the well-classified stable flux-tubes. Since the main changes on flux-tubes are caused by quark pair creations and annihilations, the first goal can be achieved by classifying flux-tubes with the operations corresponding to quark pair creations and annihilations. The second goal can be achieved by considering the connectedness of flux-tubes between given boundary points sat by quarks. The relations between classified flux-tubes allow the construction of topological spaces, and the connectedness of flux-tube can be used to define connection amplitudes which are related to the measure. Let's consider these formulations in detail.

The starting point for a systematic description of flux-tubes rests on the classification of flux-tubes, which can be done by counting the number of boundary points. Since flux-tubes start from quark and end at antiquarks, we need only count the number of quarks and antiquarks to classify flux-tubes[6]. We can represent the set of flux-tubes with a quarks and b antiquarks sitting at boundaries as $F_{a,\bar{b}}$. Omitting the number 0, except for F_0 which represents the flux-tube set with no boundary point corresponding to glueballs, mesons composed of one quark-antiquark pair can be taken to have flux-tubes $F_{1,\bar{1}}$, and simple baryon and antibaryon flux-tubes can be represented as F_3 and $F_{\bar{3}}$. For the classified flux-tube sets, we can consider relationships between them which are generated by quark pair creations and annihilations. The division of a flux-tube is generated by a quark pair creation, and the union by a pair annihilation. These relationships can be used to construct topological spaces of flux-tubes which are necessary to define physical amplitudes

The assumptions for the construction of topological spaces are as follows;
(1) Open sets are stable flux-tubes.
(2) The union of stable flux-tubes becomes a stable flux-tube.
(3) The intersection between a connected stable flux-tube and disconnected stable flux-tubes is the reverse operation of the union.

With these assumptions, we can follow the flux-tube sets that can be produced from a given flux-tube by repeating union and intersection operations. If the produced sets are closed under these operations, we can classify the constructed topological spaces and this classification process can be done by counting the numbers of incoming and outgoing 3-junctions in a given closed set. When we include the excited flux-tube set F_0, two 3-junctions can be created making it impossible to assign fixed numbers of 3-junctions to a given set, and therefore, we omit this possibility in this talk. Let's consider first the flux-tubes $F_{1,\bar{1}}$ corresponding to simple quarkonium mesons. If quark pair creations are repeated disconnecting the flux-tubes, we get in general n quarkonium meson states represented by $F_{1,\bar{1}}^n$. The inverse process does not change the situation, so we get the simplest non-trivial topological space

$$T_0 = \{\phi, F_{1,\bar{1}}, F_{1,\bar{1}}^2, \cdots, F_{1,\bar{1}}^n, \cdots\}. \tag{1}$$

Since there exists only one kind of flux-tube $F_{1,\bar{1}}$ in this space, we may reduce the notation as

$$T_0 = \{\phi, F_{1,\bar{1}}\}, \tag{2}$$

where it is assumed that $F_{1,\bar{1}}$ can be multiplied repeatedly without violating the law of baryon number conservation. Then, the topological space for baryon-meson system can be represented as

$$T_1 = \{\phi, F_3\}, \tag{3}$$

and the baryon-meson-baryon space becomes

$$T_2 = \{\phi, F_3^2\}. \tag{4}$$

When outgoing 3-junctions exist, we need another index to represent the topological space. For example, the space with two incoming 3-junctions and one outgoing 3-junction is represented as

$$T_{2,\bar{1}} = \{\phi, F_3^2 F_{\bar{3}}, F_3 F_{2,\bar{2}}, F_{4,\bar{1}}\}. \tag{5}$$

In general, we can write down the spaces as

$$T_{i,\bar{j}} = \{\phi, F_3^i F_{\bar{3}}^j, F_3^{i-1} F_{\bar{3}}^{j-1} F_{2,\bar{2}}, \cdots\}, \tag{6}$$

where i is the number of incoming 3-junctions and j that of outgoing 3-junctions. For a given space, the baryon number $B = i - j$ is fixed and the number of boundary points is reduced by (1,1) pair by one union operation.

355

Now let's try to define physical amplitudes related to the measures on flux-tubes. Since the flux-tubes can be classified into different topological spaces, it is better to consider the physical amplitudes in a given space. The formation of a space is generated by repeated union and intersection operations, and therefore, the physical amplitudes can be taken to be closely related to the connection and disconnection of flux-tubes. Then, it is natural to define the amplitudes A for a quark to be connected to another quark or antiquark through given flux-tube open set. In order to quantify A, we can assume the existence of a measure M of A satisfying the conditions

(1) M(A) decreases as A increases,

(2) $M(A_1) + M(A_2) = M(A_1 A_2)$ when A_1 and A_2 are independent. (7)

The first condition states that the connection probability increases for smaller measure of flux-tube which is physically acceptable. The other condition states the relation between two flux-tubes that can be joined to form single flux-tube. Thus, the amplitude A can be used to estimate the connectedness of a flux-tube quantitatively, and it can be named as connection amplitude. The measure M of A can be solved as functions of A from the above two conditions;

$$M(A) = -k \ln \frac{A}{A_0},$$ (8)

where A_0 is a normalization constant and k is an appropriate parameter.

The connection amplitude A has to be represented in terms of some physically measurable quantities. One reasonable method to convert the amplitude A into a concrete form is to consider the measure M as a metric function defined on the flux-tube. For the simplest flux-tubes $F_{1,\bar{1}}$, a metric function can be introduced between the two boundary points $\mathbf{x}$ and $\mathbf{y}$ corresponding to the positions of a quark and an antiquark. A general form of distance function between the two points $\mathbf{x}$ and $\mathbf{y}$ can be written down as $|\mathbf{x} - \mathbf{y}|^\nu$ with ν being an arbitrary number. This distance function can be made metric for the points $\mathbf{z}$ satisfying

$$|\mathbf{x} - \mathbf{z}|^\nu + |\mathbf{z} - \mathbf{y}|^\nu \geq |\mathbf{x} - \mathbf{y}|^\nu.$$ (9)

The set of points $\mathbf{z}$ not satisfying this triangle inequality can be taken as forming the inner part of the flux-tube where it is impossible to define a metric from boundary points with given ν. With this metric condition, we can figure out the shape of flux-tube, and we may take $|\mathbf{x} - \mathbf{y}|^\nu$ as an appropriate measure to deduce a concrete form for the connection amplitude A. The lower limit of ν can be fixed to be 1 because there exists no point $\mathbf{z}$ satisfying the triangle inequality with $\nu < 1$. In order to account for various possibilities, we need to sum over contributions from different ν's. For a small increment $d\nu$, the product of the two probability amplitudes for $|\mathbf{x} - \mathbf{y}|^\nu$ and $|\mathbf{x} - \mathbf{y}|^{\nu+d\nu}$ to satisfy the metric conditions can be accepted as the probability amplitude for the increased region to be added to

356

the inner connected region which is out of the metric condition. Then the full connection amplitude becomes

$$A = A_0 \exp\{-\frac{1}{k}\int_1^\alpha F(\nu)r^\nu d\nu\}, \tag{10}$$

where all possibilities from the line shape with $\nu = 1$ to the arbitrary shape with $\nu = \alpha$ have been included. The weight factor $F(\nu)$ has been introduced in order to account for possible different contributions from different ν's, and the variable r is

$$r = \frac{1}{l}\mid x - y \mid \tag{11}$$

with l being a scale parameter. When we consider the case of $\alpha = 2$, which corresponds to a spherical shape flux-tube, and the case of equal weight $F(\nu) = 1$, we get

$$A = A_0 \exp\{-\frac{1}{k}\frac{r^2 - r}{\ln r}\}. \tag{12}$$

We will use this form of connection amplitude with slight changes. The changes can be made if we replace

$$A \longrightarrow r^\beta A, \tag{13}$$

for which the conditions on M still hold with $\beta > 0$. Then general A becomes

$$A = \frac{A_0}{r^\beta}\exp\{-\frac{1}{k}\frac{r^2 - r}{\ln r}\}. \tag{14}$$

III. GLUONIC STRUCTURES IN HADRONS

The connection amplitude A for a quark to be connected to another quark through a given gluonic flux-tube is introduced to describe quark pair creations. In order to predict the positions of the created quark pair, the probability amplitude to have a quark pair at a given position is taken to be equal to the overlap of connection amplitudes calculated for the initial and the final boundary points. The overlap function γ can be written as

$$\gamma = A_i A_f, \tag{15}$$

where A_i and A_f represent the connection amplitudes before and after quark pair creation.

The definition of gluonic densities has to be closely related to physically observable quantities. Although these quantities can be measured, it is impossible to observe directly the gluonic densities because gluons are confined. During the formation process of final hadrons, quark pairs are created by confined gluons. The probability amplitude to have a quark pair at a specified position is proportional to the amplitude that can be used to describe gluonic behaviors. If we take the latter amplitude as representing gluonic densities, then the gluonic structures can be deduced from the amplitude for quark pair creations. Since we have introduced connection amplitude to describe quark pair creations, the connection amplitude can be used to predict gluonic densities. We will take this point of view in this talk, and the gluonic densities in hadrons will be estimated by calculating the overlap of connection amplitudes for various configurations of boundary points.

Let's consider the case of a proton accelerated in the positive direction of z-axis. We can assume that electric fields are applied along positive **z** direction during the acceleration. Then the three boundary quarks will be oriented in such a way that the two positively charged up quarks go ahead pulling the other negatively charged down quark. If the flux-tubes connecting the three boundary quarks do not break, the distance between the up quarks and the down quark will be contracted along the direction of z-axis. Now the relative positions of three boundary quarks will change according to the Lorentz boosts, and the corresponding gluonic structures will change. If we take different connection amplitudes as given in Eq.(14) with different values of β, the gluonic structures turn out to be deformed from one to another. These deformations are shown in Fig.1 where the three boundary quarks are taken to be at three vertices of a equilateral triangle which corresponds to a proton at rest. We have taken the values of β as 0, 0.2, 0.3, 0.5, 1, and 2.

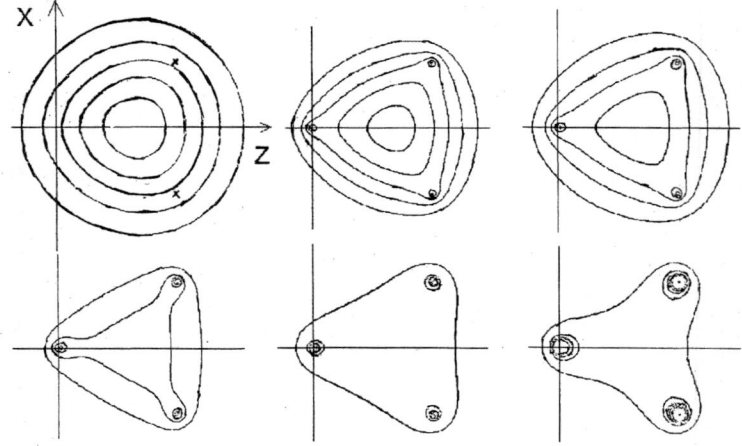

FIGURE.1. Gluonic structures of a proton at rest with different values of β. β values are taken as 0, 0.2, 0.3, 0.5, 1, and 2 respectively.

In cases of motion along **z** direction, the contracted gluonic structures appear as shown in Fig.2, where only one case with β equal to 0.5 is considered and 4 contracted cases are shown. The down quark sits at origin and the up quarks in contracted cases are taken to be at z positions 0.5, 0.3, 0.2, and 0.01 respectively. The x positions are fixed at 0, 0.5 and -0.5.

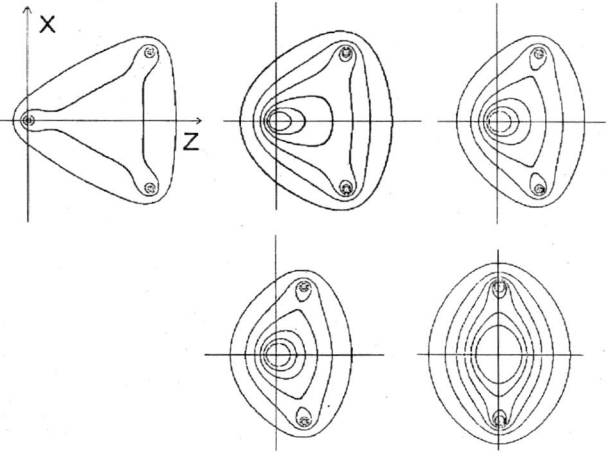

FIGURE.2. Contracted gluonic structures with β equal to 0.5. The contracted distances along z-axis are chosen arbitrarily.

The changes of γ values along axes parallel to z-axis are shown in Fig.3. The 5 figures correspond to those in Fig.2, and the different curves in each figure represent γ values for different values of x.

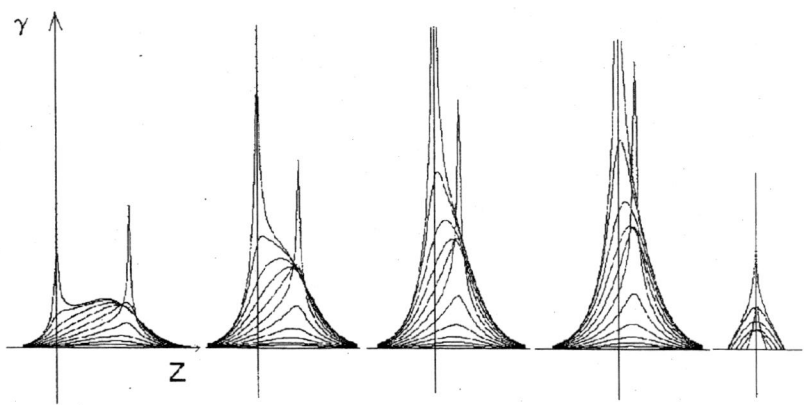

FIGURE.3. γ values along axes parallel to z-axis. Scales of γ are arbitrary.

359

IV. MOMENTUM SPACE FORMULATION AND JETS

In scattering problems, the final data can be obtained by analyzing particle trajectories. Since quarks cannot exist alone, hadronization processes are necessary to form particles that can generate such trajectories. These hadronization processes cannot be described by perturbative QCD, and we need to construct models to account for the non-perturbative features of QCD. In our flux-tube model, we can account for these non-perturbative features which are related to quark pair creations leading to final jets observed in high energies. Because the created quark pairs have to be specified in momentum space for scattering states, we need to formulate our model in momentum space[7].

Since the connection amplitudes have been defined between two boundary points, it is possible to consider the same amplitudes in momentum space. The connection amplitude for two boundary points with momenta $\mathbf{p_1}$ and $\mathbf{p_2}$ can be written down as

$$A = A_0 \exp\{-\frac{1}{\tau} \int_1^\alpha G(\nu) \mid \mathbf{p_1} - \mathbf{p_2} \mid^\nu d\nu\}, \tag{16}$$

where $G(\nu)$ and α are an appropriate weight factor and the boundary value of ν respectively. In case of $G(\nu) = 1$, and $\alpha = 2$, we get

$$A(p) = A_0 \exp\{-\frac{1}{\tau} \frac{p^2 - p}{\ln p}\} \tag{17}$$

with $p = \mid \mathbf{p_1} - \mathbf{p_2} \mid$. We can apply this amplitude to describe fragmentation processes resulting in jets. We choose the example of 3-jet events formed from the fragmentation processes of Z^0's produced by e^+e^- collisions.

For Z^0 fragmentations, no initial state uncertainties exist, and the final quark jets and antiquark jets can be identified by lepton-tagging methods. For non-trivial 3-jet events, the other gluon jet can be identified clearly and we can compare different properties between the quark jet and the gluon jet. The gluonic effects appear as the probability amplitudes for creation of quark pairs which generate many hadrons resulting in the formation of jets. For the $A(p)$ in Eq. (17), the probability amplitude for a created quark pair to be connected in momentum to the quark and the gluon boundaries is proportional to

$$C_1 = A_0^3 \exp\{-\frac{1}{\tau}(\frac{a^2 - a}{\ln a} + \frac{b^2 - b}{\ln b} + \frac{c^2 - c}{\ln c})\}, \tag{18}$$

where a and b are the distances in momentum space from the created quark pair to gluon and quark boundaries, and c is the distance between gluon and antiquark boundaries. The other amplitude C_2 for the created quark pair to be connected to the antiquark and the gluon boundaries has the same form with the interchanged roles of quark and antiquark boundaries. The probability to get particles in a given direction is

$$P = \int^h \mid C_1 + C_2 \mid^2 dR. \tag{19}$$

360

where h is determined by phase space configurations. The shape of momentum phase space can be determined by using parton model assumptions. Longitudinal components are taken to be proportional to the total parton energy which becomes generally the total jet energy. On the other hand, transverse components are assumed to be small resulting from uncertainties in momentum specifications. The simplest choice is that the transverse momentum components increase with the values of the longitudinal components. One way to put this assumption into a concrete form of phase space is to draw a trapezoid along the direction of the jet momentum[8]. The base line of this trapezoid along highest energy quark jet is given a length parameter 2d which is divided equally by the tip of the jet momentum vector. Another length parameter e is assigned to the upper side of the trapezoid. Moreover, additional parameter f has to be introduced to account for larger transverse component in case of gluon jet than in case of quark jet. When the phase boundary lines are crossed near the origin, we simply take the outer lines as a first approximation.

For experimental data, we used 3-jet events from the OPAL group[9]. In our model, we have five parameters A_0, τ, d, e, and f. The normalization constant A_0 has been determined by the peak of the highest quark jet for each data sample representing particle multiplicities. The widths of the jets are controlled by the parameter d, and the depths of the inter-jet regions are related to the value of e. For gluon jet region, f has to be multiplied to the value of e. Finally the parameter k has been fixed by fitting the peak value of secondary quark jet. In this way, we

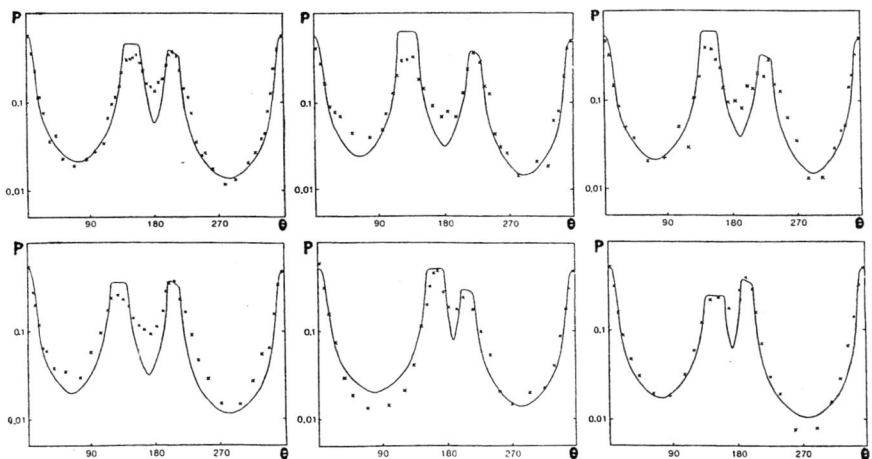

FIGURE.4. Six data samples for 3-jet events with different energy configurations. Parameters were fixed to the first data sample, and the other five samples predicted.

have obtained the values for the first data sample as

$$d = 0.15, \quad e = 0.015, \quad f = 1.5, \quad \tau = 0.7. \tag{20}$$

With these parameters, we can predict the multiplicity distributions for the other five data samples. The six data samples with predicted curves and data points are shown in Fig.4.

V. SUMMARY AND OUTLOOK

In this talk, we have shown that a new approach to non-perturbative QCD is possible. Gluonic flux-tube picture can be changed into a systematic formalism by constructing topological spaces, and quantitative calculations can be made by introducing connection amplitude. We can apply these formalisms to estimate the gluon densities in hadrons, which can be Fourier transformed into momentum space resulting in non-perturbative gluon distribution functions. In momentum space formalism, we can describe fragmentation procedures making it possible to predict particle multiplicities in jet events.

Further works are needed to settle down some problems. We need to study more on small-x physics where the distinction between intrinsic and extrinsic gluon contributions is not so clear. It is also interesting to try to calculate the properties of glueballs and pomeron in our new formalism. For future experiments, we need to analyze many jet systems discriminating gluon jets from quark jets. Of course, applications to larger systems such as nuclei need more works and are expected to provide new insights about the really strong interactions.

ACKNOWLEDGMENTS

This work is supported in part by the Ministry of Education of Korea through the Basic Research Institute Program under project No. 1998-015-D00071.

REFERENCES

1. J.Kwiecinski, A.D.Martin, and P.J.Sutton, Phys. Rev. **D52**, 1445(1995).
2. J.Ellis, K.Geiger, and H.Kowalski, Phys. Rev. **D54**, 5443(1996).
3. N.G.Hyun and J.B.Choi, J. Korean Phys. Soc. **28**, 20(1995); J.B.Choi, K.H.Cho, and N.G.Hyun, *ibid.* **31**, 735(1997).
4. N.Isgur and J.Paton, Phys. Rev. **D31**, 2910(1985); R.Kokoski and N.Isgur, *ibid.* **D35**, 907(1987).
5. J.B.Choi and S.U.Park, J. Korean Phys. Soc. **24**, 263(1991).
6. J.B.Choi and W.J.Kim, J. Korean Phys. Soc. **25**, 477(1992).
7. J.B.Choi, J. Korean Phys. Soc. **30**, 28(1997).
8. J.B.Choi, J. Korean Phys. Soc. **27**, 605(1994).
9. M.Z.Akrawy et al.(OPAL Collaboration), Phys. Lett. **261B**, 334(1991).

πNN coupling from the two-point correlation function with a pion [1]

Hungchong Kim

Department of Physics, Tokyo Institute of Technology, Tokyo 152-8551, Japan.

Abstract. We construct three different sum rules from the two-point correlation function with a pion, $i \int d^4x e^{iq \cdot x} \langle 0|T J_N(x) \bar{J}_N(0)|\pi(p)\rangle$, beyond the soft-pion limit or beyond the chiral limit. In constructing the sum rules beyond the soft-pion limit, we consider the Dirac structures, $i\gamma_5 \not{p}$ and $\gamma_5 \sigma_{\mu\nu} q^\mu p^\nu$. The $i\gamma_5$ Dirac structure is used to construct a sum rule beyond the chiral limit. Our results suggest that the sum rule beyond the chiral limit provides the πNN coupling consistent with its empirical value.

Within QCD sum rules, the πNN coupling constant, $g_{\pi N}$, is often calculated [1–4] from the correlation function,

$$i \int d^4x e^{iq \cdot x} \langle 0|T J_p(x) \bar{J}_n(0)|\pi^+(p)\rangle \ , \tag{1}$$

where J_p is the proton interpolating field [5] and J_n is the neutron interpolating field. This correlation function contains the three distinct Dirac structures, $i\gamma_5, i\gamma_5 \not{p}$ and $\gamma_5 \sigma_{\mu\nu} q^\mu p^\nu$, each of which can be used to construct a separate sum rule for the πNN coupling constant. In this work, we study each sum rule separately to suggest a reliable method of constructing a QCD sum rule for $g_{\pi N}$. A successful determination of $g_{\pi N}$ may provide a solid ground for extending the formalism to determine other meson-baryon couplings within the conventional QCD sum rules.

In constructing a QCD sum rule from the correlator Eq.(1), one should go beyond the soft-pion limit. In the soft-pion limit, Eq.(1) is not independent from the nucleon chiral-odd sum rule [1] and therefore not useful to determine $g_{\pi N}$. Thus, for an independent determination of $g_{\pi N}$, one needs to go beyond the soft-pion limit. To construct a sum rule beyond the soft-pion limit, we can consider the Dirac structure, $i\gamma_5 \not{p}$ or $\gamma_5 \sigma_{\mu\nu} q^\mu p^\nu$. These Dirac structures, as one power of the pion momentum p_μ is taken out as an overall factor, provide sum rules beyond the soft-pion limit.

The OPE side of Eq. (1) can be calculated by separating the quark-antiquark component of the pion wave function,

---

[1] This work is supported by Research Fellowships of the Japan Society for the Promotion of Science.

CP494, *New Directions in Quantum Chromodynamics*, edited by C.-R. Ji and D.-P. Min

$$D^{\alpha\beta}_{aa'} \equiv \langle 0|u^\alpha_a(x)\bar{d}^\beta_{a'}(0)|\pi^+(p)\rangle \; , \qquad (2)$$

from the rest quark propagators. This quark-antiquark component can be written in terms of the following three matrix elements,

$$\langle 0|\bar{d}(0)\gamma_\mu\gamma_5 u(x)|\pi^+(p)\rangle \; , \qquad \langle 0|\bar{d}(0)\gamma_5\sigma_{\mu\nu}u(x)|\pi^+(p)\rangle \; ,$$
$$\langle 0|\bar{d}(0)i\gamma_5 u(x)|\pi^+(p)\rangle \; , \qquad (3)$$

whose few moments are relatively well-known [8].

The first two matrix elements participate in the sum rules with the Dirac structures, $i\gamma_5\,\slashed{p}$ and $\gamma_5\sigma_{\mu\nu}q^\mu p^\nu$. Following the standard prescription of QCD sum rules, we obtain for the $i\gamma_5\,\slashed{p}$ structure,

$$g_{\pi N}\lambda^2_N(1 + AM^2)$$
$$= \frac{f_\pi}{m}M^2 e^{m^2/M^2}\left[\frac{E_1(x_\pi)}{2\pi^2}M^4 + \frac{E_0(x_\pi)}{2\pi^2}M^2\delta^2 + \frac{1}{12}\left\langle\frac{\alpha_s}{\pi}G^2\right\rangle + \frac{2\langle\bar{q}q\rangle^2}{9f^2_\pi}\right] \; . \qquad (4)$$

Here $x_\pi = S_\pi/M^2$ with S_π being the continuum threshold and $E_n(x) = 1 - (1 + x + \cdots + x^n/n!)\,e^{-x}$. δ^2 is the twist-4 contribution to the first matrix element in Eq. (3). The unknown single pole A represents the contribution of $N \rightarrow N^*$ [6] as well as the PS-PV scheme dependent $N \rightarrow N$ [3]. Thus, physical content of A is coupling-scheme dependent. In obtaining Eq. (4), we have taken out one power of the pion momentum and took the limit $p_\mu \rightarrow 0$ in the rest of the correlator. By fitting the RHS with a straight line within an appropriate Borel window, we can determine $g_{\pi N}\lambda^2_N$ and $g_{\pi N}\lambda^2_N A$. Then combining this sum rule with the nucleon mass sum rule, we can determine the coupling $g_{\pi N}$. This sum rule is however strongly sensitive to the continuum threshold we choose and the contribution from the unknown single pole term A is very large. The reason for having large continuum and large contribution from $N \rightarrow N^*$ is because the positive-parity higher resonances add up with the negative-parity higher resonances in making the continuum or A [4]. Therefore, we can not determine the coupling reliably from this sum rule.

The sum rule for $\gamma_5\sigma_{\mu\nu}q^\mu p^\nu$ can be constructed similarly [3],

$$g_{\pi N}\lambda^2_N(1 + BM^2) = -\frac{\langle\bar{q}q\rangle}{f_\pi}e^{m^2/M^2}\left[\frac{M^4 E_0(x_\pi)}{12\pi^2} + \frac{4}{3}f^2_\pi M^2\right.$$
$$\left. + \left\langle\frac{\alpha_s}{\pi}G^2\right\rangle\frac{1}{216} - \frac{m^2_0 f^2_\pi}{6}\right] \; . \qquad (5)$$

The unknown single pole term is represented by B whose physical content is independent of the coupling schemes. This can be also checked explicitly by constructing B using effective models for higher resonances. m^2_0 is a parameter associated with the dim-5 quark-gluon mixed condensate. We emphasize that this sum rule is independent of the PS and PV coupling scheme employed in the phenomenological side. When this sum rule is combined with the nucleon chiral-odd sum rule, we obtain $g_{\pi N} \sim 10$ [3]. This result is insensitive to the continuum threshold and the unknown single-pole term contributes little to the sum rule.

Our result from the $\gamma_5 \sigma_{\mu\nu} q^\mu p^\nu$ sum rule, even though the sum rule has some nice features, is not quite consistent with the empirical value for $g_{\pi N}$. The result $g_{\pi N} \sim 10$ is rather close to the one in the chiral limit. Certainly a further improvement may be needed. We seek an improvement by constructing a sum rule beyond the chiral limit. Our calculations up to now are performed beyond the soft-pion limit by taking the leading order of the pion momentum p_μ, but for the rest of the correlator the chiral limit $p^2 = m_\pi^2 = 0$ is taken. Thus, it is not clear whether the calculation is done beyond the chiral limit and this may cause the discrepancy with the empirical $g_{\pi N}$.

In constructing a sum rule beyond the chiral limit, we consider the Dirac structure $i\gamma_5$ at the order $p^2 = m_\pi^2$ [7]. Specifically, we expand the correlator Eq. (1) in terms of the pion momentum p_μ and take the terms proportional to $p^2 = m_\pi^2$. For a consistent chiral counting, the terms linear in quark mass m_q should be kept in the OPE side as well as the $p^2 = m_\pi^2$ terms since they are related by the Gell-Mann$-$Oakes$-$Renner relation. The pion matrix element contributing to the OPE in this sum rule is the third term in Eq. (3), twist-3 pion wave function. For the m_q terms, we need to take the zeroth moment of the pion wave function but for the $p^2 = m_\pi^2$ terms, we need to take the second moment. Both moments are relatively well-known [8].

The sum rule for the $i\gamma_5$ at the order $p^2 = m_\pi^2$ can be constructed straightforwardly and it is

$$g_{\pi N} \lambda_N^2 e^{-m^2/M^2} [1 + CM^2] = -M^4 E_0(x_\pi) \left[\frac{\langle \bar{q}q \rangle}{12\pi^2 f_\pi} + \frac{3 f_{3\pi}}{4\sqrt{2}\pi^2} \right]$$
$$- f_\pi \langle \bar{q}q \rangle M^2 + \frac{1}{72 f_\pi} \langle \bar{q}q \rangle \left\langle \frac{\alpha_s}{\pi} G^2 \right\rangle - \frac{1}{3} m_0^2 f_\pi \langle \bar{q}q \rangle , \quad (6)$$

where $f_{3\pi} = 0.003$ GeV^2 coming from the three-particle wave function of a pion [8]. After transferring the e^{-m^2/M^2} factor to the RHS, the LHS is again a linear function of the Borel mass M^2. The two empirical parameters, $g_{\pi N} \lambda_N^2$ and $g_{\pi N} \lambda_N^2 C$, are determined by fitting the RHS within an appropriate Borel window. We found that this sum rule is insensitive to the continuum threshold we choose. The unknown single-pole term C is small. Thus, this sum rule has similar nice features as the $\gamma_5 \sigma_{\mu\nu} q^\mu p^\nu$ sum rule. The OPE structure looks also similar. Contributions from the terms containing $f_{3\pi}$ and the gluon condensate are not so large in this sum rule. All others, which are important in our sum rule, have the same sign making the OPE strength large. In contrast, the quark-gluon mixed condensate (which contains m_0^2) in the $\gamma_5 \sigma_{\mu\nu} q^\mu p^\nu$ sum rule Eq. (5) has an opposite sign from other important OPE terms, which makes the OPE strength smaller than the present case.

By combining with the nucleon chiral-odd sum rule, we eliminate the dependence on λ_N^2 and obtain the coupling $g_{\pi N}$. Our result is

$$g_{\pi N} = 13.3 \pm 1.2 . \quad (7)$$

The errors mainly come from the uncertainty in the QCD parameter m_0^2. Our result is remarkably close to the empirical value of $g_{\pi N}$.

In summary, we have constructed the three sum rules depending on the Dirac structure from the correlator Eq. (1). The $i\gamma_5 \not{p}$ sum rule is found to be not reliable to calculate $g_{\pi N}$. The $\gamma_5\sigma_{\mu\nu}q^\mu p^\nu$ has some nice features but the result is not quite consistent with the empirical value of $g_{\pi N}$. As a further improvement of the sum rule, we have constructed the $i\gamma_5$ sum rule beyond the chiral limit. Our result shows that going beyond the chiral limit is important in achieving consistency with the phenomenology. It might be useful to apply the current framework to other meson-baryon couplings in future.

REFERENCES

1. H. Shiomi and T. Hatsuda, *Nucl. Phys.* **A 594**, 294 (1995).
2. L.J. Reinders, H. Rubinstein and S. Yazaki, Phys. Rep. **127**, 1 (1985).
3. Hungchong Kim, Su Houng Lee and Makoto Oka, *Phys. Lett.* **B 453**, 199, (1999).
4. Hungchong Kim, Su Houng Lee and Makoto Oka, *Phys. Rev. D* **60**,034007 (1999).
5. B. L. Ioffe and A. V. Smilga, Nucl. Phys. **B 234** (1984) 109.
6. B. L. Ioffe, Nucl. Phys. **B 188** (1981) 317.
7. Hungchong Kim, *nucl-th/9903040*.
8. V. M. Belyaev, V. M. Braun, A. Khodjamirian and R. Rückl, Phys. Rev. **D** 51 (1995) 6177.

Decoupling of heavy quarks from QCD and applications in Higgs-boson phenomenology

Bernd A. Kniehl

II. Institut für Theoretische Physik, Universität Hamburg, Luruper Chaussee 149, 22761 Hamburg, Germany

Abstract. If QCD is renormalized by minimal subtraction (MS), at higher orders, the strong coupling constant α_s and the quark masses m_q exhibit discontinuities at the flavour thresholds, which are controlled by so-called decoupling constants, ζ_g and ζ_m, respectively. Adopting the modified MS ($\overline{\text{MS}}$) scheme, we derive simple formulae which reduce the calculation of ζ_g and ζ_m to the solution of vacuum integrals. This allows us to evaluate ζ_g and ζ_m through three loops. We also establish low-energy theorems, valid to all orders, which relate the effective couplings of the Higgs boson to gluons and light quarks, due to the virtual presence of a heavy quark h, to the logarithmic derivatives w.r.t. m_h of ζ_g and ζ_m, respectively. We also consider the effective QCD interaction of a CP-odd Higgs boson and verify the Adler-Bardeen nonrenormalization theorem at three loops.

INTRODUCTION

It is generally believed that quantum chromodynamics (QCD) is the true theory of the strong interactions. There are still open questions, concerning the origin of confinement or as to why the quark masses m_q and the asymptotic scale parameter Λ have the values they happen to have. The answers to these questions probably lie outside the scope of perturbative QCD, which forms the basis of this presentation. In perturbative QCD, the strong coupling constant $\alpha_s = g^2/(4\pi)$, where g is the gauge coupling, is small enough to serve as a useful expansion parameter, and quarks and gluons may appear as asymptotic states of the scattering matrix.

QCD is a nonabelian Yang-Mills theory based on the gauge group SU(3). In the covariant gauge, the Lagrangian reads

$$\mathcal{L} = \sum_{q=1}^{n_f} \bar{\psi}_q^i \left(i \not{D}^{ij} - \delta^{ij} m_q \right) \psi_q^j - \frac{1}{4} \left(G_{\mu\nu}^a \right)^2 - \frac{1}{2\xi} \left(\partial^\mu G_\mu^a \right)^2 + \left(\partial^\mu \bar{c}^a \right) \nabla_\mu^{ab} c^b,$$

$$D_\mu^{ij} = \delta^{ij} \partial_\mu - ig[T^a]^{ij} G_\mu^a, \qquad \nabla_\mu^{ab} = \delta^{ab} \partial_\mu - g f^{abc} G_\mu^c,$$

$$G_{\mu\nu}^a = \partial_\mu G_\nu^a - \partial_\nu G_\mu^a + g f^{abc} G_\mu^b G_\nu^c, \tag{1}$$

CP494, *New Directions in Quantum Chromodynamics*, edited by C.-R. Ji and D.-P. Min

with $n_f = 6$ flavours of quarks ψ_q^i $(i = 1, 2, 3)$, gluons G_μ^a $(a = 1, \ldots, 8)$, Faddeev-Popov ghosts c^a, and gauge parameter ξ. The generators T^a satisfy the commutation relations $[T^a, T^b] = if^{abc}T^c$, where f^{abc} are the structure constants.

In the calculation of QCD quantum corrections, one generally encounters, among other things, ultraviolet (UV) divergences, which must be regularized and removed by renormalization. In quantum electrodynamics, it is natural to employ the on-shell renormalization scheme, where the fine-structure constant is renormalized in the limit of the photon being on its mass shell. Due to confinement, this limit cannot be taken in QCD, and it is natural to employ the most convenient renormalization scheme instead. It has become customary to use dimensional regularization [1] in connection with minimal subtraction (MS) [2]. I.e., the integrations over the loop momenta are performed in $D = 4 - 2\epsilon$ space-time dimensions, introducing a 't Hooft mass scale μ to keep the (renormalized) coupling constant dimensionless. The poles in ϵ that emerge as UV divergences in the physical limit $\epsilon \to 0$ are then combined with the bare (UV-divergent) parameters and fields in Eq. (1) so as to render them renormalized (UV finite). This is always possible because QCD is a renormalizable theory. In the modified MS ($\overline{\text{MS}}$) scheme [3], the specific combination of transcendental numbers that always appears along with the poles in ϵ is also subtracted. In the following, bare quantities will be denoted by the superscript '0.' Specifically, we have

$$g^0 = \mu^\epsilon Z_g g(\mu), \qquad m_q^0 = Z_m m_q(\mu), \qquad \xi^0 - 1 = Z_3(\xi(\mu) - 1),$$
$$\psi_q^{0,i} = \sqrt{Z_2}\psi_q^i(\mu), \qquad G_\mu^{0,a} = \sqrt{Z_3}G_\mu^a(\mu), \qquad c^{0,a} = \sqrt{\tilde{Z}_3}c^a(\mu). \qquad (2)$$

In the MS-like schemes, the renormalization constants Z may be written in the simple form

$$Z = 1 + \sum_{i=1}^\infty \sum_{j=1}^i Z_{ij} \frac{a^i}{\epsilon^j}, \qquad (3)$$

where $a = \alpha_s/\pi$ is the renormalized couplant and Z_{ij} are numbers. I.e., the Z factors do not explicitly depend on dimensionful parameters. In particular, Z_m is generic for all q. A crucial advantage of the MS-like schemes is that Z_g and Z_m are ξ independent to all orders. This property carries over to α_s and m_q, so that it makes sense to extract these parameters from experimental data.

Z_g and Z_m carry the full information on how α_s and m_q run with μ. In fact, from the μ independence of g^0 and m_q^0 it follows that

$$\beta(a) \equiv \frac{da}{d\ln\mu^2} = -a\left(\frac{d\ln Z_g^2}{d\ln\mu^2} + \epsilon\right) = -\sum_{n=0}^\infty \beta_n a^{n+2},$$
$$\gamma_m(a) \equiv \frac{d\ln m_q}{d\ln\mu^2} = -\frac{d\ln Z_m}{d\ln\mu^2} = -\sum_{n=0}^\infty \gamma_n a^{n+1}. \qquad (4)$$

The Callan-Symanzik β function and the quark mass anomalous dimension γ_m are universal in the MS-like schemes. Moreover, β_0 and β_1 are universal in the larger class of schemes which have mass-independent β functions. In the MS-like schemes, the coefficients β_n and γ_n are known through four loops, i.e., $n = 3$ [4].

To summarize, the MS-like schemes offer several advantages. They are easy to implement in symbolic manipulation programs and are tractable at high numbers of loops. Furthermore, α_s and m_q are ξ independent to all orders and may thus be regarded as physical observables. The price to pay is that heavy quarks do not automatically decouple. However, as will become clear in the following, the theoretical ambiguity associated with the matching at the flavour thresholds is negligible if higher orders are taken into account.

DECOUPLING OF HEAVY QUARKS

The decoupling theorem states that the infrared structure of unbroken, non-abelian gauge theories is not affected by the presence of heavy fields coupled to the massless gauge fields [5]. As a consequence, a heavy quark h decouples from physical observables measured at energy scales $\mu \ll m_h$ up to terms of $\mathcal{O}(\mu/m_h)$. However, the proof of this theorem relies on the use of mass-dependent β functions. Thus, this theorem does not automatically hold for the parameters and fields in MS-like schemes. The standard way out is to implement explicit decoupling by using the language of effective field theory.

As an idealized situation, consider full QCD with $n_l = n_f - 1$ light quarks q, with $m_q \ll \mu$, plus one heavy quark h, with $m_h \gg \mu$. The idea is to construct an effective theory, QCD$'$, by integrating out the h quark. The parameters and fields of the effective theory, which will be denoted by a prime, are related to their counterparts of the full theory by the decoupling relations,

$$g^{0\prime} = \zeta_g^0 g^0, \qquad m_q^{0\prime} = \zeta_m^0 m_q^0, \qquad \xi^{0\prime} - 1 = \zeta_3^0(\xi^0 - 1),$$
$$\psi_q^{0\prime,i} = (\zeta_2^0)^{1/2}\psi_q^{0,i}, \qquad G_\mu^{0\prime,a} = (\zeta_3^0)^{1/2}G_\mu^{0,a}, \qquad c^{0\prime,a} = (\tilde{\zeta}_3^0)^{1/2}c^{0,a}. \tag{5}$$

By gauge invariance, the most general form of the effective Lagrangian $\mathcal{L}'$ emerges from Eq. (1) by only retaining the light degrees of freedom and reads

$$\mathcal{L}'\left(g_s^0, m_q^0, \xi^0; \psi_q^{0,i}, G_\mu^{0,a}, c^{0,a}; \zeta^0\right) = \mathcal{L}\left(g_s^{0\prime}, m_q^{0\prime}, \xi^{0\prime}; \psi_q^{0\prime,i}, G_\mu^{0\prime,a}, c^{0\prime,a}\right), \tag{6}$$

where ζ^0 collectively denotes all decoupling constants of Eq. (5). The latter may be derived by imposing the condition that the results for n-particle Green functions of light fields in both theories should agree up to terms of $\mathcal{O}(\mu/m_h)$.

As an example, let us consider the q-quark propagator. Up to terms of $\mathcal{O}(\mu/m_h)$, we have [6]

$$\frac{i}{\not{p}[1 + \Sigma_V^0(p^2)]} = \int dx \, e^{ip\cdot x} \left\langle T\psi_q^0(x)\bar{\psi}_q^0(0)\right\rangle$$
$$= \frac{1}{\zeta_2^0}\int dx \, e^{ip\cdot x} \left\langle T\psi_q^{0\prime}(x)\bar{\psi}_q^{0\prime}(0)\right\rangle = \frac{1}{\zeta_2^0}\frac{i}{\not{p}[1 + \Sigma_V^{0\prime}(p^2)]}, \tag{7}$$

where the subscript V reminds us that the self-energy of a massless quark only consists of a vector part. Note that $\Sigma_V^{0\prime}(p^2)$ only contains light degrees of freedom, whereas $\Sigma_V^0(p^2)$ also receives virtual contributions from the h quark. As we are interested in the limit $m_h \to \infty$, we may nullify the external momentum p, which entails an enormous technical simplification because then only tadpole integrals have to be considered. In dimensional regularization, one also has $\Sigma_V^{0\prime}(0) = 0$. Thus, we obtain

$$\zeta_2^0 = 1 + \Sigma_V^{0h}(0), \tag{8}$$

where the superscript h indicates that only diagrams involving closed h-quark loops need to be computed. In a similar fashion, one obtains

$$\zeta_m^0 = \frac{1 - \Sigma_S^{0h}(0)}{1 + \Sigma_V^{0h}(0)}, \qquad \zeta_3^0 = 1 + \Pi_G^{0h}(0), \qquad \tilde{\zeta}_3^0 = 1 + \Pi_c^{0h}(0),$$

$$\tilde{\zeta}_1^0 = 1 + \Gamma_{G\bar{c}c}^{0h}(0,0), \qquad \zeta_g^0 = \frac{\tilde{\zeta}_1^0}{\tilde{\zeta}_3^0(\zeta_3^0)^{1/2}}, \tag{9}$$

where Σ_S, Π_G, Π_c, and $\Gamma_{G\bar{c}c}$ denote the scalar part of the q-quark self-energy, the gluon self-energy, the ghost self-energy, and the $G\bar{c}c$ vertex function, respectively. Typical Feynman diagrams contributing to ζ_2^0, ζ_m^0, ζ_3^0, $\tilde{\zeta}_3^0$, and $\tilde{\zeta}_1^0$ are depicted in Fig. 1. The full set of diagrams are generated and evaluated with the symbolic manipulation packages QGRAF [7] and MATAD [8], respectively. The renormalized counterparts of ζ_g^0 and ζ_m^0,

$$\zeta_g = \frac{Z_g}{Z_g'}\zeta_g^0, \qquad \zeta_m = \frac{Z_m}{Z_m'}\zeta_m^0, \tag{10}$$

are found to be UV finite and ξ independent and to satisfy the appropriate renormalization group equations, which constitutes a strong test. The resulting decoupling relations take a particularly simple form if the matching scale is chosen to be $\mu_h = m_h(\mu_h)$, namely,

$$\frac{a'}{a} = \zeta_g^2 = 1 + c_2 a^2 + c_3 a^3, \qquad \frac{m_q'}{m_q} = \zeta_m = 1 + d_2 a^2 + d_3 a^3,$$

$$c_2 = \frac{11}{72}, \qquad c_3 = \frac{564731}{124416} - \frac{82043}{27648}\zeta(3) - \frac{2633}{31104}n_l, \qquad d_2 = \frac{89}{432},$$

$$d_3 = \frac{2951}{2916} - \frac{\ln^4 2}{54} + \frac{\ln^2 2}{9}\zeta(2) - \frac{407}{864}\zeta(3) + \frac{103}{72}\zeta(4) - \frac{4}{9}\operatorname{Li}_4\left(\frac{1}{2}\right)$$

$$+ n_l\left(\frac{1327}{11664} - \frac{2}{27}\zeta(3)\right), \tag{11}$$

where ζ and Li_4 are Riemann's zeta function and the dilogarithm, respectively. c_2 and d_2 were previously calculated [9]. Three-loop expressions for ζ_2 and ζ_3, which

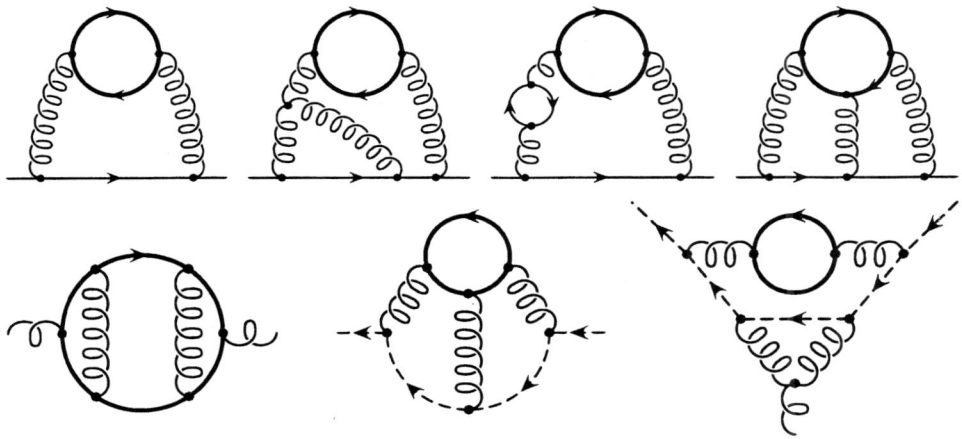

FIGURE 1. Typical Feynman diagrams contributing to ζ_2^0, ζ_m^0, ζ_3^0, $\tilde{\zeta}_3^0$, and $\tilde{\zeta}_1^0$.

may be useful for parton model calculations, are available for the covariant gauge [10].

The phenomenological implications of Eqs. (4) and (11) are illustrated in Fig. 2. For consistency, $(n + 1)$-loop evolution must be accompanied by n-loop matching. Figure 2 shows how $\alpha_s^{(5)}(M_Z)$, consistently evaluated from $\alpha_s^{(4)}(M_\tau) = 0.36$ to a given order, depends on the scale $\mu^{(5)}$, measured in units of the bottom-quark pole mass $M_b = 4.7$ GeV, where the bottom-quark threshold is crossed. In Fig. 3, the analogous study is performed for $m_c^{(5)}(M_Z)$ calculated from $\mu_c = m_c^{(4)}(\mu_c) = 1.2$ GeV using $\alpha_s^{(5)}(M_Z) = 0.118$. As expected, the dependence on the unphysical scale $\mu^{(5)}$ is gradually getting weaker as we go to higher orders.

EFFECTIVE LAGRANGIANS AND LOW-ENERGY THEOREMS

An interesting and perhaps even surprising aspect of ζ_g and ζ_m is that they carry the full information about the virtual h-quark effects on the couplings of a CP-even Higgs boson H to gluons and q quarks, respectively. To reveal this connection, starting from the bare Yukawa Lagrangian of the full theory,

$$\mathcal{L}_{\text{Yuk}} = -\frac{H^0}{v^0}\left(\sum_{q=1}^{n_l} m_q^0 \bar{\psi}_q^0 \psi_q^0 + m_h^0 \bar{\psi}_h^0 \psi_h^0\right), \qquad (12)$$

where v is the Higgs vacuum expectation value, one integrates out the h quark by taking the limit $m_h^0 \to \infty$ and so derives the effective Lagrangian,

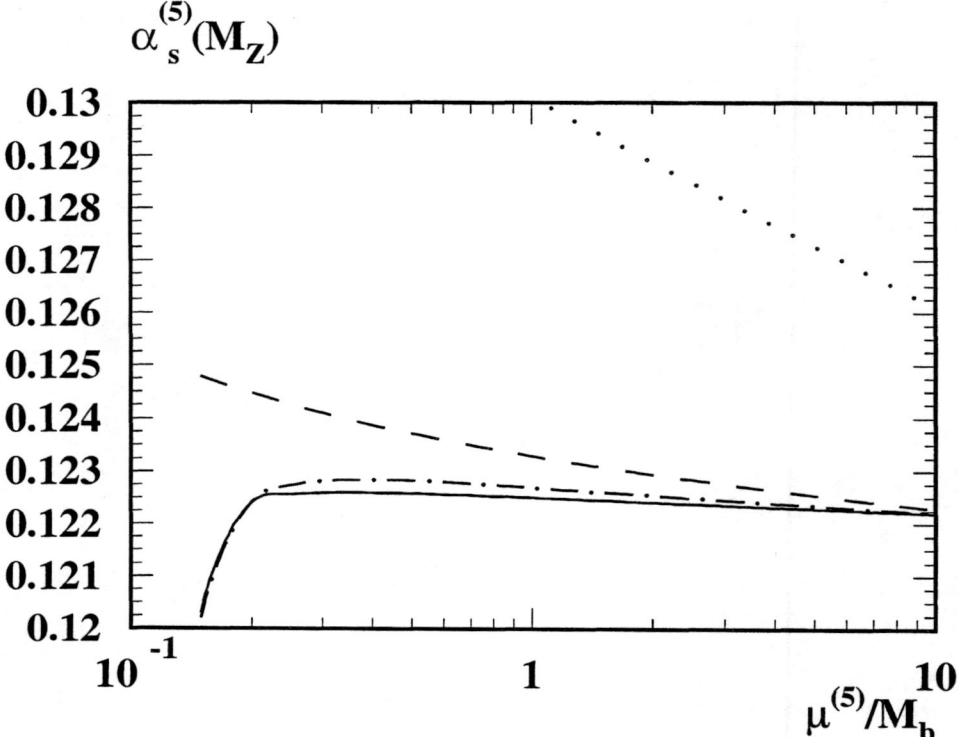

FIGURE 2. $\mu^{(5)}$ dependence of $\alpha_s^{(5)}(M_Z)$ calculated from $\alpha_s^{(4)}(M_\tau) = 0.36$ and $M_b = 4.7$ GeV with evolution at one (dotted), two (dashed), three (dot-dashed), and four (solid) loops and appropriate matching.

$$\mathcal{L}'_{\text{Yuk}} = -\frac{H^0}{v^0} \sum_{i=1}^{5} C_i^0 \mathcal{O}_i' = -2^{1/4} G_F^{1/2} H \sum_{i=1}^{5} C_i[\mathcal{O}_i], \tag{13}$$

which is spanned by a natural basis of composite scalar operators with mass dimension four [11]. The operators,

$$\mathcal{O}_1' = \left(G_{\mu\nu}^{0\prime,a}\right)^2, \qquad \mathcal{O}_2' = \sum_{q=1}^{n_l} m_q^{0\prime} \bar{\psi}_q^{0\prime} \psi_q^{0\prime}, \qquad \mathcal{O}_3' = \sum_{q=1}^{n_l} \bar{\psi}_q^{0\prime} \left(i \not{D}^{0\prime} - m_q^{0\prime}\right) \psi_q^{0\prime},$$

$$\mathcal{O}_4' = G_\nu^{0\prime,a} \left(\nabla_\mu^{ab} G^{0\prime,b\mu\nu} + g_s^{0\prime} \sum_{q=1}^{n_l} \bar{\psi}_q^{0\prime} T^a \gamma^\nu \psi_q^{0\prime}\right) - \partial_\mu \bar{c}^{0\prime,a} \partial^\mu c^{0\prime,a},$$

$$\mathcal{O}_5' = (\nabla_\mu^{ab} \partial^\mu \bar{c}^{0\prime,b}) c^{0\prime,a}, \tag{14}$$

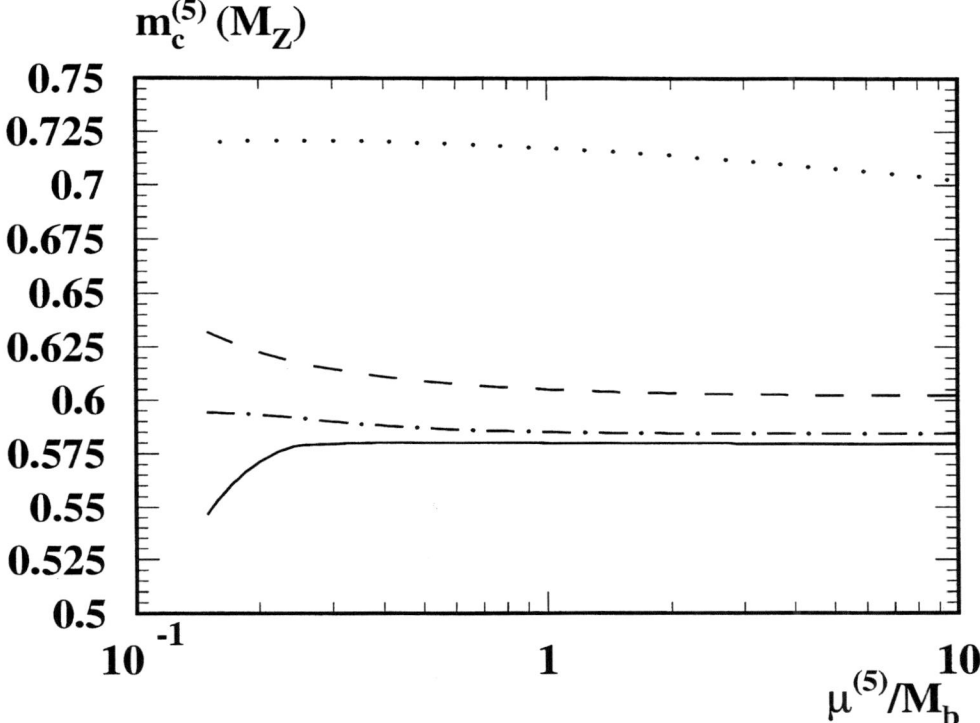

FIGURE 3. $\mu^{(5)}$ dependence of $m_c^{(5)}(M_Z)$ calculated from $\mu_c = m_c^{(4)}(\mu_c) = 1.2$ GeV, $M_b = 4.7$ GeV, and $\alpha_s^{(5)}(M_Z) = 0.118$ with evolution at one (dotted), two (dashed), three (dot-dashed), and four (solid) loops and appropriate matching.

are only constructed from light degrees of freedom, while all residual dependence on the h quark resides in the Wilson coefficients C_i^0.

The derivation of C_i^0 proceeds similarly to Eq. (9). Considering appropriate one-particle-irreducible Green functions which contain a zero-momentum insertion of $\mathcal{O}_h = m_h^0 \bar{\psi}_h^0 \psi_h^0$ in the limit $m_h^0 \to \infty$, one finds [10]

$$\zeta_3^0(-4C_1^0 + 2C_4^0) = -\frac{1}{2}\partial_h^0 \Pi_G^{0h}(0), \quad \zeta_m^0 \zeta_2^0(C_2^0 - C_3^0) = 1 - \Sigma_S^{0h}(0) - \frac{1}{2}\partial_h^0 \Sigma_S^{0h}(0),$$

$$\zeta_2^0 C_3^0 = -\frac{1}{2}\partial_h^0 \Sigma_V^{0h}(0), \quad \zeta_3^0(C_4^0 + C_5^0) = \frac{1}{2}\partial_h^0 \Pi_c^{0h}(0), \quad \tilde{\zeta}_1^0 C_5^0 = \frac{1}{2}\partial_h^0 \Gamma_{G\bar{c}c}^{0h}(0,0), \qquad (15)$$

with $\partial_h^0 = (m_h^{02}\partial/\partial m_h^{02})$, which may be solved for C_i^0. Only $\mathcal{O}_1'$ and $\mathcal{O}_2'$ contribute to physical observables. They mix under renormalization as [11]

$$[\mathcal{O}_1'] = \left[1 + 2\left(\frac{\alpha_s'\partial}{\partial\alpha_s'}\ln Z_g'\right)\right]\mathcal{O}_1' - 4\left(\frac{\alpha_s'\partial}{\partial\alpha_s'}\ln Z_m'\right)\mathcal{O}_2', \qquad [\mathcal{O}_2'] = \mathcal{O}_2', \qquad (16)$$

where the brackets denote the renormalized counterparts. C_1 and C_2 are accordingly determined from the second equation in Eq. (13). They are diagrammatically calculated through three loops [10]. Inserting Eqs. (8) and (9) into Eqs. (15), one obtains the low-energy theorems [10]

$$C_1 = -\frac{1}{2}\frac{\partial \ln \zeta_g^2}{\partial \ln m_h^2}, \qquad C_2 = 1 + 2\frac{\partial \ln \zeta_m}{\partial \ln m_h^2}, \tag{17}$$

which are valid to all orders in α_s. Fully exploiting the present knowledge of Eq. (4) [4], one may construct the four-loop terms of ζ_g and ζ_m involving $\ln m_h^2$ and so obtain C_1 and C_2 from Eq. (17) to one order beyond the diagrammatic calculation. The expansions in $a = \alpha_s^{(n_f)}(\mu_h)/\pi$ read [10]

$$\begin{aligned}
C_1 &= -\frac{a}{12}[1 + 2.7500\,a + (9.7951 - 0.6979\,n_l)a^2 \\
&\quad + (49.1827 - 7.7743\,n_l - 0.2207\,n_l^2)a^3], \\
C_2 &= 1 + 0.2778a^2 + (2.2434 + 0.2454\,n_l)a^3 \\
&\quad + (2.1800 + 0.3096\,n_l - 0.0100\,n_l^2)a^4.
\end{aligned} \tag{18}$$

Having established $\mathcal{L}'_{\text{yuk}}$, we are able to make higher-order predictions for the QCD interactions of a light H boson by just computing massless diagrams. For instance, the $H \to gg$ partial decay width at three loops is found to be [12]

$$\Gamma(H \to gg) = \frac{G_F M_H^3}{36\pi\sqrt{2}}a'^2\left[1 + 17.917\,a' + a'^2\left(156.808 - 5.708\ln\frac{m_t^2}{M_H^2}\right)\right], \tag{19}$$

where $a' = \alpha_s^{(5)}(M_H)/\pi$. The three-loop $\mathcal{O}(\alpha_s^2 G_F m_t^2)$ corrections to $\Gamma(H \to q\bar{q})$, with $q = u, d, s, c, b$, may also be obtained from Eq. (13) [13]. Analogously, the QCD interactions of a CP-odd Higgs boson A may be described by an effective Lagrangian involving composite pseudoscalar operators with mass dimension four [14]. The resulting counterpart of Eq. (19) is found to be [14]

$$\Gamma(A \to gg) = \frac{G_F M_A^3}{16\pi\sqrt{2}}a'^2\left[1 + 18.417\,a' + a'^2\left(171.544 - 5\ln\frac{m_t^2}{M_A^2}\right)\right], \tag{20}$$

where $a' = \alpha_s^{(5)}(M_A)/\pi$. As a by-product of this analysis [14], the Adler-Bardeen nonrenormalization theorem [15], which states that the anomaly of the axial-vector current is not renormalized in QCD, is verified through three loops by an explicit diagrammatic calculation.

COMPARISON WITH SCALE OPTIMIZATION PROCEDURES

It is interesting to compare the exact values of the $\mathcal{O}(\alpha_s^2)$ corrections in Eqs. (19) and (20) with the estimates one may derive from the knowledge of the $\mathcal{O}(\alpha_s)$ correction through the application of well-known scale optimization procedures, based on

Grunberg's concept of fastest apparent convergence (FAC) [16], Stevenson's principle of minimal sensitivity (PMS) [17], and the proposal by Brodsky, Lepage, and Mackenzie (BLM) [18] to resum the leading light-quark contribution to the renormalization of the strong coupling constant. These procedures lead to the generic expression

$$\Gamma(H \to gg) = \frac{G_F M_H^3}{36\pi\sqrt{2}} \left(\frac{\alpha_s^{(5)}(\xi M_H)}{\pi} \right)^2 \left(1 + \bar{K}_1 \frac{\alpha_s^{(5)}(\xi M_H)}{\pi} \right)$$
$$= \frac{G_F M_H^3}{36\pi\sqrt{2}} a'^2 \left[1 + 17.917\, a' + \bar{K}_2 a'^2 \right], \tag{21}$$

where $a' = \alpha_s^{(5)}(M_H)/\pi$, and similarly for $\Gamma(A \to gg)$. The FAC, PMS, and BLM values of ξ, $\bar{K}_1$, and $\bar{K}_2$ for $\Gamma(H \to gg)$ and $\Gamma(A \to gg)$ with $n_l = 5$ are summarized in Table 1. The values of $\bar{K}_2$ should be compared with the true coefficients of the $\mathcal{O}(\alpha_s^2)$ corrections in Eqs. (19) and (20). We observe that the sign and the order of magnitude is correctly predicted in all cases. Furthermore, the three values of $\bar{K}_2$ for $\Gamma(H \to gg)$ are indeed smaller than their counterparts for $\Gamma(A \to gg)$. Similarly to Ref. [19], the FAC and PMS results almost coincide.

TABLE 1. FAC, PMS, and BLM values of ξ, $\bar{K}_1$, and $\bar{K}_2$ for $\Gamma(H \to gg)$ and $\Gamma(A \to gg)$ with $n_l = 5$.

	$H \to gg$			$A \to gg$		
	ξ	$\bar{K}_1$	$\bar{K}_2$	ξ	$\bar{K}_1$	$\bar{K}_2$
FAC	0.097	0	263.346	0.091	0	277.601
PMS	0.087	−0.841	263.876	0.081	−0.841	278.131
BLM	0.174	4.5	242.484	0.174	5	252.547

SUMMARY

A consistent $\overline{\text{MS}}$ description of $\alpha_s(\mu)$ and $m_q(\mu)$ with μ evolution through four loops and threshold matching through three loops is now available. Effective Lagrangians and low-energy theorems are useful tools to treat the hadronic decays of light CP-even and CP-odd Higgs bosons through three loops. The sign and the order of magnitude of the resulting three-loop corrections are correctly predicted by scale optimization procedures.

ACKNOWLEDGMENTS

The author is grateful to W.A. Bardeen, K.G. Chetyrkin, and M. Steinhauser for their collaboration and to the organizers of the Eleventh International Light-Cone School and Workshop on New Directions in Quantum Chromodynamics for their excellent work.

REFERENCES

1. C.G. Bollini and J.J. Giambiagi, *Phys. Lett.* **40** B, 566 (1972); G. 't Hooft and M. Veltman, *Nucl. Phys.* B **44**, 189 (1972).
2. G. 't Hooft, *Nucl. Phys.* B **61**, 455 (1973).
3. W.A. Bardeen, A.J. Buras, D.W. Duke, and T. Muta, *Phys. Rev.* D **18**, 3998 (1978).
4. J.A.M. Vermaseren, S.A. Larin, and T. van Ritbergen, *Phys. Lett.* B **400**, 379 (1997); **405**, 327 (1997); K.G. Chetyrkin, *Phys. Lett.* B **404**, 161 (1997).
5. K. Symanzik, *Commun. math. Phys.* **34**, 7 (1973); T. Appelquist und J. Carazzone, *Phys. Rev.* D **11**, 2856 (1975).
6. K.G. Chetyrkin, B.A. Kniehl, and M. Steinhauser, *Phys. Rev. Lett.* **79**, 2184 (1997).
7. P. Nogueira, *J. Comput. Phys.* **105**, 279 (1993).
8. M. Steinhauser, Ph.D. thesis, Karlsruhe University (Shaker Verlag, Aachen, 1996); in these proceedings.
9. W. Bernreuther and W. Wetzel, *Nucl. Phys.* B **197**, 228 (1982); **513**, 758(E) (1998); W. Bernreuther, *Ann. Phys.* **151**, 127 (1983); *Z. Phys.* C **20**, 331 (1983); **29**, 245 (1985); S.A. Larin, T. van Ritbergen, and J.A.M. Vermaseren, *Nucl. Phys.* B **438**, 278 (1995).
10. K.G. Chetyrkin, B.A. Kniehl, and M. Steinhauser, *Nucl. Phys.* B **510**, 61 (1998).
11. T. Inami, T. Kubota, and Y. Okada, *Z. Phys.* C **18**, 69 (1983); V.P. Spiridonov, INR Report P–0378 (1984).
12. K.G. Chetyrkin, B.A. Kniehl, and M. Steinhauser, *Phys. Rev. Lett.* **79**, 353 (1997).
13. K.G. Chetyrkin, B.A. Kniehl, and M. Steinhauser, *Phys. Rev. Lett.* **78**, 594 (1997); *Nucl. Phys.* B **490**, 19 (1997).
14. K.G. Chetyrkin, B.A. Kniehl, M. Steinhauser, and W.A. Bardeen, *Nucl. Phys.* B **535**, 3 (1998).
15. S.L. Adler and W.A. Bardeen, *Phys. Rev.* **182**, 1517 (1969).
16. G. Grunberg, *Phys. Lett.* **95** B, 70 (1980); **110** B, 501(E) (1982); *Phys. Rev.* D **29**, 2315 (1984).
17. P.M. Stevenson, *Phys. Rev.* D **23**, 2916 (1981); *Phys. Lett.* **100** B, 61 (1981); *Nucl. Phys.* B **203**, 472 (1982); *Phys. Lett.* B **231**, 65 (1984).
18. S.J. Brodsky, G.P. Lepage and P.B. Mackenzie, *Phys. Rev.* D **28**, 228 (1983); S.J. Brodsky and H.J. Lu, *Phys. Rev.* D **51** (1995) 3652.
19. J. Kubo and S. Sakakibara, *Phys. Rev.* D **26**, 3656 (1982); A.L. Kataev and V.V. Starshenko, *Mod. Phys. Lett.* A **10**, 235 (1995); K.G. Chetyrkin, B.A. Kniehl, and A. Sirlin, *Phys. Lett.* B **402**, 359 (1997).

Multiquark picture for $\Lambda(1405)$ and $\Sigma(1620)$

Seungho Choe

Department of Physics, Yonsei University, Seoul 120-749, Korea

Abstract. We propose a new QCD sum rule analysis for the Λ (1405) and the Σ (1620). Using the I=0 and I=1 multiquark sum rules we predict their masses.

One of interesting subjects in nuclear physics is to study properties of the excited baryon states. For example, in the case of the Λ (1405) its nature is not revealed completely [1]; i.e. an ordinary three quark state or a $\bar{K}N$ bound state or the mixing state of the previous two possibilities. In the QCD sum rule approach [2] there have been several works on the Λ (1405) using three-quark interpolating fields [3,4] or five-quark operators [5]. In this work we focus on the decay modes of the Λ (1405) and the Σ (1620) and get the mass of each particle introducing multiquark sum rules.

Let's consider the following correlator:

$$\Pi(q^2) = i \int d^4x e^{iqx} \langle T(J(x)\bar{J}(0)) \rangle, \tag{1}$$

where J is the $\pi\Sigma$ (I=0) multiquark interpolating field, $J_{\pi^+\Sigma^- + \pi^0\Sigma^0 + \pi^-\Sigma^+}$.

Here, for the Σ we take the Ioffe's choice [6]; e.g. $\pi^0\Sigma^0$ means $\epsilon_{abc}(\bar{u}_e i\gamma^5 u_e - \bar{d}_e i\gamma^5 d_e)([u_a^T C\gamma_\mu s_b]\gamma^5\gamma^\mu d_c + [d_a^T C\gamma_\mu s_b]\gamma^5\gamma^\mu u_c)$, where u, d and s are the up, down and strange quark fields, and a, b, c, e are color indices. T denotes the transpose in Dirac space and C is the charge conjugation matrix.

The OPE side has two structures:

$$\Pi^{OPE}(q^2) = \Pi_q^{OPE}(q^2)\slashed{q} + \Pi_1^{OPE}(q^2)\mathbf{1}. \tag{2}$$

In this paper, however, we only present the sum rule from the Π_1 structure (hereafter referred to as the Π_1 sum rule) because the Π_1 sum rule is generally more reliable than the Π_q sum rule as emphasized in Ref. [7]. The OPE side is given as follows.

$$\begin{aligned}
\Pi_1^{OPE}(q^2) = &-\frac{7\, m_s}{\pi^8\, 2^{18}\, 3^2\, 5}q^{10}ln(-q^2) + \frac{7}{\pi^6\, 2^{15}\, 3^2}\langle\bar{s}s\rangle q^8 ln(-q^2) \\
&+ \frac{35\, m_s^2}{\pi^6\, 2^{14}\, 3^2}\langle\bar{s}s\rangle q^6 ln(-q^2) - \frac{121\, m_s}{\pi^4\, 2^9\, 3^2}\langle\bar{q}q\rangle^2 q^4 ln(-q^2)
\end{aligned}$$

CP494, *New Directions in Quantum Chromodynamics*, edited by C.-R. Ji and D.-P. Min
© 1999 American Institute of Physics 1-56396-908-4/99/$15.00

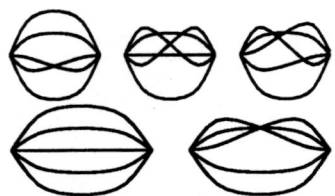

FIGURE 1. "bound" (upper) and "unbound" (lower) diagrams. Solid lines are the quark propagators.

$$+ \frac{11}{\pi^2 \, 2^6}\langle \bar{q}q \rangle^2 \langle \bar{s}s \rangle q^2 ln(-q^2) - \frac{m_s^2}{\pi^2 \, 2^6 \, 3}(14\langle \bar{q}q \rangle^3 - 33\langle \bar{q}q \rangle^2 \langle \bar{s}s \rangle)ln(-q^2)$$

$$- \frac{m_s}{2^4 \, 3^3}(140\langle \bar{q}q \rangle^4 + 3\langle \bar{q}q \rangle^3 \langle \bar{s}s \rangle)\frac{1}{q^2}, \tag{3}$$

where m_s is the strange quark mass and $\langle \bar{q}q \rangle$, $\langle \bar{s}s \rangle$ are the quark condensate and the strange quark condensate, respectively. Here, we let $m_u = m_d = 0 \neq m_s$ and $\langle \bar{u}u \rangle = \langle \bar{d}d \rangle \equiv \langle \bar{q}q \rangle \neq \langle \bar{s}s \rangle$. We neglect the contribution of gluon condensates and concentrate on tree diagrams such as Fig. 1, and assume the vacuum saturation hypothesis to calculate quark condensates of higher dimensions. Note that only some typical diagrams are shown in Fig. 1.

The contribution of the "bound" diagrams is a $1/N_c$ correction to that of the "unbound" diagrams, where N_c is the number of the colors. In Eq. (3) we set $N_c = 3$. The "unbound" diagrams correspond to a picture that two particles are flying away without any interaction between them. In the $N_c \to \infty$ limit only the "unbound" diagrams contribute to the $\pi\Sigma$ multiquark sum rule. Then, the $\pi\Sigma$ multiquark mass ($m(\pi\Sigma)$) should be the sum of the pion and the Σ mass in this limit.

Eq. (3) has the following form:

$$\Pi_1^{OPE}(q^2) = a \, q^{10}ln(-q^2) + b \, q^8 ln(-q^2) + c \, q^6 ln(-q^2) + d \, q^4 ln(-q^2)$$

$$+ e \, q^2 ln(-q^2) + f \, ln(-q^2) + g \, \frac{1}{q^2}, \tag{4}$$

where $a, b, c, \cdots, g$ are constants. Then, we parameterize the phenomenological side as

$$\frac{1}{\pi}Im\Pi_1^{Phen}(s) = \lambda^2 m\delta(s - m^2) + [-a \, s^5 - b \, s^4 - c \, s^3 - d \, s^2 - e \, s - f]\theta(s - s_0), \tag{5}$$

where m is the $m(\pi\Sigma)$ and s_0 the continuum threshold. λ is the coupling strength of the interpolating field to the physical Λ (1405) state. The Borel-mass dependence of the $m(\pi\Sigma)$ shows that there is a plateau for the large Borel mass. However, this is a trivial result from our crude model on the phenomenological side. Hence we do not take this as the $m(\pi\Sigma)$ and neither as the Λ (1405) mass. Instead, we draw the Borel-mass dependence of the coupling strength λ^2 at $s_0 = 2.789$ GeV^2

378

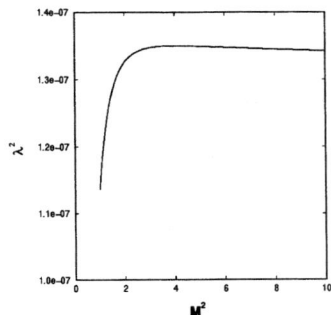

FIGURE 2. The Borel-mass dependence of the coupling strength λ^2 from the $\pi\Sigma$ multiquark sum rule at $s_0 = 2.789$ GeV^2.

as shown in Fig. 2, where the s_0 is taken by considering the next Λ particle [1]. There is the maximum point in the figure. It means that the $\pi\Sigma$ multiquark state couples strongly to the physical Λ (1405) state at this point. Then we take the Λ (1405) mass as the $m(\pi\Sigma)$ at the point. However, it would be better to determine an effective threshold s_0 from the present sum rule itself.

Thus, the steps for getting the $m(\pi\Sigma)$ are as follows. First, consider "unbound" diagrams only and choose a threshold s_0 in order that the average mass between the fiducial Borel interval becomes the $m(\pi) + m(\Sigma)$. Second, consider whole diagrams ("unbound" + "bound" diagrams) and draw the Borel-mass dependence of the coupling strength λ^2 using the above s_0. Last, determine the $m(\pi\Sigma)$ where the λ^2 has the maximum value, and thus take this as the Λ (1405) mass. Following the above steps we get the $m(\pi\Sigma) = 1.424$ GeV at $s_0 = 3.082$ GeV^2.

There is another I=0 multiquark state; i.e. the $\bar{K}^0 n + K^- p$ multiquark state. Similarly, we obtain the $m(\bar{K}N) = 1.589$ GeV at $s_0 = 3.852$ GeV^2. This corresponds to the Λ (1600) mass. It is interesting to note that the masses from two multiquark states are similar at the same threshold as shown in Table 1.

Now, we can extend our previous analysis to the I=1 multiquark states and thus get the Σ (1620) mass. There are three decay channels for the Σ (1620). Then, we can construct the following multiquark interpolating fields; $J_{\bar{K}^0 n - K^- p}$, $J_{\pi^+ \Sigma^- - \pi^- \Sigma^+}$, and $J_{\pi^0 \Lambda}$ (or $J_{\pi^\pm \Lambda}$). In Table 2 we present each multiquark mass.

We have obtained the I=0 and I=1 multiquark masses which are slightly different from the experimental values [1]. One of corrections is to include the isospin

TABLE 1. Mass of the $\bar{K}N$, $\pi\Sigma$ (I=0) multiquark states ($\langle \bar{q}q \rangle = -(0.230 \text{ GeV})^3$, $\langle \bar{s}s \rangle = 0.8 \langle \bar{q}q \rangle$, and $m_s = 0.150$ GeV).

s_0 (GeV^2)	$m(\bar{K}N)$ (GeV)	$m(\pi\Sigma)$ (GeV)
3.852	1.589	1.612
3.082	1.405	1.424

379

TABLE 2. Mass of the $\bar{K}N$, $\pi\Sigma$, and $\pi\Lambda$ (I=1) multiquark states ($\langle\bar{q}q\rangle = -(0.230\text{ GeV})^3$, $\langle\bar{s}s\rangle = 0.8\langle\bar{q}q\rangle$, and $m_s = 0.150$ GeV).

s_0 (GeV^2)	$m(\bar{K}N)$ (GeV)	$m(\pi\Sigma)$ (GeV)	$m(\pi\Lambda)$ (GeV)
3.852	1.589	1.606	1.581

symmetry breaking effects (i.e. $m_u \neq m_d \neq 0$, $\langle\bar{u}u\rangle \neq \langle\bar{d}d\rangle$, and electromagnetic effects) in our sum rules. On the other hand, one can consider the contractions between the $\bar{u}$ and u (or between the $\bar{d}$ and d) quarks in the initial state which have been excluded in our previous calculation. However, it is found that this correction is very small comparing to other $1/N_c$ corrections, i.e. the contribution of "bound" diagrams. Another possibility is the correction from the possible instanton effects [8] to the I=0 and I=1 states, respectively.

In this work we have neglected the contribution of gluon condensates and that of other higher dimensional operators including gluon components. Since we have considered the Π_1 sum rule, only the odd dimensional operators can contribute to the sum rule. Thus, for example, the contribution of the gluon condensates is given by the terms like $m_s\langle\frac{\alpha_s}{\pi}G^2\rangle$ and thus can be neglected comparing to other quark condensates of the same dimension.

In summary, the Λ (1405) and Σ (1620) masses are predicted in the QCD sum rule approach using the $\bar{K}N$, $\pi\Sigma$, and $\pi\Lambda$ multiquark interpolating fields (both I=0 and I=1).

The author thanks Prof. D.-P. Min and Prof. C.-R. Ji for their effort to make NuSS'99 successful. This work was supported in part by the Korea Science and Engineering Foundation (KOSEF).

REFERENCES

1. Particle Data Group, Eur. Phys. J. **C3**, 1 (1998).
2. Shifman, M.A., Vainshtein, A.I. and Zakharov, V.I., Nucl. Phys. **B147**, 385, 448 (1979); Reinders, L.J., Rubinstein, H.R., and Yazaki, S., Phys. Rep. **127**, 1 (1985); Narison, S., "QCD Spectral Sum Rules", World Scientific Lecture Notes in Physics, Vol. 26 (1989); and references therein.
3. Leinweber, D.B., Ann. Phys. (N.Y.) **198**, 203 (1990).
4. Kim, H. and Lee, Su H., Z. Phys. **A357**, 425 (1997).
5. Liu, J.P., Z. Phys. **C22**, 171 (1984).
6. Ioffe, B.L., Nucl. Phys. **B188**, 317 (1981); **B191**, 591 (E) (1981).
7. Jin, X., and Tang, J., Phys. Rev. **D56**, 515 (1997).
8. For a recent review, see Schäfer, T. and Shuryak, E.V., Rev. Mod. Phys. **70**, 323 (1998); Forkel, H. and Banerjee, M.K., Phys. Rev. Lett. **71**, 484 (1993); Forkel, H. and Nielsen, M., Phys. Rev. **D55**, 1471 (1997).

EFFECTIVE FIELD THEORY
IN HADRONS

Elusive Neutrons from Nuclei in Effective Field Theory

Silas R. Beane*

*Department of Physics, University of Maryland
College Park, MD 20742-4111

Abstract. We review recent computations of neutral pion photoproduction and Compton scattering on the deuteron in baryon chiral perturbation theory. Progress in extracting the neutron electric dipole amplitude, which is relevant in neutral pion photoproduction, and the neutron polarizabilities, which are relevant in Compton scattering, is discussed.

INTRODUCTION

The absence of suitable neutron targets in low-energy scattering experiments requires the use of nuclear targets like deuterium and helium in order to extract neutron scattering data. The extent to which neutron data can be reliably extracted depends on how under control the errors are in computing the nuclear corrections to free nucleon motion. Of course precise calculations of hadron processes are possible only where a small dimensionless expansion parameter is identified. This is the main motivation behind the ongoing intense effort to develop a perturbative theory of nuclear interactions [1]. The dimensionless parameters relevant to low energy QCD and therefore to nuclear physics consist of ratios of external momenta to various characteristic energy scales, like the nucleon mass. Effective field theory is the technology which develops a hierarchy of scales into a perturbative expansion of physical observables.

In this paper we describe several recent effective field theory calculations whose objective is to extract neutron properties from nuclear scattering processes in a systematic way. We first discuss a computation of neutral pion photoproduction on the deuteron and its dependence on nucleon parameters. We then describe a calculation of Compton scattering on the deuteron at photon energies of order the pion mass. Here the ultimate objective is to learn about neutron polarizabilities from nuclear Compton scattering. The basic power-counting scheme is reviewed in the first section. In the second section we discuss photoproduction on the deuteron. The third section is dedicated to Compton scattering on the deuteron.

CP494, *New Directions in Quantum Chromodynamics*, edited by C.-R. Ji and D.-P. Min

WEINBERG POWER-COUNTING

At energies well below the chiral symmetry breaking scale, $\Lambda_\chi \sim 4\pi f_\pi \sim M \sim m_\rho$, the interactions of pions, photons and nucleons can be described systematically using an effective field theory. This effective field theory, known as chiral perturbation theory (χPT), reflects the observed QCD pattern of symmetry breaking. In QCD the chiral $SU(2)_L \times SU(2)_R$ symmetry is spontaneously broken. Here we are interested in processes where the typical momenta of all external particles is $p \ll \Lambda_\chi$, so we identify our expansion parameter as p/Λ_χ. In QCD $SU(2)_L \times SU(2)_R$ is softly broken by the small quark masses. This explicit breaking implies that the pion has a small mass in the low-energy theory. Since m_π/Λ_χ is then also a small parameter, we have a dual expansion in p/Λ_χ and m_π/Λ_χ. We take Q to represent either a small momentum *or* a pion mass.

In few-nucleon systems, a complication arises in χPT due to the existence of shallow nuclear bound states and related infrared singularities in A-nucleon reducible Feynman diagrams evaluated in the static approximation [2]. The fundamental problem is that nuclear physics introduces a new mass scale, the nuclear binding energy, which is very small compared to a typical hadronic scale like Λ_χ. One way to overcome this difficulty is to adopt a modified power-counting scheme in which χPT is used to calculate an effective potential which generally consists of all A-nucleon irreducible graphs. The S-matrix, which includes all reducible graphs as well, is then obtained through iteration by solving a Lippmann-Schwinger equation [2]. This version of nuclear effective theory is known as the Weinberg formulation. There now exists a competing power-counting scheme in which all nonperturbative physics responsible for the presence of low-lying bound states arises from the iteration of a single operator in the effective theory, while all other effects, including all higher dimensional operators *and* pion exchange, are treated perturbatively [3,4]. This version of the effective theory is known as the Kaplan-Savage-Wise (KSW) formulation. This is relevant here because Compton scattering on the deuteron has been computed to next-to-leading order in the KSW formulation [5]. We will discuss this result and its relation to our calculation. A comprehensive and up-to-date review of nuclear applications of effective field theories can be found in Ref. [1].

It should be noted that typical nucleon momenta inside the deuteron are small— on the order of $\sqrt{MB}$ or m_π, with B the deuteron binding energy—and consequently, *a priori* we expect no convergence problems in the χPT expansion of any low-momentum electromagnetic or pionic probe of the deuteron. Although in principle we could use wavefunctions computed in χPT, we will consider wavefunctions generated using modern nucleon-nucleon potentials [6]. Generally we find that any wavefunction with the correct binding energy gives equivalent results to within the theoretical error expected from neglected higher orders in the chiral expansion. Presumably we are insensitive to short distance components of the wavefunction because we are working at low energies and the deuteron is a large object.

PHOTOPRODUCTION ON THE DEUTERON

A striking example of the power of effective field theory is neutral pion photoproduction on the deuteron at threshold. The $O(Q^4)$ χPT prediction for the electric dipole amplitude in neutral pion photoproduction on the deuteron is [7]

$$
\begin{aligned}
E_d &= E_d^{ss} + E_d^{tb,3} + E_d^{tb,4} \\
&= (0.36 - 1.90 - 0.25) \times 10^{-3}/M_{\pi^+} \\
&= (-1.8 \pm 0.2) \times 10^{-3}/M_{\pi^+}
\end{aligned}
\tag{1}
$$

where $E_d^{tb,3}$ and $E_d^{tb,4}$ represent various 3- and 4-body corrections and

$$
\begin{aligned}
E_d^{ss} = \frac{1 + M_\pi/m}{1 + M_\pi/m_d} &\left\{ \frac{1}{2} \left(E_{0+}^{\pi^0 p} + E_{0+}^{\pi^0 n} \right) \int d^3p \, \phi_f^*(\vec{p}) \, \vec{\epsilon} \cdot \vec{J} \phi_i(\vec{p} - \vec{k}/2) \right. \\
&\left. - \frac{k}{m} \hat{k} \cdot \int d^3p \, \hat{p} \, \frac{1}{2} \left(P_1^{\pi^0 p} + P_1^{\pi^0 n} \right) \phi_f^*(\vec{p}) \, \vec{\epsilon} \cdot \vec{J} \phi_i(\vec{p} - \vec{k}/2) \right\},
\end{aligned}
\tag{2}
$$

where ϕ represents the deuteron wavefunction, here taken from the Argonne V18 potential. The $O(Q^4)$ χPT predictions for the electric dipole amplitudes of the proton and neutron are [8]

$$
E_{0+}^{\pi^0 p} = -1.16 \times 10^{-3}/M_{\pi^+} \qquad E_{0+}^{\pi^0 n} = +2.13 \times 10^{-3}/M_{\pi^+}.
\tag{3}
$$

Note the large value of the neutron electric dipole amplitude. The corresponding neutron cross section is a factor of four larger than the proton cross section, in complete violation of classical intuition. The proton empirical value from MAMI [9] is

$$
E_{0+}^{\pi^0 p} = (-1.31 \pm 0.08) \times 10^{-3}/M_{\pi^+},
\tag{4}
$$

in agreement with the χPT prediction. Of course in order to test the neutron prediction we must consider the deuteron. The SAL prediction [10] for the deuteron is

$$
E_d^{exp} = (-1.45 \pm 0.09) \times 10^{-3}/M_{\pi^+}
\tag{5}
$$

and therefore overlaps with the χPT prediction within 1.5 σ. One may wonder about the sensitivity of the deuteron electric dipole amplitude to the neutron contribution. For instance, if we set $E_{0+}^{\pi^0 n} = 0$ the χPT prediction becomes $E_d = -2.6$, completely at odds with the experimental value. This result is a striking confirmation of the large χPT prediction for the neutron. See Fig. 1.

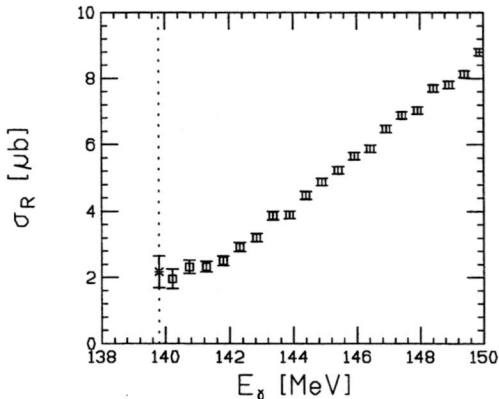

FIGURE 1. SAL data versus χPT prediction (star).

DEUTERON COMPTON SCATTERING

Nucleon Compton scattering has been studied in χPT in Ref. [11], where the following results for the polarizabilities were obtained to order Q^3:

$$\alpha_p = \alpha_n = \frac{5e^2 g_A^2}{384\pi^2 f_\pi^2 m_\pi} = 12.2 \times 10^{-4}\,\text{fm}^3;$$

$$\beta_p = \beta_n = \frac{e^2 g_A^2}{768\pi^2 f_\pi^2 m_\pi} = 1.2 \times 10^{-4}\,\text{fm}^3. \tag{6}$$

Here we have used $g_A = 1.26$ for the axial coupling of the nucleon, and $f_\pi = 93$ MeV as the pion decay constant. Note that the polarizabilities are *predictions* of χPT at this order. The $O(Q^3)$ χPT predictions diverge in the chiral limit because they arise from pion loop effects.

Recent experimental values for the proton polarizabilities are [12] [1]

$$\alpha_p + \beta_p = 13.23 \pm 0.86^{+0.20}_{-0.49} \times 10^{-4}\,\text{fm}^3,$$

$$\alpha_p - \beta_p = 10.11 \pm 1.74^{+1.22}_{-0.86} \times 10^{-4}\,\text{fm}^3, \tag{7}$$

where the first error is a combined statistical and systematic error, and the second set of errors comes from the theoretical model employed. These values are in good agreement with the χPT predictions.

On the other hand, the neutron polarizabilities are difficult to obtain experimentally and so the corresponding χPT prediction is not well tested. One way to extract neutron polarizabilities is to consider Compton scattering on nuclear targets. Consider coherent photon scattering on the deuteron. The cross section in the forward direction naively goes as:

---

[1] These are the result of a model-dependent fit to data from Compton scattering on the proton at several angles and at energies ranging from 33 to 309 MeV.

$$\frac{d\sigma}{d\Omega}\bigg|_{\theta=0} \sim (f_{Th} - (\alpha_p + \alpha_n)\omega^2)^2. \tag{8}$$

The sum $\alpha_p + \alpha_n$ may then be accessible via its interference with the dominant Thomson term for the proton, f_{Th} [13]. This means that with experimental knowledge of the proton polarizabilities it may be possible to extract those for the neutron. Coherent Compton scattering on a deuteron target has been measured at $E_\gamma = 49$ and 69 MeV by the Illinois group [14]. An experiment with tagged photons in the energy range $E_\gamma = 84.2 - 104.5$ MeV is under analysis at Saskatoon [15].

Clearly the amplitude for Compton scattering on the deuteron involves mechanisms other than Compton scattering on the individual constituent nucleons. Hence, extraction of nucleon polarizabilities requires a theoretical calculation of Compton scattering on the deuteron that is under control in the sense that it accounts for *all* mechanisms to a given order in a systematic expansion in a small parameter.

In the remainder of this paper we will review a recent computation of Compton scattering on the deuteron for incoming photon energies of order 100 MeV in the Weinberg formulation [16]. As in the computation of the electric dipole amplitude in photoproduction, baryon χPT is used to compute an irreducible scattering kernel (here to order Q^3) which is then sewn to external deuteron wavefunctions.

In Figures 2 and 3 we display our results at 69 and 95 MeV. For comparison we have included the calculation at $O(Q^2)$ where the γN T-matrix in the single-scattering contribution is given by the Thomson term on a single nucleon. It is remarkable that to $O(Q^3)$ no unknown counterterms appear. All contributions to the kernel are fixed in terms of known pion and nucleon parameters such as m_π, g_A, M, and f_π. Thus, to this order χPT makes *predictions* for Compton scattering.

The curves show that the correction from the $O(Q^3)$ terms gets larger as ω is increased, as was to be expected. Indeed, while at lower energies corrections are relatively small, in the 95 MeV results the correction to the differential cross section from the $O(Q^3)$ terms is of order 50%, although the contribution of these terms to the *amplitude* is of roughly the size one would expect from the power-counting: about 25%. Nevertheless, it is clear, even from these results, that this calculation must be performed to $O(Q^4)$ before conclusions can be drawn about polarizabilities from data at photon energies of order m_π. This is in accord with similar convergence properties for the analogous calculation for threshold pion photoproduction on the deuteron [7].

We have also shown the Illinois data points at 69 MeV [14]. Statistical and systematic errors have been added in quadrature. The agreement of the $O(Q^3)$ calculation with the 69 MeV data is very good, although only limited conclusions can be drawn, given that there are only two data points, each with sizeable error bars.

Comparing to the calculations of deuteron Compton scattering in the KSW formulation of effective field theory [5], we see that the result of Ref. [5] is significantly

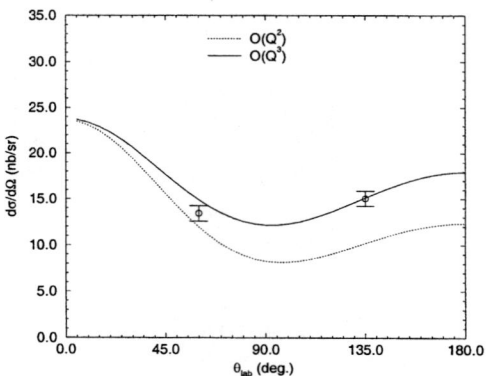

FIGURE 2. Results of the $O(Q^2)$ (dotted line) and $O(Q^3)$ (solid line) calculations at a photon laboratory energy of 69 MeV.

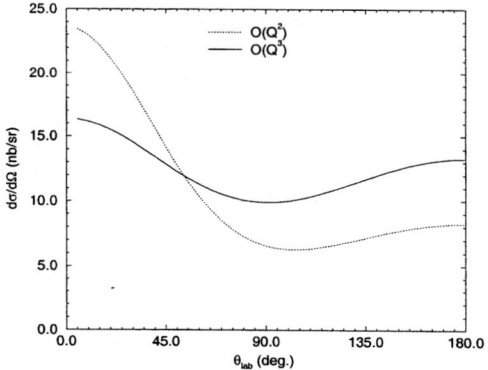

FIGURE 3. Results of the $O(Q^2)$ (dotted line) and $O(Q^3)$ (solid line) calculations at a photon laboratory energy of 95 MeV.

lower than those presented here at both 49 and 69 MeV. At 49 MeV (not shown here) the agreement of Ref. [5]'s calculation with the data is better than ours. We shall show in the next section that this is partly because 49 MeV is at the lower end of the domain of applicability of the Weinberg formulation. At 69 MeV our calculation does a slightly better job of reproducing the (two) data points available. The qualitative agreement among these calculations is a reflection of the similarities of mechanisms involved. Ours is however the only calculation to incorporate the full single-nucleon amplitude instead of its polarizability approximation. Our tendency to higher relative cross sections in the backward directions is at least in part due to this feature.

Although nominally the domain of validity of the Weinberg formulation extends well beyond the threshold for pion production, the power-counting fails at low energies well before the Thompson limit is reached. By comparing $O(Q^4)$ and $O(Q^3)$ contributions, it is straightforward to show that χPT breaks down when

$$\frac{|\vec{p}\,|^2}{\omega M} \sim 1. \tag{9}$$

Here $\vec{p}$ is a typical nucleon momentum inside the deuteron and ω is the photon energy. Since our power-counting is predicated on the assumption that all momenta are of order m_π, we find that power-counting is valid in the region

$$\frac{m_\pi^2}{M} \ll Q \ll \Lambda_\chi. \tag{10}$$

Therefore, in the region $\omega \sim B$ the Weinberg power-counting is not valid, since the external probe momentum flowing through the nucleon lines is of order Q^2/M, rather than order Q. It is in this region that the Compton low-energy theorems are derived. Therefore our power-counting will not recover those low-energy theorems. Of course the upper bound on the validity of the effective theory should increase if the Δ-resonance is included as a fundamental degree of freedom.

In Ref. [5] Compton scattering on the deuteron was computed to the same order discussed here, one order beyond leading non-vanishing order. An advantage of KSW power-counting is that the effective field theory moves smoothly between $Q < B$ and $Q > B$. KSW power-counting is valid for nucleon momenta $Q < \Lambda_{NN} \sim 300$ MeV. Thus in the KSW formulation deuteron polarizabilities and Compton scattering up to energies $\omega < \Lambda_{NN}^2/M \sim 90$ MeV can be discussed in the same framework. Here we are interested mostly in the region $\omega \sim m_\pi$, and so we regard ourselves as being firmly in the second regime. We stress that the value of Λ_{NN} is uncertain; it is conceivable that $\Lambda_{NN} \sim 500$ MeV in which case the range of the KSW formulation would extend well beyond pion production threshold.

In order to test the sensitivity of our calculation to higher-order effects we added a small piece of the $O(Q^4)$ amplitude for Compton scattering off a single nucleon. As one would expect, we find that the cross section at 95 MeV is much more sensitive to these $O(Q^4)$ terms than the cross section at 49 MeV. In our view, a full $O(Q^4)$ calculation in χPT is necessary if any attempt is to be made to extract the neutron polarizability from the Saskatoon data within this framework.

389

ACKNOWLEDGMENTS

I thank Veronique Bernard, Harry Lee, Mané Malheiro, Ulf Meißner, Dan Phillips and Ubi van Kolck for enjoyable collaborations. This research was supported by DOE grant DE-FG02-93ER-40762 (DOE/ER/40762-193, UMPP#00-016).

REFERENCES

1. R. Seki, U. van Kolck and M. J. Savage eds., *Nuclear Physics with Effective Field Theory*, World Scientific, 1998; U. van Kolck, nucl-th/9902015.
2. S. Weinberg, *Phys. Lett. B.* **251**, 288 (1990); *Nucl. Phys. B.* **363**, 3 (1991).
3. U. van Kolck, in *Mainz 1997, Chiral Dynamics: Theory and Experiment*, ed. A. Bernstein *et al*, Springer-Verlag, 1998, hep-ph/9711222; *Nucl. Phys. A.* **645**, 273 (1999), nucl-th/9808007.
4. D. B. Kaplan, M. J. Savage, and M. B. Wise, *Phys. Lett. B.* **424**, 390 (1998), nucl-th/9801034; *Nucl. Phys. B.* **534**, 329 (1998), nucl-th/9802075.
5. J.-W. Chen, H. W. Griesshammer, M. J. Savage and R. P. Springer, nucl-th/9806080; nucl-th/9809023; J.-W. Chen, nucl-th/9810021.
6. S. Weinberg, *Phys. Lett. B.* **295**, 114 (1992).
7. S. R. Beane, C. Y. Lee, and U. van Kolck, *Phys. Rev. C.* **52**, 2914 (1995), nucl-th/9506017; S.R. Beane, V. Bernard, T.S.H. Lee, Ulf-G. Meißner and U. van Kolck, *Nucl. Phys.* **618**, 381 (1997), hep-ph/9702226.
8. V. Bernard, N. Kaiser and Ulf-G. Meißner, *Phys. Lett. B.* **378**, 337 (1996), hep-ph/9512234.
9. M. Fuchs *et al*, *Phys. Lett. B.* **368**, 20 (1996).
10. J. Bergstrom *et al*, *Phys. Rev. C.* **57**, 3202 (1998).
11. V. Bernard, N. Kaiser, and Ulf-G. Meißner, *Phys. Rev. Lett.* **67**, 1515 (1991); *Nucl. Phys. B.* **383**, 442 (1992); V. Bernard, N. Kaiser, J. Kambor, and Ulf-G. Meißner, *Nucl. Phys. B.* **388**, 315 (1992).
12. J. Tonnison, A. M. Sandorfi, S. Hoblit, and A. M. Nathan, *Phys. Rev. Lett.* **80**, 4382 (1998), nucl-th/9801008.
13. D. Drechsel, *et al.*, in *Mainz 1997, Chiral Dynamics: Theory and Experiment*, ed. A. Bernstein *et al*, Springer-Verlag, 1998, p. 264, nucl-th/9712013.
14. M. Lucas, Ph. D. thesis, University of Illinois, unpublished (1994).
15. D. Hornidge, private communication; G. Feldman, private communication.
16. S.R. Beane, M. Malheiro, D.R. Phillips and U. van Kolck, nucl-th/9905023.

Effective Field Theory for Nuclei, Dense Matter, and the Cheshire Cat

Mannque Rho

School of Physics, Korea Institute for Advanced Study, Seoul 130-012, Korea
and
Service de Physique Théorique, CE Saclay, 91191 Gif-sur-Yvette, France

Abstract. In this talk I discuss three related topics based on some of the recent developments in hadron and nuclear physics: one, effective field theory approach to two-nucleon systems; two, an explanation of the flavor singlet axial charge in the proton (i.e., "proton spin problem") in terms of a Cheshire Cat phenomenon; and three, the quark-hadron duality in hadronic matter at high density and "qualitons" at high density ("superqualitons"). The principal common theme in these discussions will be the emergence of the generic feature of the Cheshire Cat Principle.

INTRODUCTION

I would like to discuss in this talk three topics which superficially would look unrelated but in essence can be connected by the general theme of Cheshire Cat. I will first look at two-nucleon systems in terms of macroscopic variables of QCD, namely hadrons; I will then go back to an elementary hadron, in particular the proton and examine its microscopic structure by "punching" a hole in the proton and then putting quarks and gluons of the appropriate quantum numbers inside and argue that there is a continuity in the descriptions in terms of hadronic variables and in terms of quark-gluon variables; finally the continuity of quarks and hadrons at high density will be described in terms of qualitons – quark solitons. Many of my collaborators have contributed to the development discussed in this talk. Among them are Gerry Brown, Bengt Friman, Deog Ki Hong, Kuniharu Kubodera, Kurt Langfeld, Hee-Jung Lee, Dong-Pil Min, Byung-Yoon Park, Tae-Sun Park, Vicente Vento and Ismail Zahed.

TWO-NUCLEON SYSTEMS IN EFFECTIVE FIELD THEORY

Consider two nucleon interactions at very low energy. We are generically interested in the energy-momentum probe much less than the pion mass $m_\pi \sim 140$

CP494, *New Directions in Quantum Chromodynamics*, edited by C.-R. Ji and D.-P. Min

MeV. At this energy, according to Weinberg's "theorem," the content of QCD can be phrased in terms of the nucleons and pions. In fact if we are probing a scale much less than the pion mass, we can even ignore the pions and work with the nucleons only. The corresponding framework is an effective field theory (EFT). Much work has been done on this EFT for two-nucleon systems [1–5] (see [6] for recent reviews). There are two classes of observables to look at. One is scattering process and the other is response function to external fields. The two are of course complementary in revealing the physics involved. Recent efforts have been put more on scattering than on response functions although more information has traditionally been gained from the latter in nuclear physics. Both need to be treated simultaneously which I will do here.

In the literature so far, there are broadly two approaches to EFT in nuclear physics. One is the original Weinberg approach [1] where a systematic power counting is made only to the "irreducible graphs," for which chiral perturbation theory (with pions figuring crucially) becomes applicable in organizing the expansion and the reducible graphs are summed to all orders with the irreducible graphs entering as vertices. This scheme used in Ref. [1–3,5] – which in spirit is close to the original Wilsonian EFT but incurs possible errors in the power counting – involves a scale $\Lambda \gtrsim m_\pi$ as the counting is applied only to the irreducible terms. I will call this the Λ counting. The other approach [4] motivated to account more transparently for the large s-wave scattering lengths in two-nucleon scattering purports to do a systematic counting for the S-matrix as a whole which amounts to summing all graphs involving leading non-derivative four-Fermi contact interactions while treating all others, including pion exchanges, as perturbation. This approach renders a more systematic accounting of the powers of Q/Λ where $Q = \sqrt{MB}$, p as well as m_π but at the expense of certain predictivity. This is referred to as Q-counting scheme.

There have been lots of hot debates as to whether one scheme is more powerful and or more consistent than the other, mainly in connection with the scattering [6]. While the situation is not completely settled, I believe that it is fair to say that the two are both consistent with the tenet of EFT and roughly equivalent in its power. Eschewing the debate which seems somewhat academic, I will simply focus on the Λ-counting approach in this talk. In fact, in the processes that I will consider, I would claim that the Λ counting is more adaptable to — and more predictive in treating – nuclear physics problems than the Q counting. One great advantage of the Λ scheme is that it allows one to calculate precisely defined corrections to what can be obtained from so-called realistic potential models (PM in short) that have been developed by nuclear theorists since a long time, thus giving the realistic potential models (PM) a first-principle justification. It allows us to study processes involving not just few-body but also many-body systems. For instance, it is possible to calculate the "hep" process in the Sun $p + {}^3\text{He} \rightarrow {}^4\text{He} + e^+ + \nu_e$ (which is currently an exciting issue after the recent Surperkamiokande neutrino data) with an accuracy that can be controlled systematically. It has also been successfully applied to calculating axial charge transitions of heavy nuclei (see [7]) with the additional ingredient of BR scaling [8].

While there is no definitive evidence that the Λ scheme is fully justified for n-body systems with $n \gg 2$, it definitely works for $n = 2$ systems. In Ref. [5], it has been shown in a cutoff regularization (with a finite cutoff as required by EFT [9]) that the EFT results of the leading order terms in all two-body observables at low energy $E \ll m_\pi$ are precisely reproduced by the potential model results. This is the case not only for scattering amplitudes but also all electroweak response functions. What EFT brings in addition to what we get from the PM is a systematic procedure to compute corrections to the leading order results. For low-energy processes, this status of the PM can be understood by the fact that the tail of the wave functions is a physical quantity and the realistic potential models which are fit to experiments have the *correct* asymptotic properties in the wave function. This point has also been stressed and clarified by Phillips and Cohen in a recent important paper [10].

Considered to order Q^n where n is the order in the Λ counting (which I will consider relative to the leading order term in the expansion of the irreducible graphs), the s-wave scattering amplitudes are accurately postdicted [5,11] up to $p \leq m_\pi$ for $n = 2$ and a cutoff appropriate to the number of pions exchanged (one or two) in the irreducible graphs. Deuteron properties are also well understood within the same scheme [5]. The scheme allowed the calculations to order Q^2 and Q^3 of the proton fusion process in the Sun [12]

$$p + p \rightarrow d + e^+ + \nu_e \tag{1}$$

and of the threshold np capture [2,13] with polarized projectile and target nucleons

$$\vec{n} + \vec{p} \rightarrow d + \gamma. \tag{2}$$

The process (1) crucial for the solar neutrino problem is given in the scheme to an accuracy of $1 \sim 3$ percents (the uncertainty here is due to the exchange current that appears at order Q^3). The unpolarized cross section for (2) has been computed to the accuracy of 1 percent in a complete agreement with the experiment. More significantly, the polarization observables P (circular polarization) and η (anisotropy) have been predicted *parameter-free* in Ref. [13][1]. This is a genuine prediction since there are no experimental data available. (They are currently being measured in ILL of Grenoble [15].)

In all these postdictions and predictions, there is very little Λ dependence as required by the tenet of EFT. This is a clear indication that the scheme is fully consistent.

One can go up in the momentum range by doing higher order calculations. Phillips and Cohen [10] discuss how the two-body EM form factors can be described in the Λ scheme. Pushing somewhat the validity of the scheme, one can calculate even the process

$$e + d \rightarrow e + n + p \tag{3}$$

---

[1] A similar prediction in the Q scheme was made by Chen, Rupak and Savage [14].

involving large momentum transfers $q \gtrsim 1$ GeV. In fact this process measured in 1980's at ALS of Saclay and elsewhere is considered to be the *unambiguous* confirmation of meson-exchange currents in nuclei (see [16]).[2]

"PROTON SPIN" AND THE CHESHIRE CAT

The nucleons that figured in the above section were color-singlet point-like fields that say nothing explicit about quarks and gluons. Let me now imagine puncturing a hole of size of radius R in the proton and populating the inside with the QCD degrees of freedom, quarks and gluons. How to do this consistently with QCD is known and the model involved is the chiral bag (this is described extensively in [18]) which consists of a quark-gluon sector inside the bag and a color-singlet hadronic sector outside, with the two sectors connected by suitable boundary conditions. When constructed with relevant degrees of freedom and in consistency with the symmetries of QCD, the model gives what is now called "Cheshire Cat." In short, the Cheshire Cat Principle [18] states that at low energy, physics involving hadrons should be independent of the bag size R. It has been shown that this principle is operative semi-quantitatively in *all* properties of the nucleon [19] *except for* the flavor-singlet axial charge (FSAC) of the proton which is related to the "proton spin." Here I will briefly describe – leaving the details to the paper by Lee et al [20] – how this Cheshire Cat property can be recovered in the FSAC when chiral symmetry and chiral anomaly are judiciously taken into account. It turns out that the interplay between the boundary conditions and Casimir effects plays a crucial role.

Since the flavor-singlet axial current is not conserved because of the anomaly, the color cannot be confined inside the bag unless a suitable boundary condition is put at the surface that cancels the outflow of the color [21]. The boundary term that does this is proportional to the Chern-Simons current on the surface, i.e., the Chern-Simons flux (which is invariant under neither small nor large gauge transformation). This influences the FSAC of the proton nontrivially. Briefly, what happens is that the FSAC contributed by the matter fields (quarks inside the bag and η' outside the bag) and the FSAC contributed by the gauge field (gluons inside the bag) more or less (or possibly exactly if treated rigorously) cancel, leaving behind only the small contribution from the (gauge field) vacuum fluctuation which is effectively a Casimir effect caused by the boundary with its color-anomalous boundary condition. The cancellation and the remnant small FSAC are shown in Fig. 1.

The upshot of the small result left over which is independent of the size R provides yet another compelling evidence that the proton could be equivalently understood

---

[2] When I suggested this process at the second INT-Caltech EFT meeting as a case for testing the EFT strategy in the Q counting, everyone (!) in the audience chuckled and said the process is completely out of reach for EFT (see [17]). I grant that this may be true in the Q scheme at least for the moment but *not* in the Λ scheme where it has worked stunningly, confirming the "chiral filter hypothesis." *Voilà* the power of the Λ scheme!

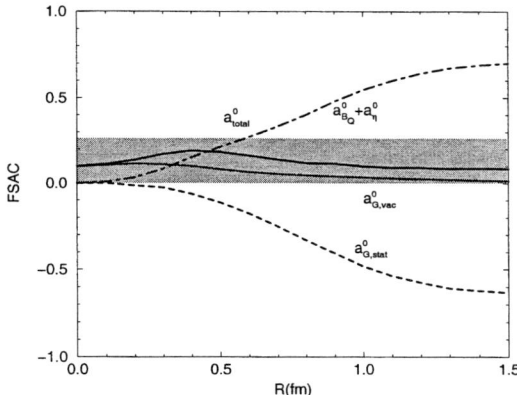

FIGURE 1. The flavor singlet axial charge of the proton as a function of the bag radius compared with the experiment; it consists of three contributions: (a) *matter field contribution*: quark plus η ($a_{B_Q}^0 + a_\eta^0$), (b) *gauge field contribution*: the static gluons coupled to the quark source ($a_{G,stat}^0$), (c) *Casimir contribution*: the gluon vacuum fluctuation ($a_{G,vac}^0$), and (d) the sum total (a_{total}^0). The shaded area represents the range admitted by experiments.

both in terms of quarks/gluons and in terms of macroscopic hadronic variables. When R is taken to be big, it is the QCD variables that take over, e.g., the MIT bag. When the size R is shrunk to a point, the proton is a skyrmion. Thus we have the equivalence of the skyrmionic proton and the quark-gluonic (QCD) proton, that is, the Cheshire Cat. Since there is no way one can exactly bosonize four-dimensional QCD, the equivalence cannot be made exact. We simply have an approximate equivalence which can be made more precise by doing more work. What determines the language to use is the kinematic condition of the process one looks at.

DENSE MATTER AND THE CHESHIRE CAT

The last topic I would like to discuss is infinite nuclear matter at large density. The old lore that at an asymptotic density, the matter can be described by perturbative QCD with weakly interacting boring quarks is simply wrong. What may be happening at high density is something a lot more exciting than that. This explains why lots of people are presently writing papers about it.

What is most surprising and in some sense unexpected is that at high density the Cheshire Cat picture becomes more prominent. In fact, at high density, there ceases to be any real distinction between quarks and hadrons. This can be best seen in terms of a quark soliton analogous to the qualiton Kaplan [22] introduced as a model for the constituent quark. The mechanism I will discuss exploits that at high density diquarks condense giving rise to color superconductivity as recently discussed in the literature [23]. Since the resulting qualiton is formed from a color superconducting ground state, it seems proper to call it superqualiton [25]. It has been argued on general symmetry and dynamical grounds [24] that at high density, hadronic matter of flavor $SU(3)$ is characterized by the condensate

$$\left\langle q_{L\alpha}^{ia} q_{L\beta}^{jb} \right\rangle = -\left\langle q_{R\alpha}^{ia} q_{R\beta}^{jb} \right\rangle = \kappa \ \epsilon^{ij} \epsilon^{abI} \epsilon_{\alpha\beta I} \tag{4}$$

where κ is some constant, i, j are $SL(2, C)$ indices, a, b are color indices, and α, β are flavor indices. Equation (4) holds for parity-even states. Such a condensate locks color and flavor so that global color and chiral symmetry are broken to the diagonal subgroup $SU(3)_{C+L+R}$. The consequence of this is that there is an invariant $U(1)$ subgroup that contains a "twisted" photon, measured with which all excitations carry integer charges reminiscent of the Han-Nambu quarks and have quantum numbers that correspond to those of the mesons and baryons present at zero density. There is then a continuity between the excitations at high density in terms of quarks and gluons and hadronic excitations at low density in terms of baryons and mesons. This clearly is a case of Cheshire Cat.

Now to see that this is the Cheshire Cat in the sense formulated in terms of the chiral bag [18], consider the excitation of a quark on top of the diquark-condensed "vacuum." In [24], such a quark is argued to behave like a baryon. Now I claim that this quark is a quark soliton, i.e., superqualiton [25].

To describe the low-energy dynamics of the color-flavor locking phase, introduce a field $U_L(x)$ which maps space-time to the coset space, $M_L = SU(3)_c \times SU(3)_L/SU(3)_{c+L}$. One can take it to be

$$U_{L a\alpha}(x) \equiv \lim_{y \to x} \frac{|x - y|^{\gamma_m}}{\kappa} \epsilon^{ij} \epsilon_{abc} \epsilon_{\alpha\beta\gamma} q_{Li}^{b\beta}(-\vec{v}_F, x) q_{Lj}^{c\gamma}(\vec{v}_F, y), \tag{5}$$

where γ_m is the anomalous dimension of the diquark field of order α_s and $q(\vec{v}_F, x)$ denotes the quark field with momentum close to a Fermi momentum $\mu\vec{v}_F$. The pairing involves quarks near the opposite edge of the Fermi surface. Similarly, we introduce a right-handed field $U_R(x)$, also a map from space-time to $M_R = SU(3)_c \times SU(3)_R/SU(3)_{c+R}$, to describe the excitations of the right-handed diquark condensate. If this field takes a vacuum expectation value as a consequence of the diquark condensation which will, owing to (4), have the form

$$\langle U_{L a\alpha} \rangle = -\langle U_{R a\alpha} \rangle = \kappa \, \delta_{a\alpha}, \tag{6}$$

then 16 Nambu-Goldstone bosons will get excited [3]. Eight of them will get eaten up by the gluons to give masses to the gluons. The massive gluons then turn into massive vector mesons whose quantum numbers are those of the light-quark

---

[3] Actually there are 17 of them, one of which has to do with spontaneous breaking of the baryon number.

vector mesons present at zero density. The remaining eight (pseudoscalar) Nambu-Goldstone bosons are the equivalents of the ones present at zero density and are represented by the interpolating field (5). In analogy to the usual skyrmion at zero density, this field supports a soliton which is a fermion, the quantum numbers of which are identical to those of the usual baryon.

The effective Lagrangian that gives rise to this soliton should in principle be derived from QCD with the help of renormalization group flows toward the Fermi surface of high quark density. Such a derivation would determine the parameters that figure in the effective Lagrangian such as the "pion decay constant" F etc which would carry information on the superconductivity gap etc. At the moment such an effective Lagrangian is not known, so a detailed study of it excitation structure cannot be discussed.

However one can venture to make a few interesting conjectures. Viewed as a superqualiton whose mass is given by the soliton mass, there is nothing that requires that the soliton mass be equal to or near the superconductivity gap Δ (which is dictated by the condensate). In fact there is nothing which would prevent the mass from being much less than the gap. Thus one could imagine that light fermions are excited *within* the gap. Correlations between light superqualitons could rearrange the ground state into a different form from that of the standard superconductivity. For this reason the phenomenon of color superconductivity in QCD at high density could be completely different from the usual BCS superconductivity. A similar point in a different context was raised in [26].

CONCLUSION

The most important outcome of the recent development of EFT in nuclear physics is that the highly successful approach to nuclear structure using realistic nuclear potentials (PM) is rendered a first-principle interpretation in that it represents the leading term in the EFT expansion with the corrections thereof systematically calculable. This confers the power of modern field theory techniques to the standard nuclear physics approach that has been practiced with success since a long time. I am suggesting that this "bridging" comes about thanks to a possible duality between QCD variables and macroscopic (color-singlet) variables that I refer to as the Cheshire Cat Principle. The "proton spin problem" is an illustration of this in the basic structure of the hadron.

In the case of high density, the picture becomes even more intriguing. There we see emerging the symbolic (approximate) equality

$$\text{"Quark"} \approx \text{"Qualiton"} \approx \text{"Baryon"}. \tag{7}$$

It is amusing that the notion of the Cheshire Cat which was conceived by the need to reconcile the traditional meson-exchange description with the modern QCD description for nuclear processes [27] (i.e., the "little bag" with pion cloud, chiral bag etc) at low density re-emerges at high density where one would have expected the bona-fide QCD to be uniquely applicable.

Acknowledgments

I would like to thank the organizers of this meeting, particularly Chueng-Ryong Ji and Dong-Pil Min for inviting me to give this talk. I would also like to acknowledge the hospitality of KIAS and its Director, C.W. Kim while this paper was being written.

REFERENCES

1. S. Weinberg, Phys. Lett. B **251**, 288 (1990) ; Nucl. Phys. B **363**, 3 (1991) ; Phys. Lett. B **295**, 114 (1992) .
2. M. Rho, Phys. Rev. Lett. **66**, 1275 (1991) ; T.-S. Park, D.-P. Min and M. Rho, Physics Reports **233**, 341 (1993) ; Phys. Rev. Lett. **74**, 4153 (1995) ; Nucl. Phys. A **596**, 515 (1996) .
3. C. Ordonez and U. van Kolck, Phys. Lett. B **291**, 459 (1992) ; C. Ordonez, L. Ray and U. van Kolck, Phys. Rev. Lett. **72**, 1982 (1994) ; Phys. Rev. C **53**, 2086 (1996) ; U. van Kolck, Phys. Rev. C **49**, 2932 (1994) .
4. D.B. Kaplan, M. Savage and M. Wise, Nucl. Phys. B **478**, 629 (1995) ; Phys. Lett. B **424**, 390 (1998) ; nucl-th/9801034; nucl-th/9802075; nucl-th/9804032.
5. T.-S. Park, K. Kubodera, D.-P. Min and M. Rho, Phys. Rev. C **58**, R637 (1998) ; Nucl. Phys. A **646**, 83 (1999) .
6. 1998 and 1999 Joint Caltech-INT Workshops: *Nuclear Physics with Effective Field Theory.*
7. M. Rho, "Chiral symmetry in nuclear physics," nucl-th/9812012.
8. G.E. Brown and M. Rho, Phys. Rev. Lett. **66**, 2720 (1991) .
9. G.P. Lepage, "How to renormalize the Schrödinger equation?", nucl-th/9706029.
10. D.R. Phillips and T.D. Cohen, "Deuteron electromagnetic properties and the viability of effective field theory methods un the two-nucleon system," nucl-th/9906091.
11. C.H. Hyun, D.-P. Min and T.-S. Park, "Next-to-leading order np scattering in cut-off effective field theory," to appear.
12. T.-S. Park, K. Kubodera, D.-P. Min and M. Rho, Astrophys. J. **507**, 443 (1998).
13. T.-S. Park, K. Kubodera, D.-P. Min and M. Rho, nucl-th/9904053, nucl-th/9906005.
14. J.-W. Chen, G. Rupak and M.J. Savage, "Suppressed amplitudes in $np \to d + \gamma$," nucl-th/9905002.
15. T.M. Müller, private communication; T.M. Müller, D. Dubbers, P. Hautle and O. Zimmer, "Measurement of the γ anisotropy in the $\vec{p}\vec{n}, \gamma d$-process," in Proceedings of the "International Workshop on Particle Physics with Slow Neutrons," ILL, Grenoble, France, 22-24 October 1998.
16. See e.g. B. Frois and J.F. Mathiot, Comments on Nucl. Part. Phys. **18**, 307 (1989).
17. R. Seki and H. Griesshammer, "Open discussion and new ideas," in the Proceedings of the 1999 Joint INT-Caltech EFT Worskshop.

18. M.A. Nowak, M. Rho and I. Zahed, *Chiral Nuclear Dynamics* (World Scientific, Singapore, 1996).

19. A. Hosaka and H. Toki, Phys. Repts. **277**, 65 (1996).

20. H.-J. Lee, D.-P. Min, B.-Y. Park, M. Rho and V. Vento, "The proton spin in the chiral bag model: Casimir contribution and Cheshire Cat Principle," Nucl. Phys. **A**, in press, hep-ph/9810539.

21. H.B. Nielsen, M. Rho, A. Wirzba and I. Zahed, Phys. Lett. **B269**, 389 (1991); Phys. Lett. **281**, 345 (1992).

22. D.A. Kaplan, Phys. Lett. **B235**, 163 (1990); Nucl. Phys. **B351**, 137 (1991).

23. M. Alford, K. Rajagopal and F. Wilczek, Phys. Lett. B **422**, 247 (1998); R. Rapp, T. Schäfer, E.V. Shuryak and M. Velkovsky, Phys. Rev. Lett. **81**, 53 (1998); J. Berges and K. Rajagopal, Nucl. Phys. B **538**, 215 (1999); N. Evans, S. Hsu, and M. Schwetz, Nucl.Phys. **B551**, 275 (1999); Phys. Lett. B **449**, 281 (1999); K. Langfeld and M. Rho, hep-ph/9811227; G. W. Carter and D. Diakonov, Phys. Rev. D **60**, 016004 (1999); M. Alford, J. Berges, and K. Rajagopal, hep-ph/9903502; T. Schwarz, S. Klevansky and G. Papp, hep-ph/9903048; D. T. Son, hep-ph/9812287, Phys. Rev. D **59**, 094019 (1999); R. D. Pisarski and D. H. Rischke, nucl-th/9811104; nucl-th/9903023; nucl-th/9906050; D. K. Hong, hep-ph/9812510; hep-ph/9905523; D. K. Hong, I. A. Shovkovy, V. A. Miransky, and L. C. R. Wijewardhana, hep-ph/9906478; T. Schäfer and F. Wilczek, hep-ph/9906512.

24. T. Schäfer and F. Wilczek, Phys. Rev. Lett. **82**, 3956 (1999), hep-ph/9811473; hep-ph/9903503.

25. D.K. Hong, M. Rho and I. Zahed, "Qualitons at High Density," hep-ph/9906551.

26. R.D. Pisarski and D.H. Rischke, "Why color-flavor locking is just like chiral symmetry breaking?," nucl-th/9907094.

27. G.E. Brown and M. Rho, Phys. Lett. **B82**, 177 (1979).

Chiral approach to antikaon dynamics in the nuclear medium

A. Ramos[*] [1] and E. Oset[†] [2]

[*]Departament d'Estructura i Constituents de la Matèria, Universitat de Barcelona,
Diagonal 647, 08028 Barcelona, Spain
[†]Departamento de Física Teórica and IFIC, Centro Mixto Universidad de Valencia-CSIC
46100 Burjassot, Valencia, Spain

Abstract. A s-wave meson-nucleon interaction in the $S = -1$ sector is built up by means of a coupled-channel Bethe-Salpeter equation, using the lowest order chiral Lagrangian and a cut off to regularize the loop integrals. The position and width of the $\Lambda(1405)$ resonance as well as the K^-p scattering cross sections at low energies are well reproduced. This model is applied to calculate the K^--meson self-energy in nuclear matter self-consistently. The in-medium effective $\bar{K}N$ interaction includes the effects of Pauli blocking on the nucleons, mean-field potentials for the baryons and the self-energy of the π and $\bar{K}$ mesons. The incorporation of the medium modified meson spectral densities, substantially spread out over energies, gives an attractive effective $\bar{K}N$ interaction which is weaker than that obtained using free meson propagators. The moderately attractive K^- nuclear potential obtained (around -40 MeV at normal nuclear matter density) seems to be compatible with existing kaonic atom data but would make the phenomenon of kaon condensation very unlikely.

INTRODUCTION

The properties of antikaons in the nuclear medium are of especial relevance in determining the possible occurence of kaon condensation in dense matter [1]. The phenomenon would occur if, at some density, the K^- energy is lowered below the corresponding electron chemical potential. Both the enhancement of the K^- yield in ion-ion collisions measured at GSI [2] and kaonic atom data [3] can be explained by assuming that the K^- meson feels an attractive potential in the nuclear medium. However, the strength of this attraction has not clearly been established yet.

In this contribution we study the properties of the K^- meson in nuclear matter from a microscopic approach that incorporates the medium effects on the model of Ref. [4], which describes the $\bar{K}N$ scattering properties in free space from the lowest-order chiral Lagrangian.

[1] Supported by DGICYT contract number PB95-1249 and EEC-TMR contract CT98-0169
[2] Supported by DGICYT contract number PB96-0753 and EEC-TMR contract CT98-0169

Since the $\bar{K}N$ system couples strongly to other meson-baryon channels and generates the $\Lambda(1405)$ resonance just below threshold, one cannot expect chiral perturbation theory to work. This is why the $\Lambda(1405)$ has to be either introduced explicitly as an elementary field [5] or generated dynamically through the Lippmann-Schwinger (LS) [6] or the Bethe-Salpeter (BS) [4] equations. These latter approaches allow for a microscopic incorporation of the medium effects on the the $\bar{K}N$ interaction [7–10]. One source of density dependence is the Pauli blocking on the nucleons in the intermediate states iterated by the LS or BS equation. The effect is to shift the resonance to higher energies [7–9] and to change the $\bar{K}N$ interaction at threshold from a repulsive to an attractive one. However, a self-consistent calculation of the K^- self-energy leaves the resonance unchanged [10], due to a compensation of the repulsive shift on the $\Lambda(1405)$ with the attraction felt by the K^- meson.

The importance of these medium effects makes it interesting to investigate other medium modifications of the particles participating in building up the $\bar{K}N$ interaction. In this contribution, we present results obtained from a model [11] that also includes the dressing of the pions in the $\pi\Lambda$, $\pi\Sigma$ intermediate states, which couple strongly to the $\bar{K}N$ state. This is done through a pion self-energy that contains one- and two-nucleon absorption, conveniently modified to include the effect of nuclear short-range correlations. The medium effects on the nucleons and hyperons is also considered via density-dependent mean-field binding potentials.

$\bar{K}N$ INTERACTION IN FREE SPACE

The meson-baryon interaction Lagrangian at lowest order in momentum is given by

$$L_1^{(B)} = \langle \bar{B} i \gamma^\mu \frac{1}{4f^2} [(\Phi \partial_\mu \Phi - \partial_\mu \Phi \Phi) B - B(\Phi \partial_\mu \Phi - \partial_\mu \Phi \Phi)] \rangle \ , \tag{1}$$

where Φ represents the octet of pseudoscalar mesons and B the octet of $1/2^+$ baryons. The symbol $\langle \rangle$ denotes the trace of SU(3) matrices.

The coupled channel formalism requires to evaluate the transition amplitudes between the different meson-baryon channels. For K^-p scattering there are ten channels, namely K^-p, $\bar{K}^0n$, $\pi^0\Lambda$, $\pi^0\Sigma^0$, $\pi^+\Sigma^-$, $\pi^-\Sigma^+$, $\eta\Lambda$, $\eta\Sigma^0$, $K^+\Xi^-$ and $K^0\Xi^0$, while in the case of K^-n scattering there are six: K^-n, $\pi^0\Sigma^-$, $\pi^-\Sigma^0$, $\pi^-\Lambda$, $\eta\Sigma^-$ and $K^0\Xi^-$. These amplitudes have the form

$$V_{ij} = -C_{ij} \frac{1}{4f^2} \bar{u}(p_i) \gamma^\mu u(p_j)(k_{i\mu} + k_{j\mu}) \ , \tag{2}$$

where $p_j, p_i (k_j, k_i)$ are the initial, final momenta of the baryons (mesons) and C_{ij} are SU(3) coefficients that can be found in Ref. [4]. At low energies the spatial components can be neglected and the amplitudes reduce to

$$V_{ij} = -C_{ij}\frac{1}{4f^2}(k_j^0 + k_i^0) \ . \tag{3}$$

The coupled-channel BS equations in the center of mass frame read

$$T_{ij} = V_{ij} + \overline{V_{il}\, G_l\, T_{lj}} \ , \tag{4}$$

where the indices i, l, j run over all possible channels and

$$\overline{V_{il}\, G_l\, T_{lj}} = i \int \frac{d^4q}{(2\pi)^4}\frac{M_l}{E_l(\vec{q})}\frac{V_{il}(k_i, q)\, T_{lj}(q, k_j)}{k^0 + p^0 - q^0 - E_l(\vec{q}) + i\epsilon}\frac{1}{q^2 - m_l^2 + i\epsilon} \ , \tag{5}$$

with M_l, E_l and m_l being, respectively, the baryon mass, baryon energy and meson mass in the intermediate state. As was shown in Ref. [4], the off-shell part of V_{il} and T_{lj} goes into renormalization of coupling constants and the problem reduces to one of inverting a set of algebraic equations with loop integrals G_l that read

$$\begin{aligned}
G_l(\sqrt{s}) &= i \int \frac{d^4q}{(2\pi)^4}\frac{M_l}{E_l(\vec{q})}\frac{1}{\sqrt{s} - q^0 - E_l(\vec{q}) + i\epsilon}\frac{1}{q^2 - m_l^2 + i\epsilon} \\
&= \int_{|\vec{q}|<q_{max}} \frac{d^3q}{(2\pi)^3}\frac{1}{2\omega_l(\vec{q})}\frac{M_l}{E_l(\vec{q})}\frac{1}{\sqrt{s} - \omega_l(\vec{q}) - E_l(\vec{q}) + i\epsilon} \ ,
\end{aligned} \tag{6}$$

where $\sqrt{s} = p^0 + k^0$ and $\omega_l(\vec{q}) = \sqrt{m_l^2 + \vec{q}^2}$. The value of the cut-off, $q_{max} = 630$ MeV, was chosen to reproduce the K^-p scattering branching ratios at threshold, while the weak decay constant, $f = 1.15f_\pi$, was taken in between the pion and kaon ones to optimize the position of the $\Lambda(1405)$ resonance. The predictions of the model for several scattering observables are summarized in Table 1.

TABLE 1. K^-p threshold ratios and K^-N scattering lengths

	This work	Exp.
$\gamma = \dfrac{\Gamma(K^-p \to \pi^+\Sigma^-)}{\Gamma(K^-p \to \pi^-\Sigma^+)}$	2.32	2.36±0.04 [12,13]
$R_c = \dfrac{\Gamma(K^-p \to \text{charged})}{\Gamma(K^-p \to \text{all})}$	0.627	0.664±0.011 [12,13]
$R_n = \dfrac{\Gamma(K^-p \to \pi^0\Lambda)}{\Gamma(K^-p \to \text{neutral})}$	0.213	0.189±0.015 [12,13]
a_{K^-p} (fm)	$-1.00 + i\,0.94$	$-0.67 + i\,0.64$ [14] -0.98 (from Re(a)) [14] $(-0.78\pm0.18) + i(0.49\pm0.37)$ [15]
a_{K^-n} (fm)	$0.53 + i\,0.62$	$0.37 + i\,0.60$ [14] 0.54 (from Re(a)) [14]

$\bar{K}$ IN THE NUCLEAR MEDIUM

The medium effects on the $\bar{K}N$ interaction are incorporated by replacing the free propagators by in-medium ones in the meson-baryon loop. For the nucleons we take mean-field propagators

$$\frac{1}{\sqrt{s} - q^0 - E_l(\vec{q}) + i\epsilon} \rightarrow \left\{ \frac{1 - n(\vec{q}_{\text{lab}})}{\sqrt{s} - q^0 - E_l(\vec{q}) + i\epsilon} + \frac{n(\vec{q}_{\text{lab}})}{\sqrt{s} - q^0 - E_l(\vec{q}) - i\epsilon} \right\} , \quad (7)$$

where $n(\vec{q}_{\text{lab}})$ is the occupation probability of a nucleon of momentum $\vec{q}_{\text{lab}}$ in the lab frame. For the hyperons (Λ and Σ) the occupation probability is simply zero. The single particle energies $E_l(\vec{q})$ now contain a mean-field potential of the type $U_0\rho/\rho_0$, with $U_0 = -50$ (-30) MeV for nucleons (hyperons). The dressed meson propagator used for the $\bar{K}$ and π mesons is

$$D_l(q^0, \vec{q}, \rho) = \frac{1}{(q^0)^2 - \vec{q}^2 - m_l^2 - \Pi_l(q^0, \vec{q}, \rho)} , \quad (8)$$

where $\Pi_l(q^0, \vec{q}, \rho)$ is the meson self-energy. The resulting loop integral becomes

$$G_l(P^0, \vec{P}, \rho) = \int_{|\vec{q}| < q_{\text{max}}} \frac{d^3q}{(2\pi)^3} \frac{M_l}{E_l(\vec{q})} \int_0^\infty d\omega\, S_l(\omega, \vec{q}, \rho)$$

$$\times \left\{ \frac{1 - n(\vec{q}_{\text{lab}})}{\sqrt{s} - \omega - E_l(\vec{q}) + i\epsilon} + \frac{n(\vec{q}_{\text{lab}})}{\sqrt{s} + \omega - E_l(\vec{q}) - i\epsilon} \right\} , \quad (9)$$

where $(P^0, \vec{P})$ is the total four-momentum in the lab frame. The function $S_l(\omega, \vec{q}, \rho) = -\text{Im}D_i(\omega, \vec{q}, \rho)/\pi$ is the meson spectral density that in the case on undressed mesons reduces to $\delta(\omega - \omega_l(\vec{q}))/2\omega_l(\vec{q})$.

From the effective $\bar{K}N$ interaction, $T_{\text{eff}}(P^0, \vec{P}, \rho)$, which is obtained by solving the coupled-channel BS equation using the dressed meson-baryon loop of Eq. (9), one can easily calculate the s-wave $\bar{K}$ self-energy ($\bar{K} = K^-$ or $\bar{K}^0$)

$$\Pi_{\bar{K}}^s(q^0, \vec{q}, \rho) = 2 \sum_{N=n,p} \int \frac{d^3p}{(2\pi)^3} n(\vec{p})\, T_{\text{eff}}^{\bar{K}N}(q^0 + E(\vec{p}), \vec{q} + \vec{p}, \rho) . \quad (10)$$

Note that a self-consistent approach is required since one calculates the $\bar{K}$ self-energy from the effective interaction T_{eff} which uses $\bar{K}$ propagators which themselves include the self-energy being calculated. A p-wave contribution to the $\bar{K}$ self-energy coming from the coupling of the $\bar{K}$ meson to hyperon-hole excitations is also included.

The pion self-energy contains the effect of one- and two-nucleon absorption, conveniently modified to include the effect of nuclear short-range correlations (see Ref. [16] for details).

403

RESULTS AND DISCUSSION

To assess the importance of dressing the mesons we show first their spectral density. In Fig. 1 the spectral density of the π meson in nuclear matter at density $\rho = \rho_0$, where $\rho_0 = 0.17$ fm^{-3}, is shown as a function of energy for several momenta. The strength is distributed over a wide range of energies and, as the pion momentum increases, the position of the peak is increasingly lowered from the corresponding one in free space. The in-medium peak positions are at 158, 179 and 212 MeV for $q = 100$, 200 and 300 MeV/c, respectively, while the corresponding free values are 172, 244 and 331 MeV. Note also, to the left of the peaks, the typical structure of the $1p1h$ excitations which give rise to $1p1h\Lambda$ and $1p1h\Sigma$ components in the effective $\bar{K}N$ interacion.

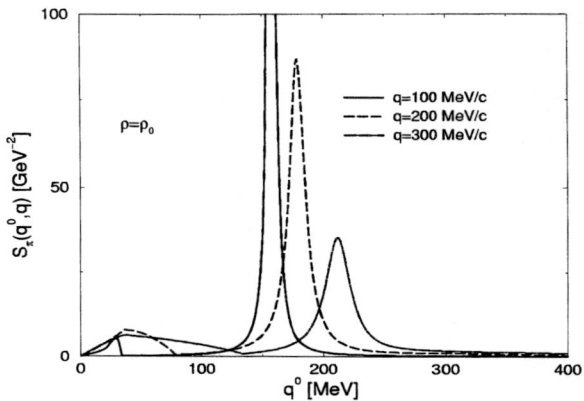

FIGURE 1. Pion spectral density at $\rho = \rho_0$ for several momenta

The spectral function of a K^- meson of zero momentum is shown in Fig. 2 for various densities: ρ_0, $\rho_0/2$ and $\rho_0/4$. The results in the upper panel include only Pauli blocking effects. The two excitation modes visible at $\rho_0/4$ correspond to the K^- pole branch, appearing at an energy smaller than the kaon mass m_K due to the attractive medium effects, and to the $\Lambda(1405)$-hole excitation, located above m_K because of the shifting of the $\Lambda(1405)$ resonance to energies above the K^-p threshold. As density increases, the K^- feels an enhanced attraction while the $\Lambda(1405)$-hole peak moves to higher energies and loses strength, a reflection of the tendency of the $\Lambda(1405)$ to dissolve in the dense nuclear medium. The (self-consistent) incorporation of the $\bar{K}$ propagator in the BS equation softens the effective interaction T_{eff}, which becomes more spread out in energies. The resulting K^- spectral function (middle panel in Fig. 2) shows the displacement of the resonance back to lower energies because the attraction felt by the $\bar{K}$ meson lowers the threshold of $\bar{K}N$ states. Finally, the K^- spectral function obtained when the dressing of the pions is incorporated in the determination of T_{eff} (bottom panel in Fig. 2) is smoother.

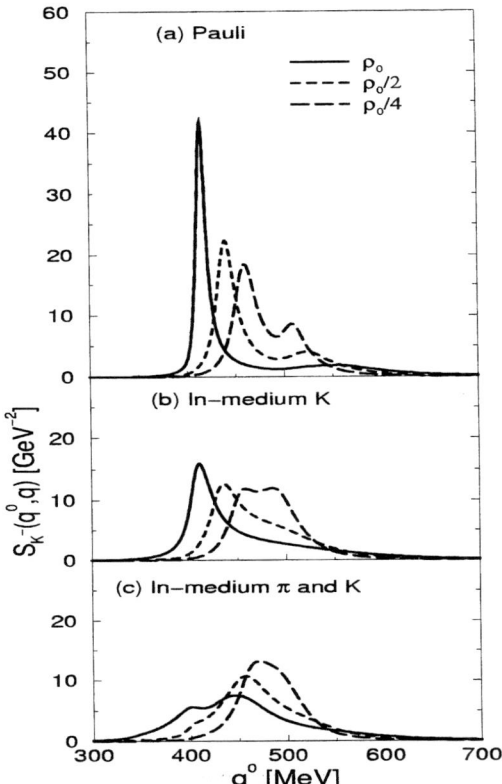

FIGURE 2. K^- spectral density for zero momentum

The K^- pole peak appears at a higher energy and, thus, the kaon experiences less attraction than in the two other approaches. However, more strength is found at very low energies, especially at ρ_0, due to the coupling of the K^- to the $1p1h$ and $2p2h$ components of the pionic strength. It is precisely the opening of the $\pi\Sigma$ channel, on top of the already opened $(1p1h)\Sigma$ and $(2p2h)\Sigma$ ones, the reason for the cusp structure which appears around 400 MeV in the spectral functions shown in the bottom panel.

The isospin averaged in-medium scattering length, defined as

$$a_{\text{eff}}(\rho) = -\frac{1}{4\pi}\frac{M}{m_K + \langle E_N \rangle}\frac{\Pi_{\bar{K}}(m_K, \vec{q} = 0, \rho)}{\rho} \, , \tag{11}$$

with $\langle E_N \rangle = M + U_N(\rho) + 3p_F^2/10M$ being the average nucleon energy in the Fermi sea, is shown if Fig. 3 as a function of the nuclear density ρ. The change of $\text{Re}\,a_{\text{eff}}$ from negative to positive values indicates the transition from a repulsive interaction

405

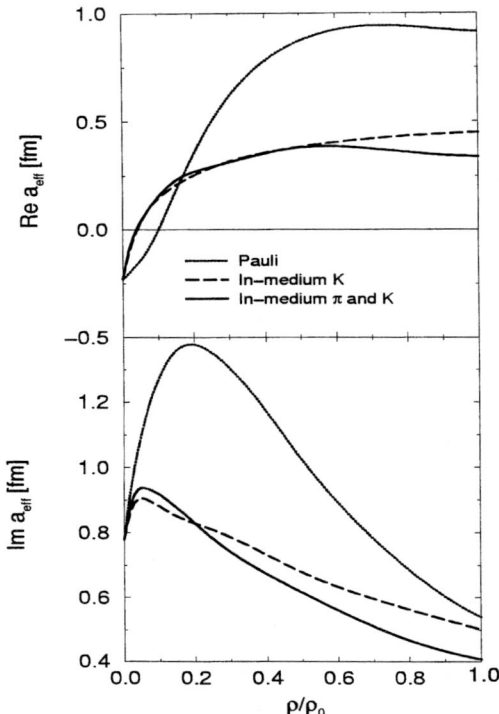

FIGURE 3. Real (top) and imaginary (bottom) parts of the K^-N scattering length as a function of density

in free space to an attractive interaction in the medium. This happens at a density of about $0.1\rho_0$ when only Pauli blocking effects are considered and at even lower densities ($\rho \sim 0.04\rho_0$) when one also considers the dressing of the mesons.

The value of Re a_{eff} at ρ_0 corresponds to a K^- nuclear potential depth of around -40 MeV. This is about half the attraction of that obtained with other theories and approximation schemes [8,9,17–20], which predict potential depths at the center of the nucleus in between -140 and -75 MeV.

The implications of our in-medium K^- self-energy on kaonic atoms is currently being analyzed [21] in the framework of a Local Density Approximation. The K^- optical potential is obtained from the self-energy, $V_{\text{opt}}(\rho) = \Pi_{\bar{K}}(m_K, \vec{q} = 0, \rho)/2m_K$, by replacing the nuclear matter density ρ by the density profile $\rho(r)$ of the nucleus. Both the energy shifts and widths of atomic kaonic states in several nuclei agree well with the experimental data. The effects of the non-localities (momentum dependence) of the kaon self-energy on kaonic atoms remain to be studied [22].

Although the consequences of our results for kaon condensation need to be explored carefully taking the full distribution of K^- strength, the position of the peak already indicates that the quasiparticle K^- energy is lowered only moderately, making the phenomenon quite unlikely.

REFERENCES

1. Kaplan, D.B., and Nelson, A.E., *Phys. Lett.* **B175**, 57 (1986).
2. Barth, R. et al., *Phys. Rev. Lett.* **78**, 4007 (1997). Laue, F. et al, *Phys. Rev. Lett.* **82**, 1640 (1999).
3. Friedman, E., Gal, A., and Batty, C.J., *Nucl. Phys.* **A579**, 518 (1994).
4. Oset, E., and Ramos, A. *Nucl. Phys.* **A635**, 99 (1998).
5. Lee, C.-H., Brown, G.E., Min, D.P., and Rho, M., *Nucl. Phys.* **A385**, 481 (1995); Lee, C.-H., *Phys. Rep.* **275**, 255 (1996).
6. Kaiser, N., Siegel, P.B., and Weise, W., *Nucl. Phys.* **A594**, 325 (1995); Kaiser, N., Waas, T., and Weise, W., *Nucl. Phys.* **A612**, 297 (1997).
7. Koch, V., *Phys. Lett.* **B337**, 7 (1994).
8. Waas, T., Kaiser, N., and Weise, W., *Phys. Lett.* **B365**, 12 (1996); *ibid.* **B379**, 34 (1996).
9. Waas, T., and Weise, W., *Nucl. Phys.* **A625**, 287 (1997).
10. Lutz, M., *Phys. Lett.* **B426**, 12 (1998).
11. Ramos, A., and Oset, E., nucl-th/9906016.
12. Tovee, D.N. et al., *Nucl. Phys.* **B33**, 493 (1971).
13. Nowak, R.J. et al., *Nucl. Phys.* **B139**, 61 (1978).
14. Martin, A.D., *Nucl. Phys.* **B179**, 33 (1981).
15. Iwasaki, M. et al., *Phys. Rev. Lett.* **78**, 3067 (1997).
16. Ramos, A., Oset, E., and Salcedo, L.L., *Phys. Rev.* **C50**, 2314 (1994).
17. Li, G.Q., Lee, C.-H., Brown, G.E., *Nucl. Phys.* **A625**, 372 (1997); *ibid*, *Phys. Rev. Lett.* **79**, 5214 (1997).
18. Mao, G., Papazoglou, P., Hofmann, S., Schramm, S., Stöcker, H., and Greiner, W., *Phys. Rev.* **C59**, 3381 (1999).
19. Schaffner, J., and Mishustin, I.N., *Phys. Rev.* **C53**, 1416 (1996); Schaffner-Bielich, J., Mishustin, I.N., and Bondorf, J., *Nucl. Phys.* **A625**, 325 (1997).
20. Tsushima, K., Saito, K., Thomas, A.W., and Wright, S.V., *Phys. Lett.* **B429**, 239 (1998).
21. Okumura, Y., Hirenzaki, S., Toki, H., Oset, E., and Ramos, A., in preparation.
22. Lutz, M., nucl-th/9802033.

U(1) Problem at Finite Temperature

Romuald A. Janik[1,2], Maciej A. Nowak[2,3], Gábor Papp[4] and Ismail Zahed[5]

[1] *Service de Physique Théorique, CEA Saclay, 91191 Gif-sur-Yvette, France*
[2] *Marian Smoluchowski Institute of Physics, Jagellonian University, 30-059 Krakow, Poland*
[3] *GSI, Planckstr. 1, D-64291 Darmstadt, Germany*
[4] *CNR Department of Physics, KSU, Kent, Ohio 44242, USA &*
HAS Research Group for Theoretical Physics, Eötvös University, Budapest, Hungary
[5] *Department of Physics and Astronomy, SUNY, Stony Brook, New York 11794, USA.*

Abstract. We model the effects of a large number of zero and near-zero modes in the QCD partition function by using sparse chiral matrix models with an emphasis on the quenched topological susceptibility in the choice of the measure. At finite temperature, the zero modes are not affected by temperature but are allowed to pair into topologically neutral near-zero modes which are gapped at high temperature. In equilibrium, chiral and U(1) symmetry are simultaneously restored for total pairing, evading mean-field arguments. We analyze a number of susceptibilities versus the light quark masses. At the transition point the topological susceptibility vanishes, and the dependence on the vacuum angle θ drops out. Our results are briefly contrasted with recent lattice simulations.

INTRODUCTION

The current theoretical resolution of the U(1) problem in QCD relies on the assumption that the QCD vacuum supports a finite topological susceptibility [1–3]. This assumption is supported by current lattice simulations [4], anomalous Ward identities [5], chiral effective Lagrangians [6] and canonical quantization [7], although there are questions in covariant quantization [8].

In this paper, we will adopt the current view and proceed to analyze what happens to the U(1) problem at finite temperature. At infinite temperature, both the anomaly and topological effects become negligible, so that the U(1) symmetry is effectively restored. In this limit there is an exact chiral and U(1) degeneracy modulo quark masses. The question then is what happens at finite temperature? Does the U(1) restoration coincide or differ from the conventional chiral restoration [9]?

Recently, a number of lattice simulations [10–13] and model calculations [14] have attempted to answer this and other questions with somewhat opposite conclusions. It is our purpose in this letter to try to address some of these issues in a lat-

CP494, *New Directions in Quantum Chromodynamics*, edited by C.-R. Ji and D.-P. Min

tice motivated matrix model, focusing on the interplay between zero and near-zero modes, the effects of light quark masses and the importance of the thermodynamical limit. In many ways our analysis will parallel instanton-like calculations in a solvable context, with interesting lessons for these calculations as well as lattice simulations. Indeed, one of the main thrust of the present letter is to provide a minimal framework for a model analysis of current lattice results.

In section 2, we motivate the use of a class of chiral matrix models by reviewing some recent lattice calculations. In section 3, we discuss the saddle point results following from the present model under some generic assumptions on the interplay between zero and near-zero modes. In section 4, we address certain aspects of the chiral and U(1) transitions, including a number of susceptibilities. In section 5, we comment on a number of recent lattice simulations in light of our results. Our conclusions are in section 6.

FORMULATION

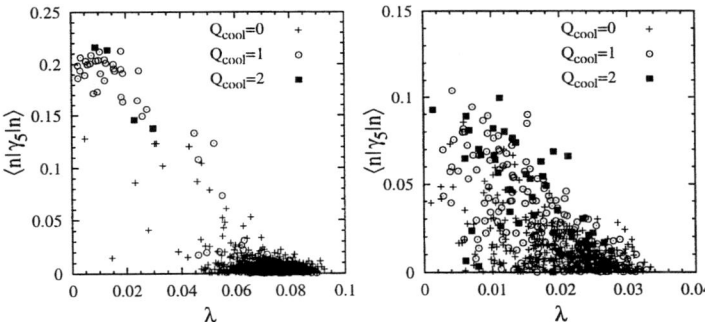

FIGURE 1. Lattice results from Ref. [10] for quenched $\beta = 6.2$ (left) and staggered, $N_f = 2$, $\beta = 5.55$, and $m_q a = 0.00625$ (right) lattices for topological sectors 0 (crosses), 1 (open circles) and 2 (full boxes).

Lattice Motivation

Recently Kogut, Lagae and Sinclair [10] have studied the chirality content $r_n \equiv \langle n|\gamma_5|n\rangle$ of the low-lying quark eigenstates λ_n for staggered fermions with $D|n\rangle = \lambda_n|n\rangle$. Their results after cooling are displayed in Fig. 1 in the (r, λ) plane for $\beta = 6/g^2 = 5.55$ and 6.2, on a $16^3 \times 8$ lattice. $|r| = 1$ in the continuum (about 1/4 on the lattice) corresponds to an eigenvalue with topological charge ± 1, while $r = 0$ corresponds to a non-topological eigenvalue. The high temperature configurations ($\beta = 5.55$) are characterized by a depletion in the zero modes ($r = 1/4$), and an enhancement in the near-zero modes ($r = 0$). Throughout we will refer to the modes with definite chirality as zero modes, and those without as near-zero modes.

We note that at high temperature and in the continuum, the near-zero modes are gapped by $\pm \pi T$.

Model

A simple way to analyze the interplay between the zero and near-zero modes around the chiral transition point is through a matrix model. A pertinent example for the fermion matrix D, was discussed in [15] (and references therein)

$$
D = \begin{pmatrix}
ime^{i\theta} & A+d & 0 & \Gamma_R^\dagger \\
A^\dagger+d & ime^{-i\theta} & \Gamma_L^\dagger & 0 \\
0 & \Gamma_L & ime^{i\theta} & B \\
\Gamma_R & 0 & B^\dagger & ime^{-i\theta}
\end{pmatrix},
\tag{1}
$$

and the partition function is

$$
Z[m, \theta] = \langle \det D \rangle.
\tag{2}
$$

For each flavor, the entries in the matrices have respectively n, n, n_+ and n_- elements corresponding to the number of right handed near-zero modes, left handed near-zero modes, right handed zero modes and left handed zero modes. Here d denotes a diagonal matrix with equal entries, d, the matrices assigned to the near-zero modes are square matrices, while the ones assigned to the zero modes are rectangular matrices. The fluctuations in the rectangularity of the matrices induce the proper U(1) breaking [16]. The hopping between zero and near-zero modes is characterized by the overlap matrices Γ.

The averaging in (2) is done with respect to the local fluctuations in the topological charge $\chi = (n_+ - n_-)$, with a Gaussian width fixed by the quenched topological susceptibility χ_*,

$$
e^{-\frac{(n_+-n_-)^2}{2\chi_* V}} \equiv e^{-\frac{\chi^2}{2\chi_* V}}
\tag{3}
$$

and a Gaussian measure for the random matrix elements A, Γ, B, with width $\Sigma = 1$. The latter is physically tied to the quark condensate $|\langle q^\dagger q \rangle| \sim (200\,\mathrm{MeV})^3$. For simplicity, we take the width of the Gaussians to be the same since the hopping between the low-lying modes may be random enough not to distinguish between zero and near-zero modes. The temperature effects on the near-zero modes are parameterized by the deterministic and off-diagonal entries d. They cause the near-zero modes to be gapped by typically $d = \pm \pi T$ at high temperature, setting the range of validity of the current assumptions [15]. The depletion in the number of zero modes caused by an increase in the temperature will be discussed below.

We observe that the columns and rows in the fermion matrix (1) may be rearranged to give instead

$$
\begin{pmatrix}
ime^{i\theta} & 0 & A+d & \Gamma_R^{\dagger} \\
0 & ime^{i\theta} & \Gamma_L & B \\
A^{\dagger}+d & \Gamma_L^{\dagger} & ime^{-i\theta} & 0 \\
\Gamma_R & B^{\dagger} & 0 & ime^{-i\theta}
\end{pmatrix}
\tag{4}
$$

where we grouped the right and left handed modes, respectively. We denote the total number of modes (size of the full matrix) by $2N = 2n + n_+ + n_-$. We note that in the limit $mV \ll 1$ the matrix model (2) yields sum-rules for the quark eigenvalues that are consistent with those discussed in [17], following the general arguments in [18] with the identification $N = \mathrm{n}V$, where n is an order 1 quantity measuring the density of modes. Physically, the latter sets the scale for the fermionic contribution to the energy density.

Distribution of Eigenmodes

The matrices D in (1) have (in the chiral limit) a continuous distribution of eigenvalues $\bar{\rho}(\lambda)$, with a superimposed Dirac delta function at zero virtuality $\lambda = 0$, for fixed topological charge χ,

$$
2N \rho(\lambda) = |\chi| \delta(\lambda) + 2N \bar{\rho}(\lambda) .
\tag{5}
$$

To assess the dependence of r_n on λ one can use the fact that λ_n and εr_n are the real and imaginary parts of the non-hermitian operator $D + \varepsilon i \mathbf{1}_5$, in first order perturbation theory. Here $\mathbf{1}_5 = \mathrm{diag}(1, -1, 1, -1)$ with each identity assigned to its pertinent subspace. Taking ε infinitesimally small and rescaling the imaginary part of the eigenvalue we obtain the abovementioned dependence[1].

One can now use the chiral structure of D (its anti-commutativity with $\mathbf{1}_5$) to show that the square

$$
(D + i\varepsilon \mathbf{1}_5)^2 = D^2 - \varepsilon^2
\tag{6}
$$

is a hermitian operator. This means that the pair $\lambda_n + i\varepsilon r_n$ is either purely real or purely imaginary. It follows that all nonzero eigenvalues ($\lambda_n^2 \gg \varepsilon$) have vanishing r_n while the $|\chi|$ topological ones have zero eigenvalue and $r_n = \pm 1$. In the limit $mV > 1$ the random matrix model (2) allows for a model dependent assessment of the distribution of the low-lying modes of the Dirac operator in the infinite volume limit using the methods discussed in [20].

---

[1] It is interesting to note that non-hermitean operators of this form with $\varepsilon = 1/\sqrt{2N}$ yield a generic distribution of non-hermitean eigenvalues [19].

PARTITION FUNCTION

Bosonization

For equal masses, the partition function (2) can be readily bosonized. The result for fixed size matrices is

$$Z_{N,n_\pm}[m,\theta] = \int dP dP^\dagger e^{-NPP^\dagger} e^{-\frac{\chi^2}{2\chi_* V}} \times \tag{7}$$
$$\left[(z+P)(\bar{z}+P^\dagger) + d^2\right]^n (z+P)^{n_+} (\bar{z}+P^\dagger)^{n_-}.$$

Here $z = me^{i\theta}$ stands for degenerate flavors. The number of near-zero modes n, and the number of zero modes $n_\pm$, fix the size of the matrices in (7). However, their distribution is partly fixed by the Gaussian distribution (3), while the remaining part is fixed by equilibrium arguments as we now discuss.

Detailed balance

At finite temperature the change in the total number of zero modes can be argued generically. Indeed, with increasing temperature the zero modes may deplete either by pairing into topologically neutral aggregates of near-zero modes [21] or screening [22]. For the simplest neutral aggregate (molecule [21]), this pairing is reminiscent of the Kosterlitz-Thouless one in four-dimensions [23].

The chemistry of small neutral aggregates can be described by the probability of formation p_f and breaking p_b. For molecular arrangements in equilibrium, detailed balance implies

$$p_b\, n = p_f \left(\frac{n_-}{N}\right) n_+. \tag{8}$$

The l.h.s stands for the number of pairs broken. The r.h.s stands for the number of pairs formed which is the formation probability, times the probability (n_-/N) to find an unpaired negative charge, times the total number of positive charges n_+. In equilibrium, the number of pairing matches the number of breaking.

We may now calculate the square of $(n_+ + n_-)$ from the constraint $2N = 2n + n_+ + n_-$ and subtract χ to obtain

$$n_+ n_- = (N-n)^2 - \frac{1}{4}\chi^2. \tag{9}$$

Therefore n satisfies

$$\frac{1}{2}n^2 - N(1+\delta)\,n + \frac{1}{2}N^2 - \frac{1}{8}\chi^2 = 0 \tag{10}$$

with $\delta = p_b/2p_f$. Since from (3) $\chi \sim \sqrt{N}$ we obtain

$$n = N \underbrace{\left(1+\delta-\sqrt{\delta^2+2\delta}\,\right)}_{\alpha} - \frac{1}{8N}\frac{1}{\sqrt{\delta^2+2\delta}}\chi^2 + \mathcal{O}(\chi^4) \tag{11}$$

and

$$n_\pm = N - n \pm \frac{\chi}{2}. \tag{12}$$

We are effectively left with a 'filling fraction' α and a contribution to the topological susceptibility χ. For $\alpha \to 1$ we have $\delta \to 0$ ($p_b \ll p_f$). In this case, practically all the zero modes are paired (the unpaired ones are of order $1/\sqrt{N}$) with U(1) effectively restored [2]. For $\alpha \to 0$ we have $\delta \to \infty$ ($p_b \gg p_f$). In this case, all the zero modes are unpaired and U(1) is broken.

Saddle point analysis

We insert (11-12) into the partition function (7), and perform a linear shift in χ,

$$\chi = \tilde{\chi} - i2N \cdot y \tag{13}$$

with the requirement that the term linear in $\tilde{\chi}$ vanishes in (7). The resulting consistency condition (saddle point) reads

$$\frac{1}{2}\log\frac{z+P}{\bar{z}+P^\dagger} + 2iay = 0 \tag{14}$$

where

$$a = \frac{1}{\chi_*}\mathbf{n} + \frac{1}{2}\frac{\alpha}{1-\alpha^2}\log\frac{|z+P|^2+d^2}{|z+P|^2}. \tag{15}$$

The parameter y is just proportional to the average topological charge

$$\langle n_+ - n_- \rangle = 2V\mathbf{n}y, \tag{16}$$

while P and $P^\dagger$ in the above equations are the saddle point solutions following from the 'action'

$$(1-\alpha_*)\log|z+P|^2+\alpha_*\log(|z+P|^2+d^2) - iy\log\frac{z+P}{\bar{z}+P^\dagger} - PP^\dagger + \frac{2\mathbf{n}}{\chi_*}y^2 \tag{17}$$

where the effective 'filling fraction' is

$$\alpha_* = \alpha\left(1 + \frac{y^2}{1-\alpha^2}\right). \tag{18}$$

---

[2] Eq. (11) holds for $1 - \alpha \gg N^{-1/4}$, hence in this paper the limit $\alpha \to 1$ is understood always *after* the thermodynamical limit $N \to \infty$.

Writing out the saddle point equations for P and $P^\dagger$ and subtracting yield

$$\bar{z}P - zP^\dagger = 2iy.\tag{19}$$

This suggests the decomposition $e^{-i\theta}P = Q + iy/m$, with Q real, being the chiral condensate $|\langle q^\dagger q \rangle|$ and satisfying the saddle point equation

$$(1-mQ-Q^2-\frac{y^2}{m^2})\,[(m+Q)^2+\frac{y^2}{m^2}+d^2] = \alpha_*d^2.\tag{20}$$

This equation will be analyzed next.

RESULTS

The model is totally specified by (1-3) and (8). The thermodynamical limit will be understood as $N, V \to \infty$, with N/V fixed. The parameters are: the width of the Gaussian $\Sigma = 1$, the current mass m, the quenched topological susceptibility χ_*, the deterministic entries d, the vacuum angle θ, the filling fraction α and the mode-density $\mathbf{n} = N/V$. Generically, the effects of temperature cause $0 < d = \pi T$ and $0 \le \alpha \le 1$.

Chiral condensate

The χ saddle point equation (14) has a trivial solution, $y = 0$ for $\theta = 0$. Inserting this back into the equation for the condensate Q and setting $m = 0$ yields

$$Q^2 = \frac{1}{2}\left(1-d^2 + \sqrt{(1-d^2)^2 + 4(1-\alpha)\,d^2}\right).\tag{21}$$

This result is similar to the one considered by [24], although our physical interpretation is different. Indeed, in our case α measures the amount of U(1) breaking and follows from the rectangular character of the matrices as opposed to the square matrices used in [24]. It is fixed by detailed balance. We have ignored the trivial solution with $Q = -m$, by maximizing the effective action (17)

$$F/N = -Q^2 + \log{(m+Q)^2} + \alpha \log \frac{(m+Q)^2 + d^2}{(m+Q)^2}.\tag{22}$$

For $d < 1$, the solution with $Q \neq 0$ sets in independently of α. For $\alpha = 1$, the zero modes pair into near-zero modes, and $Q^2 = 1 - d^2$ [15,24,25]. This is a U(1) symmetric phase with broken chiral symmetry. A qualitative assessment of the range of temperature where this can take place follows by reinstating the dimensionful constants, that is $d = \pi T < \sqrt[3]{\Sigma}$. Hence $T < 70$ MeV, which is outside the range of validity of our model (see above). However, this points to the fact that the near-zero modes are sufficiently gapped at already moderate temperatures,

414

leaving the zero modes as the only contributors to the chirally broken phase. Indeed, at $d = 1$ we have $Q = \sqrt[4]{1-\alpha}$, which is zero-mode driven. From here on, only the case with $d > 1$ will be discussed unless specified otherwise.

For $\alpha \to 1$,

$$Q(m) = \frac{d}{\sqrt{d^2-1}} \sqrt{1-\alpha} + \mathcal{O}(m) \tag{23}$$

while for $\alpha = 1$,

$$Q(m) = \frac{m}{d^2-1} \left[1 - \left(\frac{d^2}{d^2-1} \right)^3 m^2 \right] \tag{24}$$

The pairing mechanism suggests an integer 'exponent' $\delta = 1$. We recall that for $d = \alpha = 1$, $Q = m^{1/3}$ and $\delta = 3$ which is mean-field [15,24,25].

Isotriplet susceptibilities

A measure of U(1) breaking in the matrix model can be assessed by investigating the difference in the π^0 and a^0 isotriplet susceptibilities [10,11]

$$\omega = \left\langle q^\dagger i 1_5 \tau^3 q q^\dagger i 1_5 \tau^3 q \right\rangle_c - \left\langle q^\dagger \tau^3 q q^\dagger \tau^3 q \right\rangle_c \tag{25}$$

and is amenable to the quark eigenvalue distribution $\rho(\lambda)$ through [12]

$$\omega = 4m^2 \int_0^\infty d\lambda \frac{\rho(\lambda)}{(m^2 + \lambda^2)^2} . \tag{26}$$

For $\alpha \to 1$, the matrix model yields $Q \to 0$ with a gapped spectrum in the chiral limit, hence $\omega = 0$ [3]. This observation is similar to the one we made in [26] without due care to the U(1) problem as we noted. In general, ω can be related to the resolvent

$$G(z) = \frac{1}{N} \left\langle \mathrm{Tr} \frac{1}{z-D} \right\rangle . \tag{27}$$

Specifically,

$$\omega = \frac{\mathrm{Im}\ G(im)}{m} - \mathrm{Re}\ G'(im) = \frac{Q(m)}{m} - Q'(m) \tag{28}$$

where $Q(m)$ follows from (20).

For $\alpha < 1$ we have

---

[3] In fact it would suffice that $\rho(\lambda)$ vanishes as $\sim \lambda^a m^b$ with $a > -1$ and $a + b > 1$.

415

$$\omega = \frac{d}{d^2 - 1} \frac{\sqrt{1 - \alpha}}{m} + \mathcal{O}(m^2) \qquad (29)$$

which is to be compared to $\omega \sim \sqrt[4]{1 - \alpha}/m$ for $d = 1$. The $1/N$ corrections to (29) are

$$2\frac{|\chi|}{N} \frac{1}{m^2} + \frac{\chi^2}{N^2} (\dots). \qquad (30)$$

The first term is the contribution of the zero modes in (5) through (26). Since $\chi \sim \sqrt{N}$ both contributions in (30) are subleading in comparison to (29) in the thermodynamical limit. These effects may still be present in current lattice assessments of ω as we discuss below.

For $\alpha = 1$, we have

$$\omega = \frac{d^6}{(d^2 - 1)^3} m^2 + \mathcal{O}(m^4) \qquad (31)$$

implying that ω flips from $1/m$ to m^2 at the transition point. We note that $\omega \sim 1/m^{2/3}$ for $d = 1$, $\alpha = 1$ which is the mean-field result [15]. It is noteworthy that only integer 'exponents' are produced by the pairing transition, a point in support of some general arguments made in [11].

Topological susceptibility

The topological susceptibility in the matrix model is simply given by

$$\chi_{top} = -\frac{\partial^2}{\partial \theta^2} \log Z = -2\frac{\partial y}{\partial \theta}. \qquad (32)$$

Expanding the consistency equation to linear order in y, we obtain

$$y = \frac{-\theta}{2a + \frac{1}{m(m+Q)}} \qquad (33)$$

with a defined in Eq. (15), which gives

$$\frac{1}{\chi_{top}} = \frac{1}{\chi_*} + \sum_{i=1}^{N_f} \frac{1}{2m_i(m_i + Q_i)} + \frac{\alpha}{2(1 - \alpha^2)} \log \frac{(m+Q)^2 + d^2}{(m+Q)^2} \qquad (34)$$

where we have reinstated the flavor dependence. The first contribution is the quenched susceptibility, the second contribution is the screening caused by the near-zero modes and the unpaired zero modes, and the third contribution stems from the paired zero modes. Note that χ_{top} vanishes not only for massless quarks but also for maximal pairing with $\alpha = 1$, as the asymmetry of D's become minimal. This happens as $Q \to 0$, in qualitative agreement with recent lattice simulations [4].

Pseudoscalar susceptibilities

The connected and disconnected pseudoscalar susceptibilities associated with $q^\dagger 1_5 q$ may be assessed in a similar way. These susceptibilities were recently addressed on the lattice [10]. In our case, the disconnected part χ_5^{dis} reads

$$\chi_5^{dis} = \frac{1}{N} \left\langle \text{Tr} 1_5 \frac{1}{im - D} 1_5 \frac{1}{im - D} \right\rangle \tag{35}$$

and is readily amenable to (34) through

$$\chi_5^{dis} = \frac{1}{V} \left\langle \frac{(n_+ - n_-)^2}{m^2} \right\rangle = \frac{\chi_{top}}{m^2}. \tag{36}$$

In the broken phase χ_{top} is dominated by the the second term in (34) for small m, hence $\chi_5^{dis} \sim 1/m$. As $Q \to 0$, the limits $m \to 0$ (chiral) and $\alpha \to 1$ (pairing) do not commute. For fixed mass and $\alpha \to 1$, $\chi_5^{dis} \sim (\alpha-1)\log m/m^2 \sim 0$, while for $\alpha = 1$ and $m \to 0$, $\chi_5^{dis} \sim 1/(1+Q/m) \sim 1$. In both cases, χ_5^{dis} is finite. Note that for $d = 1$, $\chi_5^{dis} \sim m^{2/3}$.

The connected part χ_5^{conn} follows from the identity [16]

$$\chi_{top} = \frac{2m}{N_f^2} Q - \frac{m^2}{N_f^2} \left(\chi_5^{disc} - \chi_5^{conn} \right). \tag{37}$$

This is the random matrix version of the QCD anomalous Ward identity [5]. Hence

$$\chi_5^{conn} = (N_f^2 + 1)\chi_5^{disc} - 2\frac{Q}{m} \tag{38}$$

for $m > 0$. Again, the connected part of the susceptibility is plagued with similar ambiguities in the chiral and pairing limits. For $\alpha = 1$ and $m \to 0$, χ_5^{conn} is finite.

Scalar susceptibilities

The connected and disconnected isosinglet susceptibility associated with $q^\dagger q$ may be estimated in our case as well, following the lattice conventions [10,11],

$$\chi_S^{conn} = \frac{1}{N} \left\langle \text{Tr} \frac{1}{im - D} \frac{1}{im - D} \right\rangle \tag{39}$$

and

$$\chi_S^{disc} = \frac{1}{N} \left\langle \text{Tr} \frac{1}{im - D} \text{Tr} \frac{1}{im - D} \right\rangle - \frac{1}{N^2} \left\langle \text{Tr} \frac{1}{im - D} \right\rangle^2. \tag{40}$$

Both susceptibilities follow from (20). Specifically,

$$\chi_S^{conn} = Q'(m) = \frac{1}{d^2 - 1} \tag{41}$$

$$\chi_S^{disc} = Q^2(m) = \frac{m^2}{(d^2 - 1)^2} . \tag{42}$$

for $\alpha = 1$. This is to be compared with the mean-field result for $d = 1$, $\chi_S^{conn} = 1/m^{2/3}$ and $\chi_S^{disc} = m^{2/3}$. The factorized result for the disconnected isosinglet susceptibility follows from the absence of correlations in the number $(n_+ + n_-)$.

θ angle dependence

In the symmetric phase and for small m, the θ dependence of the free energy $\ln Z/V$ is simple. Indeed, since $y \sim m$, for $\alpha < 1$ we may neglect the last term in the consistency equation (14) and obtain

$$\sum_i \underbrace{\arctan \frac{y/m_i}{Q_i}}_{-\phi_i} = -\theta . \tag{43}$$

The saddle point equation (20) in the chiral limit can be solved. Defining $Q_i + iy/m_i \equiv |Q_{i*}|e^{i\phi_i}$, the result for each flavor is

$$\frac{y^2}{m_i^2 \sin^2 \phi_i} = Q_{i*}^2 . \tag{44}$$

where Q_{i*} follows from Q in (21) through the substitution $\alpha \to \alpha_*$ for $m = 0$. Hence

$$\sum_i \phi_i = \theta , \tag{45}$$

$$m_1 \sin \phi_1 = \ldots = m_{N_f} \sin \phi_{N_f} . \tag{46}$$

These equations are analogous to the zero-temperature equations originally derived in QCD [1–3] and more recently in a matrix model [27]. Therefore the dependence of the free energy on θ in the broken phase is the same as the vacuum one. The temperature dependence is only implicit through Q_*.

As Q_* approaches zero at the critical point and in the chiral limit, the dependence on θ changes. For small m, we may no longer neglect the last term in the consistency equation (14) as it diverges. Geometrically the line that intersects the curves of the arctan's becomes nearly vertical so that y is for all purposes 0 regardless of the value of the θ angle. This extends the result $\chi_{top} = 0$ obtained earlier at $\theta = 0$ to $\theta \neq 0$.

The fact that the free energy no longer depends on θ at the critical point and beyond, may be traced to the occurrence of a non-analytic term $|\chi|$ in the partition function. Indeed from (10) and for $\alpha = 1$

418

$$n = N - \frac{1}{2}|\chi| .$$ (47)

Inserting this into the partition function (7), we obtain

$$e^{i\theta\chi} e^{-b|\chi|} e^{-\frac{\chi^2}{2\chi_* V}}$$ (48)

with b a *positive* factor stemming from (7). Performing the integral/sum over χ gives a vanishing contribution to the free energy $\log Z/V$. Specifically, the sum over χ is for a range of parameters well approximated by

$$2\text{Re} \, \frac{1}{1 - e^{i\theta - b}}$$ (49)

which gives zero contribution to the free energy. This is a direct consequence of the total quenching of the topological fluctuations in the paired configurations of zero modes. The simultaneous restoration of chiral and U(1) symmetry at finite temperature yields a symmetric phase that preserves strong CP.

COMPARISON TO LATTICE

In a first lattice study by Bernard et al. [11], chiral symmetry restoration was found to precede the U(1) restoration. Their analysis relied on gauge configurations at fixed lattice spacing $a \sim 1/6T_c \sim 0.25$ fm [11] for $N_t = 6$. Since finite volume effects were not investigated, it may be that the small U(1) breaking effects detected in these simulations through a lattice measurement of ω for staggered fermions are of the type (30). However, simple estimates based on their numbers appear to be on the larger side of their reported results [20]. As we already noted, the pairing mechanism supports integer 'exponents' for ω, a point sought in [11].

In a second lattice study by Kogut et al. [10], the low-lying quark eigenvalues of the staggered Dirac operator where investigated. Their analysis shows that the disconnected isosinglet susceptibility χ_5^{dis}, decreases but remains finite in the high temperature phase. The finite result was shown to follow from the eigenmodes with finite chirality (topological). The conclusion was that the U(1) symmetry was not restored in the symmetric phase, although again finite volume effects were not investigated. In the present matrix model, we have observed that χ_5^{dis} remains finite in the chiral and U(1) symmetric phase for $d > 1$, when the thermodynamical limit is carried. Also, we have noted an ambiguity in the limits $m, \alpha \to 0, 1$, suggesting that the cooling procedure may be subtle while carrying the chiral limit. Indeed, lattice cooling affects the "filling fraction" α.

In a third lattice study by Chandrasekharan et al. [12], the chiral condensate and ω were calculated using also staggered fermions for fixed $\beta = 5.3$ and $N_t = 4$. Although their results were found to be consistent with those of Bernard et. al. [11], they concluded that the anomalous effects were small, hinting at the possible restoration of U(1) in the symmetric phase. Although their conclusions are closer to ours in spirit, they differ in content since their small value of ω was obtained from a linear extrapolation in the current quark mass, as opposed to a quadratic extrapolation suggested by our results. Also, we have observed that the θ-dependence drops in the symmetric phase in distinction to a general assumption they made.

In a fourth lattice study by Vranas et al. [13], lattice simulations with domain wall fermions were carried at $N_t = 4$. It was found that the high temperature phase preserves chiral symmetry with a small amount of U(1) breaking, although with a somehow heavier pion mass. The method preserves flavor symmetry and incorporates the effect of the anomaly at every stage of the simulation. It is indeed encouraging that the results of these simulations are the closest to ours.

CONCLUSIONS

We have used a simple matrix model to analyze the interplay between zero and near-zero modes at finite temperature. While the model finds its motivation in the lattice results described above, it was originally argued from an NJL model with U(1) breaking [15]. At finite temperature, the pairing mechanism at work in the zero mode sector is reminiscent of the one originally suggested in the context of instantons [21]. The present model is by no means exhaustive as additional effects, e.g. Debye screening, have been omitted. Their consideration goes beyond the scope of this work.

This notwithstanding, our results indicate that chiral and U(1) symmetry are simultaneously restored for maximum pairing of zero modes. Although the chiral condensate receives contribution from all low-lying modes, its depletion to zero requires that the zero modes are paired and the near-zero modes are gapped. A simple estimate shows that the near zero modes are substantially gapped at moderately low temperatures, suggesting their early decoupling. This rules out the possibility of a U(1) restoration prior to a chiral restoration, and suggests that both symmetry restorations occur simultaneously.

The transition by pairing the topological charges is followed by a number of observations regarding the topological, scalar and pseudoscalar susceptibilities for small current quark masses. In particular, integer 'exponents' were observed in contrast to the fractional exponents expected from general universality arguments. These susceptibilities have been extensively studied on the lattice. Our comparison with the most recent lattice simulation using domain wall fermions is very encouraging, although some improvements regarding the extrapolation to zero quark mass and finite volume effects are still warranted in the staggered simulations. In many ways, our results should benefit the more complex instanton calculations when they become available.

Finally, we have shown that in the symmetric phase the topological susceptibility vanishes in the thermodynamical limit. As a result, the partition function develops a non-analyticity in the net topological charge that causes the symmetric phase to be CP even whatever the vacuum angle. While admittedly this is a result of the present matrix model, it should be interesting to see whether it carries to QCD in the infinite volume limit.

ACKNOWLEDGMENTS

We would like to thank S. Chandrasekharan for a discussion. IZ thanks Norman Christ for several discussions, and C.-R. Ji and D.-P. Ming for the invitation to a pleasant meeting. This work was supported in part by the US DOE grants DE-FG-88ER40388 and DE-FG02-86ER40251, by the Polish Government Project (KBN) grant 2P03B00814 and by the Hungarian grant OTKA-T022931.

REFERENCES

1. R.J. Crewther, in *Field Theoretical Methods in Particle Physics*, Ed. W. Rühl, Plenum 1980; R.J. Crewther, Nucl. Phys. **B209**, 413 (1982).
2. E. Witten, Annals of Phys. **128**, 363 (1980).
3. P. Di Vecchia and G. Veneziano, Nucl. Phys. **B171**, 253 (1980).
4. B. Alles et al., Nucl. Phys. Proc. Suppl. **73**, 518 (1999).
5. R.J. Crewther, Phys. Lett. **B70**, 349 (1977); M.A. Shifman, A.I. Vainshtein and V.I. Zakharov, Nucl. Phys. **B166**, 439 (1980); D.I. Diakonov and M.I. Eides, Sov. Phys. JETP **54**, 232 (1981).
6. C. Rozenzweig, J. Schechter and G. Trahern, Phys. Rev. D **21**, 3388 (1980); P. Natt and R. Arnowitt, Phys. Rev. D **23**, 1789 (1981).
7. R. Jackiw in *Relativity, Groups and Topology*, Les Houches 1983.
8. H.Yamagishi and I. Zahed, *hep-ph/9507296*; *hep-th/9709125*.
9. E. Shuryak, Comments Nucl. Part. Phys. **21** 235 (1994).
10. J.B. Kogut, J.-F. Lagaë, D.K. Sinclair, Nucl. Phys. Proc. Suppl. **53**, 269 (1997); Nucl. Phys. Proc. Suppl. **63**, 433 (1998).
11. C. Bernard, et al., Phys. Rev. Lett. **78**, 598 (1997).
12. S. Chandrasekharan, et al., Phys. Rev. Lett. **82**, 2463 (1999).
13. P. Vranas et al., *hep-lat/9903024*.
14. T. Schaefer, Phys. Lett. **B389**, 445 (1996); T. Cohen, Phys. Rev. **D54**, R1867 (1996); S.H. Lee and T. Hatsuda Phys. Lett. **54**, R1871 (1996).
15. M.A. Nowak, M. Rho, I. Zahed, *"Chiral Nuclear Dynamics"*, World Scientific, Singapore 1996; R.A. Janik, M.A. Nowak, I. Zahed, Phys. Lett. **B392**, 155 (1997).
16. R.A. Janik, M.A. Nowak, G. Papp, I. Zahed, Nucl. Phys. **B498**, 313 (1997).
17. A.V. Smilga, Phys. Rev. **D59**, 114021 (1999).
18. R.A. Janik, M.A. Nowak, G. Papp, I. Zahed, Acta Phys. Pol. **B29**, 3957 (1998).
19. K.B. Efetov, Phys. Rev. Lett. **79** 491 (1997); Phys. Rev. **B56**, 9630 (1997).
20. M.A. Nowak, G. Papp and I.Zahed, Phys. Lett. **B389**, 137 (1996).
21. E. Ilgenfritz and E. Shuryak, Phys. Lett. **B325**, 263 (1994); T. Schaefer and E. Shuryak, Rev. Mod. Phys. **70**, 323 (1990).
22. D. Gross, R. Pisarski and L. Yaffe, Rev. Mod. Phys. **53**, 43 (1981).

23. I. Zahed, Nucl. Phys. **B 427**, 561 (1994).

24. T. Wettig, A. Schäfer and H. Weidenmüeller, Phys. Lett. **B367**, 28 (1996).

25. A.D. Jackson and J.J.M. Verbaarschot, Phys. Rev. **D53**, 7223 (1996); M. Stephanov, Phys. Lett. **B275**, 249 (1996);

26. M.A. Nowak, G. Papp and I. Zahed, Phys. Lett. **B389**, 341 (1996).

27. R.A. Janik, M.A. Nowak, G. Papp and I.Zahed, *hep-ph/9901390*.

Strange Form Factors of Octet and Decuplet Baryons

Soon-Tae Hong

Department of Physics and Basic Science Research Institute,
Sogang University, C.P.O. Box 1142, Seoul 100-611, Korea

Abstract. The strange form factors of baryon octet are evaluated, in the chiral models with the general chiral SU(3) group structure, to yield the theoretical predictions comparable to the recent experimental data of SAMPLE Collaboration and to study the spin symmetries. Other model predictions are also briefly reviewed to compare with our results and then the strange form factors of baryon octet and decuplet are predicted.

Triggered by the EMC experimental result [1] on inelastic muon-proton scattering, there have been considerable discussions concerning the strangeness in hadron physics. Beginning with Kaplan and Nelson's work [2] on the charged kaon condensation the theory of condensation in dense matter has become one of the central issues in nuclear physics and astrophysics together with the supernova collapse. The K^- condensation at a few times nuclear matter density was later interpreted [3] in terms of cleaning of $\bar{q}q$ condensates from the QCD vacuum by a dense nuclear matter and also was further theoretically investigated [4] in chiral phase transition.

Quite recently, the SAMPLE collaboration [5] reported the experimental data of the proton strange form factor through parity violating electron scattering [6]. On the other hand, McKeown [7] has shown that the strange form factor of proton should be positive by using the conjecture that the up-quark effects are generally dominant in the flavor dependence of the nucleon properties. This result is contrary to the negative values of the proton strange form factor which result from most of the model calculations [8–11] except that of Hong and Park [12] based on the SU(3) chiral bag model (CBM) and that of Meissner and co-workers [13].

Now let us consider the strange form factors of baryons in the chiral models, such as Skyrmion, MIT and chiral bag models with the general chiral SU(3) group structure. In these models, using the electromagnetic (EM) currents obtained from the model Lagrangian, the magnetic moment operator is given by the sum of isovector

CP494, *New Directions in Quantum Chromodynamics*, edited by C.-R. Ji and D.-P. Min
© 1999 American Institute of Physics 1-56396-908-4/99/$15.00

and isoscalar parts, $\hat{\mu}^i = \hat{\mu}^{i(3)} + \frac{1}{\sqrt{3}}\hat{\mu}^{i(8)}$ where

$$\hat{\mu}^{i(a)} = -\mathcal{N}D^8_{ai} - \mathcal{N}'d_{ipq}D^8_{ap}\hat{T}^R_q + \frac{N_c}{2\sqrt{3}}\mathcal{M}D^8_{a8}\hat{J}_i - \mathcal{P}D^8_{ai}(1 - D^8_{88}) + \frac{\sqrt{3}}{2}\mathcal{Q}d_{ipq}D^8_{ap}D^8_{8q}.$$

Here D^8_{ab} is the adjoint representation of SU(3) and $\hat{J}_i = -\hat{T}^R_i$ ($i = 1, 2, 3$) and $\hat{T}^R_p$ ($p = 4, 5, 6, 7$) are the right SU(3) operators along the isospin and strangeness directions, respectively. $\mathcal{M}$, $\mathcal{N}$ and $\mathcal{N}'$ are the inertia parameters obtained in the chiral symmetric limit and $\mathcal{P}$ and $\mathcal{Q}$ are the inertia parameters coming from the explicit current flavor symmetry breaking (FSB) effects in the adjoint representation where the chiral symmetry breaking mass terms cannot contribute to the magnetic moment operator.

Now, in order to take into account the missing symmetry breaking mass effects, we employ the quantum mechanical perturbative scheme where we use the SU(3) cranking and treat the symmetry breaking terms perturbatively. In this scheme, the Hamiltonian is split up into two pieces, the SU(3) flavor symmetric and symmetry breaking parts, $H = H_0 + H_{SB}$ where

$$H_0 = M + \frac{1}{2}\left(\frac{1}{\mathcal{I}_1} - \frac{1}{\mathcal{I}_2}\right)\hat{J}^2 + \frac{1}{2\mathcal{I}_2}(C_2(SU(3)) - \tfrac{3}{4}Y^2_R), \quad H_{SB} = m(1 - \hat{D}^8_{88}). \quad (1)$$

Here M is the static mass of the baryon and $\mathcal{I}_1$ and $\mathcal{I}_2$ are the moments of inertia along the isospin and strangeness directions, respectively. $\hat{J}^2$ and $C_2(SU(3))$ are the Casimir operators in the SU(2) and SU(3) group, Y^2_R the right hypercharge operator and m the inertia parameter denoting the symmetry breaking strength.

In the higher dimensional irreducible representation of SU(3) group, the baryon wave function is described as $|B\rangle = |B\rangle_8 - C^B_{\overline{10}}|B\rangle_{\overline{10}} - C^B_{27}|B\rangle_{27}$ with the representation mixing coefficients C^B_λ, to yield the implicit FSB contribution to the magnetic moment

$$\delta\mu_{2,B} = -2 \sum_{\lambda=\overline{10},27} \frac{{}_8\langle B|\hat{\mu}^i|B\rangle_{\lambda\lambda}\langle B|H_{SB}|B\rangle_8}{E_\lambda - E_8}. \quad (2)$$

Now the magnetic moments of baryon octet in the FSB case can break up into three parts

$$\mu_B = \mu_{0,B}(\mathcal{M}, \mathcal{N}, \mathcal{N}') + \delta\mu_{1,B}(\mathcal{P}, \mathcal{Q}) + \delta\mu_{2,B}(m\mathcal{I}_2) \quad (3)$$

where the first term $\mu_{0,B}$ comes from the chiral symmetric contribution, $\delta\mu_{1,B}$ is due to the explicit FSB and $\delta\mu_{2,B}$ is obtained from the implicit FSB in the representation mixing as shown in Eq. (2).

Following the above scheme we can obtain the proton magnetic moment

$$\mu_p = \frac{1}{10}\mathcal{M} + \frac{4}{15}(\mathcal{N} + \frac{1}{2}\mathcal{N}') + \frac{8}{45}\mathcal{P} - \frac{2}{45}\mathcal{Q} + m\mathcal{I}_2(\frac{2}{125}\mathcal{M} + \frac{8}{1125}(\mathcal{N} - 2\mathcal{N}')).$$

Here one notes that the coefficients are solely given by the SU(3) group structure of the chiral models and the physical informations such as decay constants and masses are included in the inertia parameters.

Now one can easily see the spin symmetries as follows. First, in the adjoint representation of the SU(3) chiral symmetric limit with $\mathcal{M}$, $\mathcal{N}$ and $\mathcal{N}'$, we have the U-spin symmetry, $\mu_{0,p} = \mu_{0,\Sigma^+}$, $\mu_{0,n} = \mu_{0,\Xi^0}$, $\mu_{0,\Sigma^-} = \mu_{0,\Xi^-}$ and $\mu_{0,\Lambda} = -\mu_{0,\Sigma^0}$. Secondly, in the implicit representation mixing FSB contributions, we can obtain the V-spin symmetry relations, $\delta\mu_{2,p} = \delta\mu_{2,\Xi^-}$, $\delta\mu_{2,n} = \delta\mu_{2,\Sigma^-}$ and $\delta\mu_{2,\Sigma^+} = \delta\mu_{2,\Xi^0} = \frac{1}{2}\delta\mu_{2,p}$. Finally, we can see that the s-flavor channel possesses the I-spin symmetry, $\mu_B^{(s)} = \mu_{\bar{B}}^{(s)}$.

Using the flavor projection operators in the EM currents of the chiral models we can obtain the strange components of the nucleon magnetic moments, which are degenerate in the isomultiplets and respect the above I-spin symmetry

$$\mu_N^{(s)} = -\frac{7}{60}\mathcal{M} + \frac{1}{45}(\mathcal{N} + \frac{1}{2}\mathcal{N}') + \frac{1}{45}\mathcal{P} + \frac{1}{90}\mathcal{Q} + m\mathcal{I}_2(\frac{43}{2250}\mathcal{M} - \frac{38}{3375}(\mathcal{N} - \frac{13}{19}\mathcal{N}')).$$

On the other hand, the form factors of the baryons, with internal structure, are defined by the matrix elements of the EM currents

$$\langle p + q|J_{EM}^\mu|p\rangle = \bar{u}(p+q)(\gamma^\mu F_{1B}(q^2) + \frac{i}{2m_B}\sigma_{\mu\nu}q^\nu F_{2B}(q^2))u(p)$$

where $u(p)$ is the spinor of the baryons. Using the flavor projection operators in the EM currents as before, in the limit of zero momentum transfer, one can obtain the strange form factors of the baryons

$$F_{1B}^{(s)}(0) = S, \quad F_{2B}^{(s)}(0) = -3\mu_B^{(s)} - S \tag{4}$$

in terms of the strange quantum number of the baryon S and the strange components of the baryon magnetic moments. In Table 1, we obtain the chiral model predictions that the CBM with $R \approx 0.6$ fm corresponding to $\theta(R) = \pi/2$ yield $F_{2p}^{(s)} = 0.30$ comparable to the experimental data [5] $F_{2p}^{(s)} = 0.23 \pm 0.37 \pm 0.15$ (exp) within about 30% errors. Here one notes that the large positive values of the proton strange form factors originate from $\delta F_{2p}^{(s),1}$ (with $\mathcal{P}$ and $\mathcal{Q}$) and $F_{2p}^{(s),2}$ (with $m\mathcal{I}_2$), the explicit and implicit FSB contributions due to $f_\pi \neq f_K$, $m_\pi \neq m_K$ and $m_u = m_d \neq m_s$.

TABLE 1. The strange form factors of the octet and decuplet baryons.

	$F_{2N}^{(s),0}$	$\delta F_{2N}^{(s),1}$	$\delta F_{2N}^{(s),2}$	$F_{2N}^{(s)}$	$F_{2\Lambda}^{(s)}$	$F_{2\Xi}^{(s)}$	$F_{2\Sigma}^{(s)}$	$F_{2\Delta}^{(s)}$	$F_{2\Sigma^*}^{(s)}$	$F_{2\Xi^*}^{(s)}$	$F_{2\Omega}^{(s)}$
CBM	−0.19	−0.12	0.61	0.30	0.49	0.25	−1.54	1.67	0.84	0.56	0.83
SM	−0.13	−0.09	0.20	−0.02	0.51	0.09	−1.74	0.04	−0.10	−0.03	0.24

Now let us briefly review other model predictions. To the dispersion theory prediction [8] $F_{2p}^{(s)} = -0.31$, the kaon loop correction is included [9] to yield $F_{2p}^{(s)} = -0.40$ where the SU(3) flavor symmetric baryon octet, for example $m_N = m_\Lambda$, is used. On the other hand, neglecting the sea-quark fluctuation effects, the nonrelativistic constituent quark model produces [10] $F_{2p}^{(s)} = -0.0324$. In the Skyrmion model, Park and collaborators [11] evaluates the proton strange form factor to yield $F_{2p}^{(s)} = -0.13$, which has the same sign but is much larger than our Skyrmion prediction due to the fact that they used the different Skyrmion parameter $e = 4.0$ and missed the contribution from the term proportional to $f_K^2 - f_\pi^2$ in the inertia parameter $m\mathcal{I}_2$. Very recently Meissner and co-workers [13] included the kaon loop corrections in the heavy baryon chiral perturbation theory to yield $F_{2p}^{(s)} = 0.18$, which is positive also.

Similarly to the baryon octet case, one can obtain the magnetic moments of Δ baryons in the s-flavor channel

$$\mu_\Delta^{(s)} = -\frac{7}{48}\mathcal{M} + \frac{1}{12}(\mathcal{N} - \frac{1}{2\sqrt{3}}\mathcal{N}') + \frac{2}{21}\mathcal{P} + \frac{5}{168}\mathcal{Q} + m\mathcal{I}_2(\frac{85}{2016}\mathcal{M} - \frac{25}{504}(\mathcal{N} - \frac{2}{5\sqrt{3}}\mathcal{N}')).$$

Substitution of the above equation into Eq. (4) yields the strange form factors of Δ baryons whose numerical values are listed in Table 1, together with the predictions for the other octet and decuplet baryons.

ACKNOWLEDGMENTS

We would like to thank B.Y. Park, D.P. Min, M. Rho and G.E. Brown for helpful discussions and constant concerns.

REFERENCES

1. Ashman,J., et al., Phys. Lett. **B206**, 364 (1988).
2. Kaplan,D.B., and Nelson,A.E., Phys. Lett. **B175**, 57 (1986).
3. Brown,G.E., Kubodera,K., and Rho,M., Phys. Lett. **B192**, 273 (1987).
4. Lee,G.Q., Lee,C.H., and Brown,G.E., Nucl. Phys. **A625**, 372 (1997).
5. Mueller,B., et al., Phys. Rev. Lett. **78**, 3824 (1997).
6. McKeown,R.D., Phys. Lett. **B219**, 140 (1989); Beise,E.J., and McKeown,R.D., Comm. Nucl. Part. Phys. **20**, 105 (1991).
7. McKeown,R.D., Los Alamos Preprint hep-ph/9607340 (1996).
8. Jaffe,R.L., Phys. Lett. **B229**, 275 (1989).
9. Musolf,M.J., and Burkardt,M., Z. Phys. **C61**, 433 (1994).
10. Koepf,W., Henley,E.M., and Pollock,S.J., Phys. Lett. **B288**, 11 (1992).
11. Park,N.W., Schechter,J., and Weigel,H., Phys. Rev. **D43**, 869 (1991).
12. Hong,S.T., and Park,B.Y., Nucl. Phys. **A561**, 525 (1993).
13. Meissner,Ulf-G., et al., Los Alamos Preprint nucl-th/9904076 (1996).

HIGH TEMPERATURE
AND DENSITY QCD

J/ψ at Finite Density in QCD Sum Rule Analysis

Arata Hayashigaki

Yukawa Institute for Theoretical Physics,
Kyoto University, Kyoto 606-8502, Japan

Abstract. For the "anomalous" J/ψ suppression in *Pb-Pb* collision announced by NA50 Collaboration I propose new mechanism accessible to the novel behavior of the data. The point of new idea is to focus on in-medium effects on the masses of mesons containing charmed quarks. In QCD sum rule analysis, I find that the *D*-meson involving one charmed quark feels more attractive force from the surrounding nucleons than the charmonium composed of two charmed quarks. As a result, it will appear that as the nucleon density increases $D\bar{D}$ threshold in turn falls below higher-lying charmonium states (ψ', χ_c) which have the energy levels below the threshold in vacuum. Finally, also for the low-lying J/ψ state, a larger decay channel to the threshold could be opened up. Therefore, this mechanism can qualitatively explain some features of the anomalous J/ψ suppression without advent of deconfinement phase.

INTRODUCTION

The NA50 Collaboration (CERN-SPS) has reported a strong suppression of J/ψ and ψ' production in *Pb-Pb* collision at 158 GeV per nucleon [1]. The suppression of J/ψ production (relative to Drell-Yan process) shows large discrepancy from conventional nuclear absorption models [2] exhibited by measurements from *p-A* up to *S-U* collisions. In particular, it seems to expose some characteristic suppression forms, which might be a first indication for color deconfinement [3]. The suppression form of $J/\psi(1S)$ shows the discontinuity around two points of the effective length L of nuclear matter over which the produced $c\bar{c}$ traverses. In contrast, the first excited state $\psi'(2S)$ seems to show only one discontinuity at smaller E_T value observed in the *S-U* collision. As is well known, the conventional attempts quantitatively encounter considerable difficulties to explain such peculiar behaviors. That is, such models allow only a gradual change from slight suppression in the *p-A* to strong suppression in the *Pb-Pb* collisions. On the other hand, once the local energy density exceeds certain threshold value in the heavy-ion collisions and onset of a new state of matter (deconfinement) is switched on, the nature of deconfinement gives more satisfactory results [4,5]. This is based on the idea that if the deconfinement

CP494, *New Directions in Quantum Chromodynamics*, edited by C.-R. Ji and D.-P. Min
© 1999 American Institute of Physics 1-56396-908-4/99/$15.00

transition were of first order, onset of the suppression will become discontinuous. But I feel that the suppression of χ_c-feeding introduced to explain double threshold structure of the data is not obvious in the plasma-based model, because from a estimate of formation time of $c\bar{c}$ resonances the hadronization of $c\bar{c}$ system like the χ_c will not occur at small central region of the collision where plasma bubbles are produced. Are such suppression phenomena peculiar to the phase transition? I suggest such phenomena could occur even in nuclear matter. Indeed, the formation time of charmonium is so long that final state interaction of the chamonium with nucleons or comovers will occur at the peripheral region of the nucleus (about a few times as large as normal matter density), although its magnitude of interaction depends on transverse momentum of the charmonium produced. Motivated by this point, it is reasonable as a first step to study matter effect at such finite density. It was found in Ref. [6] that the light quark (u, d) condensates may be substantially reduced in medium and as a result light hadron masses would also decrease. On the other hand, if heavier quarks (s, c) feel so weaker interaction than the light quarks, the masses of mesons composed of heavy-light and heavy-heavy quarks may be expected to scale differently with density. In particular, the difference becomes larger with increasing the density. If it is the case of $|\delta m_\psi| < |\delta m_D|$, charmonia below the $D\bar{D}$ threshold in vacuum may rise above the threshold in matter. Therefore, the level crossing between the charmonium and the $D\bar{D}$ threshold can cause the abrupt decrease of J/ψ survival probability due to the decay to $D\bar{D}$ state. I advocate new mechanism to explain drastic change of the suppression which has ever been ruled out in normal nuclear absorption: $J/\psi + N \to D + \bar{D} + X$. This paper is composed as follows: First, in QCD sum rule analysis I make a detailed study of the mass modification for the J/ψ at finite density and second, a rough estimate of that for D-meson in terms of the same approach. In conclusion, I discuss the J/ψ suppression through these results.

IN-MEDIUM EFFECT FOR J/ψ

I calculate the medium modification of J/ψ on the basis of the relation between J/ψ-N scattering length and the mass shift [7,8] through extending vacuum QCD sum rules (QSR's) to finite density [†]. First by applying QSR to J/ψ-N forward scattering amplitude, I evaluate the scattering length. The superposition of such elementary J/ψ-N elastic scattering at low energy affects the effective mass of J/ψ in nuclear matter. When one works in the dilute nucleon gas, one finds that the mass shift is linearly dependent on the density and the scattering length in the framework of QSR. In this approach with the Fermi gas model, in-medium correlation function (density ρ_N) is divided into vacuum part and static one-nucleon part near the normal matter density by applying OPE to the correlators at deep Euclidean region. That is, in the framework of QSR, in-medium correlation function

[†] For the detailed discussion of this section, refer to Ref. [9]. The similar QCD sum rule analysis was performed in Ref. [10].

can be approximated reasonably well to the linear density of nuclear matter that all nucleons are at rest:

$$\Pi^{\text{NM}}_{\mu\nu}(q) = i \int d^4 x e^{iq \cdot x} \langle \text{T} J_\mu(x) J^\dagger_\nu(0) \rangle_{\rho_N} \simeq \Pi^0_{\mu\nu}(q) + \frac{\rho_N}{2M_N} T_{\mu\nu}(q), \tag{1}$$

where $q^\mu = (\omega, \boldsymbol{q})$ is the 4-momentum of J/ψ ($J_\mu = \bar{c}\gamma_\mu c$). The one-nucleon part corresponds to the forward J/ψ-N scattering amplitude with the spin of nucleon averaged:

$$T_{\mu\nu}(\omega, \boldsymbol{q}) = i \int d^4 x e^{iq \cdot x} \langle N(ps) | \text{T} J_\mu(x) J^\dagger_\nu(0) | N(ps) \rangle, \tag{2}$$

where $|N(ps)\rangle$ denotes the nucleon state with $p = (M_N, \boldsymbol{p} = 0)$ and spin s normalized covariantly as $\langle N(\boldsymbol{p}) | N(\boldsymbol{p}') \rangle = (2\pi)^3 2p^0 \delta^3(\boldsymbol{p} - \boldsymbol{p}')$. By applying QSR to $T_{\mu\nu}$ directly, I can relate the scattering length extracted from the QSR for $T_{\mu\nu}$ with the mass shift, $\delta m_{J/\psi} = 2\pi \frac{M_N + m_{J/\psi}}{M_N m_{J/\psi}} \rho_N a_{J/\psi}$. For applying the dispersion relation to $T(\omega, \boldsymbol{q}) = T_\mu^\mu/(-3)$, I parametrize the spectral function such as $\rho(u, \boldsymbol{q} = 0) = a \delta'(u^2 - m^2_{J/\psi}) + b \delta(u^2 - m^2_{J/\psi}) + c \delta(u^2 - s_0)$, near the pole position of J/ψ. Here δ' is the first derivative of δ function with respect to u^2 and the parameter a is related to the spin-averaged scattering length $a_{J/\psi}$ as $a = 8\pi f^2_{J/\psi} m^4_{J/\psi}(M_N + m_{J/\psi}) a_{J/\psi}$, where the coupling $f_{J/\psi}$ and the J/ψ mass $m_{J/\psi}$ are defined by $\langle 0 | J_\mu | J/\psi^{(h)}(q) \rangle = f_{J/\psi} m^2_{J/\psi} \epsilon^{(h)}_\mu(q)$ with the polarization vector $\epsilon^{(h)}_\mu$. s_0 is the continuum threshold in vacuum. Moreover, among these parameters I impose the constraint from low energy theorem that in the low energy limit, $\omega \to 0$, $T(\omega, \boldsymbol{0})$ becomes equivalent to Born term $T^{\text{Born}}(\omega, \boldsymbol{0})$ ($= 0$ for lack of charmed quarks inside a nucleon in this system). Finally, through the QSR I determine two unknown phenomenological parameters a and b. On the other hand, I give the following OPE expression for n-th derivative of $T_{\mu\nu}$ with respect to q^2, up to dimension-4 operators.

$$\frac{1}{n!} \left(\frac{d}{dq^2} \right)^n \frac{T^{\text{OPE}}(q^2)}{q^2} = \frac{1}{3} \left[C^{(n)}_G(\xi) \left\{ \left\langle \frac{\alpha_s}{\pi} G^2 \right\rangle_N - 4 \left\langle \frac{\alpha_s}{\pi} \mathcal{ST}(G^a_{0\sigma} G^a_{0\sigma}) \right\rangle_N \right\} \right.$$
$$\left. + \{ D^{(n)}_1(\xi) - D^{(n)}_2(\xi) - D^{(n)}_3(\xi) \} \left\langle \frac{\alpha_s}{\pi} \mathcal{ST}(G^a_{0\sigma} G^a_{0\sigma}) \right\rangle_N \right], \tag{3}$$

where $\xi = -q^2/4m^2_c$ (m_c; charmed quark mass $1.3 \sim 1.35$ GeV). $\mathcal{ST}$ means making the twist-2 operators symmetric and traceless in its Lorentz indices. The explicit forms of Wilson coefficient $C^{(n)}_G$, $D^{(n)}_1$, $D^{(n)}_2$ and $D^{(n)}_3$ are given in Ref. [9]. Eventually I evaluate the parameters in terms of the moment sum rule, $\hat{T}^{(n)\,\text{ph}}(\xi \; ; \; a, b) = \hat{T}^{(n)\,\text{OPE}}(\xi)$. Here I determine both n and q^2 to reproduce the experimental value of J/ψ bare mass by applying the moment sum rule to $\Pi^0_{\mu\nu}$. After inserting the sets of ξ and n obtained thus into the moment sum rule, I can determine unknown parameters a and b simultaneously. Here the values of other parameters are given in Ref. [9]. The direct application of moment sum rule to the forward J/ψ-N scattering amplitude supplies us the fascinating results for the J/ψ-N interaction, which is consistent with Ref. [10]. That is, the J/ψ-N scattering length $a_{J/\psi}$ indicates negative value (about -0.1 ± 0.02 fm). This result suggests that J/ψ-N interaction is very weakly attractive. Moreover the result gives very slight decreasing mass (about -4 to -7 MeV), about 0.1 to 0.2 % at normal matter density.

IN-MEDIUM EFFECT FOR D-MESON

Next, as well as the above scheme I estimate the mass shift of D-meson through D-N scattering length. In this case, the pseudoscalar current ($J_5 = \bar{c}i\gamma_5 q$) is used for the D-meson, where q indicates massless light quarks. For simplicity, I don't take account of the isospin decomposition on the D-meson. Therefore, the in-medium mass of D-meson obtained by this analysis can be regarded as the isospin averaged result. The coupling f_D and the D-meson mass m_D are defined by $\langle 0|J_5 |D(q)\rangle$ $= if_D m_D^2/m_c$ ($m_D = 1.87$ GeV). After applying the Borel transformation to the forward D-N scattering amplitude, I obtain that on the OPE side with $\nu = m_c^2/M^2$ (M; Borel mass) and the Whittaker function G,

$$
B_M T^{\text{OPE}}(q^2) = e^{-\nu} \left[-m_c \langle \bar{q}q \rangle_N - 2 \left(1 - \frac{m_c^2}{M^2} \right) \left\langle q^\dagger i D_0 q \right\rangle_N \right.
$$
$$
+ \frac{1}{12} \left\langle \frac{\alpha_s}{\pi} G^2 \right\rangle_N + \frac{1}{3} \left\langle \frac{\alpha_s}{\pi} \mathcal{ST}(G_{0\sigma}^a G_{0\sigma}^a) \right\rangle_N
$$
$$
\left. \times \{ -2 + G(1,2,\nu) + G(2,2,\nu) + 2G(2,3,\nu) + 2G(1,3,\nu) \} \right] \quad (4)
$$

and on the phenomenological side, if as before one assumes $T^{\text{Born}} = 0$ [tt],

$$
B_M T^{\text{ph}}(q^2) = a \left(\frac{1}{M^2} e^{-m_D^2/M^2} - \frac{s_0}{M^4} e^{-s_0/M^2} \right) + b \left(e^{-m_D^2/M^2} - \frac{s_0}{M^2} e^{-s_0/M^2} \right). \quad (5)
$$

As before I perform two-parameter fitting by means of OPE expanded up to dimension 4 including twist-2 operators. In fact, I can derive D-N scattering length a_D ($= -m_c^2 a/8\pi (M_N + m_D)f_D^2 m_D^4$) as a function of M^2 by removing the parameter b from both the matching equation of Eqs. (4) and (5) and its first derivative with respect to M^2. For coupling f_D and continuum threshold s_0, I adopt $f_D = 0.187$ GeV read off from a stable curve in the vicinity of $s_0 = 7.5$ GeV^2 using Borel sum rule of vacuum correlation function for the D-meson. This f_D value is very close to other calculations [11,12] and experimental data (≤ 0.31 GeV). When one uses nucleon matrix elements for quark fields such as $\langle \bar{q}q \rangle_N = 5.3$ GeV and $\left\langle q^\dagger i D_0 q \right\rangle_N = 0.34$

[tt] This implies one ignores the channels of D^+ or $D^0 + N \to \Lambda_c$ or Σ_c and so on. In the case of D^- or $\bar{D}^0$, since the anti-charm quark cannot couple to three quarks inside a nucleon by OZI rule, Born term will be zero.

GeV^2, comparatively stable curves of a_D are obtained at the reliable Borel mass region. To summarize the results for D-meson analysis, the D-N scattering length is -1.25 ± 0.05 fm. This result suggests that the D-N interaction is more attractive than the J/ψ-N interaction. I apply this result to effective mass of D-meson at normal matter density in the linear density approximation as before. Then the mass reduction is -83 ± 4 MeV, about 4 % of the total mass. I find this result leads to larger decrease than the charmonium. An origin of this mechanism originates in that on the OPE side, $m_c \langle \bar{q}q \rangle_N$ term is dominant for the D-meson in contrast to only gluon operator contribution for the charmonium. In Quark-Meson-Coupling model, this contribution will correspond to quark-σ meson coupling. Indeed, this model predicts the mass shift of D-meson becomes -60 MeV for the scalar potential at normal matter density [13].

CONCLUSION

Now I can come to some important conclusions through all the above results. Namely, if the mass modification of higher-lying charmonium states is very slight as well as the J/ψ in matter, the $D\bar{D}$ threshold falls below ψ' at the normal density and χ_c at twice as the normal density. These results could produce some important features relevant to recent NA50 experimental data. First, the behavior of level crossing may lead to onset of discontinuous property for the suppression form. It is well known from p-A collision data that the J/ψ observed in nuclear collisions are directly produced only about 60% and the remainder comes from excited states ($\chi_c(1P)$, $\psi'(2S)$) with the ratio of 3 to 1 [14]. So the suppression of such feeddown effect from ψ' and χ_c to J/ψ could lead to direct decrease of J/ψ survival probability through the level crossings. First, the suppression of ψ' nearest to the $D\bar{D}$ threshold can be observed as the first very slight suppression of the J/ψ production. In fact, the ψ' data indicates a strong suppression even in cooler S-U collision. Next, three states of χ_c which is very close each other, induce the sequent 2nd suppression through the level crossings. Finally, direct suppression due to the J/ψ itself can be observed as the 3rd suppression with further increase of the matter density. Thus the stair-shaped suppression form is not necessarily a phenomenon as far as deconfinement phase. Not theoretically but also experimentally we should investigate whether in fact such level crossing could occur at least at normal matter density. To that end, I hope realization of the inverse kinematics experiments, in which the nuclear beam is incident on a hydrogen target, because this experiment can be feasible for measuring the decays of the charmonium and the D-meson [15] inside a nucleus. I can also suggest some observational consequences caused by such level crossings. One of them is to observe the change of decay width for the charmonium. In vacuum the resonances above $D\bar{D}$ threshold, for example ψ'' state have width of order MeV because of strong open charm channel. On the other hand, the resonances below the threshold have very sharp width of a few hundreds keV. So given level crossing, the decay modes will change drastically at least one order of magnitude. The another is to observe the enhancement of D-meson at intermediate mass region of dilepton ($1.5 \leq M \leq 2.5$ GeV). In fact, such a dilepton enhancement was observed [16]. Thus I expect that the matter effect gives a considerable impact on the anomalous J/ψ suppression and we can not rule that possibility out. It might be considered as one of new suppression mechanisms due to matter effect without advent of deconfinement phase. Needless to say, the investigation of finite temperature effect [17,18] to the mass modification must be performed in the future.

ACKNOWLEDGEMENTS

I would like to thank T. Hatsuda for useful discussions.

REFERENCES

1. L. Rammello et al., NA50 Collaboration, *Nucl. Phys.* **A638** (1998) 261c; M.C. Abreu et al., NA50 Collaboration, *Phys. Lett.* **450B** (1999) 456.
2. C. Gershel and J. Hüfner, *Phys. Lett.* **207B** (1988) 253; D. Kharzeev, C. Lourenco, M. Nardi and H. Satz, *Z. Phys.* **C74** (1997) 307.
3. T. Matsui and H. Satz, *Phys. Lett.* **178B** (1986) 416.
4. R. Vogt, *Phys. Rep.* **310** (1999) 197.
5. D. Kharzeev, M. Nardi and H. Satz, *hep-ph*/9707308; D. Kharzeev, *Nucl. Phys.* **A638** (1998) 279c.
6. T. Hatsuda and T. Kunihiro, *Phys. Rev. Lett.* **55** (1985) 158; *Phys. Rep.* **247** (1994) 221.
7. Y. Kondo and O. Morimatsu, *Phys. Rev. Lett.* **71** (1993) 2855.
8. Y. Koike and A. Hayashigaki, *Prog. Theor. Phys.* **98** (1997) 631.
9. A. Hayashigaki, *Prog. Theo. Phys.* **101** (1999) 923.
10. F. Klingl, S. Kim, S.H. Lee, P. Morath and W. Weise, *Phys. Rev. Lett.* **82** (1999) 3396.
11. T.M. Aliev and V.L. Eletskii, *Sov. J. Nucl. Phys.* **38**(6) (1983) 936.
12. A.X. El-Khadra, A.S. Kronfeld, P.B. Mackenzie, S.M. Ryan and J.N. Simone, *Phys. Rev.* **D58** (1998) 014506.
13. K. Tsushima, D.H. Lu, A.W. Thomas and R.H. Landau, *Phys. Rev.* **C59** (1999) 2824; A. Sibirtsev, K. Tsushima and A.W. Thomas, *nucl-th*/9904016.
14. L. Antoniazzi et al., E705 Collaboration, *Phys. Rev. Lett.* **70** (1993) 383.
15. G.A. Alves et al., *Phys. Rev. Lett.* **70** (1993) 722; M.J. Leitch et al., *Phys. Rev. Lett.* **72** (1994) 2542.
16. M. Masera et al., HELIOS-3 Collaboration, *Nucl. Phys.* **A590** (1995) 93c; E. Scomparin et al., NA50 Collaboration, *Nucl. Phys.* **A610** (1996) 331c; M. Gazdzicki and C. Markert, *hep-ph*/9904441.
17. R. Vogt and A. Jackson, *Phys. Lett.* **206B** (1988) 333.
18. A. Sibirtsev, K. Tsushima, K. Saito and A.W. Thomas, *nucl-th*/9904015.

Vector mesons in nuclear matter

Su Houng Lee*

*Department of Physics, Yonsei University [1]
Seoul 120-749, Korea

Abstract. In this talk, I will summarize the progress in understanding the changes the vector meson spectral density in nuclear medium using the constraint equations obtained from the Borel transformed dispersion relation and QCD Operator Product Expansion. We will discuss the results for the scalar mass shift and dispersion effects (three momentum dependence) for the light quark system (ρ, ω), the strange quark system (ϕ) and the heavy quark system (J/ψ) in nuclear medium. For the light quark systems, a nontrivial change in the mass and/or width are expected, while the dispersion effects are found to be small. Existing model calculations for the dispersion effects are compared to the constraint equation in detail. Very small, but accurate mass shift is obtained for the heavy quark system.

INTRODUCTION

The properties of vector meson in nuclear medium have been the focus of current interest due to their potential role to provide one with a direct observable of the nuclear medium effects, associated with chiral symmetry restoration, via dileptons in p-A or A-A reactions [1]. Indeed dileptons from Relativistic Heavy Ion Collisions (RHIC) [1] seemed to suggest a non-trivial change in the vector meson spectral density in a hot/dense environment, which can be understood in terms of model calculations based on either decreasing vector meson masses in hot/dense medium [2–4] or increased vector meson width with nontrivial momentum dependence [5,6]. In all of the approaches, the central question is, how the spectral density changes in hot/dense matter [7]. In this talk, I will discuss the model independent QCD constraints for the changes of the spectral density in nuclear matter for the light quark system (ρ, ω), the strange quark system (ϕ) and the heavy quark system (J/ψ).

[1] This work was supported in part by KOSEF through grant no. 976-0200-002-2 and the Korean Ministry of Education through grant no. 98-015-D00061

CP494, *New Directions in Quantum Chromodynamics*, edited by C.-R. Ji and D.-P. Min

Vector mesons in vacuum

The distinctive features appearing in the dilepton spectrum in p-A or A-A reactions are the vector meson peaks; the ρ, ϕ and the J/ψ, whose (mass, width) in the vacuum are (770, 150), (1020, 4.4), (3100, 0.086) MeV. It is interesting to compare the phase spaces of their decay into their corresponding two pseudo particles. For the ρ, its two pion decay, which accounts for most of the total width, has a large phase space. This is so because the pion is a Goldstone boson and has a small mass. For the J/ψ, its decay into D-mesons are forbidden, because the D mesons are not Goldstone bosons and two times its mass is greater than the mass of J/ψ. For the ϕ, the situation is something in between and its decay into two kaons has very little phase space. Hence, chiral symmetry breaking is partly responsible for the large difference in their width. As for the masses of the vector mesons, among other models, QCD sum rules provide an indirect relation to QCD condensates. For the light quark system, $\langle \bar{q}q \rangle$ is dominantly responsible for its mass, for the strange quark system, $\langle \bar{s}s \rangle$ is responsible, and for the J/ψ, the charmed quark mass and a small non perturbative contribution from the gluon condensate $\langle \frac{\alpha_s}{\pi} G^2 \rangle$ are responsible.

Therefore, if chiral symmetry gets restored at finite density or temperature, nontrivial changes will occur to the masses and widths of the vector mesons.

Quark condensate at finite density

The temperature dependence of the quark condensates have been calculated long ago on the lattice [8,9], showing that the light quark condensate goes to zero above the critical temperature. For heavier quark masses, the changes become smaller. Since there does not exit any lattice result at finite density, we will use linear density approximation in what follows.

$$\langle O \rangle_{\rho_n} = \langle O \rangle_0 + \rho_n \times \langle O \rangle_N \tag{1}$$

where O is any operator, ρ_n the nucleon density, and the subscripts $\rho_n, 0, N$ denotes the nuclear, the vacuum and the nucleon expectation values. Then, we have the following model independent result [10,11] for the quark condensate

$$\langle \bar{q}q \rangle_{\rho_n} = \langle \bar{q}q \rangle_0 + \frac{\Sigma_{\pi N}}{2\hat{m}} \rho_n \sim \langle \bar{q}q \rangle_0 \cdot \left(1 - 0.2 \frac{\rho_n}{\rho_0} \right) \tag{2}$$

where , ρ_0 is the nuclear matter saturation density and 0.2 changes to about 0.05 for the strange quark condensate. One notes that already at nuclear matter, we have partial restoration of chiral symmetry; namely, the chiral order parameter is reduced by 20%. Nuclear matter provides a stable environment with non trivial vacuum changes. Hence, if anything happens to the vector mesons at high temperature or density, the tendencies will already be apparent at nuclear matter.

Gluon condensate at finite density

The temperature dependence of the non-perturbative gluon condensates has also been calculated on the lattice [12–15]. The result is that it will stay almost constant up to the critical temperature and then reduce to about 60% of its vacuum value above the critical temperature. The leading density behavior can be obtained from the trace anomaly relation to leading order in α_s [10] $T_\mu^\mu = -\frac{9}{8}\frac{\alpha_s}{\pi}G^2 + \sum m_q \bar{q}q$, which when combined with the most recent determination of the nucleon mass in the chiral limit [16] gives,

$$\langle \frac{\alpha_s}{\pi}G^2 \rangle_{\rho_n} = \langle \frac{\alpha_s}{\pi}G^2 \rangle_0 \cdot \left(1 - 0.05\frac{\rho_n}{\rho_0} \right). \tag{3}$$

Hence, we also have a non-trivial change in the gluon condensate, but its relative change is smaller than the case of the quark condensate.

As we have seen, the condensates have non-trivial change in the nuclear medium. Using QCD Operator Product Expansion (OPE), we will derive constraints that the changes of the condensates provides on the vector meson spectral density in nuclear medium.

QCD CONSTRAINTS

There have been many model calculations to study vector meson properties in nuclear medium. Here, we will avoid any model calculation and derive a constraint equation that any model calculation should satisfy. The foundation of this approach was laid in ref. [4].

Consider the correlation function of the vector current $J_\mu = \bar{q}\gamma_\mu q$ at finite density;

$$\Pi_{\mu\nu}(\omega^2, \mathbf{q}^2) = i \int d^4 x e^{iqx} \langle T[J_\mu(x)J_\nu(0)] \rangle_{\rho_n}. \tag{4}$$

Here $q = (\omega, \mathbf{q})$. In what follows, when we give result for explicit vector meson, we will use the currents $J_\mu^{\rho,\omega} = \frac{1}{2}(\bar{u}\gamma_\mu u \mp \bar{d}\gamma_\mu d)$ for the ρ, ω mesons, $J_\mu^\phi = \bar{s}\gamma_\mu s$ for the ϕ and $J_\mu^{J/\psi} = \bar{c}\gamma_\mu c$ for the J/ψ.

In general, because the vector current is conserved, the polarization tensor in eq.(4) will have only two invariant functions [17].

$$\Pi_{\mu\nu}(\omega^2, \mathbf{q}^2) = \Pi_T q^2 P_{\mu\nu}^T + \Pi_L q^2 P_{\mu\nu}^L, \tag{5}$$

where we assume the ground state to be at rest, such that, $P_{00}^T = P_{0i}^T = P_{i0}^T = 0$, $P_{ij}^T = \delta_{ij} - \mathbf{q}_i\mathbf{q}_j/\mathbf{q}^2$ and $P_{\mu\nu}^L = (q_\mu q_\nu/q^2 - g_{\mu\nu} - P_{\mu\nu}^T)$. When $\mathbf{q} \to 0$, $\Pi_L = \Pi_T$, as in the vacuum.

We will make a small $\mathbf{q}$ expansion of the correlation function and look at its energy dispersion relation at fixed $\mathbf{q}$,

$$\mathrm{Re}\Pi_{L,T}(\omega^2, \mathbf{q}^2) = \mathrm{Re}\left(\Pi^0(\omega^2, 0) + \Pi^1_{L,T}(\omega^2, 0)\ \mathbf{q}^2 + \cdots\right)$$

$$= \int_0^\infty du^2 \left(\frac{\rho^0(u^2, 0)}{(u^2 - \omega^2)} + \frac{\rho^1_{L,T}(u^2, 0)}{(u^2 - \omega^2)}\ \mathbf{q}^2 + \cdots\right), \tag{6}$$

where $\rho(u^2, \mathbf{q}^2) = 1/\pi\,\mathrm{Im}\Pi^R(u^2, \mathbf{q}^2)$, and R denotes the retarded correlation function. We will construct a constraint equation for Π^0, Π^1_L and Π^1_T. For Π^1_L, Π^1_T, we will only look at the "non-trivial" $\mathbf{q}$ dependence [18]. A simple method to extract the "non-trivial" $\mathbf{q}$ dependence is to express the polarization function in terms of $Q^2, \mathbf{q}^2$ ($\Pi(Q^2, \mathbf{q}^2)$) and extract the linear $\mathbf{q}^2$ term.

In general for each polarization functions (Π^0, Π^1_L, Π^1_T), the OPE [19,20] looks as follows,

$$\Pi(\omega^2) = \sum_n C_n \langle O_n \rangle. \tag{7}$$

Here the O_n are operators of (mass) dimension n, renormalized at a scale μ^2, and C_n are the perturbative Wilson coefficients, which for the light quark system can be written as $C_n = \frac{c_n}{(-\omega^2)^{n/2}}$ and for heavy quark system as $C_n = \frac{c_n}{m_h^n}$ at $\omega^2 = 0$.

Let us first discuss the light quark system. After the Borel transformation, the dispersion relation for any one of the polarization function(Π^0, Π^1_L, Π^1_T) becomes

$$\mathrm{B.T.}\,\mathrm{Re}\Pi(M^2) = \int ds\,\rho(s)e^{-s/M^2}, \tag{8}$$

The left hand side, which is the Borel transform (B.T.) of the OPE, up to dimension 6 looks as follows,

$$\mathrm{B.T.}\,\mathrm{Re}\Pi(M^2) = \mathrm{B.T.}\,\Pi(M^2)_{pert} + \frac{c_4}{1!M^2}\langle O_4 \rangle + \frac{c_6}{2!M^4}\langle O_6 \rangle \tag{9}$$

The truncation is valid as long as M^2 is sufficiently large. The minimum M^2_{min} is usually determined by requiring the correction from higher dimensional operators to be less than 30% of the perturbative contribution. Now the constraints for the spectral density would be eq.(8), applied above $M^2 > M^2_{min}$.

As can be seen in eq.(8), the Borel transformation also changes the weighting factor of the spectral density to an $exp(-s/M^2)$. This has the following advantage for practical applications of our constraint. For small values of the Borel mass, the contribution of the spectral density at larger energy is exponentially suppressed. Consequently, in a model calculation, one can concentrate on the changes of the spectral density near the vector meson mass region and below and model the higher energy part with a simple pole like contribution.

The constraint equation in the vacuum are well satisfied by the spectral density in the vacuum. As we will see, in most cases the changes in the operators $\langle O_n \rangle$ are known. Hence, starting from the vacuum form of the spectral density (ImΠ), we can study what changes are consistent with the constraint equation. This provides model independent QCD constraints that any model calculation should satisfy. One can go one step further and try to parameterize the spectral density with a simple delta function type of pole and a continuum and determine the changes in the parameters.

LIGHT QUARK SYSTEM ρ, ω

Π_0

The operators that dominate the change in the OPE in eq.(8) for the light quark system (ρ, ω) are the quark operators [4]

$$\langle O_4 \rangle \rightarrow \langle \bar{q} \gamma_\mu D_\nu q \rangle \propto \int dx x [q(x) + \bar{q}(x)]$$
$$\langle O_6 \rangle \rightarrow \langle (\bar{q}q)^2 \rangle \propto \langle (\bar{q}q)^2 \rangle_0 + 2\rho_n \langle \bar{q}q \rangle_0 \langle \bar{q}q \rangle_N \quad (10)$$

The first equation of eq.(10) dominates the changes in the dimension 4 operators and is related to the well known second moment of the quark distribution function. The second equation of eq.(10), for which we have used the ground state saturation hypothesis [4], dominates the changes in the dimension 6 operators.

Using a delta function assumption for the spectral density in the constraint equation gives the following result for the scalar mass shift at $\mathbf{q} = 0$ [4],

$$\frac{m_V(\rho_n)}{m_V(\rho_n = 0)} = 1 - (0.16 \pm 0.06)\frac{\rho_n}{\rho_0} \quad (11)$$

This result is also consistent with other model calculations [22–24] or the Brown-Rho scaling argument [3].

A detailed comparison of the constraint equation in eq.(8) with a hadronic calculation, based on chiral SU(3) dynamics with explicit vector mesons were performed in [21]. The result shows a very good agreement between the OPE and the phenomenological spectral density put into the constraint equation in eq.(8). However, the hadronic calculation gave a large increase in width with a small decrease in mass for the ρ and a large decrease in mass with a small increase in width for the ω. Hence only the result for the ω is consistent with the result in eq.(11). Later it was found that this was a general result, given the uncertainty in the ground state hypothesis for the four quark condensate in the medium [25]; namely, that there exists a band in the mass vs. width plane that satisfies the constraint equation in eq.(8).

Π_T

The constraint for the nontrivial $\mathbf{q}$ dependence in eq.(8) has no $\Pi(M^2)_{pert}$ and has contributions from operators with explicit spin index [18]. The contributions from dimension 4 operators are related to the twist-2 matrix elements and are well known. The dimension 6 operators are dominated also by the twist-2 matrix elements. Hence the constraint equation has little uncertainty coming from the OPE. The operators that dominate are

$$\langle O_4 \rangle \rightarrow \langle \bar{q}\gamma_\mu D_\nu q \rangle_N \propto \int dx x [q(x) + \bar{q}(x)]$$

$$\langle O_6 \rangle \rightarrow \langle \bar{q}\gamma_\mu D_\nu D_\alpha D_\beta q \rangle_N \propto \int dx x^3 [q(x) + \bar{q}(x)] \tag{12}$$

Using a delta function assumption for the spectral density and allowing the parameters to change to leading order in density and in $\mathbf{q}^2$, we find the following non-trivial momentum dependence in the peak position [18],

$$\frac{m_\rho(\rho_n)}{m_\rho(\rho_n = 0)} = 1 - (0.023 \pm 0.007) \left(\frac{\mathbf{q}}{0.5}\right)^2 \frac{\rho_n}{\rho_0}, \tag{13}$$

where $\mathbf{q}$ is in GeV/c unit and for the ω meson, 0.023 ± 0.007 changes to 0.016 ± 0.005. This shows a very small momentum dependence compared to the expected scalar mass shift in eq.(11).

A detailed comparison of the constraint equation in eq.(8) for the momentum dependence for the transverse direction has been made [26] to the hadronic calculation, where the vector-meson nucleon scattering amplitude is obtained by resonance saturation in the s-channel. The result shows that the existing model calculations tend to overestimate the constraint. This is due to the large $\rho - N - \Delta(1232)$ coupling, which is obtained from the Bonn potential [27]. However, the existing calculations used a non-covariant monopole form factor [5,28] normalized off shell $F(\mathbf{q}) = \frac{\Lambda^2}{\Lambda^2 + \mathbf{q}^2}$. On the other hand, the large $\rho - N - \Delta(1232)$ coupling in the Bonn potential is defined with a dipole form factor normalized at the on shell point of the vector meson, $F_\rho(q^2) = \left(\frac{\Lambda_\rho^2 - m_\rho^2}{\Lambda_\rho^2 - q^2}\right)^2$. This reduces the Delta contribution to the rho meson self energy at the invariant mass around $m_\Delta - m_N$ by approximately a factor of 4. After this correction, we find that the model calculations give very good agreement with the constraint equation [26].

Π_L

The constraint for the longitudinal direction is dominated by twist-2 quark and gluon operators $\langle \bar{q}\gamma_{\mu_1} D_{\mu_2}..D_{\mu_n} q \rangle_N \propto \int dx x^{n-1}[q(x) + \bar{q}(x)]$, $\langle G^\alpha_{\mu_1} D_{\mu_2}..G_{\mu_n \alpha} \rangle_N \propto \int dx x^{n-1} g(x)$,

Using a delta function assumption for the spectral density and allowing the parameters to change to leading order in density and in $\mathbf{q}^2$, we find [18],

$$\frac{m_V(\rho_n)}{m_V(\rho_n = 0)} = 1 - (0.004 \pm 0.002) \left(\frac{\mathbf{q}}{0.5}\right)^2 \frac{\rho_n}{\rho_0}, \tag{14}$$

for both the ρ and ω. This is a very small effect and no detailed comparison with any hadronic calculations exits yet.

STRANGE QUARK SYSTEM ϕ

The OPE in the constraint equation for the ϕ meson is dominated by $\langle m_s \bar{s}s \rangle$ for the scalar mass shift and $\langle \bar{s}\gamma_\mu D_\nu s \rangle$ for the momentum dependence. Assuming a delta function ansatz for the pole, we find

$$\frac{m_\phi(\rho_n)}{m_\phi(\rho_n = 0)} = 1 - (0.03 \pm 0.015)\frac{\rho_n}{\rho_0} + (0.0005 \pm 0.0002)\left(\frac{\mathbf{q}}{0.5}\right)^2 \frac{\rho_n}{\rho_0}, \qquad (15)$$

for the transverse vector meson and the momentum dependence for the longitudinal ϕ is about a factor of two larger.

HEAVY QUARK SYSTEM J/ψ

For the heavy quark system, we look at the constraints from the moments [29,30],

$$M_n \equiv \frac{1}{n!}\left(\frac{d}{d\omega^2}\right)^n \mathrm{Re}\Pi(\omega^2)\Bigg|_{\omega^2 = -Q_0^2} = \int_{4m_c^2}^\infty \frac{\rho(s)}{(s + Q_0^2)^{n+1}}ds. \qquad (16)$$

The changes in OPE in nuclear medium for M_n is dominated by the change in the gluon condensate in eq.(3).

To study the J/ψ at rest in the nuclear matter, we will approximate the spectral density with a delta function for the lowest state. This is valid even in nuclear matter, because for a J/ψ at rest, inelastic interactions with nucleons such as $J/\psi + N \rightarrow \bar{D} + \Lambda_c$ do not occur. With this assumption, the constraint equation allows for the determination of mass shift of the J/ψ in nuclear matter. We find [31] for the mass shift

$$\Delta m_{J/\psi} \simeq -7\,\mathrm{MeV}. \qquad (17)$$

This corresponds to small J/ψ- and η_c-nucleon scattering lengths $a = -\mu_r \Delta m/(2\pi \rho_N) \simeq (0.1 - 0.2)\,\mathrm{fm}$ (μ_r is the meson-nucleon reduced mass). Our results for the mass shifts of the lowest $\bar{c}c$ states are surprisingly close to those reported in ref. [32–35].

Although the expected mass shift is small, the result has little uncertainty coming from OPE and puts reliable constraint on charmonium mass shift which should be met by further studies of heavy quark systems in dense matter.

CONCLUSION

We have derived and explored the consequences of model independent constraints for the vector meson polarization at $\mathbf{q} = 0$ and $\mathbf{q} \neq 0$ for all the vector mesons $\rho, \omega, \phi, J/\psi$. Most of them have very little uncertainty in the OPE side of the constraint equation and can be used as reliable constraints on all model calculation of the vector meson properties in medium.

REFERENCES

1. Quark matter 97, Nucl. Phys. **A 638** (1998). Quark Matter 99 proceeding.
2. G.Q.Li, C.M. Ko and G.E.Brown, Phys. Rev. Lett. **75** 4007 (1995).
3. G.E. Brown and M. Rho, Phys. Rev. Lett. **66** 2720 (1991).
4. T. Hatsuda and Su H. Lee, Phys. Rev. **C** 46R34 (1992) ; T. Hatsuda, Su H. Lee and H. Shiomi, Phys. Rev. **C 52** 3364 (1995).
5. R.Rapp, G. Chanfray and J. Wambach, Nucl. Phys. **A 617** 472(1997).,hep-ph/9907502,hep-ph/9907342.
6. F. Klingl, W. Weise, hep-ph/9802211.
7. G.E. Brown, et.al., Acta Phys. Polon. **B 29** 2309(1998).
8. J. Kogut et.al. , Phys. Rev. Lett. **50** 393 (1983).
9. M. Fukugita and A. Ukawa, Phys. Rev. Lett. **57** 503 (1986).
10. E.G. Drukarev and E.M. Levin, Prog. Part. Nucl. Phys. **27** 77 (1991).
11. T. D. Cohen, R.J. Furnstahl, D. Phys. Rev. **C 45** 1881 (1992).
12. M. Campostrini and A. Di Giacomo, Phys. Lett. **B 197** 403(1987).
13. Su H. Lee, Phys. Rev. **D 40** 2484 (1989).
14. C. Adami, T. Hatsuda, I. Zahed, Phys. Rev. **D 43** 921 (1991).
15. V. Koch, G.E. Brown, Nucl. Phys. **A 560** 345 (1993).
16. B. Borasoy and U.G. Meißner, Phys. Lett. **B 365** 285(1996).
17. J. Kapusta, Nucl. Phys. **B 148** 461 (1979).
18. Su H. Lee, Phys. Rev. **C 57**927 (1998).; Su H. Lee and H. Kim, Nucl. Phys. **A 642**165 (1998).
19. K.G. Wilson, Phys. Rev. **179** 1499 (1969).
20. T. Muta "Foundations of Quantum Chromodynamics", World Scientific, Singapore 1987.
21. F. Klingl, N. Kaiser and W. Weise, Nucl. Phys. **A 624** 527(1997).
22. H.-C. Jean, J. Piekarewicz and A, G, Williams, Phys. Rev. **C 49** 1981(1994); H. Shiomi and T. Hatsuda, Phys. Lett. **B 334** 281 (1994).
23. K. Saito and A.W. Thomas, Phys. Rev. **C 51** 2757 (1995).
24. B. Friman and M. Soyeur, Nucl. Phys. **A 600** 477 (1996).
25. S. Leupold, W. Peters and U. Mosel, Nucl. Phys. **A 628** 311 (1998).
26. B. Friman, Su H. Lee and H. Kim, Nucl. Phys. **A 653** 91 (1999).
27. R. Machleidt, K. Holinde and C. Elster, Phys. Rept. **149** 1 (1987).
28. W. Peters, M.Post, H. Lenske, S. Leupold and U. Mosel, Nucl. Phys. **A 632** 109 (1998).
29. M.A. Shifman, A.I. Vainshtein and V.I. Zakharov, Nucl.,Phys., **B 147**385 (1979); Nucl.,Phys., **B 147** 448(1979).
30. L.J. Reinders, H.R. Rubinstein and S. Yazaki, Phys. Rep. **127** 1 (1985).
31. F. Klingl, S.Kim, Su H. Lee, P. Morath and W. Weise, Phys. Rev. Lett. **82** 3396 (1999).
32. M. Luke, A.V. Manohar, M.J. Savage, Phys. Lett. **B 288** 355 (1992).
33. S.J. Brodsky, G.A. Miller, Phys. Lett. **B 412** 125 (1997).
34. G.F. de Teramond et.al., Phys. Rev. **D 58** 034012 (1998).
35. A. Hayashigaki, Prog.Theor.Phys.**101** 923 (1999)

The Center-Symmetric Phase of QCD

F. Lenz

Institut für Theoretische Physik III
Universität Erlangen–Nürnberg
Staudtstraße 7, D-91058 Erlangen, Germany

Abstract. The role of the center-symmetry and the dynamics of Polyakov loops are described within the axial gauge representation of QCD. Realization of the center symmetry is shown to result from non-perturbative gauge fixing and concomitant confinement like properties emerging even at the perturbative level are displayed. The dynamics of the Polyakov loops in the deconfined phase is shown to be severely constrained by the combined requirement of thermodynamic and magnetic stability.

I INTRODUCTION

The center symmetry [1–3] distinguishes the phases of Yang Mills theories. This symmetry is realized in the confining and spontaneously broken in the deconfined phase with the Polyakov loops serving as order parameter. Unlike in studies of lattice QCD, the center symmetry has not been the subject of systematic analytical investigations of QCD. For instance, perturbative approaches in general imply a change of the underlying gauge symmetry from $SU(N)$ to $U(1)^{N^2-1}$ when the coupling vanishes and thereby break the Z_N center symmetry. For the center symmetry to be preserved in the path integral formulation, the Faddeev–Popov determinant arising in the process of gauge fixing cannot be treated perturbatively. Likewise, for the center symmetry to be preserved in the canonical formalism, the Gauss law has to be resolved non-perturbatively. Only then the center symmetry is guaranteed to appear as the correct residual gauge symmetry. The process of non-perturbative gauge fixing is crucial for a description of the center symmetric phase and can be carried out explicitly with only few gauge choices. I shall describe QCD within the modified axial gauge; in this representation the Polyakov loop actually appears as a fundamental rather than composite degree of freedom which makes this gauge choice particularly relevant for an investigation of the center symmetry. The gauge choice also unveils another fundamental property of the formal structure of QCD. Unlike in QED, a global – for all field configurations valid – elimination of redundant variables is possible in QCD only at the expense of introducing coordinate singularities and thus of including singular gauge field configurations in such gauge fixed formulations [4]. Formation of Gribov horizons [5] or appearance of magnetic

CP494, *New Directions in Quantum Chromodynamics*, edited by C.-R. Ji and D.-P. Min
© 1999 American Institute of Physics 1-56396-908-4/99/$15.00

monopoles [6,7] represent two prominent examples of the occurrence of singular field configurations as a result of non-perturbative gauge fixing. It is thus tempting to connect the realization of the center symmetry with the emergence of singular field configurations and to identify these non-perturbative basic structures as the origin of the characteristic properties of the confining phase of QCD. I will present a study of the role of the center symmetry, display the emergence of monopole like field configurations and discuss the implications for QCD in both the confined and deconfined phases. This discussion summarizes the results of a series of investigations of QCD in axial gauge [8–11].

II   AXIAL GAUGE QCD

In order to define properly center symmetry transformations and Polyakov loops, one space-time direction has to be chosen compact. Here the 3-direction is assumed to be of finite extent L. The choice of a spatial compact direction is compatible with the canonical formulation of QCD. The order parameter which characterizes the phases of QCD appearing when compressing the system is the vacuum expectation value of the trace of the Polyakov loop operator

$$P\left(x_\perp\right) = P \exp\left\{ ig \int_0^L dx^3 A_3\left(x\right) \right\} \tag{1}$$

at finite extension $(x_\perp = (x_0, x_1, x_2))$. Under gauge transformations $U(x)$, $P\left(x_\perp\right)$ transforms as

$$P(x_\perp) \to U\left(x_\perp, L\right) P(x_\perp) U^\dagger\left(x_\perp, 0\right) . \tag{2}$$

The coordinates $x = (x_\perp, 0)$ and $x = (x_\perp, L)$ describe identical points, and we require periodicity properties imposed on the field strengths not to change under gauge transformations. This is achieved if U satisfies

$$U\left(x_\perp, L\right) = c_U \cdot U\left(x_\perp, 0\right) \tag{3}$$

with c_U being an element of the center of the group. Thus gauge transformations can be classified according to the value of c_U (± 1 in SU(2)). Therefore under gauge transformations

$$\mathrm{tr}(P(x_\perp)) \to \mathrm{tr}(c_U P(x_\perp)) \overset{SU(2)}{=} \pm \mathrm{tr}(P(x_\perp)). \tag{4}$$

Whenever gauge fixing is carried out exactly and with the help of strictly periodic gauge fixing transformations $(\Omega, c_\Omega = 1)$ the resulting formalism must contain the center symmetry

$$C: \quad \mathrm{tr}(P(x_\perp)) \to -\mathrm{tr}(P(x_\perp)) \tag{5}$$

as residual gauge symmetry. Each gauge orbit is represented by two gauge field configurations.

We pass to the axial gauge representation of QCD by applying the gauge fixing transformation

$$\Omega(x) = \Omega_D\left(x_\perp\right)\left(P^\dagger(x_\perp)\right)^{x^3/L} P\exp\left\{ig\int_0^{x^3} dz A_3\left(x_\perp, z\right)\right\}. \tag{6}$$

in which the axial gauge is reached in 3 steps [8]. In the presence of the third factor only, the gauge transformation would eliminate A_3 completely. In order to preserve the periodic boundary conditions of the gauge fields the second term reintroduces zero mode fields which in turn are diagonalized by Ω_D. Thus the gauge condition reads

$$\Omega\left(x\right)\left(A_3(x_\perp) + \frac{1}{ig}\partial_3\right)\Omega^\dagger\left(x\right) = \left(a_3\left(x_\perp\right) + \frac{\pi}{gL}\right)\tau_3. \tag{7}$$

By the gauge transformation, the 3 component of the gauge field is transformed to zero apart from the eigenvalues of the Polyakov loops. The elementary rather than composite nature of the Polyakov loop variables $a_3(x_\perp)$ in axial gauge is manifest. The gauge fixing transformation Ω is periodic and consequently field configurations which before gauge fixing are related by a gauge transformation with $c = -1$ are not identified. Therefore center symmetry transformations (cf. Eq.(5)) appear as residual symmetry transformations of the gauge fixed theory. The effect of C on an arbitrary gauge field is most conveniently written in a spherical color basis

$$\Phi_\mu(x) = \frac{1}{\sqrt{2}}(A_\mu^1(x) + i A_\mu^2(x))e^{-i\pi x^3/L} \tag{8}$$

as

$$C: \quad a_3 \to -a_3 \quad, \quad A_\mu^3 \to -A_\mu^3 \quad, \quad \Phi_\mu \to \Phi_\mu^\dagger \quad, \quad (\mu \neq 3). \tag{9}$$

The center symmetry transformation C acts as (Abelian) charge conjugation with the "photons" described by the neutral fields $A_\mu^3(x), a_3(x_\perp)$. For identification of the center symmetry with charge conjugation symmetry, the shift in the definition of the Polyakov loop variables in Eq.(7) and the phase change in the definition of the charged fields (Eq.(8)) have been introduced. This definitions will also simplify the description of the dynamics. The phase change in Eq.(8) makes the charged fields antiperiodic

$$\Phi_\mu(x_\perp, x^3 = L) = -\Phi_\mu(x_\perp, x^3 = 0). \tag{10}$$

If the center symmetry is realized $gLa_3(\mathbf{x}_\perp)$ has to be distributed symmetrically around the origin.

Apart from the discrete center symmetry transformation all other symmetries related to the gauge invariance have been used to eliminate A_3. In such a case of a global, non-perturbative gauge fixing we have to expect singular field configurations to emerge. These appear in the diagonalization of the Polyakov loops (Ω_D in Eq.(6)). This diagonalization can be viewed as choice of coordinates in color space in which the color 3-direction is identified with the direction of the Polyakov loop. This choice of coordinates becomes ambiguous if $gLa_3(x_\perp) = \pm\pi$, i.e. if the Polyakov loop is in the center of the group

$$P(x_\perp) = \pm\mathbb{1}. \tag{11}$$

This requirement determines a point on the group manifold S^3 and thus, for generic cases, fixes (locally) uniquely the position $x_\perp$. At these points, the gauge transformed field

$$A'_\mu(x) = \Omega_D(x_\perp) A_\mu(x) \Omega_D^\dagger(x_\perp) + s_\mu(x_\perp) \quad , \quad \mu \neq 3 \tag{12}$$

with

$$s_\mu(x_\perp) = \Omega_D(x_\perp) \frac{1}{ig} \partial_\mu \Omega_D^\dagger(x_\perp), \tag{13}$$

in general, is singular with Ω_D.

The expression for the axial gauge QCD partition function can be written as

$$Z = \sum_n Z_n = \sum_n \int D[a_3^n] \int \prod_{\mu \neq 3} D[A_\mu] e^{-S[A+s, a_3^n]}. \tag{14}$$

The integration variables, the unconstrained degrees of freedom, are the 3 components of the gauge field ($A_\mu(x), \mu \neq 3$) and the eigenvalues of the Polyakov loops. The partition function has been decomposed according to the number of singularities n of the field configurations. For this decomposition to be meaningful, regularization of the generating functional is required. The singular field $s(x_\perp)$ is determined by the Polyakov loop variables

$$s = s[a_3^n]. \tag{15}$$

III   DYNAMICS IN AXIAL GAUGE QCD.

I display the dynamical content of the above expression for the generating functional by discussing a hierarchy of approximations to Z with increasing complexity.

The QCD generating functional in the naive axial (or temporal) gauge is obtained if only the sector without singularities is kept and the dependence on the eigenvalues of the Polyakov loops is disregarded. As a consequence of these approximations, the generating functional becomes actually ill-defined as has been noticed by Schwinger 35 years ago [15]. In definition of propagators certain "$i\epsilon$" prescriptions have to be applied. Due to the approximations, the center-symmetry is not present anymore.

Keeping the zero singularity sector only one might proceed by accounting for the dependence of Z on a_3. The simplest form of these dynamics results, if these variables are treated as Gaussian variables, i.e. if the non-flat measure

$$d\,[a_3] = \prod_{y_\perp} \cos^2\left(gLa_3(y_\perp)/2\right) \Theta\left((\pi/gL)^2 - a_3^2(y_\perp)\right) da_3\,(y_\perp) \qquad (16)$$

is replaced by the flat measure da_3. In this way, one effectively treats the Polyakov loop eigenvalues as the zero modes in QED. It is therefore not surprising that the center-symmetry is lost again and Debye screening like in QED [17] is obtained.

First characteristic properties of QCD are encountered if, still in the absence of singular field configurations, the non-flat measure of the Polyakov loop variables is properly taken into account. These properties will be the subject of the following section. In particular, the perturbative phase reached in this way will be seen to be center-symmetric.

The role of singular field configurations in the $n \neq 0$ sectors (cf.Eq.(14)) is very poorly understood. In particular it has not been possible so far to identify those sectors which dominate the partition function nor has the dynamics of the quantum fluctuations around singular fields been studied systematically. Nevertheless, basic and well understood properties of QCD permit a certain indirect characterization of the dynamics in these sectors (cf. [11]).

A Polyakov Loop Dynamics

Here I discuss in some detail the dynamics in the sector where no singularities are present . Unlike in more standard approaches, the non-Gaussian nature of the Polyakov loop variables $a_3(x_\perp)$ is explicitly taken into account and the finite limit of integration associated with these variables is respected [10]. First the Polyakov loop dynamics in the absence of coupling to the other degrees of freedom is considered. The generating functional is, in the Euclidean, given by

$$Z_0 = \int d\,[a_3] \exp\left\{-1/2 \int d^4x (\partial_\mu a_3(x_\perp))^2\right\} \qquad (17)$$

$$= \int_{-\pi/2}^{\pi/2} \prod_{x_\perp} d\tilde{a}_3\,(x_\perp) \cos^2 \tilde{a}_3\,(x_\perp) \exp\left\{-\frac{2\ell}{g^2 L} \sum_{y_\perp, \delta_\perp} (\tilde{a}_3(y_\perp + \delta_\perp) - \tilde{a}_3(y_\perp))^2\right\} .$$

Transverse space time has been discretized with ℓ and $\delta_\perp$ denoting lattice spacing and lattice unit vectors respectively and the Polyakov loop variables have been rescaled

$$\tilde{a}_3(x_\perp) = gLa_3(x_\perp)/2 .$$

In the continuum limit,

$$\frac{\ell}{g^2 L} \sim \frac{\ell}{L} \frac{1}{\ln \frac{\ell}{L}} \to 0 , \qquad (18)$$

and therefore the nearest neighbor interaction generated by the Abelian field energy of the Polyakov loop variables is negligible. As a consequence, in the absence of coupling to other degrees of freedom, Polyakov loops are ultralocal, they do not propagate,

$$\langle \Omega | T \left(a_3 \left(x_\perp \right) a_3 \left(0 \right) \right) | \Omega \rangle \sim \left(\frac{\ell}{g^2 L} \right)^{x_\perp / \ell} \to \delta^3 \left(x_\perp \right) . \tag{19}$$

Propagation of excitations induced by $a_3(x_\perp)$ can only arise by coupling to the other microscopic degrees of freedom. Ultralocality permits the Polyakov loop variables a_3 to be integrated out. In this way, the following effective action is obtained

$$S_{\text{eff}} \left[A_\mu \right] = S_{\text{YM}} \left[A_\mu, A_3 = 0 \right] + S_{\text{gf}} \left[\int_0^L dz\, A_\mu^3 \right] + M^2 \int d^4 x\, \Phi_\mu^\dagger(x) \Phi^\mu(x). \tag{20}$$

The Polyakov loop variables have left their signature in the geometrical mass term of the charged gluons (cf. Eq.(8))

$$M^2 = (\pi^2/3 - 2)/L^2 \tag{21}$$

and in the antiperiodic boundary conditions (Eq.10). The neutral gluons remain massless and periodic. The antiperiodic boundary conditions reflect the mean value of the Polyakov loop variables, the geometrical mass their fluctuations; notice that in both of these corrections, the coupling constant has dropped out. I emphasize that periodic boundary conditions for the gluon fields are imposed in the representation (14) of the generating functional. The antiperiodic boundary conditions in (10) describe the appearance of Aharonov-Bohm fluxes in the elimination of the Polyakov loop variables. Periodic charged gluon fields may be used if the differential operator ∂_3 is replaced by

$$\partial_3 \to \partial_3 + \frac{i\pi}{2L} [\tau_3 , \quad . \tag{22}$$

As for a quantum mechanical particle on a circle, such a magnetic flux is technically most easily accounted for by an appropriate change in boundary conditions – without changing the original periodicity requirements. With regard to the rather unexpected physical consequences, the space-time independence of this flux is important, since it induces global changes in the theory. These global changes are missed if Polyakov loops are treated as Gaussian variables.

The role of the order parameter is taken over by the neutral color current in 3-direction $u\left(x_\perp\right)$ which is generated by the 3-gluon interaction

$$u \left(x_\perp \right) = i \int_0^L dx_3\, \Phi_\mu^\dagger \left(x \right) \overset{\leftrightarrow}{\partial_3} \Phi^\mu \left(x \right) . \tag{23}$$

This composite field is odd under charge conjugation (cf.(9))

$$C: \quad u(x_\perp) \to -u(x_\perp). \tag{24}$$

It determines the vacuum expectation value of the Polyakov loops

$$\langle \Omega | P(x_\perp) | \Omega \rangle \quad \propto \quad \langle \Omega | u(x_\perp) | \Omega \rangle \tag{25}$$

and the corresponding correlation function

$$\langle \Omega | T[P(x_\perp) P(0)] | \Omega \rangle \quad \propto \quad \langle \Omega | T[u(x_\perp) u(0)] | \Omega \rangle \tag{26}$$

which in turn yields the static quark-antiquark interaction energy [16]. Up to an irrelevant factor we have after rotation to the Euclidean ($r = |x_\perp^E|$)

$$\exp\{-LV(r)\} = \langle \Omega | T \left[u(x_\perp^E) u(0) \right] | \Omega \rangle, \tag{27}$$

i.e., the static quark-antiquark potential is given directly by (the $a = b = 3, \mu = \nu = 3$ component of) the vacuum polarization tensor $\Pi_{\mu\nu}^{ab}$ and not by the zero mass propagator with corresponding self-energy insertions as obtained in the standard Gaussian treatment. This remarkable consequence of the ultralocality property (19) of the Polyakov loop variables provides a direct connection between confinement and certain spectral properties of gluonic states. If, as required in the center symmetric phase, the vacuum expectation value of the Polyakov loop operator vanishes and if the spectrum of states excited by u exhibits a gap ΔE, Eq.(27) implies a linear rise in V for large separations

$$V(r) \to \sigma r = \Delta E r / L . \tag{28}$$

Thus in axial gauge, confinement is connected to a shift in the spectrum of gluonic excitations to excitation energies

$$E \geq \sigma L \tag{29}$$

which diverges with the extension L becoming infinite. Comparison with the interaction energy of adjoint static charges suggests the negative charge conjugation parity (cf.Eq.(24)) of the intermediate "2-gluon" states contributing to V in Eq.(27) to be the distinctive property which leads to infinite excitation energies.

The system described by the effective action (20) exhibits remarkable properties already at the perturbative level. Most importantly the center symmetry is realized in the perturbative vacuum, i.e. in the ground state obtained by dropping all the terms containing the coupling constant g. Geometrical mass (Eq.(21)) and Aharonov-Bohm flux (Eq.(22)) are not affected by such a perturbative treatment. The perturbative ground-state is even under charge conjugation and the expectation value of the Polyakov loop vanishes

$$\langle \Omega_{\mathrm{pt}} | P(x_\perp) | \Omega_{\mathrm{pt}} \rangle = 0, \tag{30}$$

indicating an infinite free energy of a static quark. Indeed perturbative analysis of the correlation function (27) yields a linearly rising static interaction energy. However the perturbative string tension decreases with increasing extension ($\propto L^{-2}$). The change from this value of the string tension to the physical one together with the emergence of the proper QCD scale is beyond a perturbative treatment also after elimination of the Polyakov loop variables and requires treatment of the dynamics of Polyakov loops coupled to singular fields. The perturbative vacuum shares with the QCD vacuum certain properties also after including dynamical quarks. In particular, application of perturbation theory shows the interaction energy of static quarks to cease to rise indefinitely and to be given at asymptotic separations by the non-perturbative value of twice the mass of the dynamical quarks. Unlike in QED, Fermion loops do not yield the "Uehling" potential but describe at this "perturbative" level a string breaking mechanism. For small distances, Coulomb-like behavior must emerge if the separation is small on the scale of Λ_{QCD} and small in comparison with the extension L. This is possible only, if the vacuum polarization tensor possess an essential singularity at infinite momentum

$$\int d^3x \, e^{ipx} \langle \Omega | T \left[u\left(x\right) u\left(0\right)\right] | \Omega \rangle \rightarrow e^{-\sqrt{g^2 Lp/\pi}}. \tag{31}$$

Obviously, finite order perturbation theory cannot yield such a singularity; it however can be shown that, with increasing order in g, increasingly high powers of pL appear; two loop evaluation of the short distance behavior indicates exponentiation.

B   The Deconfined Phase

The perturbative center symmetric phase with its signatures of confinement cannot be relevant for QCD at extensions smaller than $L_c = 1/T_c$ or temperatures beyond the critical temperature T_c. Not only do we expect the center symmetry to be broken at small extensions but also dimensional reduction to QCD_{2+1} to happen. Aharonov-Bohm fluxes induce anti-periodic boundary conditions and therefore yield a decoupling of the charged gluons if dimensional reduction takes place in the center symmetric phase (high temperature confining phase). The small extension or high temperature limit of the center symmetric phase is therefore QED_{2+1}. In order to reach the correct high temperature phase, the deconfinement phase-transition arising when compressing the QCD vacuum, must be accompanied by screening of the Aharonov-Bohm fluxes and simultaneously by a weakening of the increase in the geometrical mass. In the following we shall treat the Aharonov-Bohm fluxes as phenomenological space-time independent quantities (cf. Eq.(7))

$$\chi = gLa_3 + \pi \tag{32}$$

and assume for simplicity the geometrical mass M (cf. Eq.(21)) to vanish in the deconfined phase.

The strength χ is limited by the requirement of thermodynamic stability. By covariance, positive Casimir energy ϵ at finite extension corresponds to a negative pressure of the corresponding system at finite temperature. The expression for the Casimir energy of the charged gluons

$$\varepsilon(L,\chi) = -\frac{4}{\pi^2 L^4} \sum_{n=1}^{\infty} \frac{\cos(n\chi)}{n^4} = \frac{4\pi^2}{3L^4} B_4\left(\frac{\chi}{2\pi}\right), \quad B_4(x) = -\frac{1}{30} + x^2(1-x)^2 \quad (33)$$

yields the following requirement for thermodynamic stability

$$\chi \leq 1.51. \quad (34)$$

Complete screening ($\chi = 0$) of the Aharonov-Bohm fluxes, compatible with thermodynamic stability, is unlikely to take place in the deconfined phase. Such a system will exhibit, like perturbative QCD at zero temperature, a magnetic instability (cf. [18]). Spontaneous formation of magnetic fields on the other hand is prevented in the presence of sufficiently strong Aharonov-Bohm fluxes. Calculation of the Casimir energy for gluons subject to a color-magnetic homogeneous background field

$$A_\mu^a|_{\text{bg}} = \delta^{a3}\delta_{\mu 1} x_2 H . \quad (35)$$

leads, for given extension L or temperature T to a lower limit for χ. Figure 1 displays the region of thermodynamic and magnetic stability. Obviously, due to the requirement of magnetic stability, the Stefan-Boltzmann limit (corresponding to $\chi = 0$) cannot be reached for any finite temperature. Identification of the Aharonov-Bohm flux with the minimal allowed values sets upper limits to energy density and pressure. They are shown in Figure 2 and are reminiscent of lattice data ([19]) in their slow, logarithmic approach to the Stefan-Boltzmann limit. The finite value of the Aharonov-Bohm flux accounts for interactions present in the deconfined phase and specifically gives rise to values of the interaction measure $\epsilon - 3P$ which are in qualitative agreement with lattice data. Other quantities like the Debye screening mass are also affected by the non-vanishing Aharonov-Bohm flux. A one loop calculation

$$m_{el}^2 = \frac{4g^2}{L^2} B_2\left(\frac{\chi}{2\pi}\right), \quad B_2(x) = 1/6 - x(1-x) \quad (36)$$

reproduces qualitatively the approximate linear dependence of m_{el}^2 over a large temperature regime as observed in lattice calculations. Finally these results can also be used to estimate the change in energy density across the phase transition. In this phenomenological treatment the phase transition is accompanied by a change in strength of the Aharonov-Bohm flux from the center symmetric value π to a value bounded by Eq.(34). The resulting limit

$$\Delta\epsilon = \epsilon(L_c, \chi = 1.51) - \epsilon(L_c, \chi = \pi) \leq -\frac{7\pi^2}{180}\frac{1}{L_c^4} \quad (37)$$

once more compares well with the lattice result [20]

$$\Delta\epsilon = -0.45 \frac{1}{L_c^4} \ . \tag{38}$$

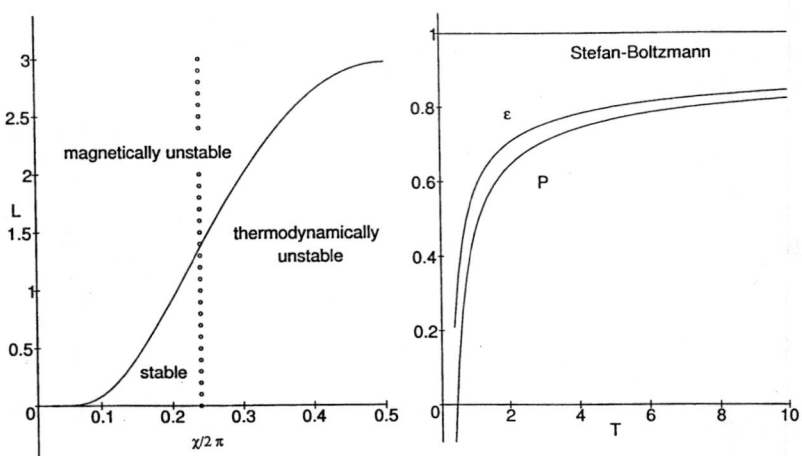

Fig.1: Regions of stability and instability in the (L, χ) plane. To the right of the circles, thermodynamic instability; above the solid line, magnetic instability.

Fig.2: Energy density and pressure normalized to Stefan-Boltzmann values vs. temperature in units of Λ_{MS}.

IV   SUMMARY

The axial gauge representation of QCD provides an appropriate framework for description of the center-symmetric confining phase of QCD. The order parameter variables associated with the realization of the center symmetry, the Polyakov loops, play a central role in this formulation. In this particular gauge, they are elementary rather than composite degrees of freedom. Ultralocality of these variables implies even at the perturbative level a confining interaction between static quarks as well as string breaking by dynamical quarks. For a proper theoretical treatment, coupling of the Polyakov loop variables to singular monopole like configurations is likely to be an essential element of the dynamics. Polyakov loop variables are also very useful for a phenomenological description of the deconfined phase and in particular for a characterization of the interaction effects at high temperatures. Thermodynamic and magnetic stability is not compatible with a system of non- or weakly interacting Polyakov loops. Stability of the deconfined high temperature phase can be achieved only by non-perturbative dynamics.

This work has been supported by the Bundesministerium für Bildung, Wissenschaft, Forschung und Technologie.

REFERENCES

1. L. Susskind, *Phys. Rev. D* **20**, 2610 (1979).
2. L. McLerran and B. Svetitsky, *Phys. Lett. B* **98**, 195 (1981).
3. J. Kuti, J. Polonyi, and K. Szlachanyi, *Phys. Lett. B* **98** , 199 (1981).
4. I. M. Singer, *Comm. Math. Phys.* **60**, 7 (1978).
5. V. N. Gribov, *Nucl. Phys. B* **139**, 1 (1978).
6. S. Mandelstam, *Phys. Rep.* **67**, 109 (1980).
7. G. 't Hooft, *Nucl. Phys. B* **190**, 455 (1981).
8. F. Lenz, H. W. L. Naus, and M. Thies, *Ann. Phys.* **233**, 317 (1994);
   F. Lenz, E. J. Moniz, and M. Thies, *Ann. Phys.* **242**, 429 (1995).
9. V. L. Eletsky, A. C. Kalloniatis, F. Lenz and M. Thies, *Phys. Rev. D* **57**, 5010 (1998).
10. F. Lenz and M. Thies, *Ann. Phys.* **268** 308 (1998).
11. O. Jahn and F. Lenz, *Phys. Rev. D* **58**, 85006 (1998)
12. A. V. Smilga, *Ann. Phys.* **234**, 1 (1994).
13. For a recent review, see K. Kanayo, *Nucl. Phys. B* (Proc. Suppl.) **47**, 144 (1996).
14. T. Reisz, *Z. Phys.* **C53**, 169 (1992).
15. J. Schwinger, *Phys. Rev.* **130**, 402 (1963).
16. B. Svetitsky, *Phys. Rep.* **132**, 1 (1986).
17. N. Weiss, *Phys. Rev.* **D24**, 475 (1981).
18. G.K. Savvidy, Phys.Lett. **71B** (1977) 133.
19. J. Engels, F. Karsch, H. Satz, I. Montvay, Nucl. Phys. **B205**, 545 (1982); Phys. Lett. **101B**, 89 (1981).
20. J. Engels, F. Karsch, and K. Redlich, *Nucl. Phys.* **B435**, 295 (1995).

Hadronic Masses at High Temperature by Lattice QCD Approach

Ph. deForcrand^a, M. García Peréz^b, T.Hashimoto^c, S.Hioki^d,
Y.Liu^e, H.Matsufuru^f, O.Miyamura^{e1} , A.Nakamura^e, T.Umeda^e,
I.O.Stamatescu^g ,T.Takaishi^h

^{a}ETH-Zürich, CH-8092, Switzerland
^{b}Universidad Autónoma de Madrid, E-28049, Spain
^{c}Fukui University, Fukui 910-8507, Japan
^{d}Tezukayama University, Nara 631-8501, Japan
^{e}Hiroshima University, Higashi-Hiroshima 739-8521 , Japan
^{f}RCNP, Osaka University, Osaka 567 , Japan
^{g}FEST,Schmeilweg 5, D-69118 Heidelberg, Germany
^{h}Hiroshima University of Economics, Hiroshima 731-01 , Japan

Abstract. Two new measurements of hadronic masses at high temperature are presented. One is an investigation for temporal hadronic mass at high temperature by anisotropic lattice with optimized hadronic operator. Temporal and screening masses for $0^\pm$ and $1^\pm$ mesons in quench approximation are presented. In addition, spatial structure of the hadronic operator is examined. The other is a study for response of hadronic masses and couplings to chemical potential at high temperature. Method and preliminary results for light-light and heavy-light 0^- mesons are presented.

INTRODUCTION

Dynamical mass of hadrons at finite temperature is one of the most interesting quantities but definitive result has not yet been given by Lattice Quantum Chromo Dynamics (Lattice QCD). Based on measurements on hadronic screening mass, restoration of chiral symmetry and liberation of quark are discussed. [1] However, relation to the dynamical mass is not well-known. A pioneering work for direct measurement of dynamical mass has given several interesting features of hadrons and deconfined quark but further extensive analysis seems difficult. [2] In principle, if we are successful to determine spectral function it gives corresponding dynamical green function. However, determination of spectral function from thermal hadronic correlators belongs to a class of ill-posed problem. There are several proposals to overcome this difficulty. Utilization of maximum entropy method which gives most

---

[1] Talk presented by O.Miyamura

CP494, *New Directions in Quantum Chromodynamics*, edited by C.-R. Ji and D.-P. Min
© 1999 American Institute of Physics 1-56396-908-4/99/$15.00

probable solution has recently been suggested. [3–5] Here we take another approach in which temporal resolution is kept fine and hadronic operator is optimized.

The other topic discussed in this work is response of hadron mass to chemical potential. As is well known, there is difficulty for simulation of finite density system by Lattice QCD approach. The reason is that fermionic determinant becomes complex and it gives oscillating behavior in quantum average. This makes simulation hard to give reliable results. It has been also known that quench approximation leads to essentially different world. [6] In spite of such difficult situation, density effects to hadrons such as mass-shift are very interesting and important subjects both theoretically and experimentally. [7,8] There are several approach to circumvent this difficulty and they seems successful in limited extent. [9,10] As for response to the chemical potential, baryon number susceptibility at zero baryon density has been studied and abrupt jump at the transition point has been reported. [11] In this paper, we examine response to the chemical potential for mass and coupling of hadrons at finite temperature with zero baryon density. [12]

STUDY OF TEMPORAL HADRONIC MASSES BY ANISOTROPIC LATTICE WITH OPTIMIZED HADRONIC OPERATORS

Anisotropic lattice

On lattices at finite temperature, temporal extension is

$$a_t N_t = \frac{1}{T}. \tag{1}$$

while thermo-dynamical limit requires $a_s N_s \gg a_t N_t$. Here a_t (a_s) and N_t (N_s) are lattice spacing and number of lattice site in temporal (spatial) direction. In this situation, if we want to keep good resolution in temporal direction, anisotropic lattice is a reasonable approach. There, spatial resolution is coarse compared with temporal one, $a_s > a_t$. Such formulation is called anisotropic lattice. [13] On an anisotropic lattice , gluonic action and fermion operator are given by

$$S_G = \beta[\gamma_G^{-1} \sum_{s>s'} (1 - \frac{1}{3} ReTr(Pss')) + \gamma_G \sum_s (1 - \frac{1}{3} ReTr(Pst))] \tag{2}$$

and

$$D = \frac{1}{2\kappa_\sigma}[1 - \kappa_\sigma(\sum_s (\Gamma_s^- U_s T_s^- + \Gamma_s^+ U_s^\dagger T_s^+) + \gamma_F(\Gamma_t^- U_t T_t^- + \Gamma_t^+ U_t^\dagger T_t^+))] \ , \tag{3}$$

where $P_{ss'}$ and P_{st} are spatial-spatial and spatial-temporal plaquettes and $\Gamma_\mu^\pm = 1 \pm \gamma_\mu$ and $(T_\mu^\pm)_{nm} = \delta_{n,m\pm\mu}$. The parameters γ_G, γ_F control ratio of lattice spacings $\xi \equiv a_s/a_t$. In classical level, they coincide but they can be different due to quantum correction. Therefore we determine ξ and γ_F to be consistent with Euclidian symmetry at given γ_G. Fig.1 shows matching method in which spatial-spatial Wilson loops are adjusted to spatial-temporal ones by rescaling: $t \rightarrow \xi t$.

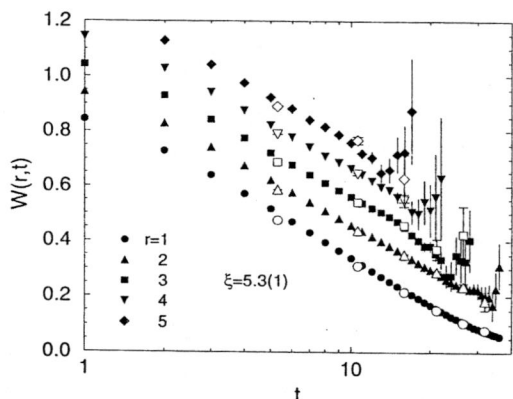

FIGURE 1. Matching method for determination of $\xi = a_s/a_t$. Filled symbols are spatial-temporal Wilson loops: $W_{st}(r,t)$. Open symbols are spatial-spatial Wilson loops: $W_{ss}(r,\xi t)$.

This method is used to fix anisotropy of the lattice at zero temperature. At finite temperature, nonperturbative determination of ξ needs more procedures. A possible way is to compare spatial and temporal quantities of both isotropic and anisotropic lattices. On the other hand, in perturbation theory , the quantum correction to lattice spacing is infra-red insensitive quantity and thus does not vary at temperatures lower than inverse of lattice spacings. In this work, we fix ξ at zero temperature and use the same value for lattices at finite temperature.

Optimization of hadronic operators

In order to find the lowest hadronic excitations, measurement of correlator of hadronic operators is a standard way. With enough temporal separation , exponential tail gives mass and coupling as

$$C(t) = \sum_{x,y,z} < H(x,y,z,t)H(0,0,0,0)^\dagger > \approx Ae^{-mt} \quad . \tag{4}$$

where H is a relevant hadronic operator. At finite temperature, temporal extension is limited by eq.(1) and we can not keep enough separation. Therefore straight-forward application of the method does not work. In order to circumvent this difficulty, hadronic operators are optimized to be dominated by the lowest state.

This can be done at $T = 0$. Then , at finite temperature, we examine the correlator precisely and detect the temperature dependent effects. Actually we use following form for the correlator after Coulomb gauge fixing,

$$G_i(r,t) = \sum_{\vec{m_1},\vec{m_2},\vec{n}} w(\vec{m_1})w(\vec{m_2}) < Tr[M(\vec{m_1},0;\vec{n},t)\Gamma_i\gamma_5 M(\vec{m_2},0;\vec{n}+\vec{r},t)^\dagger \gamma_5 \Gamma_i^\dagger >$$

(5)

where $\Gamma_i = \gamma_5, \gamma_1, 1, \gamma_1\gamma_5$ for $0^-, 1^-, 0^+, 1^+$ mesons and $M(\vec{m}, m_0; \vec{n}, n_0)$ is a quark propagator from $(\vec{n}, n_0)$ to $(\vec{m}, m_0)$. $w(\vec{m})$ is either $\delta(\vec{m})$ (point) or $exp(-\alpha|\vec{m}|^p)$ (exp) where α and p are used as optimization parameters. Fig.2 shows an example of the optimization for 0^- meson on $12^3 \times 72$ lattice where $\xi = 5.3(1)$ at $\gamma_G = 4$. The effective mass $m_{eff}(t)$ is defined through $G(0,t) = Acosh(m_{eff}(t)(t - N_t/2))$. A plateau sets on precautiously in the case of optimized operator with $\alpha = 0.379$ and $p = 1.289$ while usual point hadronic operator shows approximate plateau for $t > 18$.

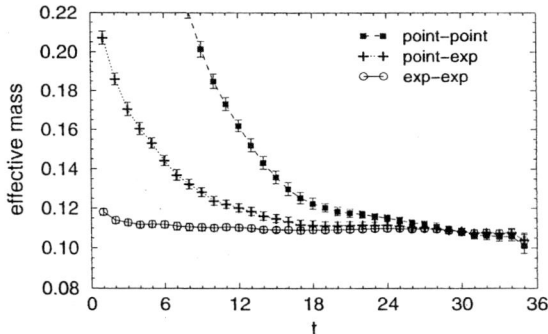

FIGURE 2. Effective mass of 0^- meson by the optimized operator. In case of "exp", $\alpha = 0.379$ and $p = 1.289$.

Simulations and results

Simulations are carried out on $12^3 \times N_t$ lattices at $\beta = 5.68$. Bare anisotropy parameter is taken as $\gamma_G = 4$. We use quench approximation in this study. On the $N_t = 72$ lattice, lattice spacings are determined as $a_s/a_t = 5.3(1)$ and $a_t = 0.044(1)fm$ based on analysis of static quark potential. At finite temperature , N_t is chosen to cover $T = 0.93$ to $T = 1.5T_c$ as

$$N_t(T/T_c) = 72(0.0), 20(0.93), 16(1.15), 12(1.5) \quad .$$

(6)

It is noted that phase transition point is slightly above $1/(18a_t)$. For the fermionic part, we tune fermionic anisotropy parameter γ_F so as to reproduce the same a_s/a_t

in the hadronic correlator. Quark mass is controlled by the κ_σ parameter and it is chosen to be 0.081 ,0.084 and 0.086 . Those values correspond to middle quark mass. At each N_t , 60 measurements per data are performed.

Results are shown in Fig.s 3-5. Effective masses (in unit of a_t) of pseudoscalar and vector mesons at $N_t = 20$, 16 and 12 is presented in Fig.3.

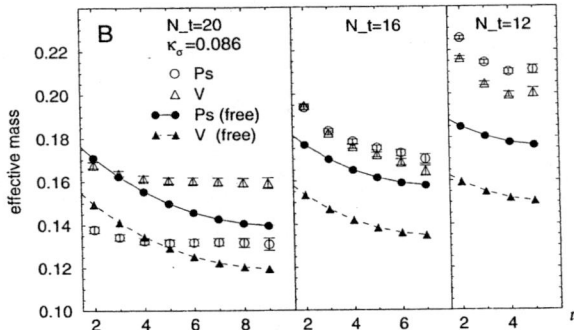

FIGURE 3. Effective masses at finite temperature by optimized hadronic operators. $\kappa_\sigma = 0.086$. Filled symbols and curves are those calculated by free quark propagators.

At $N_t = 20(T/T_c = 0.93)$, there is a plateau and good signal of ground states are found. However, above the transition ($N_t = 16$ and 12), no plateau is seen. For comparison, correlators composed of free quark are shown. The data of $N_t = 16$ and 12 look similar to the latter. This result seems consistent with deconfining of ground state mesons on this lattice. For further investigation, we examine "wave function" $G(r,t)/G(0,t)$ as a function of spatial separation between quark and antiquark, r.

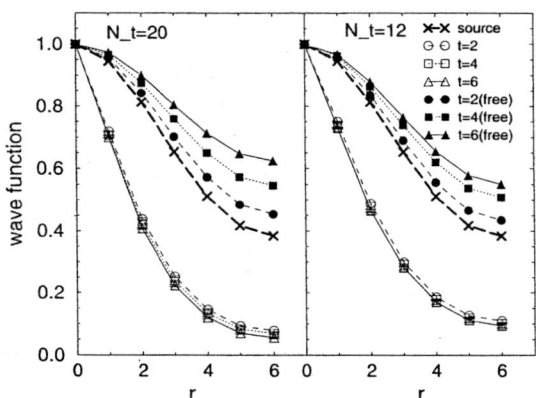

FIGURE 4. $G(r,t)/G(0,t)$ for 0^- meson. $\kappa_\sigma = 0.086$. "source" means distribution: $\int d\vec{y}w(\vec{y} + \vec{r})w(\vec{y})$.

Fig.4 shows "wave function"s of pseudoscalar meson at $t = 2,4$ and 6. At $N_t = 20$, the data of different t converge to a localized shape while free quark gives spreading of "wave function" as t increases (filled symbols). Surprisingly, similar tendency persists at $N_t = 12$. Thus, even at $T = 1.5T_c$, quark and antiquark prefer to stay together.

Difference between spatial and temporal correlators is also studied. Fig.5 shows masses in unit of lattice spacing. Here, temporal mass on $N_t = 16$ and 12 lattices is defined by averaging lowest three points in the effective mass. Chiral extrapolation is made.

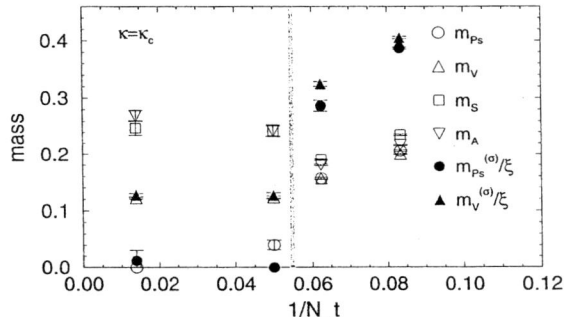

FIGURE 5. Temporal and screening masses below and above phase transition. Chiral extrapolation is made. Filled symbols are for the shieling masses.

The temporal masses clearly show feature of chiral symmetry restoration. Pseudoscalar meson becomes massive and parity partners degenerate above T_c. Although such feature has been suggested by measurement of screening mass [1], we observe it directly in temporal masses. In addition, difference between temporal mass and screening mass is manifest in the data. The latter rises strongly in the deconfined phase. It is noted that such difference has been suggested by Nambu-Jona-Lasinio model. [14] We also note that in the deconfined phase, all measured masses approximately degenerate irrespective of spin and all screening masses roughly agree with $2\pi/N_t \approx 0.39$ for $N_t = 18$. This seems consistent with liberated quarks. On the other hand, as shown before, "wave function" indicates attractive correlation between quark and antiquark. Apparently, this interesting situation requires more investigations and deeper understanding. Finally, a remark is added. Since the present lattice is quenched one, it undergoes the first order transition. In real situation, quark loop effects may soften abrupt change of hadronic masses near the transition. This point will be clarified by simulation with dynamical quarks.

CHEMICAL POTENTIAL RESPONSE OF HADRONIC MASSES AND COUPLINGS AT HIGH TEMPERATURE

As stated in the introduction, lattice study at finite density is still difficult. In this situation, a possible way is to examine response of physical quantities to chemical potential. Here we consider hadronic correlators and examine chemical potential response of them.

Basic frame work is as follows. Suppose that a hadronic correlator is dominated by a single pole,

$$G(x) = \sum_{y,z,t} < H(x,y,z,t)H(0,0,0,0)^\dagger > \approx Ae^{-mx} \quad . \tag{7}$$

Then we take derivative with respect to chemical potential μ,

$$G(x)^{-1}\frac{\partial G(x)}{\partial \mu} = A^{-1}\frac{\partial A}{\partial \mu} - \frac{\partial m}{\partial \mu}x \quad . \tag{8}$$

In eq.(8), if the left hand side is measured as a function of x , the linear term gives chemical potential response of mass while the constant term gives response of coupling.

Next problem is how to get the derivative of the correlator . For this purpose, we go back to definition of the hadronic correlator. For simplicity of explanation, we consider flavor non-singlet meson channel.

$$< H(n)H(0)^\dagger >= Z^{-1} \int [dU]Tr(M(n;0)\Gamma M(0;n)\Gamma^\dagger)det(D)exp(-S_G) \tag{9}$$

where $Z = \int[dU]det(D)exp(-S_G)$ and $M = D^{-1}$. Then, using formulae

$$\frac{\partial det(D)}{\partial \mu} = Tr(M\frac{\partial D}{\partial \mu})det(D) \text{ and } \frac{\partial M}{\partial \mu} = -M\frac{\partial D}{\partial \mu}M \quad , \tag{10}$$

we get

$$\frac{\partial < H(n)H(0)^\dagger >}{\partial \mu} = - < Tr(M\dot{D}M\Gamma M\Gamma^\dagger) > - < Tr(M\Gamma M\dot{D}M\Gamma^\dagger) >$$
$$+ < Tr(M\Gamma M\Gamma^\dagger)Tr(\dot{D}M) > - < Tr(M\Gamma M\Gamma^\dagger) >< Tr(\dot{D}M) > \tag{11}$$

where short handed notations ,

$$\dot{X} = \frac{\partial X}{\partial \mu} \quad \text{and} \quad < Y >= Z^{-1} \int [dU]det(D)exp(-S_G)Y \tag{12}$$

are used. Note that for flavor singlet scalar meson, hair-pin diagrams also contribute to eq.(11).

In the case of staggered fermion formalism, derivative of the fermion operator is

$$\frac{\partial D_{KS}}{\partial \mu} = \frac{a}{2}[U_t(n)e^{a\mu}\delta_{n+t,m} + U_t^\dagger(n-t)e^{-a\mu}\delta_{n-t,m}].$$ (13)

At $\mu = 0$, we can evaluate eq.s (9) and (11) by usual technique of lattice QCD simulation.

Here we present preliminary results for $N_f = 2$ (u and d quarks). In this case, we have two independent chemical potentials, μ_u and μ_u. Instead, following combinations are convenient,

$$\mu_S = (\mu_u + \mu_d)/2 \quad \text{and} \quad \mu_V = (\mu_u - \mu_d)/2$$ (14)

where μ_S is usual chemical potential corresponding to baryon number. Then derivatives with respect to μ_S and μ_S are

$$\frac{\partial}{\partial \mu_S} = \frac{\partial}{\partial \mu_u} + \frac{\partial}{\partial \mu_d} = \frac{\partial}{\partial \mu_u} - \frac{\partial}{\partial \mu_{\bar{d}}} \quad \text{and} \quad \frac{\partial}{\partial \mu_V} = \frac{\partial}{\partial \mu_u} - \frac{\partial}{\partial \mu_d} = \frac{\partial}{\partial \mu_u} + \frac{\partial}{\partial \mu_{\bar{d}}} \; .$$ (15)

Two remarks are noted. For degenerate system of u and d quarks,

$$\frac{\partial G_{u\bar{d}}}{\partial \mu_S} = \frac{\partial G_{u\bar{d}}}{\partial \mu_u} - \frac{\partial G_{u\bar{d}}}{\partial \mu_{\bar{d}}} = 0 \; .$$ (16)

at $\mu_u = \mu_d = 0$. On the other hand, for light l quark and heavy H antiquark system,

$$\frac{\partial G_{l\bar{H}}}{\partial \mu_S} = \frac{\partial G_{l\bar{H}}}{\partial \mu_V} = \frac{\partial G_{l\bar{H}}}{\partial \mu_l}$$ (17)

holds in the heavy quark limit of H.

Response of mass and coupling for pseudoscalar mesons

Here we present preliminary results for $N_f = 2$ staggered fermion on $16 \times 8^2 \times 4$ lattice. Measurement is done for screenings mass of hadrons. Parameters of simulation are

$$\beta(T/T_c) = 5.20(0.90), 5.26(0.97), 5.32(1.06), 5.34(1.09)$$ (18)

Staggered quark mass parameter is taken as $ma = 0.025$. Hybrid MonteCarlo algorithm is used to generate configuration and a hundred measurements per data are carried out. At present, only response with respect to μ_V is presented. Data of μ_S are noisy and require more statistics. In case of light-heavy pseudoscalar meson , light quark mass parameter is $ma = 0.025$ while $am = 0.25$ is taken for the heavy quark.

Results on response are shown in Fig.6. As seen in the figure, response of light-light mass below T_c is small while it rises above T_c. For light-heavy meson, the response is larger and reveals some increase above T_c. As for the response of coupling, feature is different. Both Light-light and heavy-light systems show similar value and it is rather insensitive to the phase.

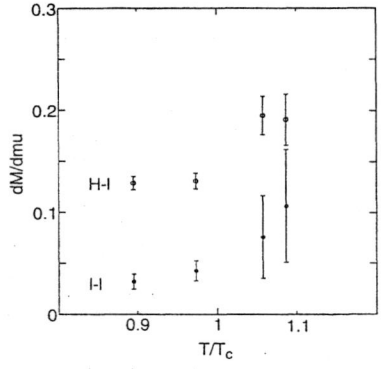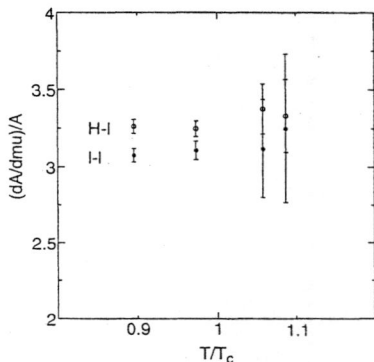

FIGURE 6. Response of mass (left) and coupling(right) to μ_V for light-light (filled circle) and heavy-light (open circle) 0^- mesons. $am = 0.025$ for light quark and $am = 0.25$ for heavy quark.

Although present results are preliminary, we find remarkable characteristics of the chemical potential response. Possible interpretations for light-light system are as follows. Weak response of mass at low βs indicates persistency of Nambu-Goldstone boson nature at least up to $T = 0.97T_c$. Growth of it above T_c is consistent with chiral restoration since the meson loses the Nambu-Goldstone boson character. For heavy-light system, response of mass is sizable strength in confinement phase. As remarked before, response to the vector chemical potential coincides with that to the scalar chemical potential in the heavy quark limit. Thus , this is an indicative result for mass-shift induced by density effect for heavy-light system.

SUMMARY AND DISCUSSIONS

In this work, two new measurements of hadronic masses at high temperature are discussed. Study for hadronic mass at high temperature by anisotropic lattice with optimized hadronic operator is carried out. Temporal and screening masses for $0^\pm$ and $1^\pm$ mesons in quench approximation show manifest difference. Chiral restoration in high temperature phase is clearly seen in temporal mass spectrum. There is an indication that strong correlation between quark and antiquark remains. A study for chemical potential response of hadronic masses and couplings at high temperature is performed. Results for light-light and heavy-light pseudoscalar mesons are presented. The response of mass is considerably different for light-light and light-heavy pseudoscalar mesons.

ACKNOWLEDGMENTS

We thank JSPS,DFG and Europian Network "Finite Temperature Phase transitions in Particle Physics" for support. This work is supported also by Grant-in-Aide for Scientific Research by Monbusho, Japan (11694085). H.M. thanks T.Kunihiro for discussions. Simulations have been carried out at Fujitsu Parallel Comp. Res. Center and INSAM, Hiroshima University.

REFERENCES

1. C.DeTar and J.Kogut, *Phys. Rev.* **D36** (1987) 2828; C.DeTar, *Phys. Rev.* **D36** (1988) 2328.
2. G. Boyd, Dourendu Gupta and F. Karsch , *Z.f. Physik* **C64** 331 (1994).
3. Ph. de Forcrand et al., QCDTARO-collaboration , *Nucl.Phys.* **B(Proc.Suppl.)63A-C** 460 (1998).
4. E.G.Klepfish, C.E.Creffield and E.R.Pike, *Nucl.Phys.* B(Proc.Suppl.)63A-C 655 (1998).
5. Y.Nakamura, M.Asakawa and T.Hatsuda, "Hadronic Spectral Functions in Lattice QCD", hep-lat/9905034.
6. M.A.Stephanov, *Phys.Rev.Lett.* **76** 4472 (1996).
7. T.Hatsuda and S.H.Lee, *Phys. Rev.* **D46** R34 (1992).
8. G.Agakishev et al., CERES collaboration, Nucl. Phys. **A638** 159 (1998).
9. I.M.Barbour, S.E.Morrison, E.G.Klepfish, J.B.Kogut, and M.Lombardo, *Nucl.Phys.* **B(Proc.Suppl.)60A** 220 (1998).
10. O.Kaczemarek, J.Engels, F.Karsch, E.Laermann, "Lattice QCD at Nonzero Baryon Number", hep-lat/9905022.
11. S.Gottlieb et al., Phys. Rev. **D55** 6852 (1997)
12. See also Ph. deForcrand et al., QCDTARO-collaboration , *Nucl.Phys.* **B(Proc.Suppl.)73** 477 (1999).
13. G.Berger, F.Karsch, A.Nakamura and I.O.Stamatescu, *Nucl.Phys.* **B304** 587 (1987)
14. See T.Hatsuda and T.Kunihiro, Phys. Rep. **C247** 221 (1994).

A picture of the Yang-Mills deconfinement transition and its lattice verification[1]

M. Engelhardt[2], K. Langfeld, H. Reinhardt[3] and O. Tennert

Institut für theoretische Physik, Universität Tübingen
Auf der Morgenstelle 14, 72076 Tübingen, Germany

Abstract. In the framework of the center vortex picture of confinement, the nature of the deconfining phase transition is studied. Using recently developed techniques which allow to associate a center vortex configuration with any given lattice gauge configuration, it is demonstrated that the confining phase is a phase in which vortices percolate, whereas the deconfined phase is a phase in which vortices cease to percolate if one considers an appropriate slice of space-time.

HEURISTICS OF THE CENTER VORTEX PICTURE

A discussion of the deconfinement transition in Yang-Mills theory presupposes a picture of the phenomenon of confinement. Conversely, any picture of confinement should be able to accomodate the deconfinement phase transition. The work presented here is concerned specifically with the so-called center vortex picture of confinement; this picture is based on the conjectured presence of center vortices in typical Yang-Mills gauge configurations. These vortices represent closed magnetic flux lines in three space dimensions, describing closed two-dimensional world-sheets in four space-time dimensions. Space-time in the following will always be considered Euclidean. The magnetic flux represented by the vortices is furthermore quantized such that a Wilson loop linking vortex flux takes a value corresponding to a non-trivial center element of the gauge group. In the case of $SU(2)$ color discussed here, the only such element is (-1). For N colors, there are $N-1$ different possible vortex fluxes corresponding to the $N-1$ nontrivial center elements of $SU(N)$.

Consider an ensemble of center vortex configurations in which the vortices are distributed randomly, specifically such that intersection points of vortices with a given two-dimensional plane in space-time are found at random, uncorrelated

---

[1] Invited talk presented by M.Engelhardt.
[2] Supported by Deutsche Forschungsgemeinschaft under DFG En 415/1-1.
[3] Supported by Deutsche Forschungsgemeinschaft under DFG Re 856/4-1.

CP494, *New Directions in Quantum Chromodynamics*, edited by C.-R. Ji and D.-P. Min

locations. In such an ensemble, confinement results in a very simple manner. Let the universe be a cube of length L, and consider a two-dimensional slice of this universe of area L^2, with a Wilson loop embedded into it, circumscribing an area A. On this plane, distribute N vortex intersection points at random, cf. Fig. 1 (left). According to the specification above, each of these points contributes a factor (-1) to the value of the Wilson loop if it falls within the area A spanned by the loop; the probability for this to occur for any given point is A/L^2.

The expectation value of the Wilson loop is readily evaluated in this simple model. The probability that n of the N vortex intersection points fall within the area A is binomial, and, since the Wilson loop takes the value $(-1)^n$ in the presence of n intersection points within the area A, its expectation value is

$$\langle W \rangle = \sum_{n=0}^{N} (-1)^n \binom{N}{n} \left(\frac{A}{L^2}\right)^n \left(1 - \frac{A}{L^2}\right)^{N-n} = \left(1 - \frac{2\rho A}{N}\right)^N \xrightarrow{N \to \infty} \exp(-2\rho A)$$

(1)

where in the last step, the size of the universe L has been sent to infinity while leaving the planar density $\rho = N/L^2$ of vortex intersection points constant. Thus, one obtains an area law for the Wilson loop, with the string tension $\sigma = 2\rho$.

This simple mechanism lies at the core of the center vortex picture of confinement. After having been proposed already in [1,2], evidence that the Yang-Mills dynamics actually favors the formation of magnetic flux tubes arose in the framework of the Copenhagen vacuum [3]. Also lattice studies were initiated with the aim to study vortices [4,5]. These studies in essence defined vortices via their effect on Wilson loops, as discussed above. While this definition has the advantage of being gauge invariant, it does not allow to easily localize vortices, i.e. associate a collection of vortex world-surfaces with any given lattice gauge configuration.

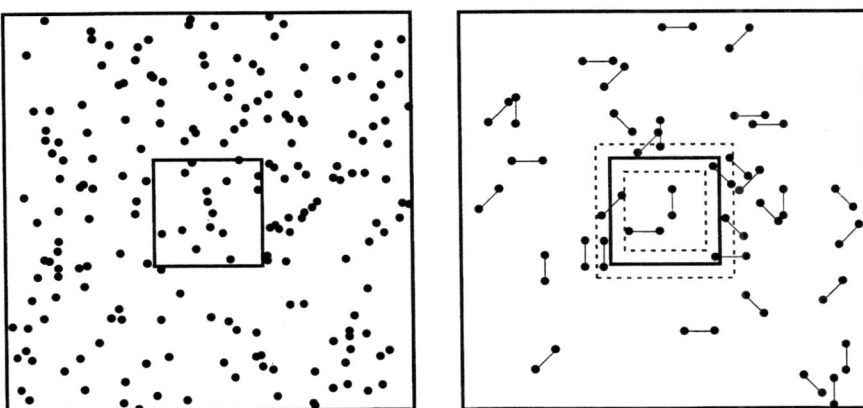

FIGURE 1. Simple models for confining (left) and deconfining (right) vortex ensembles.

The absence of techniques allowing to carry out such an identification for a long time posed a considerable obstacle to the study of center vortex physics, especially the study of their global properties. These properties, however, constitute a crucial aspect for many applications, as a closer examination of the above heuristic picture shows. Namely, for vortex intersection points to be distributed in a sufficiently random manner on a space-time plane to induce an area law for the Wilson loop, the vortices must form networks which percolate throughout space-time. To see this, consider the converse, namely that vortices can be separated into clusters of bounded extension. This implies that any vortex intersection point on a plane comes with a partner a finite distance (smaller than the bound on the cluster extension) away, because vortices are closed. For simplicity, assume the pairs of intersection points to occur with a fixed mutual distance d, and distribute N pairs on a space-time plane containing a Wilson loop of area A, cf. Fig. 1 (right), where the lines between the points in the figure are merely to guide the eye in identifying pairs of points. Now, the probability that any given pair contributes a factor (-1) to the Wilson loop is pPd/L^2, where P denotes the perimeter of the loop, since only pairs whose midpoints lie within a strip of width d around the Wilson loop are able to contribute a factor (-1), and they do this with a probability p related to the angular distribution of the pairs. Note that p is independent of the dimensions of the Wilson loop. The probability that n pairs contribute a factor (-1) is again binomial, in complete analogy to above, and one consequently obtains a perimeter law for the expectation value of the Wilson loop in the limit of an infinite universe,

$$\langle W \rangle = \left(1 - \frac{2pPd}{L^2}\right)^N \stackrel{N \to \infty}{\longrightarrow} \exp(-\rho pPd) \qquad (2)$$

where $\rho = 2N/L^2$ again denotes the density of points. Thus, in the absence of percolation, confinement disappears. This leads to the conjecture that the deconfinement phase transition in the vortex picture may take the guise of a percolation transition. However, as already indicated above, to test such global properties of vortices in lattice experiments, new techniques are needed which allow to associate a vortex world-sheet configuration with any given lattice gauge configuration. These techniques have only been furnished quite recently, sparking renewed interest in the vortex picture. The present work is one contribution to these efforts.

LOCATING VORTICES ON THE LATTICE

The abovementioned techniques, introduced in [6–8], employ a two-step procedure familiar from the dual superconductor picture of confinement. First, one uses the gauge freedom to bring a given gauge configuration as close as possible to the collective degrees of freedom under consideration; in the case of the dual superconductor, that is the Abelian degrees of freedom, in particular, the monopoles. The second step consists of projecting onto these degrees of freedom, i.e. neglecting residual deviations away from, say, Abelian configurations in the case of the dual superconductor. This second step clearly constitutes a truncation of the theory.

This idea was adapted to the case of vortex degrees of freedom as follows [6–8]. One fixes gauge configurations to the *maximal center gauge*,

$$\max \sum_i |\text{tr } U_i|^2 \tag{3}$$

where the U_i are the link variables on a space-time lattice. This procedure biases links towards elements of the center of the gauge group. Next, one performs a truncation of the configurations, namely *center projection*,

$$U \longrightarrow \text{sign tr } U \tag{4}$$

i.e. one replaces each $SU(2)$ link variable by the center element closest to it in the group. Thus, one remains with a lattice of center elements. Such a lattice can be associated in the standard fashion with a vortex configuration. One examines all plaquettes on the lattice, and if a plaquette takes the value (-1), a vortex is said to pierce that plaquette. Thus, vortices in the lattice formulation are defined on the dual lattice, i.e. the lattice shifted by the vector $(a/2, a/2, a/2, a/2)$ w.r.t. the original one, a denoting the lattice spacing. One can easily convince oneself that the vortices defined in this way have all the properties postulated further above.

Having isolated vortices on the lattice, the first question to answer is whether these degrees of freedom do indeed determine the physics of confinement, i.e. whether they furnish the full string tension found in exact calculations without any truncations. Without this basis, more detailed considerations of vortex properties run the risk of being academical. One carries out two lattice experiments, both times using the full Yang-Mills action as a weight, but in one experiment, one calculates the observable in question, such as the Wilson loop, using the full configurations; in the other experiment, one uses the center projected configurations. If the results agree, the observable is said to display *center dominance*. Center dominance for the string tension has indeed been verified in $SU(2)$ lattice gauge theory both at zero temperature [6–8] and at finite temperatures [9,10], including the so-called "spatial string tension" all the way into the deconfined regime. Furthermore, the vortex density obeys the proper scaling law as dictated by the renormalization group for physical quantities, cf. [11] (note erratum in [9]) and [8].

VORTEX PERCOLATION PROPERTIES

Given techniques allowing to locate vortex world-sheets in space-time, or vortex loops on three-dimensional slices thereof, it is possible to discriminate between different vortex clusters. In the following, three-dimensional slices of space-time, where one of the space directions is left away, will be considered, since this displays the relevant percolation properties most clearly. To define a cluster, one finds a link on the dual lattice which is part of a vortex and furthermore locates all adjacent links which are also part of the vortex. This is repeated with all new links found, until no further links exist which are connected with the cluster in question. Having detected all vortex clusters in this manner, it is possible to determine the space-time extension of each cluster, i.e. the largest distance between any pair of points on the cluster. In a percolating phase, most of the available vortex length will be organized into clusters of the maximal possible extension, whereas in a phase with no vortex percolation, most of the vortex material present in the configuration will be concentrated in clusters much smaller than the typical extension of the universe.

To generate "vortex material distributions" which allow to read off which scenario is realized, one simply measures both the extension of each cluster as well as the number of links contained in it, and adds the latter number to a bin corresponding

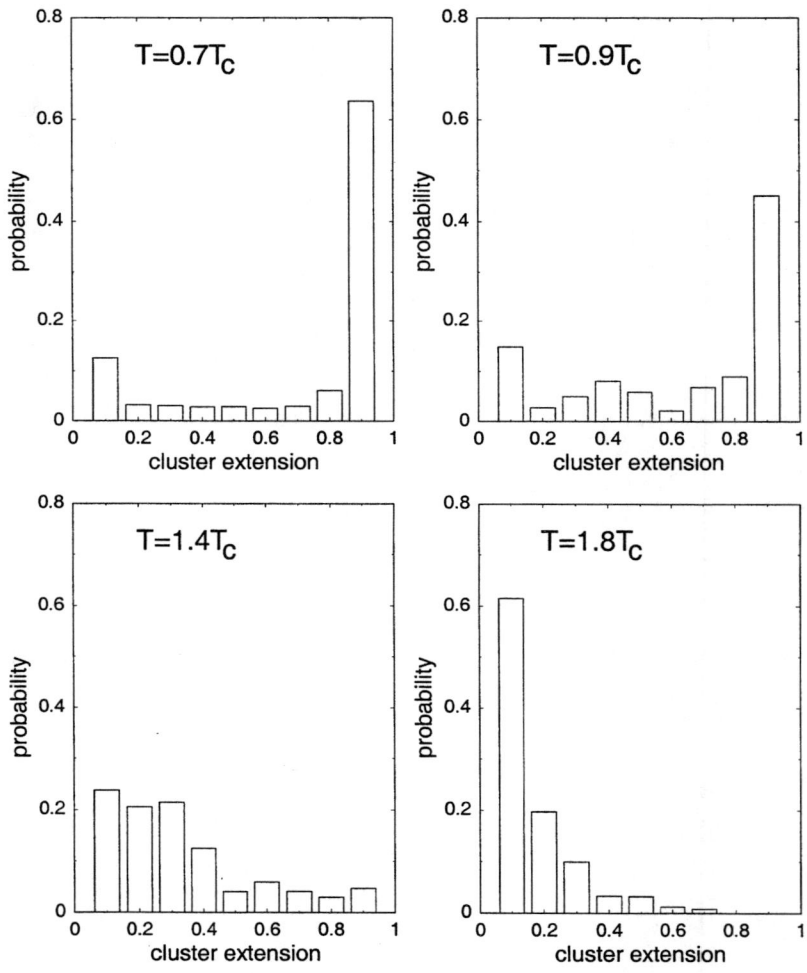

FIGURE 2. Vortex material distributions.

to the cluster extension in question. Fig. 2 displays such distributions, obtained for $\beta = 2.4$ on $12^3 \times N_t$ lattices [10], which have been normalized such that the integral over the distributions gives unity, and where the cluster extension on the horizontal axis is in units of the maximal extension possible in the universe in question. In view of Fig. 2, one indeed obtains a transition from a confining phase, in which vortices percolate, to a deconfining phase, in which they cease to percolate. This confirms the conjecture proposed above in the introductory section. If one analyzes the small vortex clusters dominating the deconfined phase in more detail, one finds that a large part of these vortices wind in the (Euclidean) temporal direction, i.e. the space-time direction whose extension is identified with the inverse temperature. Therefore, one finds that the typical configurations in the two phases can be characterized as displayed in Fig. 3 in a three-dimensional slice of space-time, where one space direction has been left away. Note that Fig. 3 also furnishes an explanation of the spatial string tension in the deconfined phase. A spatial Wilson loop embedded into Fig. 3 (right) can exhibit an area law, since intersection points of winding vortices with the minimal area spanned by the loop can occur in an uncorrelated fashion despite those vortices having small extension. Note also the dual nature of this (magnetic) picture as compared with electric flux models [12]. In such models, electric flux percolates in the *deconfined* phase, while it does not percolate in the confining phase.

OUTLOOK

While it has thus been established *how* vortices generate the confining and deconfining phases of Yang-Mills theory, it remains to be clarified what the essential features of the *dynamics* underlying their behavior are. One interesting observation in this context is that a simple model of vortices as random surfaces in

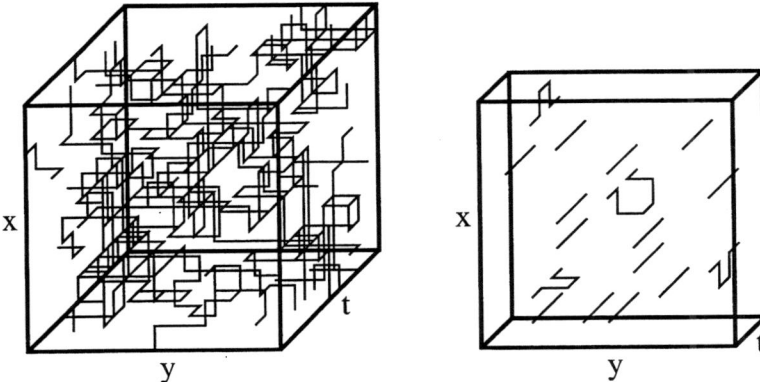

FIGURE 3. Typical vortex configurations in the confining (left) and the deconfined phase (right).

four-dimensional space-time already is able to generate the vortex phenomenology described above, i.e. a percolating confining and a non-percolating deconfining phase, separated by a transition as a function of temperature. The necessary ingredients are an action per unit vortex area (i.e. a Nambu-Goto term), and an action penalty related to the curvature of the vortex surfaces. By construction, this model can be understood in terms of the entropy associated with random surfaces in a given space-time domain; it contains no further dynamics. Evaluating the partition function of such a model amounts to counting possible vortex surface configurations given a certain vortex density (enforced by the Nambu-Goto term), and given an ultraviolet cutoff on the space-time fluctuations of the surfaces (enforced by the curvature penalty). A detailed report on a lattice investigation of this model will be given in an upcoming publication.

Further issues being, or recently having been, investigated include: The Pontryagin index associated with center vortex configurations [13,14], and the breaking of chiral symmetry [13]; the continuum meaning of the maximal center gauge [14]; generalizations to $SU(3)$ color [8,15]; and whether a random surface model for vortices can be justified in terms of a low-energy effective theory describing infrared Yang-Mills dynamics [14].

REFERENCES

1. 't Hooft, G., *Nucl. Phys.* **B138**, 1 (1978).
2. Cornwall, J.M., *Nucl. Phys.* **B157**, 392 (1979).
3. Nielsen, H.B., and Olesen, P., *Nucl. Phys.* **B160**, 380 (1979);
   Ambjørn, J., and Olesen, P., *Nucl. Phys.* **B170** [FS1], 60, 265 (1980);
   Olesen, P., *Nucl. Phys.* **B200** [FS4], 381 (1982).
4. Mack, G., and Petkova, V.B., *Ann. Phys.* (NY) **123**, 442 (1979);
   Mack, G., and Petkova, V.B., *Ann. Phys.* (NY) **125**, 117 (1980);
   Mack, G., *Phys. Rev. Lett.* **45**, 1378 (1980).
5. Tomboulis, E.T., *Phys. Rev.* **D 23**, 2371 (1981);
   Tomboulis, E.T., *Phys. Lett.* **B303**, 103 (1993).
6. Del Debbio, L., Faber, M., Greensite, J., and Olejník, Š., *Phys. Rev.* **D 55**, 2298 (1997).
7. Del Debbio, L., Faber, M., Greensite, J., and Olejník, Š., hep-lat/9708023.
8. Del Debbio, L., Faber, M., Giedt, J., Greensite, J., and Olejník, Š., *Phys. Rev.* **D 58**, 094501 (1998).
9. Langfeld, K., Tennert, O., Engelhardt, M., and Reinhardt, H., *Phys. Lett.* **B452**, 301 (1999).
10. Engelhardt, M., Langfeld, K., Reinhardt, H., and Tennert, O., hep-lat/9904004.
11. Langfeld, K., Reinhardt, H., and Tennert, O., *Phys. Lett.* **B419**, 317 (1998).
12. Patel, A., *Nucl. Phys.* **B243**, 411 (1984).
13. De Forcrand, P., and D'Elia, M., *Phys. Rev. Lett.* **82**, 4582 (1999).
14. Engelhardt, M., and Reinhardt, H., hep-th/9907139.
15. Montero, A., hep-lat/9906010.

Chiral Lagrangian with BR Scaling for Dense Nuclear Matter

Chaejun Song

Center for Theoretical Physics
Seoul National University
Seoul 151-742, Korea

Abstract. I review here some works done in collaboration with G. Brown, D.-P. Min and M. Rho. With a simple Brown-Rho (BR) scaled chiral model Lagrangian we discuss how BR scaling can be applied to nuclear matter and connected with relativistic Landau Fermi-liquid theory.

Although we believe Quantum Chormodynamics (QCD) is the fundamental theory for strong interaction, we do not deal with quarks and gluons in low energy physics. In low energy physics our degrees of freedom are hadrons; nucleons, pions, etc. Then how can we relate the hadron physics with the fundamental QCD? We have an answer. We can use the effective Lagrangian which is governed by QCD symmetry, especially chiral symmetry. We all know that the so-called chiral effective Lagrangians describe the low energy physics very beautifully in free/dilute space [1]. In this workshop such success and progress are introduced by many speakers.

However, we do not know how chiral effective theory can be applied to hadronic matter. Partly because hadronic matter gives a new energy scale. Fermi momentum, and partly because there need more massive degrees of freedom, e.g, vector mesons or higher order operator in the nucleon fields.

Brown and Rho [2] applied the strategy of the effective theory to the in-medium theory, instead of deriving the hadronic matter properties from the chiral effective Lagrangian directly. It is assumed that the the in-medium effective Lagrangian has the same structure as the chiral effective Lagrangian in free space but its parameters are changed by the modification of the vacuum. They derived the scaling relation among the in-medium parameters from the large N_c QCD-oriented effective Lagrangian;

$$\Phi(\rho) \approx \frac{f_\pi^\star(\rho)}{f_\pi} \approx \frac{m_v^\star(\rho)}{m_v} \approx \frac{m_s^\star(\rho)}{m_s} \approx \frac{M^\star(\rho)}{M}. \qquad (1)$$

Here M represents nucleon mass. Brown-Rho scaling describes the meson at high density very successfully and economically. For example, Li, Ko, and Brown [3]

CP494, *New Directions in Quantum Chromodynamics*, edited by C.-R. Ji and D.-P. Min
© 1999 American Institute of Physics 1-56396-908-4/99/$15.00

showed that a chiral Lagrangian with scaled meson masses describes the dilepton enhancement in CERES experiment.

Our work [4,5] has two targets. One is how BR scaling can describe the ground state of nuclear matter. And the other is how to bridge chiral Lagrangian and Landau Fermi-liquid theory, which is a well-known in-medium theory, via BR scaling. We will construct a simple Walecka-type model Lagrangian that implements BR scaling and deal with it using static mean field approximation.

$$\mathcal{L}_{BR} = \bar{N}(i\gamma_\mu(\partial^\mu + ig_v^\star\omega^\mu) - M^\star + h\phi)N$$
$$-\frac{1}{4}F_{\mu\nu}^2 + \frac{1}{2}(\partial_\mu\phi)^2 + \frac{m_\omega^{\star 2}}{2}\omega^2 - \frac{m_s^{\star 2}}{2}\phi^2 \tag{2}$$

The parameters of the Lagrangian satisfy BR scaling (1). Since the effective nucleon mass in medium decrease more rapidly than vector meson mass according to QCD sum rule [6], $\bar{N}h\phi N$ is put in to account for the difference. We will take the scaling in the simple form

$$\Phi(\rho) = \frac{1}{1 + 0.28\rho/\rho_0} \tag{3}$$

with normal nuclear metter density ρ_0 so as to give $\Phi(\rho_0) = 0.78$ found in QCD sum rule calculations [7], as well as from the in-medium Gell-Mann-Oakes-Renner relation [8].

We know that the proper saturation properties do not appear with BR-scaled masses only [9]. We add one more assumption that the in-medium ωNN coupling also decreases. Brockmann and Toki's calculation [10] already observed the decrease of ωNN coupling. And the recent calculation of nucleon flow data by Li, Brown, Lee, and Ko [11] also showed that $g_v^\star/m_\omega^\star$ is independent of density at low densities and decreases slightly at high densities. So we shall take the scaling vector coupling as

$$\frac{g_v^\star}{g_v} = \frac{1}{1 + z\rho/\rho_0} \tag{4}$$

with z equal to or slightly greater than 0.28. The h is assumed not to scale although it is easy to take into account the density dependence if necessary.

Note that the model has density-dependent parameters. If you fix the density first and calculate the pressure with the model, the thermodynamic contradiction appears. In this way thermodynamic pressure $-\frac{\partial E}{\partial V}$ is not the same as the hydrostatic pressure $< T_{ii} > /3$ at comoving frame. For consistent treatment of the model, it is convenient to define $\check{\rho}u^\mu \equiv \bar{N}\gamma^\mu N$ with unit fluid 4-velocity $u^\mu = \frac{1}{\sqrt{1-v^2}}(1, v)$. v is $< \bar{N}\gamma N > / < N^\dagger N >$. With it we can deal with density as bilinear fields before taking expectation value, i.e, $\rho =< \check{\rho} >$. Let's define $\check{\Sigma} \equiv \frac{\partial \mathcal{L}}{\partial \check{\rho}}$. The equations of motion become

$$[i\gamma^{\mu}(\partial_{\mu} + ig_v^{\star}\omega_{\mu} - iu_{\mu}\check{\Sigma}) - M^{\star} + h\phi]N = 0 \qquad (5)$$

$$(\partial^{\mu}\partial_{\mu} + m_s^{\star 2})\phi = h\bar{N}N \qquad (6)$$

$$\partial_{\nu}F^{\nu\mu}_{\omega} + m_{\omega}^{\star 2}\omega^{\mu} = g_v^{\star}\bar{N}\gamma^{\mu}N. \qquad (7)$$

Note that the equations of motion for bosonic fields are the same as those of Walecka model. Only the fermionic equation of motion includes the additional term $\gamma^{\mu}u_{\mu}\check{\Sigma}$. We can obtain the conserved canonical energy-momentum tensor;

$$T^{\mu\nu} = i\bar{N}\gamma^{\mu}\partial^{\nu}N + \partial^{\mu}\phi\partial^{\nu}\phi - \partial^{\mu}\omega_{\lambda}\partial^{\nu}\omega^{\lambda}$$
$$- \frac{1}{2}[(\partial\phi)^2 - m_s^{\star 2}\phi^2 - (\partial\omega)^2 + m_{\omega}^{\star 2}\omega^2 - 2\check{\Sigma}\bar{N}\!\!\!/ N]g^{\mu\nu}. \qquad (8)$$

In comoving frame, i.e, $v = 0$, the energy density is independent of Σ. It means that we can fix the density from the start when dealing with the energy. But Σ-dependent term appears when we calculate the pressure. We can easily check that (8) gives the consistent pressure.

Based on the thermodynamic consistency in mean field level, we reproduce the infinite nuclear matter properties successfully. We list three sets of parameters in Table 1. We take $m_{\omega} = 783$ MeV and $M = 939$ MeV. And we take $m_s = 700$ MeV consistent with what is argued in [8]. Nuclear matter properties predicted with the parameters of Table 1 are given in Table 2. We obtain the proper bulk properties of nuclear matter, including the compression modulus less than 300 MeV. It is encouraging that the simple model with BR scaling describes the physics of nuclear matter successfully.

Following closely Matsui's analysis of Walecka model in mean field [12], we connect the mean field theory of the BR-scaled chiral Lagrangian and Landau Fermi-liquid theory. We can obtain the quasiparticle interaction by $f_{ij} = \frac{\partial \varepsilon_i}{\partial n_j}$ in the comoving frame. The Landau parameters in our model are;

$$F_0 = \frac{3E_F}{k_F}\rho[C_v^2 + 4C_v\rho\frac{\partial C_v}{\partial\rho} + \frac{m_N^{\star}}{E_F}\frac{\partial m_N^{\star}}{\partial n_j} + \rho^2\{(\frac{\partial C_v}{\partial\rho})^2 + C_v\frac{\partial^2 C_v}{\partial\rho^2}\} \qquad (9)$$

$$+(m_N^{\star} - M^{\star})^2\left\{(\frac{\partial\tilde{C}_h}{\partial\rho})^2 + \tilde{C}_h\frac{\partial^2\tilde{C}_h}{\partial\rho^2}\right\} + 2\tilde{C}_h\frac{\partial\tilde{C}_h}{\partial\rho}(m_N^{\star} - M^{\star})\frac{\partial}{\partial n_j}(m_N^{\star} - M^{\star})$$

$$-2\tilde{C}_h\frac{\partial\tilde{C}_h}{\partial\rho}(m_N^{\star} - M^{\star})\frac{\partial M^{\star}}{\partial\rho} - \tilde{C}_h^2\frac{\partial M^{\star}}{\partial\rho}\frac{\partial}{\partial n_j}(m_N^{\star} - M^{\star}) - \tilde{C}_h^2(m_N^{\star} - M^{\star})\frac{\partial^2 M^{\star}}{\partial\rho^2}]$$

and

TABLE 1. Parameters

SET	h	g_v	z
S1	6.62	15.8	0.28
S2	5.62	15.3	0.30
S3	5.30	15.2	0.31

TABLE 2. Nuclear matter properties predicted with the parameters of Table 1. The effective nucleon mass is $m_N^\star = M^\star - h\phi_0$.

SET	$E/A - M$(MeV)	k_{eq}(MeV)	K(MeV)	$m_N^\star/M$	$\Phi(k_{eq})$
S1	-16.0	257.3	296	0.619	0.79
S2	-16.2	256.9	263	0.666	0.79
S3	-16.1	258.2	259	0.675	0.78

$$F_1 = -\frac{3(C_v^2 - \frac{\Sigma_0}{\rho})\rho}{E_F + (C_v^2 - \frac{\Sigma_0}{\rho})\rho}. \tag{10}$$

with $C_v(\check{\rho}) \equiv \frac{g_v^\star(\check{\rho})}{m_\omega^\star(\check{\rho})}$, $\tilde{C}_h(\check{\rho}) \equiv \frac{m_s^\star(\check{\rho})}{h}$ and $\Sigma_0 = <\check{\Sigma}>$. We can see that those Landau parameters satisfy the relativistic Landau Fermi-liquid relations in [13];

$$K = 3k_F \left(\frac{\partial \varepsilon_i}{\partial k_i}\right)_{k=k_F} (1 + F_0)$$

$$k_F \left(\frac{\partial k_i}{\partial \varepsilon_i}\right)_{k=k_F} = \mu(1 + F_1/3)$$

$$c_1^2 = \frac{k_F^2}{3\mu^2} \frac{1 + F_0}{1 + F_1/3}$$

where μ is chemical potential and c_1 is first sound velocity.

We construct a simple BR-scaled chiral model. We see its phenomenological success and map it into Landau Fermi-liquid. It makes a basis to describe the fluctuations under the extreme condition from what we know at normal nuclear density.

REFERENCES

1. Recent progress can be seen in *Nuclear physics with effective field theory* edited by R. Seki, U. van Kolck, and M. Savage (World Scientific, Singapore, 1998)
2. G.E. Brown and M. Rho, Phys. Rev. Lett. **66**, 2720 (1991).
3. G.Q. Li, C.M. Ko and G.E. Brown, Phys. Rev. Lett. **75**, 4007 (1995).
4. C. Song, G. E. Brown, D.-P. Min, and M. Rho, Phys. Rev. C **56**, 2244 (1997).
5. C. Song, D.-P. Min, and M. Rho, Phys. Lett. B **424**, 226 (1998).
6. R.J. Furnstahl, X. Jin, and D.B. Leinweber, Phys. Lett. B **387**, 253 (1996).
7. X. Jin and D.B. Leinweber, Phys. Rev. C **52**, 3344 (1995).
8. G.E. Brown and M. Rho, Phys. Rep. **269**, 333 (1996).
9. G.E. Brown and R. Machleidt, e-print nucl-th/9210010.
10. R. Brockmann and H. Toki, Phys. Rev. Lett. **68**, 3408 (1992).
11. G.Q. Li, G.E. Brown, C.-H. Lee, and C.M. Ko, e-print nucl-th/9702023.
12. T. Matsui, Nucl. Phys. **A370**, 365 (1981).
13. G. Baym and S. Chin, Nucl. Phys. **A262**, 527 (1976).

CRITICAL EXPERIMENTS
TESTING QCD

Measurement and Origin of the $\bar{u}, \bar{d}$ Asymmetry in the Nucleon Sea

Gerald T. Garvey

Physics Division, Los Alamos National Laboratory
MS H846, Los Alamos, New Mexico, USA

Even though physicists have established the presumed correct theory of strong interactions, QCD, calculation of the partonic structure of hadrons remains beyond our present capability. Those of us who measure these partonic distributions are anxious to learn at this workshop, of the progress being made in carrying out the theory in the light cone reference frame. In this talk I will present results from a direct measurement (FNAL/E866/NUSEA) of the relative $\bar{d}, \bar{u}$ content of the nucleon sea and discuss perspectives on the origin of the observed flavor asymmetry.

Based on charge and baryon number conservation, a few definite statements [1] can be made regarding integrals of the parton distributions. For example

$$\int_0^1 [u_p(x,Q^2) - \bar{u}_p(x,Q^2)]dx = \int_0^1 [d_n(x,Q^2) - \bar{d}_n(x,Q^2)]dx = 2 \qquad \text{1a)}$$

and

$$\int_0^1 [d_p(x,Q^2) - \bar{d}_p(x,Q^2)]dx = \int_0^1 [u_n(x,Q^2) - \bar{u}_n(x,Q^2)]dx = 1 \qquad \text{1b)}$$

There are also direct equalities coming from symmetry principles such as charge symmetry. On such grounds we believe that $u_p(x,Q^2) = d_n(x,Q^2)$, ect..

However, there is a characteristic integral of the nucleon's parton distribution whose value has ben widely but incorrectly assumed to be zero, namely;

$$\int_0^1 [\bar{u}_p(x,Q^2) - \bar{d}_p(x,Q^2)]dx = ?? \qquad \text{2)}$$

This integral was thought to be approximately zero because anti quarks were assumed to be produced via pair production by gluons. As the mass difference between up and down quarks is small (~2.5 MeV) compared to a confinement scale of 10^{-13} cm, pair production creates approximately identical numbers of up and down pairs. The role that virtual pions play in determining the character of the nucleon sea has been neglected, particularly within the high-energy community. It was always clear that mesons were required at the hadronic level to account for the long range part of the nucleon–nucleon force, to implement PCAC and as a critical degree of freedom in the realization of chiral quark models. The vanishing role virtual pions in the partonic description of the nucleon and their "rediscovery" is described in a recent issue of Science[2].

CP494, *New Directions in Quantum Chromodynamics*, edited by C.-R. Ji and D.-P. Min
© 1999 American Institute of Physics 1-56396-908-4/99/$15.00

It became necessary to reconsider the role of virtual pions in the nucleon with the 1991 report of the New Muon Collaboration's (NMC) measurement [3] of the integral of the difference of the $F_2(x,Q^2)$ structure functions of the proton and neutron. They measured F_2 using deep inelastic scattering(DIS) of muons from hydrogen and deuterium. NMC's most recent [4] value for I_G is,

$$I_G = \int_0^1 [F_2^p(x,Q^2) - F_2^n(x,Q^2)]dx / x = 0.235 \pm 0.026 \qquad 3)$$

rather than 1/3, as would be the case if the integral in eq 2 were equal to zero. I_G expressed in terms of parton distributions is referred to as the Gottfried Sum Rule (GSR)[5],

$$I_G = \frac{1}{3}\int_0^1 [u_{v,p}(x,Q^2) - d_{v,p}(x,Q^2)]dx - \frac{2}{3}\int_0^1 [\bar{d}_p(x,Q^2) - \bar{u}_p(x,Q^2)]dx \qquad 4)$$

where $u_{v,p}(x,Q^2) \equiv u_p(x,Q^2) - \bar{u}_p(x,Q^2)$ and charge symmetry is used to reference all parton distributions to the proton. Thus difference from 1/3 in eq 3 is readily cast as a relative excess of $\bar{d}$ quarks in the sea of the proton.

$$\int_0^1 [\bar{d}_p(x,Q^2) - \bar{u}_p(x,Q^2)]dx = 0.147 \pm 0.039$$

Following publication of the NMC result, the Drell-Yan process [6] was suggested [7] as a means of directly investigating the anti-quark content of the neutron and proton. Comparing the Drell-Yan yield from a nuclear target having a large neutron excess to the yield from deuterium FNAL/E-772 showed [8] that several of the parameterisations initially adopted to accommodate the violation of the GSR were incorrect. Following the suggestion of ref [7], NA51 carried out [9] a comparison of the D-Y yield from hydrogen (H) and deuterium (D) using the 450GeV proton beam at CERN. They obtained a total of ~6000 D-Y events on D and H and deduced,

$$\frac{\bar{u}(x = 0.18)}{\bar{d}(x = 0.18)} = 0.51 \pm 0.04 \pm 0.05$$

further demonstrating that $\bar{d}_p(x) > \bar{u}_p(x)$. A host of papers were published calculating the asymmetry. Two recent reviews[10,11] provide extensive accounts of the existing literature and ongoing activity.

This paper presents the results from E866 at FNAL; a high statistics measurement of the Drell-Yan (D-Y) yield from deuterium (^{2}H) relative to hydrogen (H). These results allow reliable extraction of the x dependence of $\bar{d}_p(x) / \bar{u}_p(x)$ and $\bar{d}_p(x) - \bar{u}_p(x)$ over the range $0.03<x<0.3$. Complete expressions for the D-Y cross section are readily found in the literature [10,11]. We are only concerned with the ratio of the D-Y yield from deuterium relative to hydrogen. E866 observes dilepton pairs from the kinimatic regime where the momentum fraction of the incident beam parton (x_1) is appreciably greater than the parton momentum fraction in the target (x_2). Thus for the most part an incident valance quark in the beam annihilates on a sea quark in the target producing a D-Y dilepton ($\mu^+\mu^-$) pair . In such a case the ratio of the D-Y cross sections is approximately given by,

$$\frac{\sigma_{DY}(x_1,x_2:p\to D)}{2\sigma_{DY}(x_1,x_2:p\to H)}=(\frac{1}{2})\frac{1+\dfrac{\overline{d}_p(x_2)}{\overline{u}_p(x_2)}\left(1+\dfrac{d_p(x_1)}{4u_p(x_1)}\right)}{\left(1+\dfrac{d_p(x_1)\overline{d}_p(x_2)}{4u_p(x_1)\overline{u}_p(x_2)}\right)} \qquad 5)$$

If $\overline{d}_p(x_2) = \overline{u}_p(x_2)$ the ratio is 1. In obtaining eq. 5), charge symmetry is assumed and small contributions from the sea of the incident proton are neglected. The term $d_p(x_1)$ / $4u_p(x_1)$ is small, typically the order of 1/8 and becomes smaller as $x_1 \to 1$, so that the ratio as directly measured in the experiment is roughly $\frac{1}{2}\left(1+\dfrac{\overline{d}_p(x_1)}{\overline{u}_p(x_1)}\right)$. In the actual analysis of the data the complete D-Y expression for the cross section is used.

E 866 uses the extracted 800 GeV beam from the Fermilab Tevatron, with 2×10^{12} protons per 20 sec spill to bombard 7.6 cm diameter by 50 cm thick targets of liquid hydrogen (LH) and liquid deuterium (LD). Fig.1 shows the spectrometer configuration On passing through the target the beam is intercepted in a beam dump located inside the second dipole magnet.

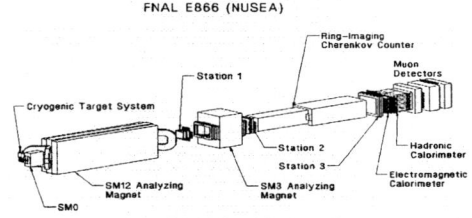

An extensive absorber in the gap of the second magnet (SM12) removes scattered hadrons and allows only muons to traverse the detection system consisting of four tracking stations and a momentum analyzing magnet. The beam spot is 8mm wide by 2mm high and its intensity is monitored by two secondary-emission detectors and a quarter wave rf cavity. The luminosity from the individual targets is

Figure 1. E866 Spectometer Layout

monitored by a multiple-element scintillation counter which also yields the DAQ and trigger live time and the beam duty factor. The targets are cycled every five beam spills with one spill per cycle devoted to collecting data from an empty target flask. The 3-dipole magnet spectrometer is similar to the one [20] employed in previous experiments (E-605, E772, and E789) except that the station 1 drift chambers have been replaced. Other improvements that have been incorporated into E866 are a programmable trigger and faster data acquisition system(DAQ)

Three different spectrometer setting were employed to optimize data collection at low, intermediate, and high masses of the dilepton pairs ($M^2=Sx_1x_2$). In a six month run completed on 3/24/97, 1.3×10^{17} protons were recorded on target and over 350,000 D-Y events written to tape. The data recorded at the low and intermediate mass setting have systematic uncertainties associated with random coincidences and rate dependent efficiencies at the several percent level and will be ready for publication soon, however the events recorded at the high mass setting are free from these effects because of the greatly reduced rates in the tracking chambers. Therefore we present results using for the most part the data from the high mass setting. Fig.2 shows the ratio of the D-Y yield

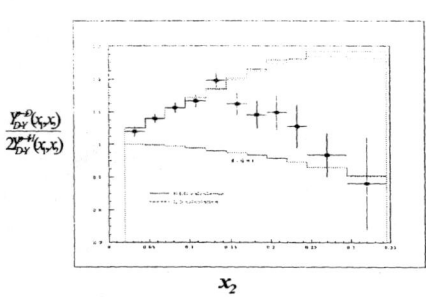

$$\frac{Y_{DY}^{p \to D}(x_1, x_2)}{2 Y_{DY}^{p \to H}(x_1, x_2)}$$

x_2

Figure 2. Ratio of the Drell-Yan yield from p→D relative to that from 2x p→H

for p→D to 2(p→H) as a function of x_2. Explicit cuts on the range of the data are $Sx_1x_2 > 4$ Gev and $x_1-x_2 > 0$. The data clearly show the D-Y yield from p→D exceeds that from 2(p→H) over an appreciable range in x_2. Referring to eq 5. this excess is immediately ascribable to $\bar{d}_p(x_2) / \bar{u}_p(x_2) > 1$. The reason for the excess in yield is that $\bar{d}_p(x_2) / \bar{u}_p(x_2) > 1$ implies by charge symmetry, $\bar{d}_n(x_2) / \bar{u}_n(x_2) < 1$, hence the mean square charge of the partons in the neutron sea is larger than that in the proton sea. The upper curve in Fig.2 shows the prediction for the ratio using the CTEQ4M [12] structure functions which have been fitted to accommodate both the NMC and NA51 results. The lower curve uses the CTEQ4M structure functions but with $\bar{d}_p(x_2) = \bar{u}_p(x_2)$. Fig. 2 shows that next to leading order (NLO) QCD corrections have a negligible effect on the predicted ratio. While our results are slightly below the CTEQ prediction for $.06 < x_2 < .2$ it is clear that $\bar{d}_p(x_2) > \bar{u}_p(x_2)$ at least out to $x_2 = .25$.

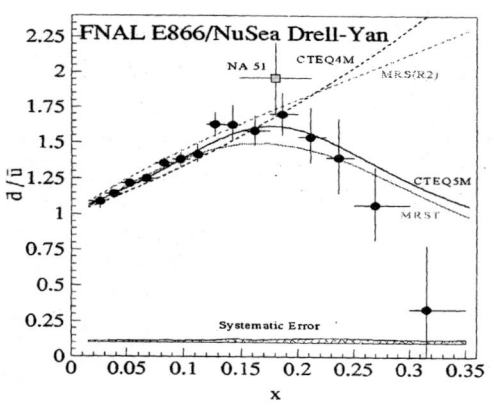

Figure 3. $\bar{d}_p(x)/\bar{u}_p(x)$ as extracted from the cross-section ratio

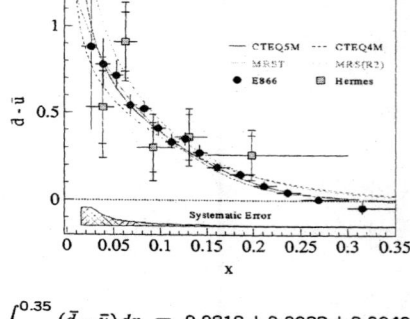

$$\int_{0.015}^{0.35} (\bar{d} - \bar{u}) \, dx = 0.0818 \pm 0.0082 \pm 0.0049$$

$$\int_0^1 (\bar{d} - \bar{u}) \, dx = 0.118 \pm 0.011$$

Figure 4. $\bar{d}_p(x) - \bar{u}_p(x)$ obtained from $\bar{d}_p(x)/\bar{u}_p(x)$

Fig 3 shows the values extracted for $\bar{d}_p(x_2) / \bar{u}_p(x_2)$ along with the predictions made by the some of the parameterized parton distributions. The data presented in Fig. 3 includes some preliminary results from our low and intermediate mass D-Y measurements. We extract $\bar{d}_p(x_2) / \bar{u}_p(x_2)$ from the ratio of the yields, by successively iterating the value for $\bar{d}_p(x_2) / \bar{u}_p(x_2)$ under the constraint that $\bar{u}_p(x_2) + \bar{d}_p(x_2) = [\ \bar{u}_p(x_2) + \bar{d}_p(x_2)]_{PD}$ until the measured ratio is obtained. The term in brackets in the previous sentence refers to the particular set of parton distributions (CTEQ, MRS, RGV) being used in the iteration. The CTEQ4M and MRS(R2) [13] distributions were created before our published work [14] while CTEQ5M [15] and MRST [16] are subsequent. A qualitative feature of our result, that was not reflected in any prediction is the rapid reduction towards 1 of $\bar{d}_p(x_2) / \bar{u}_p(x_2)$ beyond $x_2=0.2$. This suggests that the perturbatively generated symmetric sea dominates the non-perturbative sea at both small $(x<.01)$ and large $(x> .25)$ values of x. The value extracted for $\bar{d}_p(x_2) / \bar{u}_p(x_2)$ does not depend on the particular parton distribution employed or whether a LO or NLO calculation of the ratio is employed. The data beyond $x=0.2$ is statistics limited and very interesting for study at high intensity with lower proton beam energy, for example, at new Main Ring Injector nearing completion at Fermilab

Our measurement of $\bar{d}_p(x_2) / \bar{u}_p(x_2)$ in conjunction with the various parton distribution values for $u_p(x_2) + \bar{d}_p(x_2)$ is used to determine $\bar{u}_p(x_2) - \bar{d}_p(x_2)$ as shown in fig. 4. The value for integral of $\bar{u}_p(x_2) - \bar{d}_p(x_2)$ is shown at the bottom of the figure. Fig. 4 also shows a variety of fits to the distribution as well as data from a recent HERMES publication [17]. There is a negligible difference in the value for the integral if, say MRS(D), is used rather than CTEQ4M. The integral reaches a value of 0.06 at $x_2=0.03$ which means that some 60% of the violation of the GSR lies below this point if the value extracted for the GSR by NMC is assumed to be correct. Also shown are the values of the integral for the CTEQ4M, MRS(D) and GRV [18] structure functions. In every case our data is lower than the structure functions which had been adjusted to fit the NMC value for the integral and the NA51 single point at $x_2=0.18$. The issue of the compatibility of our result with the NMC value for the GSR is not resolved as significant contributions to the integral of $\bar{u}_p - \bar{d}_p$ can arise from the region $x < .03$. The low and intermediate mass data from E866 can help address this issue but careful attention will have to be paid to systematic errors effecting the relative normalization and possible nuclear effects such as shadowing.

ORIGIN OF THE $\bar{d}$, $\bar{u}$ ASYMMETRY

Thus there is now clear evidence that there is a surplus of $\bar{d}_p$ relative to $\bar{u}_p$ in the proton.. When the earliest DIS experiments indicated that the value of the Gottfried integral might be less than 1/3, Field and Feynman suggested [19] that it could be due to Pauli blocking in so far as the creation of $u\bar{u}$ pairs would be suppressed relative to $d\bar{d}$ pairs because of the presence of two valance u-quarks in proton as compared to a single valance d-quark. Ross and Sachrajda [20] questioned that this effect would be appreciable because of the large phase-space available to the created $q\bar{q}$ pairs. They also showed that perturbative QCD would not produce a $\bar{d}$, $\bar{u}$ asymmetry. Steffens and Thomas [21] recently looked into this issue, explicitly examining the consequences of Pauli blocking. They similarly concluded that the blocking effects were small, particularly when the anti-quark is in a virtual meson. Thus another, presumably non-perturbative, mechanism must be found to account for the large measured $\bar{d}, \bar{u}$ asymmetry.

Virtual Meson-Baryon States

A natural origin for this flavor asymmetry is the virtual states of the proton containing isovector mesons. This point appears to have first been made by A.W.Thomas, in a publication [22] treating SU(3) symmetry breaking in the nucleon sea. Sullivan [23] had earlier shown that virtual meson-baryon states directly contribute to the nucleon's structure function. As mentioned earlier a large number of authors have contributed to calculating the asymmetry from this perspective, so recent reviews [10,11] should be consulted for a complete list of contributors.

Conservation of electric charge and isospin naturally create a flavor asymmetry when isovector mesons are involved. For example the Fock decomposition of the proton into its πN components yields

$$|p\rangle \rightarrow \sqrt{\frac{1}{3}}|p_0\pi^0\rangle + \sqrt{\frac{2}{3}}|n_0\pi^+\rangle \qquad 6)$$

where p_0 and n_0 are regarded as proton and neutron states with symmetric seas. The $\bar{d}/\bar{u}$ ratio for the configurations shown in eq 6 is 5! Thus the presence of such virtual states can readily generate a large flavor asymmetry. Allowing for the possibility of $\pi\Delta$ configurations it is easy to show that

$$\frac{\bar{d}_p(x)}{\bar{u}_p(x)} = \frac{\frac{1}{6}a + \frac{b}{3} + S_0(x)}{\frac{1}{6}a + \frac{2b}{3} + S_0(x)} \qquad 7)$$

where $a(b)$ is the probability of $\pi N(\pi\Delta)$ components in the proton and $S_0(x)$ is the amount of symmetric sea. Note that the $\pi\Delta$ component has the opposite effect on the flavor asymmetry to the πN component. To properly characterize the $\bar{d}/\bar{u}$ ratio the relative size of the symmetric sea must be known, however if one considers the integral of $\bar{d}_p(x) - \bar{u}_p(x)$, the result is given by,

$$\int_0^1 [\bar{d}_p(x) - \bar{u}_p(x)]dx = \frac{1}{3}(2a - b) \qquad 8)$$

The value extracted for the above integral from the E866 high mass data is 0.100 ±0.018 leading to a= 0.2 ± 0.036 if one accepts $b\sim a/2$ as found in many calculations.

Many attempts have been made to calculate the flavor asymmetry due to isovector mesons [6]. Most start with the following convolution expressions:

$$|p\rangle = a_0|p_0\rangle + \sum_{MB} a_{MB}[|M\rangle|B\rangle]^P \qquad 9)$$

$$x\overline{q}_p(x,Q^2) = \sum_{MB} a_{MB}^q \int_x dy f_{MB}(y) \frac{x}{y} \overline{q}_M(x/y, Q^2) \qquad 10)$$

where

$$f_{MB}(y) = \frac{g_{MpB}^2}{16\pi^2} y \int_{-\infty} dt \frac{F(t, m_p, m_B)}{(t - m_M^2)^2} F_{MpB}^2(t) \qquad 11)$$

In the above expressions x is the fraction of proton's momentum carried by the antiquark, and y is fraction carried by the meson (M). The meson-proton-baryon couplings are characterized by coupling constants g_{MpB}, and form factors $F_{MpB}(t)$. As pions are the only mesons considered and the baryons are usually restricted to nucleons and deltas, the coupling constants are well known and the partonic structure of the pion, $\overline{q}_\pi(x,Q^2)$, is fixed by the Drell-Yan cross section using high energy pion beams. The only uncertainties are the form factors $F_{\pi pN}(t)$ and $F_{\pi p\Delta}(\tau)$. One attempts to determine these form factors by using [24,25] the measured yields from a variety of high energy hadronic reactions at small $p_\perp$ such as

$$p + p \rightarrow n + X$$
$$p + p \rightarrow \Delta^{++} + X$$

Even though there is a sizable amount of available data, employing such a procedure does not produce a precise result. The cutoff parameters used in the extracted form factors are the order of 1 GeV but the uncertainties in their values produce differences of factors of two in the predicted Drell-Yan yield.

Even though calculation of the integral of $\overline{d}(x) - \overline{u}(x)$ is often in agreement with experiment it is not possible to achieve a quantitative fit to the measured x dependence of the difference [26]. Calculations of $\overline{d}_p(x)/\overline{u}_p(x)$ are even more unsuccessful, as knowledge of the x dependence of the symmetric sea is required in this instance.

More recent attempts [24,25] to calculate the asymmetry due to isovector mesons find a smaller $\pi\Delta$ component in the nucleon than is presented following eq 8. The πN component is still about 20% while the $\pi\Delta$ piece is down around 6%. As a result the integrated asymmetry is too large, typically 0.16. However considerable progress appears to have been made by Reggeizing the virtual mesons[25]. This procedure has two interesting consequences first it shows that the principle contributions to the asymmetry come from virtual pions and secondly it reduces the contributions from these pions at large x allowing the ratio $\overline{d}_p(x)/\overline{u}_p(x)$ to approach 1 for x>0.3.

Chiral Models:

An alternative approach [27,28,29,30] also employing virtual pions to produce the $\overline{d}, \overline{u}$ asymmetry uses constituent quarks and pions as the relevant degrees of freedom.

483

Such models are usually referred to as chiral models[31]. This approach which has less dynamical freedom that the M/B expansion can be shown to predict $\bar{d}_p(x)/\bar{u}_p(x) < 11/7 = 1.57$ independent of x or Q^2. This limit is observed to be exceeded for $0.12 < x < 0.2$ as shown in fig.3. The limit on the maximum value of the ratio can be traced to the fact that there is no mechanism to suppress virtual $\Delta\pi$ like configurations which are seen in eq.2 to reduce the asymmetry. Some extensions of the simple chiral quark model can obtain agreement with certain features of the observed asymmetry but only with the introduction of additional parameters. A calculation [32] of $\bar{d}_p(x) - \bar{u}_p(x)$ in the chiral model yields too soft a distribution relative to experiment, indicating that more dynamics must be included in the chiral quark model if it is to produce more than rough qualitative agreement with the observed asymmetry in the up, down sea of the nucleon.

Charge Symmetry

The nucleon's structure functions can be investigated by the deep inelastic scattering (DIS) of neutrinos or charged leptons. The $F_2^{\mu N}(x,Q^2)$ structure function extracted from DIS muon scattering is defined as

$$F_2^{\mu N}(x,Q^2) \equiv \frac{F_2^{\mu p}(x,Q^2) + F_2^{\mu n}(x,Q^2)}{2} \qquad 12)$$

$$= \frac{5}{18}x[u + \bar{u} + d + \bar{d} + \frac{2}{5}(s + \bar{s}) + \frac{8}{5}(c + \bar{c})] \quad 13)$$

In going from eq 6) to 7), the (x,Q^2) dependence has been suppressed and charge symmetry ($u_p = d_n = u$, ect) has been employed. For neutrino DIS one has

$$F_2^{\nu N}(x,Q^2) = x[u + \bar{u} + d + \bar{d} + 2s + 2\bar{c}] \qquad 14)$$

again using CS and suppressing the (x,Q^2) dependence in the parton distributions. Neglecting the contribution of charmed quarks in the nucleon and correcting for the small difference due to strange quark contributions one expects

$$\frac{\frac{18}{5}F_2^{\mu N}(x,Q^2)}{F_2^{\nu N}(x,Q^2)} \cong 1 \qquad 15)$$

Recently Boras, Londergan and Thomas (BLT) [33] compared the parton distributions, $F_2^{\nu N}(x,Q^2)$ and $F_2^{\mu N}(x,Q^2)$. They found after making the necessary small corrections that the ratio in eq. 15 at common values of (x, Q^2) is 1 for $x > 0.1$ but for $x < 0.1$ the ratio progressively decreases below 1 as x decreases. BLT suggested that this effect might be due to a violation of charge symmetry in the sea of the nucleon. While charge symmetry has been well tested at the hadronic level, and known to hold for the nucleon to much better than 1% there is little information on the partonic level. To achieve agreement with experiment BLT found it necessary to set $\bar{d}_n(x) \approx 1.25\bar{u}_p(x)$ at small x. As the total number of sea quarks is kept equal in the neutron and proton, one has $\bar{d}_n(x) \approx \bar{d}_p(x)$ and $\bar{u}_n(x) \approx \bar{u}_p(x)$, rather than the expected relationships from CS ($\bar{d}_p(x) = \bar{u}_n(x), \bar{u}_p(x) = \bar{d}_n(x)$). The claimed violation of CS is therefore extremely large, approximately 40%. Fortunately, this radical alteration of the conventional view does not appear to be the case, as shown in a recent letter by Bodek et al [34] who show that the measured asymmetry of W+(W-) production at CDF/FNAL is consistent with CS and in strong disagreement with the suggestion of BLT. The reasons for the discrepancy of the DIS data with eq. 9 must be found elsewhere.

Spin Dependent Structure Functions

As the pion is a $J^{\pi} = 0^-$ meson it must be emitted in a p-wave requiring that the nucleon flip its spin 2/3 of the time upon emitting a pion . Thus it appears that such a process might account for the reduction of g_A from the SU96) value of 1.667 and the overall reduction in the quark contribution to the nucleon spin as observed in deep inelastic scattering . The contribution to g_A from virtual pions is

$$g_A = \Delta u - \Delta d$$
$$= \frac{5}{3} - \frac{20}{27}(2a + b) + \frac{32}{27}\sqrt{2ab}$$

16)

Fig. 5 below shows the effect of pions on the value of g_A where the values of a and b are constrained by measurements of the flavor asymmetry. In the case of the flavor asymmetry there is no interference between the N and Δ contributions; but this is not the case for g_A.

Fig. 5 shows that virtual pions appreciably reduce the value g_A from the SU(6) value of 5/3 however it can not be the whole story. Presumably relativistic effects also play a large role in the reduction [35]. Similarly virtual pions reduce the total spin carried by u and d quarks as shown in eq.17 below,

$$\Delta q = \Delta u + \Delta d = 1 - \frac{2}{3}(2a - b)$$

17)

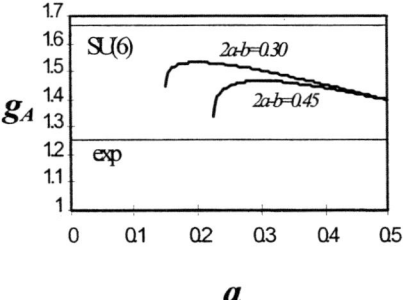

g_A

a

Figure 5. Effect on the value of g_A of virtual pions

However the reduction of Δq obtained for the range of a and b extracted from the flavor asymmetry is still lies well above the value[36] extracted ($\Delta u + \Delta d \approx 0.25$) from polarized DIS.

However, there is a aspect of spin distributions predicted by virtual pion emission that is present in the data. The spin carried by anti-quarks in the virtual pion model must be zero. Indeed the measured value [36], $\Delta \bar{q} = 0.01 \pm 0.04 \pm 0.03$ is consistent with that picture. It is very likely that meson will prove to play an important part in understanding the nucleon's spin structure but that role is not clear at present.

References

1. Hinchcliffe, I. and Kwiatkowski, A., *Ann Rev of Nuc and Part. Sci.* **46**, 609 (1996)
2. Watson, A *Science* **283** 472 (1999)
3. Amaudruz, P. et al., *Phys Rev. Lett.* **66** 2712 (1991)
4. Arneodo, M. et al., *Phys. Rev.* **D50** R1 (1994)
5. Gottfried, K., *Phys. Rev. Lett.* **18** 1174 (1967)
6. Drell S. and Yan, T., *Phys. Rev. Lett.* **25** 316 (1971)
7. Ellis S. and Sterling W., *Phys. Lett.* **B256** 258 (1991)
8. .Mcgaughey, P.L et al., *Phys. Rev. Lett.* **69** 1726 (1991)
9. Baldit, A. et al., *Phys Lett.* **B332** 244 (1994)
10. Kumano,S., Phys Rept. 303 183 (1998)
11. Speth J. and Thomas A., Adv. Nuc. Phys. 24 83 (1998)
12. Lai, H.L. et al., *Phys. Rev.* **D55** 1280 (1997)
13. Martin, A.D., Roberts, R. G. and Sterling, W. J., *Phys. Lett.* **B387** 419 (1996)
14. Hawker, E. A. et al., *Phys Rev. Lett.* **80** 3715 (1998)
15. Lai, H.L., et al hep-ph/9903282
16. Sterling, W.H. private communication
17. Ackerstaff, K., et al *Phys Rev. Lett* **81** 5519 (1998)
18. Gluck, M., Reya, E. and Vogt, A., *Z. Phys.* **C67** 433 (1995)
19. Feynman, R. P. and Field, R. D., *Phys. Rev.* **D15** 2590 (1977)
20. Ross, D. A. and Sachrajda, C. T., *Nuc. Phys.* **B149** 497 (1979)
21. Steffens, F. M. and Thomas, A. W., *Phys. Rev.* **C55** 900 (1997)
22. Thomas, A. W., *Phys. Lett.* **126B** 97 (1983)
23. Sullivan, J. D., *Phys. Rev.* **D5** 1732 (1972)
24. Holtzman, H., Szczurek, A. and Speth, J., *Nuc Phys.* **A596** 631 (1996)
25. Nikolaev, N. N., Schafer, W., Szczurek, A. and Speth, J., *Phys Rev* **D60** 014004 (1999)
26. Peng, J. C. et al (E866 NUSEA Collaboration), *Phys Rev.* **D58** 092004 (1998)
27. Eichten, E. J., Hinchcliffe, I. and Quigg, C., *Phys. Rev.* **D45** 2269 (1992), D47 R747 (1993)
28. Cheng, T.P. and Li, L.F., *Phys. Rev. Lett.* **74** 2872 (1995)
29. Cheng, T.P. and Li, L. F., *Phys. Lett.* **B366** 365 (1996)
30. Suzuki, K. and Weise, W., *Nuc. Phys.* **A634** 141 (1998)
31. Manohar, A. and Georgi, H., *Nuc. Phys._**B234** 189 (1984)
32. Szczureck, A., Buchmans, A. and Faessler, A., *Jour. Phys.* **C22** 1741(1996)
33. Boros, C., Londergan, J.T. and Thomas, A.W., *Phys.Rev.Lett.* **81** 4075 (1998) *Phys.Rev.* **D59** 074021 (1999)
34. Bodek, A. et al , hep-ex/9904022
35. Signal, A. I. and Thomas, A. W. *Phys. Lett.* **B191** 205 (1987)
36. Adeva, B. et al (SMC Collaboration), *Phys. Lett._**B420** 180 (1998)

The PHENIX Experiment at RHIC

Ju Hwan Kang

Department of Physics, Yonsei University
Seoul, 120-749, Korea
for
the PHENIX Collaboration[1]

Abstract. The PHENIX experiment at RHIC is currently under construction to start data collection in late 1999. The physics goals of PHENIX, which are mainly QGP searches and spin physics studies, are described. The design and capabilities of the PHENIX detector to address the physics goals are also discussed.

INTRODUCTION

The Relativistic Heavy Ion Collider (RHIC) at Brookhaven National Laboratory is in the final stage of its construction. Its primary purpose is to study Quark Gluon Plasma (QGP) by colliding heavy-ion beams up to Au at the maximum colliding energy of $\sqrt{s} = 200$ GeV/nucleon. RHIC is also designed to study spin physics by accelerating polarized proton beams up to $\sqrt{s} = 500$ GeV at the luminosity of 2×10^{32} cm^{-2}s^{-1} or $\sqrt{s} = 200$ GeV at the luminosity of 8×10^{31} cm^{-2}s^{-1} with large polarizations of ~ 70 %. Starting from the year 2000, there will be two months of spin physics running per year. The PHENIX detector [1,2] is one of the two large detectors at RHIC. It is designed to detect photons, leptons, and hadrons with excellent momentum resolution and particle identification so it can measure a wide range of proposed signatures for QGP. Due to its capabilities to detect prompt photons and leptons including muons, PHENIX will also undertake a very attractive spin physics program at RHIC.

PHYSICS GOALS

PHENIX will study many proposed signatures of the deconfinement transition and the restoration of chiral symmetry [3]. Since it is difficult to find a single definitive signature for the phase transition, PHENIX will simultaneously measure

---

[1] For the complete PHENIX Collaboration author list, please refer to Ref. 1. Visit *http://www.phenix.bnl.gov* for the most current PHENIX information.

CP494, *New Directions in Quantum Chromodynamics,* edited by C.-R. Ji and D.-P. Min
© 1999 American Institute of Physics 1-56396-908-4/99/$15.00

various signatures as a function of energy density on $A + A$, $p + A$, and $p + p$ collisions with different collision energies. Utilizing polarized $p + p$ collisions at RHIC, PHENIX will measure the gluon polarization and the anti-quark polarization in the nucleon. Those are not directly measured in polarized deep-inelastic scattering (DIS) experiments in which virtual photons are used to probe the nucleon, since the photon does not couple directly to the gluon and couples to quark and to anti-quark in a same way. Furthermore, flavor decomposition of quark and anti-quark will be possible by analyzing W-producing processes.

QGP Searches

PHENIX will measure the suppression of J/ψ and ψ' production relative to that of the Υ to estimate the strength of Debye Screening in the deconfined plasma. Considering the bound state radii of J/ψ, ψ', and Υ: $R_\Upsilon(0.13 \text{ fm}) < R_{J/\psi}(0.29 \text{ fm}) < R_{\psi'}(0.56 \text{ fm})$, the Υ can be used as a reference since it is unlikely to be screened at energy densities attainable at RHIC. J/ψ suppression relative to the Drell-Yan continuum will also be measured to compare with existing results from CERN. Comparison of charmonium production relative to that of open charm identified in PHENIX through the semi-leptonic decay of charm mesons $(D, \bar{D})$ will give us an estimate of the subsequent dissolving of any created charmonium assuming the known initial state effects such as gluon shadowing can be measured in $p + A$ collisions.

At the initial stage of RHIC collisions, the hard scattering of quarks and gluons is predicted to be the dominant process. By measuring high p_T photons, high p_T leptons, and high p_T jets identified by their leading particle, products of the hard scattering can be separated so that we can study the parton distribution functions with $p + p$, $p + A$, and $A + A$ collisions. Substantial energy loss and re-scattering are predicted for the outgoing quarks and gluons in a hot dense deconfined QGP, leading to enhanced acoplanarity and energy imbalance of the two back-to-back jets. This would have drastic effects on both single and di-hadron spectra at large p_T.

As the temperature or baryon density increases, the spontaneously broken chiral symmetry at zero temperature and/or baryon density is expected to be restored. The restoration of chiral symmetry is predicted to cause changes in the mass and width of some vector mesons. In particular the properties of ϕ meson will be investigated in PHENIX as those can be measured easily due to its narrow mass width. PHENIX will also measure the relative branching ratio of ϕ mesons decaying via K^+K^- or e^+e^- channels as the ratio would be sensitive to changes in its properties due to the smaller mass difference (33 MeV) between the ϕ meson and two charged kaons. The temporary restoration of chiral symmetry in nuclear collisions may result in the formation of domains of disoriented chiral condensate (DCC). Such domains would decay into pions having the pion charge ratio $N_{\pi^0}/N_{\pi^+,\pi^0,\pi^-}$ substantially different from $\frac{1}{3}$. By measuring the above ratio or fluctuations in the charged pion multiplicity, PHENIX will study the chiral symmetry restoration which complements the study of deconfinement.

The thermal radiation of the plasma will be examined by measuring emitted photons; direct or virtual ($\gamma^* \to e^+e^-$, $\mu^+\mu^-$). Unlike hadrons suffering the strong interactions, photons and leptons retain information about the early history of the collisions. The spectrum of emitted photon would be different depending on the evolution process, which can be through the formation of a plasma with many degrees of freedom, just a hot hadronic gas state without the deconfinement, or a long-lived mixed state. Enhanced production of strangeness and charm will be investigated by measuring $K^{\pm}$, ϕ meson, and D mesons since those have been proposed as QGP signatures.

Spin Physics

PHENIX will measure the gluon polarization by looking at high p_T prompt photon production, which is dominated (~ 90 %) by the gluon compton process ($gq \to \gamma q$). A double longitudinal-spin asymmetry of this process will be measured to extract the gluon polarization. The partonic level asymmetry calculated in QCD, and the quark distribution and quark polarization measured in lepton scattering experiments will be used in this measurement. The PHENIX acceptance ($|\eta| \leq$ 0.35) for photons strongly favors the samples having $x_g \approx x_q$ and greatly simplifies the relation.

The Drell-Yan process ($q\bar{q} \to \gamma^* \to l^+l^-$) will be studied to extract the anti-quark polarization which is not directly measurable in the polarized DIS experiments. Because of the photon (γ^*) in the intermediate state, this process has the maximum analyzing power so that the partonic level asymmetry becomes just -1. By neglecting contributions from heavier quarks, the measured asymmetry which is the sum (weighted by the squared charge) of $\bar{u}$ and $\bar{d}$ contributions and abundance will give information on the $\bar{u}$ polarization with the $\bar{d}$ contamination.

More detailed flavor analysis is possible with the asymmetry measurements for W productions. For these processes, the parity-violating asymmetry between the longitudinally polarized proton and unpolarized proton can be written as the linear combination of the quark and anti-quark polarizations (Δu, $\Delta \bar{d}$ for W^+; Δd, $\Delta \bar{u}$ for W^-). Thus the flavor decomposition of the anti-quark helicity distribution is possible. The $W^{\pm}$ productions can be identified by detecting muons with $p_T \geq$ 20 GeV since W decaying into muon is the dominant source of the muons in this region. The reconstruction of the parton kinematics for W productions ($W \to l\nu$) is not simple because of the undetectable neutrino. Fortunately, due to the V-A nature of the weak process, there is a good correlation between the muon energy and momentum fraction carried by the incident quark, x_q. By neglecting the p_T of W, $x_{\bar{q}}$ is related to x_q with a simple kinematics relation containing $\sqrt{s}$ and M_W.

Studies to understand the contribution of the spin of sea quarks and the polarization of gluons to the total proton spin would help understand the spin structure of the nucleon. A further goal utilizing the polarized proton beams will be the precise test of fundamental symmetries such as the parity violation.

EXPERIMENTAL OVERVIEW

PHENIX is designed to detect leptons, photons, and hadrons with excellent momentum resolution and particle identification [4]. It is also designed to handle high recorded event rates (100 Hz for $Au + Au$ central events, ~ 10 kHz for $p + p$ minimum bias events) and trigger on the rare event using exclusive leptons, photons, and high p_T hadrons. As shown in Figure 1, PHENIX has east and west arms to cover the central rapidity region ($-0.35 \leq \eta \leq 0.35$), and north and south muon arms to cover the higher rapidity regions. Each of the four arms has a geometric acceptance of approximately one steradian. An axial magnetic field will be used in the central arms which measure photons, electrons and hadrons. The radial magnetic field will be used to measure muon momentum in the muon arms. The beam-beam counter having two arrays of quartz Čerenkov telescopes surrounding the beam pipe provides the time ($\sigma_T \sim 50$ ps) and longitudinal position of the interaction. The multiplicity vertex detector near the interaction point has two concentric barrels of silicon strips and two end-plates of silicon pads. It measures the vertex position ($\sigma_Z \sim 200$ μm) and charge multiplicity, N_{ch} and $d^2N_{ch}/d\eta d\phi$, in the wide range of pseudorapidity ($-2.5 \leq \eta \leq 2.5$).

Central Arms

Tracking in a central arm uses the information provided by the drift chambers, pad chambers, and time expansion chambers (TEC). The drift chambers provide precise projective measurements of particle trajectories, with a position resolution of 150 μm. The pad chambers provide three-dimensional space points which are essential for pattern recognition. $r - \phi$ information is provided by the TEC which will be used for particle identification also. Using this tracking system the mass resolution of $\phi \rightarrow e^+e^-$ is determined to better than 0.5 % for $p_T < 2$ GeV/c.

Particle identification is also realized by utilizing several detectors. A time-of-flight wall with excellent timing resolution ($\sigma_T \sim 85$ ps) covers part ($\Delta\phi = 30°$) of the central arm acceptance to separate kaons from pions up to 2.5 GeV/c. The π/K separation up to 1.4 GeV/c for the rest of the azimuthal coverage is available by the lead-scintillators ($\sigma_T \sim 300$ ps) for the electromagnetic calorimeter. For electron identification, information from the ring-imaging Čerenkov detector, the dE/dx measurement by the TEC, and information from the electromagnetic calorimeter are combined to give a π/e rejection of 10^{-4}. The electromagnetic calorimeter consisting of ~ 28000 cells has high resolution and high granularity. Each is about 15 to 18 radiation lengths long with a cross section area of about one Molière radius square, or $\Delta\eta \sim \Delta\phi \sim 0.01$. It has an energy resolution of $6 \sim 8$ %$/\sqrt{E}$ and can reliably separate π^0 mesons from single photons up to $p_T = 25$ GeV/c.

Muon Arms

The muon arms will be used to measure the production of vector mesons decaying into dimuons in heavy-ion collisions for masses ranging from that of the φ to the Υ. Measurements of the charmonium particles (J/ψ, ψ') along with Υ are crucial for the deconfinement study by providing a systematic understanding of the Debye screening effect, while the ϕ meson can be used for the studies of the chiral symmetry restoration and strangeness enhancement. The muon arms also allow studies of the continuum dilepton spectrum in a wide region of rapidity and mass by augmenting e^+e^- pairs measured in the central arms. The high mass continuum ($m_{\mu^+\mu^-} > 4$ GeV) is expected to arise primarily from hard scattering (Drell-Yan). The measurements of the Drell-Yan process and W production using high p_T single muon will be very important for the study of the spin physics. The $e\mu$ coincidences will probe heavy-quark production and help understand the shape of the continuum dilepton spectrum. Unambiguous measurements of the $D\bar{D}$ cross section can be made through the correlation of opposite-sign $e\mu$ pairs, by utilizing both of the central and muon arms. Since none of the opposite-sign $e\mu$ pairs are expected to arise from either the Drell-Yan or thermal production, like-sign $e\mu$ pairs can be used to remove the backgrounds due to random semi-leptonic decays of pions and kaons.

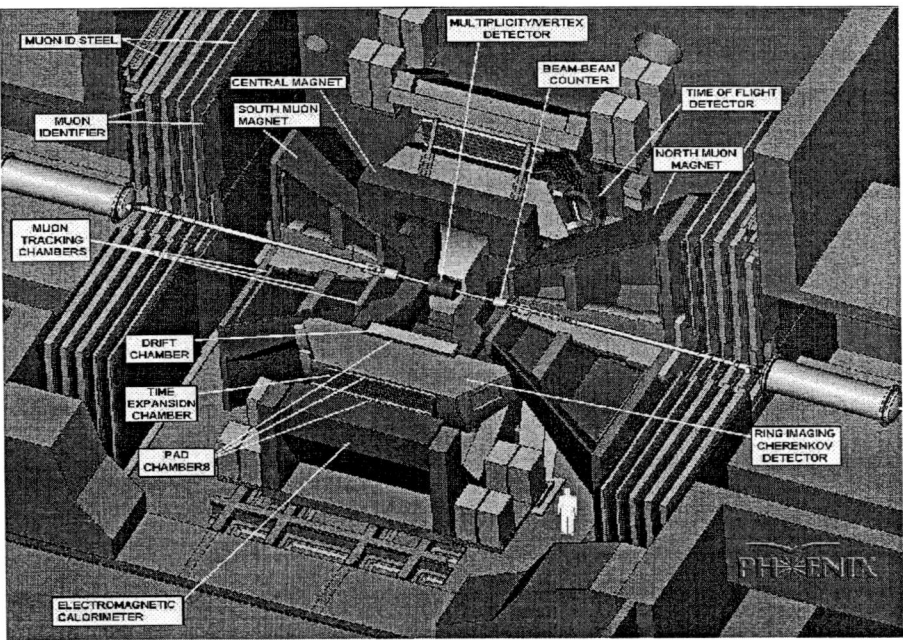

FIGURE 1. A schematic drawing of the PHENIX Experiment.

The coverage of muon arms is $-2.3 \leq \eta \leq -1.1$ and $1.1 \leq \eta \leq 2.4$. Following a hadron absorber, each has the muon tracker containing three stations of cathode strip chambers, which provide the position resolution of $100~\mu$m. The muon identifier after the tracker consists of panels of plastic proportional (Iarocci) tubes alternating with plates of steel absorber. The pion contamination of identified muons is $\sim 10^{-4}$, matching the high π/e rejection in the central arms. The excellent momentum resolution of identified tracks in the muon arms yields a mass resolution of 100 MeV/c^2 for $J/\psi \rightarrow \mu^+\mu^-$ and separates $\Upsilon(1S)$ from the $\Upsilon(2S+3S)$.

SUMMARY

The PHENIX experiment will study the deconfinement transition and the chiral symmetry restoration by measuring a wide range of proposed signatures. It is designed to detect leptons, photons, and hadrons with excellent momentum identification and particle identification. Due to its capabilities to detect prompt photons and leptons including muons, the experiment is very well suited to the measurements of the polarizations of gluon and anti-quark which greatly improve the understanding of the spin structure in the nucleon.

Acknowledgments

The author wishes to acknowledge the financial support of the Korean Research Foundation made in the program year of 1999 and the Yonsei University Research Fund (1999). Construction of the PHENIX has been supported by the Department of Energy (USA), Monbu-sho and STA (Japan), RSA, RMEA, and RMS (Russia), BMBF (Germany), FRN and the Knut & Alice Wallenberg Foundation (Sweden), and MIST and NSERC (Canada).

REFERENCES

1. Morrison D. P. *et al.*, *Nucl. Phys.* **A638**, 565c-569c (1998).
2. Saito N. *et al.*, *Nucl. Phys.* **A638**, 575c-578c (1998).
3. Harris J. W. and Müller B., *Annu. Rev. Nucl. Part. Sci.* **46**, 71-107 (1996).
4. *PHENIX Conceptual Design Report*, BNL 1993 (unpublished).

Spin and Strangeness in the Nucleon

Robert D. McKeown

W. K. Kellogg Radiation Laboratory[1]
California Institute of Technology
Pasadena, CA, USA 80307

Abstract. The static properties of the nucleon must ultimately arise from the quark and gluon constituents and the properties of Quantum Chromodynamics. The quark spin contribution to the spin of the nucleon is experimentally rather small, and there is some evidence that strange quarks are partly responsible. In addition, new efforts to address the role of gluons are in progress. The magnetic moment of the nucleon may contain contributions from strange quark-antiquark pairs, and this aspect of nucleon structure can be studied via parity-violating electron scattering. The present status of these experimental efforts and their impact on our knowledge of nucleon structure is discussed.

INTRODUCTION

In the decade since the discovery [1] by the EMC collaboration that the spin dependent structure function of the proton violated the Ellis-Jaffe (EJ) sum rule [2] we have seen tremendous interest in the study of the spin and flavor structure of the nucleon. Further experimental studies of inclusive spin-dependent deep inelastic scattering (DIS) have attained remarkable precision, and the theoretical analysis has become quite sophisticated. Nevertheless, detailed studies of the physical processes that are presumed responsible for the violation of the EJ sum rule are still in progress, including measurements of the strange quark contribution to the nucleon's quark spin content and the role of polarized gluons. The present status and future direction of these studies will be briefly addressed in this paper.

The possible role of the strange quarks in the spin structure of the nucleon and the EJ sum rule has spawned a great deal of interest in the study of the contribution of strange quarks to the neutral weak form factors of the nucleon. The axial form factor is directly related to the first moment of the nucleon spin-dependent structure function studied in DIS. However, the determination of strange quark vector matrix elements provides completely new information on the role of strange quarks in the structure of the nucleon [3]. These can be accessed through measurements of parity-violating electron scattering [4,5] and an extensive program

---

[1] Supported by the National Science Foundation.

of experiments is in progress to address this issue. The status and future directions in this subject are also discussed in this paper.

SPIN-DEPENDENT DEEP INELASTIC SCATTERING

Since the pioneering work of the EMC collaboration [1], a great deal of experimental [6–9] and theoretical [10,11] work has been done on this topic. The inclusive scattering of a charged lepton from a nucleon at large momentum transfer ($Q^2 > 1$ GeV^2) can be described by four structure functions: F_1, F_2, g_1, and g_2. Each of these functions depends on the kinematic variables Q^2 (the squared momentum transfer) and the Bjorken scaling variable x (in the laboratory frame where the nucleon is initially at rest $x = Q^2/2M\nu$, where M is the nucleon mass and ν is the energy transfer in this frame). In measurements of spin-dependent deep inelastic scattering, the two structure functions g_1 and g_2 can be determined. Of most interest for the present discussion is the leading twist-2 structure function g_1, which contains information related to the quark spin structure of the nucleon. The first moment of g_1 has a simple interpretation in the Quark Parton Model (QPM)

$$\Gamma_1(Q^2) \equiv \int_0^1 g_1(x, Q^2)dx \tag{1}$$

$$= \frac{1}{2}\sum_f e_f^2 \Delta q_f \tag{2}$$

where $\Delta q_f = \int[q_f^{++}(x) - q_f^{-+}(x)]dx$ in which $q_f^{++}(x)$ is the probability distribution corresponding to quark (flavor f) having positive helicity in a positive helicity nucleon and q_f^{-+} is the probability distribution corresponding to the quark having negative helicity in a positive helicity nucleon. The three flavor combinations that contribute to Γ_1 are (again in the QPM)

$$a_3 = \Delta u - \Delta d \tag{3}$$

$$a_8 = \Delta u + \Delta d - 2\Delta s \tag{4}$$

$$a_0 = \Delta\Sigma \equiv \Delta u + \Delta d + \Delta s . \tag{5}$$

The Ellis-Jaffe sum rule [2] predictions for Γ_1 are obtained from the above expressions by equating $a_3 = F + D$ and $a_8 = 3F - D$, where F and D are obtained from axial matrix elements in hyperon and nucleon beta decays, and also assuming $\Delta s = 0$ (so $a_8 = a_0$).

In a more detailed treatment in next-to-leading-order (NLO) QCD radiative effects and Q^2 evolution are included, with the result that Γ_1 depends upon Q^2 as does the flavor decomposition [10]. In particular, the singlet axial charge is written

$$a_0 = \Delta\Sigma - 3\frac{\alpha_s}{2\pi}\Delta G \tag{6}$$

where ΔG is the first moment of the analogous gluon spin-structure function

$$\Delta G = \int \Delta g \, dx = \int [g^{++}(x) - g^{+-}(x)] dx \ . \tag{7}$$

In practice the inclusive deep-inelastic scattering measurements of Γ_1^p and Γ_1^n can be used to yield meaningful information on the triplet and singlet combinations a_3 and a_0. The combination $\Gamma_1^p - \Gamma_1^n$ isolates the triplet and tests the Bjorken sum rule, which relates $a_3 = g_A$ where $g_A = 1.26$ from neutron beta decay experiments [11]. Experimentally, this relation is verified to better than 10%.

The fact that Γ_1^p is experimentally much smaller than the Ellis-Jaffe prediction implies that the value of a_0 is also much smaller than naively expected, $a_0 \sim 0.15 - 0.3 \ll 1$. From Equations 5 and 6, one sees several possibilities for this reduction in the singlet axial charge. The fits to the available spin-dependent DIS data favor both a negative contribution from the strange quarks (typically $\Delta s \simeq -0.1$) and a positive value of the gluon polarization $\Delta G(Q^2 = 1\,\text{GeV}^2) \sim 1.5$. The gluon polarization result is very uncertain and is favored by the apparent degree of Q^2 evolution of $g_1(x, Q^2)$ present in the experimental data. The situation with a_0 leads to the following two questions which are being addressed by further experimental study:

- What is the role of strange quarks in nucleon structure, and is $\Delta s < 0$ partly responsible for $a_0 \ll 1$?

- What is the role of gluons in the spin structure of the nucleon?

NEW RESULTS FROM HERMES

The HERMES experiment [12] at the HERA storage ring at DESY is capable of a variety of measurements to further study the spin structure of the nucleon. One method is to explore the flavor decomposition of the quark spin contributions via spin-dependent semi-inclusive deep inelastic scattering (SIDIS). In SIDIS, one observes an energetic final state hadron in coincidence with the DIS lepton. In addition to the DIS variables Q^2 and x one observes the fractional energy transfer to the hadron $z \equiv E_h/\nu$, where E_h is the energy of the detected hadron. (We generally integrate over the hadron transverse momentum.) The cross section for this process can be written

$$\frac{d\sigma^h}{dx \, dQ^2 \, dz} \propto \sum_f e_f^2 q_f(x, Q^2) D_f^h(z, Q^2) \tag{8}$$

where $q_f(x, Q^2)$ is the parton distribution function (as in DIS) and $D_f^h(z, Q^2)$ is the fragmentation function corresponding to the probability that the quark of flavor f hadronizes into hadron type h with energy fraction z.

In a similar fashion, the spin-dependent asymmetry for polarized SIDIS is

$$A_1^h \propto \frac{\sum_f e_f^2 \Delta q_f(x, Q^2) D_f^h(z, Q^2)}{\sum_f e_f^2 q_f(x, Q^2) D_f^h(z, Q^2)} \tag{9}$$

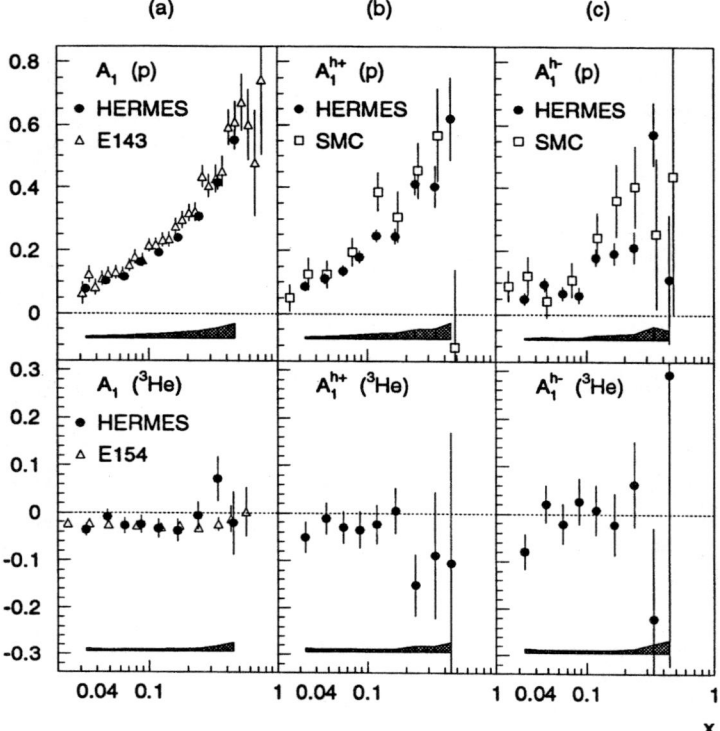

FIGURE 1. Semi-inclusive asymmetries measured in the HERMES and SMC experiments for positive (h^+) and negative (h^-) hadrons on proton and ^{3}He targets. Inclusive asymmetries measured at HERMES and SLAC are also shown for comparison.

where now we define $\Delta q_f(x, Q^2) \equiv q_f^{++}(x, Q^2) - q_f^{-+}(x, Q^2)$. The fragmentation functions are rather well-known from measurements of unpolarized SIDIS distributions, and thus this method enables the determination of additional linear combinations of the spin dependent parton distribution functions $\Delta q_f(x, Q^2)$.

Examples of the asymmetries for SIDIS measured by HERMES for proton and ^{3}He targets are shown in Figure 1. These measurements allow extraction of individual combinations of $\Delta q_f(x, Q^2)$ as shown in Figure 2. Further measurements with the enhanced particle identification due to the addition of a ring-imaging Cerenkov counter will enable K^+ SIDIS asymmetries to be determined at HERMES. Thus we can expect more detailed information in the near future, and the K^+ asymmetries should provide more direct access to the strange quark polarizations.

Another area of current study at HERMES is the spin-dependence in high p_T pion photoproduction. At large p_T, model calculations indicate that the dominant

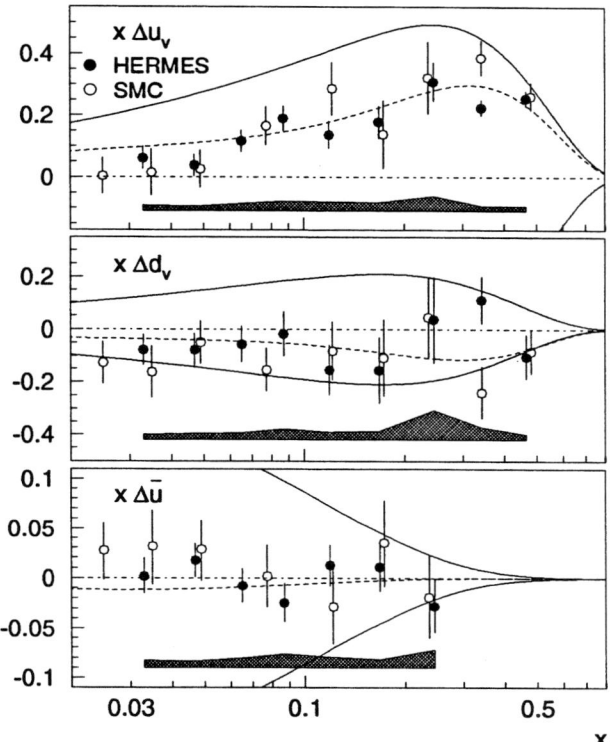

FIGURE 2. Parton distribution functions determined from semi-inclusive asymmetries measured in the HERMES and SMC experiments. The subscript V indicates valence distributions. The solid lines indicate the positivity limit and the dashed lines are a parametrization due to Gehrmann and Stirling. [13]

photoproduction process for π^+-π^- pairs is the photon gluon fusion (PGF) process. This is a variation of a method proposed by Bravar, von Harrach, and Kotzinian [14]. Thus, one could potentially obtain information on the gluon polarization $\Delta g(x, Q^2)$ by studying the spin-dependence of this process. Hadron pairs (predominantly pion pairs) are produced at HERMES by quasi-real polarized photons arising from small-angle polarized positron scattering. The observed asymmetries for this process when at least one of the two hadrons has $p_T > 1.5$ GeV/c are shown in Figure 3. There are substantial negative asymmetries observed. One should note that, for a proton target,

- the asymmetry for photon absorption by quarks is expected to be *positive* as observed in Figure 1,

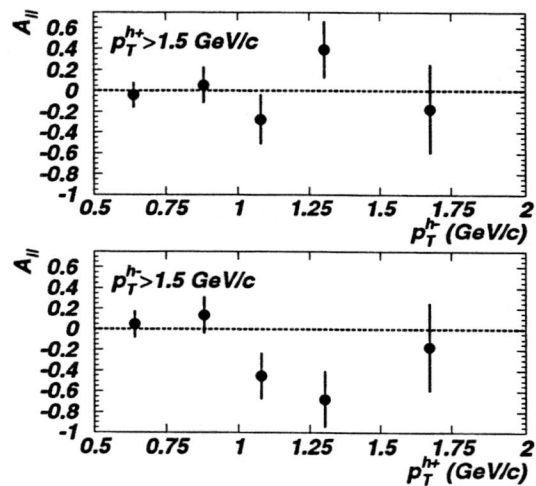

FIGURE 3. Asymmetries measured in high p_T photoproduction of hadron pairs in the HERMES experiment. The top (bottom) panel requires the positive (negative) hadron to have transverse momentum $p_T > 1.5$ GeV/c.

- the asymmetry for photon gluon fusion is expected to be substantially negative (for $\Delta g > 0$), as is the case for the photoproduced hadron pairs measured at HERMES.

Thus, we observe that, in the region where PGF is expected to dominate the measured yield of pion pairs, the asymmetry tends to be *negative*. However, the limitations of the theory and model calculations of these pairs does not allow a definitive statement of the magnitude of Δg at this time and the interpretation of these data remains qualitative. Nevertheless, it is intriguing that we seem to be observing the first direct evidence that $\Delta g > 0$ at $x \simeq 0.17$.

THE SAMPLE EXPERIMENT

As first discussed by Kaplan and Manohar [3], strange quark-antiquark pairs may contribute significantly to vector matrix elements of the nucleon (such as the magnetic moment) and the strange quark components can be accessed experimentally through study of the neutral weak current. Whereas the normal magnetic moment corresponds to the magnetic coupling to the photon, the weak magnetic moment represents the analogous coupling to the Z boson. The weak magnetic moment is equally fundamental and just as important as the electromagnetic moment first measured in 1933.

The weak magnetic moment of the proton can be measured via the small parity-violating effect in the elastic scattering of longitudinally polarized electrons from protons [4]. Using this technique, the SAMPLE experiment is providing the very first determination of the proton's weak magnetic moment and therefore the first glimpse of the role of strange quarks in the magnetic properties of the nucleon.

The magnetic moments of the proton and neutron presently provide two constraints on the flavor decomposition, but a third quantity is necessary to uniquely determine all three contributions. By utilizing the (already known) proton (μ_p) and neutron (μ_n) magnetic moments, the neutral weak magnetic moment of the nucleon can be related to the contribution from strange quarks:

$$\mu_p^Z = (\mu_p - \mu_n) - 4\sin^2\theta_w\,\mu_p - \mu_s \tag{10}$$

where θ_w is the weak mixing angle and μ_s is the strange quark contribution. In the limit $q^2 = 0$ the matrix element

$$\langle N|\bar{s}\gamma^\mu s|N\rangle_{q^2\to0} = \mu_s\bar{U}_N[\frac{i}{2M_N}\sigma^{\mu\nu}q_\nu]U_N \tag{11}$$

defines the quantity μ_s. No other experimental information on μ_s presently exists. Thus, the SAMPLE experiment provides the first experimental information on μ_s.

Since the initial suggestion by Kaplan and Manohar, there have been many theoretical calculations of the strange magnetic moment μ_s [15–32]. Most of these result in values of μ_s in the range $-0.5 \to 0.$, including a recent lattice study [32]. However, there are 2 calculations [20,31] in the chiral bag model that yield $\mu_s > 0$.

The experiment is performed using a 200 MeV longitudinally polarized electron beam incident on a liquid hydrogen target. The scattered electrons are detected in a large solid angle (~ 1.5 sr) Cerenkov detector at backward angles $130° < \theta < 170°$. This results in an average $Q^2 \simeq 0.1$ (GeV/c)^2. If one assumes very small strange quark contribution, the parity-violating asymmetry would be -7.2×10^{-6} (i.e., -7.2 ppm).

The detector signals are integrated over the $\sim 15\mu$sec of the beam pulse and digitized. The beam intensity is similarly integrated and digitized. The ratio of integrated detector signal to integrated beam charge is the normalized yield which is proportional to cross section (plus background). The goal is to measure the beam helicity dependence of the cross section, or equivalently, the helicity dependence of the normalized yield.

The helicity of the polarized electron beam is randomly chosen for each of 10 consecutive beam pulses (the beam repetition rate is 600 Hz) and then the complement helicities are used for the next 10 pulses. The normalized detector asymmetry is computed for "pulse pairs" separated by 1/60 of a second to minimize systematic errors. The electron polarization is measured using a Moller apparatus on the beamline typically once per day.

In the measurement of a small asymmetry, it is essential that the experimental conditions (i.e., beam properties) remain as identical as possible when the helicity is reversed. There are many steps taken to insure this basic goal. An especially

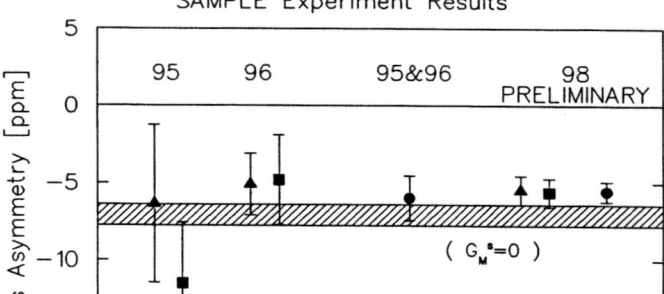

FIGURE 4. Results for the average parity-violating asymmetry measured in the 1995/6 and 1998 running periods. For each running period we display the result for the "$\lambda/2$ normal" (squares) and "$\lambda/2$ reverse" (triangles) settings of slow helicity reversal, which are in good agreement. The combined result for both helicity states and the 95/96 running periods and the 1998 period is also shown (filled circles). The error bars include statistical errors only. The hatched region indicates the asymmetry band (due to the uncertain axial radiative correction) for $\mu_s = G^s_M = 0$.

important one is to actively feedback the helicity correlated beam intensity difference to the polarized injector; this system reduces the helicity correlations in the beam intensity from typically 100-200 ppm to ~ 1 ppm.

Residual differences in the beam parameters are monitored, continuously if possible. The position and angle at the target in both transverse dimensions (x and y), the beam energy, and the "halo" of the beam are continuously monitored for every beam pulse. The detector normalized yield asymmetry can be corrected for the helicity correlated effects observed in these beam monitors. Other properties, such as the size of the beam, can be monitored during special runs to search for such effects.

Finally, we also have the capability to manually reverse the beam helicity by inserting a $\lambda/2$ plate which reverses the helicity of the light (and the beam) relative to all electronic signals. A real parity violation signal will change sign under this "slow reversal". Electronic crosstalk and other effects will not change under "slow reversal", so this is an important test that our signal is real physics rather than a spurious systematic effect in the experiment.

We have studied the composition of the detected signal by using special runs at low beam current and computer simulations. We find that 55% of the detected signal is due to elastic scattering from protons; the measured asymmetry is corrected for this dilution of the signal.

During 1995/6 the SAMPLE experiment acquired its first production data, and these results were previously published [33]. The measured asymmetry indicated that the strange magnetic form factor was positive (contrary to most theoretical calculations) although the experimental uncertainty was large.

During 1998, a more substantial data set was acquired and many improvements were implemented to reduce systematic errors. The preliminary physics asymmetries for each of the slow helicity states in this data run are displayed in Figure 4 along with the 1995/6 results. All the measurements are consistent and are well behaved under $\lambda/2$ reversal. The 1998 data show a substantial improvement in the statistical precision. The final analysis of the 1998 data is still in progress and almost complete.

During the summer of 1999, we plan to run the experiment with deuterium in the target. The deuterium asymmetry is quite sensitive to the isovector axial radiative correction, but is almost completely insensitive to the strange magnetic moment. Thus, we can reduce our uncertainty in μ_s by using deuterium and proton data together to almost eliminate the radiative correction uncertainty. (The hatched region in Figure 4 will be essentially reduced to a line.) The deuterium experiment is thus crucial to a precise quantitative interpretation of the hydrogen data.

CONCLUSIONS

The experimental study of the spin and flavor structure of the nucleon is a very active field at the present time. Substantial progress has been achieved in our understanding of nucleon structure and the role of QCD. In the area of parton distribution functions, we are beginning to see more detailed information emerge from spin-dependent semi-inclusive deep inelastic scattering experiments such as HERMES. High p_T photoproduction is a promising method of addressing the role of gluon spin. Further studies of this type will be carried out by the upcoming COMPASS experiment at CERN [34]. In addition, we can hopefully expect data at lower x from polarized collider experiments in the future.

A substantial program of experiments to explore the neutral weak elastic form factors of the nucleon in parity-violating electron scattering is presently in progress. The SAMPLE experiment is nearing completion and will provide new information on the strange quark contribution to the proton's magnetic moment. Experiments at other laboratories [35–37] are in progress or under construction, and we can expect a complete mapping of the Q^2 dependence of the strange vector form factors over the next few years.

REFERENCES

1. J. Ashman, *et al.*, Nucl. Phys. B **364**, 1 (1989).
2. J. Ellis and R. Jaffe, Phys. Rev. **D 10**, 1444 (1974).
3. D. Kaplan and A. Manohar, Nucl. Phys. B **310**, 527 (1988).
4. R. D. McKeown, Phys. Lett. **B219**, 140 (1989).
5. D. H Beck, Phys. Rev. **D39**, 3248 (1989).
6. A. Airapetian, *et al.*, Phys. Lett. **B442**, 484 (1998).
7. B. Adeva, *et al.*, Phys. Rev. **D58**, 112001 (1998).
8. K. Abe, *et al.*, Phys. Rev. **D58**, 112003 (1998).
9. K. Abe, *et al.*, Phys. Rev. Lett. **79**, 26 (1997).
10. R. D. Ball, S. Forte, and G. Ridolfi, Phys. Lett. **B378**, 255 (1996).
11. G. Altarelli, *et al.*, Nucl. Phys. B **496**, 337 (1997).
12. K. Ackerstaff, *et al.*, Nucl. Instr. Meth. **A417**, 230 (1998).
13. T. Gehrmann and W. J. Stirling, Phys. Rev. **D53**, 6100 (1996).
14. A. Bravar, D. von Harrach, and A. Kotzinian, Phys. Lett **B421** 349 (1998).
15. R. L. Jaffe, Phys. Lett. **B229**, 275 (1989).
16. M. J. Musolf and M. Burkardt, Z. Phys. **C61**, 433 (1994).
17. W. Koepf, E. M. Henley, and S. J. Pollock, Phys. Lett **B288** 11 (1992).
18. B. R. Holstein in *Proceedings of the Caltech Workshop on Parity Violation in Electron Scattering*, E. J. Beise and R. D. McKeown, Eds., World Scientific, 27 (1990).
19. N. W. Park, J. Schechter, and H. Weigel, Phys. Rev. **D43**, 869 (1991).
20. S. Hong and B. Park, Nucl. Phys. **A561**, 525 (1993).
21. S. C. Phatak and Sarira Sahu, Phys. Lett **B321** 11 (1994).
22. Chr. V. Christov, *et al.,* Prog. Part. Nucl. Phys. **37**, 1 (1996).
23. H.-W. Hammer, U.-G. Meissner, and D. Drechsel, Phys. Lett. **B367**, 323 (1996).
24. H. Ito, Phys. Rev. **C52**, R1750 (1995).
25. H. Weigel, *et al.*, Phys. Lett. **B353**, 20 (1995).
26. D. Leinweber, Phys. Rev. **D53**, 5115 (1996).
27. P. Geiger and N. Isgur, preprint hep-ph/9610445.
28. M. J. Musolf, H.-W. Hammer, and D. Drechsel, Phys. Rev. **D55**, 299 (1997).
29. M. J. Musolf and H. Ito, Phys. Rev. **C 55** 3066 (1997).
30. U.-G. Meissner *et al.*, Phys. Lett. **B408**, 381 (1997).
31. S-T. Hong, B-Y. Park, and D-P. Min, Phys. Lett. **B414**, 229 (1997).
32. S. J. Dong, K. F. Liu, and A. G. Williams, Phys.Rev. D58, 074504 (1998).
33. B. A. Mueller *et al.*, Phys. Rev. Lett. **78**, 3824 (1997).
34. CERN SPSLC 96-14, Mar. 1996.
35. K. A. Aniol, *et al.*, Phys. Rev. Lett. **83**, 1096 (1999).
36. Mainz proposal A4/1-93 94-11 (D. von Harrach, spokesperson).
37. TJNAF experiment E91-017 (D. Beck, spokesperson).

Dileptons from P-Nucleus Collisions

Jen-Chieh Peng

Physics Division, Los Alamos National Laboratory
Los Alamos, New Mexico 87545

Abstract. Recent results from fixed-target dimuon production experiments at Fermilab are presented. Various QCD tests using dilepton production data are discussed. We emphasize that clear evidence for scaling violation in the Drell-Yan process remains to be established. Further theoretical and experimental work are needed to understand the polarization of the Drell-Yan pairs. We also discuss the nuclear medium effects for dilepton productions. In particular, we discuss the use of Drell-Yan data to deduce the energy-loss of partons traversing nuclear medium.

INTRODUCTION

The experimental detection of high-mass lepton pairs produced in hadronic reactions has a long and rich history. The famous quarkonium states that revealed the existence of the charm and beauty quarks in the 1970s were discovered through their dilepton decay branches. They are superimposed on a continuum, which was anticipated theoretically in 1970 [1], and is now known as the Drell-Yan (DY) process. The DY process, electromagnetic quark-antiquark annihilation, is closely related to the deeply inelastic lepton scattering (DIS). By 1980, DY production was already a source of information about antiquark structure of the nucleon. Additionally, DY production with beams of pions and kaons yielded the structure functions of these unstable particles for the first time. Also notable in the history of the DY process were the discoveries of the $W^\pm$ and Z^0 particles in 1983, produced by a generalized (vector boson exchange) quark-antiquark annihilation mechanism.

New experimental work has been carried out in recent years by few but prolific collaborations working in the fixed-target programs at the CERN SPS accelerator and at Fermilab. A series of fixed-target dimuon production experiments (E772, E789, E866) have been carried out at Fermilab in the last 10 years. Some of the highlights from these experiments, namely the observation of pronounced flavor asymmetry in the nucleon sea and the absence of antiquark enhancement in heavy nuclei, are discussed by Garvey at this Conference. In this paper, we will focus on other areas of dilepton physics studied in these experiments. To follow the main theme of this Conference, we first discuss the status of various QCD tests using

CP494, *New Directions in Quantum Chromodynamics*, edited by C.-R. Ji and D.-P. Min
© 1999 American Institute of Physics 1-56396-908-4/99/$15.00

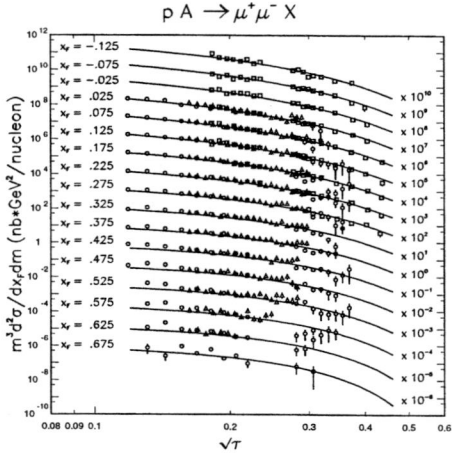

$$pA \rightarrow \mu^+\mu^- X$$

FIGURE 1. Proton-induced Drell-Yan production from experiments NA3 [2] (triangles) at 400 GeV/c, E605 [3] (squares) at 800 GeV/c, and E772 [4] (circles) at 800 GeV/c. The lines are absolute (no arbitrary normalization factor) next-to-leading order calculations for $p + d$ collisions at 800 GeV/c using the CTEQ4M structure functions [5].

dilepton productions. We will then discuss the nuclear medium effects for dilepton productions. In particular, the relevance of the Fermilab experiments on the issue of parton energy loss in nuclei will be presented.

QCD TESTS IN DILEPTON PRODUCTION

In the "Naive" Drell-Yan model, the differential cross section is given as

$$M^3 \frac{d^2\sigma}{dMdx_F} = \frac{8\pi\alpha^2}{9} \frac{x_1 x_2}{x_1 + x_2} \times \sum_a e_a^2 [q_a(x_1)\bar{q}_a(x_2) + \bar{q}_a(x_1)q_a(x_2)]. \tag{1}$$

Here $q_a(x)$ are the quark or antiquark structure functions of the two colliding hadrons evaluated at momentum fractions x_1 and x_2. The sum is over quark flavors. The Feynman-x (x_F) is equal to $x_1 - x_2$.

Although the simple parton model originally proposed by Drell and Yan enjoyed considerable success in explaining many features of the early data, it was soon realized that QCD corrections to the parton model were required. Historically, two experimental features demanded theoretical improvement: first, the experimental cross section was about a factor of two larger than the parton-model value, and second, the distribution of dilepton transverse momenta extended to much larger values than are characteristic of the convolution of intrinsic parton momenta. We now discuss several consequences of QCD corrections to the DY observables.

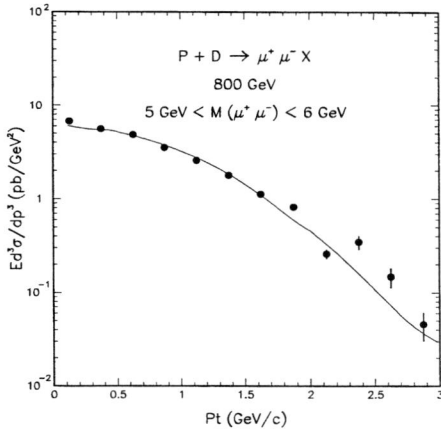

FIGURE 2. Comparison between the NLO calculation from Ref. [6] and the E772 data [4].

Absolute Cross Sections and p_T Distribution

The inclusion of the NLO diagrams for the DY process brings excellent agreements between the calculations and the data. As an example, Figure 1 shows the NA3 data [2] at 400 GeV, together with the E605 [3] and E772 [4] data at 800 GeV. The solid curve in Figure 1 corresponds to NLO calculation for 800 GeV $p+d$ ($\sqrt{s}$ = 38.9 GeV) and it describes the NA3/E605/E772 data well.

Berger et al. [6] recently compared their NLO calculations with the E772 data. As shown in Figure 2, the p_T distribution is well reproduced by the calculation. At $p_T > 2$ GeV/c, the DY cross section is shown to be dominated by processes involving gluons. This suggests the interesting possibility of probing gluon density using large p_T DY events.

Scaling Violation

The right-hand side of Eq. (1) is only a function of x_1, x_2 and is independent of the beam energies. This scaling property no longer holds when QCD corrections to the DY are taken into account.

While logarithmic scaling violation is well established in DIS experiments, it is not well confirmed in DY experiments at all. No evidence for scaling violation is seen. As discussed in a recent review [7], there are mainly two reasons for this. First, unlike the DIS, the DY cross section is a convolution of two structure functions. Scaling violation implies that the structure functions rise for $x \leq 0.1$ and drop for $x \geq 0.1$ as Q^2 increases. For proton-induced DY, one often involves a beam quark with $x_1 > 0.1$ and a target antiquark with $x_2 < 0.1$. Hence the effects of scaling

$$p\,A \rightarrow \mu^+\mu^-\,X$$

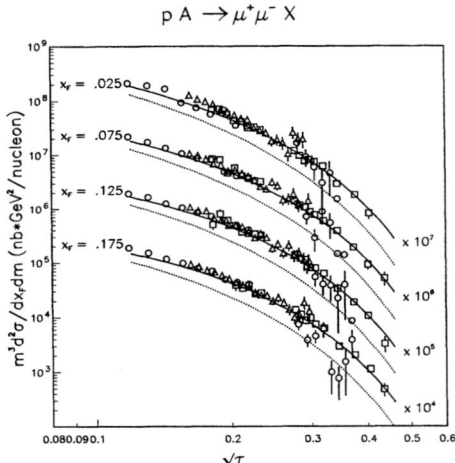

FIGURE 3. Comparison of DY cross section data with NLO calculations using MRST [10] structure functions. Note that $\tau = x_1 x_2$. The E772 [4], E605 [3], and NA3 [2] data points are shown as circles, squares, and triangles, respectively. The solid curve corresponds to fixed-target p+d collision at 800 GeV, while the dotted curve is for p+d collision at $\sqrt{s} = 500$ GeV.

violation are partially cancelled. Second, unlike the DIS, the DY experiment can only probe relatively large Q^2, namely, $Q^2 > 16$ GeV^2 for a mass cut of 4 GeV. This makes it more difficult to observe the logarithmic variation of the structure functions in DY experiments.

Possible indications of scaling violation in DY process have been reported in two pion-induced experiments, E326 [8] at Fermilab and NA10 [9] at CERN. E326 collaboration compared their 225 GeV $\pi^- + W$ DY cross sections against calculations with and without scaling violation. They observed better agreement when scaling violation is included. This analysis is subject to the uncertainties associated with the pion structure functions, as well as the nuclear effects of the W target. The NA10 collaboration measured $\pi^- + W$ DY cross sections at three beam energies, namely, 140, 194, and 286 GeV. By checking the ratios of the cross sections at three different energies, NA10 largely avoids the uncertainty of the pion structure functions. However, the relatively small span in $\sqrt{s}$, together with the complication of nuclear effects, make the NA10 result less than conclusive.

RHIC provides an interesting opportunity for unambiguously establishing scaling violation in the DY process. Figure 3 shows the predictions for $p + d$ at $\sqrt{s} = 500$ GeV. The scaling-violation accounts for a factor of two drop in the DY cross sections when $\sqrt{s}$ is increased from 38.9 GeV to 500 GeV. It appears quite feasible to establish scaling violation in DY with future dilepton production experiments at RHIC.

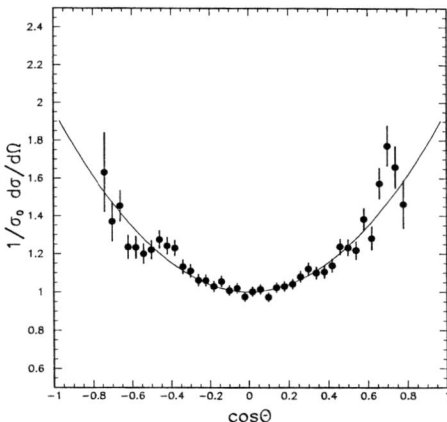

FIGURE 4. Drell-Yan angular distribution from Fermilab E772 [12]: $p + Cu$ collisions at 800 GeV/c. The dimuons cover the mass region $11 \leq M_{\mu^+\mu^-} \leq 17$ GeV/c^2 with $-0.3 \leq x_F \leq 0.8$ and $p_t \leq 6$ GeV/c. Mean values for p_t, x_F, and M are 1.4 GeV/c, 0.16, and 11.9 GeV/c^2, respectively. The solid curve is a fit to the data with the form $1 + \lambda cos^2\theta$, where λ is $0.96 \pm .04 \pm .06$.

Decay Anugular Distributions

In the parton model, the angular distribution of dileptons is characteristic of the decay of a transversely polarized virtual photon,

$$\frac{d\sigma}{d\Omega} = \sigma_0(1 + \lambda cos^2\theta), \tag{2}$$

where θ is the polar angle of the lepton in the virtual photon rest frame and $\lambda = 1$. Early experimental data from both pion and proton beams [11] were consistent with this form but had large statistical errors.

Recently, E772 has performed a high-statistics study of the angular distribution for DY events [12] with masses above the Υ family of resonances. About 50,000 events were recorded from 800 GeV/c $p + Cu$ collisions, using a copper beam dump as the target. Figure 4 shows the acceptance-corrected angular distribution, integrated over the kinematic variables. Analyzed in the Collins-Soper reference frame [13], the data yield $\lambda = 0.96 \pm 0.04 \pm 0.06$(systematic).

Including higher-order QCD corrections to the DY process [14,15] results in the more complicated form of the angular distribution,

$$\frac{d\sigma}{d\Omega} \propto 1 + \lambda cos^2\theta + \mu sin2\theta cos\phi + \frac{\nu}{2}sin^2\theta cos2\phi, \tag{3}$$

where ϕ is the azimuthal angle and λ, μ, and ν are angle-independent parameters. NLO calculations predict [16] small deviations from $1 + cos^2\theta$ ($\leq 5\%$) for below 3

GeV/c. The relevant scaling parameter for the magnitude of these deviations is p_t/Q, implying that NLO corrections become important when $p_t \simeq Q$. A relation, $1 - \lambda - 2\nu = 0$, developed by Lam & Tung [17], is analogous to the Callan-Gross relation in DIS. Measurements with pion beams at CERN [18] and at Fermilab [19] have shown that the Lam-Tung relation is clearly violated at large p_t.

Pion-induced DY experiments have unexpectedly shown that transverse photon polarization changes to longitudinal ($\lambda \simeq -1$) at large x_F [18–21]. The dependence of λ is qualitatively consistent with a higher-twist model originally proposed by Berger & Brodsky [22,23]. However, the quantitative agreement is poor. The model's basis can be described as follows. As of the muon pair approaches unity, the Bjorken-x (momentum fraction) of the annihilating projectile parton must also be near unity. Thus, the whole pion contributes to the DY process. This can be treated with perturbation theory, with the result that the transverse polarization of the virtual photon becomes longitudinal. The angular distribution at large becomes

$$\frac{d\sigma}{d\Omega} \propto (1 - x)^2(1 + \lambda cos^2\theta) + \alpha sin^2\theta, \qquad (4)$$

where α is $\propto p_t^2/Q^2$.

Eskola et al [24] have shown that an improved treatment of the effects of nonasymptotic kinematics greatly improves quantitative agreement with the λ values from the pion data. Brandenburg et al [25] have extended the higher twist model to specifically include pion bound-state effects. They predict values for λ, μ and ν that are in good agreement with the pion data at large x_F. Unfortunately, the results are quite sensitive to the choice of the pion Fock state wave functions, which are not well constrained by experimental data.

NUCLEAR MEDIUM EFFECTS OF DILEPTON PRODUCTION

From a high-statistics measurement of dilepton production in 800 GeV proton-nucleus interaction, the target-mass dependence of DY, J/Ψ, Ψ', and Υ productions have been determined in E772 [26–28]. As shown in Figure 5, different nuclear dependences are observed for different dilepton processes. While the DY process shows almost no nuclear dependence, pronounced nuclear effects are seen for the production of heavy quarkonium states. E772 found that J/Ψ and Ψ' have similar nuclear dependence. The nuclear dependences for Υ, Υ' and Υ'' are less than that observed for the J/Ψ and Ψ'. Within statistics, the various Υ resonances also have very similar nuclear dependences.

Although the integrated DY yields in E772 show little nuclear dependence, it is instructive to examine the DY nuclear dependences on various kinematic variables. Using the simple A^α expression to fit the DY nuclear dependence, the values of α are shown in Figure 6 as a function of $x_T(x_2)$, M, x_F, and p_t. Several features are observed:

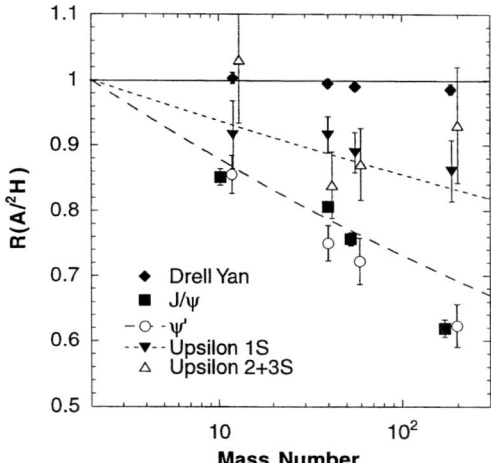

FIGURE 5. Ratios of heavy-nucleus to deuterium integrated yields per nucleon for 800 GeV proton production of dimuons from the Drell-Yan process and from decays of the J/ψ, ψ', $\Upsilon(1S)$, and $\Upsilon(2S+3S)$ states [7]. The short dash and long dash curves represent the approximate nuclear dependences for the $b\bar{b}$ and $c\bar{c}$ states, $A^{0.96}$ and $A^{0.92}$, respectively.

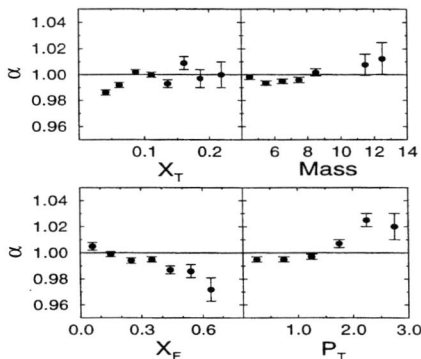

FIGURE 6. Nuclear dependence coefficient α for 800 GeV p+A Drell-Yan process versus various kinematic variables [26].

1. A suppression of the DY yields from heavy nuclear targets is seen at small x_2. This is consistent with the shadowing effect observed in DIS. In fact, E772 provides the only experimental evidence for shadowing in hadronic reactions. The reach of small x_2 in E772 is limited by the mass cut ($M \geq 4$ GeV) and by the relatively small center-of-mass energy (recall that $x_1x_2 = M^2/s$). p-A collisions at RHIC clearly offer the exciting opportunity to extend the study of shadowing to much smaller x.

2. $\alpha(x_F)$ shows an interesting trend, namely, it decreases as x_F increases. It is tempting to attribute this behavior to initial-state energy-loss effect. However, there is a strong correlation between x_F and x_2 ($x_F = x_1 - x_2$), and it is essential to separate the x_F energy-loss effect from the x_2 shadowing effect. Figure 7 shows α versus x_F for two bins of x_2, one in the shadowing region ($x_2 \leq 0.075$) and one outside of it ($x_2 \geq 0.075$). There is no discernible x_F dependence for α once one stays outside of the shadowing region. Therefore, the apparent suppression at large x_F in Figure 6 reflects the shadowing effect at small x_2 rather than the energy-loss effect.

3. $\alpha(p_t)$ shows an enhancement at large p_t. This is reminiscent of the Cronin Effect [29] where the broadening in p_t distribution is attributed to multiple parton-nucleon scatterings. It is instructive to compare the p_t broadening for DY process and quarkonium production. Figure 8 shows $\Delta\langle p_t^2 \rangle$, the difference of mean p_t^2 between p-A and p-D interactions, as a function of A for DY, J/Ψ, and $\Upsilon(1S)$ productions at 800 GeV. The DY and Υ data are from E772 [30], while the J/Ψ results are from E789 [31], E771 [32], and preliminary E866 analysis [33]. More details on this analysis will be presented elsewhere [30]. Figure 8 shows that $\langle p_t^2 \rangle$ is well described by the simple expression $a+bA^{1/3}$. It also shows that the p_t broadening for J/Ψ is very similar to Υ, but significantly larger (by a factor of 5) than the DY. A factor of 9/4 could be attributed to the color factor of the initial gluon in the quarkonium production versus the quark in the DY process. The remaining difference could come from the final-state multiple scattering effect which is absent in the DY process.

Baier et al. [34] have recently derived a relationship between the partonic energy-loss due to gluon bremsstrahlung and the mean p_t^2 broadening accumulated via multiple parton-nucleon scattering:

$$-dE/dz = \frac{3}{4} \, \alpha_s \, \Delta\langle p_t^2 \rangle. \tag{5}$$

This non-intuitive result states that the total energy loss is proportional to square of the path length traversed by the incident partons. From Figure 8 and Eq. 5, we deduce that the mean total energy loss, ΔE, for the p+W DY process is ≈ 0.6 GeV. Such an energy-loss is too small to cause any discernible effect in the x_F (or x_1) nuclear dependence. As shown in Figure 7, the dashed curve corresponds to $\Delta E = 2.0 \pm 1.7$ GeV (for p+W), and the E772 data are consistent with Eq. 5. A

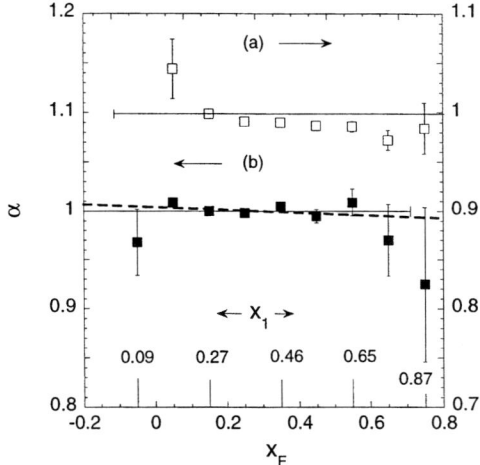

FIGURE 7. Nuclear dependence coefficient α for the Drell-Yan process [26] versus x_F for (a) $x_2 \leq 0.075$, right scale, and (b) $x_2 \geq 0.075$, left scale. The thin solid lines show $\alpha = 1$. The dashed line is a linear least-squares fit to the lower points. Also shown is the mean value of x_1 for (b).

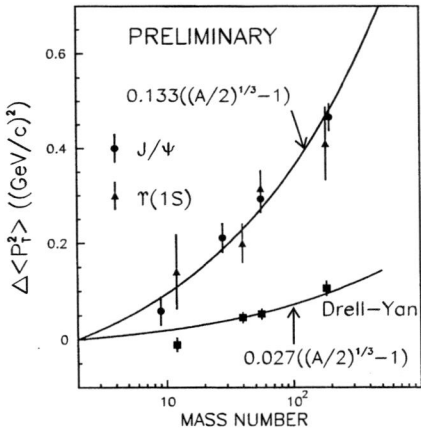

FIGURE 8. The change of mean p_t^2 for nuclear target, $\Delta\langle p_t^2 \rangle = \langle p_t^2 \rangle(A) - \langle p_t^2 \rangle(D)$, for 800 GeV p+A Drell-Yan process and J/Ψ and $\Upsilon(1S)$ productions. The solid curves are best fits to the A-dependence of $\Delta\langle p_t^2 \rangle$. The J/Ψ and $\Upsilon(1S)$ productions have identical curves for the A-dependence fits.

511

much more sensitive test for Eq. 5 could be done at RHIC, where the energy-loss effect is expected to be much enhanced in A-A collision [35].

REFERENCES

1. Drell S. D., and Yan T. M., *Phys. Rev. Lett.* **25**, 316 (1971).
2. Badier J. et al., *Z. Phys.* **C26**, 489 (1984).
3. Moreno G. et al., *Phys. Rev.* **D43**, 2815 (1991).
4. McGaughey P. L. et al., *Phys. Rev.* **D50**, 3038 (1994).
5. Lai H. L. et al., *Phys. Rev.* **D55**, 1280 (1997).
6. Berger E. L., Gordon L. E., and Klasen M., *Phys. Rev.* **D58**, 074012 (1998).
7. McGaughey P. L., Moss J. M., and Peng J. C., hep-ph/9905409, to be published in *Annu. Rev. Nucl. Part. Sci.* (1999).
8. Greenlee H. B. et al., *Phys. Rev. Lett.* **55**, 1555 (1985).
9. Freudenreich K., *Int. J. Mod. Phys.* **A5**, 3643 (1990).
10. Martin A. D. et al., *Eur. Phys. J.* **C4**, 463 (1998).
11. Kenyon I. R., *Rep. Prog. Phys.* **45**, 1261 (1982).
12. McGaughey P. L., *Nucl. Phys.* **A610**, 394c (1996).
13. Collins J. C., and Soper D. E., *Phys. Rev.* **D16**, 2219 (1977).
14. Oakes R. J., *Nuovo Cim.* **A44**, 440 (1966).
15. Collins J. C., *Phys. Rev. Lett.* **42**, 291 (1979).
16. Chiappetta P., and Le Bellac M., *Z. Phys.* **C32**, 521 (1986).
17. Lam C. S., and Tung W. K., *Phys. Rev.* **D21**, 2712 (1980).
18. Guanziroli M et al., *Z. Phys.* **D37**, 545 (1988) .
19. Conway J. S. et al., *Phys. Rev.* **D39**, 92 (1989).
20. Badier J. et al., *Z. Phys.* **C11**, 195 (1981).
21. Heinrich J. G. et al., *Phys. Rev.* **D44**, 1909 (1991)
22. Berger E. L., and Brodsky S. J., *Phys. Rev. Lett.* **42**, 940 (1979).
23. Berger E. L., *Z. Phys.* **C4**, 289 (1980).
24. Eskola K. J. et al., *Phys. Lett.* **B333**, 526 (1994).
25. Brandenburg A et al., *Phys. Rev. Lett.* **73**, 939 (1994).
26. Alde D. A., et al., *Phys. Rev. Lett.* **64**, 2479 (1990).
27. Alde D. A. et al., *Phys. Rev. Lett.* **66**, 133 (1991).
28. Alde D. A. et al., *Phys. Rev. Lett.* **66**, 2285 (1991).
29. Cronin J. W. et al., *Phys. Rev.* **D11**, 3105 (1975).
30. McGaughey P. L., Moss J. M., and Peng J. C., to be published (1999).
31. Schub M. H. et al., *Phys. Rev.* **D52**, 1307 (1995).
32. Alexopoulos T. et al., *Phys. Rev.* **D55**, 3927 (1997).
33. Leitch M. J., in Proceedings of "Quarkonium Production in Relativistic Nuclear Collisions", Institute for Nuclear Theory, Seattle, WA, May 1998.
34. Baier R. et al., *Nucl. Phys.* **B484**, 265 (1997); Baier R. et al., *Nucl. Phys.* **B531**, 403 (1998).
35. Baier R. et al., *Nucl. Phys.* **B483**, 291 (1997).

Low-x at HERA

A. De Roeck

DESY, Notkestrasse 85, 22607 Hamburg, Germany

Abstract. A selection of new low-x data from the ep collider HERA is presented. Structure function data at low x and Q^2 are discussed, as well as the extraction of the gluon distribution. Final state measurements are used to study low-x dynamics, to extract the b-quark production cross section and search for instantons.

INTRODUCTION

A new breakthrough in deep inelastic lepton-hadron scattering has been achieved by HERA, which provides for the first time collisions of electrons and protons in a collider mode, thereby increasing $\sqrt{s}$, the centre of mass system (CMS) energy of the scattering process, by an order of magnitude compared to the traditional fixed target experiments. Collisions are produced by 27.5 GeV electrons on 820 GeV protons, yielding a CMS energy of 300 GeV, and are recorded by two experiments H1 [1] and ZEUS [2].

A wealth of new results has been obtained in deep inelastic scattering (DIS: $Q^2 >$ few GeV^2) and photoproduction ($Q^2 \sim 0$) during the first years after the startup of HERA. Major highlights are measurements of the proton structure at low x in terms of the structure function F_2 and the gluon distribution $xg(x)$. Further, a perhaps unexpected large fraction of the events (about 10%) has been found to contain a large rapidity gap, and is referred to as diffraction.

An important feature of the detectors at HERA is the capability to measure the hadronic final state. Hence the production of hadrons of a low-x interaction pictured by the ladder diagram shown in Fig. 1, produced along the initial cascade, can be studied.

STRUCTURE FUNCTIONS

To probe the structure of the proton in the new kinematic region is by all means the prime task of HERA. Due to its large CMS energy, the proton structure is measured at x (Q^2) values two orders of magnitude lower (larger) than data from fixed target experiments. In the quark-parton model F_2 is a sum over the quark and

CP494, *New Directions in Quantum Chromodynamics,* edited by C.-R. Ji and D.-P. Min
© 1999 American Institute of Physics 1-56396-908-4/99/$15.00

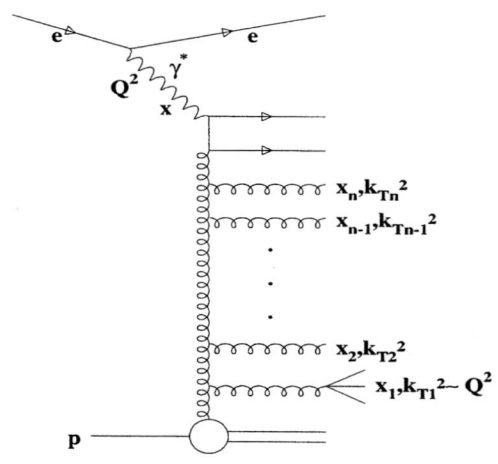

FIGURE 1. Diagram for a low-x Deep Inelastic Scattering event.

antiquark momentum fractions within a nucleon multiplied by the corresponding quark charge squared, thus directly related to the parton structure of the proton.

The first F_2 measurements in the newly reached low-x region have led to a surprise. Contrary to the tendency anticipated from measurements at fixed target energies, the structure function was found to rise steeply with decreasing x. This dramatic effect is shown in Fig. 2 for $0.6 < Q^2 < 17$ GeV^2. A lot of speculation started immediately after the first HERA results were released, mostly in terms of anticipated new QCD effects, such as the ones due to the low-x QCD evolution or the BFKL pomeron. Meanwhile it has become clear that 'traditional' QCD evolution based on the so called DGLAP [3] evolution equations can describe the data for Q^2 values larger than 1 GeV^2, as demonstrated by the fit (solid line) shown in Fig. 2, provided a suitable non-singular non-perturbative input is taken at a low starting scale Q_0^2 for the evolution ($Q_0^2 \sim 1$ GeV^2). However, also improved evolution equations which include $\ln 1/x$ resummation terms or attempts for a unified evolution equation can describe the data equally well [4,5]. Hence we conclude that the data at present do not require effects beyond DGLAP, but they can not exclude that specific low-x effects are already at work in the HERA regime [6].

The DGLAP evolution equations have been used to analyze the scaling violations of F_2 and extract the gluon density at small x. The next to leading order result in the $\overline{MS}$ scheme is shown in Fig. 3. The error bands include systematic and statistical uncertainties [7]. The knowledge of the gluon at low x has reached a precision of about 10%. A remarkable effect is seen: while the gluon grows very strongly with decreasing x for Q^2 values above a few GeV^2, it collapses completely

ZEUS 1995

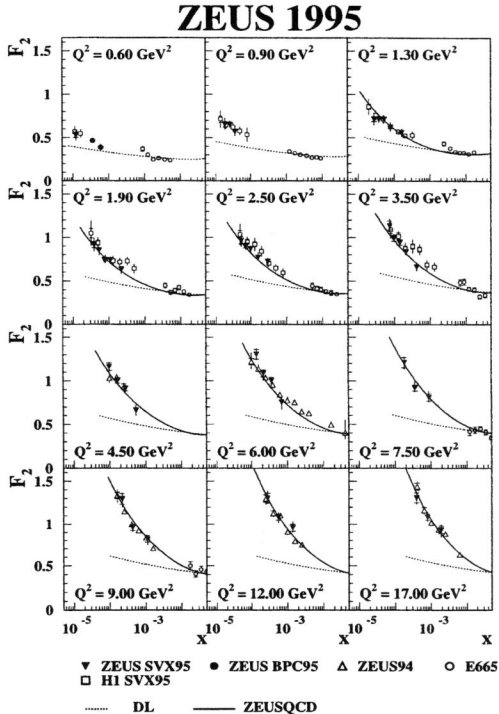

FIGURE 2. F_2 measurements at low-x and low Q^2 from H1 and ZEUS compared to a DGLAP fit (ZEUSQCD) and the Regge model prediction of Donnachie-Landshoff (DL).

for $Q^2 \sim 1$ GeV^2. Such a dramatic change seems to hint that either DGLAP is at the verge of breaking down at these scales, or other effects such as higher twists are required and/or parton screening effects are becoming important.

A region of particular interest is the low-Q^2 region, towards the photoproduction limit ($Q^2 = 0$). Improved detectors and clever experimental conditions (shifting the interaction vertex) have allowed to probe the region down to $Q^2 = 0.1$ GeV^2, and presently even down to 0.045 GeV^2, at HERA [8–10]. In this region perturbative QCD (pQCD) is not expected to be applicable, but the data is analysed with Regge phenomenology, such as the DL [11] model. The data, presented in Fig. 2, show a flattening of F_2 with decreasing Q^2. Much remains to be understood on the transition region from high Q^2 to low Q^2, or hard to soft processes, a topic for which studies just got started at HERA.

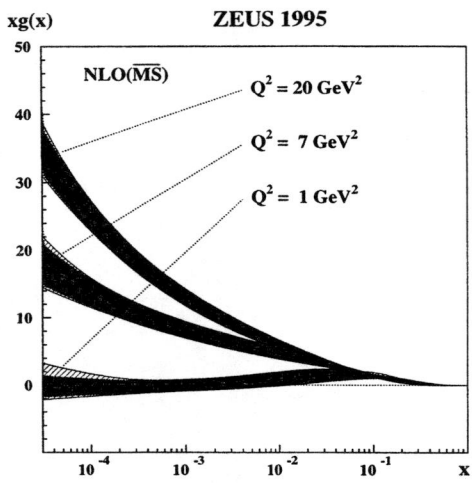

xg(x)   **ZEUS 1995**

FIGURE 3. Gluon distribution $xg(x)$ for several Q^2 values.

DIFFRACTION AND THE PION STRUCTURE

The data taken in 1992 at the ep collider HERA led to the first observation of deep inelastic scattering events with an interval in rapidity around the proton remnant direction devoid of hadron production - a "rapidity gap" [12,13]. Using the data taken in 1993 and 1994 measurements were made of the contribution of such events to the DIS cross section, quantified in terms of a structure function $F_2^{D(3)}$ [14,15]. These measurements demonstrated that the rapidity gap events could be attributed to a virtual photon–proton process which is dominantly diffractive. No substantial Q^2 dependence was observed, thus indicating that the sub–structure of diffractive exchange was of a point–like, presumably partonic, nature. This leads to the exciting possibility of measuring the partonic structure of the pomeron, an object which originates from soft hadronic scattering phenomenology and is thought to be exchanged between the incoming particles in a diffractive interaction, but is not yet understood in QCD.

New detectors in the beamline, about 100 m away from the central detector in the proton direction, to tag leading protons and neutrons have been used to measure additional processes. In the case of deep inelastic scattering with a leading neutron in the final state, the process can be viewed as the emition of a charged meson, e.g. a pion which is then probed by the virtual photon. Contributions due to the ρ and a_2 meson have been estimated to be an order of magnitude less in the kinematical region selected here [16]. Hence the partonic structure of this 'pion' is probed in the same way the partonic structure of the pomeron is probed in diffractive events. The results of such a measurement [17] are shown in Fig. 4 which tie in nicely

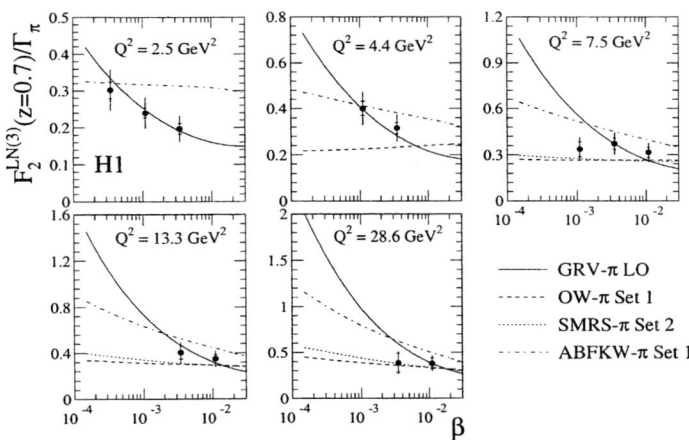

FIGURE 4. Extracted 'pion structure' as function of the fractional momentum variable β and Q^2, with various parametrizations.

with the talk on pions in the nucleon sea by G. Garvey at this conference [18]. An ambiguity in such an extraction is the assumed flux, here taken to be the one of [16].

FINAL STATES

The hadronic final state allows for measurements in the perturbative and non-perturbative sector of QCD, such as jet rate, extraction of α_s, hadronization, etc.

In DIS a parton in the proton can induce a QCD cascade consisting of several subsequent parton emissions before the final parton interacts with the virtual photon. The multiplicity and the x distribution of these emitted partons differ significantly in different approximations of QCD dynamics at small x.

At low x, pQCD evolution is complicated by the occurrence of two large logarithms in the evolution equations, namely $\ln 1/x$ and $\ln Q^2$. A complete perturbative treatment in the low-x region is not yet available, and different approximations are made resulting in different parton dynamics. At high Q^2 and high x pQCD requires the resummation of contributions of $\alpha_s \ln(Q^2/Q_0^2)$ terms, yielding the DGLAP (Dokshitzer-Gribov-Lipatov-Altarelli-Parisi) [3] evolution equations. However at small x the contribution of large leading $\ln 1/x$ terms may become important. Resummation of these terms leads to the BFKL (Balitsky-Fadin-Kuraev-Lipatov) [19] evolution equation. Hence a pertinent and exciting question is whether these $\ln 1/x$ contributions to the parton evolution can be observed experimentally.

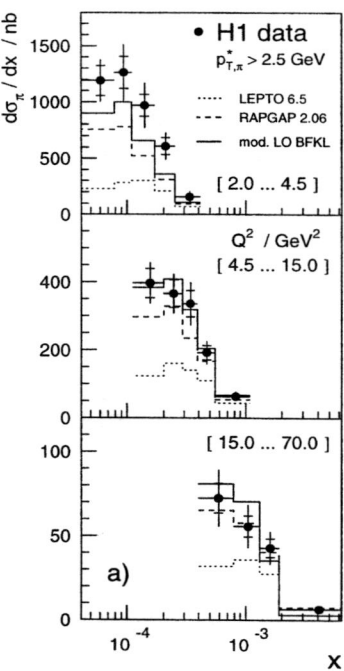

FIGURE 5. Inclusive π^0 spectrum measured in the forward region for $0.1 < y < 0.6, 5^0 < \theta < 25^0$ and $E_\pi > 8.2\text{GeV}$.

Differences between different dynamical assumptions for the parton cascade are expected to be most prominent in the phase space region towards the proton remnant direction, i.e. away from the scattered quark. In the HERA laboratory frame this corresponds to a region of small polar angles and has been generically termed "forward region". In previous analyses results have been presented on forward jet and forward inclusive charged and neutral pion production [20,21]. The data showed an excess over expectation from DGLAP (Monte Carlo model) calculations, but it appeared that this could be remedied by adding 'resolved photon' diagrams to the DIS diagram [22].

New data on forward single π^0 production, shown in Fig. 5, became available recently at larger transverse momentum, p_T. A calculation based on pQCD which uses the BFKL formalism for the perturbative part, and fragmentation functions for the hadronization, is available [23] and compares well to the data. The resolved model shows however some deficiencies. There are remaining questions on the scales to be used in all these calculations, but these forward pion data could well be the first which show evidence for BFKL effects in the hadronic final state.

Heavy flavours, in particular charm quark production, has been studied over the last 5 years at HERA. The production is generally well understood in terms

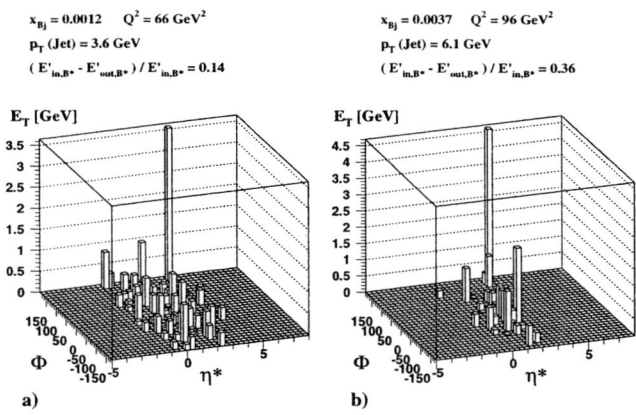

FIGURE 6. Transverse energy distribution in the $\eta - \phi$ plane for instanton induced processes (Monte Carlo).

of boson gluon fusion. Recently, beauty production as been measured and the outcome was a surprise [24]: the cross section, be it measured in a limited range of phase space, was found to be five times larger than the one predicted by the (LO) program AROMA. For $Q^2 < 1$ GeV^2, $0.1 < y < 0.8$ and $p_T^\mu > 2.0$ GeV/c, $35^0 < \theta^\mu < 130^0$, with p_T^μ, θ^μ the transverse momentum and polar angle of the muon measured in the lab frame respectively, the cross section is (preliminary) $\sigma_{vis}(ep \to b\bar{b}X) = 0.93 \pm 0.8(stat.)^{+0.21}_{-0.12}(syst.)$ nb. AROMA predicts 0.19 nb. In NLO this discrepancy is found to be reduced.

Finally, we mention the interesting possibility to measure instanton events in DIS at HERA. Instantons [25] describe the quantum tunneling between different gauge rotated classical vacua in QCD and give important information on the complicated structure of the ground state in theory of strong interaction. It has been suggested that these can have measurable effects in DIS, a nice overview of which is given in [26]. Instanton would be characterised by quark flavour democracy, a band of about 1-2 units in pseudorapidity with homogenous many-particle production and locally increased transverse energy. The H1 experiment has produced an upper limit of 300 pb in a selected kinematical region based on the analysis of the multiplicity distribution [27]. It is expected that this limit can be improved by a factor 5 to 10 soon, which would meet the expected -be it with large uncertainty- cross section of $O(100)$ pb. Two Monte Carlo events are shown in Fig. 6, a 'gold plated' one, and a 'typical' one where the homogenous instanton particle production got distorted by the high E_T current quark jet.

SUMMARY

HERA data have over the past years been pivotal to explore the QCD dynamics at low-x. New structure function analyses and data, and forward particle data hint towards phenomena beyond the simple DGLAP approaches. DIS data with leading protons can be interpreted in terms of a pion structure. The beauty quark production cross section measurement is found to be larger than expected. Non-perturbative effects such as instantons will be explored in detail in the near future.

REFERENCES

1. H1 Collab., I. Abt et al., *Nucl. Instr. and Meth.* **A386** (1997) 310 and **A386** (1997) 348.
2. ZEUS Collab., *The ZEUS Detector*, Status Report (1993).
3. Yu. L. Dokshitzer, *JETP* **46** (1977) 641.
   V. N. Gribov and L. N. Lipatov, *Sov. Journ. Nucl. Phys.* **15** (1972) 78.
   G. Altarelli and G. Parisi, *Nucl. Phys.* **B126** (1977) 298.
4. J. Kwiecinski, A. D. Martin and A. M. Stasto, *Phys. Rev.* **D56** (1997) 3991.
5. R. S. Thorne, *Phys. Lett.* **B392** (1997) 463, hep-ph/9701241 and hep-ph/9710541.
6. A. M. Cooper-Sarkar, R. C. E. Devenish and A. De Roeck, *Int. J. Mod. Phys.* **A13** (1998) 3385.
7. ZEUS Collab., J. Breitweg et al., *Eur. Phys. J.* **C7** (1999) 609.
8. H1 Collab., C. Adloff et al., *Nucl. Phys.* **B497** (1997) 3.
9. ZEUS Collab., J. Breitweg et al., *Phys. Lett.* **B407** (1997) 432.
10. ZEUS Collab., Contributed paper to EPS99, Tampere, paper 493 (1999).
11. A. Donnachie and P. Landshoff, *Phys. Lett.* **B296** (1992) 227.
12. ZEUS Collab., M. Derrick et al., *Phys. Lett.* **B315** (1993) 481.
13. H1 Collab., T. Ahmed et al., *Nucl. Phys.* **B429** (1994) 477.
14. H1 Collab., S. Aid et al., *Z. Phys.* **C76** (1997) 613.
15. ZEUS Collab., J. Breitweg et al., *Eur. Phys. J.* **C6** (1999) 43.
16. B. Kopeliovich et al., *Z. Phys* **C73** (1996) 125.
17. H1 Collab., C. Adloff et al., *Eur. Phys. J.* **C6** (1999) 587.
18. G. Garvey, these proceedings.
19. E.A. Kuraev, L.N. Lipatov, V.S. Fadin, *Sov. Phys. JETP* **45** (1972) 199;
    Y.Y. Balitsky, L.N. Lipatov, *Sov. J. Nucl. Phys.* **28** (1978) 822.
20. H1 Collab., C. Adloff et al., *Nucl. Phys.* **B538** (1999) 3.
21. ZEUS Collab., J. Breitweg et al., *Eur. Phys. J.* **C6** (1999) 239.
22. H. Jung, L. Jonsson, H. Kuster, *Eur. Phys. J.* **C9** (1999) 383.
23. J. Kwiecinski, A.D. Martin, J.J. Outhwaite, hep/ph9903439 (1999).
24. H1 Collab., Contributed paper to ICHEP98, Vancouver, paper 575. (1998).
25. N. Kochelev, these proceedings.
26. T. Carli et al., DESY preprint 99-067 (1999).
27. S. Aid et al., *Z. Phys.* **C72** (1996) 573.

POSTER SESSION

Renormalization Group Study of Thermodynamics in Large N 3D O(N) Vector Model

Taichi Itoh and Yoonbai Kim

Department of Physics and Institute of Basic Science, Sungkyunkwan University
Suwon 440-746, Korea
taichi@newton.skku.ac.kr, yoonbai@skku.ac.kr

Abstract. We show that a conformal fixed point appears in the three dimensional O(N) $(\vec{\phi}^2)^2$ theory in large-N limit and it shares renormalization group properties with the Wilson-Fisher fixed point which arises in the context of the ϵ expansion. Thermodynamic free energy density is computed and its behavior is studied along renormalization group flow lines around the conformal fixed point. Specifically, the ultraviolet free energy is not less than the infrared one except for those along the flow line corresponding to nonlinear σ model embedded in the $(\vec{\phi}^2)^2$ theory.

Introduction. About a decade ago conformal field theories (CFT) on line have attracted attention in the end of the first revolutionary era of string theory, and during recent years CFT on plane or bulk are highlighted in relation with Maldacena's conjecture. In two dimensional (2D) CFT, an intriguing idea is Zamolodchikov's C-function and C-theorem. The higher dimensional extension of C-function was proposed in Ref. [1] by a thermodynamic free energy (TFE). Though it does not necessarily show the monotonically increasing behavior as temperature increases, it is conjectured in this year [3] that the universal constraint $f_{IR} \leq f_{UV}$ might be satisfied in any renormalizable field theory where f_{IR} and f_{UV} are given by the number of lightest degrees of freedom in both low and high temperature limits, respectively. Those authors examined their constraint in various 4D gauge theories with or without supersymmetry and it showed nice agreement with the known results, however the above conjecture seems unlikely for the 3D bosonic models with a nontrivial ultraviolet (UV) fixed point due to the crossover from classical ordered phase to quantum disordered phase [2]. A specific field theory counterexample in three dimensions is O(N) nonlinear σ (NLσ) model [4].

In this note we study the fixed point structure in the 3D O(N) $(\vec{\phi}^2)^2$ theory and demonstrate the existence of a conformal fixed point (CFP) in the large-N limit. We also compute the TFE, or the C-function, and examine how the conjecture in Ref. [3] works around the CFP despite the RG flow line of O(N) NLσ model.

CP494, *New Directions in Quantum Chromodynamics*, edited by C.-R. Ji and D.-P. Min
© 1999 American Institute of Physics 1-56396-908-4/99/$15.00

Fixed Point Structure of $(\vec{\phi}^2)^2$ *Theory.* The $O(N)$ vector model with $(\vec{\phi}^2)^2$ interaction term in three (Euclidean) spacetime is described by the action

$$S = \int_0^L dt \int d^2\mathbf{x} \left\{ \frac{1}{2}(\partial_\mu \vec{\phi})^2 + \frac{1}{2}\mu^2 \vec{\phi}^2 + \frac{G}{8N}(\vec{\phi}^2)^2 \right\}, \tag{1}$$

where $\vec{\phi} = (\phi_1, ..., \phi_N)$ denotes the N-component real scalar field and the inverse of length L stands for the temperature T of the system. Linearizing the $\vec{\phi}$ field by introducing an auxiliary field λ and taking mean field approximation: $\langle \lambda \rangle = m^2$, we obtain large-$N$ free energy density

$$\frac{F(L)}{N} = \frac{1}{2\Omega} \operatorname{tr} \ln\left[-\Box + m^2\right] + \frac{\mu^2}{G} m^2 - \frac{m^4}{2G}, \tag{2}$$

where Ω denotes the volume of the spacetime. In order to derive gap equation from Eq. (2) we have to regularize UV divergence by a cutoff Λ. In terms of RG invariants R and M rewritten by dimensionless couplings $u = G\Lambda^{-1}$ and $r = \mu^2 \Lambda^{-2}$ with $(u_*, r_*) = (4\pi^2, -1)$ such as

$$R = 8\pi \left(\frac{r}{u} - \frac{r_*}{u_*}\right) \Lambda, \quad M = \frac{1}{8\pi} \left(\frac{1}{u_*} - \frac{1}{u}\right)^{-1} \Lambda, \tag{3}$$

we read the finite temperature gap equation up to $O(m/\Lambda)$

$$\frac{R}{m} + \frac{m}{M} = 1 + \frac{2}{Lm} \ln\left(1 - e^{-Lm}\right). \tag{4}$$

At zero temperature $(L = \infty)$ the second term in the right-hand side of the Eq. (4) vanishes so that the gap equation reads $m^2 - Mm + MR = 0$ of which solution should be uniquely determined as the Goldstone boson mass which reduces to zero on the critical line of second order phase transition $O(N)/O(N-1)$, to say, $R = 0$. The (u, r)-plane is separated into the symmetric phase with $R > 0$ and the broken phase with $R < 0$ as shown in Fig. 1-(a).

Both of R and M should be cutoff independent to retain the large-N theory UV finite and should obey renormalization group (RG) equations $dR/d\Lambda = 0$ and $dM/d\Lambda = 0$. Hence the Callan-Symanzik β-functions $\beta_u = \Lambda du/d\Lambda$ and $\beta_r = \Lambda dr/d\Lambda$ are determined as

$$\beta_u = -u\left(1 - \frac{u}{u_*}\right), \quad \beta_r = -r\left(2 - \frac{u}{u_*}\right) + \frac{r_*}{u_*}u. \tag{5}$$

We notice that both $u = u_*$ and $r = (r_*/u_*)u$ constitute two critical lines in the theory space spanned by u and r. We have two specific points on the critical line of the second order phase transition $r = (r_*/u_*)u$ $(R = 0)$. One is the Gaussian fixed point (GFP) at $(u, r) = (0, 0)$ $(M = 0)$ and the other is a CFP at $(u, r) = (u_*, r_*)$ $(M \to \infty)$. The CFP is located at the intersection of the critical lines so that it appears as a nontrivial ultraviolet (UV) fixed point along $u = u_*$, while it appears as a nontrivial infrared (IR) fixed point along $r = (r_*/u_*)u$ (see Fig. 1-(a)) [5]. In addition, we easily see that the critical line $u = u_*$ $(M \to \infty)$ can be mapped into the $O(N)$ NLσ model with a coupling $\alpha\Lambda$ through $(u, r) = (u_*, -\pi/2\alpha)$.

Crossover Behavior of the C-function. There is no finite temperature phase transition in 3D finite temperature bosonic systems in accordance with the Mermin-Wagner-Coleman theorem. Nevertheless the TFE of a given system shows the crossover from classical to quantum behavior. To figure out this, let us introduce the function $C(x)$, as a natural 3D extension of the Zamolodchikov's C-function, defined by $C(x) = -2\pi L^3 [F(L) - F(\infty)]/\zeta(3)$ with $x = e^{-Lm}$ [1]. Notice that the zero temperature free energy has been subtracted in the definition so that it reduces to zero at zero temperature limit. After a straightforward calculation, we obtain an expression in terms of polylogarithms [1]

$$\zeta(3) C(x) = N [Li_3(x) - Li_2(x) \ln x], \tag{6}$$

where $x = e^{-Lm}$ has to obey the finite temperature gap equation (4) so as to make the free energy density take its ground state value. Just on the CFP, the gap equation (4) provides with the golden mean $\tau = (\sqrt{5}+1)/2$ the critical scaling $m = 2\ln\tau/L$ which specifies the crossover point from classical to quantum behavior of Goldstone bosons. The C-function therefore takes the value $C(2 - \tau) \approx 0.68N$ at the CFP. [2]

Since the time scale which measures the evolution of the system is regarded as a logarithmic temperature $t = \ln(T/T_0)$, the high temperature (UV) and the low temperature (IR) limit of the C-function are given as $f_{UV} = \lim_{L\to 0} C(x)$ or $f_{IR} = \lim_{L\to\infty} C(x)$, respectively. The crossover from classical to quantum behavior of the C-function can be studied through the temperature derivative $\partial C/\partial t = -(Ldx/dL) dC/dx$ where we can confirm dC/dx is positive definite from Eq. (6) so that the sign of the derivative only depends on Ldx/dL.

The critical line $u = u_*$ has been identified with the NLσ model and the gap equation (4) in the limit $M \to \infty$ provides $Ldx/dL = -LRx(1 - x)/(1 + x)$. Therefore we see that $\partial C/\partial t < 0$ for $R < 0$ and vice versa. The C-function shows a classical decreasing behavior in the zero temperature ordered region ($R < 0$), while it shows a quantum increasing behavior in the zero temperature disordered region ($R > 0$). (f_{IR}, f_{UV}) is determined as ($N, 0.68N$) for $R < 0$ (decreasing), while ($0, 0.68N$) for $R > 0$ (increasing).

Along the critical line of the second order phase transition $r = (r_*/u_*) u$, the gap equation (4) associated with $R = 0$ tells us that Ldx/dL is expressed as a negative definite function of x multiplied by $1/LM$ which results in $\partial C/\partial t > 0$ for $M < 0$ and vice versa. The low temperature limit is controlled by the CFP so that it shows the quantum critical behavior between the zero temperature ordered and disordered phases irrespective of the sign of M. However, since the solution to the gap equation disappears in $M > 0$ at some finite temperature, the region $M > 0$ seems unphysical. Notice that for $M < 0$ the system goes to the GFP at high temperature limit and the C-function shows the increasing behavior from the CFP to the GFP. (f_{IR}, f_{UV}) are determined as ($0.68N, N$) for $M < 0$ [6].

---

[1] Polylogarithms are defined iteratively by $Li_1(x) = -\ln(1 - x)$ and $xdLi_{n+1}(x)/dx = Li_n(x)$.
[2] This value is different from the value $(4/5)N$ derived in Ref. [4] whose definition of C-function contains the zero temperature free energy partially in contrast to that in Ref. [1].

Summary and Outlook. In larger than two dimensions there arises a classical ordered phase, together with a nontrivial UV fixed point, where the thermal fluctuation dominates and the C-function shows decreasing behavior such like in the region $R < 0$ in NLσ model. However, in whole of the (u, r)-plane, we have no reason to choose the fine tuning theory just on $u = u_*$. Instead we can choose the point infinite below in $u < u_*$. The theory moves directly to the GFP without passing through the CFP along some RG flow line. The C-function decreases at first but turns to increase as the theory approaches to the critical line $R = 0$ (see Fig. 1-(b)) [7]. f_{IR} counts the number of Goldstone bosons $N - 1$ in the classical ordered phase, while f_{UV} takes the value N since all bosons behaves like massless free fields in the GFP. The conjecture in Ref. [3] therefore seems likely to work in spite of the crossover behavior of the C-function.

Acknowledgement. We are indebted to T. Appelquist, Chanju Kim, and S. Sachdev for helpful discussions. This work was supported in part by the Korea Research Foundation (1998-015-D00075).

REFERENCES

1. A. H. Castro Neto and E. Fradkin, Nucl. Phys. **B400**,525 (1993).
2. S. Chakravarty, B. I. Halperin, and D. R. Nelson, Phys. Rev. Lett. **60**, 1057 (1988).
3. T. Appelquist, A. G. Cohen and M. Schmaltz, Phys. Rev. D **60**, 045003 (1999); T. Appelquist, A. G. Cohen, M. Schmaltz and R. Shrock, Phys. Lett. **B459**, 235 (1999).
4. S. Sachdev, Phys. Lett. **B309**, 285 (1993).
5. E. Brézin and J. Zinn-Justin, Phys. Rev. B **14**, 3110 (1976).
6. A. C. Petkou and M. B. Silva Neto, Phys. Lett. **B456**, 147 (1999).
7. T. Itoh and Y. Kim, in preparation.

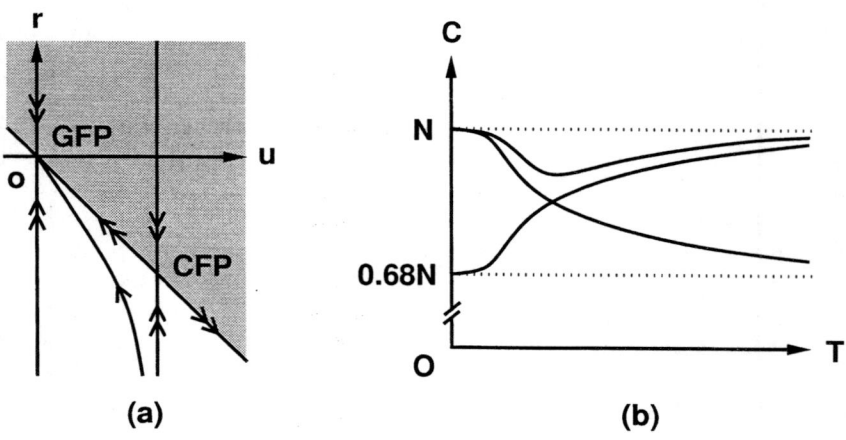

(a) **(b)**

FIGURE 1. (a) The zero temperature phase diagram in the $(\vec{\phi}^2)^2$ theory. The shaded area denotes the symmetric phase. (b) The crossover behavior of the C-function.

Form factors for B to Light meson decay
- Light Cone analysis -

Bum-Hoon Lee

Department of Physics, Sogang University
Seoul 121-742, Korea
E-mail: bhl@ccs.sogang.ac.kr

Abstract. We calculate the form factors of $B \to \pi$ and $B \to \rho$ transition matrix elements by using the light cone formalism. In the limit of $m_\pi/M_B = 0$, $m_\rho/M_B = 0$, $M_b/M_B = 1$ and $(1-x) << 1$, we show the pole types of each form factors and show that there exists relations among form factors with 2 unknown functions.

I INTRODUCTION

For the decays involving $b \to c$ transition, the heavy quark symmetry makes it possible to understand reliably through the heavy quark effective theory (HQET) [1]. However, for those involving $b \to u$ it is less likely that the heavy quark symmetry applies, and the determination of the matrix element is heavily relied on the models for the form factors. The theoretical calculation of the form factors involving $b \to u$ transition is a difficult task, since it is concerned with the non-perturbative realm of QCD. There have been active study of the form factors of $B \to \pi$ and $B \to \rho$ with such various methods as quark model, QCD sum rule and lattice calculations [2].

In this work [3], we will calculate the form factors of $B \to \pi$ and $B \to \rho$ transitions by using the method [4] based on the meson theory of Brodsky and Lepage [5]. We obtain the pole types of the form factors from q^2 dependences and also show the relations among the form factors in the limit of $m_\pi/M_B = 0$, $m_\rho/M_B = 0$, $M_b/M_B = 1$.

II FORM FACTOR RELATIONS

We can decompose the hadronic matrix element for $B \to \pi$ transition in terms of hadronic form factors [6]:

$$< \pi^-(p_\pi)|V^\mu|B^0(p_B) >$$

CP494, *New Directions in Quantum Chromodynamics*, edited by C.-R. Ji and D.-P. Min
© 1999 American Institute of Physics 1-56396-908-4/99/$15.00

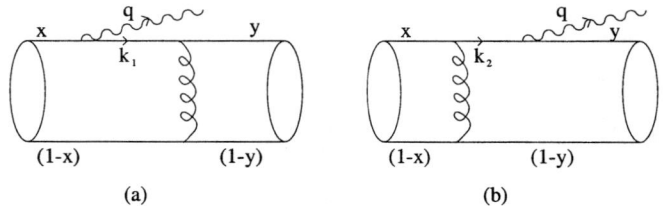

FIGURE 1. Feynman diagrams to the first order in α_s.

$$= \left(r^\mu - \frac{m_B^2 - m_\pi^2}{q^2} q^\mu\right) F_1^{B\pi}(q^2) + \frac{m_B^2 - m_\pi^2}{q^2} q^\mu F_0^{B\pi}(q^2), \tag{1}$$

where $V^\mu = \bar{u}\gamma^\mu b$, $q^\mu = (p_B - p_\pi)^\mu$, $r^\mu = (p_B + p_\pi)^\mu$. In the similar way, the hadronic matrix element for $B \to \rho$ transition is written [6] in terms of hadronic form factors V, A_0, A_1, and A_2.

We now calculate the heavy to light transition matrix element using the PQCD factorization of exclusive amplitudes. To the first order in $\alpha_s = \alpha_s(Q^2)$, two Feynman diagrams in Fig. 1 give the following amplitude:

$$< \pi^-(p_\pi)|V^\mu|B^0(p_B) > = \frac{8\pi\alpha_s}{3} \int_0^1 dx \int_0^{1-\epsilon} dy\, \phi_B(x)$$

$$\times \left[\frac{\mathrm{Tr}\{(\not{p}_\pi + m_\pi)\gamma_5\gamma^\nu \not{k}_1\gamma^\mu(\not{p}_B + g(x)M_B)\gamma_5\gamma_\nu\}}{k_1^2 Q^2}\right.$$

$$\left. + \frac{\mathrm{Tr}\{(\not{p}_\pi + m_\pi)\gamma_5\gamma^\mu(\not{k}_2 + M_b)\gamma^\nu(\not{p}_B + g(x)M_B)\gamma_5\gamma_\nu\}}{(k_2^2 - M_b^2)Q^2}\right]\phi_\pi(y), \tag{2}$$

and the similar expression for the decay into ρ. Here, $Q^\mu = (1 - x)p_B^\mu - (1 - y)p_\pi^\mu$, $k_1^\mu = -(1 - x)p_B^\mu + p_\pi^\mu$, $k_2^\mu = p_B^\mu - (1 - y)p_\pi^\mu$. The variables x and y represent the momentum fraction carried by the correspondig quarks. In (2) $g(x)$ is a phenomenological parameter for B meson wave function, and $\phi_\pi(y)$ and $\phi_B(x)$ are the distribution amplitudes for π and B mesons. In (2) we took the upper limit of the integration over momentum fraction y of a quark in the light meson as $1 - \epsilon$. The integration in the interval $1 - \epsilon \le y \le 1$ corresponds to the Drell-Yan-West [7] end-point region and is expected to be suppressed by the Sudakov form factor [4].

Now we evaluate the form factors in the limit of $m_\pi/M_B = 0$, $m_\rho/M_B = 0$, $M_b/M_B = 1$ and $(1 - x) << 1$. The approximations $m_\pi/M_B = 0$, $m_\rho/M_B = 0$ and $M_b/M_B = 1$ are reasonable. For the B meson, it is expected that the momentum fraction of the light quark, $1 - x$, is roughly given by m_{light}/M_B and small. For this dominant range of x, the approximation $(1 - x) << 1$ can be applied in the integrand safely as long as the range of q^2 not too big.

We organize the form factors in the following form:

$$F_i(q^2) = \frac{32\pi\alpha_s}{3M_B^2} \int_0^1 dx \int_0^{1-\epsilon} dy\, \phi_B(x)\, \phi_i(y)\frac{1}{(1-x)(1-y)^2}f_i, \tag{3}$$

528

where $F_i = F_0, F_1, V, A_0, A_1, A_2$, and $\phi_i(y) = \phi_\pi(y)$ for F_0 and F_1, and $\phi_i(y) = \phi_\rho(y)$ for V, A_0, A_1 and A_2. In (3) all f_i are given by

$$f_i = c1_i \frac{1}{z} + c2_i \frac{1}{z^2} \tag{4}$$

where ci_i and $c2_i$ are function of y only, and $z \equiv 1 - \frac{q^2}{M_B^2}$. The integration does not change the q^2 dependence. The form factors then can be summarized as

$$F_0(q^2) = (a_\pi + b_\pi)\frac{1}{z}, \quad F_1(q^2) = 2b_\pi \frac{1}{z} + (a_\pi - b_\pi)\frac{1}{z^2}, \tag{5}$$

$$A_1(q^2) = a_\rho \frac{1}{z}, \quad A_2(q^2) = -V(q^2) = a_\rho \frac{1}{z^2}, \quad -A_0(q^2) = (a_\rho + b_\rho)\frac{1}{z^2},$$

where

$$a_i = \frac{32\pi\alpha_s}{3M_B^2} \int_0^1 dx \int_0^{1-\epsilon} dy\, \phi_B(x)\, \phi_i(y) \frac{2g-1}{(1-x)(1-y)^2}, \tag{6}$$

$$b_i = \frac{32\pi\alpha_s}{3M_B^2} \int_0^1 dx \int_0^{1-\epsilon} dy\, \phi_B(x)\, \phi_i(y) \frac{1}{(1-x)(1-y)}$$

with $i = \pi$ or ρ.

We emphasize that the q^2 dependences of the form factors given in (5) are independent of the shapes of the distribution amplitudes $\phi_B(x)$, $\phi_\pi(y)$, $\phi_\rho(y)$ and the value of the parameter ϵ. Their dependences appear only in the values of the constants a_i and b_i in (6), which affect only the normalizations of the form factors. From (5) we find that $F_0(q^2)$ and $A_1(q^2)$ have the simple pole q^2 dependence, and $A_2(q^2)$, $V(q^2)$ and $A_0(q^2)$ have the dipole q^2 dependence. $F_1(q^2)$ has the mixture of the simple pole and dipole q^2 dependences, but the dipole q^2 dependence is dominant.

From (5) we find the relations among the form factors:

$$F_1(q^2) = F_0(q^2)(2 - \frac{1}{z}) + 2\frac{a_\pi}{a_\rho} A_1(q^2)(-1 + \frac{1}{z}) \tag{7}$$

$$F_0(q^2)\frac{1}{z} = -\frac{a_\pi + b_\pi}{a_\rho + b_\rho} A_0(q^2) \tag{8}$$

$$A_1(q^2)\frac{1}{z} = A_2(q^2) = -V(q^2). \tag{9}$$

For $q^2 = 0$, this can be checked by measuring the differential branching ratios at $q^2 = 0$.

III   CONCLUSION

We evaluated the form factors of $B \to \pi$ and $B \to \rho$ heavy to light transition matrix elements using the factorization formalism of perturbative QCD. We obtained the q^2 dependences of the form factors and the relations among them in the limit of $m_\pi/M_B = 0$, $m_\rho/M_B = 0$, $M_b/M_B = 1$ and $(1 - x) \ll 1$. $F_0(q^2)$ and $A_1(q^2)$ have the simple pole q^2 dependence, and $A_2(q^2)$, $V(q^2)$ and $A_0(q^2)$ have the dipole q^2 dependence. $F_1(q^2)$ has the mixture of the simple pole and dipole q^2 dependences. The pole types and the relations among the form factors are independent of the shapes of the distribution amplitudes $\phi_B(x)$, $\phi_\pi(y)$, $\phi_\rho(y)$ and the value of the parameter ϵ.

ACKNOWLEDGMENTS

This work was supported in part by the Basic Science Research Institute Program, Ministry of Education, Project No. BSRI-98-2414.

REFERENCES

1. N. Isgur and M.B. Wise, Phys. Lett. B232 (1989) 113; B237 (1990) 527
2. H. Wittig, OUTP-97-59-P, hep-lat/9710088 (1997)
3. Dae Sung Hwang and Bum-Hoon Lee, Eur. Phys. J. C 6, 663 (1999)
4. A. Szczepaniak, E.M. Henley and S.J. Brodsky, Phys. Lett. B243 (1990) 287; S.J. Brodsky, in *QCD 20 Years Later*, P.M. Zerwas and H.A. Kastrup, Ed. (World Scientific, 1993)
5. G.P. Lepage and S.J. Brodsky, *Phys. Lett.* B **87**, 359 (1979) ; Phys. Rev. D22 (1980) 2157; S.J. Brodsky and G.P. Lepage, Phys. Rev. D24 (1981) 1808; in *Perturbative Quantum Chromodynamics*, A.H. Mueller, Ed. (World Scientific, 1989)
6. M. Wirbel, B. Stech and M. Bauer, Z. Phys. C29 (1985) 637; C34 (1987) 103; M. Bauer and M. Wirbel, Z. Phys. C42 (1989) 671
7. S.D. Drell and T.M. Yan, Phys. Rev. Lett. 24 (1970) 181; G. West, Phys. Rev. Lett. 24 (1970) 1206
8. V.L. Chernyak and I.R. Zhitnitsky, Phys. Rept. 112 (1984) 173; V.L. Chernyak and A.R. Zhitnitsky, Nucl. Phys. B345 (1990) 137

APPENDIX

LECTURE SCHEDULE

27 May (Thu)   (APCTP seminar room 1414)
13:00 - 14:00: G. Miller (I)

28 May (Fri)   (APCTP seminar room 1414)
 13:00 - 14:30: S. Brodsky (I)   14:30 - 14:45: break
14:45 - 15:45: G. Miller (II)

29 May (Sat)   (APCTP seminar room 1414)
 09:30 - 11:00: S. Brodsky (II)   11:00 - 11:15: break
11:15 - 12:15: G. Miller (III)

1 June (Tue)   (APCTP seminar room 1414)
16:30 - 18:00: S. Brodsky (III)

8 June (Tue)   (APCTP seminar room 1414)
 16:30 - 17:30: S. Dalley (I)

9 June (Wed)   (APCTP seminar room 1414)
 16:30 - 17:30: S. Dalley (II)

10 June (Thu)   (APCTP seminar room 1414)
 15:00 - 16:00: S. Dalley (III)

11 June (Fri)   (APCTP seminar room 1414)
 15:00 - 16:00: S. Dalley (IV)   16:00 - 16:15: break
16:15 - 18:15: H.-C. Pauli (I)

14 June (Mon)   (APCTP seminar room 1414)
 14:00 - 16:00: H.-C. Pauli (II)   16:00 - 16:15: break
16:15 - 18:15: S. Pinsky (I)

15 June (Tue)   (Seoul National Univ., Sangsan MathScience Bldg
seminar room 205)
  14:00 - 16:00: H.-C. Pauli (III)   16:00 - 16:15: break
16:15 - 18:15: S. Pinsky (II)

16 June (Wed)   (Seoul National Univ., Sangsan MathScience Bldg.
seminar room 205)
 13:00 - 14:00; N. Kochelev: Instantons and spin physics
 14:00 - 16:00: H.-C. Pauli (IV)   16:00 - 16:15: break
16:15 - 18:15: S. Pinsky (III)

17 June (Thu)   (Seoul National Univ., Sangsan MathScience Bldg.
seminar room 205)
  14:00 - 16:00: H.-C. Pauli (V)   16:00 - 16:15: break
16:15 - 18:15: S. Pinsky (IV)

18 June (Fri)   (APCTP seminar room 1414)
  10:00 - 12:00: S. Pinsky (V)
 14:00 - 15:00; N. Kochelev: Unusual properties of the central
production of mesons and structure of QCD vacuum

WORKSHOP PROGRAM

June 21 (Mon)
09:00 - 09:10 Opening -- D.P. Min (Seoul)
 Chair: C.-R. Ji (North Carolina)
09:10 - 10:10   G. Garvey (LANL)   The Role of Pions in Nucleon Sea;
Neutrinos and Nucleon Structure
10:10 - 10:35   Break (Registration)
10:35 - 11:25   M. Rho (KIAS)   Effective Field Theory and Nuclei
11:25 - 12:00   H.-C. Pauli (MPI, Heidelberg)   On the Renormalized
Effective (Light-Cone) Hamiltonian in QED and QCD
Lunch
Chair: S.H. Lee (Yonsei)
13:30 - 14:30   F. Lenz (Erlangen)   The Center-Symmetric Phase of QCD
14:30 - 15:30   O. Miyamura (Hiroshima)   Hadronic Masses at High
Temperature by Lattice QCD Approaches
15:30 - 15:45   Break
Chair: F. Lenz (Erlangen)
15:45 - 16:20   J.-H. Kang (Yonsei)   The PHENIX Experiment at RHIC
16:20 - 17:05   S.H. Lee (Yonsei)   Hadrons at Finite Temperature and
Density
17:05 - 17:50   M. Engelhardt (Tuebingen)   A Picture of the
Yang-Mills Deconfinement Transition and Its Lattice Verification
17:50 - 18:25   A. Hayashigaki (Kyoto)   J/psi at Finite Density in
QCD Sum Rule Analysis
Banquet (from 19:00)

June 22 (Tue)
Chair: B.H. Lee (Sogang)
08:30 - 09:30   R. McKeown (Caltech)   Color Transparency; Neutral
Current Formfactors of the Nucleon
09:30 - 10:20   I. Zahed (Stony Brook)   Disordered Phenomena in QCD
10:20 - 10:35   Break
10:35 - 11:25   M. Lutz (GSI)   Chiral Nuclear Theory
11:25 - 12:15   S. Beane (Maryland)   Elusive Nucleons from Nuclei in
Effective Field Theory
12:15 - 12:35   S.T. Hong (Sogang)   Strange Form Factors of Octet and
Decuplet Baryons
Lunch
Chair: H. Jung (Sookmyung)
14:00 - 14:35   B.Y. Park (Chungnam)   The Proton Spin in the Chiral
Bag Model
14:35 - 15:20   M. Burkardt (New Mexico)   The Color Dielectric
Formulation of Transverse Lattice QCD
15:20 - 15:55   S. Dalley (Cambridge)   Glueballs on a Transverse
Lattice
15:55 - 16:30   B. van de Sande (Erlangen)   The Glueball Spectrum
16:30 - 16:45   Break
Chair: W. Namgung (Dongguk)
16:45 - 17:30   J. Terning (Berkeley)   Glueball Mass Spectrum from
Supergravity
17:30 - 18:05   S. Hyun (KIAS)   On DLCQ M Theory
18:05 - 18:50   Y. Nakawaki (Setsunan)   Gauge Theories in the
Light-Cone Representation
20:30 - 21:30   Poster Session (T. Itoh & Y. Kim, B.-H. Lee & D.S.

Hwang, G.R. Shin)

June 23 (Wed)
Chair: K. Yamawaki (Nagoya)
08:30 - 09:30   J.-C. Peng (LANL)   Dileptons from P-Nucleus
Collisions
09:30 - 10:20   A. Ramos (Barcelona)   Kaons in the Medium within a
Chiral Non-Perturbative Approach
10:20 - 10:40   T. Sugihara (Nagoya)   Variational Calculation of
the Effective Action
10:40 - 10:55   Break
10:55 - 11:40   F. Sannino (Yale)   Chiral Phase Transition via an
Effective Lagrangian Approach
11:40 - 12:25   Y. Kikukawa (Nagoya)   Chiral symmetry and chiral
gauge theories on the lattice (domain-wall fermion, the
Ginsparg-Wilson relation and a lattice implementation of the
eta-invariant)
Lunch & Sightseeing
Chair: J.B. Choi (Chonbuk)
19:00 - 19:20   H. Kim (TIT)   Two-Point Correlation Function with
Pion in QCD Sum Rules
19:20 - 19:40   S. Choe (Yonsei)   Multiquark picture for
Lambda(1405) & Sigma(1620)
19:40 - 20:00   C. Song (Seoul)   Chiral Lagrangian with BR Scaling
for Dense Nuclear Matter

June 24 (Thu)
Chair: M. Lutz (GSI)
08:30 - 09:05   A. de Roeck (DESY)   Low-x at HERA
09:05 - 10:05   C. Carlson (William & Mary)   Semi-Inclusive
Processes: A Different Way to Probe Hadron Structure
10:05 - 11:05   B.A. Kniehl (Hamburg)   Decoupling Relations,
Effective Lagrangians and Low-Energy Theorems
11:05 - 11:20   Break
11:20 - 12:20   J.B. Choi (Chonbuk)   Flux-tube Model for Gluonic
Structures
Lunch
Chair: C. Carlson (William & Mary)
14:00 - 14:45   K.-I. Kondo (Chiba)   Color Confinement in QCD due to
Topological Defects
14:45 - 15:20   Y. Kim & T. Itoh (Sungkyunkwan)   Fixed Point
Structure of Abelian Gauge Theories in $2 < D < 4$
15:20 - 15:55   E. Goubankova (North Carolina)   Hamiltonian
Renormalization for Bound State Problem in Gluodynamics
15:55 - 16:10   Break
Chair: G. McCartor (Southern Methodist)
16:10 - 16:45   B. Bakker (Vrije)   Embedded Bound States in
Light-Front Dynamics
16:45 - 17:20   H.M. Choi (North Carolina) Exploring Timelike
Region of QCD Exclusive Process in Relativistic Quark Model
17:20 - 17:55   D.S. Hwang (Sejong)   Hadronic Electroweak Transition
Matrix Elements in Light-Cone Field Theory
17:55 - 18:15   M.A. Taniguchi (Nagoya)   Physical Role of the
Light-Front Zero Mode
18:15 - 18:35   T. Maslowski (Warsaw)   The Construction of the
Angular Momentum Operator on the Light-Front in Perturbation Theory

June 25 (Fri)
Chair: D.-P. Min (Seoul)
08:30 - 09:15   R. Soldati (Bologna)   Consistent Light-Cone
Quantization in the Anti-Light-Cone Gauge
09:15 - 09:50   J. Hiller (Minnesota)   Pauli-Villars Regularization
in DLCQ
09:50 - 10:25   T. Heinzl (Jena)   Causality in Light-Cone Field
Theory
10:25 - 11:10   G. McCartor (Southern Methodist)   Continuum
Schwinger Model in the Light-Cone Representation
11:10 - 11:20   Closing -- C.R. Ji (North Carolina)

CONFERENCE PARTICIPANTS

Bakker, Bennard   (Vrije Univ.) blgbkkr@nat.vu.nl

Beane, Silas   (Univ. of Maryland) sbeane@quark.umd.edu

Brodsky, Stanley J.   (SLAC) sjbth@slac.stanford.edu

Burkardt, Matthias   (New Mexico State Univ.) burkardt@weizen.nmsu.edu

Carlson, Carl   (Coll. of William & Mary) carlson@physics.wm.edu

Choe, Seungho   (Yonsei Univ.) schoe@phya.yonsei.ac.kr

Choi, Ho Myeong   (North Carolina State Univ.) hmchoi@unity.ncsu.edu

Choi, Jong Bum   (Chonbuk Nat'l Univ.) jbchoi@moak.chonbuk.ac.kr

Dalley, Simon   (Cambridge Univ.) sd214@damtp.cam.ac.uk

de Roeck, Albert   (DESY) deroeck@mail.desy.de

Engelhardt, Michael   (Univ. Tuebingen) engelm@pthp1.tphys.physik.uni-tuebingen.de

Garvey, Gerald   (LANL) garvey@lanl.gov

Goubankova, Elena   (North Carolina State Univ.) egoubank@unity.ncsu.edu

Han, Sang Uk   (Sungkyunkwan Univ.)

Hayashigaki, Arata   (Kyoto Univ.) arata@yukawa.kyoto-u.ac.jp

Heinzl, Thomas   (Univ. Jena) heinzl@tpi.uni-jena.de

Hiller, John   (Univ. of Minnesota, Duluth) jhiller@d.umn.edu

Hong, Soon-Tae   (Sogang Univ.) sthong@phya.snu.ac.kr

Hwang, Dae Sung   (Sejong Univ.) dshwang@kunja.sejong.ac.kr

Hyun, Chang-Ho   (Seoul Nat'l Univ.) hch@zoo.snu.ac.kr

Hyun, Seungjun   (KIAS) hyun@ns.kias.re.kr

Itoh, Taichi   (Sungkyunkwan Univ.) taichi@newton.skku.ac.kr

Jang, Wonseok   (Kyungpook Nat'l Univ.) wsjang@nuclear.kyungpook.ac.kr

Ji, Chueng-Ryong   (North Carolina State Univ.) crji@unity.ncsu.edu

Jung, Hong   (Sookmyung Women's Univ.) jung@sookmyung.ac.kr

Kang, Ju Hwan   (Yonsei Univ.) jhkang@phya.yonsei.ac.kr

(Kyoto Univ.)

Kikukawa, Yoshio
kikukawa@gauge.scphys.kyoto-u.ac.jp

Kim, Gwang-Ho (Seoul Nat'l Univ.)
ghkim@phya.snu.ac.kr

Kim, Hungchong (Tokyo Inst. of Tech.)
hckim@th.phys.titech.ac.jp

Kim, Myungjin (Chungnam Nat'l Univ.)
myungjin@hanbat.chungnam.ac.kr

Kim, Yoonbai (Sungkyunkwan Univ.)
yoonbai@cosmos.skku.ac.kr

Kim, Wooyoung (Kyungpook Nat'l Univ.)
wooyoung@kyungpook.ac.kr

Kniehl, Bernd (Hamburg Univ.)
kniehl@vxdesy.desy.de

Kochelev, Nikolai (JINR)
kochelev@mail.desy.de

Kondo, Kei-Ichi (Chiba Univ.)
kondo@cuphd.nd.chiba-u.ac.jp

Lee, Bum Hun (Sejong Univ.)
bhl@ccs.sogang.ac.kr

Lee, Hee-Jung (Seoul Nat'l Univ.)
hjlee@fire.snu.ac.kr

Lee, Jeonghan (Pusan Nat'l Univ.)
citadel@cheerful.com

Lee, Jun-Hyoung (Sungkyunkwan Univ.)
june@newton.skku.ac.kr

Lee, Su Houng (Yonsei Univ.)
suhoung@phya.yonsei.ac.kr

Lenz, Frieder (Univ. Erlangen)
flenz@theorie3.physik.uni-erlangen.de

Lutz, Matthias (GSI)
m.lutz@gsi.de

Maslowski, Tomasz (Warsaw Univ.)
maslo@fuw.edu.pl

McCartor, Gary (Southern Methodist Univ.)
mccartor@pascal.physics.smu.edu

McKeown, Robert (California Inst. Tech.)
bmck@krl.caltech.edu

Miller, Gerald A. (Univ. of Washington)
miller@nucthy.phys.washington.edu

Min, Dong-Pil Min (Seoul Nat'l Univ.)
dpmin@phya.snu.ac.kr

Miyamura, Osamu (Hiroshima Univ.)
miyamura@fusion.sci.hiroshima-u.ac.jp

Nakawaki, Yuji (Setsunan Univ.)
nakawaki@mpg.setsunan.ac.jp

Namgung, Wuk (Dongguk Univ.)
ngw@cakra.dongguk.ac.kr

Oh, Yongseok (Seoul Nat'l Univ.)
yoh@phya.snu.ac.kr

Park, Byung-Yoon Park (Chungnam Nat'l Univ.)
bypark@chaosphys.chungnam.ac.kr

Park, Dong Hyun (Sungkyunkwan Univ.)
particle@newton.skku.ac.kr

Pauli, Hans-Christian   (MPI, Heidelberg)
                          pauli@mpi-hd.mpg.de
Peng, Jen-Chieh   (LANL)
                          peng@lanl.gov
Pinsky, Stephen S.   (Ohio State Univ.)
                          pinsky@mps.ohio-state.edu
Ramos, Angels   (Univ. Barcelona)
                          ramos@ecm.ub.es
Rho, Mannque   (KIAS)
                          rho@spht.saclay.cea.fr
Sannino, Francesco   (Yale Univ.)
                          francesco.sannino@yale.edu
Shim, Myoung-Ho   (Seoul Nat'l Univ.)
                          heisenbg@shinbiro.com
Shin, Ghi-Ryang   (Andong Nat'l Univ.)
                          gshin@andong.ac.kr
Shin, Gwansoo   (Seoul Nat'l Univ.)
                          shin@zoo.snu.ac.kr
Soldati, Roberto   (Univ. Bologna)
                          soldati@bo.infn.it
Song, Chaejun   (Seoul Nat'l Univ.)
                          chaejun@phya.snu.ac.kr
Song, Young-Ho   (Seoul Nat'l Univ.)
                          singer@phya.snu.ac.kr
Sugihara, Takanori   (Nagoya Univ.)
                          sugihara@eken.phys.nagoya-u.ac.jp
Suk, Jooyeop   (Chungnam Nat'l Univ.)
                          mars@hanbat.chungnam.ac.kr
Taniguchi, Masa-Aki   (Nagoya Univ.)
                          mass@eken.phys.nagoya-u.ac.jp
Terning, John   (Univ. of California, Berkeley)
                          terning@alvin.lbl.gov
van de Sande, Brett   (Univ. Erlangen)
                          bvds@theorie3.physik.uni-erlangen.de
Yamawaki, Koichi   (Nagoya Univ.)
                          yamawaki@eken.phys.nagoya-u.ac.jp
Yang, Gil-Seok   (Pusan Nat'l Univ.)

Yu, Cheol   (Sungkyunkwan Univ.)

Zahed, Ismail   (State Univ. of New York, Stony Brook)
                          zahed@nuclear.physics.sunysb.edu

AUTHOR INDEX